1 MONTH OF
FREE
READING

at

www.ForgottenBooks.com

By purchasing this book you are eligible for one month membership to ForgottenBooks.com, giving you unlimited access to our entire collection of over 1,000,000 titles via our web site and mobile apps.

To claim your free month visit: www.forgottenbooks.com/free903370

ISBN 978-0-265-87815-6
PIBN 10903370

SESSIONAL PAPERS

VOLUME 27

FIFTH SESSION OF THE TWELFTH PARLIAMENT

OF THE

DOMINION OF CANADA

SESSION 1915

VOLUME L.

ALPHABETICAL INDEX

TO THE

SESSIONAL PAPERS

OF THE

PARLIAMENT OF CANADA

FIFTH SESSION, TWELFTH PARLIAMENT, 1915.

See also Alphabetical List, Page 1.

LIST OF SESSIONAL PAPERS

Arranged in Numerical Order, with their titles at full length; the dates when Ordered and when presented to the Houses of Parliament: the name of the Senator or Member who moved for each Sessional Paper, and whether it is ordered to be Printed or Not Printed.

CONTENTS OF VOLUME D.

Fifth Census of Canada, 1911,—Agriculture, Volume IV: Presented by Hon. Mr. Foster, February 8, 1915.. .. ..*Printed for distribution and sessional papers.*

CONTENTS OF VOLUME 1.

(This volume is bound in three parts).

1. Report of the Auditor General for the year ended 31st March, 1914, Volume I, Parts A, B and A to L; Volume II, Parts M to U; Volume III, Parts V to Z.   Presented by Hon. Mr. White, February 9, 1915.. .. .. ..*Printed for distribution and sessional papers.*

CONTENTS OF VOLUME 2.

2. The Public Accounts of Canada, for the fiscal year ended 31st March, 1914.   Presented by Hon. Mr. White, February 9, 1915 .. .. ..*Printed for distribution and sessional papers.*

3. Estimates of sums required for the service of the Dominion for the year ending on 31st March, 1916.   Presented by Hon. Mr. White, February 8, 1915.
Printed for distribution and sessional papers.

4. Supplementary Estimates of sums required for the service of the Dominion for the year ending on the 31st March, 1915.   Presented by Hon. Mr. White, March 9, 1915.
Printed for distribution and sessional papers.

5. Further Supplementary Estimates of sums required for the service of the Dominion for the year ending on the 31st March, 1915.   Presented by Hon. Mr. White, March 27, 1915.
Printed for distribution and sessional papers.

5a. Further Supplementary Estimates for year ending 31st March, 1916.   Presented by Hon. Mr. White, March 31, 1915.. .. ..*Printed for distribution and sessional papers.*

CONTENTS OF VOLUME 3.

6. List of Shareholders in the Chartered Banks of the Dominion of Canada as on 31st December, 1914.   Presented by Hon. Mr. White, February 9, 1915.
Printed for distribution and sessional papers.

CONTENTS OF VOLUME 4.

7. Report on certified cheques, dividends, unclaimed balances and drafts or bills of exchange remaining unpaid in Chartered Banks of the Dominion of Canada, for five years and upwards prior to 31st December, 1913.   Presented by Hon. Mr. White, April 10, 1915.
Printed for distribution and sessional papers.

CONTENTS OF VOLUME 5.

(This volume is bound in two parts).

8. Report of Superintendent of Insurance for year 1914.   Presented by Hon. Mr. White, 1915.
Printed for distribution and sessional papers.

9. Abstract of Statement of Insurance Companies in Canada for year ended 31st December, 1914.   Presented by Hon. Mr. White, 1914.
Printed for distribution and sessional papers.

CONTENTS OF VOLUME 6.

CONTENTS OF VOLUME 13.

19. Report of the Minister of Public Works on the works under his control for the fiscal year ended 31st March, 1914, Volume I.   Presented by Hon. Mr. Rogers, February 8, 1915.
Printed for distribution and sessional papers.

CONTENTS OF VOLUME 14.

20. Annual Report of the Department of Railways and Canals, for the fiscal period from 1st April, 1913, to 31st March, 1914.   Presented by Hon. Mr. Cochrane, March 12.
Printed for distribution and sessional papers.

20a. Canal Statistics for the season of navigation, 1914.   Presented by Hon. Mr. Cochrane. 9th April, 1915.. .. ..*Printed for distribution and sessional papers.*

20b. Railway Statistics of the Dominion of Canada, for the year ended 30th June, 1914.   Presented by Hon. Mr. Cochrane, March 12, 1915.
Printed for distribution and sessional papers.

CONTENTS OF VOLUME 15.

20c. Ninth Report of the Board of Railway Commissioners for Canada, for the year ending 31st March, 1914.   Presented by Hon. Mr. Cochrane, February 8, 1915.
Printed for distribution and sessional papers.

20d. Telephone Statistics of the Dominion of Canada, for the year ended 30th June, 1914.   Presented by Hon. Mr. Cochrane, March 17, 1915.
Printed for distribution and sessional papers.

20e. Express Statistics of the Dominion of Canada for year ended 30th June, 1914.   Presented by Hon. Mr. Cochrane, 1915... .. ..*Printed for distribution nad sessional papers.*

20f. Telegraph Statistics of the Dominion of Canada, for the year ended 30th June. 1914.   Presented by Hon. Mr. Cochrane, March 17, 1915.
Printed for distribution and sessional papers.

CONTENTS OF VOLUME 16.

21. Forty-seventh Annual Report of the Department of Marine and Fisheries, for the year 1913-1914—Marine.   Presented by Hon. Mr. Hazen, February 8, 1915.
Printed for distribution and sessional papers.

21b. Report and evidence in connection with the Royal Commission appointed to investigate the disaster of the *Empress of Ireland.*   Presented by Hon. Mr. Hazen, 1914.
Printed for distribution and sessional papers.

CONTENTS OF VOLUME 17.

22. List of Shipping issued by the Department of Marine and Fisheries, being a list of vessels on the registry books of the Dominion of Canada on 31st December, 1914.   Presented by Hon. Mr. Hazen, 1915... .. ..*Printed for distribution and sessional papers.*

23. Supplement to the Forty-seventh Annual Report of the Department of Marine and Fisheries for the fiscal year 1913-14—Steamboat Inspection Report.   Presented by Hon. Mr. Hazen, March 3, 1915 .. ..*Printed for distribution and sessional papers.*

CONTENTS OF VOLUME 18.

24. Report of the Postmaster General for the year ended 31st March, 1914.   Presented by Hon. Mr. Casgrain, February 8, 1915.. .. .. ..*Printed for distribution and sessional papers.*

CONTENTS OF VOLUME 19.

(This volume is bound in two parts).

25. Annual Report of the Department of the Interior, for the fiscal year ending 31st March, 1914.—Volume I.   Presented by Hon. Mr. Roche, March 8, 1915.
Printed for distribution and sessional papers.

CONTENTS OF VOLUME 20.

25*a.* Report of Chief Astronomer, Department of the Interior for year ending 31st March, 1911. Presented by Hon. Mr. Roche, 1915.. .. .. ..*Printed for distribution and sessional* papers.

25*b.* Annual Report of the Topographical Surveys Branch of the Department of the Interior, 1912-13. Presented by Hon. Mr. Roche, 1914.
Printed for distribution and sessional papers.

25*c.* Report of progress of stream measurements for calendar year of 1914. Presented by Hon. Mr. Roche, 1914.. .. ..*Printed for distribution and sessional papers.*

CONTENTS OF VOLUME 21.

25*d.* Thirteenth Report of the Geographic Board of Canada for the year ending 30th June, 1914. Presented by Hon. Mr. Roche, 1915.
Printed for distribution and sessional papers.

25*e.* Report on Bow River Water Power and Storage Investigations, seasons 1911-1912-1913. Presented by Hon. Mr. Burrell, 1915.. ..*Printed for distribution and sessional papers.*

25*f.* Report of the British Columbia Hydrographic Survey for the calendar year 1913. Presented by Hon. Mr. Burrell, 1915.. ...*Printed for distribution and sessional papers.*

CONTENTS OF VOLUME 22.

26. Summary Report of the Geological Survey, Department of Mines, for the calendar year 1913. Presented, 1915.. .. ..*Printed for distribution and sessional papers.*

26*a.* Summary Report of the Mines Branch for the calendar year 1913. Presented, 1914.
Printed for distribution and sessional papers

CONTENTS OF VOLUME 23.

27. Report of the Department of Indian Affairs for the year ended 31st March, 1914. Presented by Hon. Mr. Roche, 11th February, 1915.
Printed for distribution and sessional papers.

28. Report of the Royal Northwest Mounted Police, 1914. Presented by Hon. Sir Robert Borden, 8th February, 1915 .. ..*Printed for distribution and sessional papers.*

CONTENTS OF VOLUME 24.

29. Report of the Secretary of State of Canada for the year ended 31st March, 1914. Presented by Hon. Mr. Coderre, 9th February, 1915.
Printed for distribution and sessional papers.

29*b.* Report of the work of the Public Archives for the year 1913. Presented, 1915.
Printed for distribution and sessional papers.

30. The Civil Service List of Canada, 1914. Presented by Hon. Mr. Coderre, 9th February, 1915.. .. ..*Printed for distribution and sessional papers.*

CONTENTS OF VOLUME 25.

31. Sixth Annual Report of the Civil Service Commission of Canada for the year ended 31st August, 1914. Presented by Hon. Mr. Coderre, 19th March, 1915.
Printed for distribution and sessional papers.

32. Annual Report of the Department of Public Printing and Stationery for the year ended 31st March, 1914. Presented by Hon. Mr. Coderre, 6th April, 1915
Printed for distribution and sessional papers.

33. Report of the Secretary of State for External Affairs for the year ended 31st March, 1914. Presented by Sir Robert Borden, 18th February, 1915.
Printed for distribution and sessional papers.

34. Report of the Minister of Justice as to Penetentiaries of Canada, for the fiscal year ended 31st March, 1914. Presented, 1915.. .. ..*Printed for distribution and sessional papers.*

35. Report of the Militia Council for the Dominion of Canada, for the fiscal year ending 31st March, 1914. Presented by Hon. Mr. Hughes, 10th February, 1915.
Printed for distribution and sessional papers

CONTENTS OF VOLUME 26.

CONTENTS OF VOLUME 27.

CONTENTS OF VOLUME 28.

CONTENTS OF VOLUME 28—*Continued.*

CONTENTS OF VOLUME 28—*Continued.*

CONTENTS OF VOLUME 28—*Continued.*

CONTENTS OF VOLUME 28—*Continued.*

CONTENTS OF VOLUME 28—*Continued.*

CONTENTS OF VOLUME 28—*Continued.*

CONTENTS OF VOLUME 28—*Continued.*

CONTENTS OF VOLUME 28—*Continued.*

CONTENTS OF VOLUME 28—*Continued.*

CONTENTS OF VOLUME 28—*Continued*.

CONTENTS OF VOLUME 28—*Continued.*

CONTENTS OF VOLUME 28—*Continued.*

CONTENTS OF VOLUME 28—*Continued.*

CONTENTS OF VOLUME 28—*Continued.*

CONTENTS OF VOLUME 28—*Continued.*

180. Return to an Order of the House of the 8th March, 1915, for a return showing:—1. The number of customs officers employed at the customs port of Abercorn, Quebec, on 20th September, 1911. 2. The names of these officers. 3. The salary each one received. 4. The total amount of salaries paid the officers at this port. 5. The number of customs officers employed at the port of Abercorn at the present time. 6. The names of these officers. 7. The salary each one receives. 8. The total amount of salaries paid to the officers at this port. Presented 18th March, 1915.—*Mr. Kay*... ..*Not printed.*

181. Return to an Order of the House, of the 1st March, 1915, for a copy of all petitions, letters, communications and other documents relating to or bearing upon the dismissal of Leonard Hutchinson, chief keeper at Dorchester penitentiary. Presented 18th March, 1915.—*Mr. Copp*.. .. ..*Not printed.*

182. Return to an Order of the House of the 22nd February, 1915, for a copy of all letters, telegrams and papers generally concerning the proposed construction of a bridge to connect Isle Perrot with the mainland at Vaudreuil. Presented 18th March, 1915.— *Mr. Boyer*... .. ..*Not printed.*

182a. Return to an Order of the House of the 22nd February, 1915, for a copy of all letters, telegrams and papers generally concerning the proposed construction of a bridge between the Island of Montreal and the Mainland at Vaudreuil. Presented 18th March, 1915.—*Mr. Boyer* .. ..*Not printed.*

183. Return to an Order of the House of the 22nd February, 1915, for a return showing:— 1. What properties have been acquired by the Government in the City of Regina since 21st September, 1911? 2. The descriptions of such properties by metes and bounds? 3. For what purposes such properties were acquired? 4. From whom such properties were purchased? 5. The total price and the price per foot paid for each property? 6. If any such property was acquired by expropriation, what tribunal determined the price to be paid for any property so expropriated? 7. The dates on which any such properties were acquired? Presented 18th March, 1915.—*Mr. Martin (Regina).*

184. Return to an Order of the House of the 19th February, 1915, for a copy of all letters, telegrams, memoranda, pay-lists, recommendations and any other documents whatsoever in any wise appertaining to the construction of a wharf at Lower Burlington, in the County of Hants. Presented 18th March, 1915.—*Mr. Chisholm (Inverness).*
                         Not printed.

185. Return to an Order of the House of the 24th February, 1915, for a copy of pay-rolls and all correspondence and vouchers in connection with the repairs to Jordan breakwater, Shelburne county, for which Leander McKenzie was contractor of works or foreman. Presented 18th March, 1915.—*Mr. Law* .. ..*Not printed.*

186. Return to an Order of the House of the 24th February, 1915, for a copy of all letters, telegrams, correspondence and pay-rolls in connection with repairs and extension of breakwater at Bluff Head, Yarmouth county, N.S., during year 1914. Presented 18th March, 1915.—*Mr. Law*.. .. ..*Not printed.*

187. Return to an Order of the House of the 22nd February, 1915, for a return showing the amounts expended by the Public Works Department in the County of Inverness each year from 1896 down to 1915. Presented 18th March, 1915.—*Mr. Chisholm (Inverness)*.. .. ..*Not printed.*

188. Return to an Order of the House of the 24th February, 1915, for a copy of all letters, telegrams, correspondence and pay-sheets in connection with the repairs and other work on the breakwater at Sandford, Yarmouth County, N.S., during the year 1914. Presented 18th March, 1915.—*Mr. Law*.. .. ..*Not printed.*

189. Return to an Order of the House of the 1st March, 1915, for a copy of all papers, letters, petitions and other documents relating to a mail contract with David D. Heard & Sons, between Whitby and Grand Trunk Railway station, or with one John Gimblet, Whitby. Presented 19th March, 1915.—*Mr. Pardee*.. . .. .*Not printed.*

190. Copies of Reports of the Committee of the Privy Council, approved by His Royal Highness the Governor General, relating to certain advances made to the Canadian Northern Railway Company and the Grand Trunk Pacific Railway Company, respectively, together with copies of agreements made between the said companies and His Majesty. Presented by Hon. Mr. White, 19th March, 1915.. .. ..*Not printed.*

191. Return to an Order of the House of the 11th February, 1915, for a copy of all tenders received by the Post Office Department for the mail service between Caraquet and Tracadie, Gloucester County, N.B., on the 15th day of January last, with the names of the tenderers, the respective amounts of the tenders, and the name of the new contractor. Presented 19th March, 1915.—*Mr. Turgeon*.. . .. .*Not printed.*

CONTENTS OF VOLUME 28—*Continued.*

192. Return to an Order of the House of the 8th March, 1915, for a return showing :—1. The fractional areas of homestead lands or otherwise in the province of Saskatchewan sold in the year 1914.   2. The name of the purchaser, and the price paid in each case. Presented 22nd March, 1915.—*Mr. Martin (Regina)* ... ...*Not printed.*

193. Return to an Order of the House of the 25th February, 1915, for a return showing, in reference to the answer to question No. 6 of 9th February, and answered 15th February as per page 101 unrevised *Hansard*, the cost of furnishing the Government offices in each of the said buildings.   Presented 22nd March, 1915.—*Mr. Turriff.*
Not printed.

194. Return to an Order of the House of the 1st March, 1915, for a return showing the amount of railway subsidies paid in the county of Inverness since 1896, to date, and the dates on which such subsidies were paid.   Presented 22nd March, 1915.—*Mr. Chisholm (Inverness)* ... ...*Not printed.*

195. Return to an Order of the House of the 1st March, 1915, for a copy of all letters, papers, telegrams and other documents relating to the purchase or lease of the railway from New Glasgow to Thorburn, in the county of Pictou, known as the Vale Railway, from the Acadia Coal Company, since January, 1911, to date.   Presented 22nd March, 1915. —*Mr. Macdonald.* ... ...*Not printed.*

196. Return to an Order of the House of the 1st March, 1915, for a copy of all papers, letters, telegrams, correspondence, contracts, etc., in connection with the sale of the hay grown or the lease of certain tracts of land belonging to the Intercolonial Railway, upon which hay is grown, and which are contiguous to the properties of Charles Lavoie, Cléophas Leclerc and Joseph Parent of the Parish of Bic, county of Rimouski.   Presented 22nd March, 1915.—*Mr. Lapointe (Kamouraska)* ... ...*Not printed.*

197. Return to an Order of the House of the 3rd March, 1915, for a copy of all letters, papers, telegrams, evidence taken at investigations, reports and all other documents relating to the suspension or other action in regard to the charge of drunkeness against Newton Hopper, conductor on the Intercolonial Railway, and to his subsequent reinstatement. Presented 22nd March, 1915.—*Mr. Macdonald.* ... ...*Not printed.*

198. Return to an Order of the House of the 1st March, 1915, for a copy of all letters, telegrams and other papers relating to the dismissal of Bruce Wiswell, as sectionman on the Intercolonial Railway at Stellarton, Nova Scotia.   Presented 22nd March, 1915.— *Mr. Macdonald.* ... ...*Not printed.*

199. Return to an Order of the House of the 22nd February, 1915, for a return showing :— 1. The inward tonnage freight, and also the outward tonnage freight respectively, at Loggieville station of the Intercolonial Railway for each month of 1914, and also for the month of January, 1915.   2. The inward tonnage freight, and the outward tonnage freight at Chatham station, on the Intercolonial Railway for each month of 1914, and also for the month of January, 1915.   3. The inward tonnage freight, and the outward tonnage freight at Newcastle station on the Intercolonial Railway for each month of 1914, and also for the month of January, 1915.   4. The local and through passenger traffic to and through each of the above stations, respectively, during each of the months above mentioned.   Presented 22nd March, 1915.—*Mr. Loggie.*
Not printed.

200. Return to an Order of the House of the 15th February, 1915, for a copy of all letters, telegrams and correspondence had by Margaret Lynch, or any person representing her, with reference to the expropriation of certain land beolnging to the said Margaret Lynch in the city of Fredericton, province of New Brunswick, by the Intercolonial Railway, and also of all letters, telegrams and correspondence had with F. P. Gutelius or any other official of the Intercolonial Railway with reference thereto.   Presented 22nd March, 1915.—*Mr. Carvell.* ... ...*Not printed.*

201. Return to an Order of the House of the 3rd March, 1915, for a copy of all documents bearing on the payment made to C. R. Scoles, New Carlisle, Quebec, in July, 1914, of balance of subsidy voted to the Atlantic and Lake Superior Railway on the recommendation of the Financial Comptroller.   Presented 22nd March, 1915.—*Mr. Marcil.*
Not printed.

202. Return to an Order of the House of the 1st March, 1915, for a copy of all letters, telegrams, correspondence and reports relating to the purchase of the New Brunswick and Prince Edward Island Railway, extending from Sackville to Cape Tormentine, county of Westmorland.   Presented 22nd March, 1915.—*Mr. Copp.* ... ...*Not printed.*

203. Return to an Order of the House of the 1st March, 1915, for a copy of the tariff on flour shipments now in force on the Quebec, Oriental Railway and the Atlantic. Quebec and Western Railway.   Presented 22nd March, 1915.—*Mr. Marcil.* ... ...*Not printed.*

CONTENTS OF VOLUME 28—*Continued.*

CONTENTS OF VOLUME 28—*Continued.*

CONTENTS OF VOLUME 28—*Continued.*

CONTENTS OF VOLUME 28—*Continued.*

CONTENTS OF VOLUME 28—*Continued.*

CONTENTS OF VOLUME 28—*Continued.*

CONTENTS OF VOLUME 28—*Continued.*

CONTENTS OF VOLUME 28—*Continued.*

CONTENTS OF VOLUME 28—*Continued.*

FORTY-SEVENTH ANNUAL REPORT

OF THE

Department of Marine and Fisheries

1913-14

FISHERIES

PRINTED BY ORDER OF PARLIAMENT

OTTAWA

PRINTED BY J. DE L. TACHÉ, PRINTER TO THE KING'S MOST

EXCELLENT MAJESTY

1914

[No. 39—1915]

To Field Marshal His Royal Highness Prince Arthur William Patrick Albert Duke
of Connaught and of Strathearn, K.G., K.T., K.P., etc.,etc., etc., Governor General
and Commander-in-Chief of the Dominion of Canada.

MAY IT PLEASE YOUR ROYAL HIGHNESS :

I have the honour to submit herewith, for the information of Your Royal Highness
and the legislature of Canada, the forty-seventh Annual Report of the Department of
Marine and Fisheries, Fisheries Branch.

I have the honour to be,

Your Royal Highness's most obedient servant,

J. D. HAZEN,

Minister of Marine and Fisheries.

DEPARTMENT OF MARINE AND FISHERIES,
OTTAWA, NOVEMBER, 1914.

39—A½

ERRATA.

On page 4, line 10: $959,492 should read $913,217.

On page 4, line 11: $38,592 should read $84,867.

On page 7, line 27: $4,292,657 should read $4,294,657, and $799,164 should read $797,164.

On page 141, the value of clams canned should read $51.984 instead of $19,494. The total value for Charlotte county should read $1,386,462.

TABLE OF CONTENTS.

ALPHABETICAL INDEX

TO THE

FISHERIES REPORT

1913-1914

A

B

5 GEORGE V., A. 1915

C

D

E

F

G

H

N

O

P

Q

SESSIONAL PAPER No. 39

R

5 GEORGE V., A. 1915

Y

DEPUTY MINISTER'S REPORT

To the Honourable
    J. D. HAZEN,
        Minister of Marine and Fisheries.

SIR,—I have the honour to submit the forty-seventh annual report of the Fisheries Branch of this Department, which is for the fiscal year ended March 31, 1914.

The following nineteen appendices are included:—

No. 1. Nova Scotia Fisheries.
    2. New Brunswick Fisheries.
    3. Prince Edward Island Fisheries.
    4. Quebec Fisheries.
    5. Ontario Fisheries.
    6. Manitoba Fisheries.
    7. Saskatchewan and Alberta Fisheries.
    8. Yukon Fisheries.
    9. British Columbia Fisheries.
    10. Imports and Exports of Fish.
    11. The Fisheries Patrol Service.
    12. Oyster Culture.
    13. Fish Breeding.
    14. Canadian Fisheries Museum.
    15. Fisheries Expenditure and Revenue.
    16. Fishing Bounty.
    17. United States Fishing Vessels Entries.
    18. The Outside Fisheries Staff.
    19. Report on the Biological Stations.

INTERNATIONAL FISHERIES COMMISSION.

Since my last report was submitted, the United States representative on the Commission, Mr. Job E. Hedges, of New York, resigned, and was succeeded by Dr. Hugh M. Smith, the United States Commissioner of Fisheries at Washington.

For a time, there appeared reason to hope that after more than four years' delay, the regulations as drawn up by the Commissioners would be approved by Congress. Unexpected difficulties have, however, arisen, and it is now doubtful whether they will be.

With a view to the conservation of the food resources in the boundary waters, which experience shows could best be achieved by united action by the Federal

5 GEORGE V., A. 1915

Governments of both countries, and in the hope of arriving at a satisfactory agreement, the Canadian Government has shown its willingness to prolong the consultations between its Commissioner and the successive Commissioners appointed by the United States Government. If, however, the United States authorities are unable to see their way to approve the regulations, and to co-operate with the Canadian Government in the work of conservation, obviously the Canadian Government will, to its regret, be forced, by circumstances for which it must disclaim all responsibility, to reserve liberty of action.

BIOLOGICAL STATIONS.

An interesting report, by the Secretary Treasurer of the Biological Board on the work done at and in connection with its stations during the year, forms Appendix 19 to this Report.

TRANSPORTATION OF FRESH AND MILDLY CURED FISH.

The nature of the assistance given by the Department during the past few years to aid in rapid development and expansion of the fresh and mildly cured fish business, has been fully explained in previous reports. This assistance has been continued without interruption during the year just closed, and has been extended by the inauguration of an express refrigerator car service one day each week from Mulgrave to Montreal. Shipments from Halifax, and west thereof, are consolidated in this car at Truro.

As was pointed out in my last annual report, experience has shown that to assure fresh fish packed in ice reaching distant points with the ice still unmelted, refrigerator cars are needed, even by express, and the extent to which this additional service was availed of, warrants the hope that the time is not far distant when all shipments for considerable distances will go forward in such cars.

This service was started on the 9th August, and was continued until the 24th January, when the harbours about the eastern portion of Nova Scotia were closed by ice. It is the intention to again start the service when fishing is resumed in the spring.

The condition on which the service was started was that the Department guaranteed the earnings on these cars on each trip west to the extent of those on 10,000 pounds; but with the exception of nine occasions out of the twenty-four made, the shipments were over ten thousand pounds. The occasions on which they fell short and the amounts then carried are as follows:—

	Lbs.
September 6	9,427
September 27	9,625
October 4	9,600
October 11	9,965
October 18	7,540
December 27	9,584
January 3	9,273
January 17	8,425
January 24	6,964

From this statement it will be observed that the extra cost of this service was $143.96.

It was arranged to have this car leave the coast on Saturdays, more with the object of serving the Toronto than the Montreal markets, as shipments in ordinary express cars have a much better chance of reaching Montreal than Toronto in good condition, and as shipments in this car reach Montreal with the ice thereon unmelted, they will stand transportation from there in an ordinary express car much better than they otherwise would. To best serve the Montreal markets, a car should leave the coast on Tuesdays; but Toronto dealers desire the larger shipments to reach there on Mondays. Hence, an endeavour was made to have the express companies operate two cars per week, one to Montreal and the other to Toronto; but the railway had not sufficient refrigerator cars available to enable this to be done. It is anticipated however, that by the opening of next season cars will be available, and that a bi-weekly refrigerator express service from the Atlantic coast will be in operation. With such cars running direct to Toronto, it would be practicable to distribute from there much farther west than it is now.

How the business from the east and the west, by express alone, and in less than carload—as the Department bears no share of the cost, when shipments reach carload proportions—has been expanding, will be gathered from the following figures, which show the amount paid by the Department as its share of the express charges :—

Year.	On shipments from East.	On shipments from West.
	$ cts.	$ cts.
1909-10..........	15,162 20	13,541 76
1910-11..........	16,898 13	21,896 73
1911-12.	19,620 62	35,315 10
1912-13..........	29,969 48	39,277 13
1913-14.	37,818 85	44,114 47

FISHERIES EXHIBIT AT THE TORONTO FAIR.

As was intimated in the Report for 1912-1913, it was decided that the Department would this year give a Fisheries Exhibit at the Canadian National Exhibition at Toronto. This Fair is yearly visited by people from practically all over Central Canada, so that probably by no other means could such an effective advertisement of fish be carried on for the cost involved.

To enable frozen fish to be properly displayed, a refrigerator of twenty tons capacity, with a glass front, was installed.

It was felt that the Exhibit could be made most effective if the co-operation of some of the larger fish dealers were secured, and the Department, after correspondence with the various important dealers, obtained the co-operation of the Maritime Fish Corporation of Montreal, the North Atlantic Fisheries Limited of Halifax, and the F. T. James Company, Limited, of Toronto. The first two companies above mentioned made a speciality of Atlantic fish, which were shown in frozen, smoked, pickled and other attractive ways, while the F. T. James Company, Limited, undertook the rather ambitious task of giving an Exhibit of fresh water fish, in a fresh condition, as well as of the various other lines handled by them. The Department, with the assistance of Mr. F. J. Hayward, of Vancouver, gave an exhibit of the fisheries of British Colum- bia. The Department also gave an attractive display of specimens of mounted fish from the different parts of Canada.

The Exhibit as a whole, and in detail, was a splendid success, and proved to be one of the leading features at the Fair. Almost continuously throughout the whole time at the Fair it attracted crowds of people, many of whom were inquirers as to where they could procure different kinds of the fish displayed.

As a further evidence of the high regard in which the Exhibit was held, it may be mentioned that it was awarded a gold medal.

There seems little room for doubt that it did much to increase the demand for fish, and that the public interest will be well served by making an even more compre-. hensive exhibit next year. This it is intended to do.

DAILY BAIT REPORTS.

For the purpose of assisting masters of fishing vessels to locate bait supplies during the cod-fishing season, and thus avoid the great loss of time annually spent in searching for bait from harbour to harbour, the Department, in the course of the year, 1913, put into operation a system whereby definite information concerning supplies along certain stretches of the Atlantic seaboard was collected by the local officer of the Department and despatched, daily, by telegram, to certain important sea ports, and there posted up.

The number of ports selected as receiving stations had necessarily to be limited, but, through the courtesy of the daily papers in Nova Scotia in which the telegrams were published each day, the smaller fishing communities derived the benefit of direct advice as to available supplies of bait.

During the spring ninety-three telegrams were sent from Magdalen Islands, Souris, P.E.I., and Queensport, N.S. to Canso, N.S., Halifax, N.S., Lunenburg, N.S. and Riverport, N.S. During July and August five hundred and fourteen telegrams were sent from Grand Mira, N.S., Little Bras D'Or, N.S., Petit de Grat, N.S., Lower L'Ardoise, N.S., Canso, N.S., Wine Harbour, N.S., Tangier, N.S. and Musquodoboit Harbour, N.S., to North Sydney, N.S., Canso, N.S., Halifax, N.S., Lunenburg, N.S., Riverport, N.S. and Shelburne, N.S.; also from Lockeport, N.S. to Canso, N.S., Hali- fax, N.S., Lunenburg, N.S., and Riverport, N.S.; from Shag Harbour, Middle West Pubnico and Digby, N.S., to Halifax, Lunenburg, Shelburne and Lockeport, N.S.; from Bedeque and Leoville, P.E.I., to Caraquette and Shippigan, N.B.

The service from Bedeque and Leoville, P.E.I.,, to Caraquette and Shippigan, N.B., was found to be useless, and will not be continued next season.

During September, October, November and December one hundred and seventy- five telegrams were sent from Grand Manan, N.B., Pennfield, N.B., and St. John, N.B., to Digby, N.S., Yarmouth, N.S., Pubnico, N.S. and Clarke's Harbour, N.S. Each tele- gram sent out contained specific information as to bait supplies at all the important points within the district of the officer who despatched the message. Copies of all telegrams were mailed to the Department at the end of each week, and the work closely followed.

The benefits derived from the first year's operation of this service may be gathered from the following synopsis of the reports from the officers directly concerned:—

The officer at Grand Manan, N.B., reported that the service considerably helped masters of vessels to locate bait supplies. A number of those vessels called at Louisburg during August and generally obtained bait.

The officer at Guysborough reported that masters and owners of fishing vessels benefited greatly by the information posted up at Canso.

The officer at Wine Harbour, N.S., reported that, as a result of the telegraphic information sent to Canso, Halifax, etc., during July and August, several vessels baited at Port Beckerton and other harbours in his district. All the net fishermen much appreciated the fact that the telegrams sent out brought buyers to them who paid fair prices for their herring.

The officer at Musquodoboit Harbour, N.S., reported that during July and August five vessels and a number of boats baited at Owl's Head. The telegraphic information benefited herring fishermen by bringing twenty-five sail of boats from Tancook to Eastern Passage, where they averaged about one hundred barrels per boat.

The officer at Lunenburg, N.S., reported that the bait telegrams were very beneficial to the managing owners of vessels at Lunenburg. As soon as they knew where bait was they telegraphed the information to their captains.

The officer at Allendale, N.S., reported that the reports sent out were the means of bringing a number of vessels to Lockeport, and other points in his district, for bait, which was a decided benefit not only to the vessel fishermen, but to the local net fishermen as well. All were greatly pleased with the service.

The officer at Lower Shag Harbour reported that the service was very satisfactory to trap owners and net fishermen as they readily sold their bait during the months of July and August.

The officer at Middle West Pubnico, N.S., reported that the masters and owners of vessels fresh fishing off Yarmouth made much use of the bait reports from Grand Manan, N.B. Several vessels went from Pubnico to Grand Manan for lobster bait, and were enabled to load and make their trip within two weeks; previously, owing to the lack of definite information regarding catches of bait, such a trip usually occupied four or five weeks' time.

The officer at Digby, N.S., reported the local fishermen as having declared that by knowing exactly where to get bait they caught much more fish than they would have caught without such knowledge. The captain of the schooner *Cora May* depended altogether on the reports in order to get his bait. The captain of a vessel buying fresh fish, and who had to keep the fishermen supplied with bait in order to buy their fish, also depended entirely on the telegraphic reports. In securing bait supplies another captain said that he had been able in many instances to gain several days' fishing and at times a whole trip by the information contained in the bait telegrams. The gasoline boats also benefited greatly by the information. In good fishing weather one boat would be sent directly to where bait was reported as obtainable for enough

39—B

5 GEORGE V., A. 1915

to supply several boats; thus fishing went on without interruption. This officer received many more communications from captains of vessels telling of the benèfits derived from the bait reports.

The officer at Pennfield, N.B., reported that, owing to the uncertainty of locating bait supplies during the past season, due to the fact that bait and sardine herring had never before been so scarce in Charlotte County, this service was of an especial benefit to Nova Scotia fishermen as it enabled them to go directly to where bait was available. The service also proved very beneficial to shore fishermen in this officer's district, by informing them as to where bait could be obtained, even in small quantities.

The officer at Grand Manan, N.B., reported that the service proved very beneficial as many fishing vessels from Nova Scotia got bait there for immediate use; also large quantities of bait were secured for lobster fishing purposes. The captain of a Digby vessel informed them that the Daily Bait Reports had been of great service to their fishing fleet, by saving them a lot of time and trouble, as they knew just where they could obtain their bait.

GENERAL REVIEW.

EXTENT OF FISHERIES.

To say that Canada possesses the most extensive fisheries in the world is no exaggeration; moreover, it is safe to add that the water in and around Canada contain the principal commercial food fishes in greater abundance than the waters of any other part of the world. The extraordinary fertility of what may be called our own waters is abundantly proved by the fact that, apart from salmon, all the lobsters, herring, mackerel and sardines, nearly all the haddock, and many of the cod, hake, and pollock landed in Canada are taken from within our territorial waters.

The coast line of the Atlantic provinces, from the Bay of Fundy to the Straits of Belle Isle, without taking into account the lesser bays and indentations, measures over 5,000 miles; and along this great stretch are to be found innumerable natural harbours and coves, in many of which valuable fish are taken in considerable quantities with little effort.

On the Pacific coast, the province of British Columbia, owing to its immense number of islands, bays and fiords, which form safe and accessible harbours, has a sea-washed shore of 7,000 miles.

Along this shore and within the limits of the territorial waters, there are fish and mammals in greater abundance, probably than anywhere else in the whole world.

In addition to this immense salt-water fishing area, we have in our numerous lakes no less than 220,000 square miles of fresh water, abundantly stocked with many species of excellent food fishes. In this connection, it may be pointed out that the area of the distinctly Canadian waters of what are known as the Great Lakes— Superior, Huron, Erie and Ontario—forms only one-fifth part of the total area of the larger fresh-water lakes of Canada.

The fisheries of the Atlantic coast may be divided into two distinct classes: the deep-sea and the inshore or coastal fisheries.

The deep-sea fishery is pursued in vessels of from 40 to 100 tons, carrying crews of from 12 to 20 men. The fishing grounds worked on are the several banks, which lie from 20 to 90 miles off the Canadian coast. The style is that of "trawling" by hook and line. The bait used is chiefly herring, squid and capelin; and the fish taken are principally cod, haddock, hake, pollock and halibut.

The inshore or coastal fishery is carried on in small boats with crews of from two to three men; also in a class of small vessels with crews of from four to seven men. The means of capture employed by boat fishermen are gill-nets, hooks and lines, both hand-line and trawl; and from the shore are operated trap-nets, haul seines, and weirs. The commercial food fishes taken inshore are the cod, hake, haddock, pollock, halibut, herring, mackerel, alewife, shad, smelt, flounder and sardine. The most extensive lobster fishery known is carried on along the whole of the eastern shore of Canada, whilst excellent oyster beds exist in many parts of the Gulf of St. Lawrence notably on the north coast of Prince Edward Island, and in the Northumberland strait.

The salmon fishery is, of course, the predominant one on the Pacific coast; but a very extensive halibut fishery is carried on in the northern waters of British Columbia in large, well-equipped steamers and vessels. The method of capture is by trawling, dories being used for setting and hauling the lines, as in the Atlantic deep-sea fishery. Herring are in very great abundance on the Pacific coast, and provide a plentiful supply of bait for the halibut fishery.

In the inland lake fisheries, the various means of capture in use are gill-nets, pound-nets, seines and hook-and-line to a great extent. The principal commercial fishes caught are whitefish, trout, pickerel, pike, sturgeon and fresh water herring—the latter in the lakes of Ontario only.

VALUE OF THE FISHERIES.

The total marketed value of all kinds of fish, fish products and marine animals taken by Canadian fishermen from the sea and inland lakes and rivers, during the fiscal year ended March 31, 1914, amounted to $33,207,748.

This value falls short of that for the preceding year by $181,716. This, is accounted for by the sockeye salmon run in Northern British Columbia being smaller than usual and the decrease in the value of halibut.

Of this total value the sea fisheries contributed $29,472,811; while the inland fisheries contributed $3,734,937. The former being an increase of $157,039 over that of last year, while the value of the inland fisheries decreased $338,755.

There was a total of 71,776 men employed in fishing, on 1,992 vessels, tugs and carrying smacks, and 37,686 boats; while 26,893 persons were engaged on shore in canneries, freezers, fish-houses, etc. Of this number 86,486 were engaged in the sea fisheries and 12,183 in the inland fisheries. The number of gasoline boats used in the industry was 8,700, or an increase of 2,789 over that for the preceding year.

39—B½

5 GEORGE V., A. 1915

The following table shows the value produced from the fisheries of each province in its respective order of rank, with the increase or decrease, as compared with the year 1912-13 :—

Provinces.	Value produced.	Increase.	Decrease.
	$	$	$
British Columbia	13,891,398	....	564,090
Nova Scotia	8,297,626	913,571	....
New Brunswick	4,308,707	44,653	....
Ontario	2,674,685	....	168,193
Quebec	1,850,427	....	137,814
Prince Edward Island	1,280,447	....	99,458
Manitoba	606,272	....	193,877
Saskatchewan	148,602	36,763	....
Alberta	81,319	29,703	....
Yukon	68,265	....	42,974
Totals	33,207,748	1,024,690	1,206,406
Net Decrease	....	....	181,716

The above table shows that British Columbia again produced the greatest value. although this was much lower than for the previous year. The decrease is wholly due to the comparative smallness of the sockeye salmon run in Northern British Columbia, and to a smaller halibut catch, with a much lower value.

Nova Scotia was the only province to show a substantial increase, while New Brunswick, Saskatchewan and Alberta showed slight increases over the previous year.

The following table shows the quantity of the chief kinds landed in the whole of Canada during 1913-14, and during the two preceding years :—

Kinds of fish.	1913-14.	1912-13.	1911-12.
	Cwt.	Cwt.	Cwt.
Salmon	1,551,411	1,253,997	1,136,732
Lobsters	514,646	555,138	589,141
Cod	1,664,599	1,729,070	2,097,260
Haddock	405,633	503,822	530,221
Hake and cusk	353,598	349,395	275,755
Pollock	150,094	143,324	250,881
Halibut	256,096	282,658	245,609
Herring	2,484,219	2,484,673	2,251,278
Mackerel	215,442	107,964	90,141
Sardines	141,384	281,548	404,383
Alewives	61,768	117,614	75,567
Smelts	88,728	102,360	81,748
Whitefish	137,887	140,404	131,515
Trout	73,164	73,664	80,638
Pickerel	61,603	64,839	79,610
Pike	64,925	62,492	80,328
Sturgeon	4,811	10,035	9,145
Oysters	20,828	23,377	31,746
Clams and quahaugs ....(brl)	121,335	105,303	103,347

SESSIONAL PAPER No. 39

The following table shows the relative values of the chief commercial fishes return-ing $100,000 and upwards, in their order of rank for the year under review, with the amount of increase or decrease, when compared with the values for the year 1912-13 :—

Kinds of fish.	Value.	Increase.	Decrease.
	$	$.	$
Salmon	10,833,713	808,190	....
Lobsters	4,710,062	139,048	....
Cod	3,387,109	18,359	....
Herring	3,173,129	....	177,417
Halibut	2,036,400	....	683,216
Mackerel	1,280,319	645,026	... .
Whitefish	929,962	....	124,963
Haddock	841,511	....	224,025
Smelts	810,392	....	172,408
Trout	682,619	....	26,459
Sardines	676,668	....	12,084
Hake and cusk	490,979	90,799	....
Pickerel	449,539	....	15,923
Pike	372,868	53,392	....
Clams and quahaugs	368,325	54,278	. ...,..
Pollock	187,723	9,429	....
Oysters	173,763	31,151	. ...

In the table which follows, the total results of the sea and inland fisheries are given separately. In the first two columns are shown the catch of all kinds of sea fish, and its value as realized at the vessel's or boat's side; while in the third and fourth columns are shown the various modes in which the catch was marketed, and the market value of each kind of fish. In the fifth and sixth columns are shown the quan-tity and value of all kinds of fresh-water fish caught and marketed. Such fish being practically all marketed by the fishermen in its fresh state, no distinction is made between the value of the catch as landed, and its marketed value. In the outer columns are shown the total marketed quantities of the various kinds of both sea and fresh-water fish and the market values of the same.

5 GEORGE V., A. 1915

RECAPITULATION.

Values of all Fish caught and landed in a Green State, and of the Quantities and Values of all Fish and Fish ed in a fresh, dried, pickled, canned, etc., state, for the **Whole of Canada**, during the year 1913-14.

Kinds of Fish.		Sea Fisheries. Caught and Landed. Quantity.	Value.	Sea Fisheries. Marketed. Quantity.	Value.	Inland Fisheries. Caught and Marketed. Quantity.	Value.	Both Fisheries. Total Marketed. Quantity.	Value.	Total Marketed Value.
			$		$		$		$	$
Salmon. used fresh.	cwts.	1,548,609	7,696,476	155,131	2,031,950			157,93	2,065,030	
„ canned.	cases			1,400,276	7,743,582	2,802	33,080	1,400,26	7,743,582	
„ salted (dry).	cwts.			125,021	661,210			125,01	661,210	
„ mild cured.	„			25,202	215,386			25,02	215,386	
„ smoked.				13,549	148,605			13,80	148,605	10,833,713
Lobsters. canned	cases	514,646	3,498,192	165,679	3,227,779			165,679	3,227,779	
„ shipped in shell.	cwts.			100,879	1,482,283			100,879	1,482,283	4,710,062
Cod. used fresh.	„	1,664,599	2,723,891	102,575	389,169			102,575	399,169	
„ green salted	„			91,852	302,129			91,852	302,129	
„ smoked.	„			1,128	6,640			1,128	5,640	
„ dried.	„			458,721	2,680,171			458,721	2,680,171	3,387,109
Haddock. used fresh.		405,633	779,903	146,207	337,534			146,207	337,534	
„ smoked.				27,563	171,123			27,563	171,123	
„ canned.	cases			6,947	41,662			6,347	41,662	
„ dried.	cwts.			64,312	290,792			64,312	290,792	441,511
Hake and Cusk. used fresh.	„	353,588	307,929	22,131	38,813			22,131	38,813	
„ „ dried.	„			110,405	452,165			110,405	452,165	490,979

Pollock	cwts.	150,094	147,667							
" used fresh	"			41,396	42,323			41,396	42,323	
" dried	"			36,200	145,400			36,200	145,400	187,723
Herring	"	2,352,605	1,907,754							
" used fresh	"			178,886	503,273	131,614	659,830	310,500	1,163,103	
" canned	cases.			4,936	19,274			4,936	19,274	
" smoked	cwts.			116,874	422,365			116,874	422,365	
" dry salted	"			313,178	470,379			313,178	470,379	
" pickled	brls.			131,275	525,190			131,275	525,190	
" used as bait	"			279,173	459,432			279,173	459,432	
" used as fertilizer				226,524	113,386			226,524	113,386	3,173,129
Mackerel	cwts.	215,442	999,269							
" used fresh	"			107,339	780,703			107,339	780,703	
" salted	brls.			36,015	496,072			36,015	496,072	
" canned	cases.			443	3,544			443	3,544	1,280,319
Shad	cwts.	3,865	25,842							
" used fresh	"			3,273	25,971	848	4,570	4,121	30,541	
" salted	brls.			199	2,985	51	255	250	3,240	33,781
Alewives	brls.	57,958	54,621							
" used fresh	"			14,809	19,284	3,810	7,620	18,619	26,904	
" salted	brls.			14,380	58,541			14,380	58,541	85,445
Sardines	"	141,384	282,768							
" canned	cases.			85,700	428,500			85,700	428,500	
" sold fresh	brls.			124,084	248,168			124,084	248,168	676,668
Halibut, used fresh	cwts.	255,096	1,407,052	256,096	2,036,400			256,096		2,036,400
Flounders	"	8,115	18,186	8,115	25,029			8,115		25,029
Smelts	"	88,273	521,423	88,273	809,027	455	1,365	88,728		810,392
Trout	"	4,673	43,764	4,673	50,677	68,491	631,942	73,164		682,619
Soles	"	216	324	216	1,080			216		1,080
Albacore	"	2,954	5,252	2,954	11,809			2,954		11,809
Oulachons	"	14,732	73,428	14,732	77,106			14,732		77,106
Sturgeon	"	1,229	9,755	1,229	17,385	3,582	45,105	4,811		6990
Bass	"	2,454	20,531	2,454	26,162	835	10,086	3,289		36,248
Eels	"	3,578	19,635	3,578	24868	5,048	32,032	8,626		56,900
Tom Cod	"	19,167	30,026	19,167	40,440			19,167		40,440
Swordfish	"	13,322	46,658	13,322	61,140			13,322		61,140
Whitefish	"					137,887	929,962	137,887		929,962
Pickerel	"					61,603	449,539	61,603		449,539
Perch	"					14,497	72,985	14,497		72,985
Pike	"					64,925	372,868	64,925		372,868
Tullibee	"					20,157	63,910	20,157		63,910
Maskinongé	"					130	1,659	130		1,659
Catfish	"					6,109	46,340	6,109		46,340

5 GEORGE V., A. 1915

RECAPITULATION—*Concluded*

Of the Quantities and Values of all Fish caught and landed in a Green State, and of the Quantities and Values of all Fish and Fish Products marketed in a fresh, dried, pickled, canned, etc., state, for the Whole of Canada, during the year 1913–14—*Concluded.*

Kinds of Fish		Sea Fisheries — Caught and Landed Quantity	Sea Fisheries — Caught and Landed Value $	Sea Fisheries — Marketed Quantity	Sea Fisheries — Marketed Value $	Inland Fisheries — Caught and Marketed Quantity	Inland Fisheries — Caught and Marketed Value $	Both Fisheries — Total Marketed Quantity	Both Fisheries — Total Marketed Value $	Total Marketed Value $
Octopus	cwts.	211	1,069	211	2,329			211		2,329
Gold eyes	"					5,089	12,721	5,089		12,721
Carp	"					6,721	33,606	6,721		33,606
Mixed Fish	"	19,731	57,530	19,731	76,822	99,161	316,629	118,892		393,452
Squid	brls.	2,197	7,001	2,197	9,187			2,197		9,187
Oysters	"	29,828	159,885	29,828	173,753			29,828		173,753
Clams, Quahaugs and Scallops	"	121,335	238,519							
" used fresh	cases			95,004	246,001			95,004	246,001	
" canned				26,323	122,324			26,323	122,324	368,325
Capelin (bait fd, etc.)	brls.	25,100	6,275	25,100	6,275			25,100		6,275
Dulse, C etc.	cwts.	10,755	31,267	7,473	51,795			7,473		51,795
Tongues and Sounds	"			4,041	49,811			4,041		49,811
Caviare	No.			2	532	954	8,561	571		9,093
Sturgeon Livers						453	272	453		272
Whales		792	272,400		8,122					8,122
Hair seal Skins				7,560	375			7,560		375
Beluga Skins				75	12,120			75		12,120
Fur seal Skins				404	296,119			404		296,119
Whale oil	gals.			452,666	149,022			452,666		149,022
Fish d				468,251	68,486			468,251		68,486
Fertilizer					7,478					7,478
Glue material	toms.			3,122				3,122		
Whale Bone and Meal	cwts.			10,094	16,003			10,094		16,003
Totals			21,385,192		29,472,811		3,734,937			33,207,748

REVIEW OF THE FISHERIES OF EACH PROVINCE.

NOVA SCOTIA.

The total marketed value of the fish and fish products of this province for the year 1913-14 amounted to $8,297,626.

This value shows a substantial increase over that for the preceding year of $913,571. Several kinds which showed greater values this year were: salmon, lobsters, cod, hake and cusk, mackerel, shad, halibut and clams. The values of mackerel and shad were nearly double that of last year.

There was an increase in the amount of capital invested in the fisheries of $578,620. This is shown chiefly in the increased value of gasoline boats, freezers, and ice-houses, smoke and fish-houses, and piers and wharves.

There were 882 vessels and carrying smacks, manned by 6,664 men; while on 12,908 sail and gasoline boats there were 15,648 men. On shore in the fish-houses, freezers, canneries, some 6,567 persons were employed; thus making a total of 28,879 persons engaged in the fisheries of this province.

District No. 1.

This district, which comprises the whole of the island of Cape Breton, shows a slight increase in the value of its fisheries for 1913-14, the total marketed value being $998,084 against $913,217 for the preceding year, an increase of $84,867.

Owing to unfavourable weather conditions and the dogfish pest, the catches were not so large.

An increase was shown in the catches of the following: salmon, cod, mackerel, herring and halibut.

The totals of the chief kinds landed in the district during the year, and those landed during the two preceding years were as follows:—

Kinds of fish.	1913-14.	1912-13.	1911-12.
	Cwt.	Cwt.	Cwt.
Salmon	2,406	1,903	2,690
Lobsters	51,426	53,221	49,250
Cod	114,043	101,696	146,440
Haddock	64,949	70,220	95,708
Hake,	7,388	6,541	6,384
Pollock	5,245	7,141	10,244
Herring	54,947	47,886	33,621
Mackerel	36,772	19,882	8,883

There were sixteen more men employed on vessels and smacks, 682 more on boats, and 1,088 more on shore in canneries, freezers, fish-houses, etc., making a total increase of 1,754 persons employed in the fisheries.

5 GEORGE V., A. 1915

A greater number and value of boats, lobster canneries, smoke and fish-houses, and piers and wharves are recorded.

This district shows a total increase in capital invested in the fisheries of $351,560.

District No. 2.

This district comprises the counties of Cumberland, Colchester, Pictou, Antigonish, Guysborough, Halifax and Hants. For the year under review a slight increase was recorded in the marketed value of the fish caught, the figures for 1913-14 being $2,207,721 against $2,176,181 for the previous, an increase of $31,540.

The catch of mackerel shows an increase from 19,441 cwts. to 59,225 cwts., with a corresponding higher value of $368,034 for the present year. Halibut also shows a substantial increase.

In the following table the catches of the chief kind of fish are shown, together with those for the two previous years:—

Kinds of fish.	1913-14.	1912-13.	1911-12.
	Cwt.	Cwt.	Cwt.
Lobsters	93,258	101,075	97,682
Mackerel	59,225	19,441	48,970
Cod	147,694	137,314	181,439
Halibut	21,962	13,692	17,794
Haddock	101,375	162,172	192,774
Herring	111,165	110,156	161,698

Fishing material, boats, vessels, etc., were valued this year at $1,971,321, against $1,993,889 for last year. The decrease of $22,568 is due chiefly to the fact that piers and wharves depreciated in value to quite an extent.

There were 771 men employed on vessels and smacks, 4,469 on boats, and 1,983 persons in freezers, fish-houses, canneries, etc., against 872, 4,608 and 2,033, respectively, last year. This gives a decrease of 290 persons engaged in the industry.

District No. 3.

The fisheries of this district, which comprises the counties of Lunenburg, Queens, Shelburne, Yarmouth, Digby, Annapolis and Kings, shows a considerable increase in the marketed value of the fisheries, the value this year being $5,091,821 compared with $4,294,657, or an increase of $797,164.

Lobsters, cod, herring, mackerel, hake and cusk, and halibut were caught in greater abundance than in the previous year.

One of the interesting features in the fishing industry in this part of the province is the development of the canned fish trade. Amongst the kinds being put up in tins are herring, mackerel, halibut and albacore.

SESSIONAL PAPER No. 39

The following table shows the landings of the chief kinds of fish during the year, as well as for the two preceding years:—

Kinds of fish.	1913-14.	1912-13.	1911-12.
	Cwt.	Cwt.	Cwt.
Lobsters	157,577	129,222	175,316
Cod	709,133	689,095	1,021,493
Haddock	221,062	239,880	217,876
Hake and cusk	203,838	167,998	135,218
Herring	220,361	218,105	180,043
Mackerel	66,610	45,263	8,899

The amount of capital invested in the fisheries in this district increased from $3,818,163 in 1912-13 to $4,066,791 for this year. The most important increase is shown in the value of freezers and fish-houses.

The number of persons employed in the industry, when compared with the previous year shows an increase of 877.

In Appendix No. 1 will be found full details of the Nova Scotia fisheries.

NEW BRUNSWICK.

The total marketed value of the fisheries for the province during the year 1913-14 was $4,308,707, or an increase of $44,653 over that for the previous year. Of this total value the sea fisheries contributed $4,266,759, and the inland fisheries $41,948. Each showing a slight increase.

There was a total capital investment in the industry in this province of $3,600,547, as compared with $3,508,889 for the previous year.

The value of gear, vessels, etc., in the sea fisheries is $3,491,334, while that for the inland section is $109,213.

The number of persons employed in the fisheries was 21,876, an increase of 201 during the year. Of this number 1,488 were employed on vessels and smacks, 14,052 on boats and 6,336 in fish-houses, canneries, freezers, etc.

District No. 1.

The total marketed value of the fisheries of this district, which comprises the counties of Charlotte and St. John, amounted to $1,572,119; a decrease of $40,480.

The following table shows the chief kinds landed during the year, and those landed during the two preceding years:—

Kinds of fish.	1913-14.	1912-13.	1911-12.
	Cwt.	Cwt.	Cwt.
Lobsters	11,751	12,410	8,539
Herring	197,297	189,200	190,660
Sardines (brl)	141,384	280,282	403,103
Pollock	70,802	47,954	58,210
Hake	65,180	97,524	79,412
Salmon	3,998	3,295	3,353
Cod	18,832	25,253	18,169

5 GEORGE V., A. 1915

There was $65,030 more invested in the fisheries in this district. It took 389 men to man the vessels and carrying ·smacks, 2,344 fishing in boats, and 1,034 persons employed in canneries, freezers, etc., making a total of 3,767 persons employed. This falls short of the previous total, by 143; the decrease being due to fewer persons being employed on shore in the fish-houses, canneries, etc.

District No. 2.

This district, which comprises the counties of Albert, Westmorland, Kent, Northumberland, Gloucester and Restigouche, shows a total marketed value of fish and fish products of $2,694,640. This shows a slight increase of $83,307 over the value for the preceding year. This increase was due to the higher price of lobsters, and to greater catches of herring, mackerel, clams and quahaugs.

The chief kinds of fish landed during the year, as compared with those landed during the two previous years, are shown in the following table:—

Kinds of fish.	1913–14.	1912–13.	1911–12.
	Cwt. ·	Cwt.	Cwt.
Salmon	13,090	10,004	9,144.
Lobsters	66,426	71,768	83,343
Cod	221,603	218,683	180,400
Herring	670,829	565,482	552,729
Mackerel	16,831	6,010	5,671
Smelts	66,059	79,854	64,179
Clams and quahaugs..................(brl.)	29,214	22,416	33,674

The value of fishing gear, boats, and other material was $1,567,460, as compared with $1,549,310 for the year previous.

There were 16,940 persons engaged in the industry, divided as follows: 1,050 men on vessels, 10,539 men on boats, 49 on carrying smacks, and 5,302 persons on shore in the fish-houses, freezers, canneries, etc. This gives an increase of 102 persons employed.

District No. 3.—(Inland.)

The total marketed value of the fisheries of this district, which includes the counties of Kings, Queens, Sunbury, York, Carleton, Victoria and Madawaska, amounted to $41,948, or an increase of $1,816 over the value for the preceding year. The fishermen in this district had a good season for salmon fishing and bass; the latter appearing in fairly large quantities.

SESSIONAL PAPER No. 39

The following table shows the catches of the chief kinds during the year, and during the two preceding years:—

Kinds of fish.	1913–14.	1912–13	1911–12.
	Cwt.	Cwt.	Cwt.
Salmon...	897	578	520
Trout....	728	574	579
Pickerel...	528	897	658
Alewives...	3,810	....	3,760

There was an increase in the capital investment of $8,468, and 237 more men were engaged on boats than during the previous year.

In Appendix No. 2 there will be found fuller details of the fisheries for New Brunswick.

PRINCE EDWARD ISLAND.

During the year 1913-14 the fisheries of this province were valued at $1,280,447, as compared with $1,379,906 for the preceding twelve months. This shows a decrease of $99,459, due to the falling off of the lobster, hake and smelt fishing. Big increases will be noted, however, in the return of cod, mackerel, clams and quahaugs, and oysters.

The following table shows the catches of the chief kind during the year under review, and the two preceding years:—

Kinds of fish.	1913-14.	1912-13.	1911-12.
	Cwt.	Cwt.	Cwt.
Lobsters........	92,898	136,992	118,090
Cod.............................	59,022	49,876	49,653
Herring..	85,295	83,391	79,178
Mackerel	11,496	5,448	5,005
Oysters..	12,951	8,631	8,835
Smelts	9,777	10,545	5,688
Clams and quahaugs............................	18,966	4,985	8,083

The capital invested in the fisheries increased from $851,070 to $948,667. The number of gasoline boats in use was increased by the addition of 361. To carry on the work of this industry 108 men were employed on the vessels and carrying smacks, 3,656 men on boats and 2,500 helpers on shore in the canneries, fish-houses, etc. This gives a total of 6,264 persons employed; as compared with 5,703 during the preceding year.

In Appendix No. 3 will be found more complete information on the fisheries of Prince Edward Island.

QUEBEC.

The total marketed value of the fisheries of this province amounted to $1,850,427, to which the sea fisheries contributed $1,736,581, and the inland fisheries $113,846. This shows a decrease in value of $137,814, due to the poor fishing season.

5 GEORGE V., A. 1915

Gulf Division—Sea Fisheries.

The value of the fisheries of this division, which comprises the counties of Bonaventure, Gaspé (including the Magdalen Islands), Rimouski, Chicoutimi and Saguenay (including the Island of Anticosti), shows a decrease of $135,810. This large decrease was caused by rough weather during which time the fishermen were unable to engage in their occupation, and the short season. Fishing started very late in the season and was practically over in September. Nearly all of the principal kinds of fish show decreased catches, with the exception of mackerel, which was more than double the amount landed last year.

There were 87 whales landed, this being three short of the number captured last year. Owing to higher prices which were prevalent, however, the results of this branch of the industry was about equal to that of the previous season.

In the following table there is shown the catches of the chief kinds for the year under review, together with those for the two years previous:—

Kinds of fish.	1913-14.	1912-13.	1911-12.
	Cwt.	Cwt.	Cwt.
Salmon	12,676	8,946	8,278
Lobsters	41,310	50,450	56,927
Cod	365,052	478,573	474,610
Herring	363,649	358,709	393,982
Mackerel	23,598	11,786	12,713
Smelts	12,146	4,019	3,540

The total capital invested in the fisheries of this district shows decrease from $1,379,689 to $1,331,656. The number of gasoline boats shows an increase in number of 70, and in value of $21,625.

There were 9,929 persons engaged in the industry, divided as follows: 149 men on vessels, 7,985 men on boats, 19 men in carrying smacks and 1,776 persons on shore in the canneries, fish-houses, etc. The total shows a decrease of 517 from last year.

Inland Fisheries.

The total value of what is known as the inland fisheries was $113,846, or a decrease of $2,006 from that of the previous year.

The following table shows the landings of the chief kinds of fish, compared with that for the two previous years:—

Kinds of fish.	1913-14.	1912-13.	1911-12.
	Cwt.	Cwt.	Cwt.
Pickerel	1,229	1,423	1,175
Trout	967	1,240	1,000
Eels	2,496	3,167	4,428
Sturgeon	977	1,742	2,095
Pike	935	855	914
Perch	1,823	1,722	1,726

In this division there were 1,024 men employed on boats, and 20 persons in fish-houses, freezers, canneries, etc.

There will be found fuller details of the fisheries of Quebec in Appendix No. 4.

ONTARIO.

The fisheries of this province are administered by the Provincial Government, this Department having three inspectors who exercise a general supervision only.

The Department is, therefore, indebted to the Provincial Superintendent of Game and Fisheries for the summary of the fisheries of the province, contained in this Report.

The value of the fisheries was $2,674,685, as compared with $2,842,877 for the previous year.

A table showing the catches of the chief kinds of fish taken for the past three years is given for the sake of comparison:—

Kinds of fish.	1913-14.	1912-13.	1911-12.
	Cwt.	Cwt.	Cwt.
Trout	62,204	63,707	65,120
Whitefish	52,263	58,897	44,540
Herring	130,718	170,677	131,020
Pickerel	26,564	26,656	20,225
Pike	34,547	24,732	20,985
Perch	12,427	13,931	9,572

In Appendix No. 5 will be found details of the fisheries of the province, together with reports by the three federal officers. Those who desire fuller information, however, should consult the report of Provincial Game and Fisheries Department at Toronto.

MANITOBA.

The fisheries of the province this year show a decrease in value from $800,149 in 1912-13 to $606,272. This difference was caused by the lessened catch of whitefish, owing to the stormy weather during the greater part of the summer fishing season.

The following table shows the catches of the chief kinds for the year under review, and the two preceding years:—

Kinds of fish.	1913-14.	1912-13.	1911-12.
	Cwt.	Cwt.	Cwt.
Whitefish	38,243	48,439	51,844
Pickerel	31,024	33,044	54,274
Pike	18,756	29,770	32,890
Tullibee	13,844	8,470	7,129

5 GEORGE V., A. 1915

There was an increase in the capital invested in the industry of $30,233. The number of persons employed was 1,448, distributed as follows: 92 men on vessels, 1,070 men on boats and 286 persons in freezers and fish-houses. This total falls short of that for the previous year by 208.

Appendix No. 6 gives fuller details of the fisheries of the province.

SASKATCHEWAN.

The marketed value of the fisheries of this province was $148,602; as compared with $111,839 for the previous year.

The catch of whitefish was slightly greater than during the previous year, but owing to the unfavourable condition of the roads, large quantities were held up until quite late, and consequently the prices paid to the fishermen were not so good as they otherwise might have been.

The following table shows a comparison between the catches of the chief kinds this year and for the two previous years:—

Kinds of fish.	1913-14.	1912-13.	1911-12.
	Cwt.	Cwt.	Cwt.
Whitefish..	30,993	23,120	30,856
Pike	7,936	5,197	5,975
Pickerel	1,710	2,193	2,656
Mixed fish..	4,984	2,915	3,195

There was a total of $30,941 invested in the fisheries as compared with $12,920 in 1912-13. The number of men licensed also increased from 484 to 645.

In Appendix No. 7 will be found interesting reports by the Chief Inspector and Inspector for the province, as well as statistics of the fisheries of the various districts.

ALBERTA.

The value of the fisheries of this province, for the year under review, show an increase of $29,703 over that for the previous year. The catch of whitefish increased from 8,048 cwts. to 14,012 cwts., while trout shows an increase of 1,188 cwts.

There was $15,878 invested in the fisheries this year compared with $9,744 for 1912-13. Sail boats show an increase of 154.

The number of men in boats was 4,130. In 1912-13 there were only 1,589.

Fuller details of the fisheries of this province will be found in Appendix No. 7, together with an interesting report on the *Inconnu*.

YUKON TERRITORY.

The value of the fisheries in the Yukon has decreased since 1912-13, when it was $111,239, to $68,265. Salmon, whitefish, pike and mixed fish show decreased values of $4,210, $12,124, $3,830 and $18,380, respectively.

There were $11,798 invested in the industry as compared with $10,975 for the preceding year.

Ninety-one more men were engaged in this occupation than in the previous year.

Appendix No. 8 contains fuller information of the fisheries of the Yukon.

BRITISH COLUMBIA.

The total marketed value of the fisheries of this province amounted to $13,891,398 for the year 1913-14. This was a decrease of $564,090 from the record established for the previous year.

The salmon catch in the southern part of British Columbia shows an increased value of $2,474,014; this was due to the year 1913 being a "big run" year in the Fraser river. In Northern British Columbia, or District No. 2, there was a decrease in the value of salmon of $2,040,037. In the Vancouver Island District the value shows an increase of $259,325.

Herring shows a falling off in value of $61,834.

The greatest decrease was in the halibut fishery, which has a value of $1,734,200 compared with $2,461,208 for the year previous.

The number of whales captured was 705, compared with 1,107 for 1912-13, which of course gives a corresponding lower market value to these mammals.

The catches of the chief kinds of fish are shown in the following table, also the catches of the same kinds for the two previous years :—

Kinds of fish.	1913-14.	1912-13.	1911-12.
	Cwt.	Cwt.	Cwt.
Salmon	1,509,354	1,221,057	1,103,666
Cod	29,220	28,580	25,065
Herring	649,062	729,567	545,442
Halibut	223,465	253,283	196,486

The capital invested in the fisheries increased from $9,941,049 in 1912-13 to $12,489,613.

The number of gasoline boats used increased from 1,334, with a value of $705,900, to 2,434, with a value of $1,018,150.

The number of persons employed was 20,707, or an increase of 5,079. They were divided as follows: 1,193 men on vessels, 10,055 men on boats, 68 men on carrying smacks, and 9,391 persons in fish-houses.

District No. 1.

The fisheries of this district show a substantial increase of $749,467, due to this being the year for the big run of salmon on the Fraser river. There were 732,059 cases of salmon packed, compared with 173,921 cases for 1912-13. The increase in value of the salmon fishery was $2,474,014.

5 GEORGE V., A. 1915

When the salmon are so plentiful the fishermen pay more attention to it than the other varieties, with the result that there is a decrease in the catches of the other kinds.

Herring was valued at $209,202, a decrease of $225,713 from the value of the previous year.

Halibut shows a large decrease, falling from $2,102,495 to $929,160.

In the following table there are shown the catches of the chief kinds compared with the two years previous :—

Kinds of fish.	1913-14.	1912-13.	1911-12.
	Cwt.	Cwt.	Cwt.
Salmon	797,524	410,000	445,355
Herring	29,502	46,800	19,822
Halibut	93,677	211,274	158,541
Cod	12,690	14,750	14,155
Sturgeon	1,090	5,051	5,168
Smelts	1,835	1,864	2,530

The value of gear, vessels, etc., increased from $3,895,938 in 1912-13 to $6,130,484. A very large increase of $1,516,410 is shown in the value of canneries.

There were 8,778 persons employed, as compared with 4,743 for the previous year. Of this number 143 were on vessels and carrying smacks; 5,142 on boats and 3,493 in fish-houses, freezers, canneries, etc.

District No. 2.

This district, which comprises the northern part of British Columbia and the Queen Charlotte Islands, shows a decrease in the value of the fisheries of $1,850,503. The value for the year under review being $3,230,788.

The salmon run was almost a failure; only 417,453 cases being packed, compared with 663,368 for the season of 1912-13. The value of salmon was $2,462,000, compared with $4,502,037 for the previous year. .

The value of herring fell off from $11,539 to $36,712; the catch decreasing from 166,787 cwts. to 62,240 cwts. The reason for this decrease being that the price paid for this fish, especially the salted herring, does not sufficiently pay the fishermen for their trouble.

Halibut increased in value from $203,553 to $537,440.

There were only 219 whales captured this year compared with 526 for the year preceding.

The following table gives the catches of the chief kinds during the past year, compared with the two previous years :—

Kinds of fish.	1913–14.	1912–13.	1911–12.
	Cwt.	Cwt.	Cwt.
Salmon	414,380	589,647	491,989
Halibut	107,488	29,079	27,945
Oulachons	13,950	13,800	15,000
Herring	62,240	166,787	26,410
Whales ....(Number.)	219	526	309

There was a slight increase in the capital invested in the fisheries of Northern British Columbia of $135,403.

Six hundred and seventy-five men were employed on vessels, 3,692 on boats, and 3.950 persons on shore in freezers, canneries, etc. This total number of persons engaged in the industry, 8,317, is greater than that for the preceding year by 318.

District No. 3.

This district, which comprises the Island of Vancouver and a portion of the mainland opposite thereto, had a marketed value of $3,647,823 for its fisheries for the year under review, compared with $3,110,877 for the previous year; making an increase of $536,946. The increase being mainly due to larger catches of salmon, herring, and halibut, with correspondingly greater values.

There were 486 whales captured this year, compared with 583 for the previous season.

The following table shows the catches of the chief kinds landed during the year, compared with the two previous years :—

Kinds of fish.	1913 14.	1912–13.	1911–12.
	Cwt.	Cwt.	Cwt.
Salmon	207,450	221,410	166,322
Cod	15,825	12,230	16,900
Herring	557,320	515, 80	499,210
Halibut	22,300	12,930	16,000
Clams and quahaugs ....(brl.)	10,000	8,865	4,630

There was $1,884,050 invested in the fisheries; compared with $1.705,435 for 1912-13, an increase of $178,615.

The number of persons employed in the industry in this part of the province was 3,612, an increase of 726 over last year. Four hundred and forty-three men were on vessels, 1,221 on boats and 1,948 persons in canneries, freezers, etc.

Appendix No. 9 contains statistics of the fisheries for this province, together with interesting reports by the fishery officers.

RECAPITULATION

By Provinces of the Quantities and Values of all Fish and Fish Products Marketed during the year 1913-14.

Number	Kinds of Fish	Nova Scotia Quantity	Nova Scotia Value	New Brunswick Quantity	New Brunswick Value	Prince Edward Island Quantity	Prince Edward Island Value	Quebec Quantity	Quebec Value	Ontario Quantity	Ontario Value	Number
1	Salmon, used fresh ...... cwts.	9,341	138,772	17,985	269,775	90	1,080	10,397	145,793			1
2	canned .... cases	24	183					1,576	15,760			2
3	salted (dry) .... cwts.		480									3
4	mild cured .... "											4
5	smoked .... "											5
6	Bass, canned .... cases	24	564									6
7	Cod, used fresh .... cwts.	87,449	1,280,393	24,586	491,720	37,150	743,180	16,485	313,215			7
8	shipped in shell ... "	84,063	2655	16,716	201,090			100	800			8
9	green salted .... "	58,345	292,070	11,387	22,774	3,989	7,933	250	375			9
10	smoked .... "	0,677	5,640	12,385	37,959	4,923	19,232	13,686	41,058			10
11	dried .... "	1,128	1,586									11
12	Haddock, used fresh . "	263,040	320,837	68,694	344,036	15,036	88,763	112,473	674,838			12
13	smoked .... "		7073	6,682	16,625	201	402	35	70			13
14	canned .... cases	26,833	41,662	730	3,650							14
15	Hake and Cusk used fresh cwts.	6,947	278,910	1,516	4,786	232	952	1,536	6,144			15
16	dried .... "	61,028	32,084	5,361	6,707	15	22					16
17	Pollock, used fresh . "	16,755	3285	23,750	81,815	8,379	33,516	800	2,400			17
18	dried .... "	77,476	4,576	37,747	37,747							18
19	Herring, used fresh . "	349	106,774	11,036	38,636							19
20	canned .... cases	2964	1651	78,847	78,847	4,151	6,056	2,221	3,779	130,718	658,038	20
21	smoked .... cwts.	52,549	12,614	1,332	6,660			1,300	8,600			21
22	dry salted .... "	3,604	49,454	91,025	265,961							22
23	pickled .... brls.	13,611										23
24	used as bait .... "											24
25	used as fertilizer .. "	49,240	8,927	69,177	277,316	519	2,256	10,696	42,784			25
26	Mackerel, used fresh "	78,140	149,246	72,620	116,556	39,789	73,945	61,780	92,670			26
27	canned .... brls.	306	298	126,890	63,870			99,038	49,518			27
28	Shad, canned ...... cases	87,229	581,103	17,067	169,070	2,978	29,780	75	750			28
29	salted .... "	443	3,544	232	2,736	2,848	51,264	7,841	125,456			29
30	Alewives, canned ... cases	25,094	646									30
31	salted .... brls.	943	9,338	3,007	20,405			60	630			31
32	used fresh .... cwts.	19	285	981	2,955	60	120					32
33	salted .... brls.	8,363	12,778	10,196	14,006	177	364					33
		3,743	11,807	10,460	46,380							

No.	Item	Unit											
34	Sardines, canned	cases	31,521		291,874	85,700	428,500			387	5,096		
35	Sardines, sold fresh	brls.	1,174		5,267	124,084	248,168	9,777	51,279	500	250		
36	Halibut, used fresh	cwts.	4,043		37,510	4,261	8,032	122	1,112	12,601	98,533	62,204	579,832
37	Flounders	"	1,005		12,645	60,117	601,170			1,677	19,841		
38	Smelts	"	216		1,080	2,843	28,430						
39	Trout	"	2,934		11,809								
40	Soles	"											
41	Albacore	"											
42	Oulachans	"	34		510	175	1,225			977	6,383	2,635	38,022
43	Sturgeon	"	198		1,915	1,850	21,882	242	2,258	676	8,496		
44	Bass	"	1,111		5,665	2,122	16,430	45	90	2,781	18,326	2,370	14,221
45	Eels	"	300		518	18,084	36,168			320	320		
46	Tom Cod	"	13,322		61,140								
47	Swordfish	"				26	390			514	5,140	53,263	520,123
48	Whitefish	"				598	5,280			1,229	14,330	26,564	265,645
49	Pickerel	"				4	16			1,823	9,360	12,427	62,137
50	Perch	"								935	6,179	34,547	276,378
51	Pike	"										5,738	34,429
52	Tullibee	"								115	1,284		
53	Maskinonge	"								197	985	5,264	42,115
54	Catfish	"											
55	Octopus	"											
56	Gold Eyes	"											
57	Carp	"	5,566		4,903	542	542	170	170	8,963	37,755	6,721	33,606
58	Mixed Fish	brls.	2,167		9,047	30	120					28,291	141,436
59	Squid	"	3,397		14,064	10,800	64,800	12,951	85,509				
60	Oysters	"											
61	Clams, Quahaugs and Scallops used fresh.	"	27,913		49,941	38,070	81,753	18,671	74,684	1,111	4,444		
62	" canned.	cases	175		788	18,530	90,484	290	1,740	25,100	6,275		
63	Capelin (bait fish)	brls.											
64	Dulse, kelp, etc	cwts.	597		2,673	2,788	11,122	50	1,500	63	315	84	8,411
65	Tongues and Sounds	"	2,874		28,026	1,054	19,970					453	272
66	Caviare	No.	2		632	12	150						
67	Sturgeon Livers	"											
68	Whales	"	168		184					4,872	7,398		
69	Hair Seal Skins	"								75	375		
70	Beluga Skins	"											
71	Fur Seal Skins	"											
72	Whale Oil	gals.	172,941		56,895	50,242	15,073	10,618	3,245	147,560	44,268		
73	Fish Oil	"	1,229		9,147					90,400	27,119		
74	Fertilizer	tons			7,478					244	85		
75	Glue Material	"											
76	Whale Bone and Meal	cwts.								240	4,800		
	Totals		8,297,626				4,308,707		1,280,447		1,850,427		2,674,685

RECAPITULATION—Concluded.

By Provinces of the Quantities and Values of all Fish and Fish Products Marketed during the Year 1913-14—Concluded.

Number	Kinds of Fish		Manitoba Quantity	Manitoba Value	Saskatchewan Quantity	Saskatchewan Value	Alberta Quantity	Alberta Value	Yukon Quantity	Yukon Value	British Columbia Quantity	British Columbia Value	Number
				$		$		$		$		$	
1	Salmon, used fresh	cwts.							1,820	18,200	118,300	1,491,410	1
2	" canned	cases									1,400,252	7,743,399	2
3	" mild (dry)	cwts.									123,445	645,460	3
4	" mild used	"									25,202	215,386	4
5	" smoked	"									13,525	148,025	5
6	Lobsters, canned	cases											6
7	" shipped in shell	cwts.											7
8	Cod, used fresh	cwts.									28,624	256,027	8
9	" green salted	"									181	1,810	9
10	" smoked	"											10
11	" dried	"									78	1,048	11
12	Hake, used fresh	"											12
13	" smoked	"											13
14	" canned	cases											14
15	" dried	cwts.											15
16	Hake and Cusk used fresh	dried											16
17	" dried												17
18	Pollock, used fresh												18
19	" dried												19
20	Herring, used fresh	cwts.									42,014	355,732	20
21	" canned	cases											21
22	" smoked	cwts.									7,038	98,350	22
23	" dry salted										313,178	470,379	23
24	" pickled	brls.									1,643	4,107	24
25	" used as bait										26,935	27,015	25
26	" used as fertilizer												26
27	Mackerel, used fresh	cwts.											27
28	" salted	brls.											28
29	" canned	cases											29
30	Shad, used fresh	cwts.											30
31	" salted	brls.									11	168	31
32	Alewives, used fresh	cwts.											32
33	" salted	brls.											33

No.	Item	Unit	1	2	3	4	5	6	7	8	9	10
34	Sardines, canned	cases										
35	Sardines sold fresh	brls.									223,465	1,734,200
36	Halibut, used fresh	cwts.									2,180	11,480
37	Flounders	"									2,190	21,900
38	Smelts	"	1,505	7,525	388	1,615	2,428	16,209	271	8,160	721	7,210
39	Trout	"										
40	Sole	"									14,732	77,106
41	Albacore	"									1,099	16,350
42	Oulachans	"									565	3,965
43	Sturgeon	"										
44	Bass	"										
45	Eels	"									418	3,314
46	Tom Cod	"										
47	Swordfish	"										
48	Whitefish	"	38,243	223,391	30,923	102,817	14,012	51,201	836	20,900		
49	Pickerel	"	31,024	155,020	1,710	6,941	543	2,203	5	100		
50	Perch	"	243	972			2,749	9,371	2			
51	Pike	"	18,756	56,362	7,936	24,622	290	770	15	50		
52	Tullibee	"	13,844	27,696	285	1,015				375		
53	Maskinonge	"										
54	Catfish	"	648	3,240							211	2,329
55	Cods	"										
56	Gold Eyes	"	5,089	12,721								
57	Carp	"	57,576	113,439	1,984	11,592					10,475	61,549
58	Mixed Fish	"					1,302	1,565	1,024	20,480	2,680	9,380
59	Squid	"										
60	Oysters	brls.									9,239	35,179
61	Clams, Quahaugs and Scallops, used fresh	"									7,328	29,212
62	Capelin (bait fish), canned	cases										
63	" "	brls.									4,088	38,000
64	Dulse, Crabs, Cockles, etc.	cwts.										
65	Tongues & Sounds	"										
66	Caviare	"										
67	Sturgeon Livers	No.										
68	Whales	"									2,520	630
69	Hair Seal Skins	"										
70	Beluga Skins	"									404	12,120
71	Fur Seal Skins	"									395,096	251,901
72	Whale Oil	galls.									144,050	4,690
73	Fish Oil	"									1,649	59,254
74	Fertilizer	tons.										
75	Glue Material	"									9,854	11,203
76	Whale Bone and Meal	cwts.										
	Totals			606,272		148,602		81,319		68,265		13,891,398

NUMBER OF PERSONS EMPLOYED AND AMOUNT OF CAPITAL INVESTED IN THE FISHERIES.

There was a total number of 98,669 persons engaged in the fisheries of the Dominion during the year under review. Of this number 86,486 were engaged in the sea and 12,183 in the inland fisheries.

This total number is greater than that for the previous year by 10,261, the sea fisheries engaging 9,542 more persons and the inland fisheries 719.

Of the total, 9,927 men were employed on vessels, 61,251 on boats, 598 on carrying smacks, and 26,893 in canneries, freezers, fish-houses, etc.

The capital invested increased from $24,388,459 in 1912-13 to $27,464,033. Of this amount there was $25,371,480 invested in the sea fisheries, and $2,092,553 in the inland fisheries.

Of the total capital invested $13,866,780 represents the value of vessels, boats, gear, etc., while $13,597,253 is the amount invested in canneries, freezers, wharves, fish-houses and other fixtures necessary to the carrying on of the industry.

The following tables show the details of the number and value of boats, vessels, gear, etc., and the number of persons employed in the fisheries for the whole of Canada, table No. 1 giving the sea and inland fisheries separately while table No. 2 shows the totals by provinces.

SESSIONAL PAPER No. 39

TABLE No. 1.

RECAPITULATION

Of the Number of Fishermen, etc., and of the Number and Value of Fishing Vessels, Boats, Nets, Traps, etc., used in the Sea and Inland Fisheries in the whole of Canada, for the year 1913-14.

	Sea Fisheries.		Inland Fisheries.		Total Both Fisheries.	
	Number.	Value.	Number.	Value.	Number.	Value.
		$		$		$
Steam fishing vessels............	75	1,177,575	199	518,180	274	1,695,755
Sailing and gasoline vessels............	1,247	2,504,759	....	....	1,247	2,504,759
Boats, (sail)............	24,952	1,077,453	4,034	171,136	28,986	1,248,589
" (gasoline)............	8,222	2,376,644	478	208,945	8,700	2,585,589
Carrying smacks.	471	244,745	....	.... ..	471	244,745
Gill nets, seines, trap and smelt nets, etc.	144,896	2,626,396	69,342	796,536	214,238	3,422,932
Weirs.......	720	376,170	114	28,000	834	404,170
Trawls.............................	18,913	192,221	....	....	18,913	192,221
Spears............................	.... ..	......	103	260	103	260
Skates of gear....	1,888	37,760	....	... .	1,888	37,760
Hand lines........................	64,266	51,669	7,976	13,233	72,242	64,902
Lobster traps......................	1,617,195	1,464,920	.... ..	....	1,617,195	1,464,920
" canneries..............	722	685,325	....	....	722	685,325
Salmon "	81	4,115,410	....	....	81	4,115,410
Clam "	19	29,950	....	......	19	29,950
Fish "	2	2,800	....	. ...	2	2,800
Sardine "	6	362,100	....	....	6	362,100
Freezers and ice houses......	817	2,399,560	507	199,020	1,324	2,598,580
Smoke and fish-houses....	7,496	1,533,712	209	37,090	7,705	1,570,802
Fishing piers and wharves.......... .	2,552	3,501,561	141	38,775	2,693	3,540,336
Salteries..........................	12	1,200	....	... .	12	1,200
Whaling stations..	5	550,000	....	....	5	550,000
Oil factories......................	1	40,000	....	....	1	40,000
Fishing huts and cottages....	......	......	102	81,200	102	81,200
Scows, pile drivers, etc..............	501	19,550	....	....	501	19,550
Eel traps...	....		98	178	98	178
Totals.....	... .	25,371,480	....	2,092,553	....	27,464,033
Number of men employed on vessels...	9,091	....	836	....	9,927	
" " boats.....	50,227	....	11,024	....	61,251	
" " carry. smacks	598	....	....	....	598	
" persons employed in fish houses, freezers canneries, etc......	26,570	....	323	... .	26,893	
Total...................	86,486	....	12,183	.... . .	98,669	

5 GEORGE V., A. 1915

TABLE No. 2.

RECAPITULATION

By Provinces of the number and value of Fishing Implements, Vessels, Boats, etc., used in the Fishing Industry of Canada during the year 1913–14, and of the number of persons employed.

Provinces	Persons employed — No. in Vessels, etc.	Persons employed — No. Boats	Persons employed — No. in Canneries, Fish-Houses	Vessels, Tugs and Carrying Smacks — Number	Vessels, Tugs and Carrying Smacks — Value $	Boats — Gasoline	Boats — Sail	Boats — Total Value $	Value of gill-nets, seines, trap and smelt nets, etc. $	Value of hand line, weirs, and trawls, etc. $	Value of lobster plant. $	Approximate value of salmon and other canneries, freezers, fish houses and fixtures. $	Total value. $
Nova Scotia	6,664	15,648	6,567	882	1,801,914	3,481	9,427	1,101,171	687,189	212,075	1,015,634	2,292,227	7,110,210
New Brunswick	1,488	14,052	6,336	436	305,830	1,186	7,965	628,345	625,946	365,055	403,231	1,268,140	3,600,547
Prince Edward Island	108	3,636	2,560	30	13,680		1,110	220,964	44,923	9,293	516,025	143,762	948,667
Quebec	168	9,009	1,796	41	81,250	970	5,102	331,007	270,442	66,672	215,355	481,145	1,445,871
Ontario	744	2,767	286	190	433,180	247	1,224	268,291	645,353	887		158,870	1,506,581
Manitoba	92	1,070		9	85,000	366	408	28,750	89,490	112		100,575	303,927
Saskatchewan		645				3	351	12,053	16,573	1,330			30,941
Alberta		4,130				10	205	4,682	8,796			2,400	15,878
Yukon		219	17			3	118	3,140	3,274	159		5,225	11,798
British Columbia	1,261	10,055	9,391	404	1,720,405	2,434	3,076	1,235,755	1,031,124	43,730		8,456,599	12,489,613
Totals	10,525	61,251	26,993	1,992	4,445,259	8,700	28,986	3,834,178	3,423,110	699,313	2,150,245	12,911,928	
Grand total value													27,464,033

COMPARATIVE TABLE showing the total Value of the Fisheries in the respective Provinces of Canada, from 1870 to 1913-14 inclusive, as compiled from the Annual Reports of the Department of Marine and Fisheries.

Year	Nova Scotia	New Brunswick	Prince Edward Island	Quebec	Ontario	British Columbia	Manitoba, Saskatchewan, Alberta and Yukon	Total for Canada
1870	$ 10,025	$ 1,131,433	$ No data	$ 1,161,551	$ 264,982	$ No data	$ No data	$ 8,577,391
1871	5,101,030	1,185,033		1,093,612	193,524	"	"	7,353,B9
1872	69,635	1,965,459	207,595	1,390,189	267,633	"	"	9,570,116
1873	67,385	2,285,662	288,863	1,391,564	293,091	"	"	10,754,997
1874	29,02	2,685,794	298,927	1,608,660	446,267	"	"	11,681,886
1875	5,573,451	2,427,654	494,967	1,586,759	453,194	"	"	10,350,385
1876	690	1,953,388	763,036	2,097,068	437,229	104,697	"	1,000
1877	79,558	2,133,237	840,344	2,560,147	438,223	583,433	"	12,005,934
1878	6,131,600	2,305,790	1,402,301	2,664,055	348,122	925,767	"	13,215,678
1879	21,37	2,554,722	1,675,089	2,920,395	367,133	631,746	"	13,529,254
1880	6,291,061	2,744,447	1,955,290	2,631,556	444,491	713,335	"	14,499,979
1881	4,82	2,930,904	1,855,687	2,751,962	9003	1,454,321	"	15,817,162
1882	7,131,418	3,192,339	1,272,468	1,976,516	5857	1,842,675	"	16,824,092
1883	7,689,374	3,185,674	1,085,619	2,158,997	1233	1,644,646	"	16,958,192
1884	8,763,779	3,730,454	1,293,430	1,694,561	3624	1,358,267	"	17,766,404
1885	8,283,922	4,005,431	1,141,991	1,719,460	1,342,692	1,078,038		17,722,973
1886	8,415,362	4,180,227	1,037,426	1,741,382	1,435,998	1,577,348	186,980	18,679,288
1887	8,379,782	3,559,507	876,862	1,773,567	1,531,850	1,974,887	129,084	18,386,103
1888	7,817,030	2,941,863	886,430	1,860,012	1,839,869	1,902,195	180,677	17,418,510
1889	6,346,722	3,067,039	1,041,109	1,876,194	1,963,123	3,348,067	167,679	17,655,236
1890	6,636,444	2,699,055	1,238,733	1,615,119	2,009,637	3,481,432	232,104	17,714,902
1891	7,011,300	3,571,050	1,179,856	2,008,678	1,806,389	3,9055	332,969	18,977,878
1892	6,340,724	3,203,922	1,133,368	2,236,732	2,042,198	2,849,483	1, 8954	18,941,171
1893	6,407,279	3,746,121	1,119,738	2,218,905	1,694,930	3,9963	1,042,093	20,686,661
1894	6,547,387	4,351,526	976,836	2,303,386	1,659,968	3,950,478	787,087	20,719,573
1895	6,213,131	3,868	976,126	1,867,920	1,584,473	4,401,354	752,466	20,199,338
1896	6,070,495	4,799,423	954,949	2,025,754	1,605,674	4,183,999	745,543	20,407,425
1897	8,090,346	3,934,135	1,070,202	1,737,011	1,289,822	4,205	638,418	22,783,546
1898	7,347,604	4,3068	1,043,645	1,761,440	1,433,682	3,713,101	613,355	19,667,121
1899	7,226,034	93,891	1,050,9	1,953,134	1,590,447	5,214,074	3911	21,891,706
1900	7,980,548	3,769,742	1,050,623	1,989,279	1,333,294	4,878,820	718,159	21,557,639
1901	7,909,152	4,193,364	887,024	2,174,459	1,428,078	7,942,971	958,410	25,737,153
1902	7,351,753	3,912,514	1,099,510	2,059,175	1,265,706	5,284,824	1,158,437	21,959,433
1903	7,841,602	4,186,800	1,077,546	2,211,792	1,535,144	4,5865	1,478,666	23,101,878
1904	7,287,099	4,16064	1,077,546	1,751,397	1,793,229	5,219,107	1,716,977	23,516,439

COMPARATIVE TABLE showing the total Value of the Fisheries in the respective Provinces of Canada, from 1870 to 1913-14 inclusive, as compiled from the Annual Reports of the Department of Marine and Fisheries—*Concluded.*

Year.	Nova Scotia.	New Brunswick.	Prince Edward Island.	Quebec.	Ontario.	British Columbia.	Manitoba, Saskatchewan, Alberta and Yukon.	Total for Canada.
1905	$ 8,259,085	$ 4,847,090	$ 998,922	$ 2,003,716	$ 1,708,963	$ 3,850,216	$ 1,811,570	$ 29,479,562
1906	7,798,160	4,965,225	1,168,939	2,175,035	1,734,856	7,003,347	1,492,923	26,279,485
1907-08	7,632,330	5,300,564	1,492,695	2,047,390	1,935,025	6,122,923	965,422	25,499,349
1908-09	8,069,838	4,754,298	1,378,624	1,881,817	2,100,078	6,465,038	861,392	25,451,085
1909-10	8,081,111	4,676,315	1,197,556	1,908,436	2,177,813	10,314,735	1,373,181	29,629,169
1910 11	10,119,243	4,134,144	1,153,708	1,692,475	2,025,121	9,163,235	1,676,607	29,965,433
1911-12	9,367,550	4,886,157	1,196,396	1,868,136	2,205,436	13,677,125	1,467,072	34,067,872
1912-13	7,384,055	4,264,054	1,379,905	1,988,241	2,842,878	14,455,488	1,074,843	33,389,461
1913-14	8,297,626	4,308,707	1,280,447	1,850,427	2,674,685	13,891,398	904,468	33,207,748

SESSIONAL PAPER No. 39

COMPARATIVE TABLE showing Number and Value of Vessels and Boats engaged in the Fisheries of Canada, together with the Value of Fishing Material used, since 1880.

Year.	Vessels.			Boats.		Value of Nets and Seines.	Value of other Fishing Material.	Total Capital Invested.
	Number	Tonnage.	Value.	Number	Value.			
			$		$	$	$	$
1880...	1,181	45,323	1,814,688	25,266	716,352	985,978	419,564	3,936,582
1881...... ..	1,120	48,389	1,765,870	26,108	696,710	970,617	679,852	4,113,049
1882.........	1,140	42,845	1,749,717	26,747	833,137	1,351,193	823,938	4,757,985
1883.........	1,198	48,106	2,023,045	25,825	733,186	1,243,366	1,070.930	5,120,527
1884.......	1,182	42,747	1,866,711	24,287	741,727	1,191,579	1,224,646	5,014,663
1885...	1,177	48,728	2,021,633	28,472	852,257	1,219,284	2,604,285	6,697,459
1886..	1,133	44,605	1,890,411	28,187	850,545	1,263,152	2,720,187	6,814,295
1887.........	1,168	44,845	1,989,840	28,092	875,316	1,499,328	2,384,356	6,748,840
1888.........	1,137	33,247	2,017,558	27,384	859,953	1,594,992	2,390,502	6,863,005
1889...	1,100	44,936	2,064,918	29,555	965,010	1,591,085	2,149,138	6,770,151
1890.........	1,069	43,084	2,152,790	29,803	924,346	1,695,358	2,600,147	7,372,641
1891.......	1,027	39,377	2,125,355	30,438	1,007,815	1,644,892	2,598,124	7,376,186
1892.........	988	37,205	2,112,875	30,513	1,041,972	1,475,043	3,017,945	7,647,835
1893.........	1,104	40,096	2,246,373	31,508	955,109	1.637,707	3,174,404	8,681,557
1894.........	1,178	41,768	2,409,029	34,102	1,009,189	1,921,352	4,099,546	9,439,116
1895.........	1,121	37,829	2,318,290	34,268	1,014,057	1,713,190	4,208,311	9,253,848
1896.........	1,217	42,447	2,041,130	35,398	1,110,920	2,146,934	4,527,267	9,826,251
1897..... ..	1,184	40,679	1,701,239	37,693	1,128,682	1,955,304	4,585,569	9,370,794
1898.........	1,154	38,011	1,707,180	38,675	1,136,943	2,075,928	4,940,046	9,860,097
1899.........	1,178	38,508	1,716,973	38,538	1,195,856	2,162,876	5,074,135	10,149,840
1900...	1,212	41,807	1,940,329	38,930	1,248,171	2,405,860	5,395,765	10,990,125
1901.........	1,231	40,358	2,417,680	38,186	1,212,297	2,312,187	5,549,136	11,491,300
1902... ...	1,296	49,888	2,620,661	41,667	1,199,598	2,103,621	5,382,079	11,305,959
1903	1,348	42,712	2,755,150	40,943	1,338,003	2,305,444	5,842,85	12,241,454
1904	1,316	43,025	2,592,527	41,938	1,376,165	2,189,666	6,198,584	12,356,942
1905.........	1,384	41,640	2,813,334	41,463	1,373,337	2,310,508	6,383,218	12,880,897
1906... ..	1,439	40,827	2,841,875	39,634	1,462,374	2,426,341	7,824,975	14,555,565
1907–08......	1,390	36,902	2,731,888	38,711	1,437,196	2,266,722	8,374,440	14,826,592
1908–09......	1,441	40,818	3,571,871	39,965	1,696,856	2,283,127	7,957,500	15,508,275
1909–10......	1,750	37,662	3,303,121	41,170	1,855,629	2,572,820	9,626,362	17,357,932
1910–11..	1,680	38,454	3,028,625	38,977	2,483,996	2,786,548	10,720,701	19,019,870
1911–12.. ...	1,648	....	3,502,928	36,761	2,695,650	2,453,191	12,281,135	20,932,904
1912–13......	1,669	....	4,671,923	34,501	3,072,115	4,154,880	12,489,541	24,388,459
1913–14.......	1,992	... .	4,445,259	37,686	3,834,178	3,423,110	15,761,486	27,464,033

5 GEORGE V., A. 1915

COMPARATIVE TABLE showing the Number of Persons employed in the Fishing Industry since 1895.

Year.	Number of Persons in Canneries and Fish-houses.	Number of Men in Vessels.	Number of Men in Boats.	Total Number of Fishermen.	Total Number of Persons in Fishing Industry.
1895	13,030	9,804	61,530	71,334	84,364
1896	14,175	9,735	65,502	75,237	89,412
1897	15,165	8,879	70,080	78,959	94,124
1898	16,548	8,657	72,877	81,534	98,082
1899	18,708	8,970	70,893	79,893	98,601
1900	18,205	9,205	71,859	81,064	99,269
1901	15,315	9,148	69,142	78,290	93,605
1902	13,563	9,123	68,678	77,801	91,364
1903	14,018	9,304	69,830	79,134	93,152
1904	13,981	9,236	68,109	77,345	91,326
1905	14,037	9,366	73,505	82,871	96,908
1906	12,317	8,458	67,646	76,104	88,421
1907-08	11,442	8,089	63,165	71,254	82,696
1908-09	13,753	8,550	62,520	71,070	84,823
1909-10	21,694	7,931	60,732	68,663	90,357
1910-11	24,978	8,521	60,089	68,610	93,588
1911-12	25,206	9,056	56,870	65,926	91,132
1912-13	23,327	9,076	56,005	65,081	88,408
1913-14	26,893	10,525	61,251	71,776	98,669

OTHER APPENDICES.

IMPORTS AND EXPORTS OF FISH.

Statements showing the quantities of the chief commercial fish and fish products imported into Canada for home consumption, and the quantities of the chief commercial fish and fish products, the produce of Canada, exported during the fiscal year 1913-14 will be found in Appendix No. 10.

The quantities of the different kinds exported in any one year do not necessarily bear any relation to the quantities caught in that year, for the reason that the products may not be all exported during the year in which the fish are caught.

The figures in this appendix are taken from the report of the Customs Department and are reproduced in a convenient form merely for the purpose of showing to what countries the various products of the fisheries are sent.

THE FISHERIES PATROL SERVICE.

For the prevention of illegal fishing and for the general enforcement of the fisheries regulation in the inland waters, and for the prevention of illegal lobster and other fishing on the sea coast, the Fisheries Branch has under its control in the various provinces a number of motor launches and small steamers. Reports on the work of these craft during the year 1913-14 will be found in Appendix No. 11.

OYSTER CULTURE.

In Appendix No. 12 will be found a report by the Department's Oyster Expert, on his work during the season of 1913.

FISH BREEDING.

The annual report on the work carried on in connection with the breeding of fish in the various establishments throughout the Dominion during 1913-14 will be found in Appendix No. 13.

FISHERIES MUSEUM.

In Appendix No. 14 will be found a report on the condition of the Fisheries Museum at Ottawa by the Department's Naturalist.

FISHERIES EXPENDITURE AND REVENUE.

A statement of the total expenditure and revenue in connection with the fisheries of Canada during the fiscal year ended March 31, 1914, forms Appendix No. 15 of this report.

The expenditure amounted to $1,070,857.94 divided amongst the various services as follows: Salaries and disbursements of fishery officers, $229,547.16; fish breeding, $354,675.13; miscellaneous expenditure, $486,635.65; and $158,661.25 distributed as fishing bounty.

5 GEORGE V., A. 1915

The total revenue from fishing licenses, fines, etc., in the different provinces was $110,994.63 which includes the sum of $11,728.50, paid by United States fishing vessels, as "modus vivendi" fees.

FISHING BOUNTY.

The fishermen of the Maritime Provinces received the sum of $158,661.25 as bounty on their respective catches of sea fish during the year 1913.

The number of claims received during the year was 13,412; being greater than that for the preceding year by 441. The number of claims paid was 13,533, which includes a number held over from 1912, and makes an increase of 569.

The sum of $60,887.10 was paid 910 vessels and their crews; a decrease of 55 compared with the preceding year..

To boats and boat-fishermen was paid the sum of $97,774.15, the number of boats being 12,623, and of boat-fishermen 21,557, an increase of 625 boats and 1,146 men.

The amount of bounty expended in each province for 1913 was as follows:—

Nova Scotia, $93,456; New Brunswick, $16,385.05; Prince Edward Island, $11,081.85; Quebec, $37,738.35.

Since the inception of the system in 1882, the sum of $5,058,861.62 has been paid to fishermen and vessel and boat owners, with a view to encouraging them in the development of their industry.

The regulations governing the payment of the bounty as well as the particulars respecting its distribution from appendix No. 16.

UNITED STATES FISHING VESSEL ENTRIES, ETC.

In Appendix No. 17 will be found lists of United States fishing vessels which made use of Canadian ports, and of United States fishing vessels to which "modus vivendi" licenses were issued during the year 1913-14.

On the Atlantic coast an aggregate of 1,349 entries were made by 219 vessels against 1,890 entries by 300 vessels during the preceding year.

"Modus vivendi" licenses were issued to 94 United States vessels during 1913-14, the revenue from which amounted to $11,728.50, there being a decrease of 18 in the number of vessels and of $1,771.50 in the amount of revenue.

THE OUTSIDE STAFF.

The names of the various inspectors of fisheries and fishery overseers, with the districts over which they have jurisdiction, as well as a list of officers in charge of fish hatcheries, and of officers in charge of fisheries patrol boats, will be found in Appendix No. 18.

I have the honour to be, sir,
Your obedient servant,

A. JOHNSTON,
Deputy Minister of Marine and Fisheries.

PHOTOGRAPHS

A portion of mounted fish exhibit, Toronto Exhibition

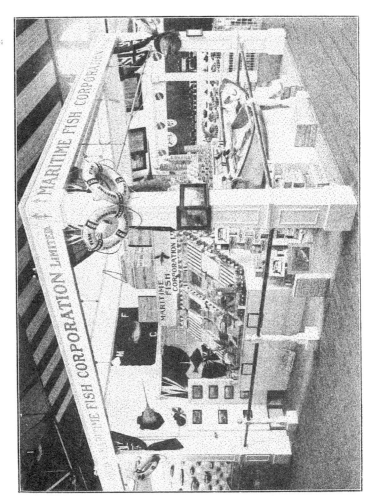

Fisheries exhibit, Toronto Exhibition.

Fisheries Exhibit, Toronto Exhibition.

Fisheries exhibit, Toronto Exhibition.

Fisheries exhibit, Toronto Exhibition.

A portion of the Lunenburg fishing fleet.

6.

Hauling in a halibut.

Deep sea fishing —Transferring the fish from the dory
to the schooner.

Deep sea fishing—Full decks.

APPENDIX No. 1.

NOVA SCOTIA.

District **No. 1.**—Comprising the four counties of Cape Breton Island. Inspector, A. G. McLeod, Whitney Pier.

District **No. 2.**—Comprising the counties of Cumberland, Colchester, Pictou, Antigonish, Guysboro, Halifax and Hants. Inspector R. Hockin, Pictou.

District **No. 3.**—Comprising the counties of Kings, Annapolis, Digby, Yarmouth, Shelburne, Queens and Lunenburg. Inspector Ward Fisher, Shelburne.

REPORT ON THE FISHERIES OF DISTRICT No. 1.

To the Superintendent of Fisheries,
       Ottawa.

SIR,—I have the honour to. submit my second annual report as fishery inspector for District No. 1 (the Island of Cape Breton), province of Nova Scotia, for the year ended March 31, 1914, together with tabulated data indicating the quantities and values of fish caught in the four counties, and in the several sections of each county, within this district, the materials used, and the persons employed in these fisheries.

Cape Breton county.—On the whole the fishery in this section of the Island was rather below the average, the lessened catch being due to an extended period of stormy weather, scarcity of bait, and the fact that the fishermen were mostly unable to set their nets on account of the presence of dogfish.

A slight increase in the lobster catch was reported from, and one new cannery was established and put in operation at, Port Morien.

Main-a-Dieu reports indicate a slight increase in the returns from cod and haddock. Dogfish appeared inshore about June 10, and remained on the coast until the latter part of November; because of this pest the mackerel and herring fisheries in this vicinity proved almost a total failure, the fishermen being prevented from putting out their nets. The salmon in this locality showed an increase over that of the previous season. Ten motor boats were added to the fishing fleet at this place.

Bad weather prevailed in the vicinity of Gabarus and occasioned losses to both gear and catch. Five new gasolene boats were added to the fleet operating from this point. A severe storm developed on the evening of May 29, and continued in unabated force until the following midnight. Destruction to fishing gear and property to the extent of $7,500 resulted. Similarly in June the loss was $370.

From the Florence station the lobster catch was reported below the average, due to the month of June being too blustery for operations. Cod fishing was satisfactory, herring bait being always available. During the past two years July herring have been a negligible quantity. Four gasolene boats were added to the fleet from this section. At Scatterie island the fishermen had a satisfactory season, the catch being well up to the average and the market price being higher than the 1913. During the severity of the storm the latter part of May about $500 damage was caused to fishing

5 GEORGE V., A. 1915

property here. Between Big Lorraine and Mira bay the property loss amounted to about $1,800.

The codfishery, as reported by the overseer at Jacksonville, was better than during the previous season, while the mackerel and herring catches were practically nil.

At Little Bras D'Or the season opened most auspiciously. From April 15 to June 1 herring was caught in abundance. From July to the close of the season both herring and squid bait were very scarce, due to the presence of dogfish. Of the earlier catch of herring a large quantity was sold to United States bank fishermen and the balance cured for use as lobster bait. The firm of T. & W. Moulton retired from business during the year, causing a reduction in the local fishery fleet of five vessels.

From Grand Mira, which section includes practically all of the coast fro 4 Gabarus bay to Lingan head, the total catch of all kinds of fish compares favourably with that of the previous 12 months. During the months of May and June lobsters were fairly plentiful, and in the latter month mackerel were schooling in large numbers. However, a severe southerly gale destroyed a large number of traps and nets, and a few days subsequently a second storm, from the north, completely demolished all of the lobster traps on the eastern shore, with the result that two canneries were compelled to close down. There was one vessel less this year.

Victoria county.—Information from all points throughout this division of the district tends to show somewhat less than an average result for the year.

The yield of lobsters at Cape North showed a decrease, rough weather interfering with attention to traps. Mackerel, salmon and halibut catches were better than during 1912-13. Codfish kept out in the deeper waters and the weather conditions being adverse the smaller boats were compelled to lie idle for a large part of the time. Swordfishing showed a decline. A regrettable feature of the season's operations in this vicinity was the loss during November, by drowning, of two fishermen, father and son, off Long point, by the sinking of their boat. During the gale in June lobster gear and salmon nets were badly wrecked and trap nets all destroyed at Cape North. In July a few lobster traps were also destroyed. In August a boat was lost at Bay St. Lawrence, being pierced by a swordfish. The crew was saved by other boats in the vicinity. Eight boats were lost at Meat cove in December, bait nets destroyed at White point, and a motor boat lost at Meat cove. At New Haven, in January, seven sailboats with all fishing tackle were lost.

Ingonish statistics show an average general catch. The spring haul of haddock was excellent, but the autumn catch fell off badly. During July, November and December large quantities of green-salted cod and haddock were shipped direct to the United States market.

The North Shore fishery records an increase in lobsters and salmon, but a decrease in the returns from codfish, mackerel and herring.

From the Bras d'Or lake, or inland water sections, the tenor of the report is about the same as that received from the coastal fisheries. The Upper Middle River stations indicate the year's catch as below the average, due mainly to high winds and cold weather during the spring and autumn months. Codfishing particularly was below par. Market prices ruling high made the lobster industry a remunerative one, although the catch was comparatively light. Preparations were made at the close of the season for a large increase in the number of traps for next year's fishing. The herring fishery was a complete failure. Oysters were exceedingly plentiful in the inland-water beds. Several parties have taken out leases under the new regulations entered into between the two governments for the propagation of this mollusk. The salmon fishery was equal to that of the previous year, but owing to prolonged periods of low water in the rivers and smaller streams angling was well-nigh impossible. One new gasolene boat and a number of smaller boats were put into commission. During the year some fishermen from the mainland of the province purchased small land areas and will next year engage in the fisheries in this locality. During the month of April the fishermen off the bay shore made the largest hauls known for many years.

Richmond county.—Spring opened earlier than for many seasons and the fishermen confidently looked forward to a banner year. At the end of May the most destructive gale on record swept the entire coast, causing a loss of from $35,000 to $40,000 in nets, traps and trap nets. Twenty-five sailboats were lost from L'Ardoise. Two and three crews to a boat spent the immediately succeeding days in picking up the wreckage. In June some further damage, though not material, was done. Had the weather been propitious the month of October would have been established as a record fishing period. There will be a big increase in the number of gasolene boats next season.

The catch of lobsters at L'Ardoise was above that of the previous season. Two vessels of the local fleet were disposed of by the owners. Although the loss to the fishermen through the May gale was very severe, still, owing to the high market prices obtained, the total year's results have been encouraging.

At Petit de Grat there was a decrease in herring, mackerel, swordfish and smelt. Six new gasolene boats and three vessels were added to the fleets at Petit de Grat and D'Escouse, respectively.

Inverness county.—The season opened with highly promising catches of lobsters, which continued well into the season. Prices ruled high on all markets and the total result gave the fishermen fair compensation. The lobsters caught last year were the largest in size for many years. Spawn lobsters were not so plentiful as in previous seasons. Herring and cod were far below the average. The fishermen generally are paying more attention to lobsters, to the almost utter neglect of the cod fishery. Two hundred cases more of lobsters were packed at one point, owing to there being more traps to each boat. The catch of mackerel during the season of 1912-13 exceeded that of any season for a number of years, and the fish graded higher than usual. This branch of the fisheries was the most profitable for a number of years. Codfish as a rule have kept to the deeper waters of the gulf. Lobster, mackerel, haddock, hake and salmon fishing was remarkably good.

Four motor boats were disabled in the gulf, three being towed in by sailing craft and one by the life-saving station boat. Two gasolene vessels and twelve gasolene boats were added to the Eastern Harbour fleet. One schooner was added and four boats removed from the fleet at Seaside.

A COMPARISON.

	1912-13.		1913-14.	
	No.	Value.	No.	Value.
Sail and gasolene vessels..	105	$ 63,620	96	$ 60,300
Sail and gasolene boats...................... .	2,571	123,360	2,909	165,839
Gill nets, seines, traps nets, etc................. ...	11,291	95,297	12,542	106,444
Trawls..	1,909	10,592	3,470	20,965
Hand Lines...........	5,583	3,482	8,040	5,852
Lobster traps......	148,675	124,007	130,937	111,026
Lobster canneries	58	36,450	70	57,270
Freezers and ice house............................	32	45,950	39	279,720
Smoke and fish house....	710	42,005	852	54,872
Fishing piers and wharves.................	187	155,645	231	182,060

—	1912-13.	1913-14.
Number of men employed on vessels......	645	491
Number of boats and smacks	3,757	4,577
Number of persons employed on shore........	1,170	2,258

The fact that there was an increase in the number of large gasolene-driven craft, gasolene and sail boats, gill nets, seines, trap nets, etc., trawls, hand lines, lobster canneries, freezers and ice houses, smoke and fish houses, and fishing piers and wharfs is clearly indicative that the fishing industry of this district is in a most flourishing condition. Power boats are gradually displacing sailing craft, and more traps, nets and lines are being handled by each individual boat and crew, with the result that under favourable conditions during any season this industry will be prosecuted with greater zeal than formerly.

The total value of the Cape Breton Island fisheries for the season of 1913-14 was $998,084, as against $959,492 for the corresponding twelve months of 1912-13, or an increase of $38,592. Increases are shown in the catches of salmon, cod, herring, mackerel, alewives, halibut, flounders, smelts, swordfish and oysters.

In the earlier part of the season fishermen, as a general rule, are very reticent about giving exact data concerning their catches, whilst at the close of the year they more readily supply any such information sought. In this district steps have already been taken to remedy this, and the fishermen themselves are now showing every disposition to assist the overseers in making full and complete statements covering the fisheries from month to month.

The fact that no fines were imposed during the past year for infractions of the fisheries regulations shows that the overseers had given the closest possible attention to their duties and so prevented violations of the fishery laws. The river guardians have, as a rule, been very faithful in the discharge of their duties, and that very little poaching indeed was done last year is due to the vigilance of those officers.

A matter worthy of note in this report is that the fishermen appear to be taking a more lively interest in their calling, due, doubtless, to the greatly improved market conditions both as to the price secured and the steady demand. The industry, although as hazardous as ever, has of late years become highly profitable, and fewer men are now attracted to the industrial centres. A fishermen's union was organized in Inverness town last year, this action having been stimulated, no doubt, by the prospect of the harbour in that locality being opened up in the immediate future.

HISTORICAL NOTES.

The fisheries along the Cape Breton coast were among the earliest in America developed for commercial purposes. Away back in the early days of the French occupation of this island the local fishery was in a very prosperous condition. The Archives of Coloniales de la Marine, Paris, contain official records of the catch, number of vessels engaged, number of men employed, and the market returns as far back as 1745. These records cover the coast from the Gut of Canso around the southern shore of Louisburg, and to the northeast extremity of the island, in which district 500 shallops and 2.500 men were yearly employed, as well as 60 brigantines. schooners and sloops carrying crews of 15 men each, making a total of 3,400 men then engaged in the industry, substantially a larger number than are to-day employed both on and off shore in the whole Cape Breton Island district.

It was computed by these old time officials that each shallop secured an annual catch of 500 quintals of cod, and the sixty larger vessels 600 quintals each, making a total season's catch of 186,000 quintals. Of course this was practically all codfish. As will be seen by the annual returns of the present year the total season's catch of cod for the entire island was 102,796 cwts.

Again, from these ancient records the information is gleaned that it required 93 sail of larger craft, each carrying 2,000 quintals of fish, and having a crew of twenty men each (a total of 1,860 seamen) to transport the year's product to France. Adding the crews of these transport ships gives a total of 5,260 persons directly or indirectly engaged in the Cape Breton fishery. Besides, 200 fishing vessels from France were annually engaged in the bank fishery, each craft having from 16 to 24 of a crew, thus adding 3,000 men more to the industry operating in the waters immediately adjacent. All of these vessels made their headquarters during the fishing season at Cape Breton ports, coming here for shelter, water and supplies.

Another fact adding importance to the Cape Breton Island fisheries in those early days was the disposition of the oil obtained from the cod livers, producing as they then did what was called a train oil, which was sent over to France for use by the manufacturers of woollen goods, for lighting purposes, and also supplied to the French sugar colonies which could not operate their local industry without it.

Much other detail covering the operations of the French fishermen at that time could be given, data showing the number of smaller vessels fishing from Ingonish, Bras d'Or, Ste. Anne's, Scatterie, Baliene, Lorraine, Louisburg, Gabarus, Ste. Esprit, Petit de Grat, and L'Ardoise, but this is sufficient for the present.

Comparisons between that early time fishing industry and the present day prosecution of our home fisheries might well suggest the possibility of a far more intense development of our fishing grounds than actually now obtains.

I am, sir, your obedient servant,

A. G. McLEOD,
Inspector of Fisheries.

REPORT ON THE FISHERIES OF DISTRICT No. 2.

To the Superintendent of Fisheries,
    Ottawa.

Sir,—I have the honour to submit my twenty-fourth annual report on the fisheries of District No. 2 of the province of Nova Scotia for the year ended March 31, 1914, together with tabulated statements showing the quantities and values of fish caught in the several counties of the district, and the material used and persons employed in the fisheries.

The aggregate value of the catch for the year is $2,207,721; as compared with the estimated value of last year's catch, which was $2,176,181, shows a slight increase of about one and one-half per cent.

Of the deep-sea fish the catch of cod shows an increase of seven per cent; haddock a decrease of about thirty-six per cent; hake an increase of about fifty per cent; pollock a decrease of about thirty-eight per cent; halibut an increase of about sixty per cent. The catch of herring was about the same as last year.

MACKEREL.

I noted in my report last year that the quantity of mackerel caught had been the smallest catch reported since 1890. The returns for this year show an increase over that for last of about sixty per cent; even with this increase it was not up to the average catch of the past twenty-four years.

There is no fishery subject to greater fluctuations in the quantity caught than this. Some' years as many as 118,087 cwts. have been reported as landed in a green state, while the quantity landed in that state this year was 59,225 cwts.

LOBSTERS.

I noted in my report for last year that this fishery has been gradually shrinking in the quantity caught, that while there were 68,352 cases packed and 5,810 cwts. exported in shell in the year 1896, in 1912 only 34,372 cases were packed and 15,141 cwts. shipped in shell.

The reports for this year show a further shrinkage; the quantity packed being 32,873 cases and 11,491 cwts. shipped in shell, or a decrease of about five per cent.

SALMON.

The reported catch is the largest during the past twenty-four years, and shows an increase of about forty per cent over that of last year.

On the Atlantic coast, in the counties of Guysboro and Halifax, the increase was about sixty per cent, while on the strait of Northumberland the increase was about twenty-four per cent.

On the Bay of Fundy division, viz., Cumberland, Colchester and Hants, the returns show a decrease of about ten per cent.

During the time the fish were ascending the rivers for spawning purposes the conditions were favourable as there was a good supply of water.

SHAD.

The reported catch is slightly over that of last year, but there were only 558 cwts. landed in a green state.

Twenty years ago the average catch was 2,000 cwts.

The close season, when these fish are in the rivers for spawning purposes, is only from Friday evening at sunset to sunrise Monday morning. The season should be made to cover the months of May and June.

ALEWIVES.

The quantity taken was about the same as last year, or about 1,200 barrels. This is about one-third of the average catch between the years 1889 and 1899.

SMELTS.

The quantity taken was less than that of last year—a decrease of about thirty-five per cent, largely owing to unfavourable ice conditions during the fishing season.

I am, sir, your obedient servant,

ROBERT HOCKIN,
Inspector of Fisheries.

REPORT ON THE FISHERIES OF DISTRICT No. 3.

To the Superintendent of Fisheries,
    Ottawa.

Sir,—I have the honour to submit the annual statistical report for District No. 3 for the year ended March 31, 1914.

The system for the gathering of the statistics has been greatly improved, particular care being taken to have the catch of the various districts reported regularly.

The statistics show that the general condition of the fisheries has been satisfactory, notwithstanding that the operations of the year were much broken. The great scarcity of ice the latter part of 1913 prevented greater success in the fresh fish trade, and the unusual ice conditions of February and March blocked several of the fishing ports, suspending operations and preventing shipments. These conditions were so unusual that for the first time in many years the Yarmouth-Boston service was suspended for a short time. Heavy gales were frequent the latter part of the season, and caused great loss, the lobstermen, particularly, being heavy sufferers. The shores were strewn with traps and gear. Many of the fishermen lost seventy-five per cent of their traps. At Flat Mud island, Shelburne, the twelve boats lost 1,400 traps in one storm alone. Fully fifty thousand traps were destroyed, and many nets and trawls lost. Three schooners, one sloop and a motor boat were lost in Shelburne county, and many boats damaged throughout the district, and five lives lost. The crew of four of the Lockeport sloop *Dollie Gray*, lost off the Shelburne coast, was saved in an exhausted condition by the American fishing vessel *Mary*, and carried to Boston. It can be seen, therefore, that the operations of the year were extremely hazardous.

On the other hand the statistics show that the year has been a prosperous one in many respects. Fish have been plentiful, and prices have run the gamut from the lowest for some years to the highest known for fresh fish. The market has been fitful, often ranging above Boston and Gloucester prices. The ports of Lockeport, Yarmouth and Digby had a greatly increased business.

The total marketed value of the catch, including by-products, was $5,091,821, as compared with $4,292,657 the preceding year, being an increase of $799,164. The following summary may, perhaps, be of value:—

LOBSTERS.

The total lobster catch was 157,577 cwts., an increase of 28,355 cwts. over the previous year. The counties reporting increases are Lunenburg, Queens, Shelburne, Yarmouth, Annapolis, and Kings. Digby reports a decrease. The total value of the marketed catch was $1,880,111, as compared with $1,394,273 the preceding year, an increase of $485,838. Canned increased from 29,269 cases to 35,194 cases, with a marketed value of $739,074. Shipped in shell increased from 56,141 cwts. to 69,597 cwts., with a marketed value of $1,141,037. Notwithstanding the increased catch over the preceding year, the shortage as compared with 1911-12 is 17,733 cwts.

The industry was more vigorously prosecuted than in any previous year. The season on the south shore opened under most favourable conditions and a record catch was secured. The latter part of the season generally was disastrous, many of the fishermen hardly paying expenses. Catch prices were high. "Shack," for packing purposes, ranged from 8½ cents per pound on the south shore to 10 cents in Digby. The marketed price secured for canned goods was higher than in previous years. These high prices have developed a system of adulteration that must seriously affect the market standard of the pack. In some instances the percentage of water added

5 GEORGE V., A. 1915

to the meat is fully 50 per cent. Regulations are needed to meet the situation. A large proportion of the catch for packing purposes is of lobsters less than seven inches in length, and the meat does not pay the expense for packing full weight meat.

COD AND HADDOCK.

The total catch of cod and haddock was 930,195 cwts., as compared with 928,975 cwts. the preceding year, an increase of 1,220 cwts. Cod increased 20,038 cwts., and haddock decreased 18,818 cwts. The marketed value of the catch was: Cod, $1,404,826; haddock, $512,043; making a total of $1,916,869, as compared with $1,810,310 the preceding year. The increase in price, notwithstanding the decrease in catch, is due to the great development of the smoked fish business in Digby county, where 2,550,900 pounds finnan haddies and 415,980 pounds smoked fillets were put up. In Shelburne and Digby counties a considerable pack of canned finnan haddies is put up. The Digby smoked fish business is especially noteworthy, not only on account of its great growth, but also for the high prices paid the fishermen for the catch. This specially prepared product finds a ready market in Montreal and other western points.

The counties showing increased catches were Yarmouth, Kings, Annapolis, and Digby.

HERRING.

The catch of herring was 220,361 cwts., with a marketed value of $262,195, as compared with 218,105 cwts. and $281,644 the preceding year. Digby and Annapolis counties show a decreased catch of nearly 50 per cent, while the only two counties showing increases were Lunenburg and Queens.

During the past three years herring has practically disappeared, and for the first time no smoked herring is reported in the statistics. For the four years 1906-1910 the average quantity of smoked herring reported was over 14,000 cwts. In 1911-12 the business dropped to 3,694 cwts., and since that time the famous "Digby Chicken" has disappeared. For the same period the total catch for Digby averaged 54,000 cwts., as compared with an average of 3,450 the past three years.

The decrease in the herring catch is a matter for serious consideration, as the success of the fresh fish business depends largely on the supply of fresh herring bait. When this bait is not available the boats and vessels are unable to continue fishing. On the south shore where the run is great and the fish of the largest and finest quality, it is to be deplored that the fishing as a staple industry has been neglected. Little provision is made for the systematic prosecution of the industry, doubtless due to the small returns for pickled herring. The catch could be very greatly increased if a satisfactory market were available. Net fishermen depend largely on selling the catch direct to the vessels seeking bait. If vessels do not happen on the grounds the catch is sold to cold storage concerns at 25 to 30 cents per bushel, and if no cold storage is within convenient distance, the business is fitful and wasteful. The matter of bait supply is becoming more acute each year. The past year unlimited quantities could have been taken in Yarmouth county, but there was no market, except when a vessel was in the district for bait. It is probable that unless action is taken to increase the market value for pickled fish this fishery will continue to be neglected. Better curing methods, with a more suitable barrel, and efficient inspection is greatly needed.

MACKEREL.

The total catch was 66,610 cwts., with a marketed value of $365,203, as compared with 45,263 cwts. and $235,533 the preceding year, an increase of 21,347 cwts. and $129,670. Lunenburg and Digby counties show an increased catch, while there is a decrease in the catch reported from Shelburne county. On several occasions the mackerel were extraordinarily plentiful in Yarmouth, but of small size.

HAKE AND CUSK.

The total catch was 203,838 cwts., with a marketed value of $308,019, as compared with 167,998 cwts. and $208,771 the preceding year. The big catch of the Lunenburg banking fleet, the spring trip having a large proportion of hake and cusk, is chiefly accountable for this increase. Digby shows a substantial increase. This fish is prepared principally for southern markets, being shipped dried.

POLLOCK.

The catch of pollock was 54,073 cwts. as against 55,144 cwts. the preceding year. Kings and Shelburne counties show an increase in the catch. The marketed value of the catch was $78,605, as compared with $67,184 the preceding year.

CANNED FISH.

The canned fish business is gradually developing. In Digby, 6,947 cases of finnan haddie, 2,013 cases kippered herring, 1,591 cases herring prepared with tomato sauce, and 450 cases mackerel. In Shelburne county and other points a considerable pack of specially prepared canned fish, including a fine quality of halibut, finds ready sale. Albacore, until several years ago a refuse fish, has been canned, and found an appreciative market. The canned fish business needs to be safeguarded by suitable regulations and inspection.

It may be noted that albacore, frequently caught in considerable numbers in mackerel and herring traps, are shipped to the New York markets, netting as high as nine cents per pound. These fish average about 600 pounds in weight each.

MEN AND PROPERTY.

The total number of men directly employed was 14,330, an increase of 877 as compared with the preceding year. Lunenburg reports a much greater number of men employed on vessels and boats than in the preceding year.

The value of vessels, boats, gear and other property is $4,066,791, as compared with $3,818,163 reported last year, an increase of $248,628. The number of gasolene boats under 10 tons is 2,027, valued at $511,990. The increase in number is 331 and in value, $77,310. For small boat fishing Lunenburg depends largely on the fine fleet of 1,243 sail boats, and employs only 58 gasolene boats. In other districts, however, the gasolene boats are in great demand.

The number of lobster traps reached the enormous total of 328,472, an increase of 18,692 over the preceding year. The number of traps have greatly increased each year since 1906, when only 160,147 were reported. The increase is accounted for by the growing scarcity of the fish, taking twice the number of traps, and many more fishermen, with greater labour and risk, to catch the quantity caught in former years.

PATROL SYSTEM.

The past year the patrol boat system has been very successful and of great value to the lobster industry. Boat "A" in Digby county and boat "B" in Yarmouth county, together with a number of smaller boats, covering the other districts, prevented much of the lobster poaching prevalent in former years. The risk of detection, carrying with it the loss of catch and gear, and prosecution of offenders, tended greatly to discourage many from attempting any poaching. As evidence of the success of the patrol system, the shipments from Yarmouth to the American markets for the first three days of the open season amounted to only 195 crates, and there is every reason to believe that these shipments were all legally caught fish. In 1911, the last year before the inauguration of the patrol system, 2,500 crates were shipped the first four

5 GEORGE V., A. 1915

days, the first shipment two days after the season opened being 1,400 crates. It is evident that the close season laws were well observed. In only one district of the seven western counties were traps put out before the morning of the opening day of the season. As the season opened on Monday, some of the fishermen in the district put out their traps on Saturday, only to have about 300 destroyed by one of the patrol boats. It should be said that the fishermen are highly pleased with the successful enforcement of the close season laws, and are actively supporting the officers in suppressing violations. Desultory summer fishing will doubtless continue, as the tourist and hotel trade offer inducements of the most tempting kind.

The fishery officers and men in charge of patrol boats gave ample evidence of zeal and resourcefulness.

RIVER AND INLAND FISHERIES.

The value of the river and inland fisheries cannot be estimated from a study of the statistics. Only 635 cwts. of trout is reported, valued at $8,985, and 1,360 cwts. of salmon with a marketed value of $26,458. As trout are not exported for sale, it is impossible to secure statistics of any value. Vast quantities are taken in practically every county in the district. Thousands of visiting anglers fish the waters each year, besides great numbers of native sportsmen. The waters of Queens alone are fished annually by some 3,000 visiting sportsmen. In Kings the salmon catch shows a decrease, due chiefly to the small size of the fish as compared with former years. In Annapolis, notwithstanding that netting was prohibited, the salmon catch was fourfold that of the preceding year. The high water in the rivers last spring, and the exceptionally low waters of the summer months, interfered with the fishermen. Great quantities of alewives and salmon went up during the high water. Some of the rivers have been alive the present year with " slink " salmon, coming from the spawning grounds.

The importance of safeguarding and developing the river fisheries cannot be too strongly pressed. The coast is most advantageously situated, and possesses a remarkable system of bays, rivers and lakes, constituting most valuable fish breeding grounds. For years past the general conditions have been bad. The pollution of the rivers and streams from sawdust and other mill refuse has been universal. Many of the best rivers have been closed to fish ascending to the spawning grounds.

For the past several years efforts have been made to improve conditions and save a most valuable asset to the people. During the past year much pollution has been prevented and the following results achieved in the way of direct improvements:—

LUNENBURG COUNTY.

One hundred dollars was expended in the removal of obstructions from the West branch of East river. Thorough work was done, and the river is now in good condition.

On the Gold river, a fish pass has been installed in the Mosher's falls dam of the Kent Lumber Company. This pass will be improved by making the lower section removable for safety during the winter months, otherwise ice conditions would destroy it.

One hundred dollars was granted for removing obstructions from Martin's river, and the completion of a fish pass at the Ezekiel Langille dam. The obstructions were removed at a cost of $31.58. The forest fires of the summer of 1913 destroyed the Langille mill, which will not be rebuilt. A wide opening has been made in the dam with the owner's consent, and therefore there was no need of any further expenditure.

The fish-pass at the dam of W. B. Langille & Co. has been made efficient, and a tight floor laid in the mill to prevent sawdust pollution.

Fifty dollars was expended in removing obstructions from the stream running into Common lake.

On the Mush-a-mush river, Mr. T. G. Nicol has installed a fish-pass at the Robar dam. Mr. Edward Ernst has taken the advice of the fishery officer, and will construct a natural fish-pass round the end of his dam, instead of installing a Hockin pass. Timothy Spidell has completed the improvements in connection with the dam owned by him. The entrance to Big lake has also been improved. There should be little difficulty in fish having free access to the lakes at the head of this river.

One hundred and nine dollars and eighty cents was expended in removing obstructions from Jodrey falls, and one hundred dollars was granted for removing obstructions from New Germany lake. The conditions in respect to these two expenditures were the same. Blasting was necessary to make the falls passable. The ascents to the spawning grounds were choked with rubbish.

The conditions on the La Have river and branches have been much improved. On this fine river the regulations were flagrantly violated. The Davidson Lumber Company have repaired the fish-passes at the two dams in the district of Bridgewater. The sawdust and other pollutions have been somewhat improved and the company has promised to erect a burner to take care of the refuse.

The Mackie mill and dam, at Upper Northfield, on the Keddy river, a branch of the La Have, has been put in good condition by the installation of a fish-pass, and provision for the care of mill refuse.

QUEENS COUNTY.

The Medway river district, above tidal waters, is in good condition. Two hundred dollars has been expended in removing fourteen piers, greatly improving the river and facilitating the ascent of fish.

The Mersey river for years has been in a deplorable condition. The salmon were becoming very scarce, as it was impossible for the fish to ascend the river. The fish-pass in the first dam was obsolete, and broken down, and on the wrong side of the river. A new pass has been installed on the west side of the river, and a new sluice gate built, which can be operated in favour of the fisheries. The cost to the mill owners was $268. At the second dam, situated at Potonac falls, the fish-pass has been extended and repaired, and the dam put in condition to assist fish-pass efficiency. At the third dam, situated at Cowies falls, there is a good natural pass, constructed by the Department some few years since at a cost of $1,100. This was the only pass in condition on the river, but was of no value, as the three dams below were closed. The fourth pass, situated at Rapid falls, has been repaired, but is not satisfactory. A new pass will be constructed the coming summer. At the fifth dam, situated at the "Guzzle," a sufficient opening has been made to provide free passage for fish. The river has been greatly improved.

SHELBURNE COUNTY.

Seventy-eight dollars and thirty cents was expended in improving the fish-pass at the Bower dam, and a portion of a grant of $60 has been expended in the removal of obstructions from the river.

One hundred dollars has been expended in removing obstructions from Round Bay brook. Also a considerable sum has been expended by the Public Works Department in improving the shore entrance to the brook.

One hundred dollars has been expended in removing obstructions from Purneys brook. This will give free access from Jordan bay and river to the lake at the head of the brook.

Jordan river has been improved by the installation of two fish-passes in the dam owned by Miller Bros., and provision made at the two new upper dams for the free passage of fish.

At Little Harbour a considerable expenditure has been made by the Public Works Department opening the canal from the sea to Matthews lake. In former years this lake was the spawning grounds for great quantities of trout, gaspereau and bass. This lake is one of the few known places in the province where bass has been abundant. For some few years, however, it was not accessible, and the appearance of the bass has not been noteworthy.

On the Clyde river a pass has been provided in the dam of the Pulp Company at "Queens." Much difficulty has been found in keeping this pass open.

YARMOUTH COUNTY.

A Hockin fish-pass has been installed in the dam of the Yarmouth Electric Light Company at Carleton, and in the Adolphus Pothier dam at Herring brook, and in the Maurice Prosser dam on the Tusket river at Kemptville. The Department paid one-half the cost of this latter pass, amounting to $84.90. The pass in the Howard Crosby dam at Carleton has been rebuilt.

One hundred dollars was expended in removing obstructions from the Salmon river. A channel ten feet wide, from tidal waters to Harpers lake, a distance of about seven miles, was cleared.

A number of the residents, under Guardian Sweeney, voluntarily cleared a valuable branch of the Tusket river from obstructions existing for many years.

DIGBY COUNTY.

A dam closed for twenty years at Salmon River lake, Maxwellton, has been opened, with the result that large quantities of alewives and other fish have access to the lake. The dam at Corberrie has also been opened.

The conditions on the Salmon river, in connection with the eel-weir obstructions, have been considerably improved, in that the fishermen below the upper stone wall and dam will have a better opportunity of catching fish. Guardian Aymar has succeeded in carrying out instructions sufficiently to provide a fairly satisfactory condition.

The Campbell Lumber Company at Weymouth has built an abutment at the end of the canal from the mill to the river and will use the canal bottom for retaining the ground pulp waste. Complaints were made, not only in respect to the pollution of the water from this waste, but also on account of the waste filling the nets of the fishermen.

ANNAPOLIS COUNTY.

A fine fish-pass has been constructed around the dam of the Annapolis Electric Light Company on the Lequille river.

A canal has been cut from the Dargie dam to the main branch of the Lequille river, which has greatly improved the conditions, and the dam at Alpena, owned by the Davidson Lumber Company, has been opened. An opening has also been made in the dam at Lake Mulgrave, owned by Clark Bros. of Bear River.

KINGS COUNTY.

The rivers and streams in this county were in very bad condition, but have been greatly improved during the year. The only improvement on the Gaspereau river has been the partially completed fish-pass at White Rock dam. During the summer an opening was made in the dam for temporary use. The pass will be completed in the spring. At the dam at large lake, about five miles above White Rock, a passage was made at the west side of the dam. The young fish came through without injury.

The conditions on the Creamer river at Millville, and on the Fales river, and on the Cornwallis river, and on the Pines brook at Waterville, in respect to sawdust pollution, have greatly improved.

I am, sir, your obedient servant,

WARD FISHER,
Inspector of Fisheries.

5 GEORGE V., A. 1915

RETURN showing the Number of Fishermen, etc., the Number and Value of Vessels,
Industry in the County of **Richmond**, Province

Number	Fishing Districts.	Vessels, Boats and Carrying Smacks.												
		Sailing and Gasolene Vessels.					Boats.					Carrying Smacks.		
		(40 tons and over).	(20 to 40 tons).	(10 to 20 tons).	Value.	Men.	Sail.	Value.	Gasolene.	Value.	Men.	Number.	Value.	Men.
	Richmond County.	No.	No.	No.	$			$		$			$	
1	Fourchu, Framboise and vicinity .	....	....	....	.....	....	160	3500	9	2900	273	6	3200	12
2	Grand River and vicinity...	....	..	....	.....	....	128	2250	7	2350	160	2	900	3
3	Point Michaud and L'Ardoise..............	....	....	2	1600	7	372	16950	8	2800	370	2	500	4
4	Rockdale and Grand Grève...	....	....	....	..	.....	150	5700	2	900	200	2	700	2
5	St. Peter's and River Bourgeois..................	2	3	5	11700	71	42	500	6	2100	82	7	1500	16
6	Louisdale and River Inhabitants.............	....	1	....	600	3	24	226	....	.....	48	..	......	....
7	Ports Malcolm and Richmond.	....	....	....	....	....	20	240	....	....	40	..	......	....
8	West Bay......	..	....	....	.....,.	....	7	70	....	....	14	..	....	....
9	Ile Madame.....	3	6	10	10800	116	313	3680	20	4800	358	8	750	13
	Totals...	5	10	17	24700	197	1216	33116	52	15850	1545	27	7550	50

SESSIONAL PAPER No. 39

Boats, and the Quantity and Value of all Fishing Gear, etc., used in the Fishing of Nova Scotia, during the Year 1913-14.

Fishing Gear										Canneries		Other Material						Persons employed in Canneries, Freezers and Fish Houses	
Gill Nets, Seines, Trap & Smelt Nets		Weirs.		Trawls.		Hand Lines.		Lobster Traps.		Lobster Canneries.		Freezers and Ice Houses.		Smoke and Fish Houses.		Fishing Piers and Wharves			
Number.	Value.	Number.	Value.	Number.	Value.	Number.	Value.	Number.	Value.	Number.	Value.	Number.	Value.	Number.	Value.	Number.	Value.		Number.
$		$		$		$		$		$		$		$		$		$	
210	1680	...	.....	...	......	180	72	13500	5550	..2	3000	..	......	22	800	6	1200	100	1
580	4650	...	.....	...	......	105	42	3400	1700	1	1000	..	......	30	1000	2	284	65	2
2800	22400	..	.....	17	155	740	296	4050	4050	1	1200	..	......	140	5340	6	1110	440	3
1100	1800	..	..	4	35	290	116	4800	3500	1	1120	..	......	35	1200	3	376	150	4
50	350	..	.....	75	280	140	67	2150	1075	1	1000	1	2650	26	4000	3	7550	35	5
90	720	2	20	...	......	20	10	....	....	..	....	..	....	2	130	1	100	.....	6
30	240	..	.....	...	......	....	....	....	....	..	....	..	....	3	40	..	....	.....	7
10	50	..	.....	10	40	20	10	....	....	..	....	..	....						8
2694	13350	..	.....	795	3975	520	260	9850	9850	6	2300	3	750	155	1850	31	11100	165	9
7565	46740	2	20	901	4485	2015	873	37750	25725	12	9620	4	3400	413	14410	52	21720	955	

5 GEORGE V., A. 1915

RETURN showing the Number of Fishermen, etc , the Number and Value of Vessels
Industry in the County of **Cape Breton**, Province

Number	Fishing districts.	Vessels, Boats and Carrying Smacks.												Fishing	
		Sailing and Gasolene Vessels.				Boats.					Carrying Smacks.			Gill Nets, Seines, Trap and Smelt Nets, etc.	
		20 to 40 tons	10 to 20 tons	Value.	Men.	Sail.	Value.	Gasolene.	Value.	Men.	Number.	Value.	Men.	Number.	Value.
		No.	No.	$			$		$			$			$
	Cape Breton County.														
1	Sydney Glace-Bay, Lingan and vicinity..................					82	2780	3	650	136	4	1100	7	122	840
2	Louisburg and vicinity............					83	3790	2	600	120	3	1700	7	235	1645
3	Upper North Sydney, Long Island and Leitches Creek					16	235	..	......	32	..	......	..	32	160
4	Port Morien and vicinity..........		1	250	3	29	810	31	7130	112	7	4350	14	493	3959
5	Main-à-Dieu and vicinity		3	1300	12	50	2000	20	4000	140	5	2500	10	366	2928
6	Gabarus and vicinity..............		1	500	5	101	6060	8	2400	267	6	4800	14	396	3500
7	Scatarie Island.					20	1640	13	3250	48	..	......	..	120	900
8	Little Bras d'Or District.........	2	6	1850	30	72	1120	5	750	96	1	500	2	228	1100
	Totals.......	2	11	3900	50	453	18435	82	18780	951	26	14950	54	1992	15032

SESSIONAL PAPER No. 39

and Boats, and the Quantity and Value of all Fishing Gear, etc., used in the Fishing of Nova Scotia, during the Year 1913-14.

Gear.						Canneries.		Other Material.						Persons employed in Canneries, Freezers and Fish-houses.	
Trawls.		Hand Lines.		Lobster Traps.		Lobster Canneries.		Freezers and Ice-houses.		Smoke and Fish-houses.		Fishing Piers and Wharves.			
Number.	Value.	Number.	Value.	Number.	Value.	Number.	Value.	Number.	Value.	Number.	Value.	Number.	Value.		Number.
	$		$		$		$		$		$		$		
48	240	150	75	5800	5800	2	2300	....	......	.....	.....	4	450	52	1
50	250	190	95	4800	4425	2	3200	....	......	.....	.....	6	1050	39	2
14	140	40	20	....	....	....	....	....	......	2	3000	....	...	10	3
29	290	205	200	8150	10187	4	5200	....	......	....	....	6	300	55	4
150	500	300	300	6000	6000	2	3000	....	...	70	2100	10	1000	190	5
....	....	130	65	13740	8740	4	5800	....	...	25	500	5	640	66	6
20	200	150	60	1200	720	1	600	....	......	20	200	....	....	13	7
94	282	230	115	3050	3800	1	1500	1	2000	40	1000	25	1290	36	8
405	1902	1395	930	42740	39672	16	21600	1	2000	157	6800	56	4730	461	

5 GEORGE V., A. 1915

RETURN showing the Number of Fishermen, etc., the Number and Value of Vessels and Industry in the County of **Victoria**, Province of

Number	Fishing Districts.	Sailing and Gasoline Vessels.			Boats.					Carrying Smacks.			Gill Nets, Seines, Trap and Smelt Nets, etc.	
		10 to 20 tons. No.	Value.	Men.	Sail.	Value.	Gasoline.	Value.	Men.	Number.	Value.	Men.	Number.	Value.
	Victoria County.		$			$		$			$		-	$
1	Iona, Washabuck & Little Narrows				66	1720	2	300	14				101	909
2	Wreck Cove to Cape Smokey				40	800	2	300	40				120	1200
3	Breton Cove and vicinity to English-town				70	1400	5	750	120				158	3500
4	Baddeck and vicinity				5	125							20	200
5	Ingonish	12	9600	60	118	10960	10	4625	275				316	8100
6	Neil's Harbour and New Haven				88	5192	4	900	116	2	500	4	152	2900
7	White Point, Dingwall & Sugar Loaf	2	1000	3	49	1225	4	900	80	2	250	4	149	2670
8	Meat Cove, Bay St. Lawrence and Vicinity				53	1060	3	450	96	4	600	8	100	1500
9	Big Bras D'Or District				4	80	2	200	6				6	48
	Totals	14	10600	63	493	22562	32	8425	837	8	1350	16	1122	21027

SESSIONAL PAPER No. 39

Boats, and the Quantity and Value of all Fishing Gear, etc., used in the Fishing
Nova Scotia, during the Year 1913-14.

Fishing Gear.						Canneries.		Other Material.						Persons employed in Canneries, Freezers and Fish-Houses.	
Trawls.		Hand Lines.		Lobster Traps.		Lobster Canneries.		Freezers and Ice-houses.		Smoke and Fish-houses.		Fishing Piers and Wharves.			
Number.	Value.	Number.	Value.	Number.	Value.	Number.	Value.	Number.	Value.	Number.	Value.	Number.	Value.		Number.
	$		$		$		$		$		$		$		
92	368	138	69	180	144	....	.....	...	.....	.	.....	....	.....	....	1
20	200	200	100	700	700	1	400	....	....	18	720	....	.....	12	2
30	300	300	150	3400	3400	2	1000	3	365	120	4809	1	3800	21	3
5	40	6	18	.....	.....									....	4
568	3976	136	115	3000	1500	5	1300	6	4150	29	9250	13	80950	200	5
40	600	390	390	3725	3725	7	2550	1	800	35	4400	3	2800	63	6
29	290	237	237	4300	4300	3	1900	2	600	13	3600	11	2600	43	7
18	216	192	192	5000	5000	2	2500	3	600	12	2400	....	.....	39	8
6	30	12	6	.....	.....	..	.....	..	.....	...	.....	....	. . .	....	9
808	6020	1611	1277	20305	18769	20	9650	15	6515	227	25170	28	90150	378	

5 GEORGE V., A. 1915

RETURN showing the Number of Fishermen, etc., the Number and Value of Vessels and
Industry in the County of **Inverness**, Province

Number	Fishing Districts.	Sailing and Gasoline Vessels.					Boats.					Carrying Smacks.			Gill Nets, Seines, Trap and Smelts Nets, etc.	
		(40 tons and over.)	(20 to 40 tons.)	(10 to 20 tons.)	Value.	Men.	Sail.	Value.	Gasoline.	Value.	Men.	Number.	Value.	Men.	Number.	Value.
	Inverness County.	No.	No.	No.	$			$		$			$			$
1	Meat Cove, Poulet Cove and Pleasant Bay	....	...	....	....	....	4	40	21	2505	48	.....	......	..	44	2830
2	Cap Rouge, Eastern Harbour, Cheticamp and Grand Etang	1	3	28	17000	159	67	3696	15	3010	204	....	......	:	317	1585
3	Friar's Head, Margaree Harbour to Smith's Cove	..	4	....	3600	16	160	8426	24	6110	406	6	1100	6	547	12980
4	Broad Cove, Port Ban to Mabou Harbour	....	....	....	....	....	34	881	14	2100	121	1	180	2	125	1000
5	West Lake Ainslie and Whycocomah Bay.....	....	...	....	....	...	25	500	....	....	40	.....	......	..	30	150
6	Little Mabou and Port Hood to Hawkesbury...	....	....	1	500	6	....	...	100	20000	175	7	2000	10	500	4500
7	West Bay, Malagawatch and Deny's Basin.	....	...	....	.. ..	....	108	1400	....	.....	112	.....	......	..	300	600
	Totals	1	7	29	21100	181	398	14946	174	33725	1106	14	3280	18	1863	23645

SESSIONAL PAPER No. 39

Boats, and the Quantity and Value of all Fishing Gear, etc., used in the Fishing of **Nova Scotia**, during the Year 1913-14.

Gear.						Canneries.		Other Material.						Persons Employed in Canneries, Freezers and Fish-houses.	
Trawls.		Hand Lines.		Lobster Traps.		Salmon Canneries.		Freezers and Ice-houses.		Smoke and Fish-houses.		Piers and Wharves.			
Number.	Value.	Number.	Value.	Number.	Value.	Number.	Value.	Number.	Value.	Number.	Value.	Number.	Value.		Number.
	$		$		$		$		$		$		$		
1	10	56	56	3950	1975	2	1600	6	180	11	130	.....	......	25	1
32	800	618	371	12792	10910	8	5150	5	2025	16	5300	14	10300	160	2
1120	5600	1515	1680	5900	5575	5	1150	4	3100	14	2910	67	6120	54	3
5	50	363	363	4500	4500	1	2500	.....	......	.. ..	......	3	14000	25	4
...	...	32	32	.....	.....	....	.....	......	......	......	.....	......	......		5
100	2000	200	200	3000	4500	6	6000	4	262500	....	......	7	35000	200	6
98	98	235	70	...	. ...	......	.... .	.....	....	14	152	4	40		
1356	8558	3019	2772	30142	27460	22	16400	19	267805	55	8492	95	65460	464	

5 GEORGE V., A. 1915

THE CATCH.

RETURN showing the Quantities and Values of all Fish caught and landed in a Green State in the County of **Richmond**, Province of **Nova Scotia**, during the Year 1913-14.

Number	Fishing Districts	Salmon, cwts.*	Salmon, value. $	Lobsters, cwts.	Lobsters, value. $	Cod, cwts.	Cod, value. $	Haddock, cwts.	Haddock, value. $	Hake and cusk, cwts.	Hake and cusk, value. $	Pollock, cwts.	Pollock, value. $	Herring, cwts.	Herring, value. $	Mackerel, cwts.	Mackerel, value. $	Shad, cwts.	Shad, value. $	Alewives, cwts.	Alewives, value. $	Number
	Richmond County.																					
1	Fourchu, Framboise and vicinity	30	270	3102	21341	1768	2652	64	77			33	33	24	31	254	795					1
2	Grand River and vicinity			926	6370	163	244	6	7			9	9	54	70	483	1455					2
3	Point Michaud and L'Ardoise	17	153	72	495	5851	97	4809	5879			843	843	2098	2727	3989	12020					3
4	Rockdale and Grand Grève			920	6329	468	92	468	562			93	93	1749	2274	2587	7795			3	3	4
5	St. Peters and River Bour geois			635	3175	8600	100	100	150			10	10	100	120	100	200					5
6	Louisdale and River in habitants					340	30					88	88	600	600	500	1000					6
7	Ports Malcolm and Rich mond					150	90							400	400	700	2840					7
8	West Bay					440	60							10	10							8
9	Ile Madame	76	761	3290	16450	9165	95	19316	24145	1793	1142	364	182	1390	1300	1047	6749					9
	Totals	123	1184	8945	54160	26935	80	24853	30820	1793	1142	1440	1258	6425	7622	9670	32854			3	3	

*Cwt = 100 lbs.

THE CATCH.

RETURN showing the Quantities and Values of all Fish caught and landed in a Green State in the County of **Richmond**, Province of **Nova Scotia**, during the Year 1913-14.—*Concluded.*

Number	Fishing Districts	Halibut, cwts.*	Halibut, value.	Flounders, cwts.	Flounders, value.	Smelts, cwts.	Smelts, value.	Trout, cwts.	Trout, value.	Eels, cwts.	Eels, value.	Tom-cod, cwts.	Tom-cod, value.	Swordfish, cwts.	Swordfish, value.	Mixed fish, cwts.	Mixed fish, value.	Squid, brls.	Squid, value.	Clams, brls.	Clams, value.	Number
	Richmond County.	$	$	$	$	$	$	$	$	$	$	$	$	$	$	$	$	$	$	$	$	
1	Fourchu, Framboise and vicinity	12	60																	70	210	1
	Grand River and vicinity																			2	6	2
3	Point Michaud and L'Ardoise																	25	75			3
4	Rockdale and Grande Grève	70	350											179	358							4
5	St. Peters and Ri er Bourgeois	8	40			70	420							68	136							5
6	Louisdale and River Inhabitants					200	1200			80	160					20	20	10	20	10	20	6
7	Port Malcolm and Richmond																					7
8	West Bay	12	80			150	900															8
9	Ile Madame			51	26									730	2478			145	290			9
	Totals	102	530	51	26	120	2520			80	160			977	2972	20	20	180	385	82	236	

*Cwts. = 100 lbs.

THE CATCH MARKETED.

RETURN showing the Quantities and Value of all Fish and Fish Products Marketed in a fresh, dried, pickled, canned, &c., state fc. the County of **Richmond**, Province of **Nova Scotia**, during the year 1913-14.

Number	Fishing Districts	Salmon, used fresh and frozen, cwts.	Salmon, canned, cases	Salmon, salted, cwts.	Lobsters, canned, cases	Lobsters, shipped in shell, cwts.	Cod, used fresh, cwts.	Cod, shipped green salted, cwts.	Cod, dried, †qtls.	Haddock, used fresh, cwts.	Haddock, smoked, cwts.	Haddock, dried, quintals	Hake and cusk, used fresh, cwts	Hake and cusk, dried, quintals	Pollock, used fresh, cwts.	Pollock, dried, quintals	Herring, used fresh, cwts.	Herring, smoked, cwts.	Number
	Richmond County.																		
1	Fourchu, Framboise and vicinity				1142	247		113	514	4		20				11			1
2	Grand River and vicinity	30			352	46	1		24			2				3			2
3	Point L'Ardoise and Grand Grève	17			29		4		1949	6		16				281			3
4	Rockdale and Grand Grève				368		183		95			34				31	276		4
5	St. Peter's and River Bourgeois				254		7		2867	6		33					30		5
6	Louisdale and River Inhabitants						1		113						10	30	20		6
7	Ports Richmond and Malcolm						1		50										7
8	West Bay	76			1304	30	419	700	1488	12791		2175	983	241	10	121			8
9	Ile Madame						3291		7										9
	Totals	123			3449	323	3807	813	7137	12807		4015	983	241	10	477	326		
	Rates	10 00			18 00	10 00	1 50	2 50	5 50	2 00		4 50	1 00	3 00	1 00	4 00	1 00		
	Values	1230			62082	3230	5860	2032	39253	25614		18067	983	723	10	1908	326		

*Cwt. = 100 lbs.   †Quintals = 112 lbs.

THE CATCH MARKETED

RETURN showing the Quantities and Values of all Fish and Fish Products Marketed in a fresh, dried, pickled, canned, &c., state, for the County of Richmond, Province of Nova Scotia, during the year 1913-14—*Concluded.*

Number	Fishing Districts	Herring, pickled, brls.	Herring, used as bait, brls.	Herring, used as fertilizer, brls.	Mackerel, used fresh, cwts.*	Mackerel, salted, brls.	Alewives, used fresh, cwts.	Halibut, used fresh, cwts.	Flounders, used fresh, cwts.	Smelts, used fresh, cwts.	Eels, used fresh, cwts.	Tom-cod, used fresh, cwts.	Swordfish, used fresh, cwts.	Mixed fish, used fresh, cwts.	Squid, used as bait, brls.	Clams and quahaugs, used fresh, brls.	Tongues and sounds, pickled and dried, cwts.	Fish oil, gals.	Number
	Richmond County.																		
1	Fourchu, Framboise and vicinity	7	1			88										70		350	1
2	Grand River and vicinity	14	6			161												21	2
3	Point Michaud and L'Ar...	51	222		95	1038							179		25	2		1350	3
4	Rockdale, and Grand Grève	62	43		106	20		70					68					79	4
5	St. Peter's and River Bourgeois	24	125		80	43	3	8		70							10	1500	5
6	Louisdale and River Inhabitants	100	15		90	77				200	80			20	10			50	6
7	Ports Malcolm and ...	101	42		20	30													7
8	Bay West	350	168		817	85	3	12	51	150			730		145			60	8
9	Ile Madame														180				9
	Totals	1609	632		4442	1751		102	51	420	80		977	20	180	82		3410	
	Rates $	4 50	2 50		5 00	11 00	2 00	7 00	1 00	6 00	2 00		4 00	1 00	3 00	3 00	3 00	.30	
	Values $	7240	1580		22210	19261	6	714	51	2520	160		3908	20	540	246		1023	

Total value.......................

*Cwts. = 100 lbs.

5 GEORGE V., A. 1915

THE

RETURN showing the Quantities and Values of all Fish caught and landed in a the year

Number	Fishing Districts.	Salmon, *cwts.	Salmon, value.	Lobsters, cwts.	Lobsters, value.	Cod, cwts.	Cod, value.	Haddock, cwts.	Haddock, value.	Hake and Cusk, cwts.	Hake and Cusk, value	Pollock, cwts.	Pollock, value.
	Cape Breton County.		$		$		$		$		$		$
1	Sydney, Glace Bay, Lingan and vicinity..................			2170	13020	1440	2160						
2	Louisburg and vicinity	18	180	2712	16272	5490	8235	2520	2520			214	175
3	Upper North Sydney, Long Island and Leitches Creek					510	1530						
4	Port Morien and vicinity...........	80	880	4065	24390	425	1104	11	25			3	3
5	Main-à-Dieu and vicinity..........	154	1625	4166	24996	5445	5445	2902	3607			246	295
6	Gabarus and vicinity.............			2647	15882	2975	5504	800	800			685	685
7	Scatarie Island.....................	12	129	617	3702	2561	4899	617	963			57	74
8	Little Bras-d'Or District	17	177	2775	16650	7533	15066	1100	1650	12	18	367	550
	Totals	281	2991	19152	114912	26379	43943	7950	9565	12	18	1572	1782

*Cwt.—100 lbs.

SESSIONAL PAPER No. 39

CATCH.

Green State in the County of **Cape Breton**, Province of **Nova Scotia**, during 1913-14.

Herring, cwts.	Herring, value.	Mackerel, cwts.	Mackerel, value.	Shad, cwts.	Shad, value.	Alewives, cwts.	Alewives, value.	Halibut, cwts.	Halibut, value.	Smelts, cwts.	Smelts, value.	Sword-fish, cwts.	Sword-fish, value.	Squid, brls.	Squid, value.	Oysters, brls.	Oysters, value.	Number.
	$		$		$		$		$		$		$		$		$	
4500	2250	39	117	....	....	....	....	....	....	21	105	40	160	....	....	....	....	1
2135	1067	3564	10692	....	....	....	....	....	....	..	....	....	....	....	....	....	....	2
568	410	....	....	....	....	....	....	....	....	....	....	....	....	....	....	....	....	3
2744	4116	37	216	..	....	....	....	37	259	30	300	....	....	....	....	5	20	4
1550	1550	525	1595	46	230	24	24	79	454	....	....	4	16	9	36	....	....	5
3940	3940	1600	4800	....	....	4	4	12	60	73	292	....	....	20	69	..	....	6
184	184	110	689	....	....	....	....	103	581	....	....	51	138	....	....	....	....	7
3742	7251	152	304	....	....	..	....	219	2106	....	....	26	104	....	....	....	....	8
19363	20768	6027	18413	46	230	28	28	450	3460	124	697	121	418	29	105	5	20	

5 GEORGE V., A. 1915

THE CATCH MAKETED.

RETURN showing the Quantities and Values of all Fish and Fish Products Marketed in a fresh, dried, pickled, canned, &c., state for the County of **Cape Breton**, Province of **Nova Scotia**, during the year 1913-14.

Number	Fishing Districts	Salmon, used fresh and frozen, cwts	Lobsters, canned, cases	Lobsters, shipped in shell, cwts	Cod, used fresh, cwts	Cod, shipped green salted, cwts	Cod, dried, qtls	Haddock, used fresh, cwts	Haddock, smoked, cwts	Haddock, dried, qtls	Hake and Cusk, dried, qtls	Pollock, used fresh, cwts	Pollock, dried, qtls	Herring, used fresh, cwts	Herring, pickled, brls	Herring, used as bait, brls	Number
	Cape Breton County.																
1	Sydney, Glace Bay, Lingan and vicinity	..	868	..	693	..	249	..		..			71	425		2037	1
2	Louisburg and vicinity	18	1085	..	11	..	1826	..		939				10	575	200	2
3	Upper North Sydney, Long Island and Leitches Creek	..	..	1030	510	..	..	..		..						265	3
4	Port Morien and vicinity	80	1214	994	24	103	65	5		2			1	22	41	1320	4
5	Main-à-Dieu and vicinity	154	1263	..	510	1248	813	106		932			82	37	433	17	5
6	Gabarus and vicinity	..	1050	..	209	..	922	..		266			228	20	773	900	6
7	Scatarie Island	12	30	342	1407	501	51	401	90	12		57				92	7
8	Little Bras-d'Or District	17	1100	25	1905	1000	1185	..		366	3		101		15	1856	8
	Totals	281	6625	2591	5269	2852	5111	512	90	2417	3	57	483	514	1897	6587	
	Rates	12 00	18 00	8 00	1 50	3 00	5 50	1 50	3 00	4 50	3 00	1 00	3 50	1 00	4 50	2 00	
	Values $	3372	119250	20728	7903	8556	29110	768	270	10876	9	57	1690	514	8536	13174	

* Cwt. = 100 lbs.   †Quintal = 112 lbs.

THE CATCH MARKETED.

RETURN showing the Quantities and Values of all Fish and Fish Products Marketed in a fresh, dried, pickled, canned, &c., state for the County of Cape Breton, Province of Nova Scotia, during the year 1913-14—*Concluded.*

Number	Fishing Districts.	Mackerel, used fresh, cwts.	Mackerel, salted, brls.	Shad, used fresh, cwt.	Alewives, used fresh, cwts.	Alewives, salted, brls.	Sardines, canned, cases.	Sardines, used brls.	Halibut, used fresh, cwts.	Flounders, used fresh, cwts.	Smelts, used fresh, cwts.	Eels, used fresh, cwts.	Tom-cod, used fresh, cwts.	Sword-fish, used fresh, cwts.	Squid, used as bait, brls.	Clams and Quahaugs, used fresh, brls.	Fish oil, gals.	Number
	Cape Breton County.																	
1	Sydney, Glace Bay, Lingan and vicinity	15									21			40			380	1
2	Louisburg and vicinity	64	1166														1100	2
3	Upper North Sydney, Long Island and Leitches Creek																	3
4	Port Morien and vicinity	7	10	46					37					4				4
5	Main-à-Dieu and vicinity	134	130		4	8			79		30				9	5		5
6	Gabarus and vicinity	160	480						12					51				6
7	Scatarie Island	110							103		73			26	20			7
8	Little Bras-d'Or District	152							219								400	8
	Totals	642	1786	46	4	8			450		124			121	29	5	1880	
	Rates $	5 00	11 00	8 00	1 50	3 00			9 00		8 00			5 00	4 00	4 00	0 30	
	Values $	3210	19646	568	6	24			4050		992			605	116	20	564	
	Total value																$ 253,414	

THE CATCH.

RETURN showing the Quantities and Values of all Fish caught and Landed in a Green State in the County of **Victoria**, Province of **Nova Scotia**, during the year 1913-14.

Number.	Fishing Districts.	Salmon, cwts.	Salmon, value.	Lobsters, cwts.	Lobsters, value.	Cod, cwts.	Cod, value.	Haddock, cwts.	Haddock, value.	Hake and Cusk, cwts.	Hake and Cusk, value.	Pollock, cwts.	Pollock, value.	Herring, cwts.	Herring, value.	Mackerel, cwts.	Mackerel, value.	Shad, cwts.	Shad, value.	Number.
	Victoria County.	$	$		$		$		$		$		$		$		$		$	
1	Iona, Washabuck and Little Narrows..	48	576	55	275	476	952							20	40					1
2	Cape Smokey to Wreck Cove.........	35	240	625	3125	400	400	90	67			250	175	820	410	90	180			2
3	Breton Cove and vicinity to Englishtown	55	440	1362	6810	460	460	104	77			332	222	7105	3552	48	96			3
4	Baddeck and vicinity...............	30	320			238	616	9	21					350	386	1	9			4
5	Ingonish..........................	77	770	1282	5771	8401	12901	14892	14892			532	479	1312	1624	1312	3317			5
6	Neil's Harbour and New Haven......	20	140	1640	7380	8200	12565	4150	3950	30	27	160	144	990	687	240	729			6
7	White Point, Dingwall and Sugar-Loaf.	173	1557	1102	4969	3100	4850	1200	1112	18	16	300	270	1040	705	120	480			7
8	Meat Cove, Bay St-Lawrence and vicinity					1125	1688	164	164			48	43	826	589	87	348			8
9	Big Bras-d'Or.....................	129	1032	1531	6889	270	405							105	90					9
	Totals........ " "	567	5115	7597	35209	22670	34538	20609	20283	48	43	1622	1333	12568	8083	1898	5150			

*Cwt. = 100 lbs.

THE CATCH.

RETURN showing the Quantities and Values of all Fish caught and landed in a Green State in the County of Victoria, Province of Nova Scotia, during the year 1913-14.—*Concluded.*

Number.	Fishing Districts.	Alewives, cwts.	Alewives, value. $	Sardines, brls.	Sardines, value. $	Halibut, cwts.	Halibut, value. $	Smelts, cwts.	Smelts, value. $	Trout, cwts.	Trout, value. $	Eels, cwts.	Eels, value. $	Sword-fish, cwts.	Sword-fish, value. $	Squid, brls.	Squid, value. $	Oysters, brls.	Oysters, value. $	Clams, brls.	Clams, value. $
	Victoria County.																				
1	Iona, Washabuck and Little Narrows	10	10															446	1338		
2	Cape Smokey to Wreck Cove																				
3	Breton Cove and vicinity to Englishtown																				
4	Baddeck and vicinity	5	10					9	53												
5	Ingonish					5	25	26	156												
6	Neil's Harbour and New Haven					86	430	1	4	2	16	8	24	25	125	3	9				
7	White Point, Dingwall and Sugar Loaf					90	450	8	32					120	420	5	15				
8	Meat Cove, Bay St. Lawrence and vicinity					59	177	2	8	1	8	2	6	124	434	1	3				
9	Big Bras-d'Or																	15	45		
	Totals	15	20			240	1082	46	253	3	24	10	30	269	979	9	27	461	1383		

*Cwt. = 100 lbs.

5 GEORGE V., A. 1915

THE CATCH MARKETED.

RETURN showing the Quantities and Values of all Fish and Fish Products Marketed in a fresh, dried, pickled, canned, &c, state, for the County of Victoria, Province of Nova Scotia, during the year 1913-14.

Number.	Fishing Districts.	Salmon used fresh and frozen, *cwt.	Salmon, canned, cases.	Lobsters, canned, cases.	Lobsters, shipped in shell, cwts.	Cod, used fresh, cwts.	Cod, shipped green salted, cwts.	Cod, dried, cwts.	Haddock, used fresh, cwt.	Haddock, smoked, cwts.	Haddock, dried, cwt.	Hake and Cusk, dried, cwts.	Pollock, used fresh, cwts	Pollock, dried, cwts.	Herring, used fresh, cwts.	Herring, pickled, brls	Herring, as bait, brls.	Number.
	Victoria County.																	
1	Iona, Washabuck and Little Narrows	48						84						83	8	60	4	1
2	Wreck Cove to Cape Smokey	35		30				133			30			110		56	160	2
3	Breton Cove and vicinity to Englishtown	55		85		223		153			34					80	1734	3
4	Baddeck and vicinity	30			55			36			3					104	5	4
5	Ingonish	77				40	40	81							320	35	500	5
6	Neil's Harbour and New Haven	10		53		750	3704	186	13189	520	231	10	12	173	3409	13	412	6
7	White Point, Dingwall and Sugar Loaf	167	12	66		120	3750	155	180		1323			53	100	38	462	7
8	Meat Cove, Bay St. Lawrence and vicinity	129	7	41		145	1236	103	174		343		1	98	60	35	334	8
9	Big Bras-d'Or			60	6	81	361	90	98		22	6	4	16	74			9
	Totals	551	19	3015	61	1371	9091	1021	13611	520	1976	16	17	533	4077	421	3611	
	Rates	12 00	7 00	18 00	8 00	2 00	4 00	6 00	2 00	3 00	5 00	4 00	1 50	3 25	1 50	5 00	2 00	
	Values	6612	133	54270	488	2742	36364	6126	27282	1560	9880	64	25	1732	6116	2105	7222	

*Cwt = 100 lbs.

THE CATCH MARKETED.

RETURN showing the Quantities and Values of all Fish and Fish Products Marketed in a fresh, dried, pickled, canned, &c., state, for the County of Victoria, Province of Nova Scotia, during the year 1913-14.—*Concluded.*

No.	Fishing Districts.	Herring, smoked, brls.	Herring, fertilizers, brls.	Mackerel, used fresh, cwts.	Mackerel, salted, brls.	Alewives, used fresh, cwts.	Alewives, salted, brls.	Halibut, used fresh, cwts.	Smelts, used fresh, cwts.	Trout, used fresh, cwts.	Eels, used fresh, cwts.	Sword-fish, used fresh, cwts.	Squid, used as bait, brls.	Oysters, used fresh, brls.	Clams and Quahaugs, used fresh, brls.	Tongues and Sounds, pickled or dried, cwts.	Hair Seal Skins, number.	Fish Oil, galls.	No.
	Victoria County.																		
1	Iona, Washabuck and Little Narrows				30	10								446				55	1
2	Wreck Cove to Cape Smokey				16													90	2
3	Breton Cove and vicinity to Englishtown					5								15				69	3
4	Baddeck and vicinity																	9	4
5	Ingonish			1	437			5	9			25						3000	5
6	Neil's Harbour and New Haven			20	73			86	26			120	5			4	4	2900	6
7	White Point, Dingwall and Sugar Loaf			25	25			90	1		8	124	5				3	900	7
8	Meat Cove, Bay St. Lawrence and vicinity			45	18			59	8	2	2		1						8
9	Big Bras-d'Or			33					2	1							12	350	9
	Totals			125	599	15		240	46	3	10	269	9	461		4	19	7373	
	Rates..............$			7 00	10 00	2 00		8 00	6 00	10 00	5 00	6 00	3 00	3 00		4 00	1 00	30c.	
	Values.............$			875	5990	30		1920	276	30	50	1614	27	1383		16	19	2212	

Total Value $177,162

*Cwt.=100 lbs.

THE CATCH.

RETURN showing the Quantities and Value of all Fish caught and landed in a Green State in the County of **Inverness**, Province of **Nova Scotia**, during the Year 1913-14.

Number	Fishing Districts	Salmon, cwts*	Salmon, value	Lobsters, cwts	Lobsters, value	Cod, cwts	Cod, value	Haddock, cwts	Haddock, value	Hake and Cusk, cwts	Hake and Cusk, value	Pollock, cwts	Pollock, value	Herring, cwts	Herring, value	Mackerel, cwts	Mackerel, value	Alewives, cwts	Alewives, value	Number
	Inverness County.																			
1	Meat Cove, Poulet Cove and Pleasant Bay	253	2277	1125	5072	225	225	1	1			1	1	924	487	1407	5160			1
2	Eastern Harbour, Cap Rouge, Chéticamp Point and Grand-Etang			4432	22160	15137	22705	1226	964	1350	877	610	397	2506	1503	5670	22650			2
3	Friar's Head Margaree Harbour to Smith's Cove	362	3620	1740	8706	17807	26710	2145	2145	225	225			1511	1511	8200	32850	1500	1530	3
4	Broad Cove, Port Ban to Mabou Harbour	780	9360	1685	7582	920	1840	165	247					636	636	1800	6000	50	75	4
5	West Lake Ainslie and Whycocomah Bay	40	400			3500	3500	8000	10000	3910	2932			24	24					5
6	Little Mabou and Port Hood to Hawkesbury			6750	33750									9590	3100	2100	11600			6
7	West Bay, Malagawatch and Deny's Basin					310	620							1400	700					7
	Totals	1435	15657	15732	77264	38069	55920	11537	13357	5485	4034	611	398	16591	7961	19177	78250	1550	1605	

*Cwts. = 100 lbs.

THE CATCH.

RETURN showing the Quantities and Values of all Fish caught and landed in a Green State, in the County of Inverness, Province of Nova Scotia, during the Year 1913-14—*Concluded.*

Number	Fishing Districts	Halibut, cwts.	Halibut, value.	Smelts, cwts.	Smelts, value.	Trout, cwts.	Trout, value.	Eels, cwts.	Eels, value.	Tom-Cod, cwts.	Tom-Cod, value.	Sword-fish, cwts.	Sword-fish, value.	Mixed fish, cwts.	Mixed fish, value.	Squid, bris.	Squid, value.	Oysters, bris.	Oysters, value.	Number
	Inverness County.		$		$		$		$		$		$		$		$		$	
1	Meat Cove, Poulet Cove and Pleasant Bay	8	35													3	9			1
2	Eastern Harbour, Cap Rouge, Chéticamp Point and Grand-Étang											54	270	10	10					2
3	Friar's Head and Margaree Harbour to Smith's Cove															48	144			3
4	Broad Cove, Port Ban to Mabou Harbour							3	12											4
5	West-Lake Ainslie and Whycocomah Bay			40	240	2	20	20	80	10	5									5
6	Little Mabou and Port Hood to Hawkesbury			100	600	30	150	10	30	120	290									6
7	West Bay, Malagawatch and Deny's Basin					5	25											300	900	7
	Totals	8	33	140	840	37	195	33	122	130	295	54	270	10	10	51	153	300	900	

Cwts. = 100 lbs.

39—3½

5 GEORGE V., A. 1915

THE CATCH MARKETED.

Return showing the Quantities and Values of all Fish and Fish Products Marketed in a fresh, dried, pickled, canned, etc., state, for the County of Inverness, Province of Nova Scotia, during the Year 1913-14.

No.	Fishing Districts	Salmon, used fresh and frozen, cwts.	Salmon, canned, cases.	Lobsters, canned, cases.	Cod, used fresh, cwts.	Cod, shipped green-salted, cwts.	Cod, dried, † Quintals.	Haddock, used fresh, cwts.	Haddock, dried, cwts.	Hake and Cusk, used fresh, cwts.	Hake and Cusk, dried, cwts.	Pollock, used fresh, cwts.	Pollock, dried, cwts.	Herring, used fresh, cwts.	Herring, pickled, brls.	Herring used as bait, brls.	Number.
	Inverness County.																
1	Meat Cove, Pollet Cove and Pleasant Bay	253		450	35	80	10	1	379			1			10	452	1
2	Cap Rouge, Eastern Harbor, Cheticamp and Grand Etang	357	5	1773	124	3145	2891	11	715							1273	2
3	Friar's Head, Margaree Harbour to Smith's Cove	780		696	500	7909	663				450					755	3
4	Broad Cove, Port Ban to Mabou Harbour	40		674	164		140		33		76					298	4
5	West Lake Ainslie and Wycocomah Bay				2300			127					203			12	5
6	Little Mabou and Port Hood to Hawkesbury			2700	310					3900	2			9590			6
7	West Bay, Malagawatch and Deny's Basin						400	8000								700	7
	Totals	1430	5	6293	3479	11134	4104	8139	1127	3900	528	1	203	9590	10	3490	
	Rates	$ 12.00	10.00	18.00	1.50	3.50	5.50	1.50	4.00	1.00	3.00	1.00	3.00	1.00	4.00	1.50	
	Values	$ 17160	50	113274	5218	38969	22572	12208	4508	3900	1584	1	609	9590	40	5235	

*Cwt=100 lbs.   † Quintals=112 lbs.

THE CATCH MARKETED.

RETURN showing the Quantities and Values of all Fish and Fish Products Marketed in a fresh, dried, pickled, canned, etc., state, for the County of **Inverness**, Province of **Nova Scotia**, during the Year 1913–14—*Concluded.*

Number	Fishing Districts.	Mackerel, used fresh, cwts.	Mackerel, salted, brls.	Alewives, used fresh, cwts.	Alewives, salted, brls.	Halibut, used fresh, cwts.	Smelts, used fresh, cwts.	Trout, used fresh, cwts.	Eels, used fresh, cwts.	Tom Cod, used fresh, cwts.	Swordfish, used fresh, cwts.	Mixed fish, used fresh, cwts.	Squid, used as bait, brls.	Oysters, used fresh, brls.	Tongues and sounds, pickled or dried, cwts.	Hair Seal Skins, Number.	Fish Oil, galls.
	Inverness County.																
1	Meat C ve, Poulet Cove and Pleasant Bay	7	467			8					54	10	3			5	20
2	Cap Rouge, Eastern Harbor, Cheticamp and Grand Etang	54	1864	600	300			2	3				48		2000		5175
3	Friar's Head, Margaree Harbour to Smith's Cove		2733				40		20	10							1330
4	Broad Cove, Port Ban to Mabou Harbour	75	575		17			30									
5	West Lake Ainslie and Wycocomah Bay						100	5	10	120							
6	Little Mabou and Port Hood to Hawkesbury	2100												300			
7	West Bay, Malagawatch and Deny's Basin																
	Totals	2236	5639	600	317	8	140	37	33	130	54	10	51	300	2000	5	6545
	Rates	$8.00	14.00	1.00	3.00	6.00	10.00	10.00	4.00	2.50	7.00	1.00	4.00	3.00	4.00	1.00	.25c.
	Value	$17888	78946	600	951	48	1400	370	132	325	378	10	204	900	8000	5	1636

Total value............ $8346 711

5 GEORGE V., A. 1915

RECAPITULATION.

Of the Quantities and Values of all Fish caught and landed in a Green State, and of the Quantities and Values of all Fish and Fish Products Marketed in a fresh, dried, pickled, canned, &c., state, for **District No. 1,** Province of **Nova Scotia,** during the year 1913-14.

Kinds of Fish.	Caught and Landed in a Green State.		Marketed.		Total Marketed Value.
	Quantity.	Value.	Quantity.	Value.	
		$		$	$
Salmon............................ cwts.	2,406	24,947	....	....	
" used fresh..................... "	....	....	2,385	28,374	
" canned.................. cases.	....	....	24	183	
					28,557
Lobsters .... cwts.	51,426	281,545	....	....	
" canned...................... cases.	....	....	19,382	348,876	
" shipped in shell.......... cwts.	....	....	2,975	24,446	
					373,322
Cod.......................... "	114,043	169,931	....	....	
" used fresh.................. "	....	....	14,026	21,723	
" green—salted............. "	....	....	23,890	85,921	
" dried..................... "	....	....	17,373	96,061	
					203,705
Haddock .... "	64,949	74,025	....	....	
" used fresh. .... "	....	....	35,099	65,872	
" smoked (finnans)........ "	....	....	610	1,830	
" dried.................. "	....	....	9,235	43,331	
					111,033
Hake and Cusk .... "	7,338	5,237	....	....	
" used fresh .... "	....	....	4,883	4,883	
" dried............ "	....	....	788	2,380	
					7,263
Pollock .... "	5,245	4,771	....	....	
" used fresh.............. "	....	....	85	93	
" dried................ "	....	....	1,696	5,939	
					6,032
Herring....................... "	54,947	44,431	....	....	
" used fresh.................. "	....	....	14,507	16,545	
" pickled.............. brls.	....	....	3,937	17,921	
" used as bait........... "	....	....	14,320	27,211	
					61,677
Mackerel..................... cwts.	36,772	134,667	....	....	
" used fresh................ "	....	....	7,445	44,183	
" salted.......... brls.	....	....	9,775	123,843	
					168,026
Shad........................ cwts.	46	230	....	....	
" used fresh.................. "	....	....	46	....	
					368
Alewives .... "	1,596	1,656	....	....	
" used fresh.................. "	....	....	622	642	
" salted .... brls.	....	....	325	975	
					1,617
Halibut, used fresh . ... cwts.	800	5,107	800	....	6,732
Flounders .... "	51	26	51	....	51
Smelts........................ "	730	4,310	730	....	5,188
Trout........................ "	40	219	40	....	400
Eels........................ "	123	312	123	....	342
Tom Cod.................... "	130	295	130	....	325
Swordfish.................... "	1,421	4,755	1,421	....	6,505
Mixed Fish..... .. .. "	30	30	30	....	30
Squid...................... brls.	269	670	269	....	887
Oysters...................... "	761	2,283	761	....	2,283
Clams .... "	87	256	....	....	
" used fresh............ "	....	....	87	....	266

RECAPITULATION.

Of the Quantities and Values of all Fish caught and landed in a Green State, and of the Quantities and Values of all Fish and Fish Products Marketed in a fresh, dried, pickled, canned, &c., state, for **District No. 1,** Province of **Nova Scotia,** during the year 1913-14— *Concluded.*

Kinds of Fish.	Caught and Landed in a Green State.		Marketed.		Total Marketed Value.
	Quantity.	Value.	Quantity.	Value.	
		$		$	$
Tongues and Sounds .... cwts.			2,004		8,016
Hair Seal Skins..... .. .. No.			24		24
Fish Oil.... .... galls.			19,208		5,435
Totals............		759,703	... ...		998,084

RECAPITULATION.

Of the Number of Fishermen, &c., and of the Number and Value of Fishing Vessels Boats, Nets, &c., in **District No. 1,** Province of **Nova Scotia,** for the year 1913-14.

—	Number.	Value.
		$
Steam Fishing Vessels (tonnage)...................	96	60,300
Sailing and Gasoline Vessels	2,560	89,059
Boats (sail)........	340	76,780
" (gasoline)...	75	27,130
Carrying Smacks....................................	12,542	106,441
Gill Nets, Seines, Trap and Smelt Nets, etc.......	2	20
Weirs......	3,470	29,965
Trawls	8,040	5,852
Hand Lines	130,937	111,626
Lobster Traps................	70	57,270
" Canneries........		
Salmon Canneries....................		
Clam Canneries.............................	39	279,720
Freezers and Ice-houses....................	852	54,872
Smoke and Fish-houses...............	231	182,060
Fishing Piers and Wharves....		
		1,072,098

Number of men employed on Vessels.... 491
     "   . "   Boats 4,439
     "   "   Carrying Smacks 138
     "   persons employed in Fish-houses, Freezers, Canneries, &c.... 2,258

        Total.... 7,326

5 GEORGE V., A. 1915

DISTRICT

RETURN showing the Number of Fishermen, etc., the Number and Value of Vessels and
Industry in the County of **Cumberland**, Province

Number.	Fishing Districts.	Steam vessels.				Boats.					Carrying Smacks.			Gill Nets, Seines, Trap and Smelt Nets, etc.	
		Number.	Tons.	Value.	Men.	Sail.	Value.	Gasoline.	Value.	Men.	Number.	Value.	Men.	Number.	Value.
	Cumberland County.			$			$		$			$			$
1	Malagash, East Wallace and Fox Harbour....	..	..	..	..	33	2450	84	15744	143	1	200	1	....	....
2	Pugwash and Gulf Shore.........	..	..	..	..	21	1052	55	8250	80	1	300	1	64	750
3	Port Philip, Northport and Amherst Shore....	..	..	..	..	11	440	13	1900	29	..	..	..	29	435
4	Wallace River..........	..	..	..	..	28	610	....	......	28	..	..	..	11	166
5	River Philip....	..	..	..	..	4	80	1	400	6	..	..	..	39	880
6	Maccan and Nappan....	..	..	..	..	....	....	2	600	6	..	..	..	....	....
7	Minudie to Apple River....	..	..	..	..	2	60	2	600	12	..	..	..	....	....
8	Advocate..	..	..	..	..	10	300	3	1200	34	..	..	..	12	96
9	Spencer Island.............	..	..	..	..	4	120	2	800	16	..	..	..	6	48
10	Port Greville...	..	..	..	..	6	180	..	......	18	..	..	..	5	40
11	Parrsboro and Two Islands......	..	..	..	..	....	....	3	750	9	..	..	..	2	25
	Totals...............	..	..	..		119	5292	165	30244	381	2	500	2	168	2440

RETURN showing the Number of Fishermen, etc., the Number and and Value of Vessels
Industry in the County of **Colchester**, Province

Number.	Fishing Districts.	Vessels, Boats and Carrying Smacks.					Fishing Gill Nets, Seines, Trap and Smelt Nets, etc.	
		Boats.						
		Sail.	Value.	Gasoline.	Value.	Men.	Number.	Value.
	Colchester County.		$					$
1	Sterling........	7	200	12	1850	17	9	225
2	Stewiacke..	....	......	90	900	165	100	1375
3	Five Islands and Economy....	1	75	...	....	1	1	60
4	Little Bass River to Highland Village.....	5	375	...	....	5	5	300
5	Great Village to Queens Village	13	975	...	......	13	13	780
	Totals.................	26	1625	102	2750	201	128	2690

No. 2.

Boats, and the Quantity and Value of all Fishing Gear, etc., used in the Fishing of **Nova Scotia,** during the Year 1913-14.

Fishing Gear.								Canneries		Other Material.				Persons Employed in Canneries, Freezers and Fish-houses.	
Weirs.		Trawls.		Hand Lines.		Lobster Traps.		Lobster Canneries.		Freezers and Ice-houses.		Smoke and Fish-houses.			
Number.	Value.	Number.	Value.	Number.	Value.	Number.	Value.	Number.	Value.	Number.	Value.	Number.	Value.	Number.	
	$		$		$		$		$		$		$		
...	...	...	...	...	...	32270	28660	14	30100	...	...	...	...	226	1
...	...	...	...	...	...	19948	19948	11	12175	...	...	...	...	125	2
2	100	...	...	...	...	6700	5150	7	1300	...	...	6	2600	32	3
...	...	...	...	...	...	...	...	...	...	...	...	4	110	...	4
															5
2	150	...	...	...	...	...	...	...	...	...				...	6
2	150	...	...	...	...	50	50	...	...	...				...	7
2	150	...	...	34	68	50	50	...	...					...	8
...	...	...	...	16	32	50	50	...	...					...	9
...	...	...	...	18	36	...	...	...	...			...	...	...	10
6	450	...	...	9	18	25	25	...	...	1	1000	...	...	3	11
14	1000	...	...	77	154	59093	53933	32	43575	1	1000	10	2710	386	

and Boats, and the Quantity and Value of all Fishing Gear, etc., used in the Fishing of **Nova Scotia,** during the Year 1913-14.

Gear.								Canneries.				Other Material.		Persons Employed in Canneries, Freezers, and Fish-Houses.	
Weirs.		Trawls.		Hand Lines.		Lobster Traps.		Lobster Canneries.		Clam Canneries.		Smoke and Fish-houses.			
Number.	Value.	Number.	Value.	Number.	Value.	Number.	Value.	Number.	Value.	Number.	Value.	Number.	Value.		Number.
	$		$		$		$		$		$		$		
...	...	...	...	...	...	4625	4625	2	1900	...	...	...	...	17	1
...	...	...	...	...	...	...	...	...	...	...	...	2	40	...	2
...	...	4	40	...	...	...	...	...	...	...	...	...	...	...	3
...	...	...	...	...	...	...	...	...	...	...	...	...	...	...	4
...	...	...	...	...	...	...	...	...	...	...	...	...	...	...	5
...	...	4	40	...	...	4625	4625	2	1900	...	...	2	40	17	

5 GEORGE V., A. 1915

RETURN showing the Number of Fishermen, &c., the Number and Value of Vessels and Industry in the County of **Pictou**, Province of

Number.	Fishing Districts.	Vessels, Boats and Carrying Smacks.								Gill Nets, Seines, Trap and Smelt Nets, &c,	
		Boats.					Carrying Smacks.				
		Sail.	Value.	Gasoline.	Value.	Men.	Number.	Value.	Men.	Number.	Value.
	Pictou County.		$		$			$			$
1	West Pictou.............	38	1270	164	26590	206	....	....	....	103	1145
2	Pictou Island	6	290	58	10300	126	1	150	2	98	830
3	Pictou Harbour...........	9	360	....	......	12	3	11500	7	50	365
4	Little Harbour and East Branch St. Mary's River....	20	700	....	......	15	....	....	....	45	2410
5	Merigomish Island....	13	520	3	600	26	1	400	2	16	1300
6	Ponds........................	5	200	18	3600	36	....	....	....	76	1762
7	Lismore....	5	200	I	200	7	....	....	....	14	900
	Totals.	96	3540	244	41290	428	5	12050	11	402	8772

RETURN showing the Number of Fishermen, &c., the Number and Value of Vessels and Industry in the County of **Antigonish**, Province of

Number.	Fishing Districts.	Vessels, Boats and Carrying Smacks.											Gill Nets, Seines, Trap and Smelt Nets, etc.	
		Sailing and Gasoline Vessels.			Boats.					Carrying Smacks.				
		(10 to 20 tons) No.	Value.	Men.	Sail.	Value.	Gasoline.	Value.	Men.	Number.	Value.	Men	Number.	Men.
	Antigonish County.		$			$		$			$			
1	Harbour au Bouche, Linwood and Cape Jack....	1	300	3	50	955	8	1510	82	3	1000	3	357	2420
2	Tracadie, Bayfield, Monk's Head & South Side Antigonish Harbour	....	....	....	45	680	13	2520	76	1	125	1	170	5200
3	North Side Antigonish Harbour, Lakevale and South Side of Cape George	...	....	....	38	670	8	1310	71	2	400	2	142	2370
4	North Side of Cape George, Georgeville and Malignant Cove to Knoidart........	...	....	....	28	560	8	1230	38	3	600	3	90	1510
	Totals............ ...	1	300	3	161	2865	37	6570	267	9	2125	9	759	11500

SESSIONAL PAPER No. 39

Boats, and the Quantity and Value of all Fishing Gear, &c, used in the Fishing Nova Scotia, during the year 1913-14.

Fishing Gear.						Canneries.		Other Material.						Persons Employed in Canneries, Freezers and Fish Houses.	
Trawls.		Hand Lines.		Lobster Traps.		Lobster Canneries.		Freezers and Ice Houses.		Smoke and Fish Houses.		Fishing Piers and Wharves.			
Number.	Value.	Number.	Value.	Number.	Value.	Number.	Value.	Number.	Value.	Number.	Value.	Number.	Value.		Numbers.
	$		$		$		$		$		$		$		
...	....	170	85	45175	45175	13	27300	....	....	....	....	....	....	250	1
....	......	32	16	24700	24700	3	14500	....	....	....	....	....	.....	131	2
...	......	18	9	....	....	...	...	...	....	....	....	....	...	15	3
...	.....	18	9	1800	1800	1	250	....	....	....	.....	....	....	14	4
2	14	16	5	2800	2500	1	1200	4	1400	4	80	....	....	30	5
4	28	18	9	5700	4485	1	1800	10	1180	10	200	...	...	27	6
1	7	6	3	800	680	...	.....	2	50						7
7	49	272	136	80975	79340	19	45050	16	2630	14	280	....	.....	467	

Boats, and the Quantity and Value of all Fishing Gear, &c., used in the Fishing Nova Scotia, during the Year 1913-14.

Fishing Gear.						Canneries.		Other Material.						Persons Employed in Canneries, Freezers and Fish Houses.	
Trawls.		Hand Lines.		Lobster Traps.		Lobster Canneries.		Freezers and Ice Houses.		Smoke and Fish Houses.		Fishing Piers and Wharves.			
Number.	Value.	Number.	Value.	Number.	Value.	Number.	Value.	Number.	Value.	Number.	Value.	Number.	Value.		Number.
	$		$		$		$		$		$		$		
53	324	131	65	7600	3800	1	1000	1	1000	43	497	2	4000	36	1
25	148	90	45	4000	2000	1	800	2	1800	31	452	....	......	24	2
56	336	51	28	9000	4500	2	2400	1	900	18	210	....	....	58	3
30	180	36	18	6400	3200	3	2600	2	2000	17	284	....	....	46	4
164	988	308	156	27000	13500	7	6800	6	5700	109	1443	2	4000	164	

RETURN showing the Number of Fishermen, etc., the Number and Value of Vessels and Boats, and the Quantity and Value of all Fishing Gear, etc., used in the Fishing Industry in the County of **Guysboro**, Province of **Nova Scotia**, during the Year 1913-14.

| | Sailing and Gasoline Vessels | | | | Boats | | | | | Carrying Smacks | | | Gill Nets, Seines, Trap and Smelt Nets, etc. | | Trawls | | |
Fishing Districts	Number (40 tons and over)	Number (10 to 20 tons)	Value $	Men	Sail	Value $	Gasoline	Value $	Men	Number	Value $	Men	Number	Value $	Number	Value $	Number
Guysboro County.																	
1 Ecum Secum					28	560	10	1800	40	1	150	2	90	450	16	96	1
2 Marie Joseph					36	1080	16	2250	46				110	500	41	246	2
3 Liscombe and Spanish Ship Bay					40	800	22	4450	55	2	400		120	700	30	300	3
4 Gegoggin					20	400	11	1670	25				57	185	10	100	4
5 St. Mary's Bay and River					30	440	8	1180	20				50	250	9	50	5
6 Wine Harbour					22	160	1	150	18				60	500	4	40	6
7 Port Hilliford					22	100	1	150	30	1	100	1	100	700	19	100	7
8 Holland's Harbour and Indian River					4	100		200	6				20	150	4	40	8
9 Port Beckerton		1	700	2	30	800	14	5400	45	1	350	2	230	2900	70	700	9
10 Fisherman's Harbour					25	500	11	2000	32	1	600	1	150	900	15	150	10
11 Country Harbour					5	100			5				10	60			11
12 Isaac's Harbour		1		1	12	600	14	4200	33	2	400	10	100	1000	25	250	12
13 Drum Head		5	5000	10	13	650	12	3600	44	5	5000		185	1850	86	840	13
14 Seal Harbour					7	350	9	3000	26				124	1240	60	600	14
15 Goldfiles Harbour					4	200	10	3400	25				150	1500	24	240	15
16 New Harbour					5	250	8	8400	23		250	3	450	450	43	430	16
17 Tor Bay					21	1560	2	325	29	2			170	1700	16	160	17
18 Larry's River		5	4000	16	65	3200	2	450	38				800	8000	75	75	18
19 Charlo's Cove		10	5000	53	56	2760	2	450	31				665	6650	98	984	19
20 Cole Harbour		2	900	6	38	1900	4	850	40				425	4250	64	640	20
21 Port Félix		1	500	5	85	4300	10	2000	95		150		820	8200	146	1460	21
22 Whitehead	1	6	5000	26	76	3275	15	3300	79	8	1600	8	1000	10000	225	2250	22
23 Raspberry and Dover		3	2000	17	43	1000	35	750	34	6	500	6	175	1750	69	280	23
24 Canso and Canso Tittle	3	17	27000	90	170	5750	2	1040	137	3	1500		2100	21000	565	5650	24
25 Fox Island Maine					22	650	2	200	17				180	1800	35	350	25
26 Half Island Cove		1	450	3	44	2900	6	1000	38	3	1015	4	860	8500	90	900	26

27	Philip's Harbour		1		850	3	32	1300	4	1010	31				575	5750	64	640
28	Queensport		1		375	5	48	2800	6	1000	42				330	3300	142	1020
29	Peas Brook						35	1060			24				325	3250	51	510
30	Halfway Cove						51	2100			56				620	6200	82	890
31	Sandy Cove and Cook's Cove						35	1240	3	500	36				390	3900	50	500
32	Guysboro and Manchester						18	600	2	300	11				285	2850	20	200
33	Port Shoreham						19	675			19				355	3550	55	550
34	St. Francis						36	1000			45				420	4200	30	300
35	Oyster Ponds						32	975			57				395	3950	28	280
36	Sand Point						25	1400	4	1200	11				569	5690	37	370
37	Middle Melford						38	2130	3	8600	47				872	8720	35	330
38	Mulgrave and Auld's Cove						7	330			13	3	8600	7	100	1000	9	90
	Totals	4	53		51675	236	1292	56485	271	64395	1431	38	20615	54	14391	140635	2402	23262

5 GEORGE V., A. 1915

RETURN showing the Number of Fishermen, etc., the Number and Value of Vessels and Boats, and the Quantity and Values of all Fishing Gear, etc., used in the Fishing Industry in the County of **Guysboro**, Province of **Nova Scotia**, during the Year 1913–14.—*Concluded.*

No.	Fishing Districts	Fishing Gear — Hand Lines Number	Hand Lines Value	Lobster Traps Number	Lobster Traps Value	Lobster Canneries Number	Lobster Canneries Value	Freezers and Ice-houses Number	Freezers and Ice-houses Value	Smoke and Fish-houses Number	Smoke and Fish-houses Value	Fishing Piers and Wharves Number	Fishing Piers and Wharves Value	Persons employed in Canneries, Freezers and Fish houses Number
	Guysboro County.													
1	Kenn Secum	70	35	2500	2500			2	100	25	400	4	200	1
2	Marie Joseph	50	25	2500	2500			1	50	25	500	6	300	2
3	Liscombe and Spanish Ship Bay	102	50	5000	5000	1	1200	3	100	40	1000	6	200	18
4	Gegoggin	40	20	1600	1600					14	180	2	100	3
5	St. Mary's Bay and River	35	18	1200	1200	1	100	6	50	12	200	1	50	4
6	Wine Harbour	30	15	1200	1200					12	300	1	50	5
7	Port Hilford	40	20	1600	1000			2	100	15	300			6
8	Holland's Harbour and Indian River	8	4	800	800								40	7
9	Port Beckerton	80	40	3500	3500	1	800	10	1000	20	800	8	400	2
10	Fisherman's Harbour	60	30	1600	1500					8	200	3	150	9
11	Country Harbour	10	5	100	100					6	100			4
12	Isaacs Harbour	42	42	1650	1650	1	1000	3	7000	20	3000	9	270	26
13	Drum Head	68	68	2200	2200	2	2500	1	2800	15	2000	8	800	23
14	Seal Harbour	80	86	1300	1300			1	900	17	1900	2	450	18
15	Goldfies Harbour	138	138	1300	1300							2	500	6
16	New Harbour	333	333	3000	3000	1	300	1	1800	37	9200	3	2500	15
17	Tor Bay	115	115	1400	1400					13	1300	3	700	12
18	Larry's River	260	260	1900	1900					31	3000	20	800	4
19	Charlo's Cove	195	195	1600	1600					20	2200	2	2300	19
20	Cole Harbour	114	114	2000	2000					18	340	15	350	20
21	Port Felix	200	290	4800	4800	1	1000			42	4500	27	1290	14
22	Whitehead	275	275	3900	3900	3	3450			38	9700	2	1270	31
23	Raspberry and Dover	108	108	1700	1700	1	800			11	1800	5	420	14
24	Canso and Canso Tittle	860	860	6800	6800	2	3400	8	77400	57	17020	30	13000	115
25	Fox Island Maine	28	28	800	800				7150					5
26	Half Island Cove	92	92	1900	1900	1	400	2	8000	20	6500	1	800	826

27 Philip's Harbour	75	75	1500	1500			5	17600	14	2600	1	800	2
28 Queensport	100	100	2100	2100	1	2500			25	6535	2	5300	
29 Peas Brook	69	69	1200	1200					14	1400			
30 Halfway Cove	114	114	2800	2800					25	3350			
31 Sandy Cove and Cook's Cove	51	51	1800	1800			1	400	16	2800	1	75	1
32 Guysboro and Manchester	29	29	500	500			2	100	10	5900	5	10000	3
33 Port Shoreham	21	21	900	900					16	2000			
34 St. Francis	60	60	2200	2200					22	3250			
35 Oyster Ponds	51	51	1800	1800					18	2750	2	800	
36 Sand Point	28	28	500	500					15	2245			
37 Middle Melford	56	56	2300	2300					25	3850			
38 Mulgrave and Auld's Cove	20	20	600	600	2	5000	3	10000	10	6750	3	7500	19
Totals	4167	3904	75350	75350	17	23460	54	129050	740	111670	200	90885	854

5 GEORGE V., A. 1915

RETURN showing the Number of Fishermen, etc., the Number and Value of Vessels and Boats, and the Quantity and Value of all Fishing Gear, etc., used in the Fishing Industry in the County of Halifax, Province of Nova Scotia, during the year 1913-14.

Number	Fishing Districts	Sailing and Gasoline vessels					Boats					Carrying Smacks			Number
		Number (40 tons and over)	Number (20 to 40 tons)	Number (10 to 20 tons)	Value $	Men	Sail	Value $	Gasoline	Value $	Men	Number	Value $	Men	
	Halifax County.														
1	North Shore	1			3000	14	130	2000	2	600	80	1	200	2	1
2	East St. Margaret's	1			2500	18	230	9100	4	800	140	1	300	2	2
3	Indian Harbour		9		9700	39	150	9000	56	12500	95	2	500	4	3
4	Peggy's Cove						100	3000	2	400	40				4
5	Dover		13		8200	48	376	11200	4	1500	135	1	200	2	5
6	Prospect				1200	9	100	3300	4	1000	90				6
7	Terrence Bay	2	8		8400	48	200	800	17	4240	140	1	400	3	7
8	Pennant	2	8		4300	32	20	600	3	900	21				8
9	Sambro	4	5		5000	32	60	4500	9	2850	19				9
10	Ketch Harbour	1	1		1500	12	60	4000	5	1500	69	1	5000	4	10
11	Portuguese Cove								10	3000	30				11
12	Herring Cove	1	3		4200	30	70	3100		300	84				12
13	Ferguson's Cove						6	200		200	14				13
14	Bedford and Grand Lake						16	425	1		16				14
15	Halifax			3	900	3	10	200			20				15
16	Dartmouth, Eastern Passage and Devil's Island			1			62	1350	26	5000	56	1		1	16
17	Cow Bay and ...						18	450	1	150	16				17
18	Seaforth and Three Fathom Harbour						35	520	1	175	30				18
19	West Chezzetcook	3			6775	51	112	1500		300	66				19
20	East Chezzetcook						14	288			13				20
21	... Harbour						31	575	6	800	31				21
22	... Harbour						58	1310	22	1050	50	1	200	1	22
23	Jeddore		2	3	1800	18	78	1150	18	3850	72	1	75	1	23
24	Ron Harbour and Owl's Head	1		1	2000	20	70	1150	15	2305	59	1	120	1	24
25	West Ship Harbour				1000	4	16	480	5	500	17		100		25
26	Ket Ship Harbour	1			600	5	6	330	4	675	21				26
27	Pleasant Harbour and Tangier								19	570	41				27

28	Pope's Harbour and Gerrard's Island	2	300	2	18	2155	11	150	3	3	800	1			
29	Spry Bay, Mor Head and Mushaboom	2	350	2	81	2790	12	2165	28	9	1300	3	1		
30	Sheet Harbour and Sober Island	1	100	1	39	1625	8	560	11	16	2050	3			
31	Beaver Harbour and Port Dufferin	1	150	1	17	700	4	140	5						
32	Quoddy and Harrigan Cove	10	1300	8	21	955	5	160	5	4	400	1		1	
33	Moser River and Smith's Cove	3			10	200	7	171	5						
34	Mitchell's Bay and Ecum Secum	3	500	3	24	985	7	280	4						
		44	10095	28	1683	60575	293	71554	2090	412	65625	60	20	7	

RETURN showing the Number of Fishermen, etc., the Number and Value of Vessels and Boats, and the Quantity and Value of all Fishing Gear, etc., used in the Fishing Industry in the County of Halifax, Province of Nova Scotia, during the year 1913-14.

Number	Fishing Districts	Gill Nets Number	Gill Nets Value $	Trawls Number	Trawls Value $	Hand Lines Number	Hand Lines Value $	Lobster Traps Number	Lobster Traps Value $	Lobster Canneries Number	Lobster Canneries Value $	Freezers and Ice-Houses Number	Freezers and Ice-Houses Value $	Smoke and Fish-Houses Number	Smoke and Fish-Houses Value $	Piers and Wharves Number	Piers and Wharves Value $	Persons employed in Canneries, Freezers and Fish-houses	Numbers
	Halifax County.																		
1	North Shore	920	24000	45	1250	300	150	3500	2800			6	1200	60	8500	60	4500	16	1
2	East St. Margaret's	2140	21720	80	2000	300	150	750	5840			3	1000	72	11000	72	5600	5	2
3	Indian Harbour	2271	29025	210	5760	212	125	4000	3200			3	450	35	5000	30	2000		3
4	Peggy's Cove	520	6400	45	850	90	45	2000	1600					23	5600	14	1050		4
5	Dover	3120	50000	300	4000	300	150	6000	4800			3	1400	61	8000	60	3600		5
6	Prospect	1840	33250	70	1750	210	105	4500	3600	1	400	2	1200	35	10500	36	4700	1	6
7	Terrence Bay	768	11500	250	3200	220	110	7000	5600		700	3	1250	50	7500	50	2500	7	7
8	Pennant	359	5250	34	720	30	15	1450	1100			3	30	12	2400	12	900		8
9	Sambro	912	9600	125	1980	240	120	950	760	1	2000	3	1300	20	3000	21	2500	20	9
10	Ketch Harbour	243	3810	69	690	150	75	3450	2750			2	1200	23	3000	14	1500	20	10
11	Portuguese Cove	611	6080	40	560	76	38	1885	1508					19	2850	20	500		11
12	Herring Cove	126	4460	110	2300	150	75	1300	1040					20	3000	20	500		12
13	Ferguson's Cove	46	275	4	20	14	7	300	400					7	1400	7	700		13
14	Bedford and Grand Lake	26	1188	3	15	10	5	50	25					6	300				14
15	Halifax	30	1200	10	50	20	10	50	350			6	30000	6	20000	6	6200	260	15
16	Dartmouth, Eastern Passage and Devil's Island	248	985	70	1050		100	2000	250					35	6500	27	800	12	16
17	Cow Bay and Lawrencetown	84	344			34	17	1000	450					13	260	5	250		17
18	Seaforth & Three Fathom Harbour	95	382	5	100	80	40	1000	500					53	1275	5	500		18
19	West Chezzetcook	432	1730			246	124	1500	750					10	320	3	300		19
20	East Chezzetcook	44	176			30	15	1000	500					25	500	10	600		20
21	Petpeswick Harbour	73	283	5	80	76	38	1000	500	1	1300			16	375	7	300	8	21
22	Musquodoboit Harbour	146	1294	8	140	100	50	1750	875	1	1000			26	550	12	350	12	22
23	Jeddore	250	1020			175	88	2500	1250	2	3000			46	950	14	600	21	23
24	Clam Harbour & Owl's Head	227	1710	2	40	180	90	3500	1750	2	1900	2	275	24	900	12	1000	19	24
25	West Ship Harbour	73	339	2	56	32	16	1000	500					11	350	9	700		25
26	East Ship Harbour	113	532	4	66	56	28	1675	1200			1		10	310	7	135	10	26
27	Pleasant Harbour and Tangier	182	728	5	65	97	48	3000	2400			1		21	410	14	350	16	27

28	Pope's Harbour and Gerrard's Island	201	804	2	16	81	41	1350	1080	1	500	1	20	14	455	11	285	23	25
29	Spry Bay, Taylor Head and Mushaboon	492	123	6	70	179	89	6500	5200	1	300	1	3.	36	780	20	387	40	29
30	Sheet Harbour and Sober Island	203	812	18	184	112	56	2900	2520				50	21	422	15	230	20	30
31	Beaver Harbour and Port Dufferin	48	92	4	35	25	13	1275	1000	2	800	1	50	10	227	6	95	35	31
32	Quoddy and Harrigan Cove	43	172	6	30	41	29	1600	1280	1	2000			7	90	3	95	25	32
33	Moser River and Smith Cove	21	84	1	5	16	8	750	590					6	79	1	10	5	33
34	Mitchell's Bay and Ecum Secum	78	312	8	45	30	15	1600	1280	2	2100			10	163	3	55	43	34
	Totals	16885 210700	1539 27360		4113	2077	80935		60033	16 160 0		39 39455		848 116966		606 99592		595

39—4½

5 GEORGE V., A. 1915

RETURN showing the Number of Fishermen, &c., the Number and Value of Vessels and Boats, and the Quantity and Value of all Fishing Gear, &c., used in the Fishing Industry in the County of Hants, Province of Nova Scotia, during the year 1913-14.

Number.	Fishing Districts.	Fishing Vessels, Boats, Tugs, &c.					Fishing Gear.			
		Boats.					Gill Nets.		Weirs.	
		Sail.	Value.	Gasoline.	Value.	Men.	Number.	Value.	Number.	Value.
	Hants County		$		$			$		$
1	Hantsport to Windsor..............	6	250	2	280	9	13	500	....	....
2	Windsor to Noel...	5	240	....	......	5	10	350	3	110
3	Maitland to Shubenacadie..........	20	290	....	....	30	40	390	....	....
4	Shubenacadie to Grand Lake	33	395	....	....	34	60	596	....	....
		64	1175	2	280	78	123	1836	3	110

THE CATCH.

RETURN showing the Quantities and Values of all Fish caught and landed in a Green State in the County of Cumberland, Province of Nova Scotia, during the year 1913-14.

Number	Fishing Districts.	Salmon, value.	Lobsters, cwts.	Lobster, value.	Cod, cwts.	Cod, value.	Haddock, cwts.	Haddock, value.	Hake & Cusk, cwts.	Hake & Cusk, value.	Pollock, cwts.	Pollock, value.	Herring, cwts.	Herring, value.	Mackerel, cwts.	Mackerel, value.	Shad, cwts.	Shad, value.	Alewives, cwts.	Alewives, value.	Halibut, cwts.	Halibut, value.	Flounders, cwts.	Flounders, value.	Smelts, cwts.	Smelts, value.	Trout, cwts.	Trout, value.	Eels, cwts.	Eels, value.	Oysters, brls.	Oysters, value.	Number
	Cumberland County.																																
1	Malagash, East Wallace and Fox Harbour		11775	58875									300	150											45	270							1
2	Pugwash and Gulf Shore		7062	35310									800	400											202	1212					100	500	2
3	Port Philip, Northport and Amherst Shore	30	1607	8035									13000	3900					16	20					265	1564	1	8			500	2500	3
4	Wallace River	46																	300	375					120	600	2	16	1	5			4
5	After Philip																		188	235					97	485							5
6	Maccan and Nappan												500	550			74	444															6
7	Middle to Apple River				800	1200	177	222	20	20	1823	1823	500	650							26	250											7
8	Advocate	70	400	2000	400	600	100	125	20	20	1000	1000	1500	1650							20	200											8
9	Spencers Island	80			325	487	100	125					1500	1650							10	100											9
10	Port Greville	80											200	400																			10
11	Parrsboro and Two Islands	270	100	500	200	300	100	125	125		800	800	400	812	3	21					10	100	12 00	12 00									11
		270	20944	104720	1725	2587	477	597	40	40	3623	3623	18706	10062	3	21	74	444	504	630	66	960			729	4151	3	24	1	5	600	3000	
		570																															

*Cwts.=100 lbs.

5 GEORGE V., A. 1915

THE CATCH MARKETED.

RETURN showing the Quantities and Values of all Fish and Fish Products Marketed in a fresh, dried, pickled, canned, &c., state, for the County of Cumberland, Province of Nova Scotia, during the year 1913-14.

No.	Fishing Districts.	Salmon, used fresh and frozen, cwts.	Lobsters, canned, cases.	Lobsters, shipped in shell, cwts.	Cod, used fresh, cwts.	Cod, shipped green-salted, cwts.	Cod, dried, † quintals.	Haddock, used fresh, cwts.	Haddock, dried, quintals.	Hake and Cusk, used fresh, cwts.	Hake and Cusk, dried, quintals.	Pollock, used fresh, cwts.	Pollock, dried, quintals.	Herring, used fresh, cwts.	Herring, smoked, cwts.	Herring, pickled, brls.	No.
	Cumberland County.																
1	Malagash, East Mace and Fox Harbour		4710														1
2	Pugwash and Gulf Shore		2823	5											5000		2
3	Port Philip, Northport and Amherst Shore	3	634	22													3
4	Wallace River	4															4
5	River Philip	7															5
6	Man and ...	8												300		10	6
7	Minudie to Apple River	8			336	110	86	55	43	4	6	331	520	300		20	7
8	Advocate				168	55	43	25	25	4	6	166	260	860		12	8
9	Spencers Island				112	70	27	25	25			26	200	1175		8	9
10	Port Greville				56	35	17	25	25					125		2	10
11	Parrsboro and Two Islands	27												370		2	11
	Totals	57	8167		672	270	173	130	118	8	12	523	1040	3130	5000	54	
	Rates	$ 5.00	18.00	10.00	2.00	3.00	6.00	2.00	4.50	1.50	4.00	1.50	4.00	2.00	3.00	4.50	
	Values	$ 855	147006	20	1344	810	1038	260	531	12	48	784	4160	6260	15000	243	

*Cwt. = 100 lbs.   † Quintal = 112 lbs.

THE CATCH MARKETED.

RETURN showing the Quantities and Value of all Fish and Fish Products Marketed in a fresh, dried, pickled, canned, &c., state for the County of Cumberland, Province of Nova Scotia, during the year 1903 -14—Concluded.

Number	Fishing Districts	Herring, used as bait, brls.	Herring, used as fertilizer, brls.	Mackerel, used fresh, cwts.	Shad, used fresh, cwts.	Shad, salted, brls.	Alewives, used fresh, cwts.	Alewives, salted, brls.	Sardines, canned, cases.	Sardines, sold fresh or salted, brls.	Halibut, used fresh, cwts.	Flounder, used fresh, cwts.	Smelts, used fresh, cwts.	Trout, used fresh, cwts.	Eels, used fresh, cwts.	Oysters, used fresh, brls.	Number
	Cumberland County.																
1	Malagash, East Wallace and Fox-Harbour	150											45			100	1
2	...sh and Gulf Shore	400	500										202				2
3	Port Philip, ...port and Amherst Shore	1000											265	1		500	3
4	Wallace River						16	100					120	2	1		4
5	River Philip	80					65	41					97				5
6	...an and Nappan	60									26						6
7	Minudie to ...	300			22	19					20						7
8	Advocate	150									10						8
9	Spencers Island	20									10						9
10	Port Grevil...	12															10
11	Parrsboro and ...o Islands			3								12					11
	Totals	2172	500	3	22	19	81	141			66	12	729	3	1	600	
	Rates	1.75	.50	8.00	10.00	15.00	2.00	4.00			10.00	5.00	7.00	10.00	5.00	6.00	
$	Values	3801	250	24	220	285	162	564			660	60	5103	30	5	3600	$

Total Value $198,385

5 GEORGE V., A. 1915

THE

RETURN showing the Quantities and Values of all Fish caught and landed in a
the year

Number.	Fishing Districts.	Salmon, *cwts.	Salmon, value.	Lobsters, cwts.	Lobsters, value.	Cod, cwts.	Cod, value.	Pollock, cwts.	Pollock, value.	Herring, cwts.	Herring, value.	Shad, cwts.
	Colchester County.		$		$		$		$		$	
1	Sterling................	....	.....	1031	5155	....	......	.....	.....	.....	.....	
2	Stewiacke........	88	880	. ..	....	.....	......	.....	.....	.....	.....	80
3	Five Islands........	......	.. ..	. .	.....	85	170	20	20	44	44	6
4	Economy......	55	550	...	......	40	80	15	15	41	41	.....
5	Little Bass River to Highland Village............	75	750	.....	.. ..	......	.....	....	.....	.....	.....	52
6	Great Village to Queens Village.	175	1750	....	......	......	.....	.....	.....	.....	....	84
	Totals............	393	3930	1031	5155	125	250	35	3	85	85	222

*Cwt. = 100 lbs.

SESSIONAL PAPER No. 39

CATCH.

Green State in the County of **Colchester,** Province of **Nova Scotia,** during 1913–14.

Shad, value.	Alewives, cwts.	Alewives, value.	Halibut, cwts.	Halibut, value.	Smelts, cwts.	Smelts, value.	Trout, cwts.	Trout, value.	Bass, cwts.	Bass, value.	Oysters, brls.	Oysters, value.	Clams, brls.	Clams, value.	Number.
$		$		$		$		$		$		$		$	
....	....	....	....	....	80	400	2	16	....	....	135	675	....	....	1
640	550	825	....	....	....	....	7	56	3	15	....	....	....	....	2
30	....	....	10	100	....	....	2	16	....	....	....	....	....	....	3
....	....	....	5	50	....	....	7	56	....	....	....	....	....	....	4
260	....	....	....	....	....	....	1	8	....	....	....	....	....	....	5
420	....	....	....	....	....	....	2	16	....	....	....	....	....	....	6
1350	550	825	15	150	80	400	21	168	3	15	135	675			

5 GEORGE V., A. 1915

THE CATCH

RETURN showing the Quantities and Values of all Fish and Fish Products Marketed in
Nova Scotia, during

Number.	Fishing Districts.	Salmon, used fresh and frozen, *cwts	Lobsters, canned, cases.	Lobsters, shipped in shell, cwts.	Cod, used fresh, cwts.	Pollock, used fresh, cwts.
	Colchester County.					
1	Sterling		412	1		
2	Stewiacke	88				
3	Five Islands				85	20
4	Economy	55			40	15
5	Little Bass River to Highland Village	75				
6	Great Village to Queens Village	175				
	Totals	393	412	1	125	35
	Rates	15	18	10	3	1.50
	Values	5895	7416	10	375	53
	Total value					

* Cwts.=100 lbs.

SESSIONAL PAPER No. 39

MARKETED.

a fresh, dried, pickled, canned, &c., state, for the County of **Colchester**, Province of the year 1913-14.

Herring, used fresh, cwts.	Herring, smoked, cwts.	Herring, used as bait, brls.	Shad, used fresh, cwts.	Shad, salted, brls.	Alewives, used fresh, cwts.	Alewives, salted, brls.	Halibut, used fresh, cwts.	Smelts, used fresh, cwts.	Trout, used fresh, cwts.	Bass, used fresh, cwts.	Oysters, used fresh, brls.	Number.
.....	....	....	80	....	550	....	....	80	2	....	135	1
.....	....	....	....	....	.....	....	....	.	7	3	....	2
10	10	7	6	....	.....	....	10	.....	2	.....	....	3
9	10	6	52	....	.....	....	5	.....	7	.....	....	4
.....	.....	....	84	.....	....	....	....	.....	1	.....	....	5
.....	.....	....	....	.....	....	....	....	.....	2	.....	....	6
19	20	13	222	... ...	550	....	15	80	21	3	135	
2	3	1.75	10	....	2	....	10	7	10	10	6	
38	60	23	2220	....	1100	....	150	560	210	30	81(	

.... $18,950

5 GEORGE V., A. 1915

THE.

RETURN showing the Quantities and Values of all Fish caught and landed in a Green state

Number.	Fishing Districts.	Salmon, *cwts.	Salmon, value.	Lobsters, cwts.	Lobsters, value.	Cod, cwts.	Cod, value.	Haddock, cwts.	Haddock, value.	Hake and cusk, cwts.	Hake and cusk, value.
	Pictou County.		$		$		$		$		$
1	West Pictou	....	....	12500	62500	115	230	70	105	100	125
2	Pictou *Island*..............	....	....	6930	34650	54	108	12	18	....	. .
3	Pictou Harbour........	40	400	10	50	102	204	6	9	3	4
4	Little Harbour and East Branch, St. Mary's River............	238	2380	1000	5000	66	132	....	....	12	15
5	Merigomish *Island*.............	328	3280	1120	5600	7	14	....	....	....	....
6	Ponds.......................	57	570	1775	8875	15	30	....	....	....	....
7	Lismore	90	900	....	......	5	10	....	....	....	....
	Totals..............	753	7530	23335	116675	364	728	88	132	115	144

*Cwt. ± 100 lbs.

SESSIONAL PAPER No. 39

CATCH.

in the County of **Pictou**, Province of **Nova Scotia**, during the year 1913-14.

Herring, cwts.	Herring, value.	Mackerel, cwts.	Mackerel, value.	Alewives, cwts.	Alewives, value.	Smelts, cwts.	Smelts, value.	Trout, cwts.	Trout, value.	Bass, cwts.	Bass, value.	Eels, cwts.	Eels, ale.	Mixed fish, cwts.	Mixed fish, value.	Oysters, brls.	Oysters, value.	Number.
	$		$		$		$		$		$		$		$		$	
2849	2849	5	30	20	40	78	390	2	16							60	300	1
1950	1950	45	225															2
400	400	14	70	410	615	57	285	2	16	15	75							3
						25	125											
																		4
360	360	3	15			120	600					20	80					5
600	600	220	1100			91	455	10	80									6
960	960							5	40									7
7119	7119	287	1440	430	655	371	1855	19	152	15	75	20	80			60	300	

5 GEORGE V., A. 1915

THE CATCH

RETURN showing the Quantities and Values of all Fish and Fish Products Marketed in
Nova Scotia, during

Number.	Fishing Districts.	Salmon, used fresh and frozen, *cwts.	Lobsters, canned, cases.	Lobsters, shipped in shell, cwts.	Cod, used fresh, cwts.	Cod, dried, †Quintals.	Haddock, used fresh, cwts.
	Pictou County.						
1	West Pictou.	. . . .	4996	. . . .	85	10	70
2	Pictou *Island*	. . . . .	2772	. . . .	10	15	4
3	Pictou Harbour.	40	. . . .	10	102	. . . .	6
4	Little Harbour and East Branch St. Mary's.	238	396	10	30	12	. . . .
5	Merigomish *Island*	328	448	. . . .	7	. . . .	. . . .
6	Ponds.	57	711	. . . .	15	. . . .	. . . .
7	Lismore.	90	. . . .	. . . .	5	. . . .	. . . .
	Total.	753	9323	20	254	37	80
	Rates. . . . .$	15.00	18.00	10.00	2.00	6.00	2.00
	Values. . . . .$	11295	167814	200	508	222	160
	Total value	. . . .	. . . .	. . . .	. . . .	. . . .	. . . .

* Cwt. = 100 lbs,   † Quintal = 112 lbs.

MARKETED.

a fresh, dried, pickled, canned, etc., state, for the County of Pictou, Province of the year 1913-14.

Haddock, dried, quintals.	Hake and Cusk, dried, quintals.	Herring, used fresh, cwts.	Herring, pickled, brls.	Herring, used as bait, brls.	Mackerel, used fresh, cwts.	Mackerel, salted, brls.	Alewives, used fresh, cwts.	Smelts, used fresh, cwts.	Trout, used fresh, cwts.	Bass, used fresh, cwts.	Eels, used fresh, cwts.	Oysters, used fresh, cwts.	Number.
	34	649	......	1100	5	......	20	78	2	......	...	60	1
3	.....	150	185	625	......	15	410			15			2
....	1	400	......	......	14	......		57	2				3
.....	12							25					4
.....		185	25	50	3	......		120			20		5
.....		50	50	180	100	40		91	10				6
.....		465	50	180	......				5				7
3	47	1899	310	2135	122	55	430	371	19	15	20	60	
4.50	4.00	2.00	4.50	1.75	7.00	12.00	2.00	7.00	10.00	10.00	5.00	6.00	
13	188	3798	1395	3737	854	660	860	2597	190	150	100	360	

.... $195,101

5 GEORGE V., A. 1915

7HE

RETURN showing the Quantities and Values of all Fish caught and landed in a
the year

Number.	Fishing Districts.	Salmon, *cwts.	Salmon, value.	⬧s, cwts.	Lobsters, value.	Cd, cwts.	Cd, value.	Haddock, cwts.	Haddock, value.	Hake and t&k.	⬧e and Cusk, ⬧ne.	Pollock, ⬧.	Pollock, value.	Herring, cwts.
	Antigonish County.	$		$		$		$		$		$		
1	Harbour au Bouche, Linwood and Cape Jack.................. ...	52	520	2732	13660	380	380	21	21	250	262	34	22	3193
2	Tracadie, Bayfield Monk's Head and South Side Antigonish Harbour	739	7390	1605	8025	147	147	12	12	26	19	....	..	1684
3	North Side Antigonish Harbour Lakevale and South Side Cape George......	184	1840	5135	25675	579	579	185	185	940	470	...	....	1365
4	North Side Cape George, Georgeville and Malignant Cove to Knoydart.....................	133	1330	1757	8755	316	316	178	178	1120	560	....	....	1185
	Totals	1108	11080	11229	56145	1422	1^{122}	396	396	2436	1311	34	22	7427

*Cwt=100 lbs.

SESSIONAL PAPER No. 39

CATCH.

Green State in the County of **Antigonish**, Province of **Nova Scotia**, during 1913-14.

Herring, value.	Mackerel, cwt.	Mackerel, value.	Smelts, cwts.	Smelts, value.	Trout, cwts.	Trout, value.	Bass, cwts.	Bass, value.	Eels, cwts.	Eels, value.	Tom cod, cwts.	Tom cod, value.	Mixed fish, cwts.	Mixed fish, value.	Squid, brls.	Squid, value.	Oysters, brls.	Oysters, value.	Number.
$		$		$		$		$		$		$		$		$		$	
1596	32	128	22	110	1	10	....	...	6	30	12	12	200	100	43	86	....	......	1
842	11	44	102	510	2	20	72	288	62	310	6	6	40	20	9	18	490	1960	2
682	10	40	79	395	4	40	....	...	10	50	..	...	30	15	6	12	....	....	3
592	112	448	....	......	1	10	34	136	....	....	...	....	20	10	18	36	....	......	4
3712	165	660	203	1015	8	80	106	424	78	390	18	18	290	145	76	152	490	1960	

5 GEORGE V., A. 1915

THE CATCH

RETURN showing the Quantities and Values of all Fish and Fish Products Marketed in
Nova Scotia, during

Number.	Fishing Districts.	Salmon, used fresh and frozen, cwts.*	Lobster, canned, cases.	Cod, used fresh, cwts	Cod, shipped green lted, cwts.	Cod, dried, fquintals.	Haddock, used fresh, cwts.	Hak, dried, qtls.	Hake and Gk, used fish, cwts.	Hake and Gk, dried, quintals.	Pollock, used fish, lbs.	Pollock, dried, qtls.
	Antigonish County.											
1	Harb. au Bouche, Linwood and Cape Jack	52	1093	80	60	60	21	.....	50	97	34	....
2	Tracadie, Bayfield, Monk Head and South Side Antigonish Harbour.	739	642	35	20	21	12	......	26	......	....	....
3	North Side Antigonish, Harbour, Lakevale and South Side Cape George.... ...	184	2054	250	40	83	45	47	30	300	....	....
4	North Side Cape George, Georgeville and Malignant Cuve to Knoydart........	133	703	96	20	60	30	49	90	340	....	....
	Totals	1108	4492	461	140	224	108	96	196	737	34	....
	Rates........... ....$	15.00	18.00	2.00	3.00	6.00	2.00	4.50	1.50	4.00	1.50	....
	Value. ....$	16620	80856	922	420	1344	216	432	294	2948	51	....
	Total Value...											

*Cwt. = 100 lbs.   †Quintal = 112 lbs.

SESSIONAL PAPER No. 39

MARKETED.

a fresh, dried, pickled, canned, &c., state, for the County of **Antigonish**, Province of the year 1913-14.

Herring, used fresh, cwts.	Herring, pickled, brls.	Herring, used as bait, brls.	Herring, used as fertilizer, brls.	Mackerel, used fresh, cwts.	Mackerel, salted, brls.	Shad, used fresh, cwts.	Smelts, used fresh, cwts.	Trout, used fresh, cwts.	Bass, used fresh, cwts.	Eels, used fresh, cwts.	Tom Cod, used fresh, cwts.	Mixed Fish, used fresh, qtls.	Squid, used as bait, brls.	Oysters, used fresh, brls.	Fish Oil, galls.	Number.
70	44	1399	96	32	....	....	22	1	....	6	12	200	43	....	337	1
24	20	800	....	11	....	....	102	2	72	62	6	40	9	490	80	2
15	50	600	....	10	....	....	79	4	....	10	....	30	6	....	144	3
17	36	530	....	1	37	....	....	1	34	....	....	20	18	....	346	4
126	150	3329	96	54	37	....	203	8	106	78	18	290	76	490	907	
2.00	4.50	1.75	·50c	7.00	12	....	7.00	10.00	10.00	5.00	3.00	1.50	4.00	6.00	·40c	
252	675	5825	48	378	444	....	1421	80	1060	390	54	435	304	2940	363	

....$118,772

5 GEORGE V., A. 1915

THE CATCH

RETURN showing the Quantities and Values of all Fish caught and landed in a Green State in the County of **Guysboro**, Province of **Nova Scotia**, during the Year 1913-14.

Number	Fishing Districts	Salmon, cwts.	Salmon, value.	Lobsters, cwts.	Lobsters, value.	Cod, cwts.	Cod, value.	Haddock, cwts.	Haddock, value.	Hake and Cusk, cwts.	Hake and Cusk, value.	Pollock, cwts.	Pollock, value.	Herring, cwts.	Herring, value.	Mackerel, cwts.	Mackerel, value.	Alewives, cwts.	Alewives, value.	Number
	Guysboro County.																			
1	Ecum Secum	8	80	514	2570	1361	2041	68	102	3	3	135	135	662	662					1
2	Marie Joseph	4	40	963	4815	862	1293	10	15	194	194	8	8	1066	1066					2
3	Liscomb and Spanish Ship Bay	10	100	560	2800	1537	2305	65	98	23	23	51	51	2227	2227			15	15	3
4	Gegoggin	28	280	201	1005	667	1000	33	50	3	3	20	20	625	625					4
5	St. Mary's Bay and River	154	1540	205	1040	68	225	24	36					846	846					5
6	Wo.	36	360	210	1050	342	102					3	3	262	262					6
7	Port	20	200	282	1410	24	513	22	33	9	9	9	9	387	387					7
8	Island's Harbour & Indian River	8	80	102	510	735	36							500	500	21	126			8
9	Port Beckerton	15	150	610	3050	357	1102	176	264	54	54	33	33	2065	2065	176	1069			9
10	Fisherman's Bour	12	120	275	1375		536	31	47	34	34	12	12	1703	1703	100	600			10
11	Harbour	32	320	10	50															11
12	Isaac's Harbour	10	100	1643	8415	681	851	86	129	20	15	47	28	1911	1911	247	803			12
13	Drum Head			2420	12100	901	1126	179	268	104	78	226	135	730	730	782	2541			13
14	Seal Harbour					939	1174	130	195	2	1	283	169	280	380	200	650			14
15	Coddle's Harbour					520	650	101	161	4	3	169	101	414	414	35	114			15
16	New Harbour	4	40	433	2165	1440	1800	399	598	55	41	673	403	677	677	163	530			16
17	Tor Bay					450	563	30	45			276	165	180	180	85	276			17
18	Larry's River	2	20			1654	2067	650	975	30	22	452	271	840	840	674	2190			18
19	Charlo's Cove					1824	2290	860	1290	676	507	450	270	1059	1059	375	1219			19
20	Cle Har	6	60			920	1150	479	718	65	48	102	61	518	518	142	461	8	8	20
21	Port Hic			134	670	1972	2465	1000	1500	106	79	205	123	1577	1677	1073	3487	25	25	21
22	Whitehead	3	30	2393	11965	4573	5716	2500	3750	215	161	671	402	732	732	1549	5034	30	30	22
23	Raspberry and Dover			949	4645	1476	1845	350	525	87	65	192	61	100	100	76	245	60	60	23
24	Canso and Canso Tittle	320	3200	3930	19650	28384	354800	46814	70221	1918	1438	1092	655	2444	2444	4253	13822			24
25	Fox Island Maine	4	40			500	625	376	564	32	24	465	39	150	150	676	1872	75	75	25
26	Half Island Cove			371	1856	2034	2543	1256	1384	320	240	472	283	471	471	375	1219			26
27	Philip's Har					1003	1254	1046	1644	276	207	327	196	230	230	280	910			27
28	Queensport	28	280	1576	7880	3092	3965	1870	2805	656	492	759	455	1140	1140	2076	6747			28
29	Peas Brook					735	919	150	225	80	60	163	97	225	225	54	175			29
30	Half Way Cove					782	977	370	555	116	87	125	75	936	936	276	894			30

SESSIONAL PAPER No. 39

		34	32	33	34	35	36	37	38											
31 Sandy Cove and Cook's Cove	60	600			375	469	140	210	56	40	210	126	358	358	300	975		
32 Guysboro and Manchester	100	1000			256	320	20	30	10	7	150	90	152	152	91	306		
33 Port Shoreham	218	2180			206	257	98	147	20	15	150	90	190	190	82	266		
34 St. Francis					325	406	172	258	15	11	41	24	240	240	136	507		
35 Oyster Ponds					106	134	75	113	8	6	23	13	500	500	35	114		
36 Sand Point					73	92	30	45	6	4	17	10	235	235	145	471		
37 Middle Melford	90	800			32	40	98	147	26	19	93	55	362	362	2078	67753		
38 Mulgrave and Auld's Cove			2797	13985	10	13	25	38	4	3	22	13	20	20	59	192		
Totals	1172	11720	20581	102905	61366	78234	59783	86 75	5340	4007	7645	4691	26824	26824	16535	54551	213	213

*Cwt.=100

THE CATCH.

RETURN showing the Quantities and Values of all Fish caught and landed in a Green State in the County of Guysboro, Province of Nova Scotia, during the year 1913-14—Concluded.

Number	Fishing Districts	Halibut, cwts.	Halibut, value.	Soles, cwts.	Soles value.	Smelts, cwts.	Smelts, value.	Trout, cwts.	Trout, value.	Eels, cwts.	Eels, value.	Swordfish, cwts.	Swordfish, value.	Squid, brls.	Squid, value.	Clams, brls.	Clams, value.	Number.
	Guysboro County.																	
1	Ecum Secum	11	55													6	6	1
2	Marie Joseph	56	280							25	100	13	26					2
3	...ard Spanish Ship Bay	66	330					2	20									3
4	Gegoggin											4	8					4
5	Mary's Bay and Ri...er							2	20	10	40					35	35	5
6	Wine Harbour															12	12	6
7	Port Hilford									25	100							7
8	Holland's Harbour and Indian River																	8
9	Port Beckerton															10	10	9
10	Fisherman's Harbour	160	800			1	5	2	4			51	102					10
11	...ey Harbour	14	70			1	5					90	180			8	8	11
12	Isaac's Harbour	10	60			7	35					132	396					12
13	Drum Head	39	234					1	2	15	19	437	1311					13
14	Seal Harbour	12	72			6	30					194	582					14
15	...'s Harbour	14	84									59	177					15
16	New Harbour	88	528									313	939					16
17	Tor Bay	20	120					3	6	3	4							17
18	Larry's River	30	180			651	325			50	62							18
19	Charlo's Cove	75	450					4	8			360	1080					19
20	Cole Harbour	50	300									6	18					20
21	Port Felix	74	444									476	1428					21
22	...	700	4200					6	12			1052	3156					22
23	Raspberry ...and Dover	4515	27090	156	234							302	906					23
24	...o and Canso Tittle		480									2367	7101	851	3404			24
25	Fox Island Maine	3	18	60	90							472	1416					25
26	Mill I...nd Cove	10	60									150	450					26
27	Philip's Harbour	50	300					1	2									27
28	...	4	24							3	4							28
29	Peas Brook													75	300			29
30	Half Way Cove																	30

		31	32	33	34	35	36	37	38	
31	Sandy Cove and Cook's Cove									71
32	Guysboro and Manchester									71
33	Port Shoreham							200		3904
34	St. Francis						50			976
35	Oyster Ponds									19276
36	Sand Point									6478
37	Middle Melford		6	1	1	3	4			344
38	Mulgrave and Auld's Cove		5	1	1	2	3			153
			16	4	6		4			104
			8	2	3		2			36
		125	225							750
		25	45							150
										824
										216
										36179
	Totals									6081

*Cwt. = 100 lbs.

THE CATCH

RETURN showing the Quantities and Values of all Fish and Fish
the County of **Guysboro**, Province of

Number.	Fishing Districts.	Salmon, used fresh and frozen, cwts.	Salmon, smoked, lbs.	Lobsters, cases.	Lobsters, shipped in shell, cwts.	Cod, used fresh, cwts.	Cod, shipped green salted, cwts.	Cod, dried, †Quintals.	Haddock, used fresh, cwts.	Haddock, smoked, cwts.	Haddock, dried, quintals.
	Guysboro County.										
1	Ecum-Secum	8	..	......	514	....	....	455	......	....	22
2	Marie Joseph	4	...	......	603	....	....	237			3
3	Liscomb and Spanish Ship Bay	10	....	354	200	....	150	408	.....	....	21
4	Gegoggin	28	....	31	24	.....	150	222	.....	..	11
5	St. Mary's Bay and River	142	6	......	143	....		50	...	..	8
6	Wine Harbour	36	...	......	60	....	.....	22	...	....	
7	Port Hilford	20	...	......	84	....	..	116	....	....	7
8	Holland's Harbour & Indian River	8	...	......	40	....	...	12	....	....	
9	Port Beckerton	15	...	......	260	....	268	66	60	......	35
10	Fisherman's Harbour	12	...	384	75	....	113	55	....	....	10
11	Country Harbour	32	...	......	10	....	..		....		
12	Isaac's Harbour	10	....	502	428	..	170	101	....	..	25
13	Drum Head		....	692	600	.....	225	131	....	...	53
14	Seal Harbour		...				233	131	....	...	35
15	Coddle's Harbour		...				130	77	....	..	30
16	New Harbour	4	...	......			360	214	......	....	118
17	Tor Bay		....	157	41	.....	112	67	......	....	9
18	Larry's River	2	....			....	413	245	......	....	194
19	Charlo's Cove	...	....			.....	456	271	700	......	48
20	Cole Harbour	6	....			.....	230	128	700	......	48
21	Port-Félix	....	.	48	14	30	493	292	900	.....	27
22	Whitehead	3	....	917	103	747	1044	680	2200	....	89
23	Raspberry and Dover		....	323	102	.....	369	219	.....	....	104
24	Canso and Canso Tittle	320	....	1422	375	2987	6096	4224	22206	3872	5623
25	Fox Island Main	4	....			.....	125	74	.....	....	111
26	Half Island Cove		....	75	186	1000	408	303	1020	.....	70
27	Philip's Harbour		....			.....	251	149	900	.....	59
28	Queensport	28	....	604	69	750	673	459	1700		51
29	Peas Brook		....			.....	184	109	.....	.....	45
30	Half Way Cove		....			...	195	116	.....	.....	109
31	Sandy Cove and Cook's Cove	60	....			.....	93	55	.....	.....	42
32	Guysboro and Manchester	100	....			.....	64	38	.....	.....	6
33	Port Shoreham	218	....			.....	52	30	.....	.....	28
34	St. Francis		....			.....	81	48	.....	.....	52
35	Oyster Ponds		....			.....	27	16	.....	.....	22
36	Sand Point		....			.....	18	10	.....	...	9
37	Middle-Melford		....			.....	8	4	.....	.....	28
38	Mulgrave and Auld's Cove	90	....	1115	10	460	2	1	.....	.....	8
	Totals	1160	6	6624	4031	5974	13195	9835	30386	3872	7163
	Rates............$	15.00	.20	18.00	10.00	2.00	3.00	6.00	2.00	3.00	4.50
	Values.............$	17400	120	119232	40310	11948	39585	59010	60772	11616	32234

* Cwt. = 100 lbs.   † Quintal = 112 lbs.

MARKETED.

Products Marketed in a fresh, dried, pickled, canned, &c., state, for **Nova Scotia**, during the year 1913-14.

Hake and Cusk, dried, quintals.	Pollock, dried, quintals.	Herring, pickled, brls.	Herring, used as bait, brls.	Mackerel, used fresh, cwts.	Mackerel, salted, brls	Alewives, salted, brls	Halibut, used fresh, cwts.	Soles, used fresh, cwts.	Smelts, used fresh, lbs.	fish, used fresh, cwts.	Eels, used fresh, cwts.	Swordfish, used fresh, cwts.	Squid, used as bait, brls.	Clams and quahaugs, used fresh, brls.	Hair seal skins, No.	Fish oil, gals.	Number.
1	45	194	40	....	....	....	11	....	....	....	....	6	....	....	4	100	1
65	2	305	70	....	....	....	56	....	....	....	25	....	....	....	50	60	2
8	17	642	150	....	....	5	66	....	....	1	....	13	....	....	2	116	3
1	7	181	40	....	....	....	....	....	1	....	....	....	....	....	....	25	4
3	3	250	50	....	....	....	7	2	10	4	....	35	....	12	....	12	5
....	1	67	30	....	....	....	....	....	6	....	25	....	....	....	....	12	6
3	3	100	44	....	7	....	....	....	6	....	25	....	....	10	3	12	7
....	....	100	100	....	7	....	....	....	2	....	....	....	....	10	....	....	8
18	11	355	500	....	58	....	160	....	....	....	....	51	....	....	....	162	9
12	4	334	350	....	33	....	14	....	....	....	....	....	....	....	....	86	10
....	....	....	....	....	....	....	....	....	....	....	....	90	....	8	4	....	11
7	16	300	537	100	49	....	10	....	....	2	....	132	....	....	....	170	12
35	75	200	137	200	39	....	39	....	....	....	....	437	....	....	....	250	13
1	94	100	67	50	57	....	12	....	....	....	....	194	....	....	....	102	14
2	56	125	103	10	8	....	14	....	....	1	15	59	....	....	....	115	15
18	224	175	130	10	51	....	88	....	65	....	....	313	....	....	....	104	16
1	92	50	33	....	28	....	20	....	....	....	....	....	....	....	....	493	17
10	151	250	123	....	225	....	30	....	....	3	....	....	....	....	....	202	18
225	150	250	176	100	92	2	75	....	....	....	3	360	....	....	....	510	19
55	34	163	70	10	44	8	50	....	....	4	50	6	....	....	....	726	20
36	69	300	356	522	187	10	74	....	....	....	....	476	....	....	....	59	21
72	224	....	337	1478	27	20	700	....	....	6	....	1052	....	....	....	765	22
29	34	33	10	38	13	....	80	....	....	....	....	302	....	....	....	673	23
639	365	50	946	4076	59	25	4515	156	....	....	....	2367	851	....	....	1092	24
11	22	25	42	400	59	....	3	....	....	....	....	472	....	....	....	62	25
107	158	55	155	272	34	....	10	60	....	....	....	150	....	....	....	450	26
92	109	....	103	200	27	....	50	....	....	....	....	....	....	....	....	149	27
215	253	50	465	2076	....	....	4	....	....	1	3	....	75	....	....	573	28
27	54	....	10	54	....	....	....	....	....	....	....	....	....	....	....	296	29
38	42	....	325	275	....	....	....	....	....	....	....	....	....	....	....	310	30
19	72	....	159	300	....	....	....	....	25	....	....	....	....	....	....	270	31
4	50	....	65	94	....	....	....	....	45	8	5	....	....	....	....	79	32
7	50	....	83	82	....	....	....	....	....	2	....	....	....	....	....	49	33
5	14	....	103	156	....	....	....	....	....	3	1	....	....	....	....	173	34
3	7	....	90	35	....	....	....	....	....	....	1	....	....	....	....	92	35
2	6	....	105	145	....	....	....	....	....	....	....	....	....	....	....	42	36
9	31	....	158	2078	....	....	....	....	....	....	2	....	....	....	....	....	37
2	7	....	7	59	....	....	....	....	....	2	3	....	50	....	....	....	38
1782	2552	4654	6359	12820	1252	79	6081	216	150	36	143	6478	976	71	64	8379	
4.00	4.00	4.50	1.75	7.00	12.00	4.00	10.00	5.00	7.00	10.00	5.00	4.00	5.00	3.00	1.25	.40	
7128	10208	20943	11128	89740	15024	280	60810	1080	1050	360	715	25912	4880	142	80	3351	

Total Value $645,058

THE CATCH.

RETURN showing the Quantities and Values of all Fish caught in a Green State in the County of **Halifax**, Province of **Nova Scotia**, during the year 1913-14.

No.	Fishing Districts	Salmon cwts.	Salmon value	Lobsters cwts.	Lobsters value	Cod cwts.	Cod value	Haddock cwts.	Haddock value	Hake and cusk cwts.	Hake and cusk value	Pollock cwts.	Pollock value	Herring cwts.	Herring value	Mackerel cwts.	Mackerel value	Shad cwts.	Shad value	No.
	Halifax County.																			
1	North Shore	6	72	200	1600	2900	5600	402	804	4000	4000	400	600	240	480	2325	11625			1
2	…St …ets	75	940	500	4000	5300	10600	2500	5000	2500	2500	460	690	1000	2000	2325	11625			2
3	Indian Harbour	125	1300	700	5600	7000	14000	1000	2000	5500	5500	420	630	1200	2400	3880	19400			3
4	…s Cove	75	900	160	1280	420	840			200	200	150	225	100	200	16.20	8000			4
5	…br	600	7200	1250	10000	7350	14700	6000	12000	3500	3500	425	637	1800	3600	11640	58200			5
6	Prospect	500	6000	420	3360	1300	2600	900	1400	760	760	235	353	1500	3600	7750	38800			6
7	Terence Bay	20	240	350	2800	6200	12400	4025	8050	3500	3500	360	540	1800	3600	5430	27150			7
8	Pennant	20	240	150	1200	4200	8400	4100	8200	1550	1550	235	354	700	1400	775	3875			8
9	Sambro	20	240	210	1640	3900	7800	4200	8400	2400	2400	160	240	700	700	775	3875			9
10	Ketch Harbour	300	3600	25	200	6340	13880			150	150	250	375	800	1600	870	4350			10
11	Portuguese Cove	126	1512	150	1200	290	400	10	20	2300	2300	150	225	800	1600	700	3500			11
12	…ring Cove	2	24	160	1280	5800	11600	10	20			170	255	900	1800	3200	16000			12
13	Fergusons Cove	14	168			30	60					9	14	5	10	40	200			13
14	Bedford and Grand Lake					35	70					9	13	7	14	40	200			14
15	Halifax					2000	4000	1520	3040	50	50	110	165	20	40	25	125	222	1332	15
16	Dartmouth Eastern Passage and Devils Island	1	12	73	5840	1784	2676	976	1464			174	174	1990	1990	191	1337			16
17	Cow Bay and Lawrencetown	7	84			175	263	20	30			27	27	913	913	20	140			17
18	Seaforth and Three Fathom Harb.	5	60			120	180	26	39			36	36	803	893	25	175			18
19	West Chezzetcook					7443	11765	450	675			210	210	2606	2606	79	553			19
20	East Chezzetcook					164	246	35	53			27	27	494	494	4	28			20
21	Petpeswick Harbour			129	1373	843	1268	211	316			744	744	776	776	12	84			21
22	…t Harbour	84	1008	261	1742	1958	2952	421	631			1188	1188	1503	1503	20	140			22
23	…re	4	48	1024	7458	4256	6384	483	724	30	30	1323	1323	4239	4239	44	308			23
24	Clam Harbour and …ie Head	10	120	1856	15342	1782	2673	204	306	231	231	111	111	2442	2442	101	707			24
25	West Ship Harbour	1	12			499	748	57	85	51	51	8	8	482	482	6	42			25
26	East Ship					567	850	78	78	27	27	81	81	1765	1767					26
27	…l					1716	2574	348	348	15	16	451	461	3840	3840	21	84			27
28	Pope's Harbour & Gerrard's Island	6	60	917	4585	1725	2587	273	273			207	207	3939	3939	39	156			28
29	Spry Bay, Taylors Head & Mushaboom	17	170	1029	5145	2850	4275	282	282	219	219	333	333	5886	5886	90	360			29

30	Sheet Harbour and Solvor Island	25	250	75	375	1137	1705	384	384	1758	1758	21	21	3308	3308	186	744				30
31	Beaver Harbr. and Port Dufferin	2	20	1954	9770	459	688	24	24	24	24	63	63	972	972	6	24				31
32	Quoddy and Harrigan Cove	5	50	2221	11105	438	657	18	18	29	29	6	6	387	387	3	12				32
33	Mower River and Smith's Cove	22	220	.....	.....	156	234	3	3	3	3	6	6	465	465	3	12				33
34	Mitchell's Bay & Ecum Secum	.....	.....	1667	8335	723	1084	21	21	33	33	18	18	2440	2440						34
	Totals	2096	24998	16138	105270	82682	150759	40631	78388	30890	30890	8577	10350	51004	62573	42235	211831	222	1332	

Cwt. = 100 lbs.

5 GEORGE V., A. 1915

THE CATCH.

RETURN showing the Quantities and Values of all Fish caught and landed in a Green State in the County of **Halifax**, Province of **Nova Scotia**, during the year 1913-14—*Concluded.*

No.	Clams, value $	Clams, brls.	Oysters, value $	Oysters, brls.	Squid, value $	Squid, brls.	Mixed fish, value $	Mixed fish, cwts.	Swordfish, value $	Swordfish, cwts.	Eels, value $	Eels, cwts.	Albacore, value $	Albacore, cwts.	Trout, value $	Trout, cwts.	Smelts, value $	Smelts, cwts.	Flounders, value $	Flounders, cwts.	Halibut, value $	Halibut, cwts.	Alewives, value $	Alewives, cwts.	Fishing Districts.	No.
																									Halifax County.	
1	150	75			340	136	150	300	900	300	60	10	400	320	280	40			4	3	6300	900	200	100	North Shore	1
2	120	60			250	100	100	200	300	100	72	12	385	310	210	30			3	2	15680	2240			East St. Margarets	2
3	40	20			125	25	50	100	600	200	18	3	250	200	3						11550	1650			Indian Harbour	3
4	50	25			63	25	25	50	450	150	12	2	93	70	7	1			1140	760	210	30			Peggys Cove	4
5	100	50			250	100	50		600	200	66	11	250	200	35	5					12600	1800	100	50	Dover	5
6	90	40			50	20			300	120	60	10	250	200	21	3					8750	1250			Prospect	6
7	54	27			50	20			360	120	78	13	250	220	70	10					9800	1400			Terrence Bay	7
8	20	10			100	20	50		270	100	12	2	125	100	56	8					7000	1000			Pennant	8
9	12	6			50	40			300	100	9	1	225	180	21	3					5950	850	170	85	Sambro	9
10	10	5			75	20			720	240	6	1	160	180	7	1					9800	1400			... Harbour	10
11	10	5			63	25			150	100	24	4	225	80	14	2					7000	1006			Portuguese Cove	11
12	40	15			37	15			450	150	36	6	137	80	7	1					9100	1300			Herring Cove	12
13	10	20									12	2	100		70	10									Fergusons Cove	13
14	50	5									54	9			70	5									Bedford and Grand Lake	14
15	15	25									12	2			35	5					3150	450			Halifax	15
16									345	115			33	11											Dartmouth, Eastern Passage and ...'s Island	16
17	80	80									16	4							50	50	850	170			Cow Bay and Lawrencetown	17
18	40	40									4	1			27	3			10	10	5	2			Seaforth and Three Fathom Harb.	18
19	180	180									16	4			18	2	50	50	20	20	40	8			West Chezzetcook	19
20	1300	1300									12	3			36	4	50	50	18	18	5	1			East ...	20
21	1400	1400									32	6			36	7	30	30	20	20	60	12			... Harbour	21
22	909	900									36	7			63	5	10	10	20	20	165	21			... Harbour	22
23	740	740							3	3	24	5	15	5	45	7	1000	200	20	20	215	43			Jeddore	23
24	40	40							13	13	20	5			45	5	16	3	22	22	225	45	4	4	Clam Harbour and Owls Head	24
25	1975	1975									16	4	15	5	63	7	30	8	22	22	45	9	2	2	West Ship Harbour	25
26	20	20															40	6	13	13	156	39			East Ship Harbour	26
27	5	5													40						120	30			Pope's Harbour and Tangier	27
28	24	8							5	5											64	16			... Harbour & Gerrard's Island	28
29	15	5							7	7			30		30						144	36			Spry Bay, ... & Musha-boom	29
30		10																								30

30	Sheet Harbour and Sober Island				16		64																	
31	Beaver Harbor, and Port Dufferin				15		60			15	6	60												
32	Quoddy and Harrigan Cove				12		48			105	3	30												
33	Moser River and Smith's Cove				3		12			60	5	50						2	2			1		
34	Mitchell's Bay and Ecum Secum				51		204			420	4	40										5		
	Totals	249	484	15800	109322	978	1360	400	2150	184	1437	2166	2740	139	732	1952	5828	850	425	561	1403	6	30	7096	755

Cwt.=100 lbs.

THE CATCH MARKETED.

RETURNS showing the Quantities and Value of all Fish and Fish Products Marketed in a fresh, dried, pickled, canned, &c., state for the County of Halifax, Province of Nova Scotia, during the year 1913–14.

Halifax County.

No.	Fishing District	Herring, used as bait, brls.	Herring, pickled, brls.	Herring, smoked, cwts.	Herring, used fresh, cwts.	Pollock, dried, quintals	Pollock, used fresh, cwts.	Hake and Cusk, dried, quintals	Hake and Cusk, used fresh, cwts.	Haddock, dried, quintals	Haddock, smoked, cwts.	Haddock, used fresh, cwts.	Cod, dried, ½ quintals	Cod, used fresh, cwts.	Lobsters, shipped in shell, cwts.	Lobsters, canned, cases	Salmon, smoked, cwts.	Salmon, used fresh and frozen, cwts.
1	North &c.	10	60		40	134		1340		101		100	930	100	50		2	6
2	East St. Margarets	320	83		100	153		835		450		500	1733	100	200			75
3	Indian Harbour	415	107		50	140		1850		850		1000	2000	100	275			125
4	Peggy's &c.	10	24		10	50		63		170		500	107	500	50			7
5	Dover	200	435		100	142		1170		1500		1500	2117	600	500			600
6	Prospect	19	350		150	78		255		630		500	1570	600	100			500
7	Terrence Bay	290	468		25	120		1150		231		200	261	15	100	9		20
8	Pennant	6	215		10	89		520		1000		1000	570	5	70	140		20
9	Sambro	50	198		10	53		617		1035	300	1000	1132	500	1360			20
10	&c. Harbour	25	192		25	84		800		1030		1000	1980	1000	20	476		300
11	Portuguese Cove	100	178		25	9		50		468		100	60	1000	40			126
12	Herring Cove	120	240		20	57		780		400		3000	1600	1000	40			2
13	Ferguson's &c.	75		490	5		9											14
14	Bedford and Grand Lake				7		9											
15	Halifax				20		110		50			600		2000	730			
16	Dartmouth, Eastern Passage and Devil's Island	356	95	510		58				42		850	378	650				
17	&c. Bay and Laurencetown	10	296		20	9				6		2	56	10		12	4	1
18	Seaforth and &c. Fathom Harbour	12	289		3	12		10		8		6	38	6	104	95	3	7
19	West	100	800		3	70		77		157		2	2610	13	72	345		5
20	East Chezzetcook	10	157		6	9		17		11		25	54	2	334	496		
21	Petpeswick Harbour	24	240		2	248		9		62		10	273	26	866			76
22	&c. Harbour	44	500		8	396				137		8	1412	12				4
23	Jeddore	117	1334		5	441			50	158		6	592	20	247	275		4
24	Clam Harbour and &c.'s Head	110	740		6	37				66		3	165	4				1
25	West Ship Harbour	29	140		4	3				13			189					
26	East Ship Harbour	10	589			27				26			572					6
27	Pleasant &c. and Tangier	20	1280			151		9		116								
28	Pope's Harbour and Gerrard's Island	30	1313			69		5		91			575					

29	Spry Bay, Taylor's Head and Mushaboom	17		313	244		950		94	73		111			1926	30
30	Sheet Harbour and Saber Island	20	5		75		379		128	586					1076	40
31	Beaver Harbour and Port Dufferin			629	369		153		8	8		7			394	10
32	Quoddy and Harrigan Cove	2		745	536		146		6	9		21			129	4
33	Moser River and Smith's Cove	5	4				52		1	1		2			132	3
34	Mitchell's Bay and Ecum Secum	18		460	530		241		1	11		6			794	24
	Totals	2069	18	3835	6912	11949	23546	810	9020	10136	118	2820	650	490	14714	2652
	Rates $	15	20	18	10	2	6	4	4.50	4	1.50	4	3	3	4.50	1.75
	Values $	31035	360	69300	69120	23898	141276	3240	40590	40644	11280	1318	1470	66123	4641

* Cwts. = 100 lbs.   † Quintal = 112 lbs.

THE CATCH MARKETED.

Returns showing the Quantities and Values of all Fish and Fish Products, Marketed in a fresh, dried, pickled, canned, &c., state, for the County of **Halifax**, Province of **Nova Scotia**, during the year 1913-14—*Concluded.*

Numbers	Fishing District	Mackerel, used fresh, cwts.	Mackerel, salted, cwts.	Shad, used fresh, cwts.	Alewives, used, cwts.	Halibut, used, cwts.	Flounders, used, cwts.	Smelts, used fresh, cwts.	Trout, used fresh, cwts.	Albacore, used, cwts.	Eels, used fresh, cwts.	Sword Fish, used fresh, cwts.	Mixed Fish, used fresh, cwts.	Squid, used as bait, brls.	Oysters, used fresh, brls.	Clams and Quahaugs, used fresh, brls.	Clams and Quahaugs, canned, cases.	Fish Oil, galls.	Numbers
	Halifax County.																		
1	North Shore	825	500		100	900	3		40	320	10	300	300	135		75		2400	1
2	East St. Margarets	1125	400			2240	2		30	310	12	300	290	100		60		3550	2
3	Indian Harbour	2400	450			1650			3	200	11	200	100	50		26		6500	3
4	Peggy's Cove	1375	75		50	30			1	70	23	150	50	55		95		600	4
5	do	8040	1200			1800	760		5	200	10	200	100	100		50		3400	5
6	Prospect	7100	200			1250			3	300	13	200				50		4795	6
7	Terrence Bay	4530	300			1400			10	220	18	125	100	20		40		3650	7
8	Pennant	760				1000			8	100	2	90		40		27		3300	8
9	Sambro	655	5		85	850			3	160	1	125		20		10		3200	9
10	Ketch Harbour	510	120			1400			1	160	4	100		40		15		700	10
11	Portuguese Cove	520	60			1000			2	170	6	240		20		30		4500	11
12	Herring Cove	2600	200			1300				110	9	50		25		25		700	12
13	Ferguson's Cove	40								80	2	150		15				4240	13
14	Bedford and Grand Lake	25							10		9							16	14
15	Halifax			222					5		2							2914	15
16	Dartmouth, Eastern Passage and Devil's Island					450			10									43251	16
17	Cow Bay and Laurencetown	38	51		8	170	50	50	3	11	4		115			80		710	17
18	Seaforth and Three Fathom	6	6			1	10	60	2							40		33	18
19	West Chezzetcook	4	5			3	20	6	4		4							28	19
20	East Chezzetcook	4	7			8	20	2	4	6	3					1300		1266	20
21	do		1			12	18	200	4		8	3				1400		30	21
22	Musquodoboit	3	1			21	20	3	5		9					900		290	22
23	Jeddore	2	3			43	20	8	6	5	7	13				740		596	23
24	Clam Harbour and Owl's Head	2	6		4	45	22		5		5	7			6	40		900	24
25	West Ship Harbour	1	33		2	9	13		6		4					1800		398	25
26	East Ship Harbour		2			39										20	175	90	26
27	Pleasant Harbour and Tangier		7									5				5		300	27
28	Pope's Harbour and Gerrard's Island		13			16						7				8		609	28

29	Spry Bay, Taylor's Head and Mushaboom	30			36				3							10		79729
30	Sheet Harbour and Sober Island	63			16				6							4		73130
31	Beaver Harbour and Port Dufferin	2			15			15	3									10031
32	Quod y and Harrigan Cove	1			12			60	5			2						6532
43	Moser River and Smith's Cove	1			3				4							1		4033
34	Mitchell's Bay and Ecum Secum				51											5		12734
	Totals	30666	3854	222	249	158000	978	400	184	2166	139	1952	850	561	6	6921	175	51041
	Rates ...... $	7	12	10	2	10	5	7	10	4	5	4	1.50	4	6	2	4 50	.40
	Values ...... $	214662	46248	2220	498	158000	4890	2800	1840	8664	695	7808	1275	2244	36	13842	788	20416

Total value $1,027,295

59—6

5 GEORGE V., A. 1915

THE CATCH.

RETURN showing the Quantities and Values of all Fish caught and landed in a Green State, in the County of **Hants**, Province of **Nova Scotia**, during the year 1913-14.

Number.	Fishing Districts.	Salmon, *cwts.	Salmon, value.	Cod, cwts.	Cod, value.	Shad, cwts.	Shad, value.	Alewives, cwts.	Alewives, value.	Trout, cwts.	Trout, value.	Bass, cwts.	Bass, value.	Clams, brls.	Clams, value.	Number.
	Hants County.		$		$		$		$		$		$		$	
1	Hantsport to Windsor............	17	170	3	6	15	120	300	450	36	288	...	....	20	30	1
2	Windsor to Noel.................	14	140	7	14	13	104	255	353	16	128	...	....	25	37	2
3	Maitland to Shubenacadie........	19	190	...	....	...	....	270	405	4	32	1	5	..	..	3
4	Shubenacadie to Grand Lake	6	60	...	....	12	96	230	345	3	24	10	50	..	..	4
	Totals....	56	560	10	20	40	320	1055	1583	59	472	11	55	45	67	

Cwts. = 1

THE CATCH MARKETED.

RETURN showing the Quantities and Values of all Fish and Fish Products Marketed in a fresh, dried, pickled, canned, &c., state, for the County of **Hants**, Province of **Nova Scotia**, during the year 1913-14.

Number.	Fishing Districts.	Salmon, fresh and frozen, *cwts.	Cod, fresh, cwts.	Shad, fresh, cwts.	Alewives, fresh, cwts.	Trout, fresh, cwts.	Bass, fresh, cwts.	Clams & Quahangs, fresh, brls.	Number.
	Hants County.								
1	Hantsport to Windsor..................	17	3	15	300	36	.....	20	1
2	Windsor to Noel.........................	14	7	13	255	16	.....	25	2
3	Maitland to Shubenacadie.............	19	.....	.....	270	4	1	.....	3
4	Shubenacadie to Grand Lake.........	6	.....	12	230	3	10	.....	4
	Totals	56	10	40	1055	59	11	45	
	Rates......................$	15 00	2 00	10 00	2.00	10 00	10 00	2 00	
	Values................ $	840	20	400	2110	590	110	90	

Total value........................ .... $4,160

* Cwt. = 100 lbs.

RECAPITULATION.

Of the Quantities and Values of all Fish caught and landed in a Green State, and of the Quantities and Values of all Fish and Fish Products Marketed in a fresh, dried, pickled, canned, etc., state, for **District No. 2**, Province of **Nova Scotia**, during the year 1913-14.

Kinds of Fish.	Caught and Landed in a Green State.		Marketed.		Total Marketed Value.
	Quantity.	Value.	Quantity.	Value.	
		$		$	$
Salmon..... .... cwts.	5,635	60,388	.....		
" used fresh....... .... "	....	....	5,596	83,940	
" smoked....... .... "	....	....	24	480	
					84,420
Lobsters .... "	93,253	490,870	... .	....	
" canned............ ......cases.	...		32,873	591,714	
" shipped in shell............. cwts.			11,491	114,910	
					706,624
Cod ....:... "	147,694	234,000	....		
" used fresh.... ... . "	....	....	19,445	39,015	
" green-salted..................... "	....	....	13,605	40,815	
" dried .... "	....	....	33,815	202,890	
					282,720
Haddock.............. .. .. "	101,375	169,188	....		
" used fresh.......... .... "	....	....	42,618	97,150	
" smoked (finnans)............ "	....	....	4,682	14,856	
" dried .... "	....	....	16,400	73,800	
					185,806
Hake and Cusk.....,.............. "	38,211	35,882	....		
" " used fresh .... "	....	....	254	381	
" " dried .... "	....	....	12,714	50,856	
					51,237
Pollock.......... .... "	19,914	18,721	....		
" used fresh.................. "	....	....	710	1,065	
" dried. ... . "	....	....	6,412	25,648	
					26,713
Herring .... "	111,165	110,375	...:		
" used fresh .... "	....	....	5,833	11,666	
" smoked..................... "	....	....	5,510	16,530	
" pickled... .... brls.	....	....	19,882	89,469	
" used as bait .... "	....	....	16,660	29,155	
" used as fertilizer.............. "	....	....	596	298	
					147,118
Mackerel............... .... cwts.	59,225	268,503	....		
" used fresh.................. "	....	....	43,665	305,658	
" salted..................... brls.	....	....	5,198	62,376	
					368,034
Shad....... ,............ .... cwts.	558	3,446	....		
" used fresh........ . ... "	....	....	506	5,060	
" salted .... brls.	....	....	19	285	
					5,345
Alewives .. .. cwts.	3,001	4,390	....		
" used fresh..... .... "	....	....	2,365	4,730	
" salted.... .... brls	....	....	211	844	
					5,574
Halibut, used fresh.................. cwts.	21,962	146,311	21,962	....	219,620
Flounders............................ "	990	1,420	990	....	4,950
Smelts........... . ... "	1,933	10,321	1,933	....	13,531
Trout .... "	330	2,437	330	....	3,300
Soles..... . ... "	216	324	216	....	1,080
Albacore..... ... . "	2,166	2,740	2,166	....	8,664
Bass..... ... . "	135	569	135	....	1,350
Eels........... .... "	381	1,531	381	....	1,905
Tom Cod..... .... "	18	18	18	....	54
Swordfish............................ "	8,430	25,104	8,430	....	33,720
Mixed fish.... .... "	1,140	570	1,140	. .. .	1,710

SESSIONAL PAPER No. 39

RECAPITULATION.

OF the Quantities and Values of all Fish caught and landed in a Green State, and of the Quantities and Values of all Fish and Fish Products Marketed in a fresh, dried, pickled, canned, etc., state, for District No. 2, Province of Nova Scotia, during the year 1913-14.

Kinds of Fish.	CAUGHT AND LANDED IN A GREEN STATE.		MARKETED.		Total Marketed Value.
	Quantity.	Value.	Quantity.	Value.	
		$		$	$
Squid............... brls.	1,613	5,459	1,613	....	7,428
Oysters............ "	1,291	5,965	1,291	.....	7,746
Clams............. "	7,212	7,693	....	....	....
" used fresh........ "	....	......	7,037	14,074	....
" canned......... cases.	....	......	175	788	
					14,862
Hair Seal Skins .... No.	....	....	64	....	80
Fish Oil .... galls.	....	....	60,327	....	24,130
Totals............	....	1,606,245	....	.....	2,207,721

RECAPITULATION.

OF the Number of Fishermen, etc., and of the Number and Value of all Fishing Vessels, Boats, Nets, etc., in District No. 2, Province of Nova Scotia, for the year 1913-14.

———	Number.	Value.
		$
Sailing and Gasoline Vessels.........	145	117,500
Boats (sail)...............	3,848	136,536
" (gasoline)	1,114	206,104
Carrying Smacks..........	82	45,385
Gill Nets, Seines, Trap and Smelt Nets, etc........	32,856	378,573
Weirs......	17	1,110
Trawls................	4,116	51,699
Hand Lines.............	8,937	6,427
Lobster Traps........	327,978	286,781
" Canneries....	93	135,785
Freezers and Ice-houses.....	116	177,835
Smoke and Fish-houses	1,723	233,109
Fishing Piers and Wharves	808	194,477
Total value	....	1,971,321

Number of men employed on Vessels.........		651
" " Boats		4,469
" " Carrying Smacks.........		120
" persons employed in Fish-houses, Freezers, Canneries, &c....		1,983
Total............		7,223

5 GEORGE V., A. 1915

DISTRICT No. 3.

RETURN showing the Number of Fishermen, &c., the Number and Value of Vessels and Boats, and the Quantity and Value of all Fishing Gear, &c., used in the Fishing Industry, in the County of Lunenberg, Province of Nova Scotia, during the year, 1913-14.

Number	Fishing Districts	Sailing and Gasoline Vessels					Boats					Fishing Gear — Gill Nets, Seines, Trap & Smelt Nets, &c.		
		Number (40 tons and over)	No. (20 to 40 tons)	No. (10 to 20 tons)	Value $	Men	Sail	Value $	Gasoline	Value $	Men	Number	Value $	Number
	Lunenburg County.													
1	Fox Point						100	2700	4	1000	50	157	5000	1
2	Mill Cove			1	600	3	125	3000			50	175	3800	2
3	Lodge and N.W. Cove			1	600	3	60	1500	2	600	30	130	3800	3
4	Aspotogan						36	1000	3	1040	20	95	3700	4
5	Bayswater, Blandford and Deep Cove			19	16500	50	170	5000	10	3500	65	310	7200	5
6	Chester Bay			3	15000		80	2500	4	1200	50	100	4800	6
7	Mahone Bay and Martin's River	9			35000	160	110	2500	5	1500	50	143	2130	7
8	Tancooks				9000	55	250	18000	30	9000	150	730	13050	8
9	Lunenburg Harbour to Kingsbury	68		18	767910	1350	163	3749			150	828	18800	9
10	Halifax For and Islands	41		20	414279	2140	143	5295			164	1156	1100	10
11	Petite Ki me to Vogler's Cove	1		8	4000	9	72	2620			108	680	6440	11
	Totals	118	1	70	1243389	3779	1243	47274	58	17800	867	4554	80500	

RETURN showing the Number of Fishermen, &c., the Gear and Value of Vessels and Boats, and the Quantity and Value of all Fishing Gear, &c., used in the Fishing 1 May, in the County of Lunenburg, Province of Nova Scotia, during the year 19.

Number	Fishing Districts	Fishing Gear						Canneries		Other Material						Persons employed in Canneries, Freezers, and Fish-houses.	Number
		Trawls.		Hand Lines.		Lobster Traps.		Lobster Canneries.		Freezers and Ice-houses.		Smoke and Fish-houses.		Fishing Piers and Wharves.			
		Number.	Value.	Number.	Value.	Number.	Value.	Number.	Value.	Number.	Value.	Number.	Value.	Number.	Value.		
			$		$		$		$		$		$		$		
	Lunenburg County.																
1	Fox Point	50	750	100	50	1000	1000	....	....	1	200	15	1000	5	500	....	1
2	Mill Cove	60	1000	100	50	1000	1000	....	....	....	....	18	720	5	500	....	2
3	Lodge and N.W. Cove	30	450	75	35	1500	1500	1	600	....	....	20	800	15	600	....	3
4	Aspotogan	5	75	40	20	2000	2000	....	....	....	....	8	400	7	350	26	4
5	Bayswater, Blandford and Deep Cove	50	1200	300	150	3500	3500	1	600	1	100	60	2000	25	1200	....	5
6	Chester Bay	3	90	25	12	1500	1500	....	....	1	200	10	400	8	890	20	6
7	Mahone Bay and Martin's River	65	2600	30	15	1000	1000	1	600	1	100	30	5300	25	6000	75	7
8	Tancooks	180	3100	600	300	4000	4000	1	600	1	140	60	5300	40	1600	20	8
9	Lunenburg Harbour to Kingsbury	367	101152	440	220	5430	5430	2	1300	....	....	199	29850	12	175600	14	9
10	LaHave River and Islands	396	10692	....	....	6480	6480	1	600	....	....	190	28500	8	52000	10	10
11	Petite Rivière to Vogler's Cove	16	432	....	....	4225	4225	....	....	....	....	42	6300	....	....	10	11
	Totals	1222	30641	1710	852	31635	31635	7	4300	5	700	652	77970	148	238550	169	

5 GEORGE V., A. 1915

RETURN showing the Number of Fishermen, etc., the Number and Value of Vessels and Industry in the County of **Queens**, Province of

Number.	Fishing Districts.	Steam Vessels.				Sailing and Gasoline Vessels.				Boats.					Carrying Smacks.		
		Number.	Tons.	Value.	Men.	(20 to 40 tons.) Number.	(10 to 20 tons.) Number.	Value.	Men.	Sail.	Value.	Gaso ine.	Value.	Men.	Number.	Value.	Men.
	Queens County.			$				$			$		$			$	
1	Port Medway	..	..	.....	..	...	6	2600	21	26	2050	16	2900	120	..	....	..
2	Mill Village	..	..	.....	..	...	...	......	..	18	216	....	....	36	..	....	..
3	Greenfield	..	..	.....	..	..	....	......	..	14	260	....	.....	....	..	....	..
4	Liverpool, Western Head and Brooklyn	..	..	.....	..	1	2	6000	18	10	500	50	7500	75	..	....	..
5	Gull Islands, White and Hunt's Point, Summerville	..	..	.....	..	....	...	......	..	11	550	26	3900	65	..	....	..
6	Port Mouton & S.W.P. Mouton	2	45	8000	6	....	3	1500	10	10	500	25	3750	75	6	18000	18
7	Port Joli, Port L'Hébert, Sandy Bay	..	..	.....	..	....	...	......	..	....	....	28	4200	55	..	....	..
8	East and West Berlin, Beach Meadows and Eastern Head	..	..	.....	..	.....	1	500	3	20	1000	10	1500	85	1	· 500	2
	Totals	2	45	8000	6	1	12	10600	52	109	5076	155	23750	511	7	18500	20

SESSIONAL PAPER No. 39

Boats, and the Quantity and Value of all Fishing Gear, etc., used in the Fishing Nova Scotia, during the year 1913-14.

Fishing Gear.								Other Material.								Persons employed in Canneries, Freezers and Fish-houses.	
Gill Nets, Seines, Trap & Smelt Nets, etc.		Trawls.		Hand Lines.		Lobster Traps.		Lobster Canneries.		Freezers and Ice-houses.		Smoke and Fish-houses.		Piers and Wharves.			
Number.	Value.	Number.	Value.	Number.	Value.	Number.	Value.	Number.	Value.	Number.	Value.	Number.	Value.	Number.	Value.	Number.	Number.
	$		$		$		$		$		$		$		$		
349	2940	20	80	250	150	4200	4200	1	1800	3	250	68	1750	20	750	20	1
50	300									4	160	20	200	12	140		2
										5	250	15	200				3
30	15000	250	1000	300	225	8000	8000	1	2000	2	6000	60	3000	6	500	3	4
		150	600	250	187	7500	7500					50	2500	4	200		5
6	3000	300	1200	350	262	10000	10000	4	8000	2	1000	25	1250	8	2000	51	6
1	500	100	400	200	150	5000	5000	1	500			10	500	2	500		7
2	1000	200	800	250	187	7500	7500	1	500	1	500	25	1250	2	500		8
438	22740	1020	4080	1600	1161	42200	42200	8	12800	17	8160	273	10650	54	4590	74	

5 GEORGE V., A. 1915

RETURN showing the Number of Fishermen, etc., the Number and Value of Vessels and Boats, and the Quantity and Value of all Fishing Gear, etc., used in the Fishing Industry in the County of Shelburne, Province of Nova Scotia, during the year 1913-14.

No.	Fishing Districts	Steam Vessels				Sailing and Gasoline Vessels					Boats					Carrying Smacks			No.
		Number	Tonnage	Value $	Men	No. (40 tons and over)	No. (20 to 40 tons)	No. (10 to 20 tons)	Value $	Men	Sail	Value $	Gasoline	Value $	Men	Number	Value $	Men	
	Shelburne County.																		
1	Woods Harbour						1	3	500	8	5	875	110	22000	198	4	1000	6	1
2	Shag Harbour and Bear Point							12	500	11	27	1050	45	11500	117	4	400	12	2
3	Cape Island								4300	67	46	5000	218	52500	545				3
4	Barrington							2	325		3	30	10	2500	20				4
5	Port La Tour and Baccaro							2	2135	46	45	575	47	8180	145	2	350	4	5
6	Cape Negro and Blanche										23	555	28	5000	82				6
7	Port Saxon, Clyde Riv. N.E. and N.W. Harb.										35	634	8	1200	44				7
8	Black Point, Red Head and Round Bay										46	700	10	1500	44				8
9	Roseway, Carleton Village and McNutt Isld					2		3	2000	7	82	1130	33	4860	115	1	200	2	9
10	Gunning Cove, Churchover and Birchtown					1		1	200	21	35	375	10	1500	45				10
11	Shelburne and Sandy Point						4	5	15500	86	95	1000	36	1100	200				11
12	Jordan East and West							1	2500	9	42	346	31	840	80	3	200	2	12
13	Lockeport	2	50	11000			5	15	45000	164	196	3735	67	10000	330		1900	9	13
	Totals	2	50	11000		8	9	54	72960	408	681	16625	653	137630	1964	15	4050	35	

SESSIONAL PAPER No. 39

RETURN showing the Number of Fishermen, etc., the Number and Value of Vessels and Boats, and the Quantity and Value of all Fishing Gear, etc., used in the Fishing Industry in the County of Shelburne, Province of Nova Scotia, during the year 1913-14—Con.

No.	Fishing Districts.	Gill nets, seines, trap and Smelt nets, etc. Number	Value $	Trawls. Number	Value $	Hand Lines. Number	Value $	Lobster Traps. Number	Value $	Lobster Canneries. Number	Value $	Freezers and Ice Houses. Number	Value $	Smoke and Fish Houses. Number	Value $	Fishing Piers and Wharves. Number	Value $	Persons Employed in Canneries, Freezers and Fish-houses. Number
	Shelburne County.																	
1	Woods Harbour........	347	3160	10	200	94	94	14795	14795	6	4000	2	250	15	3000	23	4500	96
2	Shag Harbour and Bear Point...	273	1636	11	22	93	93	8485	8485	2	700	..	....	33	2754	18	3350	27
3	Cape Island............	1865	15685	395	790	789	789	50100	50100	4	3150	7	1250	37	7500	37	7000	157
4	Barrington.............	28	150	..	....	50	50	2000	2000	1	....	3	300	..	....	..	....	..
5	Port La Tour and Baccaro....	343	3295	62	124	513	513	6180	6180	1	500	3	300	87	3500	6	1300	54
6	Cape Negro and Blanche......	310	2235	52	195	55	55	5000	500	1	150	1	30	4	800	7	1000	36
7	Port Saxon, Clyde Riv. and N. E. & N. W. Harb.	37	596	40	280	60	60	715	715	..	....	..	....	24	240	11	250	..
8	Black Point, Red Head and Round Bay...	300	2560	40	280	80	80	4000	4000	1	200	..	....	55	1550	11	1400	3
9	Roseway, Carleton Village and McNutt Iid...	400	2400	88	616	80	80	4300	4300	..	....	..	....	90	2000	34	2550	..
10	Gunning Cove, Churchover and Birchtown...	193	824	20	140	80	80	1000	1000	..	....	..	....	25	1800	15	1500	..
11	Shelburne and Sandy Point....	709	5245	544	3808	300	300	3000	3000	..	....	..	....	56	5600	40	2000	..
12	Jordan, East and West.......	355	2564	50	350	150	150	2987	2935	2	300	..	....	67	3000	41	1000	46
13	Lockeport.............	773	9019	716	4892	366	366	9000	9000	2	6500	4	140500	41	3450	69	5720	130
	Totals.........	5933	49433	2028	11617	2821	2821	111512	111512	19	15500	20	142870	534	35194	312	82160	529

5 GEORGE V., A. 1915

RETURN showing the Number and Value of Vessels and Boats, and the Quantity and
Yarmouth, Province of Nova Scotia,

Number	Fishing Districts	Steam Vessels.				Sailing and Gasoline Vessels.					Boats.					Carrying Smacks.		
		Number.	Value.	Value.	Men.	40 tons and over.) No.	20 to 40 tons.) Number.	10 to 20 tons.) Number.	Value.	Men.	Sail.	Value.	Gasoline.	Value.	Men.	Number.	Value.	Men.
	Yarmouth County.			$					$			$		$			$	
1	Port Maitland							4	1500	13	3	1200	48	14400	108			
2	Sandford										12	180	36	10800	92	1	350	3
3	Yarmouth	6	180	22000	26	4	2	6	19300	116	11	1100	83	24900	165			
4	Arcadia										5	300	30	9000	65			
5	Pinkney's Point										6	350	32	9600	62			
6	Comeau Hill							2	700	7	12	540	61	18300	146	2	600	4
7	Wedgeport	3	70	11000	18		1	2	1200	9	10	425	110	33000	240	2	600	5
8	Salmon River										20	115			23			
9	Tusket							1	300	2	101	606			105			
10	Eel Brook										11	132	3	800	16			
11	Argyle							5	1700	17	9	150	35	11500	79			
12	Pubnicoes					4	2	12	23300	143	19	285	67	20100	166	2	600	1
	Totals	9	250	33000	44	8	5	32	48000	307	219	5383	505	152400	1267	7	2150	16

SESSIONAL PAPER No. 39

Value of all Fishing Gear, etc., used in the Fishing Industry in the County of
during the year 1913–14.

Fishing Gear										Canneries		Other Material						Persons employed in canneries, Freezers and Fish-Houses.	
Gills Nets, Seines, Trap and Smelt Nets, etc.		Weirs.		Trawls.		Hand Lines.		Lobster Traps.		Lobster Canneries.		Freezers and Ice Houses.		Smoke and Fish Houses.		Fishing Piers and Wharves.			
Number.	Value.	Number.	Value.	Number.	Value.	Number.	Value.	Number.	Value.	Number.	Value.	Number.	Value.	Number.	Value.	Number.	Value.		Number.
	$		$		$		$		$		$		$		$		$		
196	3475	..	....	80	400	85	85	8100	8100	3	1800	1	150	10	1300	1	1200	65	1
187	7920	..	....	60	300	80	80	6100	6100	1	900	1	150	4	1000	1	100	50	2
419	4268	..	....	150	750	760	760	14095	14095	3	3200	4	5000	20	20000	15	220000	359	3
131	661	..	....	12	60	25	25	6450	6450	2	1200	....	....	....	.....	2	1200	35	4
170	850	.	....	15	75	45	45	4000	4000	....	.....	....	....	3	150	2	400	5	5
260	1300	1	500	40	200	260	260	12400	12400	3	3000	...	....	5	3000	3	2400	75	6
421	4100	6	2000	20	100	500	500	19100	19100	3	3600	1	400	25	2000	7	7700	100	7
42	240	1	300	....	....	....	...	. ...	..	..	....	....	....						8
379	1965	5	1500	....	....	....	...			....	....	....	...	22	1950	2	1000	36	9
92	448	..	....	...	..	...	...	350	350					2	600	...			10
207	1042	..	....	20	100	20	20	5050	5050	1	100	....		7	500	2	1200	10	11
412	3290	..	....	15	75	380	380	12900	12900	5	4500	3	1400	33	4300	6	6000	135	12
2916	29559	13	4300	412	2060	2155	2155	88545	88545	21	18300	10	7100	129	34200	43	241800	861	

MARINE AND FISHERIES

5 GEORGE V., A. 1915

RETURN showing the Number of Fishermen, &c., the Number and Value of Vessels and Industry in the County of **Digby,** Province

No.	Fishing Districts	Steam Vessels				Sailing and Gasoline Vessels				Boats					Carrying Smacks			Gill Nets, Seines, Trap and Smelt Nets, etc.	
		Number	Tons	Value	Men	Number	Number	Value	Men	Sail	Value	Gasoline	Value	Men	Number	Value	Men	Number	Value
	Digby County.			$				$			$		$			$			$
1	Digby and vicinity....					5	2	37000	140			10	3900	20	2	1500	2		
2	Bay View and Culloden....									12	600	17	3400	58				34	340
3	Gullivers Cove to Waterford.......									15	750	16	3200					74	740
4	Centerville..........	1	32	6000	5		1	1200	2	25	500	25	10000	36	1	400	2	50	400
5	Sandy Cove and Mink Cove......					1		1000	3	20	410	16	4000	50	1	200	2	55	1290
6	Little River and Whale Cove......						1	1000	2	30	860	22	6600	50	2	1700	4	51	600
7	Teddville and East Ferry.......									17	650	18	4500	53				65	680
8	Tiverton and Central Grove...........					2		2500	6	20	600	75	22500	180	2	1200	4	202	2200
9	Freeport...........					3		6000	45	75	1000	60	18000	150	4	5000	12	130	1300
10	Westport						1	700	3	50	500	65	19500	154	1	1000	2	250	2500
11	Smith's Cove and Brighton....									33	650	8	1500	37				8	80
12	Plympton to Weymouth									16	320	20	5000	50				104	1040
13	New Edinburgh									5	70	20	5000	50				110	1100
14	Belle View and White Cove........									14	280	17	4250	62				75	750
15	Grosse Coque ..									4	80	10	2500	28				15	150
16	Church Point........						3	1000	15	10	200	5	1200	30				20	200
17	Little Brook and Comeauville						1	1200	5	26	390	18	4500	88				4	40
18	Saulnierville........									12	180	8	1600	40				10	100
19	Meteghan River.....									9	320	5	1250	30				20	200
20	Meteghan					2	1	1000	15	20	400	8	1650	56	1	1800	3	40	400
21	Comeau's Cove.....									9	180	5	1250	28				10	100
22	Bear Cove									20	300	11	2750	62				20	200
23	Cap St. Mary's......						11	4300	56	10	500	43	10750	106	1	400	2	100	1000
24	Salmon and Beaver River............									15	360	5	1000	40				20	200
	Totals	1	32	6000	5	8 5	21	56900	292	467	9980	507	139800	1520	15	13200	33	1470	15610

SESSIONAL PAPER No. 39

Boats, and the Quantity and Value of all Fishing Gear, &c., used in the Fishing of Nova Scotia, during the year 1913-14.

Fishing Gear								Canneries				Other Material						Persons employed in Canneries, Freezers and Fish-houses	
Weirs		Trawls		Hand lines		Lobster Traps		Lobster Canneries		Salmon Canneries		Freezers and Ice-houses		Smoke and Fish-houses		Fishing-Piers and Wharves			
Number	Value	Number	Value	Number	Value	Number	Value	Number	Value	Number	Value	Number	Value	Number	Value	Number	Value	Number	Number
	$		$		$		$		$		$		$		$		$		
5	1125	450	3050	25	25	1500	1500	..	.....	....	....	9	11400	30	40000	12	31800	150	1
..	...	78	546	20	20	2000	2000	..	....	....	....	...	.....	6	200	....	.....	....	2
1	500	96	672	50	50	2000	2000	1	27000	..	...			7	300	...	...	....	3
..	...	200	1400	25	25	1500	1500			..	...	5	500	22	4700	1	500	75	4
2	1390	66	462	45	45	1600	1600	1	1500	...	...	8	1500	13	1800				5
..	...	150	1050	25	25	2000	2000	4	12000	...	...	4	600	20	2700	15	10000	45	6
..	...	74	518	320	320	1600	1600	..	....	...	...	1	25	13	700	3	500	....	7
		750	5250	500	500	4500	4500	1	1500	...	...	2	900	47	7500	24	20000	100	8
1	700	324	2268	155	155	3000	3000			...	...	4	750	73	8300	26	4100	15	9
..	...	275	1925	300	300	3500	3500	..	...	...	...	5	1500	24	3110	35	11600	25	10
7	2100	15	105	36	36	200	200			...	...	3	75	7	300	....	.....	....	11
2	600	80	560	106	106	1100	1100	..	..	...	...	4	205	17	500	5	2000	....	12
1	150	175	1050	50	50	2000	2000	1	300			..	...	9	1200	..	.....	10	13
1	800	150	900	50	50	400	400			1	1800			18	1140	...	....	25	14
4	600	40	240	10	10	400	400	1	800					4	200	...	...	18	15
..	...	25	150	60	60	750	750			1	1000			10	850	...	...	25	16
		12	72	100	100	2400	2400	1	500					20	800	..	....	25	17
		..	...	80	80	1800	1800	..	..					24	960	..	...		18
2	300	20	120	40	40	1000	1000	..	..					3	300	..	...		19
..	...	..	...	80	80	2100	2100	1	400					10	200	...	...	20	20
..	...	..	...	65	65	1700	1700	..	..					7	175	...	...		21
..	...	20	120	100	100	2000	2000	..						10	2.00	...	...	80	22
..	...	20	120	400	400	5000	5000	2	900					35	875	...	...		23
..	...	..	...	40	40	400	400	..	..					3	45	...	...	..	24
26	8175	3020	20578	2682	2682	44450	44450	13	44900	2	2800	45	17455	432	78855	121	80500	613	

·5 GEORGE V., A. 1915

RETURN showing the Number of Fishermen, etc., the Number and Value of Vessels and
Industry in the County of **Annapolis,**

	Fishing Districts.	Vessels, Boats and Carrying Smacks.									
		Sailing & Gasoline Vessels.					Boats.				
Number.		40 tons and over) Number.	20 to 40 tons) Number.	20 to 20 tons) Number.	Value.	Men.	Sail.	Value.	Gasoline	Value.	Men.
	Annapolis County.					$		$		$	
1	Margaretsville.........			....			7	140	8	2000	27
2	Port George.........			....			36	750	8	1400	50
3	Port Lorne....			....			6	150	2	400	16
4	Hampton................................			....			8	400	......	...	8
5	Phinney's Cove........................			....			30	600	8	2000	60
6	Parkers Cove.....			....			30	600	20	4000	66
7	Hilsburne.............................	1	1	....	2500	24	25	500	10	2000	40
8	Litchfield.........................		...				8	200	15	3000	45
9	Port Wade....	3	2	1	7400	65	10	200	7	1400	20
10	Victoria Beach			....			40	800	45	9500	110
11	Deep Brook and Clementsport..........			....			8	200	6	1400	20
12	Annapolis, Lequille and Nicteau Rivers.................			....			40	600	... .	.....	......
	Totaux...	4	3	1	9900	89	248	5140	129	27100	462

RETURN showing the Number of Fishermen, etc., the Number and Value of Vessels
Industry in the County of **Kings,**

	Fishing Districts.	Vessels, Boats and Carrying Smacks.									
		Sailing and Gasoline Vessels.				Boats.					
Number.		40 tons and over) Number.	20 to 40 Tons) Number.	Value.	Men.	Sail.	Value.	Gasoline.	Value.	Men.	
	Kings County.				$		$		$		
1	Morden and vicinity........... . ..	. .			. . .	6	120	5	1400	22	
2	Victoria Harbour and Ogilvie Wharf....					3	60	.....	.....	6	
3	Harbourville............................		13	1950	52	2	4	2	500	9	
4	Canada Creek....					5	75	2	250	15	
5	Chipman's Brook and Hunting Point....	...	12	6000	36	2	30	1	150	6	
6	Hall's Harbour....	41		6000	82	4	60	4	600	19	
7	Race Point and Sheffield Vault.					1	20	...	.....	3	
8	Baxters Harbour......................					9	135	1	250	21	
9	Whalen Beach & Wells Cove.					2	30	...	.	4	
10	Scotts Bay.............................					4	100	3	500	16	
11	Blomidon and Kingsport......					2	40	2	300	8	
12	Starr's Point to Wolfville.										
13	Upper Gaspereau and all inland waters..	....				12	150	...	....	24	
	Totals.........................	41	25	13950	170	52	824	20	3950	153	

SESSIONAL PAPER No. 39

Boats, and the Quantity and Value of all Fishing Gear, etc., used in the Fishing Province of **Nova Scotia**, during the year 1913-14.

Gill Nets, Seines, Trap & Smelt Nets, &c.		Weirs.		Trawls.		Hand Lines.		Lobster Traps.		Freezers and Ice-houses.		Smoke and Fish-houses.		Fishing Piers and Wharves.		Persons employed in canneries, freezers and fish-houses.	
Number.	Value.	Number.	Value.	Number.	Value.	Number.	Value.	Number.	Value.	Number.	Value.	Number.	Value.	Number.	Value.	Number.	
	$		$		$		$		$		$		$		$		
20	200	2	400	5	125	100	100	100	100	3	200	10	1000	2	1000	..	1
40	400	2	200	40	450	200	200	500	500	9	450	15	750	...	...	....	2
25	250	1	100	10	120	100	100	300	300	1	50	10	500	...	...	....	3
10	100	2	200	8	200	35	35	300	300	2	100	4	400	...	...	....	4
75	750	1	200	50	1200	200	200	1600	1600	....		20	600	...	...	....	5
60	600	..	....	150	1800	200	200	2000	2000	1	200	30	2500	1	1000	..	6
40	400	..	....	150	1800	100	100	600	600	2	2000	6	1000	1	500	30	7
30	300	1	200	149	1600	60	60	1200	1200	1	100	12	1200	...	...	....	8
....	......	1	200	500	12500	25	25	400	400	1	400	10	3000	3	3000	10	9
25	250	..	....	150	3700	200	206	2500	2500	1	700	40	2000	10	2000	30	10
4	40	5	800	5	125	25	25	....		1	1500	10	1500	6	1200	....	11
....	......	3	300	....		10	500	...	....	10	1000			10	12
329	3290	18	2600	1208	23620	1245	1245	9500	9500	32	6200	167	14450	33	9700	80	

and Boats, and the Quantity and Value of all Fishing Gear, etc., used in the Fishing Province of **Nova Scotia**, during the year 1913-14.

Gill Nets, Seines, Trap & Smelt Nets, &c.		Weirs.		Trawls.		Hand Lines.		Lobster Traps.		Freezers and Ice houses.		Smoke and Fish-houses.		Fishing Piers and Wharves.		
Number.	Value.	Number.	Value.	Number.	Value.	Number.	Value.	Number.	Value.	Number.	Value.	Number.	Value.	Number.	Value.	Number.
	$		$		$		$		$		$		$		$	
5	50	4	800	..	......	30	30	60	60	3	75	4	100	1	5000	1
6	60	2	400	..	......	8	8	..	...	1	15	1	20	1	7000	2
4	40	3	600	...	......	12	12	..	...	2	40	2	50	1	12000	3
10	100	2	500	...	......	20	20	220	220	3	50	6	120	1	8000	4
9	90	3	700	4	20	10	10	100	100	1	20	3	45	1	5000	5
18	180	2	500	10	50	25	25	100	100	4	80	5	100	1	10000	6
......		2	400	...		5	5	..	...	2	30	2	40	..		7
15	150	2	400	5	25	30	30	50	50	1	15	2	75	1	4000	8
......		1	200	..		8	8	..	...	..		1	20	..		9
8	80	6	1500	7	35	25	25	100	100	5	100	7	200	1	5000	10
2	20	1	200	..		12	12	..	......	1	15	2	40	1	10000	11
...	......	1	300	..		..		..						1	10000	12
35	270	3	800	..		...		..								13
112	1040	32	7300	26	130	185	185	630	630	23	440	38	810	10	76000	

5 GEORGE V., A. 1915

THE

RETURN showing the Quantities and Values of all Fish caught and landed in a
the year

Number.	Fishing Districts.	Salmon, cwts.	Salmon, value.	Lobsters, cwts.	Lobsters, value.	Cod, cwts.	Cod, value.	Haddock, cwts.	Haddock, value.	Hake and Cusk, cwts.	Hake and Cusk, value.
			$		$		$		$		$
	Lunenburg County.										
1	Fox Point			123	984	116	232	20	30	30	30
2	Mill Cove			191	1528	241	482	30	45	60	60
3	Lodge and N.W. Cove			192	1536	65	130	30	45	270	270
4	Aspotogan			626	5008						
5	Bayswater, Blandford and Deep Cove			813	6504	1505	3010	300	450	336	336
6	Chester	14	291	2835	22680	46	92	35	53	20	20
7	Martins River and Mahone Bay	6	107			30150	60300	4230	6345	3700	3700
8	Tancooks			905	7240	2220	4440	546	818	315	315
9	Lunenburg Harbour to Kingsbury			3353	26824	217452	434904	24125	36187	16350	16350
10	La Have River	91	1511	3353	26824	209952	419904	19570	29355	16350	16350
11	Petite Rivière to Vogler's Cove			2613	26130	7500	15000	4500	6750		
	Totals	111	1909	15004	125258	469247	938494	53386	80078	37431	37431

Cwt. = 100 lb.

SESSIONAL PAPER No. 39

CATCH.

Green State in the County of **Lunenburg**, Province of **Nova Scotia**, during 1913-14.

Pollock, cwts.	Pollock, value.	Herring, cwts.	Herring, value.	Mackerel, cwts.	Mackerel, value.	Alewives, cwts.	Alewives, value.	Halibut, cwts.	Halibut, value.	Albacore, cwts.	Albacore, value.	Sword-Fish, cwts.	Sword-Fish, value.	Squid, brls.	Squid, value.	Scallops, brls.	Scallops, value.
	$		$		$		$		$		$		$		$		$
93	93	370	185	6969	27207	....	......	......	......	75	75	70	275	....	..	....	....
285	285	36	18	2500	9375	....	......	......	......	...	..	...	...	....	..	....	....
.....	....	2C0	112	897	3111	..	......	....	......	6	6	32	160	....	..	....	...
.....	....	405	255	1419	4284	...	......	....	...	...	..	180	506	..	......	....	....
30	30	1815	1545	1809	5517	....	...	31	155	...	..	250	1000	..	......	....	....
...	.....	770	460	740	2220	75	75	1	5	...	..	..	....	15	60	....	....
.....	.....	90	45	495	1485	14	14	90	450	...	..	...	...	..	..	....	....
630	630	15920	15520	1650	5070	.....	...	16	80	...	..	710	3231	..	....	1345	4035
.....	.....	1926	1926	1301	6525	....	.....	616	3080	...	..	471	2355	....	..	....	....
.....	.....	1926	1926	1291	6455	....	......	616	3080	...	..	471	2355	....	..	....	....
.....	......	1926	1926	1291	6455	.....	......	....	....	......	....	....	...	....	..	....	....
1038	1038	25384	24918	20362	77704	89	89	1370	6850	81	81	2184	9882	15	60	1345	4035

THE CATCH MARKETED.

RETURN showing the Quantities and Value of all Fish and Fish Products Marketed in a fresh, dried, pickled, canned, &c., State, for the County of Lunenburg, Province of Nova Scotia, during the year 1913-14.

Number.	Fishing Districts.	Salmon, used fresh and frozen, cwts.	Salmon, canned, cases.	Lobsters, canned, cases.	Lobsters, shipped in shell, cwts.	Cod, dried, 4cwts.	Haddock, used fresh, cwts.	Haddock, dried, quintals.	Hake and Cusk, dried, quintals.	Pollock, dried, quintals.	Herring, pickled, brls.	Herring, used as bait, brls.	Number.
	Lunenburg County.												
1	Fox Point			5	101	39	20					185	1
2	Mill Cove			22	136	90	30			31		18	2
3	... and N.W. ...			14	157	22	30			95		88	3
4	Aspotogan	14		187	159							150	4
5	Bayswater, Blandford and Deep Cove	6		297	71	502	300		112	10	8	220	5
6	...			1054	200	10050		1403	1233		35	310	6
7	... Bay and ... River				250	15	35	175	7		425	45	7
8	Tancooks			22	2613	40	21	8023	105	10	50	90	8
9	Lunenburg Harbour to Kingsburg	91		96	2613	84	53	6523	5450	210	5040		
10	La Have river and Islands			97	2613	94		1500	5450		642	10	10
11	Petite Ri... ... to Vogler's Cove	8				80					642 642		11
	Totals	111		2437	8913	156416	506	17626	12477	346	7484	1458	
	Rates	25.00		21.00	15.00	6.00	2.00	5.00	3.00	3.00	4.00	1.00	
	Value	2775		51177	133695	938496	1012	88130	37431	1038	29936	1458	

THE CATCH MARKETED.

RETURN showing the Quantities and Value of all Fish and Fish Products Marketed in a fresh, dried, pickled, canned, &c., State, for the County of Lunenburg, Province of Nova Scotia, during the year 1913-14—*Concluded.*

Number	Fishing Districts	Mackerel, used fresh, cwts.	Mackerel, salted, cwts.	Shad, used fresh, cwts.	Alewives, used fresh, cwts.	Alewives, salted, brls.	Halibut, used fresh, cwts.	Albacore, used fresh, cwts.	Sword-Fish, used fresh, cwts.	Squid, used as bait, brls.	Scallops, used fresh, brls.	Fish Oil, gals.	Number
	Lunenburg County.												
1	Fox Point	4884	695					75	70			15	1
2	Mill Cove	1165	445									50	2
3	Lodge and N.W.	333	188					6	32			55	3
4	Aspotogan	603	272						180				4
5	Bayswater, Bird and Deep Cove	139	556		15		31		250			300	5
6		13	202		14	20	1						6
7	Malone Bay and Martin's River	75	140				90		710	15		6000	7
8	Tancooks		650				16					400	8
9	Lunenburg Harbour to Kingsbury	10	431				66		471		1345	32000	9
10	La Have River and Islands		430				66		471			17000	10
11	Petite Rivière to Volrer's Cove		430									65	11
	Totals	7344	4339		29	20	1370	81	2184	15	1345	55885	
	Rates	5.00	15.00		1.00	4.00	5.00	1.00	5.00	4.00	3.00	30c.	
	Values $	36720	65085		29	80	6850	81	10920	60	4035	16765	

Total value............ $1,425,773

* Cwts.=100 lbs.   † Quintal=112 lbs.

5 GEORGE V., A. 1915

THE

RETURN showing the Quantities and Values of all Fish caught and landed in a
year

Number.	Fishing Districts.	Salmon, *cwts.	Salmon, value.	Lobsters, cwts.	Lobsters, value.	Cod, cwts.	Cod, value.	Haddock, cwts.	Haddock, value.	Hake and &c., cwts.	Hake and &c., value.	Pollock, cwts.	Pol lok, value.	Herring, cwts.	Herring, value.
	Queens County.	$		$			$		$		$		$		$
1	Port Medway....	111	2626	1358	11070	1772	3544	226	369	275	344	374	374	342	342
2	Mill Village.....	61	1440	...	......	.....	....	...	...	...	...	...	......	......	.....
3	Greenfield.....	21	371	...	......	......	.. ..	...	...	...	...	...	......	......	....
4	Liverpool, Western Head, Brooklyn....	17	185	2500	22500	1000	2000	400	800	200	150	110	110	4000	4000
5	Gull Islands, White and Hunt's Point, Summerville.....	....	....	1765	15885	650	1300	300	600	100	75	435	435	4000	4000
6	Port Mouton and S. W. Port Mouton..........	....	.. ..	1205	10350	1050	2100	400	800	200	150	110	110	7000	7000
7	Port Joli, Port L'Hebert and Sandy Bay.......	.. ..	....	1150	10350	500	1000	200	400	200	150	110	110	4000	4000
8	East and West Berlin, Beach Meadows, Eastern Head.............	....	....	1825	16425	500	1000	500	1000	100	75	220	220	6000	6000
	Totals........	210	4622	9803	86580	5472	10944	2026	3969	1075	944	1359	1359	25342	25342

* Cwt. = 100 lbs.

SESSIONAL PAPER No. 39

CATCH.

Green State in the County of **Queens**, Province of **Nova Scotia**, during the 1913-14.

Mackerel, cwts.	Mackerel, value.	Alewives, cwts.	Alewives, value.	Halibut, cwts.	Halibut, value.	Smelts, cwts.	Smelts, value.	Trout, cwts.	Trout, value.	Albacore, cwts.	Albacore, value.	Eels, cwts.	Eels, value.	Sword-fish, cwts.	Sword-fish, value.	Clams, cwts.	Clams, value.	Number.
	$		$		$		$		$		$		$		$		$	
32	256	75	52	47	230	46	352	....	....	....	....	....	....	69	180	7	21	1
....	....	255	179	....	....	....	....	18	180	....	....	....	....	...	...	...	...	2
....	....	420	294	....	....	....	....	33	350	....	....	....	....	...	...	...	...	3
2000	14000	100	150	10	75	20	180	20	200	10	20	200	700	32	156	...	....	4
900	6300	..	...	12	90	....	....	....	....	20	40	....	.	....	....	....	....	5
800	5600	20	30	12	90	....	....	5	50	28	56	....	....	100	600	....	....	6
200	1400	....	....	10	75	....	....	....	....	....	....	....	....	...	...	...	...	7
400	2800	....	....	10	75	....	....	....	....	....	....	....	...	100	600	....	....	8
4332	30356	872	705	101	635	66	532	78	780	58	116	200	700	301	1536	7	21	

5 GEORGE V., A. 1915

THE CATCH

RETURN showing the Quantities and Values of all Fish and Fish Products Marketed **Nova Scotia**, during the

Number.	Fishing Districts.	Salmon, used fresh and frozen, *cwts.	Lobsters, canned, cases.	Lobsters, shipped in shell, cwts.	Cod, used fresh, cwts.	Cod, dried, †qtls.	...k, used fresh, ...	...k, dried; ...	Hake and ..., used fresh, cwts.	Hake and ..., dried, ...	Pollock, used fresh, cwts.	Pollock, dried, qtls.	Herring, used fresh, cwts.	Herring, pickled, brls.
	Queens County.													
1	Port Medway..............	111	109	1087	437	445	184	38	130	46	179	65	...	69
2	Mill Village................	61	....	....	....	....	....	....	....	....	....	....	....	....
3	Greenfield	21	....	....	....	....	....	....	....	....	....	....	....	....
4	Liverpool, Western Head, Brooklyn................	17	.....	300	400	200	200	75	....	67	60	17	100	467
5	Gull Island, White and Hunts Point, Summerville................	....	.....	200	200	150	100	35	....	35	60	125	100	467
6	Port Mouton and S.W. Port Mouton..............	....	2662	900	300	250	100	175	....	67	60	17	100	635
7	Port Joli, Port L'Hébert, Sandy Bay	....	.....	100	200	100	50	35	....	67	60	17	100	800
8	East and West Berlin, Beach Meadows, Eastern Head...................	....	.....	286	296	67	150	35	....	35	67	50	100	1000
	Totals	210	2771	2873	1833	1212	784	393	130	317	486	291	500	3438
	Rates......... ..... $	8 22.00	21.00	12.00	2.00	6.00	2.50	5.00	2.00	4.00	1.00	4.00	1.00	4.00
	Values.................$	4620	58191	34476	3666	7272	1960	1965	260	1268	486	1164	500	13752
	Total value....													

*Cwt. =100 lbs.　　Quintal—112 lbs.

SESSIONAL PAPER No. 39

MARKETED.

in a fresh, dried, pickled, canned, &c., State for the County of **Queens**, Province of year 1913–14.

Herring, used as bait, brls.	Mackerel, used fresh, cwts.	Mackerel, salted, brls.	Shad, used fresh, cwts.	Alewives, used fresh, cwts.	Alewives, salted, brls.	Halibut, used fresh, cwts.	Smelts, used fresh, cwts.	Trout, used fresh, cwts.	Albacore, used fresh, cwts.	Eels, used fresh, cwts.	Sword-fish, used fresh, cwts.	Clams and Quahangs, used fresh, brls.	Hair seal skins, No.	Number.
68	32	.....	.....	.....	25	47	46	18	.....	.....	69	7	80	1
.....	..	...	.....	75	60	.....	.....	35	.....	.....	.....	.....	....	2
.....	...	...	.....	195	75	...	.....	.....	.....	.....	.....	.....	....	3
1250	1790	70	....	50	17	10	20	20	10	200	32	.....	....	4
1250	675	75	..	.....	....	12	.....	.....	20	...	.....	.....	....	5
2534	350	150	.....	20	....	12	.....	5	28	.....	100	.....	....	6
750	200	.....	.....	.....	....	10	.....	...	..	.....	.....	.....	....	7
1450	300	35	.....	.....	....	10	.....	.....	.....	.....	100	.....	....	8
7302	3347	330	.....	340	177	101	66	78	58	200	301	7	80	
2.00	8.00	11.00	.....	1.00	4.00	7.00	9.00	10.00	2.00	3.50	7.00	3.00	1.00	
14604	26776	3630	.....	340	708	707	594	780	116	700	2107	21	80	

... . $180,743

5 GEORGE V., A. 1915

THE CATCH.

RETURN showing the Quantities and Values of all Fish caught and landed in a Green State, in the County of **Shelburne**, Province of **Nova Scotia**, during the year 1913-14.

No.	Fishing Districts.	Salmon, cwts.	Salmon, value.	Lobsters, cwts.	Lobsters, value.	Cod, cwts.	Cod, value.	Haddock, cwts.	Haddock, value.	Hake and Cusk, cwts.	Hake and Cusk, value.	Pollock, cwts.	Pollock, value.	Herring, cwts.	Herring, value.	Mackerel, cwts.	Mackerel, value.	Alewives, cwts.	Alewives, value.	Halibut, cwts.	Halibut, value.	No.
	Shelburne County.	$	$		$		$		$		$		$		$		$		$		$	
1	Wood's Harbour			10262	123144	9520	14280	312	312					1844	1844	1500	9000					1
2	Shag Harbour and Bear Point			4031	45372	1780	2670	963	963					2460	2460	84	504			39	234	2
3	Cape Island			10118	121416	13840	20760	4167	4167					4350	4350	390	2340			80	2880	3
4	Barrington			2390	29680	2230	3346	849	849			144	144	720	720							4
5	Port La Tour and Baccaro			2215	26560	12316	18477	4470	4470			399	399	7521	7521	420	2520	400	400	15	1050	5
6	Cape Negro and Blanche			770	9240	2202	3903	2801	2801			2757	2757	4630	4630	135	810			48	288	6
7	Port Saxon, Sable River, N.E. and N.W. Harbour	22	251	30	360	812	1218	230	230			77	53	164	164	250	750	506	378			7
8	Black Point, Red Head and Round Bay			332	2656	1350	2025	360	360			320	207	2225	1668	340	1020					8
9	Roseway, Carleton Village and McNutt's Island			1276	10208	3632	5448	2440	1830	9	6	2050	1393	10470	7853	1809	6740	15	11			9
10	Gunning Cove, Churchover and Birchtown	13	146	371	2968	3160	4740	1440	1377			1310	850	6246	6184	505	1515	33	36			10
11	Shelburne and Sandy Point	9	102	1318	10544	9570	14335	3000	2419	170	142	2111	1477	11830	8872	1680	3990	146	103			11
12	Jordan, East and West	27	313	1267	10136	2300	3450	1200	1660	7	5	720	540	4322	3241	500	1500	66	49			12
13	Lockeport			13177	105416	30029	45043	17210	29109	12432	10562	7400	6468	21633	20043	2278	8307	282	213	2202	13212	13
	Totals	71	812	47557	499720	92743	139114	39442	50547	12618	10735	17296	14295	96415	82550	9891	38996	1447	1190	2944	17664	

* Cwt. = 100 lbs.

THE CATCH.

RETURN showing the Quantities and Values of all Fish caught and landed in a Green State, in the County of Shelburne, Province of Nova Scotia, during the year 1913-14—*Concluded.*

Number	Fishing Districts	Smelts, cwts.	Smelts, value $	Albacore, cwts.	Albacore, value $	Sword-fish, cwts.	Sword-fish, value $	Squid, brls.	Squid, value $	Clams, brls.	Clams, value $	Dulse, Crabs, Cockles and other shell fish. Cwts.	Dulse, Crabs, Cockles and other shell fish. Value $	Number
	Shelburne County.													
1	W al's Harbour			99	297									1
2	Shag Harbour and Bear Point			15	45	252	1512							2
3	Cape Island					10	60							3
4	Barrington													4
5	Port La Tur and Baccaro			28	84									5
6	Cape Negro and Blanche													6
7	Port Saxon, Clyde River, N. E. and N.W Harbours													7
8	Black Point, Red Head and Round Bay													8
9	Roseway, Carleton Village and McNutt's Island									200	325			9
10	Gunning Cove, Churchover and	34	401											10
11	Sable and Sandy-Point	10	90											11
12	Jordan, East and West	31	477	80	320	14	70							12
13	Lockeport					633	3354	2	5	392	474	8	56	13
	Totals	75	968	222	746	909	4996	2	5	592	799	8	56	56

Cwt. = 100 lbs.

THE CATCH MARKETED.

RETURN showing the Quantities and Value of all Fish and Fish Products Marketed in a fresh, dried, pickled, canned, etc., State, for the County of Shelburne, Province of Nova Scotia during the year 1913-14.

No.	Fishing Districts.	Salmon, used fresh and frozen, cwts.	Lobsters, canned, cases.	Lobsters, shipped in shell, cwts.	Cod, used fresh, cwts.	Cod, shipped green salted, cwts.	Cod, dried, †quintals.	Haddock, used fresh, cwts.	Haddock, smoked, cwts.	Haddocks, dried, quintals.	Hake and Cusk, used fresh, cwts.	Hake and Cusk, dried, quintals.	Pollock, used fresh, cwts.	Pollock, dried, quintals.	Herring, used fresh, cwts.	No.
	Shelburne County.															
1	Wood's Harbour		2596	3772		2480	1520	150		54					185	1
2	Shag Harbour and Bear Point		829	1659	500	890				321					844	2
3	Cape Island		2532	3788			4446			1389					700	3
4	Barrington			2390		725	283			283				48		4
5	Port La Tour and Baccaro			1340		2583	2384	300		1390					2460	5
6	Cape Negro and Blanche	22	350	770	50		734	800		667				133	2970	6
7	Port Saxon, Clyde River, N. E. and N.W. Harbour			30	110	150	154	100		77	9		20	919	164	7
8	Black Point, Red Head and Round Bay	13		532	538	283	219	1300		85		41	60	17	85	8
9	Roseway, Carleton Village and McNutt's Island	9	111	924	600	800	498	610		480	50		90	86	5269	9
10	Gunning			371						274			140	651	4246	10
11	Shelburne and Sandy Point		145	1318	1200	1800	1588	535		481			200	389	4616	11
12	Jordan, East and West			965	583	400	306	330		289	7		50	636	12	12
13	Lockeport	27	3275	4590	6827	6477	3418	6577	242	3383	3846	2862	468	2207	9606	13
	Totals	71	9668	22680	10308	17394	15901	11702	242	9173	3912	2903	1028	5408	31135	
	Rates	15.00	21.00	18.00	2.00	3.00	5.00	2.00	7.00	4.00	1.50	4.00	1.50	4.00	1.00	
	Values	1065	207228	412902	20616	52182	79505	23404	1694	36692	5868	11612	1542	21632	31135	

*Cwts. = 100 lbs.

THE CATCH MARKETED.

RETURN showing the Quantities and Value of all Fish and Fish Products Marketed in a fresh, dried, pickled, canned, etc., State, for the County of Shelburne, Province of Nova Scotia, during the year 1913-14.—*Concluded.*

Number.	Fishing Districts.	Herring, smoked, cwts.	Herring, pickled, brls.	Herring, used as bait, brls.	Mackerel, used fresh, cwts.	Mackerel, salted, brls.	Alewives, used fresh, cwts.	Alewives, salted, brls.	Halibut, used fresh, cwts.	Smelts, used fresh, cwts.	Albacore, used fresh, cwts.	Swor l Fish, used fresh, cwts.	Squid, used as bait, brls.	Clams and Quahaugs, used fresh, cwts.	Dulse, Crabs, Cockerels, and Other Shell Fish, used fresh, cwts.	Fish Oil, gals.	Numbers.
	Shelburne County.																
1	Wd Harbour....................		1721	4748	1290	66				30		99					1
2	Shag Harbour and Bear Point....		353	276	84							15	242				2
3	Cape Island....................			1825	390		400			480							3
4	Barrington.....................		210														4
5	Port La Tour and Baccaro.......		725	1448	327	28				175		28	10			80	5
6	Cpe Negro and Blanche.........			830	135		130	125		48							6
7	Port Saxon, Clyde River, N.E. and N.W. Harbour		80	70	61	60	15				34						7
8	Black Point, Red Head and Round Bay.		900	100	46	45	33										8
9	Roseway, Carleton Village and McNutt's Island.		60	90	60	583	71	25			10				200		9
10	Gunning Cove, Churchover and Birchtown....				42	154											10
11	Shelburne and Sandy Point....		28	83	315	50	36	10					19				11
12	Jordan, East and West........		22	56	42	32	252			31	80	638		392			12
13	Lockeport......................	1250	25		1102			10	2202				2		8	500	13
	Totals..........	1500	7914	41012	3894	83	937	170	2914	75	222	909	2	592	8	1290	
	Rates..........	6.00	3.00	2.00	5.00	12.00	1.00	4.00	8.00	15.00	4.00	8.00	2.50	1.50	7.00	30c	
	Values..........	1500	32	41012	19470	96	937	680	23552	1125	888	7272	.5	888	56	1250	

Total Value............... $1,662,450

*Qtl. = 112 lbs.

THE CATCH.

RETURN showing the Quantities and Values of all Fish caught and landed in a Green State in the County of **Yarmouth**, Province of **Nova Scotia**, during the year 1913-14.

Number	Fishing Districts	Salmon, cwts.	Salmon, value.	Lobsters, cwts.	Lobsters, value.	Cod, cwts.	Cod, value.	Haddock, cwts.	Haddock, value.	Hake and Cusk, cwts.	Hake and Cusk, value.	Pollock, cwts.	Pollock, value.	Herring, cwts.	Herring, value.	Mackerel, cwts.	Mackerel, value.	Number.
	Yarmouth County.																	
1	Port Maitland	11	22	4837	38696	7027	10645	2900	3343	50	42	906	906	1500	1500	1400	6290	1
2	Sandford	18	330	4740	37920	5000	7573	2500	2882	40	34	840	840	1500	1600	1900	8533	2
3	Yarmouth			9858	78864	13202	20001	4490	5174	1224	1082	2163	2165	7120	7120	3500	15752	3
4	Arcadia			4445	35560	1500	2271	700	808	30	25	600	600	3000	3000	1005	4472	4
5	Pinkney's Point			4585	36680	1500	2271	600	692	20	17	360	360	1400	1400	1110	4930	5
6	Eau Hill	6	110	9774	78192	6500	9889	1914	2206	275	225	1100	1100	7050	7050	2600	11645	6
7	Wedgeport	12	220	10440	83840	900	13632	2100	2421	450	421	1200	1200	10108	10108	3150	14152	7
8	Salmon River	25	458	101	806											50	238	8
9	Tusket																	9
10	Eel Brook																	10
11	Argyle			4911	39288	2000	3029	440	507	20	17	50	50	6100	6100	1400	5833	11
12	Pub			10550	84400	15250	23199	4000	4610	1194	1005	2030	2032	6150	6150	2200	9872	12
	Totals	72	1320	64281	514246	60479	91769	15644	22543	3303	2818	9249	9253	43928	43928	18215	81727	

Cwt. =100 lbs.

THE CATCH.

RETURN showing the Quantities and Values of all Fish caught and landed in a Green State in the County of **Yarmouth**, Province of **Nova Scotia**, during the year 1913-14—*Concluded.*

No.	Fishing Districts.	Shad, cwts.	Shad, value. $	Alewives, cwts.	Alewives, value. $	Halibut, cwts.	Halibut, value. $	Smelts, cwts.	Smelts, value. $	Albacore, cwts.	Albacore, value. $	Eels, cwts.	Eels, value. $	Swordfish, cwts.	Swordfish, value. $	Mixed Fish, cwts.	Mixed Fish, value. $	Clams, brls.	Clams, value. $	Dulse, crabs, cockles and other shell fish, cwts.	Dulse, crabs, cockles and other shell fish, value. $
	Yarmouth County.																				
1	Port Maitland					15	96									500	250	70	140		
2	Sandford			60	60	12	77									550	275	60	120		
3	Yarmouth			23	23	1035	6739	2	24			4	28	17	85	400	200	215	430	6	6
4	Arcadia					3	38					4	28					40	80	8	8
5	Pinkney's Point			37	37	8	32			53	189	8	56					30	60		
6	Eau Hill			29	20	12	51		12	250	900			10	50	120	60	90	180		
7	Wedgeport			1450	1450	8	77	5	62			8	56	20	100	160	80	118	236	4	4
8	Salmon River	82	498	3890	3800			861	10633			21	147								
9	Tusket			600	600			130	1608			102	714								
10	Eel Brook			500	500			30	364			69	483					90	180		
11	Argyle			96	96	341	2202	73	898	50	180	30	210	30	150	200	100	202	404	12	12
12	Pubnicoes											60	420							52	52
	Totals	82	498	6686	6586	1434	9312	1102	13601	352	1269	306	2142	77	385	1930	965	915	1830	92	82

Cwt. = 100 lbs.

5 GEORGE V., A. 1915

THE CATCH MARKETED.

RETURN showing the Quantities and Values of all Fish Products Marketed in a fresh, dried, pickled, canned, &c., State, for the County of Yarmouth, Province of Nova Scotia, during the year 1913-14.

Number	Fishing Districts. Yarmouth County.	Salmon, used fresh and frozen, cwts.	Lobsters, canned, cases.	Lobsters, shipped in shell, cwts.	Cod, used fresh, cwts.	Cod, shipped green salted, cwts.	Cod, dried, qtl.†	Haddock, used fresh, cwts.	Haddock, smoked, cwts.	Haddock, dried, qtl.	Hake and Cusk, used fresh, cwts.	Hake and Cusk, dried, qtls.	Pollock, used fresh, cwts.	Pollock, dried, qtl.	Herring, used fresh, cwts.	Herring, smoked, cwts.	Herring, pickled, brls.	Herring, used as bait, brls.	Number.
1	Port Maitland	11	1335	1500	705	200	1971	450	20	805		15	0	270			280	560	1
2	Sandford	18	1296	1500	810	150	1396	360	10	705		12	80	250			966	450	2
3	Yarmouth		2673	3176	2602	3000	1518	2820	560	134		61	200	630		916	942	1370	3
4	Arcadia		1195	1458	125	100	380	371	20	179		8	60	179		300	342	500	4
5	Pinkney's Point		1254	1450	145	100	335			86		6		107		300	432	500	5
6	Comeau Hill		2645	3162	625	242	1634	940		179		89	100	321		1124	902	1010	6
7	Wedgeport		2846	3368	950	250	2514	950		342		149	120	357	262	1350	1041	1029	7
8	Salmon River										1018								8
9	Tusket	72			100	50	564	800	70			6							9
10	Eel Brook	6		59	203				50										10
11	Argyle	12	1334	1576					20	131			200	15	200	1217	508	529	11
12	Chicocs	25	2846	3435	1600	300	4351		25	1042		356		604		1040	712	1400	12
	Totals	72	17440	29684	7365	4392	14693	6791	755	3703	1018	702	880	2739	462	6247	5425	7348	
	Rates	$ 19.00	21.00	15.00	2.00	4.00	6.00	1.50	7.00	4.00	1.00	3.00	1.0	4.00	1.50	4.00	3.50	2.00	
	Value	$ 1368	366240	310460	15130	17568	88158	10186	5285	14812	1018	2106	880	10956	693	24988	18957	14696	

* Cwt. = 100 lbs.   † Qtl. = 112 lbs.

THE CATCH MARKETED.

RETURN showing the Quantities and Values of all Fish and Fish Products Marketed in a fresh, dried, pickled, canned, &c., State, for the County of **Yarmouth**, Province of **Nova Scotia**, during the year 1913-14—*Concluded.*

Number.	Fishing Districts.	Mackerel, used fresh, cwts.	Mackerel, salted, brls.	Shad, used fresh, cwts.	Alewives, used fresh, cwts.	Alewives, salted, brls.	Halibut, used fresh, cwts.	Smelts, used fresh, cwts.	Albacore, used fresh, cwts.	Eels, used fresh, cwts.	Sword fish, used fresh cwts.	Mixed Fish, used fresh, cwts.	Clams and Quahaugs, used fresh, brls.	Dulse, Crabs, Cockles and other Shell Fish used	Shell Fish, used fresh, cwts.	Fish Oil, Gallons.	Number.
	Yarmouth County.																
1	Port Maitland	1355	15				15			4		500	70			80	1
2	Sandford	1846	18				12	2		4	17	550	60			70	2
3	Yarmouth	2900	200		60		1035			8		400	215		6	400	3
4	Arcadia	615	130		23		6						40		8	50	4
5	Pinkney's Point	690	140		37		5						30			20	5
6	Chebogue Hill	1700	300		20		8			8	10	120	90			460	6
7	Wedgeport	2388	254		550		12	5	52	21	20	160	118		4	540	7
8	Salmon River				434	300			250	102							8
9	Tusket			82	465	1122		861		69							9
10	Eel Brook	26	8		335	45		130		30			90				10
11	Argyle	640	220		96	55	341	30	50	60	30	200	202		12		11
12	Pubnicoes	1081	373					73							52	1000	12
		13241	1658	82	2020	1522	1434	1102	352	306	77	1930	915		82	2620	
	Rates.........$	8.00	12.00	10.00	1.50	3.00	7.00	14.00	5.00	7.00	8.00	1.00	2.00		1.00	.30	
	Values........$	105928	19896	820	3030	4566	10038	15428	1760	2142	616	1930	1830		82	786	

Total value........ ....$1,072,383

* Qtls. = 100 lbs.

5 GEORGE V., A. 1915

THE CATCH.

RETURN showing the Quantities and Values of all Fish caught and landed in a Green State in the County of **Digby**, Province of **Nova Scotia**, during the year 1913-14.

Number	Fishing Districts	Salmon, cwts.	Salmon, value	Lobsters, cwts.	Lobsters, value	Cod, cwts.	Cod, value	Haddock, cwts.	Haddock, value	Hake and cusk, cwts.	Hake and cusk, value	Pollock, cwts.	Pollock, value	Herring, cwts.	Herring, value	Mackerel, cwts.	Mackerel, value	Alewives, cwts.	Alewives, value	Number
	Digby County.																			
1	Digby and vicinity	10	200	475	5700	13500	13125	36000	192520	20611	153268	1800	1800	2000	1500	180	900	10	10	1
2	Bay View and Culloden			520	6240	400	700	2000	5300	2100	1890	400	400	25	19					2
3	Gulliver's Cove to Waterford			823	9975	870	1522	4200	7350	6000	5400	2300	2300	1570	1177	1950	9750			3
4	Centreville			754	8400	2560	4480	9516	26169	8639	9775	254	254	1814	1845	402	2010			4
5	Sandy Cove and Mink Cove			500	6000	1500	2625	1300	3575	2540	2250	600	600	3290	2400	4200	21000			5
6	Lake River and Whale Cove			2015	20280	1275	2231	9250	25437	12225	11002	185	185	1000	730	600	3000			6
7	Tiverton and East Ferry			450	5400	986	1725	5900	9941	14250	3825	1330	1330	340	255	80	400			7
8	Freeport			1950	23040	13450	27037	3615	6479	37160	33435	3400	3400	3290	2100	312	1660			8
9	Westport			1730	20760	4814	8424	2356	17050	21562	21405	4200	4200	350	362	39	195			9
10	Smith's Cove and Brighton			1738	19200	11600	201235	6200	17050	15600	13850	4300	4300	1200	900	37	185			10
11	Plympton and Weymouth			15	180	85	148	50	137	30	27	180	180	1500	1125	1287	6435			11
12	New Edinburgh			215	2580	620	1090	300	825	240	216	700	700	500	375	1596	7980			12
13	Belliveau and White Cove			597	5700	70	140	1500	1500			8	8			1163	5112			13
14	Grosse Coques			42	515	32	64	4960	4960					170	125	450	2250			14
15	Ash Point			4	50	140	280	213	319					705	766	210	1050			15
16	Little Brook and Comeauville			100	1225	218	436	1223	1223					160	160					16
17	Saulnierville			1902	16190	950	1900	310	465					10	10					17
18	Meteghan River			146	1789	670	1340	295	442					24	24	12	60			18
19				122	1495	760	1520	340	360					210	210	234	1170			19
20	Comeau's Cove			872	7255	700	1400	329	329	75	75	75	75	609	609	50	250			20
21				172	2107	1153	2306	668	1002	15	15	15	15	10	10	10	50			21
22	Bear Cove			185	2366	560	1129	310	465					10	10	8	40			22
23	Cape St. Mary's	2	60	3482	31820	1486	2972	685	1026	120	120	120	120	1460	1460	404	2020			23
24	Salmon and Beaver River			163	1697	505	1010	320	480	240	216	51	51	10	10	70	350			24
	Totals	12	260	18972	199605	68804	98220	86431	308922	130867	116633	19918	19918	21194	17214	13259	65767	10	10	

*Cwt.=100 lbs.

THE CATCH.

RETURN showing the Quantities and Values of all Fish caught and landed in a Green State in the County of Digby, Province of Nova Scotia, during the year 1913-14.—*Concluded.*

No.	Fishing Districts	Halibut, cwts.	Halibut, value $	Flounders, cwts.	Flounders, value $	Smelts, cwts.	Smelts, value $	Trout, cwts.	Trout, value $	Albacore, cwts.	Albacore, value $	Bass, cwts.	Bass, value $	Eels, cwts.	Eels, value $	Tom-cod, cwts.	Tom-cod, value $	Mixed fish, cwts.	Mixed fish, value $	Squid, brls.	Squid, value $	Clams, brls.	Clams, value $	Dulse, Crabs, Cockles and other shell fish, cwts.	Dulse, Crabs, Cockles and other shell fish, value $
	Digby County.																								
1	Digby and … city	40	2800	93	186	10	50	7	70	5	20	10	13					20	10	2	8	10306	15974	979	1958
2	Bay View and Culloden	9	63																			30	46	10	50
3	Weir's Cove to Waterford	18	126	10	20									3	36			40	20	2	8	20	31		
4	Centreville	19	133					1	10																
5	Sandy Cove and Mink Cove	10	70							70	280							50	25	2	8	10	15		
6	Little River and Whale Cove	10	70			4	20							8	96	120	60					75	116		
7	Tiddville and East Ferry	10	70					3	30													30	46		
8	Tiverton and Central Grove	30	210					6	60																
9	Freeport	720	5040			123	1476											2356	1178						
10	Westport	340	2390					1	25							2	2					500	775		
11	Smith's Cove and Brighton							1	25							3	3					265	399		
12	Plympton and Weymouth							1	25											100	400	240	360		
13	New Edinburgh							2	50							3	3			50	200	160	240		
14	Belliveau and White Cove																					45	67		
15	Grosses Coques							1	20							4	4								
16	Church Point																								
17	Little Brook							1	30																
18	Saulnierville																								
19	Meteghan River																								
20	Meteghan							1	30																
21	Comeau's Cove																								
22	Bar …	5	35					1	25													25	37		
23	Cape St. Marys	4	28					1	25													3	4		
24	Salmon and Beaver River													60	240										
	Totals	1575	11025	103	206	137	1546	27	300	75	300	13	10	71	372	132	72	2466	1233	158	632	11709	18010	989	2008

THE CATCH MARKETED.

RETURN showing the Quantities and Value of all Fish and Fish Products Marketed in a fresh, dried, pickled, canned, &c., State, for the County of **Digby**, Province of **Nova Scotia**, during the year 1913-14.

Number	Fishing Districts	Salmon, used fr. and frozen, cwts.	Lobsters, canned, cases	Lobsters, shipped in shell, cwts.	Cod, used fresh, cwts.	Cod, smoked, cwts.	Cod, shipped green salted, cwts.	Cod, dried, quintals	Haddock, used fresh, cwts.	Haddock, smoked, cwts.	Haddock, canned, cases	Haddock, dried, quintals	Hake and Cusk, used fresh, cwts.	Hake and Cusk, dried, quintals	Pollock, used fresh, cwts.	Pollock, dried, cwts.	Herring, used fresh, cwts.	Herring, canned, cases	Herring, pickled, brls.	Herring, used as bait, brls.	Number
	Digby County.																				
1	Digby and vicinity	10		475	1450	1004		1347	8000	14000			2428	6061		600				1000	1
2	Bay, New and Culloden			529				133	2000				2106			133				13	2
3	Gulliver's Cove to Waterford		109	823				250	4200							767				785	3
4	Centreville			482				770	1199	1101				2000		84					4
5	Sandy Cove and ...			500				500	1300					2879		200				1600	5
6	Little River and Whale ...		650	390		124		425	3250	500	822					61				500	6
7	Tiddville and East Ferry			450				328	500					833		443		3604		170	7
8	Tiverton and Central Grove		59	1802				4483	3615		3125		2000	4975		1138				1600	8
9	Freeport			1730				1606	2356					1417		1400				175	9
10	Westport			1048				3883	4600	800				12383		1433				600	10
11	Smith's Cove and Brighton		276	15				2	50					6517		167				750	11
12	Plympton and Weymouth			215				201	300					5167		2				250	12
13	New Edinburgh		142	242	70			10	1500				30	80	180						13
14	... and White ...			42											200					828	14
15	Grosse ...			4	20			40	213								50			363	15
16	... Point			100	18			66	117	2480							23			80	16
17	Little Brook and Beaconville	2	581	450	60		156	196	10	553		100				25				12	17
18	Saulnierville			146	40			210	10			95					10			100	18
19	Meteghan River			122	60			253	340											225	19
20	Meteghan		260	222	10			230	320			122				5	9				20
21	...			172	20		646	322	302			100					10		50	5	21
22	Bear Cove			185	11		274	68	10			200				40				500	22
23	Cape St. Mary's	2	601	1980	20		250		85			18							150	5	23
24	Salmon and ... Rivers	2		163	20		160		266							17					24
	Totals	12	2678	12278	1769	1128	1396	17324	34543	19434	6347	635	6558	41412	380	6570	112	3604	200	9566	
	Rates	$22.00	2.00	18.00	2.00	5.00	4.00	7.00	3.00	7.00	6.00	4.00	3.00	5.00	1.00	5.00	1.00	3.50	3.00	2.00	
	Values	$26...	56238	221004	3538	5640	5684	121288	103629	136038	41162	2540	19674	207060	380	32850	112	12614	600	19132	

* Cwts. = 100 lbs.   † Quintal = 112 lbs.

THE CATCH MARKETED.

RETURN showing the Quantities and Values of all Fish and Fish Products Marketed in a fresh, dried, pickled, canned, &c., State, for the County of Digby, Province of Nova Scotia, during the year 1913–14—Concluded.

No.	Fishing Districts	Mackerel, used fresh, cwts.	Mackerel, canned, cwts.	Mackerel, salted, brls.	Alewives, used fresh, cwts.	Halibut, used fresh, cwts.	Flounders, used fresh, cwts.	Smelts, used cwts.	Trout, used fresh, cwts.	Albacore, used fresh, cwts.	Bass, used fresh, cwts.	Eels, used fresh, cwts.	Tom-cod, used fresh, cwts.	Mixed-fish, used fresh.	Squid, used as bait, brls.	Clams & Quahaugs, used fresh, brls.	Dulse, Crabs and other shell fish used fresh, cwts.	Sounds, pickl'd or dried, cwts.	Fish-oil, gals.
	Digby County.																		
1	Digby and vicinity	180			10	400	93	10	7	5	13			20	2	10306	397	85	2491
2	Bay View and Rhoden			417		9									2	30	10	10	
3	Cove to Water rd	700	356	20		18									2	20		30	2100
4	Sandy Cove and Mink Cove	3000	87	400		19												170	800
5	Little River and Ville			13		10	10									10		15	2500
6	le and East Ferry			27		10												30	
7	Tiverton and Coal Grove			104		30			1						2	75		240	11400
8	Freeport	687		13		720									100	30		103	1790
9	Westport	1596		12		340		4	2						50	500		85	1630
10	Smith's Cove and Brighton	484		200				123	3				8			265		96	
11	Plympton and	450												2356		240			
12	New Edinburgh	210		258									2			160			
13	Rieu and White Cove								6				3			15			
14	Grosse Coques								1			3							
15									2			60				25			
16	Little Brook and Comeauville	54		4					1				4			3			
17	Saulnierville	5		60		5													
18	River	1		15					1										
19	Meteghan	2		3						70									
20	Comeau's			2					1										15
21	Bear Cove	50		118														6	
22	Che St. Mary's	10		20		4	10												250
23	Salmon and Beaver River																		
	Totals	7872	443	1781	10	1575	103	137	27	75	13	71	132	2466	158	11709	407	870	23026
	Rates	5.00	8.00	10.00	1.00	7.00	2.00	12.00	15.00	4.00	5.00	6.00	.75	.50	4.00	2.00	5.00	23.00	27c.
	Values	39360	3544	17810	10	11025	206	1644	405	300	65	425	99	1233	632	23418	2035	2000	6217

Total Value $1,118,266

MARINE AND FISHERIES

5 GEORGE V., A. 1915

THE

RETURN showing the Quantities and Values of all Fish caught and landed in a Green State

Number.	Fishing Districts.	Salmon, cwts.	Salmon, value.	Lobsters, cwts.	lbs.	Cod, cwts.	Cod, value.	Haddock, cwts.	Haddock, value	Hake and Cusk, cwts.	Hake and Cusk, value.	Pollock, cwts.	Pollock, value.
	Annapolis County.		$		$		$		$		$		$
1	Margaretsville..............	50	1000	20	30C	1150	2300	292	584	75	75	70	70
2	Port George.................	60	1200	98	1470	580	1160	380	760	110	110	70	70
3	Port Lorne....	22	440	49	735	470	940	132	264	95	95	270	270
4	Hampton	30	600	225	3375	260	520	711	1422	320	320	102	102
5	Phinney's Cove........	....	....	265	3975	960	1920	1250	2500	660	660	110	110
6	Parker's Cove.................	....	....	355	5325	1130	2260	740	1480	2230	2230	80	80
7	Hilsburne....	.	. .	205	3075	6700	13400	3610	7220	4320	4320	....	....
8	Litchfield.................	....	..	159	2385	1010	2020	1450	2900	3990	3990	35	35
9	Port Wade.......	..	....	150	2250	2020	4040	3740	7480	1440	1440	....	....
10	Victoria Beach.....	....	..	298	4470	4130	8260	6100	12200	4830	4830	....	...
11	Deep Brook & Clementsport..	....	....	....	......	135	270	223	443	30	30	....	...
12	Annapolis, Lequille and Nicteaux Rivers..............	65	1800	....	......	......	......	......	......	....	....	....	....
	Totals.........	227	4540	1824	27360	18545	37090	18628	37256	18100	18100	737	737

*Cwt. = 100 lbs.

SESSIONAL PAPER No. 39

CATCH.

in the County of **Annapolis**, Province of **Nova Scotia**, during the year 1913-14.

Herring, cwts.	Herring, lbs. $	Mackerel, cwts.	Mackerel, value. $	Shad, cwts.	Shad, lbs. $	Halibut, cwts.	Halibut, value. $	Flounders, cwts.	Flounders, value. $	Trout, cwts.	Trout, value. $	Sturgeon, lbs.	Sturgeon, lbs.	Bass, cwts.	Bass, value. $	Eels, cwts.	Eels, value. $	Cod, cwts.	Cod, value. $	Gls, cwts.	Fish, value. $	Dulse, Crabs, etc.	Dulse, etc., value. $	Number.
60	60																							1
70	70	20	100																					2
20	20																							3
																								4
30	30																							5
50	50																							6
30	30																							7
40	40	100	500																					8
255	255					820	8200													2358	3537	300	500	9
220	220					467	4670	15	30															10
155	155							15	30	200	3000							20	40	3788	5682			11
				55	550					300	4500	34	510	50	500	30	150							12
930	930	120	600	55	550	1287	12870	30	60	500	7500	34	510	50	500	30	150	20	40	6146	9219	300	500	

5 GEORGE V., A. 1915

THE CATCH

RETURN showing the Quantities and Values of all Fish and Fish Products Marketed of **Nova Scotia**, during

Number.	Fishing Districts.	Salmon, used fresh and frozen, cwts.	Lobsters, shipped in shell, cwts.	Cod, used fresh, cwts.	Cod, dried, † quintals.	Haddock, used fresh, cwts.	Haddock, smoked, cwts.	Haddock, dried, quintals.	Hake and Cusk, dried, quintals	Pollock, used fresh, cwts.
	Annapolis County.									
1	Margaretsville	50	20	70	351	58	....	78	22	....
2	Port George	60	98	60	170	113	.....	87	36	40
3	Port Lorne	22	49	75	131	41	.....	28	29	....
4	Hampton	30	225	45	69	87	....	208	105	....
5	Phinney's Cove	......	265	10	316	342	....	336	219	......
6	Parkers Cove	......	355	285	281	83	110	145	741	.....
7	Hilsburne	......	205	......	2232	429	400.	791	1438	.....
8	Leitchfield	......	159	264	246	328	....	373	1328	....
9	Port Wade	......	150	570	482	1700	.....	679	477	....
10	Victoria Beach	......	298	1760	789	3000	600	622	1607	....
11	Deep Brook and Clementsport	....	....	......	45	85	......	41	9	.....
12	Annapolis, LeQuille and Nictaux Rivers	65	......	....	....	....	....	....	....	....
	Totals	227	1824	3139	5112	6266	1110	3388	6011	40
	Rates .... $	20.00	15.00	2.30	6.00	2.30	7.00	5.00	3.50	1.50
	Values .... $	4540	27360	7847	306,2	15664	7770	16940	21038	60
	Total value									

*Cwt. =100 lbs.   †Quintal=112 lbs.

SESSIONAL PAPER No. 39

MARKETED.

in a fresh, dried, pickled, canned, &c., State, for the County of **Annapolis**, Province the year 1913-14.

Pollock, dried, quintals.	Herring, used as bait, brls.	Mackerel, used fresh, cwts.	Shad, used fresh, cwts.	Halibut, used fresh, cwts.	Flounders, used fresh, cwts.	Trout, used fresh, cwts.	Sturgeon, used fresh, cwts.	Bass, used fresh, cwts.	Eels, used fresh, cwts.	Tom-cod, used fresh, cwts.	Clams, used fresh, brls.	Dulse, Crabs, Cockles and other shell fish, used fresh, cwts.	Caviare or sturgeon roe, cwts.	Fish oil, gals.	Number.
21	30													600	1
9	35	20												300	2
90	19													300	3
34														425	4
35	15													700	5
24	25													750	6
	15													1500	7
11	20	100												600	8
	128			820								2358	*100	800	9
	110			467	15									1600	10
	78				15	200				20	3788				11
			55			300	34	50	30				2		12
224	466	120	55	1287	30	500	34	50	30	20	6146	100	2	7575	
4	2.00	5.00	10.00	10.00	2.00	15.00	15.00	10.00	5.00	2.00	1.50	5.00	2.66	.30	
896	932	600	550	12870	60	7500	510	500	150	40	9219	500	532	2272	

....$169,022

*The Dulse is dried.

5 GEORGE V., A. 1915

THE

RETURN showing the Quantities and Values of all Fish caught and landed in a Green

Number.	Fishing Districts.	Salmon, cwts.*	Salmon, value.	Lobsters, cwts.	Lobsters, value.	Cod, cwts.	Cod, value.	Haddock, cwts.	Haddock, value.	Hake and Cusk, cwts.	Hake and Cusk, lbs.	Pollock, cwts.	Pollock, value.	Herring, cwts.
	Kings County.		$		$		$		$		$		$	
1	Morden and vicinity....................	90	1620	10	120	170	340	10	15	20	30	1205	1807	196
2	Victoria Harbour and Ogilvie Wharf .	20	360	....	....	60	120	20	30	20	30	175	262	150
3	Harbourville.	63	1134	....	....	55	110	10	15	....	....	320	480	80
4	Canada Creek......... ....,....	72	1296	45	540	450	900	80	120	60	90	622	933	2019
5	Chipman's Brook and Hunting Point..	81	1458	17	204	235	470	25	37	24	36	315	472	629
6	Hall's Harbour......	70	1260	35	420	795	1590	875	1312	235	352	735	1102	1516
7	Race Point and Sheffield Vault.......	75	1350	....	....	85	170	40	60	5	7	220	330	578
8	Baxter's Harbour.....	35	630	8	96	1010	2020	135	202	35	52	475	712	660
9	Whalen Beach and Well's Cove	25	450	....	....	30	60	....	....	30	45	100	150	305
10	Scott's Bay............,.......	69	1242	21	252	620	1242	305	457	25	37	225	337	915
11	Blomidon and Kingsport.............	15	270	....	....	10	20	5	7	....	....	50	75	120
12	Starr's Point to Wolfville...........	15	270	....	....	30	60	....	....	....	....	20	30	....
13	Upper Gaspereau and all inland waters	27	486	....	....	293	586	....	....	....	....	15	23	....
	Totals	657	11826	136	1632	3843	7686	1505	2257	454	681	4477	6713	7168

*Cwt. = 100 lbs.

CATCH.

State in the County of **Kings**, Province of **Nova Scotia**, during the year 1913-14.

Herring, value.	Mackerel, cwts.	Mackerel, value.	Shad, cwts.	Shad, value.	Alewives, cwts.	Alewives, value.	Halibut, cwts.	Halibut, value.	Trout, cwts.	Trout, value.	Squid, brls.	Squid, value.	Clams, brls.	Clams, value.	Number.
$		$		$		$		$		$		$		$	
196	80	400	....	....	....	....	10	100	....	....	27	13	....	....	1
150	25	125	....	....	....	....	15	150	....	....	5	3	....	....	2
80	20	100	....	....	....	....	5	50	....	....	5	3	....	....	3
2019	70	350	10	100	....	....	...	....	....	....	40	20	....	....	4
629	26	130	1	10	....	....	....	....	....	..	8	4	....	....	5
1516	60	300	7	70	....	....	10	100	....	....	15	7	....	....	6
578	25	125	10	100	....	....	....	....	....	....	....	....	....	....	7
560	15	75	....	....	....	....	7	70	....	....	3	2	....	....	8
305	10	50	1	10	....	....	....	....	....	....	2	1	....	....	9
915	60	300	115	1150	....	....	1	10	....	....	5	2	....	....	10
120	....	....	55	550	....	....	....	....	....	....	....	....	75	225	11
....	....	....	55	550	....	....	....	....	....	....	....	....	....	....	12
....	....	....	....	....	6000	6000	....	....	30	300	....	....	....	....	13
7168	391	1955	254	2540	6000	6000	48	480	30	300	110	55	75	225	

5 GEORGE V., A. 1915

THE CATCH MARKETED.

RETURN showing the Quantities and Values of all Fish and Fish Products Marketed in a fresh, dried, pickled, canned, &c., State, for the County of Kings, Province of **Nova Scotia**, during the year 1913-14.

Number	Fishing Districts.	Salmon, used fresh and frozen, *cwts.	Lobsters, shipped in shell, cwts.	Cod, used fresh, cwts.	Cod, dried, quintals.	Haddock, used fresh, cwts.	Haddock, dried, quintals.	Hake and Cusk, dried, quintals.	Pollock, used fresh, cwts.	Pollock, dried, quintals.	Herring, smoked, cwts.	Herring, pickled, brls.	Herring, used as bait, brls.	Mackerel, used fresh, cwts.	Mackerel, salted, brls.	Shad, used fresh, cwts.	Alewives, used fresh, cwts.	Alewives, salted, brls.	Halibut, used fresh, cwts.	Trout, used fresh, cwts.	Squid, used as bait, brls.	Clams and Quahaugs, used fresh, brls.
	Kings County.																					
1	Morden and vicinity	90	10		77	10		1		401		46	38	80					10		27	
2	Victoria, [blur]ur and Ogilvie Wharf	20			20	20		1		58		30	30	25					15		5	
3	Harbourville	63			18	10				107		13	25	20					5		5	
4	Canada Creek	72	45		150	20	65			207	709	145	40	40	10						40	
5	Chipman's Brook & Hunting point	81	17		78	251	15	20		165	76	132	145	83		10			10		25	
6	Hall's Harbour	70	35		265	585	116	8		245	138	315	25	30	10	1						
7	Race point [ard] [blur]d Vault	75			29			28		73	181	139	60	25		7			7		8	
8	[blur] Harbour	35	8		338	10	10	2		150	50	55	15	15		10					2	
9	Whalen Beach and Wells [blur]	25			10	105	10	12	25	333	100	25	32	10	10				1		5	
10	Scotts Bay	69	21		206	185		10		75	349	56	30	30		1						75
11	Blomidon and Kingsport	15			4	10		8		15		19				115						
12	Starr's point to Wolfville	15			10					7						55					30	
13	Upper Gaspereau & all inland waters	27		260	11				15							55	2040	1318		30		
	Totals	657	136	260	1194	980	175	152	40	1478	1594	960	523	301	30	254	2040	1316	48	30	110	75
	Rates	18.00	15.00	2.00	6.00	2.00	4.00	4.50	1.75	4.50	4.00	4.50	2.00	8.00	6.00	10.00	1.50	3.00	10.00	10.00	5.00	3.00
	Value	11826	2040	520	7164	1960	700	684	70	6651	6376	4320	1046	2408	180	2540	3060	3954	480	300	550	225

Total value .... $ 56,659

*Cwt. = 100 lbs.   Quintal = 112 lbs.

SESSIONAL PAPER No. 39

RECAPITULATION

Of the Quantities and Values of all Fish caught and landed in a Green State, and of the Quantities and Values of all Fish and Fish Products Marketed in a fresh, dried, pickled, canned, etc., State, for **District No. 3**, Province of **Nova Scotia**, during the year 1913-14.

Kinds of Fish.		Caught and landed in a Green State.		Marketed.		Total Marketed value.
		Quantity.	Value.	Quantity.	Value.	
			$		$	$
Salmon	cwts.	1,360	25,289			
" used fresh	"			1,360	26,458	
						26,458
Lobsters	"	157,577	1,454,493			
" canned	cases.			35,194	739,074	
" shipped in shell	cwts.			69,597	1,141,037	
						1,880,111
Cod	"	709,133	1,323,257			
" used fresh	"			21,874	51,317	
" green—salted	"			23,182	75,334	
" smoked	"			1,128	5,640	
" dried	"			211,852	1,272,535	
						1,404,826
Haddock	"	221,062	505,672			
" used fresh	"			61,572	157,815	
" smoked (finnans)	"			21,541	150,787	
" canned	cases.			6,947	41,662	
" dried	cwts.			35,093	161,779	
						512,043
Hake and Cusk	"	203,838	187,342			
" used fresh	"			11,618	26,820	
" dried	"			63,974	281,199	
						308,019
Pollock	"	54,073	53,313			
" used fresh	"			2,854	3,418	
" dried	"			17,056	75,187	
						78,605
Herring	"	220,361	202,050			
" used fresh	"			32,209	32,440	
" canned	cases.			3,604	12,614	
" smoked	cwts.			8,101	32,924	
" pickled	brls.			25,421	91,337	
" used as bait	"			47,169	92,880	
						262,195
Mackerel	cwts.	66,610	297,105			
" used fresh	"			36,119	231,262	
" canned	cases.			443	3,544	
" salted	brls.			10,121	130,397	
						365,203
Shad	cwts.	391	3,588			
" used fresh	"			391		
						3,910
Alewives	"	15,004	14,580			
" used fresh	"			5,376	7,406	
" salted	brls.			3,207	9,988	
						17,394
Halibut, used fresh	cwts.	8,759	58,836	8,759		65,522
Flounders	"	133	266	133		266
Smelts	"	1,380	16,647	1,380		18,791
Trout	"	635	8,970	635		8,985
Albacore	"	788	2,512	788		3,145

MARINE AND FISHERIES

RECAPITULATION

Of the Quantities and Value of all Fish caught and landed in a Green State, and of the Quantities and Values of all Fish and Fish Products Marketed in a fresh, dried, pickled, canned, etc., State, for **District No. 3**, Province of **Nova Scotia**, during the year 1913-14.

Kinds of Fish.		Caught and landed in a Green State.		Marketed.		Total Marketed value.
		Quantity.	Value.	Quantity.	Value.	
			$		$	
Sturgeon..	cwts.	34	510	34		510
Bass	"	63	510	63		565
Eels	"	607	3,364	607		3.418
Tom Cod	"	152	112	152		139
Sworfiish	"	3,471	16,799	3,471		20,915
Mixed Fish	"	4,396	2,198	4,396		3,163
Squid	brls.	285	752	285		752
Oysters	"	1,345	4,035	1,345		4,035
Clams	"	20,789	34,139			35,601
" used fresh	"			20,789		
Dulse, Cockles and other shell fish	cwts.	1,379	2,646	597		2,673
Tongue and Sounds	"			870		20,010
Caviare (Sturgeon roe)	"			2		532
Hair Seal Skins	No.			80		80
Fertilizer, etc	Tons.			1,229		9,147
Glue Material	"					7,478
Fish Oil	galls.			93,406		27,330
Totals			4,218,985			5,091,821

SESSIONAL PAPER No. 39

RECAPITULATION

Of the Number of Fishermen, etc., and of the Number and Value of Fishing Vessels, Boats, Nets, etc., in **District No. 3**, Province of **Nova Scotia**, for the year 1913-14.

	Number.	Value.
		$
Steam Fishing Vessels (tonnage 377)	14	58,000
Sailing and Gasoline Vessels	426	1,455,699
Boats (sail)	3,019	90,262
" (gasoline)	2,027	502,490
Carrying Smacks	44	37,900
Gill Nets, Seines, Trap and Smelt Nets, etc	15,752	202,172
Weirs	89	22,375
Trawls	8,936	92,526
Hand Lines	12,398	11,101
Lobster Traps	328,472	328,472
" Canneries	68	95,700
Fish "	2	2,800
Freezers and Ice-houses	152	182,925
Smoke and Fish-houses	2,225	252,129
Fishing Piers and Wharves	721	732,300
Totals		4,066,791

Number of men employed on Vessels	5,160
" " Boats	6,740
" " Carrying Smacks	104
" persons employed in Fish-houses, Freezers, Canneries, etc.	2,326
Total	14,330

5 GEORGE V., A. 1915

RECAPITULATION

Of the Quantities and Values of all Fish caught and landed in a Green State, and of the Quantities and Values of all Fish and Fish Products Marketed in a fresh, dried, pickled, canned, etc., State, for the **whole** Province of **Nova Scotia**, during the year 1913-14.

Kinds of Fish.		Caught and landed in a Green State.		Marketed.		Total marketed value.
		Quantity.	Value.	Quantity.	Value.	
			$		$	$
Salmon	cwts.	9,401	110,624			
" used fresh	"			9,341	138,772	
" canned	cases.			24	183	
" smoked	cwts.			24	480	
						139,435
Lobsters	"	302,261	2,226,908			
" canned	cases.			87,449	1,679,664	
" shipped in shell	cwts.			84,063	1,280,393	
						2,960,057
Cod	"	970,870	1,727,188			
" used fresh	"			58,345	112,055	
" green—salted	"			60,677	202,070	
" smoked	"			1,128	5,640	
" dried	"			263,040	1,571,486	
						1,891,251
Haddock	"	387,386	748,885			
" used fresh	"			139,289	320,837	
" smoked (finnans)	"			26,833	167,473	
" canned	cases.			6,947	41,662	
" dried	cwts.			61,028	278,910	
						808,882
Hake & Cusk	"	249,387	228,461			
" used fresh	"			16,755	32,084	
" dried	"			77,476	334,435	
						366,519
Pollock	"	79,232	76,805			
" used fresh	"			3,649	4,576	
" dried	"			25,164	106,774	
						111,350
Herring	"	386,473	356,856			
" used fresh	"			52,549	60,651	
" canned	cases.			3,604	12,614	
" smoked	cwts.			13,611	49,454	
" pickled	brls.			49,240	198,727	
" used as bait	"			78,149	149,246	
" used as fertilizer	"			596	298	
						470,990
Mackerel	cwts.	162,607	700,275			
" used fresh	"			87,229	581,103	
" canned	cases.			443	3,544	
" salted	brls.			25,004	316,616	
						901,263
Shad	cwts.	995	7,264			
" used fresh	"			943	9,338	
" salted	brls.			19	285	
						9,623
Alewives	cwts.	19,601	20,626			
" used fresh	"			8,363	12,778	
" salted	brls.			3,743	11,807	
						24,585

RECAPITULATION.

Of the Quantities and Values of all Fish caught and landed in a Green State, and of the Quantities and Values of all Fish and Fish Products Marketed in a fresh, dried, pickled, canned, etc., State, for **the whole** Province of **Nova Scotia,** during the year 1913-14.

Kinds of Fish.		Caught and landed in a Green State.		Marketed.		Total marketed value.
		Quantity.	Value.	Quantity.	Value.	
			$		$	$
Halibut, used fresh	cwts.	31,521	210,254	31,521	....	291,874
Flounders	"	1,174	1,712	1,174	....	5,267
Smelts	"	4,043	31,278	4,043	....	37,510
Trout	"	1,005	11,626	1,005	....	12,685
Soles	"	216	324	216	....	1,080
Albacore	"	2,954	5,252	2,954	....	11,809
Sturgeon	"	34	510	34	....	510
Bass	"	198	1,079	198	....	1,915
Eels	"	1,111	5,227	1,111	....	5,665
Tom Cod	"	300	425	300	....	518
Swordfish	"	13,322	16,658	13,322	....	61,140
Mixed Fish	"	5,566	2,798	5,566	....	4,903
Squid	brls.	2,167	6,881	2,167	....	9,067
Oysters	"	3,397	12,283	3,397	....	14,064
Clams	"	28,088	42,088	....		
" used fresh	"	....	....	27,913	49,941	
" canned	cases.	....	....	175	788	50,729
Dulse, Cockles & other shell fish	cwts.	1,379	2,646	597	....	2,673
Tongues and Sounds	"	....	....	2,874	....	28,026
Caviare (Sturgeon roe)	"	....	....	2	....	532
Hair Seal Skins	"	....	....	168	....	184
Fertilizer, etc	tons.	....	....	1,229	....	9,147
Glue material	"	....	....	....	....	7,478
Fish Oil	galls.	....	....	172,941	....	56,895
Totals			6,584,933			8,297,626

5 GEORGE V., A. 1915

RECAPITULATION.

Of the Number of Fishermen, etc., and of the Number and Value of Fishing Vessels, Boats, Nets, etc., in **the whole** Province of **Nova Scotia**, for the year 1913-14.

	Number.	Value.
		$
Steam Fishing Vessels (tonnage, 377)........	14	58,000
Sailing and Gasoline Vessels......................	667	1,633,499
Boats (sail).............	9,427	315,797
" (gasoline)............	3,481	785,374
Carrying Smacks	201	110,415
Gill Nets; Seines, Trap and Smelt Nets, etc	61,150	687,189
Weirs......	108	23,505
Trawls...............	16,522	165,190
Hand Lines.......	29,375	23,380
Lobster Traps.,.......	787,387	726,879
" Canneries........... ....,..... ...	231	288,755
Fish "	2	2,800
Freezers and Ice-houses...........	307	640,480
Smoke and Fish-houses........	4,800	540,110
Fishing Piers and Wharfes................................	1,760	1,108,837
Total............		7,110,210

Number of men employed on Vessels........	6,302
" " " Boats.	15,648
" " " Carrying Smacks....	362
" persons employed in Fish-houses, Freezers. Canneries, &c.	6,567
Total.	28,879

APPENDIX No. 2.

NEW BRUNSWICK.

District No. 1.—Comprising the counties of Charlotte and St. John. Inspector, John F. Calder, Campobello.

District No. 2.—Comprising the counties of Albert, Westmorland, Kent, Northumberland, Gloucester and Restigouche. Inspector, D. Morrison, Newcastle.

District No. 3.—Comprising the counties of Kings, Queens, Sunbury, York, Carleton, Victoria and Madawaska. Inspector, H. E. Harrison, Fredericton.

REPORT ON THE FISHERIES OF DISTRICT No. 1.

To the Superintendent of Fisheries,
Ottawa.

Sir,—I have the honour to present herewith my eighth annual report on the fisheries of District No. 1, province of New Brunswick, together with the statistics of the different sub-divisions.

The value of the catch for this year is $1,539,629, against $1,612,599 for last year, a decrease of $72,970. As a whole this has been a very unsatisfactory year for the fisheries of this district, yet at the same time some districts have done exceptionally well, especially the island of Grand Manan. All branches were prosperous there, more particularly the smoked herring industry. There was a fine run of medium herring in the weirs; large quantities were smoked and sold for good prices.

HERRING.

There was a slight increase in the quantity of herring caught as compared with the previous year, 197,297 cwt. being taken against 189,200 cwts. for 1912-13. There was, however, a large increase in the market value of cured herring, the figures being $196,792 for 1912-13 and $288,015 for this year. The increase in value, of course, is due to better prices being paid for the herring products.

SARDINES.

One hundred and forty-one thousand three hundred and eighty-four barrels of sardines were taken, against 280,282 barrels during the previous season. This shows an alarming falling off in the yield of this fishery, which is giving the fishermen much concern. The sardine fishery is the most extensive and profitable one in the district. In the past we have experienced a few poor seasons in this branch, but they have invariably been due to poor market conditions, which in turn were brought about by an over-supply of the raw material and the consequent over-stocking of the markets with the canned products. This year, however, we are confronted with a failure of the fishery. Good market conditions prevailed, a ready sale at remarkably good prices

39—9½

5 GEORGE V., A. 1915

always obtained, and every available fish was taken from the weirs, yet at the same time there was a falling off in the catch of more than 50 per cent. On the other hand owing to the high prices paid at the weirs, the value of the catch was nearly as much as during the past year. But such a state of affairs is not satisfying. It is much better to have a plentiful supply at fair prices than to have a poor catch at very high prices. When fish are plentiful they are most generally to be found all along the shores, and all of the fishermen get a share, but when they are scarce they are to be found in a few localities, and many places do not get any. Then again with high prices prevailing the daily expenditures of the trawl and line fishermen for bait is almost prohibitive. Since last fall sardines have been selling at the weirs at prices ranging from $15 to $35 per hogshead of five barrels; therefore the fishermen have to pay from $3 to $7 per barrel for bait. Of course, I do not know if there is any way by which your Department could come to the relief of the line fishermen in this matter, for the weirmen have the right to charge the market prices for their catches. But if this scarcity is to continue, the whole matter of the sardine fishery must receive very serious consideration.

SALMON.

There is an appreciable increase to be reported in the salmon catch. Last year's catch was 3,295 cwts., while that of this year was 3,998 cwts. The prosperous condition of this fishery in this district is no doubt in a great measure due to the splendid results attending the operations of the salmon hatcheries. I had the pleasure during the past spring to attend a meeting of the Corporations Committee of the Provincial Legislature, along with a large delegation of fishermen from St. John, to enter a protest against a charter being given a hydro-electric company which would permit them to build a dam across the St. John river, at the Meductic rapids. Our objection to this dam was that it would prevent the salmon from reaching the spawning beds on the Tobique river, to which nearly all salmon ascending the St. John go. I am pleased to be able to state that the Legislature refused to grant the charter.

LOBSTERS.

There is not much change to report in the condition of the lobster fishery. A slight falling off in the catch is to be noticed—12,410 cwts. in 1912-13 and 11,751 cwts. this season. The shortage in the catch is altogether due to the extremely severe weather conditions obtaining during last winter. It is now a matter of record that it was the worst winter in this section that there had been for a quarter of a century at least. For weeks at a time the fishermen were unable to reach the traps even. Some illegal fishing was done during the close season, and no doubt some small lobsters were sold during the open season. I am in hopes that with the *Phalarope* and *Sea Gull* in commission this year that very little illegality of this kind will occur.

HAKE.

A large decrease in the catch of hake has to be reported. It is regrettable that the supply of these fish is apparently rapidly diminishing. It is in no wise to be attributed to over-fishing on local grounds, but rather to the extensive trawling which is carried on all the year round in the mouth of the bay of Fundy by American and Nova Scotia schooners.

There is little to note in the other branches.

In conclusion I desire to again express my appreciation of the courteous treatment from both yourself and your officials during the past year.

I am, sir, your obedient servant,

J. F. CALDER,
Inspector of Fisheries.

SESSIONAL PAPER No. 39

REPORT ON THE FISHERIES OF DISTRICT No. 2.

To the Superintendent of Fisheries,
      Ottawa.

SIR,—I have the honour to submit my first annual report on the fisheries of District No. 2 in the province of New Brunswick for the fiscal year 1913-14, together with a statistical statement of the quantities and value of fish taken, the material used, and the number of persons engaged in the fisheries in my district.

These returns show the value of fish to be $2,694,640, against $2,611,333 for the preceding year, or an increase in value of $83,307, as compared with last year, notwithstanding the large falling off in shellfish.

SALMON.

The catch of salmon was good; although some of the districts show little increase, the returns for the whole district show a marked improvement, being 3,086 cwts. greater than 1912-13. The fall run was exceptionally good on the Miramichi river and with increased protection on the natural spawning beds and the assistance given through the hatcheries, there is every reason to believe this important industry will make steady progress.

LOBSTERS.

The returns show a falling off of 3,719 cases and an increased shipment in shell of 3,957 cwts. The market value of this shellfish, however, notwithstanding a falling off in quantity, shows an increase amounting to $30,320, and with the new stringent regulations recommended by the Shellfish Commissioners, it is to be hoped that this important industry will be revived.

COD.

The catch of this fish was about the same as last year.

HERRING.

This fishery shows an increase in the catch of 105,345 cwts.; while the value is $520,895, or $94,513 greater than that for the preceding year.

MACKEREL.

There was a large increase in the catch of mackerel; the marketed value of the catch this year is $168,166, against $60,100 last year.

SMELTS.

There was a falling off in the catch of smelts of about 19,795 cwts. The reason for such a large decrease is found in the fact that very unfavourable weather conditions prevailed during the early part of the season, and in some districts the fishermen were unable to set their nets until late in January. Large nets were prohibited, and the close season was rigidly enforced.

OYSTERS.

There was an increase of 1,561 barrels of oysters, with an increased value of $9,366.

CLAMS.

The increased value of clams was $23,894.

I am, sir, your obedient servant,

D. MORRISON,
      Inspector of Fisheries.

5 GEORGE V., A. 1915

REPORT ON THE FISHERIES OF DISTRICT No. 3 (INLAND).

To the Superintendent of Fisheries,
          Ottawa.

SIR,—I have the honour to submit my twelfth annual report on the inland fisheries of New Brunswick for the fiscal year 1913-14, together with a statement of the quantity and value of fish taken and the materials used therefor.

I am very pleased to be able to state that there are some encouraging particulars to be mentioned later in this report, and while there has been a falling off in the catch of only one kind of fish to any appreciable extent, that of pickerel, my overseers state that this fishery was not prosecuted to the extent that it was the previous year.

I have readjusted prices in some cases, which has affected the net financial showing, but which I believe to be nearer the true value to the fishermen.

A comparative statement of the values of fish and materials for the years 1912-13 and 1913-14 shows a slight increase in both:—

Years.	Value of Fish.	Value of Materials.
	$	$
1912-13	40,132	39,595
1913-14	41,948	45,213

SALMON.

I wish to make particular reference to the increased catch of salmon over the previous year, a most gratifying result, considering the fact that few additional licenses were issued.

The conditions in 1912-13 were not very favourable on account of heavy rises of water throughout the summer. This did not occur to so great an extent in 1913-14, and no doubt partly accounts for the satisfactory catch; but I am sure that the continual good work of the Department in placing salmon fry in the different tributaries of the St. John river, and the protection afforded the fish while ascending to their spawning grounds, is having its effect, and if a staff of good officers, not necessarily a great number, is allowed me, I feel that this fishery, barring unforeseen causes, can be improved from year to year, and with well-devised regulations such as we now have, will not be depleted by legitimate fishing.

For some reason few salmon were taken with the fly in the different pools in the St. John river this season, but for short periods at different times this sport was excellent on the Tobique river, and a fairly satisfactory season enjoyed by the members of the Tobique Salmon Club, the average weight of both salmon and grilse being good.

SHAD.

The officers in the counties of Kings, Queens and York report the fishermen as saying that shad were more plentiful this year than they were last; however, the return does not bear out the claim, which goes to show that it is difficult to get strictly reliable data. While I do not advocate it, and believe it is not wise to unnecessarily hedge about the fisheries with restrictions, I think, because of the great value of shad as a food fish, it might be well for the Department to consider the advisability of licensing the shad fishery as is the salmon fishery and others.

Overseer Worden of Queens county (north) has suggested, and requested me to ask the Department to restrict the number of nets that one person may set.

His reason for this is that one person will sometimes set several shad nets and, being unable to properly attend them, the meshed fish, or a great number of them, are destroyed by the fast increasing eels.

This restriction need not necessarily apply to the St. John river, because drift nets are used and it requires two men per net, but in the Washademoak lake and river the shad nets are set stationary, so that one person may, and does set and try to attend several, and is thus the cause of capturing much food for eels, too valuable a fish for that purpose.

I very heartily place the suggestion before the Department.

PICKEREL.

This fishery shows a very largely decreased catch compared with last year. My officers state that it was not followed up as in 1912-13, and I think it quite probable that it may have been somewhat overdone, that is, many of the large fish caught, leaving younger and smaller fish in the water for this year, and I have been led to this belief by fishermen.

I have been requested to suggest the advisability of making a close season for pickerel, as at present great quantities are taken in the summer and because of the time required to get them to the United States market, where most of them are sent, a large percentage is lost because of the warm weather, therefore it might be well to make the months of July, August and half of September a close season.

ALEWIVES.

A much smaller quantity of these fish were taken; nevertheless a large amount was marketed and financial returns were satisfactory.

BASS.

A most remarkable occurrence in connection with bass is the wonderful increase in the amount caught over that of last year and for many years previous.

A number of years ago, more than twenty, an old fisherman has said there was an immense run of these fish in the Belle Isle bay, Kings county, but for no known reason, unless because of over-fishing, they disappeared in one season and the catch has been almost nil since, not only in the Belle Isle water but the whole St. John river and tributaries.

Considering that the Belle Isle bay is only seven miles in length and about one-half mile wide, the run of bass must have been exceedingly good to catch 140 cwts., while only 20 cwts. were taken last season.

As there is a great demand for these fish, and being caught in the season when not a pound need be lost through heat, a snug sum of money was distributed amongst the farmers in that section.

I sincerely hope this will be an annual experience with these people, and think it would be interesting if some of your experts would tell us why bass act as they appear to in these waters.

EELS.

I have not a good word to speak for these " mud snakes." The return shows a larger quantity taken than in 1912-13, and fishermen state that they are increasing fast and no kind of fish that is netted is free from their attacks, often leaving but the skin of a salmon or shad for the fishermen, and even attacking large sturgeon and destroying much valuable roe of these fish.

Up to the present it has been impossible to catch eels in the winter, it being generally presumed that they bury themselves in the mud in cold weather, and the uncertainty of getting them into the United States market in the warm weather, seem to be more than enough protection for them.

I would respectfully suggest this: a plan to successfully catch eels in winter as a fit subject for the experts of the Department to consider.

I have much pleasure in reporting that, through the generous act of the Honourable the Minister, two modern fish-passes were placed in dams in the Salmon river, Victoria county, last season, and it is hoped that salmon will again ascend this small stream in considerable numbers.

A pass was also placed in a dam at the outlet of the Becaguimac stream, Carleton county, last summer by the owners.

On account of the dam at Marysville, near the mouth of the Nashwaak river, being carried away by ice last spring, a free passage was left for fish to ascend that stream, and it is hoped that, by restocking with fry, salmon will again become plentiful in those waters. I am unable to learn of any having passed up the Nashwaak river in 1913.

Trout fishing throughout the season of 1913 was reported to be the best for many years. This sport affords a deal of pleasure to hundreds of natives, and is becoming more attractive to foreign fishermen each season, and many United States people are building cottages beside our lakes and streams.

I am grateful for courteous treatment from the officials of the Department, and trust that my shortcomings have not been of a serious nature.

I am, sir, your obedient servant,

H. E. HARRISON,
Inspector of Fisheries.

RETURN showing the Number of Fishermen, etc., the Number and Value of Vessels and Boats, and the Quantity and Value of all Fishing Gear, etc., used in the Fishing Industry in the Counties of Charlotte and St. John, Province of New Brunswick, during the year 1913-14.

Fishing Districts.	Sailing and Gasoline Vessels				Boats				Carrying Smacks			Gill Nets, Seines, Trap and Smelt Nets, etc.		Weirs		Trawls		Number.
	No. 20 to 40 tons	No. 10 to 20 tons	Value $	Men	Sail	Gasoline	Value $	Men	Number	Value $	Men	Number	Value $	Number	Value $	Number	Value $	
Charlotte County.																		
1 Lepreau to Red Head	..	1	2000	3	100	27	1500	74	5	7290	15	85	4480	39	30000	6	200	1
2 Red Head to Letang	..	9	6100	26	142	7	2810	181	14	22400	31	264	8650	57	30000	70	3016	2
3 Letang to St. George	3	4	1200	14	258	30	6840	212	1	1000	2	284	5430	57	15785	61	900	3
4 St. George to St Stephen	..	9	67500	25	215	37	10630	315	1	1000	2	407	11732	90	72900	6	120	4
5 Grand Manan	3	35	33890	138	213	182	54600	325	..	..	..	830	21100	56	84000	63	1860	5
6 Campobello	..	5	4540	59	130	111	33000	280	..	..	..	115	3240	..	9000	30	2000	6
7 West Isles	3	1	1500	4	283	150	6100	420	18	10000	36	340	7000	84	42000	13	300	7
Totals	7	84	116600	269	1341	607	58900	4807	39	41700	86	2375	62138	397	302785	341	8396	
St. John County.																		
1 St. John Harbour	1	4	2400	18	230	45	7920	235	..	..	..	420	14400	25	17000	..	..	1
2 Lepreau to Chance Harbour	..	..	..	..	20	23	400	50	..	..	..	25	2500	5	500	25	25100	2
3 Chance Harbour to St. John	..	3	2400	18	100	110	33000	220	1	500	2	1130	13220	30	15000	30	730	3
4 Mispec to Tynemouth Creek	..	4	3000	8	12	12	1800	12	..	..	..	..	..	..	..	..	..	4
5 Tynemouth to Albert County	..	3	1400	..	17	11	1575	20	..	..	..	11	165	..	..	..	..	5
Totals	1	14	9200	32	369	203	10820	537	1	500	2	1586	30685	61	57000	55	175	

Return showing the Number of Fishermen, etc., the Number and Value of Vessels and Boats, and the Quantity and Value of all Fishing Gear, etc., used in the Fishing Industry in the Counties of Charlotte and St. John, Province of **New Brunswick**, during the year 1913-14. —*Concluded.*

Number	Fishing Districts	Hand Lines Number	Hand Lines Value	Lobster Traps Number	Lobster Traps Value	Lobster Canneries Number	Lobster Canneries Value	Sardine Canneries Number	Sardine Canneries Value	Clam Canneries Number	Clam Canneries Value	Freezers and Ice Houses Number	Freezers and Ice Houses Value	Smoke and Fish Houses Number	Smoke and Fish Houses Value	Fishing Piers and Wharves Number	Fishing Piers and Wharves Value	Pile Drivers and Scows Number	Pile Drivers and Scows Value	Persons Employed in Canneries, Freezers and Fish Houses	Number
	Charlotte County.																				
1	Letprean to Red Head	74	36	2000	2000									33	10620	5	6000	12	1200	190	1
2	Red Head to Letang		135	2410	2410					3	12000	5	2800	25	1730	26	29100	30	1000	324	2
3	Letang to St. George	289	155	494	794			3	96000	3	4500			2	2700	16	670	39	1850	20	3
4	St. George to St. Stephen	60	18	130	130	*3	6000	2	257100	1	2000	3	8450	434	172800	1	1000	250	4500	20	4
5	Grand Manan	1010	1010	14500	14500							1	500	90	6000	96	98500	40	4000	400	5
6	Campobello	500	250	1350	1200	*1	2500	2	150000	2	4000			75	2800	60	4000	30	1000	22	6
7	West Isles	400	400	1450	1450													100	6000	100	7
	Totals	2313	1849	22644	22944	4	8500	6.362100		9	22500	9	11730	660	196640	250	128470	501	19550	988	
	St. John County.																				
1	St. John Harbour			800	800							11	8500	80	62000	54	86000			45	1
2	Letprean to Chance Harbour			800	800									4	600	3	130			5	2
3	Chance Harbour to St. John			3000	3000							1	2500	35	3600	14	1400			16	3
4	Mispec to Tynemouth Creek			870	870																4
5	Tynemouth to Albert County	15	15	1107	1105																5
	Totals	15	15	6577	6577							12	9500	139	625900	71	86560			66	

* Not operated.

THE CATCH.

RETURN showing the Quantities and Values of all Fish caught and landed in a Green State in the Counties of **Charlotte** and **St. John**, Province of **New Brunswick**, during the year 1913-14.

Number	Fishing Districts.	Salmon, cwts.*	Salmon, value.	Lobsters, cwts.	Lobsters, value.	Cod, cwts.	Cod, value.	Haddock, cwts.	Haddock, value.	Hake and Cusk, cwts.	Hake and Cusk, value.	Pollock, cwts.	Pollock, value.	Herring, cwts.	Herring, value.	Mackerel, cwts.	Mackerel, value.	Shad, cwts.	Shad, value.	Alewives, cwts.	Alewives, value.	Number
	Charlotte County.																					
1	Lepreau to Red Head			626	9980							60	60									1
2	Red Head to L'Etang			884	13260	1440	2880	2700	6750	29000	21750	700	700	8260	8260	200	800	200	1000			2
3	L'Etang to St. George			268	4020	1500	3000	840	2000	1200	975	4335	4335	4385	4385	350	1400					3
4	St. George to St. Stephen			47	705	109	218	980	2450	468	351	70	70	174582	174582	6	24					4
5	Grand Manan			7070	106050	10093	20186	284	710	15620	11640	13563	13563	530	530	330	1920					5
6	Campo Bello			413	6195			4500	11250	15600	11850	24134	24134	6000	6000	24	96					6
7	West Isles			500	7500	150	300	148	370	300	225	28000	28000									7
	Totals			9808	147120	27792	35584	9412	23530	62888	46791	70862	70862	193757	193757	910	3640	200	1000			
	St. John County.																					
1	St. John Harbour	1250	18750	458	6670									1100	1100			990	4950	27000	27000	1
2	Lepreau to Chance Harbour	648	9720	479	7185	360	720			1700	1275							260	1300	526	526	2
3	Chance Harbour to St. John	2100	31500	378	5670	680	1360	20	50	1092	819			2400	2400							3
4	Mispec to Tynemouth Creek			301	4515									40	40							4
5	Tynemouth to Albert County			327	4905													1250	6250			5
	Totals	3998	59970	1943	29145	1040	2080	20	50	2792	2094			3540	3540			1250	6250	27526	27526	

*Cwt., 100 lbs.

THE CATCH.

RETURN showing the Quantities and Values of all Fish caught and landed in a Green State in the Counties of Charlotte and St. John, Province of New Brunswick, during the year 1913-14—*Concluded.*

Number	Fishing Districts	Sardines, brls.	Sardines, value.	Halibut, cwts.	Halibut, value.	Flounders, cwts.	Flounders, value.	Smelts, cwts.	Smelts, value.	Eels, cwts.	Eels, value.	Squid, brls.	Squid, value.	Scallops, brls.	Scallops, value.	Clams, brls.	Clams, value.	Dulse, Crabs, Cockles and other shell fish, cwts.	Dulse, Crabs, Cockles and other shell fish, value.	Number
	Charlotte County.																			
1	Lepreau to Red Head	11750	23500													13850	13850			1
2	Red Head to L'Etang	12900	24000	20	200	36	54							1150	1725	1990	1990			2
3	L'Etang to St. George	30252	60464			144	216	30	300							1875	1875			3
4	St. George to St. Stephen	24077	48154	123	1250	500	750	14	140			5	20			5847	5847	846	2117	4
5	Grand Manan	18645	37290	25	250	300	450	14	140					160	240	1400	1400	640	1700	5
6	Campo Bello	1739	3469									25	100					3750	3750	6
7	West Isles	24000	48000													1117	1117			7
	Totals	122434	244868	168	1680	980	1470	58	580			30	120	1310	1965	26079	26079	5276	7565	
	St. John County.																			
1	St. John Harbour	7000	14000																	1
2	Lepreau to Chance Harbour	3400	6900																	2
3	Chance Harbour to St. John	8550	17100																	3
4	Mispec to Tynemouth Creek									80	640									4
5	Tynemouth to Albert County																			5
	Totals	18950	37900							80	640									

* Cwt. = 100 lbs.

THE CATCH MARKETED.

RETURN showing the Quantities and Values of all Fish and Fish Products Marketed in a fresh, dried, pickled canned, etc., Slate, for Counties of Charlotte and St. John, Province of New Brunswick, during the year 1913-1914.

No.	Fishing Districts.	Salmon, used fresh and frozen, cwts.	Lobsters, shipped in shell, cwts.	Cod, used fresh, cwts.	Cod, shipped green-salted, cwts.	Cod, dried, quintals.	Haddock, used fresh, cwts.	Haddock, smoked, cwts.	Haddock, dried, quintals.	Hake and Cusk, used fresh, cwts.	Hake and Cusk, dried, quintals.	Pollock, used fresh, cwts.	Pollock, dried, quintals.	Herring, used fresh, cwts.	Herring, smoked, cwts.	Herring, canned, cases.	Herring, pickled, brls.	Herring, used as bait.	No.
	Charlotte County.																		
1	Lepreau to Red Head		696			411	130					60	234						1
2	Red Head to L'Etang		884	267		80	800					3800	180					1430	2
3	L'Etang to St. George		263	1250		18	920		390			227	24						3
4	St. George to St. Stephen		47	748	30	2722	44		86	1300	156		4440						4
5	Grand Manan		7070	3495	594	335	4500	700		300	9666	12160	3982	47650	900	1332	273	10856	5
6	Campo Bello		413	150			148	30			5176	21500	2166		9046		310	265	6
7	West Isles		500								5266				52593			3000	7
	Totals		9808	5850	624	3666	6542	730	476	1600	20264	37747	11036	47650	55341	1332	588	13551	
	Rates		15	2	4	6	2.50	5.00	3.50	.75	3.50	1	3.50	1	3.50	5	5	2	
	Values $		147120	11700	2496	21996	16365	3650	1666	1200	70924	37747	38626	47650	193793	6680	2940	31102	
	St. John County.																		
1	St. John Harbour	1250	458								566			1100					1
2	Lepreau to Chance Harbour	648	479								300			2400			20		2
3	Chance Harbour to St. John	2100	378	680	180		20			180				20			20		3
4	Mispec to Tynemouth-Creek		301																4
5	Tynemouth to Albert County		327																5
	Totals	3998	1943	680	180		20			180	866			3620			20		
	Rates	15	15	2	4		2.50			.75	3.50			1			5		
	Values $	59970	29145	1360	720		50			135	3031			3620			110		

*Cwts. = 100 lbs.   † Quintal = 112 lbs.

5 GEORGE V., A. 1915

THE CATCH MARKETED.

RETURNS showing the Quantities and Values of all Fish and Fish Products Marketed in a fresh, dried, pickled, canned, etc., State, for Counties of **Charlotte** and **St. John**, province of **New Brunswick**, during the year 1913–1914.—*Concluded.*

Numbers	Fishing Districts	Herring, used as fertilizer, brls.	Mackerel, used fresh, cwts.	Mackerel, salted, brls.	Shad, used fresh, cwts.	Shad, salted, brls.	Alewives, used fresh, cwts.	Alewives, salted, brls.	Sardines, canned, cases.	Sardines, sold fresh or salted, cwts.	Halibut, used fresh, cwts.	Flounders, used fresh, cwts.	Smelts, used fresh, cwts.	Eels, used fresh, cwts.	Squid, used as bait, brls.	Scallops, used fresh, brls.	Clams, used fresh, brls.	Clams, canned, cases.	Dulse, (Crabs, Cockles and other shell fish, used fresh, cwts.	Tongues and Sounds, pickled or dried, cwts.	Fish Oil, gals.	Numbers
	Charlotte County.																					
1	Lepreau to Red Head	250	200	112	90	36			60000	11750	20		36			1150	7838	6010	846	290	6500	1
2	Red Head to L'Etang		14		95	36				30292		36	14				590	1400			200	2
3	L'Etang to St. George		6	100					22500	19577		144	14		5	160	1875	3420			3000	3
4	St. George to St. Stephen		30							18645	123	500			25		2426		680	65	4500	4
5	Grand Manan			8		33			3200	1730	25						1400		1250	65	1000	5
6	Campobello					15				23200		300										6
7	West Isles																1117					7
	Totals	250	250	220	95	525			86700	105134	168	980	58		30	1310	15246	16830	2776	429	16100	
	Rates	1.00	4	12	5	15			5	2	10	1.50	10		41	1.50	1	4.80	4	40	.30	
	Value	250	1000	2640	475	525		7500	428500	210298	1680	1470	580		129	1965	15246	19494	1104	16800	4830	
	St. John County.																					
1	St. John Harbour						4500	7500		7000				80							600	1
2	Lepreau to Chance Harbour				990					34000												2
3	Chance Harbour to St. John				260		525			8550											500	3
4	Mispec to Tynemouth Creek																					4
5	Tynemouth Creek to Albert County				1250		5026	7500		18050				80							1100	5
	Totals				1250		5026	7500		18050				80	8						1100	
	Rates				5		1	.5		2											.30	
	Values				6250		5026	37500		375000				640							330	

Total Value, Charlotte County $1,353,972
" St. John 185,657

Grand total value $1,539,629

SESSIONAL PAPER No. 39 *

RECAPITULATION

OF the Quantities and Values of all Fish caught and landed in a Green State, and of the Quantities and Values of all Fish and Fish Products Marketed in a fresh, dried, pickled, canned, etc., State, for **District No. 1**, Province of **New Brunswick**, during the year 1913-14.

Kinds of Fish.		Caught and landed in a Green State.		Marketed.		Total Marketed Value.
		Quantity.	Value.	Quantity.	Value.	
			$		$	$
Salmon	cwts.	3,998	59,970	....	....	
„ used fresh.	„	....	....	3,998	....	59,970
Lobsters	„	11,751	176,265	....	....	
„ shipped in shell..	„	....	....	11,751	....	176,265
Cod	„	18,832	37,664	....	....	
„ used fresh	„	....	....	6,530	13,060	
„ green, salted	„	....	....	804	3,216	
„ dried	„	....	....	3,566	21,396	37,672
Haddock	„	9,432	23,580	....	....	
„ used fresh	„	....	....	6,562	16,385	
„ smoked (finnans)	„	....	....	730	3,650	
„ dried	„	....	....	476	1,666	21,701
Hake	„	65,180	48,885	....	....	
„ used fresh	„	....	....	1,780	1,335	
„ dried	„	....	....	21,130	73,955	75,290
Pollock	„	70,862	70,862	....	....	
„ used fresh	„	....	....	37,747	37,747	
„ dried	„	....	....	11,036	38,626	76,373
Herring	„	197,297	197,297	....	....	
„ used fres	„	....	....	51,170	51,170	
„ canned	cases	....	....	1,332	6,660	
„ smoked	cwts.	....	....	55,941	195,793	
„ pickled	brls.	....	....	608	3,040	
„ used as bait	„	....	....	15,551	31,102	
„ used as fertilizer	„	....	....	250	250	288,015
Mackerel	cwts.	910	3,640	....	....	
„ used fresh	„	....	....	250	1,000	
„ salted	brls.	....	....	224	2,640	3,640
Shad	cwts.	1,450	7,250	....	....	
„ used fresh	„	....	....	1,345	6,725	
„ salted	brls.	....	....	35	525	7,250
Alewives	cwts.	27,526	27,526	....	....	
„ used fresh	„	....	....	5,026	5,026	
„ salted	brls.	....	....	7,500	37,500	42,526
Sardines	„	141,384	282,768	....	....	
, canned	cases.	....	....	85,700	428,500	
„ sold fresh or salted	brls.	....	....	124,084	248,168	676,668
Halibut, used fresh	cwts.	168	1,680	168	....	1,680
Flounders	„	980	1,470	980	....	1,470

5 GEORGE V., A. 1915

RECAPITULATION

Of the Quantities and Values of all Fish caught and landed in a Green State, and of the Quantities and Values of all Fish and Fish Products Marketed in a fresh, dried, pickled, canned, etc., State, for **District No. 1**, Province of **New Brunswick**, during the year 1913-14.—*Concluded*.

Kinds of Fish.	Caught and landed in a Green State.		Marketed.		Total Marketed Value.
	Quantity.	Value.	Quantity.	Value.	
		$		$	$
Smelt............................ cwts.	58	580	58	....	580
Eels............................ "	80	640	80	....	640
Squid............................ brls.	30	120	30	....	120
Clams and Scallops .... "	27,389	28,044	....	....	
" used fresh................ "	....	....	16,556	17,211	
" canned.................. cases.	....	....	10,830	51,984	69,195
Dulse, Cockles & other shell fish.. .... cwts.	5,276	7,565	2,776	....	11,104
Tongues and Sounds.............. "	....	....	420	....	16,800
Fish Oil.. .... gallons.	....	....	17,200	....	5,169
Totals	....	·975,806	....	....	1,572,119

RECAPITULATION

Of the Number of Fishermen, etc., and of the Number and Value of Fishing Vessels, Boats, Nets, etc., in **District No. 1**, Province of **New Brunswick**, for the year 1913-14.

	Number.	Value.
		$
Sailing and Gasoline Vessels.................................	106	125,800
Boats (sail)...	1,710	49,720
" (gasoline)...	810	232,175
Carrying Smacks..	40	42,200
Gill Nets, Seines, Trap and Smelt Nets, etc.................	3,931	92,723
Weirs	458	339,785
Trawls	396	10,146
Hand Lines...	2,328	1,864
Lobster Traps..	29,521	29,521
" Canneries...	4	8,500
Sardine "	6	362,100
Clam "	9	22,500
Freezers and Ice-houses...................................	21	109,230
Smoke and Fish-houses....................................	780	265,840
Fishing Piers and Wharves................................	321	212,220
Pile Drivers and Scows...................................	501	19,550
Totals..		1,923,874

Number of men employed on Vessels............	301
" " Boats ...	2,344
" " Carrying Smacks............	88
" persons employed in Fish-houses, Freezers, Canneries, &c....	1,034
Total......	3,767

DISTRICT No. 2.

QUANTITY AND VALUE OF ALL FISHING GEAR, ETC., USED IN THE
FISHING INDUSTRY IN THE COUNTIES OF RESTIGOUCHE,
GLOUCESTER, NORTHUMBERLAND, KENT, WESTMORLAND,
AND ALBERT.

DISTRICT No. 2.

RETURN showing the Number of Fishermen, &c., the Number and Value of Vessels and Boats, and the Quantity and Value of all Fishing Gear, &c., used in the Fishing Industry in the Counties of Restigouche, Gloucester, Northumberland, Kent, Westmorland and Albert, Province of New Brunswick, during the year 1913-14.

Number	Fishing Districts	Sailing and Gasoline Vessels				Boats					Carrying Smacks			Gill Nets, Seines, Trap and Smelt Nets, etc.		Weirs		Trawls		Number
		No. (20 to 40 tons)	No. (10 to 20 tons)	Value	Men	Sail and Row	Value	Gasoline	Value	Men	Number	Value	Men	Number	Value	Number	Value	Number	Value	
	Restigouche County.			$			$		$			$			$		$		$	
1	Above Dalhousie	1		800	5	46	950	3	4500	56	3	350	6	1129	14500					1
2	Below Dalhousie		1	500	3	238	6200	7	1800	310	1	500	2	2355	21500					2
	Totals	1	1	1300	8	284	7150	10	6300	366	4	850	8	3484	36000					
	Gloucester County.																			
3	Beresford, &c.		1	540	4	455	11000	1	400	870				4100	29000			10	30	3
4	Bathurst and New Bandon, &c.	8		3900	35	312	6700	1	750	730				4100	25000			48	280	4
5	Caraquet, &c.	134		70000	554	365	9800	2		742	3	600	6	3150	23200			166	1360	5
6	Shippigan and Miscou Island	61		33600	280	384	20000	6	3200	852				2430	18300			140	1000	6
7	Tracadie, Inkerman, &c.	23		11500	155	360	11400	11	5600	753				5241	46700			40	300	7
	Totals	227		119500	1955	1880	58900	20	9550	3947	3	600	6	19044	142200			404	2870	
	Northumberland County.																			
8	Chatham, Neguac, &c.		5	1800	15	446	11900	6	3500	670	3	280	3	9830	99400			6	40	8
9	Bay-du-Vin, &c.		4	2600	12	231	4900	5	4000	370	10	2000	20	6400	98800					9
10	Southwest and Northwest Miramichi R. ver					196	2850			207				1900	13500					10
	Totals		9	4400	27	873	19750	11	7500	1247	13	2280	23	18130	212700			6	40	

Kent County.																
11 Richibucto, &c.	13	5900	38	246	16100	10	3800	437	1	2	3126	34460			8	120
12 Buctouche, &c.	4	3600	12	586	20100	11	5600	1005		2	2703	29750	500			120
13 Cocagne, &c.				440	12000	7	3600	850			1454	12500				
Totals	17	9500	50	1272	48200	28	12900	2292	1	2	7383	75710	500		8	129
Westmorland County.																
14 Shediac, &c.				510	16500	10	3000	975	1	2	3875	23375	500	2	20	120
15 Botsford				300	7000	200	40000	1020	1	2	1910	17300	400	6	600	
16 Sackville and Westmorland				250	75000	50	10000	600	2	6	810	5500	2000	1	100	
17 Dorchester				32	2500			64			160	1800				
Totals				1092	101000	260	53000	2659	4	10	6755	47973	2900	9	900	120
18 *Albert County*	11			15	500	2	300	28			200	800		2	200	
Grant totals	254	134700	1050	5436	236390	31	89850	10639	25	49	55196	513385	7130	11	438	3250

RETURN showing the Number of Fishermen, &c., the Number and Value of Vessels and Boats, and the Quantity and Value of all Fishing Gear, &c., used in the Fishing Industry in the Counties of Restigouche, Gloucester, Northumberland, Kent, Westmorland and Albert, Province of New Brunswick, during the year 1913-14—*Concluded.*

Number	Fishing Districts.	Hand lines. Number.	Hand lines. Value. $	Lobster traps. Number.	Lobster traps. Value. $	Lobster canneries. Number	Lobster canneries. Value. $	Clam canneries. Number.	Clam canneries. Value. $	Freezers and Ice-Houses. Number.	Freezers and Ice-Houses. Value. $	Smoke and Fish-Houses. Number.	Smoke and Fish-Houses. Value. $	Fishing Piers and Wharves. Number.	Fishing Piers and Wharves. Value. $	Persons employed in Canneries, Freezers and Fish-Houses.	Number.
	Restigouche County.																
1	Above Dalhousie	80	40	3800	3800	..	....	...	...	3	16000	2	800	..	....	30	1
2	Below Dalhousie	40	20	....	....	1	1500	...	...	5	8500	4	2100	1	200	50	2
	Totals	120	60	3800	3800	1	1500	...	.	8	24500	6	2900	1	200	80	
	Gloucester County.																
3	Beresford, &c.	300	150	3250	3250	4	1150	..	...	9	3000	25	1500	..	....	100	3
4	Bathurst, &c.	420	300	10650	10650	10	6000	..	...	9	3400	8	1200	2	1800	230	4
5	Caraquet &c., and New Bandon, &c.	2100	1300	2250	2250	6	6100	..	...	6	3600	31	8500	2	1500	475	5
6	Shippigan and Miscou Island	1800	800	45900	45900	49	33800	..	...	8	5000	28	8400	2	2000	840	6
7	Tracadie, Inkerman, &c.	725	375	17300	17300	15	9250	2	3000	10	15000	42	4600	2	1600	480	7
	Totals	5345	3125	78750	78750	84	56300	2	3000	42	30000	134	24200	8	6900	2125	
	Northumberland County.																
8	Chatham, Neguac, &c.	140	100	18500	18500	12	10650	..	...	22	14000	80	13000	..	....	345	8
9	Bay-du-Vin, &c.	50	30	10000	10000	4	4500	..	...	8	4200	2	800	..	....	260	9
10	Southwest and Northwest Miramichi River	....	....	....	....	..	....	..	...	7	2000	28	700	..	....	48	10
	Totals	230	130	28500	28500	16	15450	..	...	37	20200	110	14500	..	....	653	

SESSIONAL PAPER No. 39

Kent County.																
11	Richibucto, &c	220	90	27900	27900	12	13500	1	400	21	18600	8	600	15	7500	296
12	Buctouche, &c	350	150	12760	12760	16	10150	..	..	8	7000	5	500	..	..	315
13	Cocagne, &c	300	150	8300	8300	8	5100	1	800	3	1000	..	..	..	..	154
	Totals	870	390	48960	48960	36	28750	2	1200	32	26600	13	1100	15	7500	765
Westmorland County.																
14	Shediac, &c	90	40	23500	23500	18	20500	..	..	10	12000	12	8000	14	..	413
15	Botsford	100	40	39500	39500	20	19550	..	..	35	8000	50	3500	10	2800	95
16	Sackville and Westmorland	100	40	..	..	..	..	..	..	30	2200	100	8900	1	2000	280
17	Dorchester	..	..	..	..	..	..	..	..	..	..	15	300	..	500	30
	Totals	290	120	63000	63000	38	40050	..	..	75	22900	177	19800	25	5300	1679
18	Albert County	..	..	150	150	..	..	..	..	÷	..	2	120	..	..	..
	Grand totals	6855	3825	223160	223160	175	142050	4	4200	194	123900	442	62620	49	19900	5302

5 GEORGE V., A. 1915

THE CATCH.

RETURN showing the Quantities and Values of all Fish caught and landed in a Green State, in **District No. 2,** Province of **New Brunswick,** during the year 1913–14.

Number	Fishing Districts.	Salmon, *cwts.	Salmon, value.	Lobsters, cwts.	Lobsters, value.	Cod, cwts.	Cod, value.	Haddock, cwts.	Haddock, value.	Hake and Cusk, cwts.	Hake and Cusk, value.	Herring, cwts.	Herring, value.	Number
	Restigouche County.		$		$		$		$		$		$	
1	Above Dalhousie......	1125	16875	......	......	150	225	....	..					1
2	Below Dalhousie	1511	22665	680	3400	342	513	....	.	230	230	6403	3842	2
	Totals	2636	39540	680	3400	492	738	....		230	230	6403	3842	
	Gloucester County.													
3	Beresford. &c.........	1212	18180	524	2620	5826	8739	..	......	....	....	39445	23667	3
4	Bathurst, New Bandon, &c.	803	12045	1425	7125	18132	28698	..	......	....	....	17180	10308	4
5	Caraquet, &c..........	....	......	535	2675	112900	169350	1500	1500	2000	2000	70000	42000	5
6	Shippigan and Miscou Island	....	......	10460	52300	57000	85500	1620	1620	2370	2370	45220	27132	6
7	Tracadie, Inkerman, &c.	935	14025	5212	26060	20523	30787	...	......	....	....	34180	20508	7
	Totals..........	2950	44250	18156	90780	214383	323074	3120	3120	4370	4370	206025	123615	
	Northumberland County													
8	Chatham Neguac, &c..	1560	23400	4572	22860	1445	2168	..	......	....	....	17198	10319	8
9	Bay du Vin, &c.......	3475	52125	3350	17750	428	642	...	.	....	....	11359	6815	9
10	Southwest and North-west Miramichi River, &c.	525	7875	......	......									10
	Totals	5560	83400	8122	40610	1873	2810	....	....	...		28557	17134	
	Kent County.													
11	Richibucto, &c........	1756	26340	8830	44250	1874	2811	70	70	3390	3390	42900	25740	11
12	Buctouche, &c........	......	.....	3620	18100	377	566	..	......	....	....	48625	29175	12
13	Cocagne, &c........ ..	......	....	3165	15825	20	75	..	......	....	....	55154	33092	13
	Totals..........	1756	26340	15635	78175	2301	3452	70	70	3390	3390	146679	88007	
	Westmorland County.													
14	Shediac. &c......... .	100	1500	11450	57250	820	1230	50	50	3450	3450	177000	106200	14
15	Botsford	.....	.....	12293	61465	....	....	..	......	....	....	77800	46680	15
16	Sackville and Westmor-land................	15	225	.....	......	44	66	..	....	.....	....	28000	16800	16
17	Dorchester............	60	900	....	......	190	285	..	......	....		195	117	17
	Totals..........	175	2625	23743	118715	1054	1581	50	50	3450	3450	282995	169797	
18	*Albert County.........*	13	195	90	450	1500	2250	....	...	..	....	170	102	18
	Grand totals......	13090	196350	66426	332130	221603	333905	3240	3240	11440	11440	670829	402497	

*Cwt. = 100 lbs.

SESSIONAL PAPER No. 39

THE CATCH.

RETURN showing the Quantities and Values of all Fish caught and landed in a Green State, in **District No. 2**, Province of **New Brunswick**, during the year 1913-14—*Continued.*.

Number	Fishing Districts.	Mackerel, cwts.	Mackerel, value.	Shad, cwts.	Shad, value.	Alewives, cwts.	Alewives, value.	Halibut, cwts.	Halibut, value.	Flounders, cwts.	Flounders, value.	Smelts, cwts.	Smelts, value.	Number.
	Restigouche County.		$		$		$		$		$		$	
1	Above Dalhousie.... ...									1000	1500	3951	23706	1
2	Below Dalhousie.... ...	3	21	10	80					100	150	740	4440	2
	Totals..........	3	21	10	80					1100	1650	4691	28146	
	Gloucester County.													
3	Beresford, &c............									80	120	20	120	3
4	Bathurst & New Bandon,&c	65	455					135	1080	40	60	860	5160	4
5	Caraquet, &c	360	2520					200	1600			2000	12000	5
6	Shippigan & Miscou Island.	2064	14448					100	800			2520	15120	6
7	Tracadie, Inkerman, &c....	5432	38024			1975	1185	120	960			3993	23970	7
	Totals..........	7921	55447			1975	1185	555	4440	120	180	9395	56370	
	Northumberland County.													
8	Chatham, Neguac, &c.....	281	1967	201	1608	1180	708			1010	1515	15548	93288	8
9	Bay du Vin, &c...........	2476	17332	125	1000	684	410			350	525	12675	76050	9
10	Southwest and Northwest Miramichi River........			122	976	1600	960					160	960	10
	Totals.	2757	19299	448	3584	3464	2078			1360	2040	28383	170208	
	Kent County.													
11	Richibucto, &c...........	4284	29988	171	1368	2500	1500			201	302	4300	25800	11
12	Buctouche, &c...........	320	2240			500	300			140	210	3040	23840	12
13	Cocagne, &c..............	50	350			500	300			30	45	1890	11340	13
	Totals.... ...	4654	32578	171	1368	3500	2100			371	557	10130	60780	
	Westmorland County.													
14	Shediac, &c..............	1480	10360	8	64	600	360			310	465	4030	24180	14
15	Botsford.................					100	60					2780	16680	15
16	Sackville and Westmorland.			146	1168	600	360					650	3900	16
17	Dorchester...............			590	4720									17
	Totals....,......	1480	10360	744	5952	1300	780			310	465	7460	44760	
18	Albert County...........	16	112	36	288					20	30			18
	Grand totals.........	16831	117817	1409	11272	10239	6143	555	4440	3281	4922	60059	360354	

* Cwt. = 100 lbs.

5 GEORGE V., A. 1915

THE CATCH.

RETURN showing the Quantities and Values of all Fish caught and landed in a Green State, in **District No. 2**, Province of **New Brunswick**, during the year 1913–14—*Continued.*

Number.	Fishing Districts.	Trout, cwts.	Trout, value.	Sturgeon, cwts.	Sturgeon, value.	Bass, cwts.	Bass, value.	Eels, cwts.	Eels, value.	Tom-cod, cwts.	Tom-cod, value.	Number.
	Restigouche County.		$		$		$		$		$	
1	Above Dalhousie............	30	240	....	....	20	200	50	300	50	75	1
2	Below Dalhousie........	125	1000	....	....	10	100	40	240	24	36	2
	Totals	155	1240	....	....	30	300	90	540	74	111	
	Gloucester County.											
3	Beresford, &c..............	40	320	......	....	12	120	30	180	50	75	3
4	Bathurst, New Bandon, &c.......	50	400	......	....	25	250	40	240	2500	3750	4
5	Caraquet, &c.............	20	160	......	....	25	250	50	300	100	150	5
6	Shippigan & Miscou Island.........	10	80	......	....	42	420	65	390	....	....	6
7	Tracadie, Inkerman, &c.........	95	760	......	....	160	1600	544	3264	70	•105	7
	Totals..............	215	1720	......	....	264	2640	729	4374	2720	4080	
	Northumberland County.											
8	Chatham, Neguac, &c...........	65	520	....	....	60	600	96	576	14060	21090	8
9	Baie du Vin, &c.............	40	320	....	....	140	1400	60	360	650	975	9
10	Southwest and Northwest, Miramichi River......	1240	9920	....	....	280	2800	300	1800	....	....	10
	Totals	1315	10760	....	....	480	4800	456	2736	14710	22065	
	Kent County.											
11	Richibucto, &c	50	400	....	....	717	7170	70	420	130	195	11
12	Buctouche, &c....................	45	360	....	....	50	500	200	1200	60	90	12
13	Cocagne, &c	35	280	....	....	50	500	85	510	20	30	13
	Totals.............	130	1040	....	....	817	8170	355	2130	210	315	
	Westmorland County.											
14	Shediac, &c.................	60	480	....	....	40	400	40	240	280	420	14
15	Botsford........................	50	400	....	....	30	300	56	300	..	....	15
16	Sackville & Westmorland.........	35	280	45	225	20	200	45	270	....	....	16
17	Dorchester....	30	240	60	300	....	....	45	270	50	75	17
	Totals	175	1400	105	525	90	900	180	1080	330	495	
18	*Albert County................*	95	760	......	...	10	100	50	300	40	60	18
	Grand totals..............	2115	16920	105	525	1691	16910	1860	11160	18084	27126	

* Cwt. = 100 lbs.

SESSIONAL PAPER No. 39

THE CATCH.

RETURN showing the Quantities and Values of all Fish caught and landed in a Green State, in **District No. 2,** Province of **New Brunswick,** during the year 1913-14—*Concluded.*

Number.	Fishing Districts.	Mixed fish, cwts.	Mixed fish, value.	Oysters, brls.	Oysters, value.	Clams, brls.	Clams, value.	Quahaugs, brls.	Quahaugs, value.	Cockles, Crabs, &c., cwts.	Cockles, Crabs, &c., value.	Number.
	Restigouche County.		$		$		$		$		$	
1	Above Dalhousie					10	20					1
2	Below Dalhousie	100	100			25	.50					2
	Totals	100	100			35	70					
	Gloucester County.											
3	Beresford, &c					165	330					3
4	Bathurst, New Bandon, &c			40	200	80	160					4
5	Caraquet, &c			220	1100	850	1700					5
6	Shippigan & Miscou Island			20	100	800	1600					6
7	Tracadie, Inkerman, &c					7100	14200					7
	Totals			280	1400	8995	17990					
	Northumberland County.											
8	Chatham, Neguac, &c			2000	10000			2780	5560			8
9	Bay du Vin, &c			2505	12525							9
10	Southwest and Northwest, Miramichi River											10
	Totals			4505	22525			2780	5560			
	Kent County.											
11	Richibucto, &c	80	80	1373	6865	26	52			12	12	11
12	Buctouche, &c			1603	8015	125	250	4073	8146			12
13	Cocagne, &c			1282	6410			6804	13608			13
	Totals	80	80	4258	21290	151	302	10877	21754	12	12	
	Westmorland County.											
14	Shediac, &c			1562	7810	700	1400	5353	10706			14
15	Botsford			150	750	40	80	180	360			15
16	Sackville & Westmorland			45	225	33	66					16
17	Dorchester					30	60					17
	Totals			1757	8785	803	1606	5533	11066			
18	*Albert County*					40	80					18
	Grand totals	180	180	10800	54000	12804	25608	16410	32820	12	12	

* Cwt. = 100 lbs.

THE CATCH MARKETED.

RETURN showing the Quantities and Value of all Fish and Fish Products Marketed in a fresh, dried, pickled, canned, &c., State, for District No. 2, Province of New Brunswick, during the year 1913-1914.

Number	Fishing Districts	Salmon, used fresh and frozen, cwts.	Lobsters, canned, cases.	Lobsters, shipped in shell, cwts.	Cod, used fresh, cwts.	Cod, shipped green salted, cwts.	Cod, dried, †Quintals.	Haddock, used fresh, cwts.	Haddock, dried, quintals.	Hake and Cusk, used fresh, cwts.	Hake and Cusk, dried, quintals.	Herring, used fresh, cwts.	Herring, smoked, cwts.	Herring, pickled, brls.	Herring, used as bait, brls.	Herring, used as fertilizer, brls	Mackerel, used fresh, cwts.	Mackerel, salted, brls.	Shad, used fresh, cwts.	Shad, salted, brls.
	Restigouche County.																			
1	Above Dalhousie	1125			150					131	33	180			350	1000	3		10	
2	Below Dalhousie	1511	248	60		111	40							1174						
	Totals	2636	248	60	150	111	40			131	33	180		1174	350	1000	3		10	
	Gloucester County.																			
3	Beresford, &c.	1212	189	54	1196	413	1248					360		3745	225	13700	65			
4	Bathurst, New Bandon, &c.	803	558	30	700	3475	3494				667	1400		1860	2100	3800	360			
5	Caraquet, &c		206	20		5000	34300		500		790	1600		7600	4200	18000	2064			
6	Shippigan and Miscou Island		4076	270		1359	18100		·540·			1620		7200	9250	1750				
7	Tracadie, Inkerman, &c.	935	2057	70	140	312	16587					200		5670	1395	7000	5432			
	Totals	2950	7086	444	2036	10580	63729		1040		1457	5180		26075	17170	44140	7921			
	Northumberland County.																			
8	Chatham, Neguac, &c.	1560	1813	40	60	130	375					1200	50	1580	3049	2500	281		201	
9	Bay du Vin, &c.	3475	1412	20	388	20								675	2300	2337	2476		125	
10	Southwest and Northwest Miramichi River	525																	122	
	Totals	5560	3225	60	448	150	375					1200	50	2255	1349	4837	2757		448	

Kent County.																			
11 Richibucto, &c.	1756	3620	50	182	270	384	70			1130	10235		5755	3500	4200	4280	8	171		11
12 Buctouche, &c.		1406	105	177	100						2231		7910	5000	6332	320				12
13 Cocagne, &c.		1246	50	50							4531		4987	3000	14831	50				13
Totals	1756	6172	205	409	370	384	70			1130	16997		18652	11500	25363	4650	8	171		
Westmorland County.																				
14 Shediac, &c.	100	2980	4000	220	300		50	3450			1000	9000	15000	10000	46500	1480		8		14
15 Botsford	15	4875	100								1000	16000	4000	12100	4300			146		15
16 Sackville and Westmorland	60			44							200	10000	1334	500	500			155	145	16
17 Dorchester				50	70						60		45						145	17
Totals	175	7855	4106	314	370		50	3450			4060	35000	20379	22600	51300	1480		309		
18 Albert County.	13		90	1500			120	1046		2620		34	34			16		36		18
Totals	13090	24586	4965	4857	11581	64528	120	1046	3581	2620	27677	35084	68569	50969	129640	16807	8	974	145	
Rates	15	20	5	2	3	5	2	3	1.50	3	1	2	4	1.50	50	10	12	10	15	
Values	$1963350	491720	24825	9714	34743	3226640	240	3129	5872	7860	27677	70168	274276	85454	63320	168070	96	9740	2175	

*Cwt. = 100 lbs.   †Quintal = 112 lbs.

THE CATCH MARKETED.

RETURN showing the Quantities and Value of all Fish and Fish Products Marketed in a fresh, dried, pickled, canned, &c., State, for District No. 2, Province of New Brunswick, during the year 1913-14.

Number	Fishing Districts	Fish oil, gals.	Tongues and sounds, pickled or dried, cwts.	Clams and quahaugs, canned, cases.	Clams and quahaugs, used fresh, brls.	Oysters, used fresh, brls.	Mixed fish, used fresh, cwts.	Tom-cod, used fresh, cwts.	Eels, used fresh, cwts.	Bass, used fresh, cwts.	Sturgeon, used fresh, cwts.	Trout, used fresh, cwts.	Smelts, used fresh, cwts.	Flounders, used fresh, cwts.	Halibut, used fresh, cwts.	Alewives, salted, brls.	Alewives, used fresh, cwts.
	Restigouche County.																
1	Above Dalhousie	30			10			50	50	20		30	3951	1000			
2	Below Dalhousie				25		100	24	40	10		125	740	100			
	Totals	30			35		100	74	90	30		155	4691	1100			
	Gloucester County.																
3	Beresford, &c.	400			165	40		50	30	12		40	20	80			
4	Bathurst, New Bandon, &c.	11000	300		80	220			40	25		50	860	40	135		
5	Caraquet, &c.	17500	200		850	20		2600	50	25		20	2000		200		
6	Shippigan and Miscou Island	3200	100		800				65	42		10	2520		100		
7	Tracadie, Inkerman, &c.			7100				70	544	160		95	3995		120	625	100
	Totals	32100	600	7100	1895	280		2720	729	264		215	9395	120	555	625	100
	Northumberland County.																
8	Chatham, Neguac, &c.					2000			96	60		65	15648	1010		340	160
9	Bay du Vin, &c.					2505			60	140		40	12575	350		228	
10	Southwest and Northwest Miramichi River				2780				300	280		1240	160			467	290
	Totals				2780	4505			456	480		1345	28383	1360		1035	360

SESSIONAL PAPER No. 39

Kent County.																
11 Richibucto, &c.	100	800	201	4300	50		717	70	130	80	1373	26		12	34	812
12 Buctouche, &c.	50	150	140	3940	45		50	200	60		1603	4198				100
13 Cocagne, &c.	50	150	30	1890	35		50	85	20		1282	6294	600			
Totals	200	1100	371	10130	130		817	355	210	80	4258	10428	600	12	34	912
Westmorland County.																
14 Shediac, &c.		200	310	4030	60		40	40	280		1562	6053				
15 Botsford.	100			2780	50		30	50			150	220				
16 Sackville and Westmorland.	600			650	35	45	20	43			45	33				
17 Dorchester.					30	60		45	50			30				
Totals	700	200	310	7460	175	105	90	180	330		1757	6336				
18 **Albert County**	1360		20		95	105	10	50	40			40				
Totals	1360		20		95	105	10	50	40			40				
Totals	2960	555	3281	60059	2115	105	1691	1860	18084	180	10800	21514	7700	12	634	38042
Rates $	3		2	10		5	12	8	2	1	6	3	5	1.50	5	.30
Values $	8890		6562	600590	21150	525	20292	14890	36168	190	64800	64542	38500	18	3170	9913

Total value .... $2,694,640.

*Cwt.=100 lbs.   1Quintal=112 lbs.

5 GEORGE V., A. 1915

RECAPITULATION.

Of the Quantities and Values of all Fish caught and landed in a Green State, and of the Quantities and Values of all Fish and Fish Products Marketed in a fresh, dried, pickled, &c., state, for **District No. 2**, Province of **New Brunswick**, during the year 1913-14.

Kinds of Fish.		Caught and landed in a Green State.		Marketed.		Total Marketed Value.
		Quantity.	Value.	Quantity.	Value.	
			$		$	$
Salmon	cwts.	13,090	196,350			
" used fresh	"			13,090	196,350	
						196,350
Lobsters	"	66,426	332,130			
" canned	cases.			24,586	491,720	
" shipped in shell	cwts.			4,965	24,825	
						516,545
Cod	"	221,603	333,905			
" used fresh	"			4,857	9,714	
" green—salted	"			11,581	34,743	
" dried	"			61,528	322,640	
						367,097
Haddock	"	3,240	3,240			
" used fresh	"			120	240	
" dried	"			1,040	3,120	
						3,360
Hake	"	11,440	11,440			
" used fresh	"			3,581	5,372	
" dried	"			2,620	7,860	
						13,232
Herring	"	670,829	402,497			
" used fresh	"			27,677	27,677	
" smoked	"			35,984	70,168	
" pickled	brls.			68,569	274,276	
" used as bait	"			56,969	85,454	
" used as fertilizer	"			126,640	63,320	
						520,895
Mackerel	cwts.	16,831	117,817			
" used fresh	"			16,807	168,070	
" salted	brls.			8	96	
						168,166
Shad	cwts.	1,409	11,272			
" used fresh	"			974	9,740	
" salted	brls.			145	2,175	
						11,915
Alewives	cwts.	10,239	6,143			
" used fresh	"			1,360	1,360	
" salted	brls.			2,960	8,880	
						10,240
Halibut, used fresh	cwts.	555	4,440	555		5,550
Flounders	"	3,281	4,922	3,281		6,562
Smelts	"	60,059	360,354	60,059		600,590
Trout	"	2,115	16,920	2,115		21,150
Sturgeon	"	105	525	105		525
Bass	"	1,691	16,910	1,691		20,292
Eels	"	1,860	11,160	1,860		14,880
Tom Cod	"	18,084	27,126	18,084		36,168
Mixed Fish	"	180	180	180		180
Oysters	"	10,800	54,000	10,800		64,800
Clams	"	29,214	58,428			
" used fresh	"			21,514	64,542	
" canned	cases.			7,700	38,500	
						103,042
Dulse, Cockles and other shell fish	cwts	12	12	12		18
Tongues and Sounds	"			634		3,170
Fish Oil	galls.			33,042		9,913
Totals			1,969,771			2,694,640

RECAPITULATION.

Of the Number of Fishermen, etc., and of the Number and Value of Fishing Vessels, Boats, Nets, etc., in **District No. 2**, Province of **New Brunswick**, for the year 1913-14.

	Number.	Value.
		$
Sailing and Gasoline Vessels	265	134,700
Boats (sail and row)	5,416	236,390
Boats (gasoline)	331	89,850
Carrying Smacks	25	7,130
Gill Nets, Seines, Trap and Smelt Nets, etc	55,196	515,385
Weirs	11	1,100
Trawls	438	3,250
Hand Lines	6,855	3,825
Lobsters Traps	223,160	223,160
Lobsters Canneries	175	142,050
Clam Canneries	4	4,200
Freezers and Ice-houses	194	123,900
Smoke and Fish-houses	442	62,620
Fishing Piers and Wharves	49	19,900
		1,567,460

Number of men employed on	Vessels			1,050
"	"	"	Boats	10,539
"	"	"	Carrying Smacks	49
"	persons employed in Fish-houses, Freezers, Canneries, &c			5,302
	Total			16,940

DISTRICT No. 3 (Inland).

RETURN showing the Number of Fishermen, etc., the Number and Value of Tugs, Vessels and Boats, and the Quantity and Value of all Fishing Gear and other Material used in the Fishing Industry in the Counties of Kings, Queens, Sunbury, York, Carleton, Victoria and Madawaska, Province of New Brunswick, during the year 1913-14.

Number	Fishing Districts	Boats Sail	Value	Boats Gasoline	Value	Men	Gill Nets Number	Value	Eel Traps Number	Value	Rods and Lines Number	Value	Freezers and Ice-houses Number	Value	Smoke and Fish-houses Number	Value	Number
	County of Kings.		$		$			$		$		$		$		$	
1	St. John River District	35	700	3	600	60	100	2000			55	110			10	100	1
2	Kennebecasis River District	30	300	1	160	20	125	1500	78	78	370	555			25	250	2
3	Belle Isle Bay and River District	16	192			60	100	1200			75	150					3
	Totals	81	1192	4	760	140	325	4700	78	78	500	815			35	350	
	County of Queens.																
4	Northeastern Section	72	720	4	600	125	360	2880	20	100	100	250	10	100	35	280	4
5	Southwestern Section	90	900	5	1200	150	280	2240	20	100	50	125			15	150	5
	Totals	162	1620	9	1800	275	640	5120	20	100	150	375	10	100	50	430	
6	*County of Sunbury* (in all)	30	450	2	500	45	580	4640			50	200			15	250	6
	County of York.																
7	St. John River District	55	950	10	1000	110	110	2200			300	600					7
8	St. Croix, Magaguadavic and Oromocto Waters	110	1800	18	7000	135					500	1000	3	450			8
9	Southwest Miramichi Waters	25	250			35	30	200			125	250					9
	Totals	230	3000	28	8000	280	140	2400			925	1850	3	450			
	County of Carleton.																
10	St. John River and Westward	55	530			55	50	750			100	200					10
11	East of St. John River	10	100			10					50	200					11
	Totals	65	630			65	50	750			150	400					

County of Victoria.													
12 St. John River District	11	198		18	5	50			35	35	5	500	
13 Tobique River and Lakes	30	600		30					75	700			
Totals	41	798		48	5	50			110	735	5	500	
County of Madawaska.													
14 Madawaska River and Eastward	150	950	2	200					400	500			
15 West of Madawaska River	80	460	250	115					160	210			
Totals	230	1410	250	315					560	710			

THE CATCH.

RETURN showing the Quantities and Values of all Fish caught and marketed or consumed locally for the Counties of **Kings, Queens, Sunbury, York, Carleton, Victoria and Madawaska**, Province of **New Brunswick**, during the year 1913-14.

Number.	Fishing Districts.	Salmon, cwts.	Salmon, value.	Trout, cwts.	Trout, value.	Whitefish, cwts.	Whitefish, value.	Bass, cwts.	Bass, value.	Pickerel, cwts.	Pickerel, value.	Sturgeon, cwts.	Sturgeon, value.	Number.
			$		$		$		$		$		$	
	Kings County.													
1	St. John River District.	299	4485	20	200					37	370	60	600	1
2	Kennebecasis River District.	4	60	54	540			140	1400	10	100			2
3	Belle Isle Bay and River District.	5	75	15	150									3
	Totals.	308	4620	89	890			140	1400	47	470	60	600	
	Queens County.													
4	Northeastern Section.	17	255	10	100			15	150	156	1560			4
5	Southwestern Section.	13	195	8	80					140	1400			5
	Totals.	30	450	18	180			15	150	296	2960			
6	*Sunbury County* (in all)	57	855	15	150					163	1630	10	100	6
	York County.													
7	St. John River District.	315	4725	125	1250					6	60			7
8	St. Croix, Magaguadavic and Oromocto Waters.	20	300	140	1400			4	40	16	160			8
9	Southwest Miramichi Waters.	40	600	40	400									9
	Totals.	375	5625	305	3050			4	40	22	220			
	Carleton County.													
10	St. John River and westward.	67	1005	10	100									10
11	East of St. John River.			20	200									11
	Totals.	67	1005	30	300									

SESSIONAL PAPER No. 39

	12 13				14 15	
Victoria County.						
12 St. John River District..........	3 50	45 750	26 70	260 700		
13 Isue River and Lakes....						
Totals..............	53	795	96	960		
Madawaska County.						
14 Madawaska River and eastward........	7	105	125 50	1250 500	26	390
15 West of Madawaska River.....						
Totals...............	7	105	175	1750	26	390

* Cwt. = 100 lbs.

5 GEORGE V., A. 1915

THE CATCH

RETURN showing the Quantities and Values of all Fish caught and marketed or consumed locally for the Counties of **Kings, Queens, Sunbury, York, Carleton, Victoria and Madawaska,** Province of **New Brunswick,** during the year 1913-14. —*Concluded*.

Number	Fishing Districts	Eels, cwts.*	Eels, value.	Perch, cwts.	Perch, value.	Alewives, cwts.	Alewives, value.	Shad, cwts.	Shad, value.	Gaspereaux, cwts.	Gaspereaux, value.	Mixed Fish, cwts.†	Mixed Fish, value.†	Caviare, cwts.	Caviare, value.
	Kings County.														
1	St. John River District					150	300	40	200			25	25	1	100
2	Kennebecasis River District	134	670			129	258	126	630			50	50		
3	Belle Isle Bay and River District					30	60	17	85			10	10		
	Totals	131	670			309	618	183	915			85	85	1	100
	Queens County.														
4	Northwestern Section					560	1120	221	1105			30	30		
5	Southwestern Section	32	160			800	1600	330	1650			30	30		
	Totals	32	160			1360	2720	551	2755			60	60		
6	Sunbury County (in all)					2066	4132	12	60			50	50	½	50
	York County.														
7	St. John River District							75	375	25	50	5	5		
8	St. Croix, Magaguadavic and Oromocto Waters			4	16					10	20	10	10		
9	Southwest Miramichi Waters									40	80	2	2		
	Totals			4	16			75	375	75	150	17	17		
	Carleton County.														
10	St. John River and Westward							10	50			10	10		
11	East of St. John River											5	5		
	Totals							10	50			15	15		

												12 13			14 15

Victoria County.

12	St. John River District	16	80	8	40	15 10	15 10
13	Tobique River and Lakes						
	Totals	16	80	8	40	25	25

Madawaska County.

4	Madawaska River and Eastward					85 25	85 25
15	West of Madawaska River					110	110
	Totals					110	110

* Cwt. = 100 lbs.   † "Mixed Fish" includes greyling, bull-heads and ouananiche.

RECAPITULATION.

Of the Yield and Value of the Fisheries in **District No. 3**, Province of **New Brunswick**, during the year 1913–14.

Kinds of Fish.		Quantity.	Value.
			$
Salmon ....*Cwts.		897	13.455
Trout..	"	728	7,280
Whitefish..	"	26	390
Bass..	"	159	1,590
Pickerel...	"	528	5,280
Sturgeon..	"	70	700
Eels..	"	182	910
Perch...	"	4	16
Alewives	"	3,810	7,620
Shad, fresh..	"	788	3,940
" salted..	"	51	255
Mixed fish...	"	362	362
Caviare...	"	1½	150
			41,948

Quantity consumed in Canada........................ §
　　　　　" exported to U.S.A.......................... §

* Cwts. = 100 lbs.

RECAPITULATION.

Of the Number and Value of Boats, Nets, Traps, &c., used in the Fisheries in **District No. 3**, Province of **New Brunswick**, during the year 1913–14.

	Number.	Value.
		$
Boats (sail)...	839	8,900
" (gasoline)...	45	11,810
Gill-nets, seines and other nets..	1,740	17,660
Eel traps	98	178
Rods and lines..	2,455	5,085
Freezers and ice-houses..	18	1,050
Smoke and fish-houses...	100	1,030
Lodges built and used by native and foreign sport fishermen	55	64,000
		109,213

Number of men employed on boats 1,169

SESSIONAL PAPER No. 39

RECAPITULATION.

Of the Quantities and Values of all Fish caught and landed in a Green State, and of the Quantities and Values of all Fish and Fish Products Marketed in a fresh, dried, pickled, canned, &c., state, for whole of **New Brunswick**, during the year 1913-1914.

Kinds of Fish.	Sea Fisheries. Caught and landed in a Green State. Quantity.	Value.	Sea Fisheries. Marketed. Quantity.	Value.	Inland Fisheries. Marketed. Quantity.	Value.	Total, both Fisheries. Marketed. Quantity.	Value.	Total Marketed Value.
		$		$		$		$	$
Salmon...........cwts.	17088	256320					17985		
„ used fresh.... „			17088	256320	897	13,455	17985		
									269775
Lobsters........ „	78177	508395							
„ canned.......cases			24586	491720			24586	491720	
„ ship. in shell..cwts.			16716	201090			16716	201090	
									692810
Cod.......᷄......cwts.	240435	371569							
„ used fresh.... „			11387	22774			11387	22774	
„ green salted... „			12385	37959			12385	37959	
„ dried......... „			68094	344036			68094	344036	
									404769
Haddock......... „	12672	26820							
„ used fresh „			6682	16625			6682	16625	
„ smoked.. „			730	3650			730	3650	
„ dried..... „			1516	4786			1516	4786	
									25061
Hake.. ...... „	76620	60325							
„ used fresh.... „			5361	6707			5361	6707	
„ dried........ „			23750	81815			23750	81815	
									88522
Pollock..... „	70862	70862							
„ used fresh.. „			37747	37747			37747	37747	
„ dried. ..... „			11036	38626			11036	38626	
									76373
Herring .... „	868126	509794							
„ used fresh.. „			78847	78847			78847	78847	
„ canned ....cases			1332	6660			1332	6660	
„ smoked.....cwts.			91025	265961			91025	265961	
„ pickled.,....brls.			69177	277316			69177	277316	
„ used as bait „			72520	116556			72520	116556	
„ used as fertil. „			126890	63570			126890	63570	
									808910
Mackerel..........cwts.	17741	121457							
„ used fresh „			17057	169070			17057	169070	
„ salted.....brls.			232	2736			232	.2736	
									171806
Shad......... cwts.	2859	18522							
„ used fresh.... „			2319	16465	788	3,940	3007	20405	
„ salted.........brls.			180	2700	51	255	281	2955	
									23360
Alewives..........cwts.	37765	33669							
„ used fresh „			6386	6386	3810	7,620	10196	14006	
„ salted ... brls.			10460	46380			10460	46380	
									60386
Sardines............ „	141384	282768							
„ canned ....cases			85700	428500			85700	428500	
„ sold, fresh or salted....brls.			124084	248168			124084	248168	
									676668
Halibut... ......cwts.	723	6120	723	7230			723		7230
Flounders .... „	4261	6392	4261	8032			4261		8032

39—11R

5 GEORGE V., A. 1915

RECAPITULATION.

Of the Quantities and Values of all Fish caught and landed in a Green State, and of the Quantities and Values of all Fish and Fish Products Marketed in a fresh, dried, pickled, canned, &c., state, for the **whole** of **New Brunswick**, during the year 1913–1914.—*Concluded.*

Kinds of Fish.	Sea Fisheries.				Inland Fisheries.		Total, both Fisheries.		Total Marketed Value.
	Caught and landed in a Green State.		Marketed.		Marketed.		Marketed.		
	Quantity.	Value.	Quantity.	Value.	Quantity.	Value.	Quantity.	Value.	
		$		$		$		$	$
Smelts .. ..cwts.	60117	360934	60117	601170	...	...	60117	......	601170
Trout.............. "	2115	16920	2115	21150	728	7.280	2843	....	28430
Sturgeon.... .... "	105	525	105	525	70	700	175	....	1225
Bass............... "	1691	16910	1691	20292	159	1.590	1850	....	21882
Eels............... "	1940	11800	1940	15520	182	910	2122	....	16430
Tom Cod. ... "	18084	27126	18084	36168	...	...	18084	....	36168
White Fish........ "	...	...	...	...	26	390	26	....	390
Pickerel........ "	...	...	...	...	528	5.280	528	....	5280
Perch........ "	...	...	...	...	4	16	4	....	16
Mixed Fish........ "	180	180	180	180	362	362	542	....	542
Squid.... ......brls.	30	120	30	120	...	...	30	....	120
Oyster........ .. "	10800	54000	10800	64800	...	...	10800	....	64800
Clams and Scallops. "	56603	86472	...	...	...	...			
" used fresh... "	...	...	38070	81753	...	...	38070	81753	
" canned. .....cases	...	...	18530	90484	...	...	18530	90484	
Dulse, Cockles, etc..............cwts.	5288	7577	2788	11122	...	...	2788	....	172237
									11122
Tongues and Sounds. "	...	...	1054	19970	...	...	1054	....	19970
Caviare.. .... "	...	...	...	...	1½	150	1½	....	150
Fish Oil...........gals.	...	...	50242	15073	...	...	50242	....	15073
	...	2945577	...	4266759	...	41948	...	...	4308707

SESSIONAL PAPER No. 39

RECAPITULATION.

Of the Number of Fishermen, &c., and of the Number and Value of Fishing Vessels Boats, Nets, &c., in the **whole** Province of **New Brunswick**, for the year 1913–14.

	Sea Fisheries.		Inland Fisheries.		Total both Fisheries.	
	Number.	Value.	Number.	Value.	Number.	Value.
		$		$		$
Sailing and gasolene vessels...........	371	260,500	....	....	371	260,500
Boats (sail and row)................	7,126	286,110	839	8,900	7,965	295,010
" (gasoline).................... .	1,141	322,025	45	11,310	1,186	333,335
Carrying smacks	65	49,330	....	....	65	49,330
Gill nets, seines, trap and smelt nets,&c.	59,127	608,108	1,740	17,660	60,867	625,768
Weirs....................	469	340,885	....	....	469	340,885
Trawls ....·	834	13,396	....	....	834	13,396
Hand lines.................... ..	9,183	5,689	....	....	9,183	5,689
Eel traps	....	....	98	178	98	178
Rods and lines ....·.	....	....	2,455	5,085	2,455	5,085
Lobster traps....	252,681	252,681	....	....	252,681	252,681
" canneries..................	179	150,550	....	....	179	150,550
Sardine "	6	362,100	....	....	6.	362,100
Clam "	13	26,700	....	....	13	26,700
Freezers and ice houses........... ...	215	233,130	18	1,050	233	234,180
Smoke and fish-houses................	1,222	328,460	100	1,030	1,322	329,490
Fishing piers and wharves	370	232,120	....	....	370	232,120
Pile drivers and scows............. ...	501	19,550	....	....	501	19,550
Lodges used by fishermen..	....	....	55	64,000	55	64,000
Totals	....	3,491,334	....	109,213	....	3,600,547

	Sea Fisheries.	Inland Fisheries.	Total.
No. of men employed on vessels........	1,351	......	1,351
" " boats................	12,883	1,169	14,052
" " carrying smacks........	137	......	137
" persons " in fish houses, etc.....	6,336	......	6,336
Totals	20,707	1,169	21,876

5 GEORGE V., A. 1915

APPENDIX No. 3.

PRINCE EDWARD ISLAND.

REPORT ON THE FISHERIES OF THE PROVINCE

To the Superintendent of Fisheries,
      Ottawa.

Sir,—I have the honour to submit my annual report on the fisheries of the province of Prince Edward Island for the year 1913-14 ; also returns of catch, quantity and value of fish taken.

LOBSTERS.

The lobster pack was less than that of last season, being 17,639 cases short, and valued at $743,180.

COD.

The catch of cod this year was slightly greater than during 1912-13. There were 59,022 cwts. taken during 1913-14, against 49,876 cwts. the previous season.

HAKE.

Hake was plentiful on the south side, and but for the scarcity of bait a much larger catch would have been taken. The supply of bait controls the catch to a great extent.

HERRING.

The catch of this fish was about the same as in 1912, and a sufficient quantity was taken for lobster bait, for which they are principally used.

MACKEREL.

The catch of mackerel was 125 per cent in excess of 1912. This fish was very plentiful, being taken both by net and hook, and were very large in size, and not for many years have they been so plentiful along the coast, fishermen reaping a good return for their labour.

SMELTS.

The catch of smelts shows a considerable decrease from previous years. The prevailing mild weather in the early part of the season, and consequent losses made by shippers, are given as the reason for the decrease. The fishermen finding that their catches would only be boucht by the shippers during cold weather did not attempt to catch the same quantity as in former seasons.

OYSTERS.

The oyster fishing shows an increase of 50 per cent over 1912. The interest manifested in the protection given oyster fishing is bearing good results, and with the con tinuance of the long close season for a few years the yield will be greatly increased. Good prices were realized.

The Provincial Government has taken over the barren bottoms, and about five thousand acres have been leased for oyster culture. About seven hundred barrels have been taken from the leased areas, and no doubt in a few years the quantity will be largely increased.

CLAMS AND QUAHAUGS.

A large increase in the quantity of quahaugs taken is noted; the amount this season being 18,966 barrels, against 4,985 barrels in 1912-13.

The patrol boat service last season was satisfactory, and with the same results for the coming year I hope to see illegal lobster fishing stopped entirely.

I am, sir, your obedient servant,

J. A. MATHESON,
Inspector of Fisheries.

5 GEORGE V., A. 1915

RETURN showing the Number of Fishermen, &c., the Number and Value of Vessels and in the County of **Kings**, Province of **Prince Edward**

| | | Vessels Boats and Carrying Smacks. | | | | | | | | | | | | Fish | |
| | | Sailing and Gasoline Vessels. | | | | Boats. | | | | | Carrying Smacks. | | | Gill Nets, Seines, Trap and Smelt Nets, etc. | |
Fishing Districts.	Number.	(20 to 40 tons) Number.	(10 to 20 tons) Number.	Value.	Men.	Sail.	Value.	Gasoline.	Value.	Men.	Number.	Value.	Men.	Number.	Value.
Kings County.				$			$		$			$			$
1 Souris and Red Point.		....	3	1800	12	22	400	45	6750	144	....	....	..	163	2780
2 Fortune		....	....	....	....	18	340	15	2250	66	....	....	..	45	470
3 Annandale		....	....	....	....	55	1000	20	3000	125	....	....	..	141	1150
4 Georgetown		....	....	....	..	65	1170	20	3000	125	....	....	..	250	2000
5 Murray Harbour North		1	....	1200	9	84	1500	46	6900	180	1	200	2	325	2600
6 Murray Harbour South		8	....	4000	28	20	360	40	6000	85	1	200	1	260	2080
7 Morell and St. Peters		....	....	....	..	25	450	35	5250	120	....	....	..	325	2950
8 Naufrage		....	....	....	..	28	500	18	2700	84	....	....	..	120	960
9 North Lake		....	....	....	..	28	500	16	2400	70	....	....	..	105	840
10 East Lake		....	.	....	....	20	360	24	3600	70	..	..	:	80	640
Totals		9	3	7000	49	365	6580	279	41850	1069	2	400	3	1814	16470

Boats, and the Quantity and Value of all Fishing Gear, &c., used in the Fishing Industry Island, during the year 1913-14.

Gear.						Canneries.				Other Material.						Persons employed in Canneries, Freezers and Fish-Houses.	
Trawls.		Hand lines.		Lobster traps.		Lobster canneries.		Clam canneries.		Freezers and Ice-Houses.		Smoke and Fish-Houses.		Fishing Piers and Wharves.			
Number.	Value.	Number.	Value.	Number.	Value.	Number.	Value.	Number.	Value.	Number.	Value.	Number.	Value.	Number.	Value.	Number.	Number.
	$		$		$		$		$		$		$		$		
140	1120	70	70	9800	7350	4	2700	....	......	1	1000	4	200	1	500	110	1
15	120	60	60	8000	6000	3	2900	....	......	....	....	4	100	1	50	50	2
15	120	140	140	11000	8250	5	9800	1	50	....	....	4	100	2	500	90	3
25	200	140	140	19700	14400	6	7800	1	50	....	....	8	400	1	500	120	4
20	160	120	120	25700	19280	9	18300	2	100	....	....	10	500	....	......	110	5
200	1600	200	200	8100	6075	3	7000	1	50	....	....	12	600	1	500	80	6
140	1120	140	140	28100	21095	11	14800	....	......	....	....	12	300	1	200	160	7
10	80	70	70	15100	11320	6	6500	..	....	....	.	8	200	....	.....	110	8
15	120	70	70	9100	6820	8	6500	....	....	....	..	10	250	....	......	100	9
70	560	100	100	11000	8200	5	9900	....	....	....	....	12	300	....	......	100	10
650	5200	1110	1110	145600	108790	60	86200	5	250	1	1000	84	2950	7	2250	1030	

5 GEORGE V., A. 1915

RETURN showing the Number of Fishermen, &c., the Number and Value of Vessels Industry in the County of Queens, Province of

Number.	Fishing Districts.	Fishing Vessels, Boats, Tugs, &c.										
		Sailing and Gasoline Vessels.			Boats.					Carrying Smacks.		
		(10 to 20 tons) Number.	Value.	Men.	Sail.	Value.	Gasoline.	Value.	Men.	Number.	Value.	Men.
	Queens County.		$			$		$			$	
1	Tracadie....................	..	....	....	94	3384	68	10200	300	.....	......	....
2	New London....	5	2500	16	12	600	40	10000	104	.....	......	....
3	Point Prim...........	....	....	....	20	1000	47	8695	114	.....	......	....
4	Rustico.........	1	500	6	33	2475	60	15000	186	1	200	2
5	Wheatley River............	......	......	....	16	400	6	1025	33	.....	......	....
6	Pownal.........	......	......	....	11	300	4	1600	23	.....	......	....
7	Charlottetown.............	......	.. ....	....	20	800	6	1200	60	.....	......	....
8	Crapaud...................	......	......	....	9	290	11	2475	30	.....	......	....
9	Lot 65................... ..	......	......	....	35	1050	55	13750	180	.....	....	....
10	Bays and Rivers........	......	......	....	80	575	6	1500	92	...	.....	....
	Totals..................	6	3000	22	330	10874	303	65445	1122	1	200	2

SESSIONAL PAPER No. 39

and Boats, and the Quantity and Value of all Fishing Gear, &c., used in the Fishing
Prince Edward Island, during the year 1913-14.

Fishing Gear.								Canneries.		Other material.				Persons employed in Canneries, freezers and fish-houses.	Number.
Gill Nets, Seines, Trap and Smelt Nets, etc.		Trawls.		Hand lines.		Lobster Traps.		Lobster Canneries.		Smoke and fish-houses.		Fishing Piers and Wharves.			
Number.	Value.	Number.	Value.	Number.	Value.	Number.	Value.	Number.	Value.	Number.	Value.	Number.	Value.		
	$		$	$			$		$		$		$		
633	3665	10	200	80	48	23300	19150	8	8200	90	2600	9	9050	152	1
890	6750	11	220	150	90	9500	9500	7	3825	10	875	9	1000	69	2
70	346	10	200	65	38	19935	13345	26	7600	8	240	....	......	87	3
368	2343	20	400	200	120	16800	16800	4	8500	35	3500	2	1500	100	4
76	250	....	....	.....	......			....	....	....	....	1	400	......	5
43	187	....	....	100	50	3000	1000	1	500	....	....	....	....	10	6
......	.....	....	....					....	.....	....	....				7
58	440	....	....	...	....	4892	3525	6	2000	6	300	....	......	41	8
53	587	....	....	...	......	16600	14900	7	7700	17	272	2	1000	68	9
......	......									20	200			......	10
2191	1456?	51	1020	595	347	94027	78220	59	38325	186	7987	23	12950	527	

5 GEORGE V., A. 1915

RETURN showing the Number of Fisherman, &c., the Number and Value of Vessels and
in the County of **Prince**, Province of **Prince Edward**

	Fishing Districts.	Vessels, Boats and Carrying Smacks.										Gill Nets, Seines, Trap and Smelt Nets, etc.	
Number.		Sailing and Gasoline Vessels.			Boats.					Carrying Smacks. etc.			
		Number.	Value.	Men.	Sail.	Value.	Gasoline.	Value.	Men.	Number.	Value.	Men.	Number.	Value.
	Prince County.		$			$		$			$			$
1	Tignish.				30	2400	70	14000	220	2	500	4	105	820
2	Nail Pond				18	1440	8	1600	43				50	250
3	Skinner's Pond				54	4300	15	3000	72	1	300	3	80	320
4	Miminegash	1	400	5	20	1600	35	7000	97				70	280
5	Alberton	2	880	8	10	800	15	3000	45				100	800
6	Roxberry	1	400	4	4	320	5	1000	12				30	250
7	Brae				6	480	5	1000	20				40	320
8	West Point				3	240	14	2800	32				30	300
9	Narrows, Lot 11				15	525	3	500	23				36	180
10	Ellerslie, Lot 12				17	750	10	2000	42				75	225
11	Bideford				40	800	10	2500	82				42	210
12	Wellington				50	1010	4	890	101				10	60
13	Grand River				80	1700	4	800	170				50	260
14	Richmond Bay				40	400	54	10900	140				364	1688
15	Travellers' Rest				1	20	3	600	8				65	390
16	Malsèque	2	600	8	16	650	40	8000	96				450	4500
17	Fifteen Point				24	700	40	8200	120				294	1470
18	Summerside						3	900	6				32	192
19	Carleton				2	100	18	2700	57				66	320
20	Tryon						32	5400	61				148	740
21	Enmore				3	60	3	700	10				31	205
22	Indian River				2	40	2	500	8				25	125
	Totals	6	2280	25	415	18335	388	77900	1465	3	800	7	2193	13905

SESSIONAL PAPER No. 39

Boats, and the Quantity and Value of all Fishing Gear, &c., used in the Fishing Industry **Island**, during the year 1913-14.

Fishing Gear.										Canneries.		Other Material.						Persons Employed in Canneries, Freezers and Fish Houses.	Number.
Weirs.		Trawls.		Hand Lines.		Lobster Traps.		Lobster.		Freezers and Ice-Houses.		Smoke and Fish-Houses.		Fishing Piers and Wharves.					
Number.	Value.	Number.	Value.	Number.	Value.	Number.	Value.	Number.	Value.	Number.	Value.	Number.	Value.	Number.	Value.				
	$		$		$		$		$		$		$		$				
......	....	10	100	50	25	28000	28000	9	14400	....	....	9	1800	1	40000			220	1
......	....	8	80	20	10	5790	5790	3	6500	...	....	3	600	...	.....			58	2
......	....	40	400	100	50	12200	12200	4	4000	1	400	4	800	....	....			84	3
......	....	20	200	60	30	12000	12000	7	6550	1	400	7	1400	1	20000			72	4
......	....	38	380	40	20	6300	6300	8	3450	....	....	8	1600	1	20000			51	5
... ..	....	2	20	12	6	4500	4500	2	2000	....	....	2	400	...	.			30	6
......	....	....	....	....	....	2470	2470	4	2000	....	....	4	800	1	6000			18	7
......	....	..	...	16	8	5350	5350	7	3500	....	....	7	1400	1	15000			23	8
......	....	....	....	....	....	1700	1700	2	1000	....	....	....	...	....	.....			15	9
... ...	....	....	...	20	8	9100	6600	3	1850	....	....	....	...	1	800			39	10
......	....	....	....	....	....	1400	1100	1	760	..	...	....	...	....	...			10	11
......	.... .	....	....	....	....	5000	4000	1	600	....	....	....	...	....	.....			30	12
......	....	....	...	12	9	1500	1180	..	....	..	...	8	175	1	1600			8	13
......	....	....	....	....	....	12000	10000	2	1600	....	....	....	.	....	.....			51	14
... ..	.... ..	....	....	....	....	1500	1500	2	1500	....	....	....	..	....	....			12	15
......	....	40	120	150	150	6200	4000	3	2150	....	....	....	...	....	....			39	16
......	....	....	.	....	....	21310	17000	9	6500	....	....	....	...	....	....			104	17
......	....	....	....	....	..	900	585	1	500	....	....	1	1000	....	....			6	18
......	....	....	....	....	..	5100	4000	3	1600	....	....	...	...	....	.....			18	19
......	....	....	...	....	....	9800	7725	6	5560	....	....	...	...	....	....			42	20
......	....	..	..	....	..	2500	1970	1	200	....	....	...	...	2	2200			8	21
......	....	....	....	....	....	.....	......	..	....	....	....	...	...	....	....			5	22
......	....	158	1300	480	316	177120	137970	78	66320	2	800	53	9975	9	105600			943	

5 GEORGE V., A. 1915

THE

RETURN showing the Quantities and Values of all Fish caught and landed in a Green
Year

Number.	Fishing Districts.	Salmon, *cwt.	Salmon, value.	Lobsters, cwts.	Lobsters, value.	Cod, cwts.	Cod, value.	Haddock, cwts.	Haddock, value.	Hake and Cusk, cwts.	Hake and Cusk, value.	Herring, cwts.
	King's County.		$		$		$		$		$	
1	Souris and Red Point.........	....	....	1819	10914	1340	1675	400	280	10140	7098	5200
2	Bay Fortune...	....	....	1608	9648	200	312	10	7	300	210	1200
3	Annandale	....	....	3467	20802	200	268	15	10	450	315	2000
4	Georgetown................	....	..	4003	24018	200	300	60	42	350	245	3400
5	Murray Harbour North.......	....	....	5647	33882	200	300	40	28	200	140	5000
6	Murray Harbour South.......	...	....	2568	15408	850	1212	215	150	8660	6062	6015
7	Morell and St. Peters.........	90	900	7083	42498	1855	2444	20	14	160	112	1870
8	Naufrage.	....	....	3787	22722	100	160	10	7	120	84	1001
9	North Lake..........	...	....	3029	18174	100	150	15	10	110	77	900
10	East Lake........	...	....	3185	19110	740	1142	30	21	4·6	284	300
	Totals	90	900	36196	217176	5785	7963	815	569	20896	14627	26886

*Cwt = 100 lbs.

SESSIONAL PAPER No. 39

CATCH.

State in the County of **Kings,** Province of **Prince Edward Island,** during the 1913-14.

Herring, value.	Mackerel, cwts.	Mackerel, value.	Alewives, cwts.	Alewives, value.	Smelts, cwts.	Smelts, value.	Trout, cwts.	Trout, value.	Eels, cwts.	Eels, value.	Tom-cod, cwts.	Tom-cod, value.	Mixed fish, cwts.	Mixed fish, value.	Clams, brls.	Clams, value.	Quahaugs, brls.	Quahaugs, value.	Number.
$		$		$		$		$		$		$		$		$		$	
2600	2300	11500	....	....	80	400	5	40		50	10	10	20	10	340	680	....	....	1
600	50	250	....	....	130	650	10	80	18	25	....	..	10	5	20	40	....	....	2
1000	75	375	....	....	15	75	8	64		25	....	....	10	5	30	60	....	....	3
1700	40	200	....	....	30	150	5	40		250	....	....	15	7	80	160	....	....	4
2500	20	100	....	....	60	300	5	40		225	5	5	30	15	370	740	90	270	5
3007	15	75	....	....	140	700	12	96	3	150	10	10	70	35	135	270			6
935	1440	7200	20	20	290	1450	15	120	80	175	....	..	15	7	50	100	....	....	7
500	50	250	....	..	10	50	5	40		....	....	....	....	....	....	....	....	..	8
450	450	2250	....	..	43	215	15	120		....	....	....	....	....	....	....	....	..	9
150	1214	6070	....	....	....	....	15	120	35	175	....	....	....	....	....	....	..	10
13442	5654	28270	20	20	798	3990	95	760	215	1075	25	25	170	84	1025	2050	90	270	

39—12½

THE CATCH MARKETED.

RETURN showing the Quantities and Values of all Fish and Fish Products Marketed in a fresh, dried, pickled, canned, &c., State, for the County of **Kings**, Province of **Prince Edward Island**, during the year 1913-14.

Number.	Fishing Districts.	Salmon, used fresh and froz., *cwts.	Lobsters, canned, cases.	Cod, used fresh, cwts.	Cod, shipped green, salted, cwts.	Cod, dried, †quintals.	Haddock, used fresh, cwts.	Haddock, dried, quintals.	Hake and Cusk, dried, quintals.	Herring, used fresh, cwts.	Herring, pickled, brls.	Herring used as bait, brls.	Mackerel, used fresh, cwts.	Number.
	Kings County,													
1	Souris		728	115	160	300	50	113	3380	1300	80	1830	1000	1
2	Bay Fortune		643	15	10	60	10		100	30		685	5	2
3	Annandale		1387	15	10	50	10		150	20		900	10	3
4	Georgetown		1601	40	30	40	10	15	116	1400		1000	10	4
5	Murray Harbour, North		2259	25	20	40	30	13	66	20		2490	20	5
6	Murray Harbour, South		1027	50	60	220	40	56	2886	940	100	2385		6
7	Morell and St. Peters	90	2833	70	80	641	10	3	54	40		915	1000	7
8	Naufrage		1615	10	10	20			40	30		485	5	8
9	North Lake		1213	10	10	20	10		37			435	15	9
10	East Lake		1274	59	80	172	26	8	136	30		150	10	10
	Totals	90	14480	409	460	1453	201	208	6965	3810	180	11265	2075	
	Rates	12.00	20.00	2.00	3.00	5.00	2.00	4.00	4.00	1.50	5.00	1.50	10.00	
	Values	$ 1680	289600	818	1380	7265	402	832	27860	5715	900	16897	20750	$

† Quintal = 112 lbs.   * Cwts. = 100 lbs.

THE CATCH MARKETED.

RETURN showing the Quantities and Values of all Fish and Fish Products Marketed in a fresh, dried, pickled, canned, &c., State, for the County of Kings, Province of Prince Edward Island, during the year 1913-14.

Number	Fishing Districts	Mackerel, salted, brls.	Alewives, used fresh, cwts.	Smelts, used fresh, cwts.	Trout, used fresh, cwts.	Eels, used fresh, cwts.	Tom-cod, used fresh, cwts.	Mixed Fish, used fresh, cwts.	Clams and Quahaugs, used fresh, brls.	Clams and Quahaugs, canned, cases.	Tongues and Sounds, pickled or dried, Tons.	Fish Oil, gals.	Number
	Kings County.												
1	Souris	433		80	5	10	10	20	30		10	61	1
2	Bay Fortune	15		130	10	5		10	20		2	50	2
3	Annandale	22		15	5	5		10	15	15	2	50	3
4	Georgetown	10		30	8	50		15	40	40	2	50	4
5	Murray Harbour North	5		60	5	45		30	80	80	2	40	5
6	Murray Harbour South			140	12	30	5		90	35		50	6
7	Morell and St. Peters	147	20	290	15	35		70	50		12	87	7
8	Naufrage	15		1..	5						10		8
9	North Lake	145		43	15	35	5	15				20	9
10	East Lake	401½			15		10				12	30	10
	Totals	1193	20	798	95	215	25	170	825	290	50	1810	
	Rates	$ 18.00	2.00	8.00	10.00	10.00	2.00	1.00	4.00	6.00	30.00	0.30	
	Values	$ 21474	40	6384	950	2150	50	170	3300	1740	1500	543	

Total Value $41,800

5 GEORGE V., A. 1915

THE CATCH.

Return showing the Quantities and Values of all Fish caught and landed in a Green State, in the County of Queens, Province of Prince Edward Island, during the year 1913-14.

Number	Fishing Districts.	Lobsters,* cwts.	Lobsters, value. $	Cod, cwts.	Cod, value. $	Haddock, cwts.	Haddock, value. $	Hake and Cusk, cwts.	Hake and Cusk, value. $	Herring, cwts.	Herring, value. $	Mackerel, cwts.	Mackerel, value. $	Alewives, cwts.	Alewives, value. $	Number.
	Queens County.															
1	Tracadie	5625	33750	12246	24492			89	133	4651	2928	767	4602	130	65	1
2	New London	2090	17840	10642	21284			50	75	2606	1303	600	3600			2
3	Point Prim	2672	17232	226	452			519	778	605	302					3
4	Rustico	5362	32172	25298	50596	80	120	187	280	5306	2653	1289	7734			4
5	Wheatley River			1000	2000					477	238	208	1218			5
6	Pownal	350	2100							30	15					6
7	Charlottetown															7
8	Crapaud	1655	6330							1026	63					8
9	Lot 65	2542	192							1026	63					9
10	Bays and Rivers													402	201	10
	Totals	20796	124676	49412	98824	80	120	845	1266	15727	7965	2869	17154	532	266	

*Cwt. = 100 lbs.

THE CATCH.

RETURN showing the Quantities and Values of all Fish caught and landed in a Green State, in the County of Queens, Province of Prince Edward Island, during the year 1913-14.—*Concluded.*

Number	Fishing Districts	Smelts, cwts.*	Smelts, value $	Trout, cwts.	Trout, value $	Eels, cwts.	Eels, value $	Oysters, brls.	Oysters, value $	Clams, brls.	Clams, value $	Quahaugs, brls.	Quahaugs, value $	Number
	Queens County.													
1	Tracadie	2610	11745			27	108	1200	5400	20	40	50	290	1
2	New London	619	2785	4	22							261	1044	2
3	Point Prim	340	1530	6	33									3
4	Rustico	375	1687	17	93			499	2245	86	172			4
5	Wheatley River	270	1215											5
6	Pownal	9	40					12	54					6
7	Charlottetown	541	2434					410	1845					7
8	Crapaud	329	1480					373	1678			300	1200	8
9	Lot 65	400	1800											9
10	Bays and Rivers	376	1692					80	361					10
	Totals	5869	26408	27	148	27	108	2574	11583	106	212	611	2444	

*Cwts.=100 lbs.

THE CATCH MARKETED.

RETURN showing the Quantities and Value of all Fish and Fish Products Marketed in a fresh, dried, pickled, canned, &c., State, for the County of Queens, Province of Prince Edward Island, during the year 1913-14.

Number.	Fishing Districts.	Lobsters, canned, cases.	Cod, used fresh, cwts.	Cod, shipped green salted, cwts.	Cod, dried, quintals.	Haddock, dried, quintals.	Hake and Cusk, used fresh, cwts.	Hake and Cusk, dried, quintals.	Herring, used fresh, cwts.	Herring, pickled, brls.	Herring, used as bait, brls.	Mackerel, used fresh, cwts.	Mackerel, salted, brls.	Alewives, salted, brls.	Smelts, used fresh, cwts.	Trout, used fresh, cwts.	Eels, used fresh, cwts.	Oysters, used fresh, brls.	Clams and Quahangs, used fresh, brls.	Fish oil, gals.	Number.
	Queens County.																				
1	Tracadie	2250	535	815	3360			29	67	40	2292	200	189	43	2610	4	27	1200	70	640	1
2	New London	1196	408	749	2912			17	50	20	1248	115	162		619	6			261	1005	2
3	Point Prim	1149	226					168	20	5	285				346	17		499		238	3
4	Rustico	2145	1780	1007	7168	24	15	62	140	119	2404	332	319		375			12	86	6000	4
5	Bay River		220	390					20	10	213	100	34		270			410		325	5
6	Peal								4		13				9						6
7	Charlottetown	140													541						7
8	Crapaud	422							20	5	495				329			373	300		8
9	Lot 65	1017							10		508			134	400			80			9
10	Bays and Rivers														376						10
	Totals	8319	3170	2961	13440	24	15	276	331	199	7348	747	704	177	5869	27	27	2574	717	8208	
	Rates....$	20.00	2.00	4.00	6.00	5.00	1.50	4.00	1.00	4.00	2.00	10.00	18.00	2.00	5.00	6.00	4.00	5.00	4.00	.30	
	Values....$	166380	6340	11844	80640	120	22	1104	331	796	14796	7470	12672	354	29345	162	108	12870	2868	2462	

Total value.... $350,684

* Cwt. = 100 lbs.   † Quintal = 112 lbs.

THE CATCH.

RETURN showing the Quantities and Value of all Fish caught and landed in a Green State, in the County of Prince, Province of Prince Edward Island, during the year 1913-14.

Number	Fishing Districts	Lobsters, cwts.	Lobsters, value. $	Cod, cwts.	Cod, value. $	Hake and Cusk, cwts.	Hake and Cusk, value. $	Herring, cwts.	Herring, value. $	Mackerel, cwts.	Mackerel, value. $	Alewives, cwts.	Alewives, value. $	Smelts, cwts.	Smelts, value. $	Tom-cod, cwts.	Tom-cod, value. $	Oysters, cwts.	Oysters, value. $	Quahaugs, brls.	Quahaugs, value. $	Number
	Prince County.																					
1	Tignish	6268	31340	400	800	1000	750	1150	575	620	3100			30	150							1
2	Nail Pond	1462	7310	250	520	300	225	500	250	100	500											2
3	Skinner's Pond	2281	11405	380	760	600	300	700	350	100	500											3
4	Miminigash	1965	9825	600	300	800	400	1000	500	1000	5000			40	200							4
5	Alberton	1920	6450	500	250	700	350	800	400	600	3000			230	1150			40	280			5
6	Roxbury	768	3810	120	250	50	25	800	400	35	175			80	400			800	5600			6
7	Bras	524	2620					300	150					35	175							7
8	West Point	708	3540	76	34			300	150			40	40									8
9	Narrows, Lot 11	485	3152	510	1530			2400	2466					95	475			1874	13118	3290	12800	9
10	Ellerslie, Lot 12	1800	11700	420	1260			2229	2220					195	950			1000	7000	200	800	10
11	Bideford	350	2275					970	970					68	340			1250	8730	950	3800	11
12	Wellington	850	5525					1480	1480					52	280			3621	25347	3000	12900	12
13	Grand River							2400	2400					120	600			425	2975	1200	4800	13
14	Isl and Bay	2025	13162	559	1677			1060	770	528	1848			170	850			529	3640	7025	29100	14
15	Traveller's Rest	62?	4062					770	3440					420	2100			300	2100			15
16	Malpeque	220?	14332					16072	16072					196	980							16
17	Fifteen Point	6288	40872					320	320					450	2250							17
18	Summerside	312	2028					1380	1380					705	2525							18
19	Carleton	1455	9457					3210	3210					149	745	20	40	22	154			19
20	Tryon	335	22967					1290	1290					185	925			125	875	1320	6980	20
21	Enmore	820	5525					150	150					95	475			400	2800	34	136	21
22	Indian River																					22
	Totals	33936	214487	3925	7195	3450	2050	43583	39807	2983	14123	40	40	3110	15560	20	40	10377	72639	17129	68516	

THE CATCH MARKETED.

RETURN showing the Quantities and Values of all Fish and Fish Products Marketed in a fresh, dried, pickled, canned, etc., State, for the County of Prince, Province of Prince Edward Island, during the year 1913-14.

Number	Fishing Districts	Lobsters, canned, cases	Cod, fresh, cwts.	Cod, green-salted, shipped, cwts.	Cod, dried, quintals.	Hake and Cusk, dried, quintals.	Herring, fresh, cwts.	Herring, pickled, brls.	Herring, used as bait, brls.	Mackerel, fresh, cwts.	Mackerel, salted, brls.	Alewives, used fresh, cwts.	Smelts, used fresh, cwts.	Tom-cod, used fresh, cwts.	Oysters, used fresh, brls.	Clams and Quahaugs, used fresh, brls.	Fish Oil, gals.
	Prince County.																
1	Tignish	2507	30	185		333	10		500	10	203		30				
2	Nail Pool	584	15	122		90			250	15	28						
3	Skinners Pond	912	45	167		200			350	15	28						
4	Miminigash	786	40	280		296			500	50	316		40				
5	Alberton	516	75	213		233			400	40	186		230		40		
6	Roxbury	307	50	50		16			400	15	6		80				
7	Tée	209							150				35		800		
8	West Point	283	15	30				30	150		8						
9	Narrows, Lot 11	194	10	250					1200	11			95				
10	Ellerslie, Lot 12	720	10	205					1110								
11	Bideford	140							485				150		1874	3200	200
12	Wellington	340							740						1000	200	
13	Grand River							10	1190			40	68		1250	950	
14	Richmond Bay	810							500				52		3621	3000	
15	Travellers Rest	250							385				120		425		
16	Malpeque	882	130		143	143			1790		176		170		520	1200	400
17	Fifteen Point	2515						100	7936				420	20	300	7025	
18	Summerside	125							160				196				
19	Carleton	582							690				450		22		
20	Tryon	1368							1645				149		125	1620	
21	Enmore	340							630				185		400	34	
22	Indian River								75				95				
		14890	390	1502	143	1138	10	140	21126	156	951	40	3110	20	10377	17129	600
	Rates	20.00	2.00	4.00	6.00	4.00	1.00	4.00	2.00	10.00	18.00	2.00	5.00	2.00	7.00	4.00	.40
	Values	297200	780	6008	858	4552	10	560	42252	1560	17118	80	15550	40	72639	68516	240

Total value $517,963

SESSIONAL PAPER No. 39

RECAPITULATION.

Of the Quantities and Values of all Fish caught and landed in a Green State, and of the Quantities and Values of all Fish and Fish Products Marketed in a fresh, dried, pickled, canned, &c., State, for the Province of **Prince Edward Island**, during the year 1913-14.

Kinds of Fish.		Caught and landed in a Green State.		Marketed.		Total marketed value.
		Quantity.	Value.	Quantity.	Value.	
			$		$	$
Salmon	cwts.	90	900	....	....	
" used fresh	"	....	....	90	....	
						1,080
Lobsters	"	92,898	556,339	....	....	
" canned	cases.	....	....	37,159	....	
						743,180
Cod	"	59,022	113,982	....	....	
" used fresh	"	....	....	3,969	7,938	
" green—salted	"	....	....	4,923	19,232	
" dried	"	....	....	15,036	88,763	
						115,933
Haddock	"	895	689	....	....	
" used fresh	"	....	....	201	402	
" dried	"	....	....	232	952	
						1,354
Hake	"	25,191	17,943	....	....	
" used fresh	"	....	....	15	22	
" dried	"	....	....	8,379	33,516	
						33,538
Herring	"	85,295	61,214	....	....	
" used fresh	"	....	....	4,151	6,056	
" pickled	brls.	....	....	519	2,256	
" used as bait	"	....	....	39,789	73,945	
						82,257
Mackerel	cwts.	11,496	59,547	....	....	
" used fresh	"	....	....	2,978	29,780	
" salted	brls.	....	....	2,848	51,264	
						81,044
Alewives	cwts.	592	326	....	....	
" used fresh	"	....	....	60	120	
" salted	brls.	....	....	177	354	
						474
Smelts	cwts.	9,777	45,948	9,777	....	51,279
Trout	"	122	908	122	....	1,112
Eels	"	242	1,183	242	....	2,258
Tom-cod	"	45	65	45	....	90
Mixed Fish	"	170	84	170	....	170
Oysters	brls.	12,951	84,222	12,951	....	85,599
Clams	"	18,936	73,492	....	....	
" used fresh	"	....	....	18,671	74,684	
" canned	cases.	....	....	290	1,740	
						76,424
Tongues and Sounds	cwts.	....	....	50	....	1,500
Fish Oil	gals.	....	....	10,618	....	3,245
Totals		....	1,016,842	....	....	1,280,447

5 GEORGE V., A. 1915

RECAPITULATION.

Of the Number of Fishermen, &c., and of the Number and Value of Fishing Vessels, Boats, Nets, &c., in the Province of **Prince Edward Island,** for the year 1913-14.

	Number.	Value.
		$
Sailing and Gasoline Vessels................................	24	12,280
Boats (sail)...	1,110	35,789
Boats (gasoline)	970	185,195
Carrying Smacks.......................................	6	1,400
Gill Nets, Seines, Trap and Smelt Nets, etc....	6,198	44,923
Trawls...	859	7,520
Hand Lines...	2,185	1,773
Lobsters Traps..	416,747	324,980
Lobster-canneries.....................................	197	191,045
Clam Canneries.......................................	5	250
Freezers and Ice-houses................................	3	1,800
Smoke and Fish-houses................................	325	20,912
Fishing Piers and Wharves.............................	39	120,800
Totals.......................................		948,667

Number of men employed on Vessels......................		96
" " Boats............................		3,656
" " Carrying Smacks.................		12
" persons employed in Fish-houses, Freezers, Canneries, &c....		2,500
		6,264

APPENDIX No. 4.

QUEBEC

GULF DIVISION, SEA FISHERIES DISTRICT: COMPRISING THE COUN-
TIES OF BONAVENTURE, GASPE, SAGUENAY AND RIMOUSKI.
INSPECTOR, WM. WAKEHAM, M.D., GASPE BASIN.

GULF DIVISION, INLAND FISHERIES DISTRICT: COMPRISING THE
COUNTIES OF TEMISCOUATA, KAMOURASKA, L'ISLET, MONT-
MAGNY, CHICOUTIMI, CHARLEVOIX, MONTMORENCY AND QUEBEC.
INSPECTOR, WM. WAKEHAM, M.D., GASPE BASIN.

THE EASTERN TOWNSHIPS, INSPECTOR, C. A. BERNARD, ST. CESAIRE.

THE ST. LAWRENCE RIVER FROM HUNTINGDON COUNTY TO BELLE-
CHASSE, AND THE COUNTIES FROM PORTNEUF TO SOULANGES.
INSPECTOR, JOSEPH RIENDEAU, LONGUEUIL.

N.B.—In the inland portions of Quebec, including the inland section of the Gulf Division, the fisheries are administered by the Provincial Government. The department merely
exercises a general supervision.

To the Superintendent of Fisheries,
  Ottawa.

SIR,—I beg to submit the statistical returns, giving the quantities and value of
the various Fisheries of the Gulf Division for the season of 1913-14. These returns
show a decrease in the total value, as compared with the previous season, of $135,810.
The summer of 1913, like that of 1912, was cold and dull and fishing was late in beginning. On certain sections of the coast fishing was poor all through the season, fewer
people being engaged and the season practically ended in September.

The demand for labour at the enormous pulp works now being erected at Chandler,
in the County of Gaspe, together with the greatly increased efforts of the various
Rossing and other mills all along the coast to extend the output of pulpwood, drew
all hands away from the fishery.

SALMON.

The returns show a decided increase in the catch of salmon all over the Division,
the prices paid were in excess of any previously reached. As a consequence we have
everywhere a demand for new licenses to fish. The best salmon net fishing berths
have long ago been taken up, and, it is extremely doubtful if most of the new stations
now being applied for, will pay those who intend fishing them.

LOBSTERS.

The lobster pack again shows a decrease, only a little over 16,000 cases having
been packed, as compared with 20,000 in 1912. This falling off is generally attributed
to a poor fishing season, the weather in May and June being rough; we had no heavy
gales, and the loss of traps, due to rough weather, was below the average, but all the
same the number of days during which traps could be raised, fished and rebaited was
fewer than usual. The price of the canned product continues to rise, and the tendency to put the meat up in smaller tins is everywhere apparent. Complaints are

5 GEORGE V., A. 1915

being made by purchasers that too large a quantity of water is being put in the tin; the pound tin in some cases has been shown to contain only twelve ounces of lobster meat. If this practice extends, some remedy will have to be found by your department, with the view of protecting the consumer from so deliberate a fraud. We have raised the price of the 48-lb. case to $19, but we could very justly have put it at $20.

COD.

The cod fishery shows a considerable shrinkage, only 112,000 cwt. of dried cod having been prepared for export, as compared with 150,000 cwt. in 1912. This was due to the failure of the fishery all along the coast from Cape de Rosier West, and on the Labrador from Natashquan to the Strait of Belle Isle. On the Gaspe Coast, and at the Magdalen Islands the catch was good, but the number of hands engaged in the fishery was considerably less than usual, and as I have said before, we really have no fall fishery. The price fortunately ruled high, outside buyers, notably from Boston and Gloucester, were everywhere along the coast with schooners, competing for fish, taking it with very little cull, and in any condition. The dog fish were back in increased numbers, and were all over the gulf, even down to the Strait of Belle Isle. For the fourth season in succession, we had a failure in the inshore or summer fishing, mostly made with trap-nets, on what is known as the Canadian Labrador, or that part of the coast between Natashquan and Blanc Sablons. The fishermen insist that their failure is due to the cleaning out of the whales, by the new wholesale methods of hunting them. The capelin not being driven inshore by the whales, the cod consequently remains off shore. Trawlers and gill netters, who fish off shore, report plenty of fish in deep water.

HERRING.

The principal item of decrease for the season now being dealt with, is found in connection with the herring, the total return from all branches of this fishery being only $195,559, as against $279,614 in 1912, and practically all of this decrease was due to the falling off in the pack of pickled herring. Owing to the disease which showed itself among the summer herring, the demand for pickled herring stopped abruptly. Large quantities of these fish were found in a dull and helpless condition floating on the surface. Externally they showed conjested and even ulcerated spots towards the tail end of the fish; on opening them a conjested condition is found along the sides of the back bone, and here again the congestion seems to be confined to the after part of the fish. On exposure to the air these diseased fish decompose more rapidly than healthy fish should, a condition which usually follows death from congestion. The fishermen tell us that the herring taken well off shore are sound and good, that the diseased condition seems to be confined to the herring taken inshore. Last spring (1913) and at the time of writing (June, 1914) no disease was remarked among the schools of spring herring. Fishermen, who are ever ready with a reason for things, say that this disease is due to the gasolene escaping from the boats and floating on the surface. Whatever the cause may be, the matter is a very serious one, and it should occupy the attention of our biological branch.

MACKEREL.

The returns from the mackerel fishery show a decided gain, and there is now no doubt that the mackerel are coming back to their old haunts in the gulf. Our own people, outside of the Magdalen Islands, made no special fishery for mackerel, but a number of smacks from Nova Scotia carry on a fishery with drift nets in the Bay Chaleur, and between Gaspe and Anticosti.

WHALES.

Eighty-seven whales were killed by the two steamers operating in connection with the whaling station at Seven Islands. The returns show that the resulting product of their operations in oil, guano and bone was about the same as that of 1912, when ninety whales were killed.

BAIT.

Bait was plenty in the spring and again towards the fall, but as usual it was scarce during the summer. The bait most sought for during the late summer fishery is the squid but the fishermen say it is driven away by the dog-fish. Capelin, which for many years, had not been found along the south coast, has returned during May and the end of June in abundance. It would look as though the killing off of the whales might have something to do with the increase and return of the capelin.

INLAND SECTION OF THE GULF DIVISION.

The returns from the fishing of this section are about the same as usual. More fish have been taken in the upper section of the counties of Montmagny and Montmorency, but in the eastern part of the river from Rimouski to River Ouelle, where a considerable quantity of herring is usually taken in the Pêches, a great portion of which is usually pickled and sold in the parishes, and in Quebec, nothing whatever was done; the idea that all the herring was diseased had got abroad, and the owner of the Pêches had been warned not to send any of this herring to the usual markets. The consequence was that little or no attempt was made to take them, and whatever was taken went to the land as fertilizer.

The usual number of tourists and sportsmen visited the region between Quebec and Lake St. John, where many fishing and hunting clubs are established.

I am, sir,
Your obedient servant,

WM. WAKEHAM,

(Translation.)

To the Superintendent of Fisheries,
Ottawa.

Sir,—I have the honour to submit my report for the fiscal year ending the 1st April, 1914, as well as the statistics relating thereto. The following is the result of my personal observations, and information gathered from the fishermen and local officers.

On the Richelieu River throughout the Counties of Chambly, Verchères and St. Johns, fishing was good this year, especially at the lower end of the river, where pickerel and bass are abundant. Eel fishing at Iberville, which is becoming more difficult since the Chambly dam was built, was better and more remunerative this year.

The fishery officers are very painstaking and appear to discharge their duties. There are, however, contraventions to the law, especially in the spring when certain strangers fish in Canadian waters at the mouth of the river, and the electric company

5 GEORGE V., A. 1915

use dynamite to break the ice and open the channel, for the use of its workshops at Chambly.

In the Counties of Iberville and Mississquoi, in Rivière aux Brochets and Mississquoi Bay, fishing was better this year. Pickerel, perch and pike, especially, were very abundant. The fishery officers are diligent and the law is carefully observed.

The numerous lakes situated in the Counties of Brome, Stanstead and Sherbrooke, are very well stocked and sought after by amateurs, and the public are interested in having the law observed, thereby facilitating the task of the local officers. Salmon and trout are gradually increasing, more especially in Lake Memphremagog, and are very fine.

In the Counties of St. Hyacinthe, Bagot, Rouville and Shefford, on the Yamaska River and its tributaries, fishing was better this year and the regulations adhered to. There does not seem to have been any contravention, as complained of in past years.

I am pleased to state as regards the Counties of Richmond and Wolfe that there has been an improvement there, especially in Lake Aylmer, where the fish are increasing and the regulations better observed than in the past.

There seems to be no improvement in the Counties of Megantic and Beauce. This is the part of my district where the least progress is seen and which is the least interesting. There are very few fish there and what there is, is of poor quality. The public do not bother about fishing and do not seem to take any steps to improve the conditions. The Provincial Fishery Officers do not look after their duties and the regulations are completely ignored.

In general, fishing was better this year and the law more strictly observed. In the southeastern part of the district, particularly where the fish are more abundant, an improvement has been noticed during the past two years. The public seem to appreciate more the efforts of the Government in preserving and protecting the fish, and derive a benefit in the greater number of fishermen who come to the fishing places, as well as from the increase in the price of fish. Hence, the Provincial Officers find it easier to fulfil their duties and to have the regulations carried out.

I am sorry I cannot say as much for the lower part of the district where fish is found only in small quantities and where the regulations are not observed.

With a view to affording better protection to the fisheries in that district, I beg to repeat the recommendations made in my former reports:—

To completely forbid the use of minnow nets and of all nets in the tributaries of the St. Lawrence; if the same are not to be completely done away with they should only be allowed from the 1st October to the 31st December, and the mesh of the nets should not be less than 3 inches.

The use of seines should be prohibited during the winter, or at least, the greatest restrictions made as regards their use, and night lines should be prohibited from the 15th October to the 15th June.

I am, sir,

Your obedient servant.

(Sgd.)   C. BERNARD,

Inspector of Fisheries.

To the Superintendent of Fisheries,
Ottawa.

SIR,—I beg to submit to you my report for the fiscal year ending March 31, 1914, together with the statistical returns.

I had the opportunity, during last year, to visit the different sections of my district. In some parts of the district the yield has notably decreased in quantity as well as in quality, while elsewhere it has improved. This can be explained by the fact that all depends upon whether nets are used or not.

The great cause of trouble is the fishing with seines and nets of all descriptions. In some parts of my district fishermen have been notified, by myself and the overseers, of the harm done to the fish by using illegal implements during closed seasons when the fish is in an immature condition, or when it is not even able to ascend the small rivers or tributaries to spawn. It has had no good effect however upon them.

Lake St. Peter, Counties of Laprairie, Chambly, Verchères, Nicolet, Maskinonge, Berthier, L'Assomption, where the abuses are the worst, are fished freely on Sundays, as well as on any other day, with nets of small mesh. The result of this is shown in the decreased yield.

In other sections like Lake St. Louis, Lake St. Francis, County of Two Mountains, Richelieu, Jacques Cartier, Vaudreuil, Soulanges, owing to a great decrease in the number of seines and nefarious nets, there is a marked improvement in the fishing grounds. Though there are still nets for minnows, it must be acknowledged that the law is better observed than previously.

Where licenses prevail, the fishery officers are more active and the public more careful to comply with the regulations and protect the fish.

In some small rivers, such as Lachenaie and L'Assomption, where game fish go to spawn and in other streams in the southern part of my district, licenses are still being granted to fish with nets, which, in my opinion, should be forbidden, both in the tributaries of the St. Lawrence and in other small streams.

The law should not allow fishing with hoop nets or seines of less than three inches extension measure; while gill nets and trap nets should not be permitted in small rivers.

Nets of any kind, especially in small streams, constitute a general blockade and prevent the fish from ascending during the spawning season; the immature fish descending to the deeper waters are also entangled in the meshes.

Eels have been noticeably more abundant this last year, especially in Bellechasse County, than in previous years. Mixed fish are also more numerous.

I respectfully suggest that fishing through ice in the winter, if not prohibited, should at least be closely watched, as large quantities of immature and sometimes game fish are destroyed by such fishing.

During the year, I have visited a number of sawmills, fined the proprietors who permitted sawdust and refuse to be thrown into the streams, ordering them, at the same time, to take the necessary steps to completely stop the nuisance.

The only remedy I can see against the abuse of net fishing, and to replenish our rivers, would be to prohibit the use of nets from the County of Champlain on the north shore and the County of Nicolet on the south shore of the St. Lawrence River and its tributaries for a certain number of years, along both shores up to the Canadian Pacific Railway Bridge at Lachine.

Night lines should also be prohibited from the 15th of October to the 15th of June of each year.

Fish-ways that have been built recently work satisfactorily.

I am, sir,
Your obedient servant,

JOS. RIENDEAU,
Inspector of Fisheries, Longueuil.

5 GEORGE V., A. 1915

RETURN showing the Number of Fishermen, &c., the Number and Value of Vessels and Boats, and the Quantity and Value of all Fishing Gear, &c., used in the Fishing Industry in the Gulf Division, Province of Quebec, during the year 1913-14.

COUNTY OF BONAVENTURE.

| | | Steam Vessels. | | | | Vessels, Boats and Carrying Smacks. | | | | | | | | | | | | | | | Fishing Gear. | | |
| | | | | | | Sailing and Gasoline Vessels. | | | | | Boats. | | | | | Carrying Smacks. | | | Gill Nets, Seines, Trap and Smelt Nets, etc. | | |
Fishing Districts.	Number.	Number.	Tons.	Value. $	Men.	Number (40 tons and over).	Number (20 to 40 tons).	Number (10 to 20 tons).	Value. $	Men.	Sail.	Value. $	Gasoline.	Value. $	Men.	Number.	Value. $	Men.	Number.	Value. $	Number.
1 Subdiv. of Restigouche											53	1600		1800	60				63	3000	1
2 " Bonaventure						5			10000	25	150	5000	2	1800	220	2	500		350	3400	2
3 " Port Daniel								1	500	3	220	11500	3	1500	420				390	5700	3
Totals						5		1	10500	28	423	18100	5	3300	700	2	500	5	803	12100	

COUNTY OF GASPÉ.

| | | Steam Vessels. | | | | Vessels, Boats and Carrying Smacks. | | | | | | | | | | | | | | | Fishing Gear. | | |
Fishing Districts.	Number.	Number.	Tons.	Value. $	Men.	Number (40 tons and over).	Number (20 to 40 tons).	Number (10 to 20 tons).	Value. $	Men.	Sail.	Value. $	Gasoline.	Value. $	Men.	Number.	Value. $	Men.	Number.	Value. $	Number.
4 Subdiv. of Grand River											483	36180	9	2700	1057	2	450	4	1076	20812	4
5 " Gaspe Bay											832	68250	56	16400	1724				1913	39990	5
6 " Monts Louis and St. Anns						6			3750	30	500	12370	20	9750	730				1350	16374	6
7 Magdalen Islands S						3			2400	23	501	20040	30	6000	1566	4	3000	8	4312	52250	7
8 " N											318	9930	40	10500	725				960	17550	8
Totals						9			6150	53	2634	147070	155	45560	5498	6	3450	12	9611	149977	

SESSIONAL PAPER No. 39

COUNTY OF SAGUENAY.

	Subdivision	C1	C2	C3	C4	C5	C6	C7	C8	C9	C10	C11	C12	C13	C14	C15	C16	C17	C18
9	Subdiv. of Godbout	2	300	40000	20		11	4150	24	200	5900	3	1200	210				363	12290
10	Moisie									35	2500	3	750	60				60	7500
11	Mingan									184	16800			450				65	3429
12	Natashquan									70	5600	3	2000	160				60	4525
13	Romaine									31	1870	2	1200	53				68	910
14	St. Augustin					3		4500	18	176	7790	4	960	255	1	4000	2	180	12600
15	Bonne Esperance	1	60	8000	6					286	13350	16	6550	456				165	47493
16	Island of Anticosti									45	2600	1	1200	60				44	2500
	Totals	3	360	48000	26	3	11	8650	42	1027	55110	32	13860	1704	1	4000	2	1011	90550

COUNTY OF RIMOUSKI.

	Subdivision	C9	C10	C11	C12	C13	C17	C18
17	Subdiv. of Rimouski County	129	1872	4	3400	85	131	2425

39—13½

RETURN showing the Number of Fishermen, &c., the Number and Value of Vessels and Boats, and the Quantity and Value of all Fishing Gear, &c., used in the Fishing Industry in the Gulf Division, Province of Quebec, during the year 1913-14.

COUNTY OF BONAVENTURE.

Number	Fishing Districts.	Weirs Number	Weirs Value	Trawls Number	Trawls Value	Haul lines Number	Haul lines Value	Lobster Traps Number	Lobster Traps Value	Lobster Canneries Number	Lobster Canneries Value	Whale Factory Number	Whale Factory Value	Freezers and Ice-houses Number	Freezers and Ice-houses Value	Smoke and Fish-houses Number	Smoke and Fish-houses Value	Fishing piers and wharves Number	Fishing piers and wharves Value	Persons employed in Canneries, Freezers and Fish-houses.	Number
1	Subdiv. of Restigouche	20	230											20	500						1
2	" Bonaventure			23	110	360	180	1500	1500	3	900			8	250	70	7500			100	2
3	" Port Daniel			135	700	780	390	7500	7500	7	2500			10	650	40	3500	2	18000	130	3
	Totals	20	230	158	810	1140	570	9000	9000	10	3400			38	1400	110	11000	2	18000	230	

COUNTY OF GASPÉ.

Number	Fishing Districts.	Weirs Number	Weirs Value	Trawls Number	Trawls Value	Haul lines Number	Haul lines Value	Lobster Traps Number	Lobster Traps Value	Lobster Canneries Number	Lobster Canneries Value	Whale Factory Number	Whale Factory Value	Freezers and Ice-houses Number	Freezers and Ice-houses Value	Smoke and Fish-houses Number	Smoke and Fish-houses Value	Fishing piers and wharves Number	Fishing piers and wharves Value	Persons employed in Canneries, Freezers and Fish-houses.	Number
4	Subdiv. of Grand River			102	2600	2575	2250	10540	10540	13	1300			17	1350	109	81400	5	3000	402	4
5	" Gaspe Bay					4515	5645	4250	4250	6	1900			49	2850	468	76675	16	31545	366	5
6	" Mont Louis and St. Anns					1350	2700	350	350	1	200			5	500	25	7500	5	2000	30	6
7	Magdalen Islands South			170	800	3580	1430	50450	50450	15	1650			11	4300	87	9300	6	3000	294	7
8	" North			135	680	750	375	69800	69800	25	15600			2	2800	9	3000	9	4900	97	8
	Totals			407	3580	12372	12400	141360	141360	60	38950			84	11800	698	177875	41	44445	1189	

COUNTY OF SAGUENAY.

9	Subdiv. of Godbout.	80	8000	16	95	400	200	200	200	1	350	....		50000	92	3500	10	210	1	170	30	9
10	"	....	....	10	50	120	75	....	....	....	....	1	50000		20	1000	25	670	1	150	60	10
11	" Mingan	....	....	....	..	1800	1169	40	400	1	250	....	....		30	1500	50	30000	6	3000	10	11
12	" Natashquan	....	....	....	....	494	247	1675	1675	8	2375	....	....		2	850	91	14775	10	2050	40	12
13	" Romaine	....	....	....	....	63	63	2380	2380	7	550	....	....		1	300	15	2500	17	800	56	13
14	" St. Augustin	....	....	....	....	1408	704	2340	2340	26	1500	....	..		....	....	50	1000	130	2500	57	14
15	" Bonne Esperance	....	....	85	480	1188	299	25	25	1	100	....	....		....	....	60	15000	80	1000	53	15
16	" Island of Anticosti	....	....	....	...	50	30	3000	3000	1	7500	....	....		2	1600	6	1500	2	50000	25	16
	Totals	80	8000	111	625	5523	2787	10020	10020	45	12625	1	50000		147	8150	307	65655	247	68670	331	

COUNTY OF RIMOUSKI.

17	Subdiv. of Rimouski County	43	3556	....	....	200	200	....	..	....	....	....	....		8	800	5	1400	....	....	26	17

THE CATCH

RETURN showing the Quantities and Values of all Fish caught and landed in a Green State in the **Gulf Division**, Province of **Quebec**, during the year 1913-14.

COUNTY OF BONAVENTURE.

Number.	Fishing Districts.	Salmon, cwts.	Salmon, value.	Lobsters, cwts.	Lobsters, value.	Cod, cwts.	Cod, value.	Haddock, cwts.	Haddock, value.	Hake and Cusk, cwts.	Hake and Cusk, value.	Herring, cwts.	Herring, value.	Mackerel, cwts.	Mackerel, value.	Halibut, cwts.	Halibut, value.	Flounders, cwts.	Flounders, value.	Number.
1	Subdivision of Restigouche	1620	19440									11800	5900							1
2	" Bonaventure	550	6600	750	3750	11500	11500	1000	750	1800	900	69250	34625					500	250	2
3	" Port Daniel	830	9960	1380	6900	22250	22250	1650	1237	600	300	23725	11862	661	4305					3
	Totals	3000	36000	2130	10650	33750	33750	2650	1987	2400	1200	104775	52387	661	4305			500	250	

COUNTY OF GASPÉ.

Number.	Fishing Districts.	Salmon, cwts.	Salmon, value.	Lobsters, cwts.	Lobsters, value.	Cod, cwts.	Cod, value.	Haddock, cwts.	Haddock, value.	Herring, cwts.	Herring, value.	Mackerel, cwts.	Mackerel, value.	Halibut, cwts.	Halibut, value.	Number.
4	Subdivision du Grand River	372	4464	2245	11225	70100	70100	930	697	29760	14380					4
5	" Gaspé	1200	14400	770	3850	78005	78005	1100	825	30810	15405					5
6	Subdiv. of Monts Louis and St. Anne	390	4680	50	250	13500	13500			17400	8700					6
7	" Magdalen Islands South			11238	56190	47558	47558			66700	33330	17529	87645	154	1232	7
8	" North			20825	104125	1800	1800			72206	36103	5268	26040			8
	Totals	1962	23544	35128	175640	211563	211563	2030	1522	216876	107938	22735	113685	154	1232	

COUNTY OF SAGUENAY.

		9	10	11	12	13	14	15	16
9	Subdivision of Golbout	1632	19584	10	50	594	594	1690	849	52	416
10	Moisie	2051	24613			1540	1540	108	54	12	96
11	Mingan	718	6616	33	175	32380	32380	60	30	18	144
12	Natashquan	547	6364	231	1155	3000	3000	630	345	20	160
13	Romaine	501	6048	1097	5035	951	951	1200	600	12	96
14	St. Augustin	1236	14832	1140	5700	14904	14904	3270	1635		
15	Bonne Espérance	624	7488	9	45	65400	65400	600	300	24	192
16	Island of Anticosti	106	1272	1620	8100	810	810	16940	7970	50	400
	Totals	7418	89016	4652	20260	119539	119539	23548	11774	188	1504

COUNTY OF RIMOUSKI.

								17
17	County of Rimouski	296	3552		200	200	19450	19725	45	360

*Cwt. = 100 lbs.

THE CATCH.

RETURN showing the Quantities and Values of all Fish caught and landed in a Green State in the Gulf Division, Province of Quebec, during the year 1913-14.—*Concluded.*

COUNTY OF BONAVENTURE.

Number.	Fishing Districts.	Smelts, cwts.	Smelts, value.	Trout, cwts.	Trout, value.	Capelin, cwts.	Capelin, value.	Eels, cwts.	Eels, value.	Tom-cod, cwts.	Tom-cod, value.	Mixed fish, cwts.	Mixed fish, value.	Oysters, cwts.	Oysters, value.	Clams, cwts.	Clams, value.	Whales, number.	Whales, value.	Number.
1	Subdivision of Restigouche....	6000	36000	100	1000					250	250									1
2	" Bonaventure........	2000	12000	60	600	12000	3000	120	600	70	70					230	600			2
3	" Port-Daniel........	1750	10500	30	300											500	1500			3
	Totals	9750	58500	190	1900	12000	3000	120	600	320	320					730	2190			

COUNTY OF GASPE.

Number.	Fishing Districts.	Smelts, cwts.	Smelts, value.	Trout, cwts.	Trout, value.	Capelin, cwts.	Capelin, value.	Eels, cwts.	Eels, value.	Tom-cod, cwts.	Tom-cod, value.	Mixed fish, cwts.	Mixed fish, value.	Oysters, cwts.	Oysters, value.	Clams, cwts.	Clams, value.	Whales, number.	Whales, value.	Number.
4	Subdivision of Grand River........	870	5220	50	500											50	150			4
5	" Gaspé	1200	7200	75	750															5
6	Subdiv. of Monts Louis and St. Anns			50	500															6
7	Magdalen Islands South.....							110	550							100	300			7
8	" " North.....															30	90			8
	Totals	2070	12420	175	1750			110	550							180	540			

COUNTY OF SAGUENAY.

No.	Subdivision													9	10	11	12	13	14	15	16	—
9	Subdivision of Godbout	36	216	60	600								65	195	87						
10	" Moisie			40	400											60900					
11	" Mingan	15		40	400	5000	1250		2500	5000			75	225							
12	" Natashquan		90	35	350	6000	1500						30	90							
13	" Romaine			21	210	1000	250														
14	" St. Augustin			43	430	540	125		540	1080			31	93							
15	" Bonne Espérance			26	260	600	150														
16	" Island of Anticosti			30	300																
	Totals	51	306	295	2950	13100	3275		3040	6080			201	603	87	60900					

COUNTY OF RIMOUSKI.

No.											17
17	County of Rimouski	275	1650	50	500	55	275	300	600		

THE CATCH MARKETED.

RETURN showing the Quantities and Value of all Fish and Fish Products Marketed in a fresh, dried, pickled, canned, &c., State for the Gulf Division, Province of Quebec, during the year 1913-14.

COUNTY OF BONAVENTURE.

Number	Fishing Districts	Salmon used fresh and frozen, cwts.	Salmon salted, cwts.	Lobsters, canned, cases.	Lobsters shipped in shell, cwts.	Cod used fresh, cwts.	Cod shipped green-salted, cwts.	Cod dried, †quintals.	Haddock used fresh, cwts	Haddock dried, quintals.	Hake and Cusk dried, quintals.	Herring used fresh, cwts.	Herring smoked, cwts.	Herring pickled, brls.	Herring used as bait, brls.	Herring used as fertilizer, brls.	Mackerel used fresh, cwts.	Mackerel salted, brls.	Number
1	Subdiv. of Restigouche	1620		300			500	3500	10	330	600	100		100		5700			1
2	" Bonaventure	550		540	3.		2000	6083	25	530	200	150	300	200	500	33450		262	2
3	" Port Daniel	830			30							75		200	400	11125	75		3
	Totals	3000		840	30		2500	9583	35	860	800	325	300	500	900	50275	75	262	
	Rates	14		19	8		3	6	2	4	3	1.50	2	4	1.50	.50	10	16	
	Values.......$	42000		15960	240		7500	57498	70	3440	2400	487	600	2000	1350	25137	750	4192	

COUNTY OF GASPÉ.

Number	Fishing Districts	Salmon used fresh and frozen, cwts.	Salmon salted, cwts.	Lobsters, canned, cases.	Lobsters shipped in shell, cwts.	Cod used fresh, cwts.	Cod shipped green-salted, cwts.	Cod dried, †quintals.	Haddock used fresh, cwts	Haddock dried, quintals.	Hake and Cusk dried, quintals.	Herring used fresh, cwts.	Herring smoked, cwts.	Herring pickled, brls.	Herring used as bait, brls.	Herring used as fertilizer, brls.	Mackerel used fresh, cwts.	Mackerel salted, brls.	Number
4	Subdiv. of Grand River	572		886	30	100	3000	20000		310		100		400	4000	9730			4
5	" Gaspé	1200		301	10	150	2200	24684		366		150		500	4500	10080			5
6	" Mont Louis and Ste. Anne.	390		20			1000	3833				100	400	500	1000	980			6
7	" Magdalen Islands South			4496			1000	15186				200		500	22500	6000		5843	7
8	" Magdalen Islands North			8330			1090	600				150		5200	21000	7228		1736	8
	Totals	1962		14035	40	250	9290	64303		676		700	400	7100	53000	39938		7579	
	Rates	14		19	8	1.50	3	6		4		1.50	2	4	1.50	.50		16	
	Values.......$	27468		266665	320	375	27000	385818		2704		1050	8000	28400	79500	19969		121264	

COUNTY OF SAGUENAY.

	9	10	11	12	13	14	15	16						
9	Subdiv. of Godbout..								1632				297	
10	" Moisie..								2051				400	
11	" Mingan..								718	14			1189	232
12	" Natashquan..								547	93				10000
13	" Romaine..									403		336		1000
14	" St. Augustin..									456		824		317
15	" Bonne Esperance..									4		416		4868
16	" Island of Anticosti..	10							106	640	20			21800
														270
	Totals	30	1610	1576	5074	1896	38587		2596	7880				
	Rates.. $	8	19	10	14	3	6		4.	1.50				
	Values.. $	240	30590	15760	70756	5658	231522		10384	11820				

COUNTY OF RIMOUKI.

	17						
17	County of Rimouski..			206	100	300	500
	Rates.. $			14	3	1.50	4.
	Values.. $			4144	300	450	2000

	17
	8825
	.50
	4412

*Cwt 100 lbs.   †Quintal = 112 lbs.

5 GEORGE V., A. 1915

THE CATCH MARKETED.

RETURN showing the Quantities and Values of all Fish and Fish Products Marketed in a fresh, dried, pickled, canned, &c., State for the Gulf Division, Province of Quebec, during the year 1913-14—Con.

COUNTY OF BONAVENTURE.

Number	Fishing Districts	Whale oil, galls.	Fish oil, galls.	Beluga-skins, number.	Hair seal skins number.	Whale bone, cwts.	Tongues and Sounds pickled or dried, cwts.	Clams and Quahaugs used fresh, brls.	Whale fertilizer used fresh, tons.	Capelin, used as bait or fertilizer, brls.	Mixed-fish used fresh, cwts.	Tom-cod used fresh, cwts.	Eels, used fresh cwts.	Trout used fresh, cwts.	Smelts used fresh, cwts.	Flounders used fresh, cwts.	Halibut, used fresh, cwts.
1	Subdiv. of Restigouche		3200				20	230		12000		250	120	100	6000	500	
2	Bonaventure		6300				15	500				70		60	2000		
3	Port Daniel													30	1750		
	Totals		9700				35	730		12000		320	120	190	9750	500	
	Rates $		.30				5	4		25		1	5	12	8	.50	
	Values $		2910				175	2920		3000		320	600	2280	78000	250	
	Total value		$ 254,079														

COUNTY OF GASPÉ.

Number	Fishing Districts	Whale oil, galls.	Fish oil, galls.	Beluga-skins, number.	Hair seal skins number.	Whale bone, cwts.	Tongues and Sounds pickled or dried, cwts.	Clams and Quahaugs used fresh, brls.	Whale fertilizer used fresh, tons.	Capelin, used as bait or fertilizer, brls.	Mixed-fish used fresh, cwts.	Tom-cod used fresh, cwts.	Eels, used fresh cwts.	Trout used fresh, cwts.	Smelts used fresh, cwts.	Flounders used fresh, cwts.	Halibut, used fresh, cwts.
4	Subdiv. of Grand-River		17560				13	50						50	870		
5	Gaspé Bay		18750					100					110	75	1200		
6	Mont Louis & Ste. Anne		2300				15	30						50			
7	Magdalen Islands South		12875		275			180						175	2070		154
8	Magdalen Islands North		400					4.					5.	12.	8.		
	Totals		51775		275		28	180					110	175	2070		154
	Rates $		.30		1.50		5.	4					5	12	8		8
	Values $		15532		413		140	720					550	2100	16560		1232
	Total value		$ 1,006,380														

COUNTY OF SAGUENAY.

	9	10	11	12	13	14	15	16
9	Subdiv. of Godbout	52	36	60				2963
10	" Moisie	12		40	2300	5000	244	658	147560
11	" Mingan	18		40		6000		9195
12	" Natashquan	20	15	35		1000		1731
13	" Romaine	12		21		500	65	410
14	" St. Augustin			53	540	600		11943
15	" Bonne Esperance	24		26				2025
16	" Anticosti	50		30			31	
	Totals	188	51	295	3040	13100	201	28925	147560
	Rates $	8.	8.	12.	3.	.25	4	.30	.30
	Values $	1504	408	3540	9120	3275	804	8677	44268

Total value $400,481

COUNTY OF RIMOUSKI.

					17
17	County of Rimouski	45	275	50	55	300
	Rates $	8.	8.	12	5.	3.
	Values $	360	2200	600	275	900

Total value $15,641

5 GEORGE V., A. 1915

RECAPITULATION.

Of the Quantities and Values of all Fish caught and landed in a Green State, and of the Quantities and Values of all Fish and Fish Products Marketed in a fresh, dried, pickled, canned, &c., state, for the **Gulf Division** (Sea Fisheries District) Province of **Quebec**, during the Year 1913-14.

Kinds of Fish.		Caught and landed in a green state.		Marketed.		Total marketed value.
		Quantity.	Value.	Quantity.	Value.	
			$		$	$
Salmon................ cwts.		12,676	152,112	....	......	
" used fresh............., ... "		....	....	10,312	144,368	
" salted (dry)................... "		....	....	1,576	15,760	
						160,128
Lobsters................ "		41,310	206,550	....	....	
" canned.... cases.		....	....	16,485	313,215	
" shipped in shell .... cwts.		....	....	100	800	
						314,015
Cod.... .... "		365,052	365,052	....	....	
" used fresh ...... "		....	....	250	375	
" green—salted........... "		....	....	13,686	41,058	
" dried................ "		....	....	112,473	674,838	
						716,271
Haddock............ "		4,680	3,509	....	....	
" used fresh........... "		....	....	35	70	
" dried.... .... "		....	....	1,536	6,144	
						6,214
Hake.... .... "		2,400	1,200	....	....	
" dried........... "		....	....	800	.. ...	
						2,400
Herring........ "		363,649	181,824	....	... .	
" used fresh........ "		....	....	1,325	1,987	
" smoked........... "		....	....	4,300	8,600	
" pickled........ brls.		....	....	10,696	42,784	
" used as bait...... "		....	....	61,780	92,670	
" used as fertilizer........ "		....	....	99,038	49,518	
						195,559
Mackerel........... cwts.		23,598	117,990	....	......	
" used fresh........ "		....	....	75	750	
" salted................ brls.		....	....	7,841	125,456	
						126,206
Halibut, used fresh........ cwts.		387	3,096	387	....	3,096
Flounders.............. "		500	250	500	....	250
Smelts........ "		12,146	72,876	12,146	....	97,168
Trout.............. "		710	7,100	710	....	8,520
Eels........ "		285	1,425	285	....	1,425
Tom Cod .... "		320	320	320	....	320
Mixed Fish......... "		3,340	6,680	3,340	....	10,020
Clams .... brls.		1,111	3,333	....	... .	
" used fresh........... "		....	....	1,111	... .	
						4,444
Capelin........ cwts.		25,100	6,275	25,100	....	6,275
Tongues and Sounds .... "		....	....	63	....	315
Hair Seal Skins .... No.		....	....	4,872	....	7,308
Beluga Skins... .... "		....	....	75	....	375
Whales........ "		87	60,900	....	....	27,119
Fish Oil........ galls.		....	....	90,400	....	27,119
Whale Oil........ "		....	....	147,560	....	44,268
Whale Bone .... cwts.		....	. ....	240	....	4,800
Whale Fertilizer. .... tons.		....	....	244	....	85
Totals........		....	1,190,492	....	....	1,736,581

RECAPITULATION.

OF the Number of Fishermen, etc., and of the Number and Value of Fishing Vessels, Boats, Nets, etc., in the **Gulf Division** (Sea Fisheries District), Province of **Quebec**, for the year 1913-14.

—	Number.	Value.
		$
Steam Fishing Vessels (tonnage 360)	3	48,000
Sailing and Gasoline Vessels	29	25,300
Boats (sail)	4,213	222,152
" (gasoline)	196	65,900
Carrying Smacks	9	7,950
Gill Nets, Seines, Trap and Smelt Nets, etc	11,556	255,052
Weirs	143	11,780
Trawls	676	5,015
Hand Lines	19,235	15,957
Lobster Traps	160,380	160,380
" Canneries	115	54,975
Freezers and Ice-houses	277	22,150
Smoke and Fish-houses	1,120	255,930
Fishing Piers and Wharves	290	131,115
Whaling Station	1	50,000
Totals	. . . .	1,331,656

Number of men employed on Vessels	149
" " Boats	7,985
" " Carrying Smacks	19
" persons employed in Fish-houses, Freezers, Canneries, &c	1,776
	9,929

5 GEORGE V., A. 1915

INLAND FISHERIES.

RETURN showing the Number of Fishermen, etc., the Number and Value of Tugs, Vessels and Boats, and the Quantity and Value of all Fishing Gear and other Material used in the Fishing Industry, in the Inland Section of the Gulf Division, Province of Quebec, during the Year 1913-14.

Number	Fishing Districts.	Boats. Sail and row.	Boats. Value.	Boats. Gasoline.	Boats. Value.	Men.	Gill-Nets. Number.	Gill-Nets. Value.	Lines. Number.	Lines. Value.	Weirs. Number.	Weirs. Value.	Freezers and Ice-Houses. Number.	Freezers and Ice-Houses. Value.	Fishing Clubs. Number.	Fishing Clubs. Value.	Piers and Wharves. Number.	Piers and Wharves. Value.	Persons employed in Freezers, Fish-houses, etc. Number.	Number.
			$		$			$		$		$		$		$		$		
1	Temiscouata County	12	250	2	1500	21					13	1800	2	500	4	800			12	1
2	Kamouraska "	19	1400	3	2300	43					19	2800			8	1400			8	2
3	L'Islet "	24	1800	1	950	15					14	3200								3
4	Montmagny "	19	950	4	2300	55	7	200			18	2400								4
5	Charentini	20	500	3	1800	26														5
6	Charlevoix	12	3000	2	1200	19	10	300			14	1500	10	1000	10	5000				6
7	Montmorency "	1	300	2	900	4					7	1400			25	10000	30	750		7
8	Quebec	100	2500	10	9000	100			300	2400										8
	Totals	207	10700	27	19850	247	17	500	300	2400	84	13100	12	1500	47	17200	30	750	20	

INLAND FISHERIES

RETURN showing the Quantities and Values of all Fish caught and marketed or consumed locally, for the Inland Section of the Gulf Division, Province of Quebec, during the Year 1913-14.

Numbers	Fishing Districts	Salmon, cwts*	Salmon, value	Trout, cwts	Trout, value	Whitefish, cwts	Whitefish, value	Striped Bass, cwts	Striped Bass, value	Pickerel, cwts	Pickerel, value	Pike, cwts	Pike, value	Sturgeon, cwts	Sturgeon, value	Eels, cwts	Eels, value	Herring, cwts	Herring, value	Shad, cwts	Shad, value	Smelt, cwts	Smelt, value	†Mixed Fish, cwts	Mixed Fish, value	Numbers
1	Temiscouata County	12	180	12	180									5	25	15	75			10	100	20	60			1
2	Kamouraska	14	210	4	60									39	195	45	225	180	360	12	120	30	90			2
3	L'Islet	2½	30			12	120	4	72	15	180			32	160	39	195	516	1032	2	20	10	30	50	100	3
4	Montmagny	5	75			18	180	13	234	20	240			253	1245	195	975			14	140	50	150	130	260	4
5	Chicoutimi			13	195			82	1476	25	300															5
6	Charlevoix	17	255	20	300	5	50					50	250	48	240	5	25			12	120	20	60	50	500	6
7	Montmorency	5	75	500	5000	19	190	25	450	30	360			124	620	245	1225	200	400	10	130	25	75	150	3000	7
8	Quebec																					300	900			8
	Totals	53	825	549	5735	54	540	124	2232	90	1089	50	250	501	2485	544	2720	896	1792	60	630	455	1365	380	6200	

*Cwt. = 100 lbs.   †"Mixed Fish" includes greyling, bull-head, ouananiche, suckers, perch and tom-cod.

5 GEORGE V., A. 1915

RETURN showing the Number of Fishermen, etc., the Number and Value of Tugs, Vessels and Boats, and the Quantity and Value of all Fishing Gear and other Material, used in the Fishing Industry in the **Eastern Townships,** Province of **Quebec.** during the year 1913-14.

Number.	Fishing Districts.	Boats.					Seines.		Hoop-Nets.		Lines.	
		Sail.	Value.	Galo ine.	Value.	Men.	Number.	Value.	Number.	Value.	Number.	Value.
	Eastern Townships.		$		$			$		$	$	$
1	Counties of Chambly and St. John	42	630	2	250	42	5	200	54	810	20	100
2	Counties of Missisquoi and Iberville ..	40	800	3	375	57	28	1120	....		12	60
3	Counties of Stanstead and Sherbrooke....	25	500	5	625	35	....	....	..........'......		30	150
4	Counties of Shefford and Brome........ .	20	400	1	175	25	....	....	....	......	18	90
5	Counties of Yamaska, St. Hyacinthe, Bagot and Rouville	50	750	6	750	50	12	180	65	650	25	125
6	Counties of Richmond, Wolfe, Megantic and Beauce	25	250	1	150	25	....	....	... .		20	100
	Totals.....	202	3330	18	2325	234	45	1500	119	1460	125	625

RETURN showing the Quantities and Values of all Fish marketed or consumed locally, for the **Eastern Townships**, Province of Quebec, during the year 1913-14.

Number	Fishing Districts	Salmon, *cwts.	Salmon, value. $	Trout, cwts.	Trout, value. $	Whitefish, cwts.	Whitefish, value. $	Bass, cwts.	Bass, value. $	Pickerel, cwts.	Pickerel, value. $	Pike, cwts.	Pike, value. $	Sturgeon, cwts.	Sturgeon, value. $	Eels, cwts.	Eels, value. $	Perch, cwts.	Perch, value. $	Maskinongé, cwts.	Maskinongé, value. $	Mixed Fish, cwts.	†Mixed Fish, value. $	Number.
	Eastern Townships.																							
1	Counties of Chambly and St. John							55	550	47	470	180	1440	25	250	160	1440	460	2300	18	180	672	3360	1
2	Counties of Missisquoi and Iberville							45	450	600	7200	70	560					500	3000			775	4650	2
3	Counties of Stanstead and Sherbrooke	30	600	158	1896	88	880	65	650	890	8960	90	720			42	578	155	930			320	1920	3
4	Counties of Shefford and Brome			60	720	70	700	78	936	72	720	142	994	20	200	35	315	90	540			120	600	4
5	Counties of Yamaska, St. Hyacinthe, Bagot and Rouville					10	100	44	528	50	500	75	525			43	344	130	650	30	300	395	1975	5
6	Counties of Richmond, Wolfe, Megantic and Beauce			10	120			15	150	30	300	50	300			15	105	50	250			75	375	6
	Totals	30	600	228	2736	168	1680	302	3264	879	10150	607	4539	45	450	295	2782	1385	7670	48	480	2357	12880	

* Cwt. = 100 lb.   † "Mixed Fish" includes greylings, bull-heads and ouananiche.

39—14½

RETURN showing the Number of Fishermen, etc., the Number and Value of Tugs, Vessels and Boats, and the Quantity and Value of all Fishing Gear and other Material used in the Fishing Industry, from the boundary line, County of Huntingdon to Bellechasse on the South Shore, and from the County of Portneuf to County of Soulanges, inclusive, on the North Shore, Province of Quebec, during the year 1913-14.

Number	Fishing Districts	Boats Sail No.	Boats Sail Value	Boats Gasoline No.	Boats Gasoline Value	Men	Gill-Nets Number	Gill-Nets Value	Seines Number	Seines Value	Lines Number	Lines Value	Hoop-Nets Number	Hoop-Nets Value	Weirs Number	Weirs Value	Ice-Houses Number	Ice-Houses Value
1	Huntingdon, Soulanges, Beauharnois, Vaudreuil and Lake St. Francis	60	600			60					400	400						
2	Lake St. Louis	20	200			20			2	50	300	300						
3	Laprairie	5	50			10	10	30	12	300	20	20	1	5				
4	Chambly	20	200			40			5	125	50	50						
5	Verchères	10	100	2		29			3	75	75	75						
6	Richelieu	25	250	2	400	34					200	200	30	150				
7	Yamaska	70	700	2	60	75					100	100	300	1600				
8	Nicolet	50	500		600	54	150	150	10	250	300	300	1000	5000				
9	Champlain and St. Maurice Counties	40	400			50	30	90	15	375	300	300	150	750				
10	Maskinongé and Berthier	60	540			50	30	60	5	125	300	300	50	250				
11	L'Assomption and Terrebonne	25	250			32	10	20	5	125	250	250	500	2500				
12	Laval and Lake of Two Mountains Counties	30	300			30					400	400						
13	Huchelaga and Jacques Cartier Counties	25	250			25					700	300						
14	Bellechasse County	50	750			50									30	14900	23	2500
	Totals	480	5050	6	1600	543	230	350	57	1425	2895	2895	2031	10150	30	14900	23	2500

RETURN showing the Quantities and Values of all Fish caught and marketed or consumed locally, from the boundary line, County of Huntingdon to Bellechasse on the South Shore, and from the County of Portneuf to County of Soulanges, inclusive, on the North Shore of the Province of Quebec, during the year 1913-14.

Number	Fishing Districts	Trout, *cwts.	Trout, value	Whitefish, cwts.	Whitefish, value	Bass, cwts.	Bass, value	Pickerel, cwts.	Pickerel, value	Pike, cwts.	Pike, value	Sturgeon, cwts.	Sturgeon, value	Eels, cwts.	Eels, value	Perch, cwts.	Perch, value	Maskinongé, cwts.	Maskinongé, value	Catfish, cwts.	Catfish, value	Mixed Fish, cwts.	Mixed Fish, value	Number
1	Lake St. Francis and tributaries					50	600	30	360	20	100	75	600	550	3850	45	225	20	240	20	100	75	225	1
2	Chateauguay and Lake St. Louis			250	2500	100	1200	15	180	21	105	25	200	55	385	40	200	5	60	8	40	35	105	2
3	Laprairie County					25	300	15	180	5	25	3	24	15	105	15	75	2	24	2	10	15	45	3
4	" (hly)					5	60	5	60	6	30	5	40	15	105	8	40	2	24	2	10	700	2100	4
5	Verchères "			6	60	1	12	5	60	12	60	7	56	12	84	6	30	3	36	5	25	250	750	5
6	Richelieu "			3	30	3	36	18	216	25	125	13	120	50	350	35	175	3	24	10	50	200	600	6
7	Yamaska "					5	60	20	240	45	225	30	240	75	525	75	375	4	48	25	125	1000	3900	7
8	Nicolet "					10	120	50	600	40	200	60	400	80	560	75	375	4	48	30	150	150	450	8
9	Champlain and St. Maurice Counties	20	300			2	24	15	180	15	75	60	480	40	280	60	300	5	48	20	150	150	450	9
10	Maskinongé and Ber	20	300			6	72	15	180	55	275			90	630	25	125	6	60	15	75	125	3510	10
11	L'Assomption and Terrebonne "	150	2250			20	240	15	180	10	50	5	40	15	105	75	375	2	24	5	25	45	135	11
12	Laval and Lake of Two Mountains Counties					6	72	3	180	8	30	25	200	20	140	12	60	7	72	40	200	20	45	12
13	Hochelaga and Jacques-Cartier "					12	144	12	144	8	40	6	48	10	70	12	60	3	36	5	50	25	60	13
14	Vaudreuil "							30	360					30	210	10	50	5	60	10		15	75	14
15	Bellechasse			30	300									600	4200	20	100					80	240	15
	Totals	190	2850	292	2920	250	3000	293	3120	278	1300	431	3448	1667	11599	438	2190	67	804	197	963	2585	8655	

*Cwt. = 100 lbs.   † "Mixed Fish" includes greyling, bull-heads and onananiche.

5 GEORGE V., A. 1915

RECAPITULATION.

OF the Yield and Value of the Fisheries of the **Inland Fisheries** of the Province of **Quebec**, during the year 1913-14.

Kinds of Fish.		Quantity.	Value.
			8
Salmon.. *cwts.		85	1,425
Trout.. "		967	11,321
Whitefish....................................... "		514	5,140
Herring.. "		896	1,792
Bass.. "		676	8,496
Pickerel....................................... "		1,229	14,350
Pike.. "		935	6,179
Sturgeon....................................... "		977	6,383
Eels.. "		2,496	16,901
Perch... "		1,823	9,860
Maskinongé..................................... "		115	1,284
Smelt... "		455	1,365
Cat-fish....................................... "		197	985
Shad.. "		60	630
Mixed Fish .... "		5,622	27,735
Total.......................................		....	113,846

Quantity Consumed in Canada ⅚.
Quantity exported to U. S. A. ⅑.

*Cwts. = 100 lbs.

RECAPITULATION.

OF the Number and Value of Vessels, Boats, Nets, Traps, &c., used in the **Inland Fisheries** of the Province of **Quebec**, during the year 1913-14.

—	Number.	Value.
		$
Boats (sail)....	889	19,080
" (gasoline).....	51	23,875
Gill-nets, Seines and other nets........	2,499	15,390
Weirs........	114	28,000
Lines.....	3,320	5,920
Freezers and Ice-houses	35	4,000
Club Houses........	47	17,200
Piers and Wharves (private)	30	750
Total........	....	114,215

Number of men employed in Vessels or Tugs............	1,024
" " Boats........	20
	1,044

5 GEORGE V., A. 1915

RECAPITULATION

On the Quantities and Values of all Fish caught and landed in a Green State, and of the Quantities and Values of all Fish and Fish Products Marketed in a fresh, dried, pickled, canned, &c., state, for the Whole Province of Quebec, during the Year 1913-14.

Kinds of Fish	Sea Fisheries. Caught and landed in a Green State. Quantity.	Value.	Sea Fisheries. Marketed. Quantity.	Value.	Inland Fisheries. Marketed. Quantity.	Value.	Marketed, Both Fisheries. Quantity.	Value.	Total Marketed Value.
		$		$		$		$	$
Salmon............ ....cwts.	12,676	152,112	......	......	....	....	......	......	
" used fresh..... "	....	......	10,312	144,368	85	1,425	10,397	145,793	
" salted (dry).... "	....	....	1,576	15,760	....	....	1,576	15,760	
									161,553
Lobsters............ "	41,310	206,550	......	......	....	....			
" canned......cases.	......	......	16,485	313,215	....	....	16,485	313,215	
" shipp. in shell.cwts.	......	....	100	800	....	....	100	800	
									314,015
Cod .... "	365,052	365,052	..	..	..	..			
" used fresh........ "	......	...	250	375	....	....	250	375	
" green salted....... "	......	....	13,686	41,058	....	....	13,686	41,058	
" dried.... .. ... "	......	....	112,473	674,838	....	....	112,473	674,838	
									716,271
Haddock............ ... "	4,680	3,509	....	....					
" used fresh... "	......	......	35	70	....	....	35	70	
" dried .... "	......	....	1,536	6,144	....	....	1,536	6,144	
									6,214
Hake and Cusk....... "	2,400	1,200	....	....					
" " dried.. "	......	....	800	2,400	...	....	800	......	
									2,400
Herring............... "	363,649	181,824	......	......					
" used fresh...... "	......	....	1,325	1,987	896	1,792	2,221	3,779	
" smoked.......... "	......	....	4,300	8,600	....	....	4,300	8,600	
" pickled ...,.. .brls.	......	....	10,696	42,784	....	....	10,696	42,784	
" used as bait.... "	......	....	61,780	92,670	....	....	61,780	92,670	
" " fertilizer "	......	....	99,038	49,518	....	....	99,038	49,518	
									197,351
Mackerel. .. ..cwts.	23,598	117,990	......	......	..	..			
" used fresh.... "	......	....	75	750	...	....	75	750	
" salted........brls.	......	....	7,841	125,456	...	....	7,841	125,456	
									126,206
Shad . ...cwts.									
" used fresh....... "			....	....	60	630	60	......	630
Halibut, used fresh... "	387	3,096	......	......	387	3,096	387	......	3,096
Flounders . ... "	500	250	500	250	.	....	500	......	250
Smelts... .... "	12,146	72,876	12,146	97,168	455	1,365	12,601	......	98,533
Trout............. .... "	710	7,100	710	8,520	967	11,321	1,677	......	19,841
Eels............... "	285	1,425	285	1,428	2,496	16,901	2,781	......	18,326
Tom-cod.............. "	320	320	320	320	....	....	320	......	320
Whitefish............. "	. .	....	. ..	....	514	5,110	514	......	5,140
Bass........ ....,..... "	......	....	..	....	676	8,496	676	......	8,496
Pickerel............. "	......	....	....	....	1,229	14,350	1,229	......	14,356
Pike........ .... "	....	....	....	....	935	6,179	935	......	6,179
Sturgeon.. .. .. "	....	....	....	....	977	6,383	977	......	6,383
Perch.... ... . "	....	....	....	....	1,823	9,860	1,823	......	9,860
Maskinongé. .... "	....	....	....	....	115	1,284	115	......	1,284
Catfish.. .... "	....	....	....	....	197	985	197	......	985
Mixed fish.... ...... "	3,340	6,680	3,340	10,020	5,622	27,735	8,962	......	37,755
Clams............... .brls.	1,111	3,333	....	....	. ..	..			
" used fresh....... "	....	....	1,111	4,444	... .	....	1,111	... .	4,441
Tongues and Sounds...cwts.	......	....	63	315	..	....	63	......	315
Capelin (bait or fertili.) "	25,100	6,275	25,100	6,275	....	....	25,100	......	6,275

SESSIONAL PAPER No. 39

RECAPITULTION

OF the Quantities and Values of all Fish caught and landed in a Green State, and of the Quantities and Values of all Fish and Fish Products Marketed in a fresh, dried, pickled, canned, &c., state, for the Whole Province of Quebec, during the Year 1913-14—*Continued.*

Kinds of Fish.	Sea Fisheries.				Inland Fisheries.		Marketed, Both Fisheries.		Total Marketed Value.
	Caught and landed in a Green State.		Marketed.		Caught and Marketed.				
	Quantity.	Value.	Quantity.	Value.	Quantity.	Value.	Quantity.	Value.	
		$		$		$		$	$
Hair seal skins....... No.	......	......	4,872	7,308	......	......	4,872	......	7,308
Beluga skins.......... "	......	......	75	375	......	......	75	......	375
Whales............ "	87	60,900	......	......	......	......	......	......	
Fish oil............ gals.	......	......	90,400	27,119	......	......	90,400	......	27,119
Whale oil............ "	......	......	147,560	44,268	......	......	147,560	......	44,268
Whale bone.......... cwts.	......	......	240	4,800	......	......	240	......	4,800
Whale fertilizer. .....tons.	......	......	244	85	......	......	244	......	85
Totals.................	......	1,190,492	......	1,736,581	......	113,846	....	......	1,850,427

5 GEORGE V, A. 1915

RECAPITULATION.

Of the Number of Fishermen, etc., and of the Number and Value of Fishing Vessels, Boats, Nets, etc., in the **Whole** Province of **Quebec** for the Year 1913-14.

	Sea Fisheries.		Inland Fisheries.		Total–Both Fisheries.	
—	Number.	Value	Number.	Value.	Number.	Value.
		$		$		$
Steam Fishing Vessels (tonnage 360)..	3	48,000	....	... .	3	48,000
Sailing & Gasoline Vessels.......... ..	29	25,300	....	... .	29	25,300
Boats (sail)...................	4,213	222,152	889	19,080	5,102	241,232
" (gasoline)...................	196	65,900	51	23,875	247	89,775
Carrying Smacks.....	9	7,950	....	......	9	7,950
Gill Nets, Seines, Trap & Smelt Nets, &c	11,556	255,052	2,499	15,390	14,055	270,442
Weirs..........	143	11,780	114	28,000	257	39,780
Trawls............................	676	5,015	....	......	676	5,015
Hand Lines..............	19,235	15,957	3,320	5,920	22,555	21,877
Lobster Traps.	160,380	160,380	....		160,380	160,380
Lobster Canneries..	115	54,975	....	......	115	54,975
Freezers and Ice-houses..............	277	22,150	35	4,000	312	26,150
Smoke and Fish-houses.....	1,120	255,930	....	......	1,120	255,930
Fishing Piers and Wharves............	290	131,115	30	750	320	131,865
Club Houses..........................	....		47	17,200	47	17,200
Whaling Stations............	1	50,000	....	......	1	50,000
Totals	. . ..	1,331,656	....	114.215	... .	1,445,871

	Sea Fisheries.	Inland Fisheries.	Total.
Number of men employed on Vessels...........................	149	....	149
" " Boats,......	7,985	1,024	9,009
" " Carrying Smacks.................	19	.. ..	19
" persons employed in Fish-houses, Freezers, Canneries, &c	1,776	20	1,796
Totals.	9,929	1,044	10,973

APPENDIX No. 5.

ONTARIO

DISTRICT No. 1.—COMPRISING RAINY RIVER, THUNDER BAY, PARRY SOUND, &c., DISTRICTS. INSPECTOR, T. J. FOSTER, SAULT STE. MARIE, ONT.

DISTRICT No. 2.—COMPRISING PART OF THE COUNTY OF BRUCE, THE COUNTIES OF HURON, LAMBTON, ESSEX, KENT, ELGIN, &c., AND LAKES HURON, ST. CLAIR AND ERIE. INSPECTOR, O. B. SHEPPARD, TORONTO, ONT.

DISTRICT No. 3.—COMPRISING LAKE ONTARIO AND THE EASTERN COUNTIES OF THE PROVINCE. INSPECTOR, J. S. HURST, BELLE-VILLE, ONT.

N.B.—The Fisheries of Ontario are administered by the Provincial Government. This Department merely exercises a general supervision.

REPORT ON THE FISHERIES OF DISTRICT No. 2.

To the Superintendent of Fisheries,
          Ottawa, Ont.

Sir.—I have the honour to submit my report on the Fisheries of my District for the season 1913-14.

Commercial fishing has been about an average with former years excepting for herring in Lake Erie which shows an increase.

The angling sportsman Fisheries has been very good in most waters especially in waters removed from the thickly populated Districts.

In my visits to the various localities in my District I have heard many comments on the exhibit given by the Department at the Canadian National Exhibition last year and nothing but words of praise were expressed in regard to the same and I am satisfied that it will be the means of doing a great deal to introduce fish as a more general food through Ontario and am sure it will be a lasting benefit and a great saving to the people in general and will have the effect of fish taking the place to some considerable extent of meat which has lately been so high priced (and is continually advancing) that it has kept many of the poor and medium classes from using it to the same extent as formerly. I am pleased that the Exhibit is being continued by the Department this year and am sure it will have very beneficial and lasting effects.

I find that the law has been generally well observed and that American tourists are very much more careful in regard to the size and number of fish caught than they were in former years.

The price received by fishermen this season is practically the same as last with the exception of blue pickerel which has been considerably lower.

All of which is respectfully submitted.

                    Your obedient servant.

                              O. B. SHEPPARD,
                                Inspector of Fisheries.

5 GEORGE V., A. 1915

REPORT ON THE FISHERIES OF DISTRICT No. 1.

To the Superintendent of Fisheries,
Ottawa, Ont.   .

SIR,—I have the honour to submit my Report on the Fisheries of the North-West Division of the Province of Ontario for the fiscal year 1913-14.

I am pleased to report a continued improvement in the Fisheries of this District. Commercial fishing was equal to, and in many cases exceeded that of previous years, especially that of white fish. Game fishing also was better than hitherto and tourists report good catches of Brook Trout and Bass. No serious infractions of the Laws have been reported except in the St. Mary's River, and in the vicinity of St. Joseph Island where American poachers continue to trespass when not closely watched.

I regret that no close season has yet been made for Rainbow Trout which afford so much sport and abound in the St. Mary's River Rapids. Grey Trout also now quite numerous should be protected.

I find the fishery laws generally are being efficiently enforced.

I am, Sir,
Your obedient servant,

T. J. FOSTER,
Fisheries Inspector.

REPORT ON THE FISHERIES OF DISTRICT No. 3.

To the Superintendent of Fisheries,
Ottawa.

SIR,—I have the honour to submit my report on the fisheries in my district for the fiscal year 1913-14. Commercial fishing in my district has been very good. There was a very good catch of white fish in Lake Ontario and Bay of Quinte also a large catch of herring.

White fish are on the increase in Bay of Quinte; which is due largely to restocking. Coarse fish such as carp, catfish, bullheads, and perch show an increase over last year for which fishermen found a ready market at fair prices.

I have visited most of the fishing stations and found the law well respected: The provincial officers with their patrol boats are doing good work.

I would be very much pleased to see more fishways in this district as bass and pickerel naturally go up stream to spawn.

Anglers have enjoyed excellent sport after bass and trout which were plentiful during the past season.

The new hatchery erected by your department on Bay of Quinte will be ready to receive spawn this fall and I feel satisfied that great benefit will be derived in a few years.

I am, Sir,
Your obedient servant,

J. S. HURST,
Inspector of Fisheries.

WM. A. FOUND, Esq.,
Superintendent of Fisheries,
Ottawa.

SUMMARY

Of the Yield and Value of the Fisheries in the Province of **Ontario**, during the
year 1913-1914.

Kind of Fish.		Quantity.	Value.
			$
Salmon............cwts.		....	....
Trout.	"	62,204	579,832
Whitefish	"	52,263	520,123
Herring	"	130,718	658,038
Bass	"	....	....
Pickerel	"	26,564	265,645
Pike	"	34,547	276,378
Sturgeon	"	2,535	38,022
Eels	"	2,370	14,221
Perch	"	12,427	62,137
Maskinongé	"	....	....
Tullibee	"	5,738	34,429
Cat-fish	"	5,264	42,115
Carp	"	6,721	33,606
Mixed Fish	"	28,291	141,456
Caviare	"	84	8,411
Sturgeon Livers	No.	453	272
Total		....	2,674,685

* Cwts. = 100 lbs.

RECAPITULATION

Of the number and Value of Vessels, Boats, Nets, Traps, etc., used in the Fisheries in
the Province of **Ontario**, during the year 1913-1914.

—	Number.	Value.
		$
Steam Vessels or Tugs	190	433,180
Boats sail)	1,224	101,631
" gasoline)	366	166,660
Gill-nets, Seines and other nets	52,613	645,353
Spears	103	260
Lines (of 100 books)	463	627
Freezers and Ice-houses	333	130,500
Piers and Wharves (private)	91	28,370
Totals	....	1,506,581

Number of men employed in Vessels or Tugs............ 744
"   "   "   2,767
  3,511

5 GEORGE V., A. 1915·

APPENDIX No. 6.

MANITOBA

REPORT ON THE FISHERIES OF THE PROVINCE.

To the Superintendent of Fisheries,
   Ottawa, Ont.

Sir,—I have the honour to submit my Annual Report on the fisheries of the Province of Manitoba, and a part of the Province of Saskatchewan, for the fiscal year ended March 31, 1914, together with the statistical returns showing the yield and value of the catch.

Heavy storms caused a falling off in the quantity of whitefish taken from Lake Winnipeg during the summer. The fish taken were, however, of a fair size. Winter fishing this lake was not so successful as in the previous year. The lake was late in freezing up, and then storms broke the ice after the fishermen had their nets set.

The Fishery Overseer of Lake Winnipegosis reports that whitefish and pickerel were not nearly so plentiful as they were in the preceding year. There was also a diminished catch of whitefish and pickerel in Lake Manitoba; while in Lake Waterhen there was a small decrease in the quantity of whitefish, but an increase in pickerel. The catch of whitefish in Clearwater and Cormorant lakes was slightly greater.

From a small lake called Red Deer, fished this year for the first time, there were taken 42 cwts. of whitefish. Fishing on Lake St. Martin was much better this year than in the preceding year.

The Saskatchewan river was fished during the summer and fall with seines, chiefly as a test of whether it would pay to fish commercially in this way; also to see what effect this style of fishing would have. Fishery Overseer Stevenson reports that so far as can yet be known there are no bad effects from the operations of seines.

Beaver lake which lies ninety miles north-west from The Pas was fished this year for the first time for commercial purposes. There were fourteen licenses issued, and about eleven car-loads of fish taken from it.

Fishery Overseer Stevenson reports that the whitefish in Cormorant and Clearwater lakes are very small, and he would recommend the closing of these lakes.

The law has been very well observed, with a few exceptions. I seized a quantity of whitefish on the Little Saskatchewan spawning grounds last winter; also a quantity of jackfish that were caught in Oak lake in Southern Manitoba. The Overseers report that the ice was well cleaned on the lakes after the winter fishing was over.

There has been no addition to the number of vessels or boats operating in the Provinces, but there has been a development in fish culture. A large hatchery was built on Dauphin river last year, and the weekly reports show that a large amount of fry was hatched out. Under orders from the Department I took from the Selkirk hatchery thirty-two cans of young whitefish and placed them in Lac du Bonnet; also sixteen cans which I placed in Shoal lake. near Shoal lake station,—each can contained 8,000 whitefish.

    I am. Sir,
      Your obedient servant,

       J. A. HOWELL,
        Inspector of Fisheries.

RETURN showing the Number of Fishermen, &c, the Number and Value of Tugs, Vessels and Boats, and the Quantity and Value of all Fishing Gear and other Material, used in the Fishing Industry in the Province of Manitoba, during the year 1913-14.

Number	Fishing Districts	Steam Vessels or Tugs Number	Tons	Value $	Men	Boats Skiff and sail boats	Value $	Gasoline	Value $	Men	Gill-Nets Number	Value $	Seines Number	Value $	Lines Number	Value $	Freezers and Ice-houses Number	Value $	Smoke and Fish-houses Number	Value $	Piers and Wharves Number	Value $	Persons employed in Freezers, Fish-houses, &c.	Number
1	Lake Winnipeg	8	577	75000	80	275	22010			438	3838	38380			56	112	50	42500	50	25000	12	9100	200	1
2	Red River					69	345			93	93	930							13	4500	3	125	75	2
3	Lake Winnipegosis	1	60	10000	12	63	3340			196	2262	22620					6	4500	1	100			1	3
4	Lake Waterhen									21	225	2250												4
5	" Dauphin									6	68	684												5
6	" Manitoba					1		5		160	1000	10000							5	750				6
7	" St. Martin									88	880	8900												7
8	Moose Lake									3	30	300												8
9	Clearwater Lake									15	150	1500												9
10	Cormorant Lake									7	84	840												10
11	Sturgeon Lake									6	60	600												11
12	Rocky Lake									5	56	560												12
13	Big Sask. River, Northern Man.								33000	17	17	170	3	210			1	10000	2	4000			10	13
14	W. ... er									1	5	50												14
15	Beaver Lake									11	126	1960												15
16	Lake Red Deer									3	34	340												16
	Totals	9	637	85000	92	408	25750		33000	1070	8928	89280	3	210	56	112	57	57000	71	34350	15	9225	286	

RETURN showing the Quantities and Values of all Fish caught and marketed or consumed locally, for the County of Selkirk, Province of Manitoba, during the year 1913-14.

No.	Fishing Districts	Trout, cwts.	Trout, value $	Whitefish, cwts.	Whitefish, value $	Pickerel, cwts.	Pickerel, value $	Pike, cwts.	Pike, value $	Perch, cwts.	Perch, value $	Tullibee, cwts.	Tullibee, value $	Cat-fish, cwts.	Cat-fish, value $	Goldeyes, cwts.	Goldeyes, value $	Mixed fish, cwts.	Mixed fish, value $	No.
1	Lake Winnipeg, (summer)			19237	116022	8081	40405	731	2193	67	268	1975	3958			1205	3012	4600	9230	1
2	" " (winter)			2079	12471	8591	42955	1809	5427	176	704	11535	23070			3715	9287	1430	2890	2
	Red River (summer)					70	350	110	330									4200	8400	3
	" " (winter)					45	225	70	210									69	1200	4
3	Lake Winnipegosis (summer)			743	4458	4623	23115	361	1083					625	3125			10144	17716	5
	" " (winter)			8345	50070	6988	33440	19402	37296					23	115			4200	8400	6
4	Lake Waterhen			691	4146	252	4695	223	609									1000	2000	7
5	" Dauphin					629	1160	476	1428							59	17	500	1000	8
6	" Manitoba			68	405	625	3145	564	1512			314	628					9074	18111	9
7	" St. Martin			1020	6120	65	1250	175	525									5009	10000	10
8	Moose Lake			500	3000		275	55	165									80	160	11
9	Cedar Lake			2453	14718			206	618											12
10	Cormorant Lake	30	650	280	1618	70	350	350	1050									500	1000	13
11	Sturgeon Lake	84	1920	418	2505	80	150	46	138									60	120	14
12	Clay Lake, Northern Man.	23	615	147	882	200	400	150	450									20	400	15
13	Big Stack River	60	800	200	1200		1100	100	300									10000	20000	16
14	Beaver Lake (winter)	708	3540	1920	11520	25	125	900	2700			20	40					800	1600	17
15	Lake Red Deer			42	252	376	1889	88	264							110	275	928	1642	18
16	Rock Lake																	200	400	19
17	Turtle M (ntain District																	2500	5600	20
18	Riding "																	400	800	21
19	Cedar Lake																	500	1000	
20	Dog "																	900	1800	
21	Lakes Cross, Falcon, Hawke, Forbes																	300	600	
	M. 10																			
	Totals	1505	7525	38243	229391	31024	154020	18756	56268	243	972	13844	27696	648	3240	5089	12721	57576	113439	

† "Mixed Fish" includes greyling, bull-heads and ouananiche.

RECAPITULATION.

Of the Yield and Value of the Fisheries in the Province of **Manitoba**, during the year 1913-14.

Kinds of Fish.		Quantity.	Value.
			$
Trout	*cwts.	1,505	7,525
Whitefish	"	38,243	229,391
Pickerel	"	31,024	155,020
Pike	"	18,756	56,268
Perch	"	243	972
Tullibee	"	13,844	27,696
Cat-fish	"	648	3,240
Goldeyes	"	5,089	12,721
Mixed Fish	"	57,576	113,439
Totals			606,272

Quantity Consumed in Canada ¼.
Quantity exported to U. S. A. ¾.

* Cwts. = 100 lbs.

RECAPITULATION.

Of the Number and Value of Vessels, Boats, Nets, Traps, etc., used in the Fisheries in the Province of **Manitoba**, during the year 1913-14.

	Number.	Value.
		$
Steam Vessels or Tugs	9	85,000
Boats (sail)	403	25,750
Boats (gasoline)	3	3,000
Gill-nets, Seines and other nets	8,931	89,490
Lines	56	112
Freezers and Ice-houses	57	57,000
Smoke and Fish-houses	71	34,350
Piers and Wharves (private)	15	9,225
Totals		303,927

Number of men employed in Vessels or Tugs.......................... 92
" " Boats.. 1,070
" persons employed in Fish-houses, Freezers, etc................... 286

1,448

APPENDIX No. 7.

SASKATCHEWAN AND ALBERTA

REPORT ON THE FISHERIES OF THE PROVINCES, BY CHIEF INSPECTOR E. W. MILLER, FORT QU'APPELLE.

The Superintendent of Fisheries,
Ottawa.

SIR,—I have the honour to submit the following reports on the fisheries of the Provinces of Saskatchewan and Alberta for the fiscal year 1913-14, together with the statistical returns showing yield and value of fish and amount, etc., and value of material used.

There was an increase of no less than forty three per cent in the numbers of net licenses issued, over the figures for the previous year, and while this increase is partly due to the greater efficiency with which the fishery regulations have been carried out it affords ample proof of the development of the fishery resources of the two provinces.

At least eighty per cent of the catch of fish by nets is taken in the winter season through the ice, and it will be noted that the yield of whitefish is mainly responsible for the increase in the total output of fish; though there was also a good increase in the catch of pike.

There has been a considerable extension of the field of operations and the greater catch does not mean that the waters in the older settled districts are being overtaxed. In fact the shortening of the fishing season in the southern portion of the province has been most beneficial and the catch per net per setting shows that lakes in which the effect of over fishing had become evident, are now recovering owing to the fuller measures of protection at spawning seasons.

The steady growth in angling as carried on in the more accessible waters is very marked both for the coarser fish as well as for trout. For the first mentioned no license or permit is required and it is impossible to do more than give an approximate estimate of the catch made in that way as formed by the local guardians doing their rounds. In the Trout waters where an 'Angler's Permit' is required, the issue of such was nearly three hundred per cent larger than in 1913 but here again allowance must be made for the far greater efficiency with which it has been possible to enforce that part of the Fishery Regulations affecting the taking of sporting fish. At present it would appear that the increased protection at spawning time, was enabling the streams to sustain the inroads of the ever increasing army of anglers, but it is evident that the establishment of hatcheries of which there are two now in operation, was fully warranted by the extremely large amount of angling which is now done in our trout streams.

LICENSES ISSUED 1913-1914.

	Saskatchewan.	Alberta.
Domestic	231	192
Fisherman's	405	259
Commercial	9	5
Anglers' permits	Nil.	3,674

That a much closer oversight of the fisheries is gradually being secured is clearly evidenced by the statement of prosecutions for the year, no less than one hundred and thirty convictions having been secured.

The offences charged were as follows:—

Fishing in close season.. .. .. ''	73
Fishing without license.. .. .. ''	36
Spearing fish.. .. .. '' .. .. .. '.	4
Using fish traps.. .. .. ''	5
Fishing with illegal nets.. .. .. '' ..	5
Fishing excess of nets.. .. .. ''	5
Fishing in closed waters..	2

Ninety-five of these cases arose in Saskatchewan and thirty-five in Alberta. The total amount of fines inflicted with value of fish confiscated and sold amounted to $1,590.

A very large number of applications has been received for the stocking of waters with fish by the Department, and examination of these lakelets is being made by the officers of the Department as time and circumstances permit.

Of such as have been visited, only a small proportion offer any fair promise of success and they only for fish of the coarser species, whereas the applicant generally desires Trout or Whitefish. There are, however, in the aggregate a fairly large number of lakes the extent and quality of whose waters warrant an experimental planting of fish and I would strongly recommend that an appropriation be made to enable such to be made, preference to be given to such cases as where the faith of the applicant is evinced by a readiness to give practical assistance in the transport of the fish, etc.

The Indian and Halfbreed problem still remains unsolved, in some districts, where they have been accustomed to take fish by any means and at any season. As pointed out in former years so long as the lakes are not accessible to white fishermen they are perhaps able to withstand the results of such fishing, the total demand being but limited. When such districts become settled or even brought within a hundred miles of some railway shipping point the demand for fishery privileges so as to supply the local and provincial markets cannot be reasonably restricted. To enable the lakes to stand such fishing they must be protected during the spawning seasons and wasteful methods of fishing must give way to the more orthodox. It seems but right that the special claims of the native residents shall not be allowed to stand in the way of the development of the fisheries to the best advantage and general welfare, even if their necessities have to be provided for in other ways.

The conditions affecting the fisheries generally have changed but little since last year and having been very fully dealt with in my last report, it is unnecessary to repeat the same on this occasion.

SOUTH SASKATCHEWAN

The main fishing lake in this part of the province is Long or Last Mountain lake where 148 licenses were issued. Though nearly all the men fishing here take Fishermen's licenses and sell the larger portion of their catch, they are mostly farmers who fish in the winter only and then but intermittently. Under these conditions the total catch made is not in excess of the capacity of the lake.

In the lower Qu'Appelle lakes whitefish form only a small part of the catch and the use of nets is to be confined to the winter season in future on account of the greater amount of angling done.

At Lac Pelletier the applications for licenses are so numerous that none but Domestic licenses are issued and they only for the winter season.

5 GEORGE V., A. 1915

The numerous small lakes along the Canadian Northern Railway north of the Qu'Appelle contain but coarse fish and are not capable of sustaining much net fishing. Licenses are granted for the use of nets on such as contain Mullet only but the Pike and Pickerel waters are reserved for angling.

For a month before the close of the winter fishing season on February 15, the weather was extremely severe and but little fishing was done. It was felt by the fishermen that while the closing of the season at that date may be necessary at remote lakes where a long haul to market is required, the season could be safely extended to the end of February for lakes in settled districts where an immediate market is secure and no waste of fish to be apprehended.

The selection of a site for a Whitefish hatchery and assurance of its being in running order for next season is viewed with much satisfaction; while the intention to experiment with the planting of Black Bass in ponds in connection with the hatchery is heartily welcomed by all those interested in the introduction of game fish to Saskatchewan waters.

From some of the lakes, representations have been received that the proportion of Whitefish in their waters does not warrant the maintenance of the 5½-inch mesh standard of nets. There is some truth in these statements and it will be desirable perhaps to allow the use of smaller mesh nets in some instances, though the general permission of such in lakes suitable for Whitefish cannot be recommended.

NORTH SASKATCEHWAN

The outstanding features in the fishery operations in this district was the very mild weather and absence of snow which prevailed well into January.

This bid fair to bring disaster upon the season's work as though the catch was excellent, it was impossible to convey it to the market until sleighing came and in the meantime the fish deteriorated in quality and contracts for supplies could not be fulfilled. Eventually the whole catch was marketed without the actual wastage at one time feared, but both fishermen and wholesalers suffered by the delay.

Ile la Crosse and Dore lakes, which are the two districts giving the principal output of fish at present, were particularly affected by this unwonted failure of snow.

Jackfish and Turtle lakes both showed signs of returning fecundity, and a good catch was made in proportion to the fishing done on them.

Lac des Iles was fished for market purposes practically for the first time. The catch was good but the transport difficulties prevented the fishermen reaping much reward.

A detailed report on this district by Inspector MacDonald is appended.

SOUTH ALBERTA

The desire of the Department to secure fuller protection of the Trout streams which form the great fishery asset of this part of the province was evinced by the appointment of six additional guardians whose . work has unquestionably had an excellent effect.

Some difficulties arose from the fact that the upper waters of most of these streams are within the forest reserves controlled by the Department of the Interior, but it is now agreed that the fishery regulations are to have equal force within the reserves as without.

Note has already been made of the great increase in the number of Angler's Permits that were issued, which reached a total of 3,074.

Overseer Hoad of Calgary furnishes the following detailed report:—

SESSIONAL PAPER No. 39

639 6th Avenue West,
Calgary.

E. W. MILLER, Esq.,
.Chief Inspector of Fisheries,
Fort Qu'Appelle.

SIR,—I herewith present to you my report for 1913 of Southern Alberta.

The season just closed has been the most successful one for a great many years, fish being very plentiful in all open waters; which I believe should be attributed to the close season having been well observed the last two years. I am looking forward to next year being equally as good, as the anglers themselves are very interested in seeing the regulations properly carried out.

The number of anglers taking out permits increased from 1,250 in 1912, to 3,500 in 1913, and I have very little doubt that next year it will increase to 5,000.

I estimate the number of anglers fishing in the different streams as follows:— North Fork of Old Mans river 600, South and Middle Forks of Old Mans river 300, Pincher and Beaver creeks 200, Lees and Boundary creeks 100, High river Flat and Sullivan creeks 300, Sheep creek 300, Elbow river 300, Fish creek 300, Bow river 400 and Jumping Pound creek 200. The number of fish caught in the different streams are approximately as follows:—Pincher and Beaver creeks 10,000 cut throat trout, 1,500 grayling, North Fork of Old Mans river 30,000 cut throat, 2,000 grayling, South and Middle Fork of Old Mans river 10,000 cut throat, 1,000 grayling, Lees and Boundary creeks 5,000 cut throat, 1,000 grayling, High river 25,000 cut throat, 10,000 grayling, Sheep creek 20,000 cut throat 3,000 grayling, Fish creek 20,000 cut throat, 1,000 grayling, Elbow river 15,000 cut throat, 2,500 grayling, Jumping Pound creek 15,000 cut throat, 2,000 grayling, Bow River 25,000 cut throat, 10,000 grayling.

Good pike fishing was obtainable in the different lakes during the season, but exceptionally good fishing was to be had in Chestermere-lake, eight miles east of Calgary, where there were about 4,000 fish taken, averaging three pounds in weight.

There were 15 convictions during the year, for violations. Three fishing during closed season, ten fishing without a permit and two for fishing in closed waters.

I am able to report that there is a plentiful quantity of fish in Trout and Willow creeks and the Middle and South Forks of High river; which have been closed for the last two seasons, and with proper protection will provide good angling in the future.

During the month of August Mr. L. C. Orr of Banff gave me half a dozen Nippigon trout, which he caught in the Bow river near Banff. I deemed it advisable to report this as I am informed that they are getting plentiful, and it may be of value for the department to know this.

The different guardians appointed during the summer have given very efficient service and I am pleased to report that the majority of anglers seem willing to give them every assistance possible. The officers and men of 'D' division R. N. W. M. P. under Supt. P. C. H. Primrose have given a great deal of assistance during the year.

I have the honour to be, Sir,
Your obedient servant,

N. J. HOAD,
Fishery Overseer.

5 GEORGE V., A. 1915

NORTH ALBERTA

In this part of Alberta considerable difficulty is being experienced in satisfying the claims of those desiring to procure licenses on those lakes which are within easy access, as the applications are now in excess of the capacity of the lakes. The catch of fish made is probably not far from the limit of that possible without danger of exhaustion, except at the remoter points.

Towards the close of the season railway facilities were available for the transport of fish from the eastern end of Lesser Slave lake and there is reason to think that this lake too will soon be fished to its limit.

Wabamun, Calling, Primrose and Cold lakes all gave a largely increased catch and the minor lakes are shown to be in a sound condition.

A further extension of guardianship has become necessary in the new western districts recently opened up which contain many small lakes of local value though none of such importance as to promise any supply for other demands.

Applications were received for fishery privileges at Athabasca lake, but these were evidently with the future in view, as at present there is no means by which fish could be brought out at a profit. This great lake promises to become a big source of supply as soon as railway transport is available within a reasonable distance.

A detailed report on this district by Inspector Willson is appended.

I am, Sir,
Your obedient servant,

E. W. MILLER,
Chief Inspector of Fisheries.

REPORT ON THE FISHERIES OF NORTHERN SASKATCHEWAN, BY THE INSPECTOR OF FISHERIES.

To E. W. MILLER, Esq.,
Chief Inspector of Fisheries,
Fort Qu'Appelle, Sask.

SIR,—I have the honour to submit my report on the fisheries for the northern portion of the Province of Saskatchewan for the year ending thirty-first of March, nineteen hundred and fourteen, together with Statistical returns showing the yield and value of fish, etc., amount and value of equipment.

The weather conditions were most favourable for the fishermen on the ice, but not at all advantageous for hauling fish to the railroads for shipping, on account of the lateness of the arrival of the snow, which did curtail operations after January tenth considerably, as there was grave doubts if the first catch could be hauled to the railroads. However, the total catch on some lakes did exceed the previous year.

There were some eighty licenses issued more this year than the preceeding year, due no doubt to the scarcity of other employment throughout the Province, and especially the winter months when work was not as readily procured in the lumber camps as on previous winters.

During this year, there were eighty-three convictions and 15,669 pounds of fish seized, and 850 fathoms of net seized, the principal offence being, fishing without permits, and during closed season, the principal offenders being experienced fishermen. Fines were imposed and collected to the amount of $780 and a further $368.55 was realized from the sale of the confiscated fish and nets. The offences charged were as follows:—

Fishing in closed season··	45
Fishing without licenses..	28
Use of illegal nets.. .. .. ··	5
Use of over allowance of nets..	5

Some of the defendants were fined on two and three separate charges and no doubt will have a good effect on fishermen in future seasons. I may say, with few exceptions, fishermen seem to be satisfied with the fishing regulations as they are at present with the exception of a few lakes, where the white fish are of a poor quality, and average about 1½ pounds each. On these lakes, I would recommend a smaller mesh of net be allowed say 4 or 4½ inch mesh. This applies particularly to La Ronge lake. The Indian or half breed is no doubt taking the advantage of this ready market for their fish. Especially is this noticeable on Isle La Crosse lake as only some four domestic licenses have been issued where some fifty fisherman's licenses were taken out, and no doubt the Indians would fish for commercial purposes on Lac La Ronge, if a market were available and shorter winter roads opened up. This I did expect and as a result some eight Indians took out fishermen's license for that lake.

The total catch of white fish for this season was practically the same as 1912-1913, but, on account of the climate conditions, did not return the same amount of money to the fishermen as in the early part of the winter. The mild weather brought prices down considerably.

Dore lake produced by far the largest quantity of fish that was shipped to other points. On Isle La Crosse lake there was a larger increase in the catch than previous years, while only about half the winter catch was sent to outside markets, the balance being consumed by the settlers living around this large body of water.

Jack Fish and Murray lakes shows considerable increase over last year with fewer licensed fishermen, due principally to the large amount of angling done on this lake during the summer season.

Candle lake shows a small increase in the white fish catch but a decrease in the pike with some five fishermen less than the previous year.

The Saskatchewan river shows a large decrease in pike, but a larger increase in coarse fish.

The total catch for this portion of the Province shows an increase of 1,800 cwts., over last year, partly due to more accurate returns being sent in.

On only two lakes were any fish caught for commercial purposes during the summer season, namely Jack Fish lake and Stoney lake. On the former some 168 cwt., of white were sold in the vicinity of Battleford and on the latter lake 520 cwt., of white fish and 35 cwt. pike were shipped to the fish dealers at Prince Albert and Saskatoon, with a few odd shipments to smaller points.. The demand for fresh caught white fish did exceed this supply to a great extent.

Our great difficulty regarding summer fishing is the great irregularity in shipments, caused by various reasons, but its principal reason being the bad conditions of the trails leading to the lakes. One instance I might mention, Stoney lake situated some six miles from the railroad at Big river and the freighters could only haul 400 pounds of fish per team of horses per trip, and very often it required the greater part of a day to make the trip, more especially during wet weather. It is a very regretable fact, owing to the beautiful white fish we have in this Province that it is necessary to pay carriage on fish from other Provinces.

Regarding fish caught during closed season for the purpose of hanging, I may say I did not see or hear of any being hung. On my trip to Lac La Ronge via Montreal lake during the month of February, I called at the homes of a great majority of Indians and half breeds, but failed to see any fish being treated so.

The Indians are no doubt put to a great disadvantage in being disallowed this privilege, particularly for hunting and travelling long distances as they depend entirely

5 GEORGE V., A. 1915

upon dogs as a means of transportation during the winter months. On a trip of say, eight days they require ninety-six fish for a team of four dogs, three fish for each dog per diem. Ninety-six fresh fish weigh about three hundred pounds, while ninety-six hung fish weigh about one hundred pounds. The reason is obvious. This is undoubtedly the most serious question to be considered in the far northern parts of this Province, the hanging and drying of white fish during the spawning season. It is impossible with the present staff of fishery officers to watch more than a fringe of the northern lakes. Outside that fringe white fish are being hung every Fall as they have been for hundreds of years. By this I mean of course the lakes lying north of the Churchill river, such as Cree lake and hundreds of other lakes throughout that section. So far as Portage La Lock and Isle La Crosse and the Lac La Ronge districts are concerned, we are in a position to see that the law is enforced.

Re the stocking of inland lakes; during the year I inspected five lakes reporting favourably on four and unfavourable on one, and from the present indications this work will require considerable time and expense in the near future, and no doubt will be of great benefit to districts isolated from waters containing fish life.

I would strongly recommend some suitable fish being put into the Saskatchewan rivers, as this water flows through a portion of the country to a great extent not inhabited with any great quantity of fish and no doubt a suitable fish could be selected that would do well in its waters, and moreover, fish could migrate up other smaller streams to lakes which would be a great asset to the particular district such fish would inhabit.

In this district there are six dams, four containing fishways and two without situated as follows:—

On Turtle river four miles from Turtle lake with fishway.

On La Plonge river at Beaver river no fishway.

On Red river at Anglin lake, with fishway.

On Red river, south of Anglin lake, with fishway.

On Sturgeon river at Sturgeon lake, with fishway.

On Carrot river near Village of Kinistino, no fishway.

I may say regarding dams with no fishway, at La Plonge river the Roman Catholic mission has had instructions to construct a proper fishway there. At Kinistino village on the Carrot river the Canadian Northern railroad I presume, have had instructions to construct one there and no doubt in the course of a very short time all dams in this district will contain proper fishways.

Return figures show that only some 575 cwt. of fish were exported to the U.S.A., markets, but no doubt larger quantities are re-shipped by the Winnipeg dealers.

Some 400 cwt. or two cars were shipped to eastern Canada markets, the balance of shipments were marketed in Manitoba, Saskatchewan and Alberta. Pike and even suckers were shipped to some isolated points through the Province demanding practically the same price as whitefish.

In regard to the Statistical returns of boats, my report shows 254 sail or row boats and 9 gasoline boats. On Wakaw lake boats are used for pleasure and for fishing and cannot be called fishing boats while on Stoney lake and Candle lake boats may be called exclusive fishing boats.

The returns show all boats which might be used exclusively or partly for fishing purposes. Of the nine gasoline boats four are used exclusively for fishing purposes. The great majority of boats on the Saskatchewan rivers are used exclusively for fishing purposes. On Isle La Crosse lakes the boat or canoe is the only means of travelling. It is also the only means of fishing during the summer season as there is practically no angling done on that lake and a boat is absolutely necessary in order to set a net.

I am, Sir,

Your obedient servant,

G. C. MACDONALD,

Inspector of Fisheries.

SESSIONAL PAPER No. 39

REPORT ON THE FISHERIES OF NORTHERN ALBERTA, BY THE INSPECTOR OF FISHERIES.

To E. W. MILLER, Esq.,
    Chief Inspector of Fisheries,
        Fort Qu'Appelle, Sask.

SIR,—I have the honour to submit my report of the fisheries of Northern Alberta for the year 1913 and 1914, with statistics.

By comparison of these statistics with those of the previous year, it will be seen; that more men have been employed, and that a larger capital has been invested in boats, nets, and buildings, than in the year 1912 and 1913, in these proportions. For 1912 and 1913, $7,271. For 1913 and 1914, $15,978.

The earnings of fishermen for the past year have increased over those of the previous year proportionally, the catch being for 1912 and 1913, 11,061 cwt., valued at $40,231. For 1913 and 1914, 20,143 cwt., valued at $65,707, an increase of 9,082 cwt., valued at $25,476.

The increased attention to fisheries in Northern Alberta may be attributed:—

(1st) To a larger demand for fish, created by increased population, especially of towns, as well as the high prices for other foods; prevailing.

(2nd) Somewhat improved railway facilities, and improvement of country roads.

(3rd) Lack of employment which induced men who had not been fishermen, to take up the work.

(4th) Greater skill acquired by fishermen, better equipment and more knowledge of fishing conditions of some lakes; have also conduced to more successful fishing.

The Red Deer District, comprising Gull, Sylvan and Burnt lakes and their tributaries, all emptying into the Red Deer river; which only contain jackfish, pickerel, and suckers; is but little fished by nets. In Gull lake one fisherman's license was held, though a net was not used, the holder using baited hooks. This license will not be renewed for the ensuing year. These waters being situated in a well settled country, in which are several towns and villages, should be generally reserved for angling; particularly is this so of Gull lake and Sylvan lake, which have become summer resorts.

The Pigeon Lake District, comprising Pigeon, Battle, Buck, and Conjuring lakes, with their tributaries and effluent streams; the two former emptying into the Battle river, the two latter into the North Saskatchewan; is of considerable importance. Pigeon, Battle, and Buck lakes contain whitefish of superior quality; and Conjuring lake is well stocked with jackfish.

Because of the comparatively small area of these waters, even with the present settlement, they can only supply the local demand of Wetaskawin, and other surrounding settlements. The whitefish of Pigeon lake have been for several years greatly depleted, as compared with earlier years, but statistics obtained, and Overseer L. Ingram Wood's annual report show that decrease of whitefish is not going on now.

Although fewer licenses were issued for Pigeon lake during the past year than during the previous one, an increased quantity has been caught. No fishing operations other than that under domestic licenses and by angling, are carried on during summer in the Pigeon lake district.

Buffalo lake is situated on the watershed between the Battle river and Red Deer river, emptying into the latter stream, and is of considerable area spreading over Townships 40 and 41, Ranges 20 and 21 west of the fourth principal Meridian. It contains only jackfish and mixed fish. Although during the past year there appears to have been caught in it, rather more than double the quantity of the previous year, this does not indicate that the lake is better stocked; but is rather to be attributed to the fact, that it was fished by twice as many men as in the previous year. The smaller size of fish caught, and poorer catch reported by anglers, indicate a diminution of fish. From my investigation made in this district in March last, I am convinced that there is some truth in reports which have reached the Department of illegal fishing in

5 GEORGE V., A. 1915

Buffalo lake; though I was unable to procure evidence to justify prosecutions. Fishing during winter was largely carried on within shacks or tents, with hooks, through holes in the ice. Many of these shacks were so constructed, as to easily conceal illegal practices, such as the use of lights, snares, or spears. It may be necessary in future either to forbid the use of shacks on the ice, within which fishing is done; or if allowed, they should be so built, as to be open on one side while fishing is being done, so that illegal fishing may not be so easily concealed. As Buffalo lake is in a well settled country, within easy reach of thriving towns, and provided with railway service; the market is good. The recent order that for the summer of 1914, fishing be restricted to angling will have a good effect.

The Beaver Hills District, comprising Beaver, St. Joseph, Oliom, Ministick and Cooking lakes, all emptying through Beaver lake and Beaver creek into the North Saskatchewan river, is a fairly well settled country, where the local demand is equal to if not greater than the supply these lakes are capable of. They only contain coarse fish, though whitefish are occasionally caught in Beaver lake, where they are said to have been plentiful formerly. Though Beaver lake and Cooking lake are somewhat depleted, compared with former years, I do not think the stock is now decreasing. Angling in Beaver lake, and Cooking lake which is a summer resort near Edmonton, is of growing importance, as a sport.

The Wabuman lake district is of great present importance, situated on the G. T. P. railway about forty five miles west of Edmonton. Wabuman lake is the only lake of Northern Alberta producing whitefish, and provided with railway service, and consequently the only source of local supply of whitefish to the towns of Alberta in summer. The importance of its fisheries may be judged from the fact, that during the fishing seasons of 1913 and 1914, 3,372 cwt. were caught, of which 2,896 cwt. were whitefish, an increase of 1,119 cwt. more than the catch of 1912 and 1913. The catch reported by Guardian Bennett, my own occasional examination of shipping bills at Wabuman, and my observation of catches and size of fish convince me that there is now no decrease of whitefish in Wabuman lake. I have to confess that the opinion expressed in my report of last year to the effect that Wabuman lake was being over-fished, has not been verified by the results of fishing during 1913-14. I think there is an improved condition, chiefly due to the longer close seasons, and to closer super-vision by Guardian A. G. Bennett. It has also been suggested to me by fishermen that the level of Lake Wabuman has been lowered some two feet and the water cleaned by the opening and clearing away of obstructions in the different streams from the Goose Quill bay, thus either improving conditions for whitefish, or making it more easy to catch them.

During the past year I endeavoured to restrict the number of fishermen's licenses for Wabuman lake, but found this most difficult, without depriving old fishermen who lived on the shores of or in close vicinity to the lake, and who had for years been to some extent dependent on the industry.

As the lake is now so near the market, and the demands on me for licenses so great; I feel obliged not only in the future interests of fisheries, but in justice to the older and resident fishermen, to refuse my recommendation of many applications for licenses made by men not resident in the district. As the shores of Wabuman lake have become a summer resort angling is of increasing importance as sport.

Island lake also situated on the G. T. P. railway, though it contains only jackfish and pickerel, is becoming of more importance, because of the increased demand for coarse fish, its closeness to Edmonton which is within sixty miles, its daily rail service; and the increasing settlement about its shores.

Lac Ste. Anne which is connected with Island lake by the Sturgeon·river, which is also its effluent into the Saskatchewan river, is with very few exceptions still fished by half-breeds and Indians resident in the district.

Although without the skill of white fishermen, I have reason to think that during the past year these people have acquired more knowledge of the craft, and displayed greater energy than formerly. Lac La Nonne is also fished generally by half-breeds. The catch reported for Lac Ste. Anne and Lac La Nonne exceeds that of last year by about one-third, from information obtained from the local Guardian, and from my own observation, and that of older settlers whitefish were more plentiful than for ten previous years.

In the close season of 1913 and 1914 half-breeds of Lac Ste. Anne and Lac La Nonne again enjoyed the privilege of fishing for two nights each week for domestic use. I do not think that this privilege was seriously abused. In any case there is no evidence that there is recent decrease of white fish.

Though the fishermen of 'Lac Ste. Anne had made some preparations for the shipping of fish, they were again disappointed in not having railway service, though steel has been laid for many months. The catch was consumed in neighbouring villages and settlements.

Shiningbank lake, of about three and a half miles in length by a mile in width is situated in Township 57, Range 14 west of the 5th Meridian, about five miles from the McLeod river, and twenty-five miles north of the Grand Trunk Pacific railway.

This lake is very well stocked with whitefish, but is so small, as to be only of local importance, and is only fished by a few resident half-breeds.

Chip lake, which is situated on the Grand Trunk Pacific about eighty miles west of Edmonton, fed by a number of small streams from the west and north, empties into the Pembina, through Lobstick river. The lake is about ten miles long by four miles at its greatest width. In it 500 cwt. jackfish were caught during last winter, by hooks, through the ice, and marketed in Edmonton and the neighbourhood, the fishermen receiving $2\frac{1}{2}$ cents per pound on the ice, and affording employment to some ten homesteaders.

The Saddle lake district, comprising Mosquito, Pinehurst, Trout, and Egg lakes emptying into the Beaver river; Whitefish, Goodfish, Saddle, Island and Hollow lakes situated within Townships 57 and 62, and between Ranges 9 and 13 west of the 4th Meridian, is one of the most important fishing districts in northern Alberta, both as regards area of water, and the fishing capacity of its lakes. The most northerly of these lakes, Mosquito, Pinehurst, Trout, and Egg lakes, besides others which have not, at least recently been fished, are abundantly stocked with whitefish, trout, and tullibee, but of all the districts of Northern Alberta, now fished, the most difficult of access in winter, and inaccessible for summer fishing. The lakes of the more southerly group, which contain white fish, have been greatly depleted; probably by persistent fishing in close seasons for many years, by Indians. The lakes containing only coarse fish show no signs of depletion.

During the summer fishing seasons a considerable quantity of coarse fish is marketed in Vegreville and Edmonton, from Island and Hollow lakes, and notwithstanding the difficulties of transportation from the more northern and white fish producing lakes, 595 cwt. of trout, whitefish and tullibee, were caught and shipped to the markets. The catch of all kinds of fish in the whole district amounted to 832 cwt. I am of opinion that as railway facilities are extended this district may rank in importance with Cold lake district.

The Moose Lake District, in which are Moose, Keehewin, Muriel and Wolf lakes situated within Townships 58 and 65 and between Ranges 3 and 6 west of the 4th principal Meridian contain whitefish and tullibee, excepting Muriel lake which only produces jackfish and pickerel. Of this group Moose lake appears to have been seriously depleted of whitefish, by fishing in close season several years ago.

The quantity of fish caught in this district for the past year appears to be one third less than that of the previous year, though this difference may be attributed to the fact that though fishermen's licenses were formerly in force in Moose lake, they were not issued for the year 1913 and 1914.

5 GEORGE V., A. 1915

The comparatively small area of these lakes, and considerable settlement will forbid the shipping of fish from the district to any great extent.

Cold lake district, which includes Cold and Primrose lakes as well as Trout lake, and a group of small lakes known as Marie and Jackfish lakes, a few miles west of Cold lake; all empty into the Beaver river. A considerable portion of these waters are in the Province of Saskatchewan.

Comparison of the past years fishing operations of this district, with those of last year shows a very much larger investment in tackle than that of the previous year. Though there was some increase both of investment in fishing plant, in the catch of 1913 and 1914, over that of the previous year, I do not think the difference is as great as appears and may be partly explained by a closer and more intelligent annual report from Overseer John M. Whitley, than I was able to obtain for the year 1912 and 13.

The Lac La Biche district, embracing Lac La Biche and Heart lake, emptying northerly into the Athabaska river; and Beaver lake one of the sources of the Beaver river, is fished mostly by half-breeds and Indians. During summer fish are only caught for daily domestic use. Many of the poorer families fishing under the privilege granted half-breed and Indians, of fishing for daily use without license.

On my visit to the district in August last, I observed that nets used under this privilege, were very short, placed near the shore, and the daily catch being so small as to be readily consumed.

Because of the plea made by the old settlers that fish cannot be successfully caught in Lac La Biche in winter, the half-breeds and Indians were again allowed to fish for two nights weekly, during close season, for domestic use.

Twenty-three licenses were taken out for this district during the past year, and a fair catch reported, which was consumed by fishermen's families or found a ready market in the neighbourhood. Heart lake contains white fish and tullibee, but is only fished by Indians and trappers.

Beaver lake is only fished by the Indians of the Reserve on its shores. It is a good whitefish lake.

The whitefish of Lac La Biche are large, the fishermen using mesh of six to seven inches.

Notwithstanding that for the past two years, the Lac La Biche half-breeds and Indians have been permitted limited fishing for domestic use, during the close seasons, and that this practice was probably customary for many years. I do not think white fish have decreased recently in Lac La Biche, judging by information obtained from old settlers, and from my own observation of the ease with which whitefish seemed to be caught with inferior nets, of small size, used carelessly in August. As an offset to the objectionable fishing in the close seasons, it is to be remembered, that by all accounts few fish have been caught in winter, that fishermen are but poorly equipped with boats and nets, and have made no serious attempts to reach the markets.

The Athabaska district in which are Calling lake, in Townships 72 and 73, Ranges 21 and 22 west of the 4th Meridian, emptying into the Athabaska river; the group of Buck, Flat, Skeleton, and Horse lakes, between fifteen and thirty miles east of Athabaska landing; the Baptiste lake group from twelve to eighteen miles west and north of the landing; and Moose lake ten miles north of the Athabaska river in Ranges 24, 25 and 26 west of the 4th Meridian, and in Range 1 west of the 5th Meridian; is of considerable importance, its waters being well stocked with whitefish, tullibee, jackfish and pickerel, and provided with railway service from Athabaska landing.

Calling lake has an area of about thirty square miles, and was fished most successfully last winter, the fish being shipped to Edmonton and other points. When visiting this lake in February last, I was informed by fishermen, that the winter fishing of Calling lake had been the most successful in their experience, and that they had found the coarse fish in unusually small proportions. Their statements are confirmed by statistics which show the quantity caught during the past season to be more than double that of the previous year; the amount realized for the catch of 1914 being $10,050, or $4,712 greater than for 1913.

The fishing operations of other lakes of this group are of minor importance, the catch being generally marketed in the neighbourhood.

Lesser Slave lake is the largest of the lakes of Northern Alberta for which licenses are granted. It contains whitefish a few lake trout, pickerel, tullibee of too small size to be of value, and a few perch of very small size.

Lesser Slave lake has always been fished by Indians and half-breeds, who have persistently done most of their fishing in the spawning season. As there are some sixty families now subsisting on this privilege, and considering that in earlier years probably a larger number did so, there can be no doubt of the evil effect of this practice, and that the supply of whitefish in Lesser Slave lake is very much less than it would be if the close season were better observed. During the year 1913 and 1914 the privilege was extended by the Department to the Indians of Lesser Slave lake, of fishing for the first ten days of November, for domestic use. Mr. Robert Potts the Fishery Overseer of that time for the district reported to me that the Indians had not abused the privilege, so far as the time specified was concerned, but that undoubtedly, instead of catching enough for immediate consumption, very large quantities were hung for winter use. A considerable proportion of these fish were fed to sleigh dogs by Indians in their journeys further north when trapping and hunting.

White fish lake in Townships 78 and 79, and Ranges 9, 10 and 11 west of the 5th Meridian is of considerable area, covering about thirty-five square miles, emptying northerly by the Loon river, into the Peace river. It is uncommonly well stocked with whitefish, though only fished by some twenty-five Indian and half-breed families, who do not take out licenses, but habitually fish in the narrows, which is a running stream between the eastern and greater part of the western or smaller part of White-fish lake. This fishing is done in the close season, when whitefish fall an easy prey to fishermen. When I visited this district in February last, no fishing was being done, but I learned that some 1,000 cwt. had been caught in the close season, my information being confirmed by my own observation of considerable quantities still on hand, the condition of fish indicating that they had been caught in the close season.

As the fishing at Whitefish lake is only done for local consumption, no harm has yet been done, but fishery interests require that this lake should be soon placed under the restrictions of the close season.

The question as to the privileges which half-breeds are specifically entitled by the fishery regulations, as regards fishing for daily family consumption without fees, and in close season, has become a most difficult one.

For generations, especially in the more northern districts they have fished without licenses and in close seasons, but as settlements extend, these privileges cannot be longer conceded without injury to fishing interests, and injustice to legitimate fishermen and consumers.

But it is generally found, that as settlements extend, the means of livelihood of these people are rather curtailed than increased. I can only suggest that reasonable enforcement of the close seasons among Indians, can only be brought about by the co-operation of the Indian Department, who might exercise more authority, and perhaps find it necessary to expend more money, in assisting Indians to carry on legitimate fishing, farming, or other industries.

In many of the more thickly settled districts, especially those settled by half-breeds, the claim is made that licenses should only be issued to persons whose properties or homes are in the vicinity of fishing lakes, though the fishery regulations specify that all British subjects or homesteaders are eligible for licenses. In practice with regard to many districts in which there appear to be as many applicants for

5 GEORGE V., A. 1915

licenses, by persons resident in the vicinity, as lakes will bear. I have discouraged the issue of licenses to other than those who have long fished in these lakes, and who live in the neighbourhood.

Though there appears to be an increased demand for whitefish, the price has not increased generally.

Fishermen are seldom well enough provided with ice houses, or storage to hold fish, nor are they in such financial circumstances as to justify their waiting for any improvement in prices. But few men earn more than a living wage.

The market for jackfish has generally improved, especially for this class of fish caught in districts where there are no whitefish, and in districts well settled, and enjoying railway service.

I am, Sir,
Your obedient servant,

J. D. WILLSON,
Inspector of Fisheries.

NOTES ON THE DISTRIBUTION AND ECONOMIC IMPORTANCE OF THE 'INCONNU' (STENODUS MACKENZIE) IN THE MACKENZIE RIVER VALLEY.

By J. C. D. MELVILLE, F.R.G.S.

Member of the Canadian Fisheries Advisory Board.

The 'Inconnu' or 'Connie' (the name by which this fish is more generally known throughout the north) is a large, coarse salmonoid inhabiting, as far as the first rapids, most of the large rivers and streams of Alaska and Arctic Canada from the Yukon to the Anderson river.

This fish is nearly identical with the Russian 'Stenodus Leucicthys' found in the Caspian sea and many of the rivers of Russia and Siberia.

The general appearance of the 'Inconnu' when freshly caught is somewhat like an Atlantic salmon; the head and mouth being larger, and the scales uniformly, of a bright silver colour. The flesh is white, very rich and oily, too rich in fact for a long diet.

It must be admitted that its edible qualities are very much a matter of personal taste—some people liking it, others claiming it to be too oily and coarse. Personally, I believe, as in many other kinds of fish, the cooking plays a most important part. However, there certainly can be no disputing the 'Inconnu's' great value as dog-feed, or dried and smoked for human consumption.

In weight the 'Inconnu' will average 15-20 lbs., but 48 lbs. or even 50 lbs. is by no means uncommon.

'Baik-huli,' the name by which this fish is known to the Slave and Hare Skin Indians of the Mackenzie river, translated means 'No-tooth' which is not quite accurate.

'Stenodus' (literally 'Short-toothed') the name of Greek derivation given to this genus by Sir John Richardson better describes their condition.

'Inconnu' is a French word meaning 'Unknown' and the name was doubtless bestowed on this fish by the French half-breed 'Voyageurs' of the old Northwest or Hudson's Bay Companies when first they penetrated the remote shores of Great Slave lake and vicinity. The fish to them was new and unlike the familiar whitefish or jackfish, or anything of the kind ever seen before.

The 'Inconnu' is found on the Slave and Mackenzie rivers below the rapids at Fort Smith, as far as the Arctic ocean. They have never been taken above these rapids. On Great Slave lake they have been found as far east as the Narrows (Sir George Back reported taking one there in 1833). They also ascend most of the rivers flowing in from the north. But, the Rivière de Rochers, Stony Island, Buffalo river and Slave river, all tributaries of Great Slave lake, are the localities where the 'Inconnu' are, at spawning time, probably most plentiful. Sergeant Mellor, R.N.W.M.P., reports 'that he was nearly able to walk across Buffalo river on their backs,' and the same has been said of Rivière de Rochers (some 50 miles east of Fort Resolution). This latter is a wide shallow stream. In the upper Mackenzie river the 'Inconnu' are probably not as plentiful as on Slave river, but some are taken in gill nets every summer at all the trading posts. They do not ascend the Liard or Bear lake rivers (two of the largest tributaries) and they do not occur in Great Bear lake, although Thomas Simpson (Dease and Simpson Expedition 1836) reported that one was caught in a whitefish net near Fort Confidence (situated at the eastern extremity of the lake). In the lower Mackenzie the fish ascend from the Arctic ocean in great quantities as far as the Swift river (known as the Sans Sault rapids) above Fort Good Hope.

The Indians establish fisheries below these rapids every year. The fish leave the delta of the Mackenzie river and Great Slave lake (at both of which places they undoubtedly winter) about June and begin to return in October. They also ascend Peel's river and many natives inhabiting that section of country depend very largely on them for their food supply. Concerning other tributaries of the Mackenzie river, I regret I have no data, and can, therefore, give no information.

The 'Inconnu' spawns at Smith from about the middle of September to the beginning of October and it is at this time that the trading companies and others establish fisheries, for they are to be found in the eddies below the Smith rapids in great quantities. The fish are generally split and hung on a stage out of reach of the dogs. On account of the large size of the 'Inconnu' it is not necessary to 'put up' the great quantity that a whitefish fishery necessitates,—one good sized fish being nearly a days ration for a team of dogs. The price asked by the Indian and half-breed fishermen at Fort Smith is or was twenty-five cents a piece (for a fair-sized fish.) they are usually caught in gill nets (5½-inch mesh) set in the eddies of the rivers, but they also readily take a spoon or bait, the latter being a piece of fresh meat or fish. The Eskimo at the mouth of the Mackenzie river catch them through the ice by means of a hook and line baited with a thin strip of bone or ivory.

The 'Inconnu' is a fish of no great vitality; it being generally found dead in the nets, as compared to the whitefish or trout which caught at the same cast are still full of life. This apparent lack of vitality or sluggishness may possibly account for this species, although predominating in great numbers in the districts to which it is native, being very noticeably stopped by rapids or swift water occurring in the streams which they frequent.

The first mention we have of the 'Inconnu' is to be found in Samuel Hearn's account of his journey from Fort Prince of Wales (Churchill) to the Coppermine river. He records taking one of the fish in Great Slave lake in 1772. After mentioning all the other well known varieties, he writes, 'Besides these we caught also another kind of fish which is said by the Northern Indians to be peculiar to this lake; at least none of the kind have ever been met with in any other. The body of the fish much resembles a ·pike in shape, but the scales, which are very large and stiff, are of a beautiful bright silver colour; its mouth is large, although not provided with any teeth, and takes a bait as ravenously as a pike or trout. The sizes we caught were from two feet to four feet long.'

Sir Alexander Mackenzie in the account of his journey of 1789 down the great

5 GEORGE V., A. 1915

river which bears his name, mentions the 'Inconnu,' but gives no description. The fish was definitely described by Sir John Richardson after his first journey as naturalist and surgeon to the Franklin Expedition.

Little or nothing is known concerning the food of the 'Inconnu," it being probably small aquatic animaculæ or small fish such as minnows. An Indian told me it is also a great scavenger, devouring the carrion carried down by all rivers in a greater or less degree, but this statement needs corroboration.

It will be readily understood that the 'Inconnu' may be of vast importance and value if the Arctic and Sub-Arctic districts of the Mackenzie valley develop in the future, as is fully expected.

It is definitely known that these fish retire back from their summer haunts and spawning grounds in the rivers to Great Slave lake, the Mackenzie river delta and bays of the Arctic ocean. The sixteen miles of rapids at Fort Smith are the only impediment to them proceeding up stream to lake Athabasca, Peace river, etc. Lake Athabasca and the delta of the Athabasca river are exactly similar to Great Slave lake and the delta of the Slave river, in which waters the 'Inconnu' now flourish. This leads to the following reflection which, while of no great economic importance at present, might perhaps in the future be worthy of consideration, and at any rate is, I venture to believe, of interest.

It would be an interesting experiment to transfer some of these fish over the 'rapids'—or otherwise stock the Upper Slave river. Peace river is not very plentifully endowed with fish life, and if it were possible to introduce the 'Inconnu' it would confer a great boon on the inhabitants of this at present sparsely populated country. In the years to come the value of a great fishery in the heart of what is destined to be one of the finest farming districts in Canada can scarcely be estimated.

Little is known of the habits or life of the 'Inconnu' and that little is, I believe, all expressed in the foregoing notes. I have not the fatuity to pretend these to be a scientific or even a full practical report,—but perhaps some attention may be drawn towards a fish which in the future may be of great value and importance, but which is now indeed well named " Inconnu.'

RETURN showing the Number of Fishermen, &c., the Number and Value of Tugs, Vessels and Boats, and the Quantity and Value of all Fishing Gear and other Material, used in the Fishing Industry in the Province of Saskatchewan, during the year 1913-14.

Number	Fishing Districts	Boats				Men Licensed	Gill-Nets		Hoop-Nets		Lines		Freezers and Ice-houses		Smoke and Fish-houses		Piers and Wharves		Number
		Row.	Value. $	Gasoline.	Value. $		Number.	Value. $	Number.	Value. $	Number.	Value. $	Number.	Value. $	Number.	Value. $	Number.	Value. $	
1	Qu'Appelle Valley	75	2250			48	110	880			600	600	8	80					1
2	Long Lake	12	300	1	700	148	470	2820					10	500					2
3	Lac Pelletier	10	200			12	12	72											3
4	Fishing and Devil's Lakes					11													4
5	Wakaw Lake	36	765	2	400	82	9	45			260	260							5
6	N. and S. Saskatchewan River	15	260			4	24	144			175	175							6
7	Sturgeon and Whitefish Lakes	18	180			10	24	120	76	380	80	80							7
8	Candle Lake	3	150			25	32	288							1	60			8
9	Trout and Red Deer and Montreal Lakes	11	110			15	72	576			33	33			1	30			9
10	Stony and Ladder Lakes	12	300	2	500	38	28	224			35	35	2	160					10
11	Doré Lake					29	262	2620									2	80	11
12	Green and Hen Lakes	16	192			5	102	714			50	50							12
13	La Plonge Lake	1	20			77	5	56											13
14	Ile la Crosse Lakes	120	3600			16	540	4320											14
15	Lac des Iles					64	60	600			25	25							15
16	Lac La Ronge					47	64	512			10	10							16
17	Turtle and Loon Lakes	7	126			32	140	1120											17
18	Jackfish and Murray Lakes	15	700	5	1300	40	136	1088			62	62	1	25	2		2		18
	Totals	351	9153	10	2900	645	2090	16193	76	380	1330	1330	21	765	2	90	4	130	

RETURN showing the Quantities and Values of all Fish caught and marketed or consumed locally, for the Province of Saskatchewan, during the year 1913-14.

Number	Fishing Districts	Salmon, cwts.	Salmon, value	Trout, cwts.	Trout, value	Whitefish, cwts.	Whitefish, value	Pickerel, cwts.	Pickerel, value	Pike, cwts.	Pike, value	Tullibee, cwts.	Tullibee, value	Mixed fish, cwts.	Mixed fish, value	Number
1	Qu'Appelle Valley					109	872	160	960	650	3250	140	700	186	930	1
2	Long Lake					1094	6564	370	1850	766	3830			97	588	2
3	Lac Pelletier					20	160	33	160	105	525			92	368	3
4	Fishing and ... Lakes							50	300	385	1900			180	720	4
5	Wakaw and Lenore Lakes							100	500	390	1950			1030	3090	5
6	N. and 3. Saskatchewan River					60	300	12	120	36	288			386	1930	6
7	Sturgeon and ... Lakes					373	1865	15	75	177	898			14	28	7
8	Candle Lake					370	1850	8	40	72	288			36	168	8
9	Trout and Red Deer and Montreal Lakes			52	260	751	3755	51	256	96	384			157	314	9
10	Stony and Ladder Lakes					7853	23499			389	1167			41	82	10
11	Dore Lake					728	2912	87	348	131	262	25	75	1100	1100	11
12	Green and Waterhen Lakes			25	100	186	744	103	309	222	444			188	188	12
13	Lac La Plonge					10000	30000			10	20	120	240	5	10	13
14	Ile la ...					1000	3000	315	945	3100	6200			700	1400	14
15	Lac des Isles			70	350	580	1740	65	195	110	229			10	10	15
16	Lac la ...			91	455	1074	4296	36	72	101	202			134	268	16
17	Turtle and Loon Lakes					815	3250	12	36	296	592			98	98	17
18	Jackfish and Murray Lakes					6000	18000	94	376	405	1215			30	60	18
19	Northern Lakes			150	450			200	400	500	1060			500	500	19
	Totals			388	1615	30693	102817	1710	6941	7936	24622	285	1015	4984	11592	

* Cwts. = 100 lbs.   † "Mixed Fish" includes greylings, bull-heads and ouananiche.

RECAPITULATION.

Of the Yield and Value of the Fisheries in the Province of **Saskatchewan**, during the year 1913–14.

Kinds of Fish.		Quantity.	Value.
			$
Trout	*Cwts.	388	1,615
Whitefish	"	30,993	102,817
Pickerel	"	1,710	6,941
Pike	"	7,936	24,622
Tullib e	"	285	1,015
Mixed Fish	"	4,984	11,592
Total			148,602

* Cts. = 100 lbs.

RECAPITULATION.

Of the number and Value of Vessels, Boats, Nets, Traps, etc., used in the Fisheries in the Province of **Saskatchewan**, during the year 1913–14.

—	Number.	Value.
		$
Boats (sail)	351	9,153
" (gasoline)	10	2,900
Gill-nets, Seines and other nets	2,166	16,573
Lines	1,330	1,330
Frezers and Ice-houses	21	765
Smoke and Fish-houses	2	90
Piers and Wharves (private)	4	130
		30,941

Number of men employed in Boats.............................. 645

5 GEORGE V., A. 1915

RETURN showing the Number of Fishermen, etc., the Number and Value of Tugs, Vessels and Boats and the Quantity and Value of all Fishing Gear and other Material used in the Fishing Industry in the Province of **Alberta**, during the year 1913-14.

Number	Fishing Districts.	Boats.				Licensed Men.	Gill Nets.		Freezers and Ice-houses,		Smoke and Fish-houses.		Number.
		Sail.	Value.	Gasoline.	Value.		Number.	Value.	Number.	Value.	Number.	Value.	
			$		$			$		$		$	
1	Lee Creek and St. Mary's River	...	...	...	...	120	...	...			...	...	1
2	Pincher and Beaver Creeks	...	...	...	...	474	...	...			...	...	2
3	Old Man River	...	...	...	...	1080	...	...			...	...	3
4	High and Sheep Rivers	...	...	...	...	600	...	...			...	...	4
5	Bow and Elbow Rivers	...	...	...	...	1400	...	...			...	...	5
6	Buffalo and Gull Lakes	18	65			28	2	12	10	256	15	60	6
7	Pigeon and Buck Lakes	49	637			92	106	530					7
8	Beaverhill and Hastings Lakes	20	599			35	29	290	13	260	6	140	8
9	Wabamun and lsle Lakes	34	625	3	1200	55	169	1014	15	375	3	150	9
10	Chip and Shining Bank Lakes	2	30			19	12	72	3	120			10
11	Ste., Anne and La Nomme Lakes	30	450			48	57	399					11
12	Floating Stone and Trout Lakes	5	50			28	45	375					12
13	Cold and Primrose Lakes					65	280	2800					13
14	La Biche and Beaver Lakes	22	440			23	40	240					14
15	Calling and Baptiste Lakes					23	120	1500			3	125	15
16	Moose and Keehewin Lakes	1	20			23	34	214			2	20	16
17	Lesser Slave Lake	21	575			17	135	1350	1	200	3	700	17
	Totals	205	3482	3	1200	4130	1029	8796	42	1205	32	1195	

SESSIONAL PAPER No. 39

RETURN showing the Quantities and Values of all Fish caught and marketed or consumed locally, for the Province of **Alberta**, during the year 1913-14.

Number	Fishing Districts.	Trout.		Whitefish.		Pickerel.		Pike.		Tulli-bee.		†Mixed Fish.		Number
		Cwts.	Value.	Cwts.	Value.	Cwts.	Value.	Cwts.	Value.	Cwts.	Value.	Cwts.	Value.	
		$		$		$		$		$		$		
1	Lee Creek and St. Mary's River. ...	30	300	....	....	...	....	10	50	...	....	....	....	1
2	Pincher and Beaver Creeks.	55	550	....	....	....	....	....	....	....	....	....	....	2
3	Old Man River......	215	2150	....	....	....	....	....	....	....	....	....	....	3
4	High and Sheep Rivers.......... ...	290	2900	....	....	....	....	....	....	....	....	....	....	4
5	Bow and Elbow Rivers.	450	4500	....	....	....	....	120	600	....	....	....	....	5
6	Buffalo and Gull Lakes..............	....	....	....	....	....	....	590	2950	....	....	10	20	6
7	Pigeon and Buck Lakes....	...	.....	361	2166	13	78	111	333	...	1	40	80	7
8	Beaverhill and Hastings Lakes.......	...	.....	....	....	15	105	130	910	...	....	105	210	8
9	Wabamun and Isle Lakes.............	...	.....	2896	11584	110	550	354	1062	...	....	12	24	9
10	Chip and Shining Bank Lakes.......	...	.....	10	50	....	....	500	1500	...	....	....	....	10
11	Ste. Anne and La Nonne Lakes.....	...	.....	649	3245	3	15	43	129	...	....	9	18	11
12	Floating Stone & Trout Lakes........	257	1285	238	1190	48	240	130	380	80	240	74	148	12
13	Cold and Primrose Lakes............	1111	4444	2606	7818	201	603	191	382	...	....	183	183	13
14	La Biche and Beaver Lakes...........	...	.....	500	1500	....	....	15	30	100	200	50	50	14
15	Calling and Baptiste Lakes..... .. .	...	.....	3360	10080	80	320	200	500	72	216	426	426	15
16	Moose and Keehewin Lakes...... ...	...	.....	224	896	18	72	90	270	38	114	13	26	16
17	Lesser Slave Lake...................	20	80	3168	12672	55	220	265	265	...	....	380	380	17
	Totals............	2428	16209	14012	51201	543	2203	2749	9371	290	770	1302	1565	

*Cwt. = 100 lbs.   † "Mixed fish" includes greyling, bull-heads and ouananiche.

5 GEORGE V., A. 1915

RECAPITULATION

Of the Yield and Value of the Fisheries in the Province of **Alberta,** during the year 1913-1914.

Kinds of Fish.	Quantity.	Value.
		$
Salmon..*Cwt.		
Trout.. "	2,428	16,209
Whitefish.. "	14,012	51,201
Herring.. "		
Bass.. "		
Pickerel... "	543	2,203
Pike.. "	2,749	9,371
Sturgeon... "		
Eels.. "		
Perch... "		
Maskinongé.. "		
Tullibee... "	290	770
Cat-fish... "		
Goldeyes... "		
Mixed Fish.. "	1,302	1,565
Caviare.. "		
Total...		81,319

*Cwt.=100 lbs.

RECAPITULATION

Of the Number and Value of Boats, Nets, Traps, etc., used in the Fisheries in the Province of **Alberta,** during the year 1913-1914.

—	Number.	Value.
		$
Boats (sail)..	205	3,482
" (gasoline)..	3	1,200
Gill-nets, Seines and other nets..	1,029	8,796
Weirs..		
Lines..		
Freezers and Ice-houses..	42	1,205
Smoke and Fish-houses...	32	1,195
Piers and Wharves (private)...		
		15,878

Number of men employed in Boats................................ 4,130

APPENDIX No. 8.

YUKON TERRITORY

To the Superintendent of Fisheries,
Ottawa.

SIR,—I have the honour to submit herewith the annual report on the fisheries of the Yukon territory, for the fiscal year ending March 31, 1914.

In remote parts of this vast territory where it is almost impossible to visit I have to base my estimates on what information I can gain from traders and trappers who, alone, visit those isolated parts inhabited solely by Indians. By reference to the returns I have forwarded to you, you will observe that the catch by Indians and that by white are under two different heads.

SALMON.

Many were led to believe that the run of salmon was becoming less year by year in the Yukon and its tributary waters, owing, it was claimed, to the use of crude oil used as fuel by the steamers plying on the lower Yukon river between Dawson and St. Michaels. But this seems to have no foundation for the year 1913 saw the largest run since the year 1898.

WHITE FISH AND GRAYLING.

Grayling, which abound in all the side-streams are as plentiful as ever, excepting the Klondike river where the dredges are working and keeping the water in a muddy state. The lakes from which the whitefish are taken, La Barge, Tatleman, and Thadsun, are well stocked and do not seem to have suffered from fishing operation.

CLOSED SEASONS.

The closed season for the past year was well observed; no violations coming under my observation, or any reported by guardians.

FINES AND FORFEITURES.

I also beg leave to report that though I secured no convictions I destroyed eight nets of illegal size, also three dams which I found placed in the mouth of streams. It was impossible to locate the guilty party or parties.

I am, Sir,
Your obedient servant,

C. C. PAYSON,
Inspector of Fisheries.

RETURN showing the Number of Fishermen, etc., the Number and Value of Tugs, Vessels and Boats, and the Quantity and Value of all Fishing Gear and other Material, used in Fishing Industry in the Whole Yukon Territory, during the year 1913-14.

Number	Fishing Districts	Boats Sail.	Boats Value. $	Boats Gasoline.	Boats Value. $	Men.	Gill Nets Number.	Gill Nets Value. $	Lines Number.	Lines Value. $	Freezers and Ice-houses Number.	Freezers and Ice-houses Value. $	Smoke and Fish-houses Number.	Smoke and Fish-houses Value. $	Piers and Wharves Number.	Piers and Wharves Value. $	Persons employed in Freezers, Fish-houses, &c. Number.	Number
	Owned by Whites.																	
1	Dawson	9	300			14	40	480	45	22	1	4500	1	100			14	1
2	Pelly District	4	120			8	20	200	20	10			1	125			2	2
3	Forty-Mile	6	180			12	14	140	40	20			1	100	1	300	1	3
4	Lake La Barge	5	150			10	16	160	60	30								4
5	Carcross	6	180			14	36	432	45	22								5
6	Klondike River	9	270			11	24	300	26	12			1	100				6
7	Thistle & Sixty-Mile	5	150			5		240	16	8								7
8	All other agts of Yukon Territory.	8	240			8	22	220	60	15								8
	Owned by Indians.																	
9	Salmon River	8	150			16	20	120										9
10	Teslin Lake	6	90			12	9	90										10
11	Tagish	5	80			15	11	77										11
12	Big Lake	2	40			6	8	56										12
13	McPherson	4	80			8	7	49										13
14	Selkirk and Pelly	12	300			20	45	315	40	20								14
15	Duncan	4	80			8	12	84										15
16	Porcupine	5	100			12	9	63										16
17	Peel River	6	180			12	11	70										17
18	Rampart	8	250			16	14	98										18
19	Hootchi	6	200			12	12	80										19
	Totals	1180	31400			219	364	3274	352	159	1	4500	4	425	1	300	17	

RETURN showing the Quantities and Values of all Fish caught and marketed or consumed locally, in the whole Yukon Territory, during the year 1913-14.

Number	Fishing Districts	Salmon,* cwts.	Salmon, value. $	Trout, cwts.	Trout, value. $	Whitefish, cwts.	Whitefish, value. $	Pickerel, cwts.	Pickerel, value. $	Pike, cwts.	Pike, value. $	Maskinongé, cwts.	Maskinongé, value. $	†Mixed Fish, cwts.	Mixed Fish, value. $	Number
	Caught by Whites.															
1	Dawson	300	3000	6	180	40	1000	5	100	2	50			100	2000	1
2	Pelly District	25	250	24	750	100	2500					3	75	20	440	2
3	Forty-Mile	180	1800	4	120	5	125					4	100	40	800	3
4	Lake LaBarge			7	210	115	2875					1	25	30	600	4
5	Caross			45	1350	35	875							20	400	5
6	Klondike River	55	550	12	360	10	250							85	1700	6
7	Thistle and Sixty-Mile	45	450	5	150	8	200							70	1400	7
8	All other parts of Yukon Territory	150	1500	30	900	40	1000					5	125	110	2200	8
	Caught by Indians.															
9	Salmon River	160	1600	10	300	40	1000					2	50	70	1400	9
10	Teslin Lake			14	420	45	1125							90	380	10
11				10	300	54	1350							70	1400	11
12	Big			6	180	10	250							35	700	12
13		140	1400	5	150	27	675							40	800	13
14	Sel.irk and Pelly	300	3000	25	750	80	2000							55	1100	14
15	Duncan	50	500	10	300	22	550							30	600	15
16	Porcupine	40	400	2	60	25	625							20	400	16
17	Peel River	130	1300	20	600	50	1250							25	500	17
18	Rampart	145	1450	14	420	70	1750							64	1289	18
19	Klotchi	100	1000	22	660	60	1500							50	1000	19
	Totals	1820	18200	271	8160	836	20900	5	100	2	50	15	375	1024	29480	

* Cwt. = 100 lbs.   † "Mixed Fish" includes greyling, bull-heads and ouananiche.

5 GEORGE V., A. 1915

RECAPITULATION.

Of the Yield and Value of the Fisheries in the whole **Yukon Territory**, during the year 1913–14.

Quantity consumed in Canada.	Quantity consumed in U.S.A.	Kinds of Fish.		Quantity.	Value.
					$
......All.......	....None.......	Salmon........................... *cwts.		1,820	18,200
"	... "	Trout .. . . "		271	8,160
"	... "	Whitefish.... . ... "		836	20,900
....	....	Pickerel...... .... "		5	100
....	....	Pike........ . ... "		2	50
....	....	Maskinongé........................... "		15	375
......All	....None.	Mixed fish...................... ..... "		1,024	20,480
		Total................................		....	68,265

*Cwts. = 100 lbs.

   Quantity consumed in Canada............................. .... *all*

RECAPITULATION

Of the Number and Value of Vessels, Boats, Nets, Traps, etc., used in the Fisheries in the whole **Yukon Territory**, during the year 1913–14.

—		Number.	Value.
			$
Boats (sail)...		118	3,140
Gill-nets, Seines and other nets		364	3,274
Lines........		352	159
Frezers and Ice-houses...		1	4,500
Smoke and Fish-houses...		4	425
Piers and Wharves (private).. .		1	300
Total........		....	11,798

   Number of men employed in boats.......................... 219
    "   persons employed in Fish-houses, Freezers, etc..... ... 17
                        236

APPENDIX No. 9.

BRITISH COLUMBIA

CHIEF INSPECTOR FOR THE PROVINCE, F. H. CUNNINGHAM, NEW WESTMINSTER.

DISTRICT No. 1.—COMPRISING THE SOUTHERN PORTION OF THE PROVINCE. ASSISTANT INSPECTOR, A. P. HALLADAY, NEW WESTMINSTER.

DISTRICT No. 2.—COMPRISING THE NORTHERN PORTION OF THE PROVINCE. INSPECTOR, J. T. C. WILLIAMS, PORT ESSINGTON

DISTRICT No. 3.—COMPRISING VANCOUVER ISLAND AND PART OF THE MAINLAND ADJACENT THERETO. INSPECTOR, E. G. TAYLOR, NANAIMO.

REPORT ON THE FISHERIES OF THE PROVINCE BY THE CHIEF INSPECTOR.

To the Superintendent of Fisheries,
    Ottawa, Ont.

SIR,—I have the honour to report on the fisheries of the Province of British Columbia for the fiscal year ended 31st of March, 1914, as follows:—

COMMERCIAL VALUE OF THE FISHERIES.

The total value of the fisheries for the whole Province for the past fiscal year amounts to $13,891,398, as against $14,455,488 for the preceding fiscal year, showing a decrease of $564,090. A reference to the statistical returns will show that practically one-half of this decrease is due to the falling off in the whale fishery, as there were only 705 of these mammals captured as compared with 1,107 for the previous year. There is also a decrease of nearly 30,000 cwts. of halibut caught as compared with the previous year, and which, coupled with the low market price for several months during the season, has detracted to the extent of $727,000 from the value of this species for this year.

On the other hand, however, there is an increase of $703,302 in the value of salmon, which is explained, of course, by the fact that this was a 'big year' for the salmon run on the Fraser river. The total number of cases of pinks and cohoes packed is comparatively small as compared with the season of 1912, which is due no doubt to the fact that a poor price prevailed for these species in 1912, and as 1913 was the big sockeye run on the Fraser, less demand prevailed for the cheaper varieties.

It is very gratifying to note that whilst there is a falling off in the value of some varieties, the big run of sockeye salmon to the Fraser river keeps up to the average, as there was a total of 684,596 cases packed as compared with the previous four year cycle

5 GEORGE V., A. 1915

in 1909 when there were 567,203 cases put up, giving an increase of over 100,000 cases in favour of the past year. There were also one and a half million cases of sockeye packed in Puget Sound, the fish being caught when passing through the American waters on their way to the Fraser river. These figures would tend to show that the fish cultural operations as conducted on the Fraser river watershed are yielding results of great value to the fisheries of this Province.

The total value of the fishing vessels, boats, nets, etc., is $12,489,613, and 20,707 persons were employed in the fishing industry during the period covered by this report.

DISTRICT NO. 1.

This district comprises the Fraser river, a portion of Howe Sound, including the inland lakes of the southern part of the Province.

The quantity of salmon canned of all species in this district amounted to 732,059 cases, and the total commercial value for the whole district is $5,590,660. There were 2,560 salmon gill-net licenses issued during the year, and 35 cannery licenses. All the cannery licenses and a large proportion of the gill-net licenses were for the Fraser river. Of the operators of the 2,560 licenses, 1,071 were whites, 408 Indians and 1,081 Japanese.

In my last year's report, reference was made to the department's approval for the building of a suitable gasoline launch for patrol work on the Fraser river, and yet of sufficient sea-going qualities to enable her to visit any part of the coast of the Province; and I am pleased to say the authority thus given resulted in the construction by the Westminster Marine Railway Company, New Westminster, of the launch 'Fispa.' This boat is twin screw, 85 feet long, 14½ feet beam, equipped with modern gasoline engines having an equivalent of 96 horse-power each. This launch, during the past season, has travelled 5,912 miles without mishap, and has given good satisfaction in every way. She is used by the Chief Inspector for visiting all parts of the coast, and her services have been utilized during the past winter in special inspections of various inlets, thus enabling the captain, who is well versed in fish life, to inspect and report on the various spawning streams up which salmon ascend for propagating purposes. The information thus obtained, is valuable, and affords information of the various species of salmon frequenting the rivers and also information as to obstructions which bar the ascent of salmon to their natural spawning beds.

The protection of the fisheries of the Fraser river is looked after by the gasoline boats 'Swan,' 'Foam' and 'Elk.' These boats are in command of efficient patrolmen, and have rendered valuable service. During the past year there were 633 prosecutions in this district for violation of the Fisheries Act, and fines aggregating $5,416.75 were collected.

DISTRICT NO. 2.

The supervision of this district is under Inspector of Fisheries J. T. C. Williams, and who, during the season, visits all the fishing centres in this large area. He has his main office at Port Essington, on the Skeena river, and during the winter season, is to be found in Vancouver, where most of the fishing companies operating in the north maintain their head offices, and where their business is transacted during these months.

The salmon operations were not a success during the past season, only 417,453 cases being packed as against 663,668 cases for the previous year. The greatest decrease in the number of cases packed is on the Skeena river, which falls off to the extent of 90,000 cases; and Rivers Inlet which shows a decrease of 69,000 cases. This large decrease is somewhat difficult to account for, as a very fair proportion of fish reached the Skeena river spawning grounds and in Owekayno lake, the spawning area for Rivers Inlet, there were to be found myriads of spawning fish. The climatic condi-

SESSIONAL PAPER No. 39

tions were, however, very bad for fishing. Southeast winds accompanied by rain, were quite frequent, and during such weather, the fish invariably swim deep, thus passing under the nets. These climatic conditions applied also to Rivers Inlet.

It might also be remarked that on reference to satistics of 1909, the year of the 'big run' on the Fraser, the pack in the north was correspondingly small. For instance, Skeena river only produced 140,739 cases in that year as against 222,035 cases in 1910. In 1905 Skeena river only produced 114,000 cases as against 162,000 cases for 1906. The same remarks will apply to Rivers Inlet. I am not prepared to say that the 'big run' on the Fraser has any bearing on the northern waters, but it is a strange coincidence that in the years of plentitude on the Fraser, there is a dearth in the north.

Whilst there was a decrease of 17,000 cases on the Naas river as compared with the previous year, it is expected that the additional spawning grounds in Medziaden lake made accessible by the fishway built last season, thus enabling the salmon to surmount the falls at this point, will prove of great value to this area.

The salmon fisheries of the north coast and Queen Charlotte islands were practically a total failure with the exception of spring salmon. The spring salmon in these areas are caught by trolling. The run was a large one; but this species are all mild cured, and packed in tierces. Certain recommendations are made in Overseer Harrison's report as to the fisheries of this area, and it is suggested that consideration be given the same at the next meeting of the Fisheries Advisory Board in October.

During the past season the Fisheries Patrol Launch 'Gannet' was put in commission for the purpose of protecting the fisheries around Queen Charlotte Islands. She has performed good service under the command of Captain Haan, and for her size, she is a particularly good sea boat, strongly built, and having been especially constructed to meet the conditions prevailing around these islands.

The new Central division referred to in my report of last year was efficiently protected by Overseer James Boyd, with a chartered boat which, although somewhat slow, rendered good service. This is a large and important division, and the necessity for a special boat capable of being out all winter if necessary, was felt; consequently, a new boat was designed and built, and will be in commission for the season of 1914.

The Fisheries Regulations were well enforced. Inspector Williams has the whole district well in hand, and he received loyal support from his staff of Fishery Overseers.

DISTRICT NO. 3.

This district is under the supervision of Inspector E. G. Taylor, with office at Nanaimo, and comprises the whole of Vancouver Island and the mainland adjacent.

The statistical returns show that there is again this year an increase in the commercial value; the total being $3,647,823 as compared with $3,110,877 for the previous year. There is also an increase in the quantity of herring caught, being 557,320 cwts. for this year as compared with 515,980 cwts. for the previous season, and the value is $709,669 as against $462,963. As in district No. 2, there is a falling off in the number of whales caught in this district, being 809 captured during the fiscal year ended the 31st of March, 1913, as against 486 for a similar period ending 31st of March, 1914. Whilst the statistics show an increase of over 51,000 cases of sockeye salmon packed during the year, it must be remembered that quite a lot of these fish were purchased on the Fraser river and canned at the canneries in this district; but the pack of fish at the canneries located at Quatsino and Clayoquot was very disappointing. This is difficult to explain, unless by some freak of nature, the 'big run' to the Fraser river has some general effect on the run in the northern waters.

The cod fishery of this district has during the past season received a great impetus, and I agree with Inspector Taylor's remarks on this important fishery, that either a close season or reserve areas in which no fishing can be done at certain seasons of the

5 GEORGE V., A. 1915

year should become law. This question has, however, been referred to the biological board of Canada for their consideration and advice as to the best action to take.

The Fishery Overseers appointed to the new divisions in this district, given in detail in my report of last year, have been supplied with suitable gasoline boats. and the launches 'Cohoe,' 'Raven' and 'Gull' were commissioned and the 'Egret' and 'Heron' completed later in the year, and are now available for service. These boats were all designed by a Naval Architect for the special services to be performed.

OBSTRUCTIONS.

The removal of obstructions to the free ascent of salmon to their natural spawning grounds is one of the most important works in connection with the preservation of fish life. Unless the parent fish have free access to spawning areas, the species must grow less, and the asset of the fisheries decline.

The Department, having in view the value to be derived from systematic and economical removal of such obstructions, appointed a resident engineer, whose services are entirely devoted to this class of work coming under the Fisheries Branch of the Department. Mr. J. McHugh, an engineer with 14 years experience, was appointed to this position, and during his term of office, has rendered valuable service.

The most serious obstructions which have occurred in the Province of British Columbia, and probably in the whole of the Dominion of Canada, for many years, were those which existed in the Fraser river at China Bar, Skuzzy Rapids, White's creek and Hell's Gate. The three first named were due to the construction of the railway on the left bank of the river, the bays which existed and served as resting places for the fish, being destroyed. Whilst these obstructions were of a serious nature, the most serious one existed at Hell's Gate, which was greatly increased by a slide which took place on the 24th February last, bringing down part of the mountain and carrying with it about 20 feet of the railway tunnel which exists at this place, and greatly reducing the width of the river there.

Some preliminary work was done under the supervision of the Provincial Fisheries Department at all of these points last season, at the time when the sockeye were running, enabling large numbers to pass these points. The cost of this work was assumed by the Dominion Department.

The slide above referred to was a menace to the salmon fishery of the Fraser river, and both the Dominion and Provincial Departments were seized with the fact that strenuous work was necessary to make this point accessible to fish during the run of 1914. With this object in view, a conference of engineers was held, when it was decided that heroic action was necessary, and which resulted in the work being placed in the hands of the Pacific Dredging Company, of Vancouver, on a force contract basis, the work to be done under the supervision and to the satisfaction of the Department's resident engineer. The amount of work involved, with the short time available in which to perform it, presented a most serious problem, but fortunately the Company in question were in possession of modern machinery and skilled mechanics, and were thus able to commence the work with little delay, and to take advantage of the low stage of the water in the river at that time, thus removing large quantities of rock below the normal height of the water in the river at the time the sockeye run. It might.be stated that the intrinsic value of the machinery used is contained in the patent carrier and remodelled cable engine. To enable this machinery to work, a 700 foot span of 2¼ inch cable is stretched from bank to bank across the river. and about 170 feet above last year's low level water mark, each end being anchored in the bank and tested to carry a swiftly moving load equal in weight to 20 tons. The carriage moves on this cable with its load at the rate of 20 feet per second. and could be geared to move at a greater rate of speed if conditions permitted. The rock has to be carried by this machine from the left to the right bank of the river, where it is piled, there being no room on the left side for this purpose.

The work is proceeding very satisfactorily, and it is hoped that by the time this season's's run of sockeye reaches this part of the river, conditions will have so improved that they will have no serious difficulty in passing.

The work of constructing a passageway for fish at the Medziaden Falls, Naas river district, which was performed under the supervision of officers of the Provincial Department, has been completed, and paid for by the Dominion Department. This is a successful piece of work, and opens up large additional spawning areas in this locality. Obstructions have also been removed from the Kimsquit, Atnarco and Nicomekl rivers. Arrangements were made too, for the removal of the obstructions from Ellerslie channel, Finlayson channel, and Mary Cove.

When the work now going on in the Fraser river is completed; other rivers in the Province in which obstructions exist, will be looked after, and the work undertaken in accordance with their value as spawning streams.

REGULATIONS.

. The changes in the regulations referred to in my report of last year, as affecting the northern part of the Province, came into effect during the fishing season of 1913. These changes were in line with the desires of the Federal and Provincial Governments to encourage white fishermen to become permanent settlers in the north, and operate the fisheries. A number of licenses were reserved for this purpose, with the result that there were 167 white fishermen operating in the various fishing centres of the northern part of the Province.

Whilst it is not possible to effect such a radical change as this in one season, the results were satisfactory from a Departmental standpoint, and from the number of enquiries already made, it is certain a greater number of white men intend operating in the north during the season of 1914.

A departure was also made in connection with the protection of the herring close season being abolished and areas reserved in which no fishing was allowed during the spawning season. This was effected, and is in the greater interests of this species, as the herring on this coast spawn at different seasons of the year.

SEA LIONS AND HAIR SEALS.

Grave complaints have been made to this Department about the destruction of salmon by these mammals. Whilst these complaints emanate from all the principal fishing centres throughout the Province, the damage done by the hair seal is especially noticeable in the Fraser river.

With the object of decreasing the number of seals, the Department authorized the payment of a bounty equal to $3.50 per seal destroyed, and limited the number on which bounty would be paid for the first year to 1,000. The bounty on the first thousand seals, amounting to $3,500, was claimed in a short time, and an additional amount was granted, the details of which will appear in the annual report for next year.

Whilst the amount allowed for the destruction of each seal may seem large, it might be explained that the carcasses of the seals are very difficult to obtain after being shot, as unless hit in a vulnerable spot, they immediately sink and the carcasses cannot be secured. From information I have collected, only about one seal in every five killed, is secured; consequently, whilst bounty was paid on only one thousand, the actual destruction on this basis would amount to five thousand seals.

The question of reducing the number of sea lions does not present such a difficult problem as the hair seal, as they herd and produce their young on the islands and if the islands were raided during the month of June, practically all the pups could be destroyed and no doubt many of the old sea lions as well. The islands frequented are, however, situated in the ocean where rough weather may be expected, and it is only a

5 GEORGE V., A. 1915

substantial boat that could be used for the purpose. These lions are looked upon by those engaged in the fishing business as a serious menace to the industry, and it is recommended that the Department consider action towards reducing the herds. It is quite possible that the cannerymen and fishermen of the north would favourably consider assisting the Department in this direction.

HEAD OFFICE.

It might be of interest to refer to the head administrative office of the Fisheries Branch of the Department located in New Westminster. The office was established three years ago, and its usefulness and work have increased with the development of the fisheries throughout the Province.

During the past year 3,381 letters were received; 3,722 letters written and 1,846 accounts were approved and passed for payment. Altogether, there were 4,860 licenses issued, together with all the Revenue statements, &c., required to carry on the business of the Department. The undersigned, as Chief Inspector for the Province, visited all the fishing centres in the Province. The office staff consists of five clerks and a resident engineer, and I am pleased to testify to the unselfish and satisfactory manner in which they have performed their duties at all times.

GENERAL REMARKS.

It was a great pleasure to those engaged in the fishing industry and to the officers of the Department, to welcome to this Province last year, the Honourable Mr. Hazen, Minister of Marine and Fisheries. He visited most of the great fishing centres and canneries, making himself familiar with conditions as they existed.

The usual fishery exhibit was held in connection with the Annual Exhibition in New Westminster. This exhibit is becoming of greater attraction each year, and illustrates, to a limited extent, the species of fish indigenous to the waters of this Province, as well as the fish cultural operations as conducted by the Department throughout the Province. Such an exhibit, located practically on the banks of the mighty Fraser, is of great value from an educational and practical standpoint, and the question of a larger building and a greater display well merits the favourable consideration of the Department.

I am pleased to be able to again refer to the many profitable discussions which have been held between the Provincial Fisheries Department and myself. The results have been satisfactory and the pleasant relationship of the past still continues.

It is with great satisfaction that I refer to the assistance and loyalty rendered to me by the District Inspectors of Fisheries and Fishery Officers. Their duties have been well performed, very often under very adverse circumstances.

Careful consideration has been given to all matters which would in any way tend to benefit the fisheries of this coast, and it is hoped the decisions reached have been in the best interests of the same.

I am, Sir,
Your obedient servant,

F. H. CUNNINGHAM,
Chief Inspector of Fisheries.

SESSIONAL PAPER No. 39

REPORT ON THE FISHERIES OF DISTRICT No. 1.

F. H. CUNNINGHAM, Esq.,
    Chief Inspector of Fisheries,
        New Westminster, B.C.

SIR,—I have the honour to hand you herewith the Annual Statistical Report for District No. 1 for the fiscal year 1913-14.

The great increase in the total value of the products of the fisheries in the district over former years is very satisfactory and encouraging, showing an enormous development in this important industry.

This was the 'big run' year for the sockeye variety of salmon, and the results of the season's operation and the pack of the canners, show a marked improvement in the run over the last 'big run' year, namely, 1909-10. The following is a comparative statement:—

Salmon. Year.	Sockeye. Cases.	Other Varieties. Cases.	Total. Cases.
1909 and 1910	542,248	24,955	567,203
1913 and 1914	684,596	47,463	732,059

The increase in the pack of sockeye, you will observe, is very marked, and surely demonstrates the excellent results accruing from the operation of the hatcheries, and amply justifies the policy of the Department in their annual expenditure for building and maintaining these establishments, as well as in clearing the various streams of obstructions, and making it possible for the parent fish to reach their natural spawning grounds. The salmon in quality, also, this year, was excellent throughout.

HALIBUT.

The catch of halibut this year was very satisfactory, totalling 9,367,700 pounds, and commanding a good price, averaging $5\frac{3}{4}$ cents per pound wholesale.

HERRING.

The herring fishing in this district is rapidly increasing in importance. The means of capturing this variety of fish is, as you know, by use of gill-nets, principally in deep water, and consequently a larger and better quality of herring is taken. They are practically all used for kippering, outside of the quantity sold fresh in the local markets. The average price paid wholesale, this year, was 3 cents per pound.

OTHER VARIETIES.

During the 'big run' years for sockeye, the fishermen naturally devote more attention to that particular class of fishing, and consequently, a smaller return of other varieties is to be expected. This year, however, it will be observed that the quantity and price of these varieties are well up in average with those of former years.

BY-PRODUCTS.

The work of the 'Canada Fish Products, Limited,' was this year, under the management of Mr. Williamson, conducted very satisfactorily. Considering the large amount of offal to be handled from the various canneries and the occasional adverse circumstances, splendid work was done, as the following statement (giving the output of the establishment) will show:—

39—17

5 GEORGE V., A. 1915

Refined oil produced.. .. .. .. ·· .. ..gals. 97,650
Finished guano produced.. .. .. ·· .. ..tons.   589¼

This establishment is a very important factor in the industry of this district, as its operation not only relieves the various canners of the expense and trouble of taking the offal out to sea, but saves (which otherwise would be lost) and places on the markets, the two very valuable commodities above mentioned.

The policy of the Department in encouraging this industry is a very wise one.

VALUE OF THE FISHERIES.

The value of the product of the fisheries of this district for this year, totals $7,012,787.

HAIR SEALS.

The question of the destruction to, especially, the spring salmon, by hair seals, is still a very serious one. At certain seasons these mammals become such a pest that the fishermen have great difficulty in saving the fish caught in their nets from being either mutilated and spoiled for the market, or taken entirely. They appear to possess almost human intelligence, as immediately a fish strikes a gill-net, the seal is usually watching his opportunity, and will capture the fish before the operator of the net. can reach the spot. It will be seen that this is a very great loss to the fishermen and dealers.

It is estimated by those who are familiar with the habits of the hair seal and sea lion, that they will devour and destroy from four to ten salmon each per day, and the enormity of this loss is emphasized when it is remembered that it is upon the spring salmon that they appear to mostly prey; and this variety is always in demand and commands a good price.

The Department this year, in an endeavour to overcome this menace, appropriated a considerable sum of money for bounty, which resulted, before the close of the year, in the destruction of several thousand hair seals in the Province. This is no doubt a wise policy, and if continued, will have a very beneficial effect.

Respectfully submitted,

A. P. HALLADAY,
Assistant Inspector of Fisheries.

REPORT ON THE FISHERIES OF DISTRICT No. 2.

To F. H. CUNNINGHAM, Esq.,
Chief Inspector of Fisheries,
New Westminster, B.C.

SIR,—I have the honour to enclose my statistical report on the fisheries of the northern coast of British Columbia, District No. 2, for the fiscal year ended March 31, 1914, including the salmon packs of the different divisions. These returns show a decrease in the aggregate of the total value of fish and fish products for 1913-1914, being $3,230,788 as against $5,081,291 for 1912-1913.

This decrease is accounted for to a great extent by a reduction of 245,915 cases of salmon and the failure of the sockeye run, in all the divisions of my district in a more or less degree, and also by the fact that the lower grade of salmon, notably the humpback, though in as large if not larger quantities than ever, were quoted at so low a figure on the market, that it did not recompense the canners for packing them. The herring fisheries also were somewhat neglected on account of the exceedingly low figure they brought on the market, especially the salted article, which did not leave a margin of profit for the fishermen, and until the prices obtained for the raw product increases these fisheries will receive very little attention. There is always a certain demand for herring bait in a fresh or frozen state and this will probably increase in the near future, which will greatly assist those interested in our herring fisheries. No new canneries or salteries have been erected in 1913-1914 in my district.

The total pack of salmon for the season of 1913-1914 is as follows:—

1913 -14.	Cases.
Sockeye	183,731
Spring	24,458
Cohoe	41,169
Humpback and dog	168,095
Total	417,453

1912- 13.	Cases.
Sockeye	301,063
Spring	39,814
Cohoe	98,202
Humpback and dog	224,289
Total	663,368

Approximate decrease in detail:—

	Cases.
Skeena Rivers	90,203
Rivers Inlet	69,601
Naas River	17,739
North Coast and Q. C. Islands	68,372
Total	245,915

SKEENA RIVER.

The run of sockeye on the Skeena river was a failure, in fact it was the worst run that I have known during my sixteen years of service in the Department on the northern fisheries. It is somewhat difficult to account for this failure as my reports for four and five years back show large quantities of sockeye and other varieties of salmon on the spawning grounds, even this year reports from our officers all along the Skeena, show larger quantities of sockeye arriving on the spawning grounds, than for the last two or three years, and as salmon are not in the habit of reaching the spawning grounds either by aerial or land navigation, we must conclude they proceeded up the Skeena as usual, and on account of the climatic conditions being worse than for many seasons, notably south east winds accompanied by rain, the fishermen were unable to catch them, as they invariably swim deep and those taken are usually on the lead line while those escaping keep close on the bottom, and allow the nets to scrape over their backs. After careful investigation I have come to the conclusion that salmon like other animals are gradually becoming educated and avoid the nets when seen in clear water, recognizing the element of danger to themselves. The run of spring salmon was exceptionally heavy also that of humpback, and the cohoe was in fair numbers, also steelheads. I enclose Overseer Norrie's report in connection with his division in which he mentions the spawning grounds and other matters of interest.

39—17½

5 GEORGE V., A. 1915

RIVERS INLET.

Like the Skeena river there is a large decrease to report in the sockeye fisheries, which is practically the only variety of salmon canned on the Inlet.

The climatic conditions referred to on the Skeena were applicable to Rivers Inlet, which enabled the sockeye to escape the nets. This was due also to the fact that the sockeye were extremely small, smaller in fact than on any other season previously recorded, running all through the season fifteen to the case, the usual average being twelve to twelve and a half per case, consequently they were able to pass through the meshes of the nets, as well as avoid them in the clear water. Our officer stationed at the head of the Inlet in the mouth of the Wharnock river states that on Saturday and Sunday July 26, 27, the most marvellous sight was witnessed by himself and others, thousands upon thousands of sockeye were jumping in the rivers mouth on their way to the spawning grounds. He stated that you could see hundreds out of the water at the same time, and this continued all Saturday and part of Sunday, the fishermen were greatly excited on Sunday night, but their expectations were not realized as they came in on Monday morning with only average catches, proving that the salmon had escaped. Fishery Overseer Saugstad reports the spawning grounds on Oweekayno lake and tributary streams well stocked with salmon of all varieties. The fishery regulations were well observed throughout the season. No violations being reported.

NAAS RIVER.

The pack of salmon on this river shows the smallest decrease in any of the divisions in my district, being only some 17,000 cases, behind last year. The sockeye run was fair, we call it about a three quarter pack, I feel confident that within the next three or four years the pack of sockeye will increase considerably, as the department has opened up large additional spawning grounds on Meziaden lake, the fishway built last season enabled the salmon to surmount the falls, and proceed to the new spawning grounds. The fisheries generally are in a satisfactory condition, and Overseer Adamson reports very few violations of the fishery regulations.

NORTH COAST AND QUEEN CHARLOTTE ISLANDS.

There is also a notable decrease to record in the number of cases packed in this division, this is accounted for to some extent by the total failure of the salmon fisheries of the Queen Charlotte islands, with the exception of the spring salmon fishing. The two canneries only packing some 2,000 cases between them.

It should be remembered that the catch of spring salmon by trolling was phenomenally large, but these salmon are all mild cured and put up in tierces. This whole matter is exhaustively dealt with in Overseer Harrison's report which I beg to enclose. I may here state that I am in hearty accord with all his suggestions contained therein relative to regulating these fisheries and have submitted a report to the Chief Inspector making certain recommendations in the premises. The sockeye run at Bella Coola was small but Kimsquit was above the average. The canneries in the central division which is under the control of Overseer Boyd, put up fair packs. With the exception of Lowe inlet, these canneries packed almost exclusively fall fish, consequently their seasons work was not remunerative, the small pack of sockeye not being sufficiently large to defray their heavy expenses.

The department is removing certain obstructions in the ascent of salmon to their spawning grounds in the vicinity of East Bella Bella Cannery which I trust will materially improve these fisheries. The fisheries regulations were well observed and Overseer Boyd reports no infringements. The run of herring in Rupert Harbour was as heavy as usual, but with the exception of the Canadian Cold Storage and a few men fishing gill-nets, very little attention was paid to this branch of the industry, as explained in a previous portion of my report there is no profit in the salted article, the price on the Oriental market being so low, and until there is a fair margin of profit these fisheries will not be exploited further. The herring that are taken in the seines are placed in cold storage and used for bait, and those in gill-nets are principally sold locally. I should recommend closer attention to the sockeye salmon in my district, the removal of obstructions in sockeye streams, especially on the Skeena river, at Tatcha creek, Babine lake, and more efficient patrol of the spawning grounds especially on the Naas river, and when possible the opening of additional areas of spawning grounds which sockeye frequent.

The whaling stations operated at Naden harbour and Rose harbour on the Queen Charlotte islands, had another successful season, the number of whales captured was two hundred and nineteen including four sperm whales.

I am, Sir,
Your obedient servant,

JOHN T. C. WILLIAMS,
Inspector of Fisheries.

J. T. WILLIAMS, Esq.,
   Inspector of Fisheries,
      Vancouver, B.C.

SIR,—During the month of April the Indians made the final preparations necessary for the spring salmon fishing. The spring salmon was used principally for mild curing and was kept as fresh as possible. They had to be collected quickly and were kept on ice up to the very time of curing. There was a substantial increase in the catch of these fish during the past season. Early in the spring it was apparent that there would be a large demand for spring salmon, and the Haidas prepared to make a record catch. They built boats averaging about sixteen feet in length, and obtained a large number of trolling lines during the winter months. On May 1, all the families from the reservation left for Langara island. A number of Icelanders also arrived and took up the work. Hundreds of fishermen from the south and from Prince Rupert including many Zimshian Indians also decided to try their fortune in this new industry. When the season commenced there were over four hundred boats and canoes on the fishing grounds. Three companies namely the Wallace Fisheries, the B. C. Fisheries and the Prince Rupert Cold Storage Company sent out launches and steamers to gather the harvest. The fishing commenced in earnest on May 20. Each fisherman carried a line from one hundred to three hundred feet in length; some had 'spoons' which revolved through the water and flashed as they turned, and thus resembled a small fish in motion. Other fishermen used the herring bait. Their boats were rowed hither and thither, about one half mile off shore and only stopped when a fish was hooked. It has not been unusual to see one fisherman haul in ten salmon during a day that averaged thirty pounds each. One of the largest caught this year was a white spring salmon that weighed one hundred and ten pounds. Several tourists this year engaged in the sport, including Seton Ker the noted traveller and writer. Indian women also went out fishing, and one woman during the past season had one hundred fish to her credit. During the early days of the fishing season the fish were very large, and some trouble originated with the fishermen over the price of the fish. The Indians asked for five cents per pound, an increase of two cents over the price paid last year. The purchasers considered this demand too high, and fishing was suspended for a few days. A number of the white fishermen held meetings and

5 GEORGE V., A. 1915

decided to demand five cents also. The purchasers offered four cents per pound and a compromise was effected which continued until the end of the season. The fishermen did not leave the fishing grounds to hand over their catch. A mosquito fleet of gasoline launches dodged in and out along the coast collecting and weighing the fish, and each fisherman had an account book in which the catch was entered. That the fish were plentiful was evidenced by the fact that one of the Indians turned in $300 worth of fish after twenty days fishing during the month of June.

It has until this last season been generally thought that the only haunt of these fish in the neighbouring waters of Hecate Straits was around Langara island, but the prospectors sent out by the Canneries and the Cold Storage plants have ascertained that the spring salmon are plentiful all around Graham island. Late in the season those captured were not so large as in the early part, and many of the fishermen were of the opinion that they were of a species known as the 'Blue-backs' and weighed from ten to twenty pounds each. My opinion is that they are a distinct species of the salmon family entirely.

It is a further problem to be yet explained where these fish have their spawning grounds. The small rivers of the Queen Charlotte Islands are visited by very few of the spring salmon variety. It is probable that they come from the rivers of Alaska, and the Naas and Skeena rivers in British Columbia. In the waters of no other part of the Pacific coast are they so plentiful as around the Queen Charlotte Islands. Each year they return about the middle of May and disappear about the end of July. I am sorry to say that the Companies this year were not prepared for such a heavy run as took place, and many hundreds of fish had to be thrown away as being unfit for curing owing to the lapse of time that took place between the time they were caught and the time they reached their destination.

During the run of spring salmon the sockeyes ascend the streams, but as the work is easier catching the spring salmon and the remuneration better, the fishermen do not care to catch the sockeyes, and never make the attempt until the spring salmon run is over, and then also the sockeye run in these waters is also practically at an end; consequently whether or not the sockeyes frequent our streams in marketable numbers is yet not definitely known.

I stated in one of my letters during the month of August that I did not expect there would be a good run on humpbacks this season, as during the past thirty years I have noticed a heavy run only every alternate year, and last year these fish were very numerous, and the canneries obtained all they could handle. My opinion proved to be correct for only very few were caught, and those only in Naden Harbour at the north end of Graham Island, and at Copper river down at the south end. There is also a difference in the run of salmon between the north end and the south end of these islands. For instance, around Massett and Virago Sound when the humpback run is over the cohoes appear, and when they ascend the streams, the dog salmon appear. At the south end, when the humpback run is over the dog salmon appear, and when they ascend the streams the cohoes appear The dog salmon had commenced to run up Skidegate Inlet when I was there in August, and yet the fishermen were trolling for cohoes between Miagwun and Yatza Point in the vicinity of Virago Sound and Naden Harbour.

All the canoes, boats and gasoline launches were in good condition during the past season, and no possible fault could be found with any of them, with the exception of the gasoline launches that carried so many poles with baited hooks. I have already referred to this abuse in my former letters under the heading of suggestions for the regulation of the spring salmon fisheries. These suggestions are as follows:—A close season for trolling for spring salmon should be enforced from Saturday noon to six o'clock on Sunday evening of each week. This close season in my opinion should be inaugurated next year, as the spring salmon although plentiful at present are by no means inexhaustible. The hook and line fishermen never catch these fish more than one mile off shore and in water no stormier than the waters of the Skeena and Naas rivers. When the water is too stormy on one side of Langara island they can fish on the other and vice versa. Last year most of the Indians quit fishing generally on Saturday noon and were towed by either the cannery steamer or gasoline boat to the cannery to spend Sunday with their friends. Sunday evening they were again towed to the fishing camps and did not as a rule commence to fish before Monday morning, so practically making for themselves thirty-six hours more or less for the weekly close season. It was only the white fishermen and a few of the Indians who remained behind at the camps that fished last year on Sunday with hook and line. The leading Indians complained to me several times last year about the Whites fishing on Sunday as they themselves rested on that day and desired to see the white fishermen do the same. Taking all these points into consideration there can be very little harm done or loss caused by enforcing the weekly close season for the hook and line fishermen similar to that now in force for those who fish with nets and seines This year a large number of Whites, Zimshians, Haidas and Skidegate Indians fished every Sunday, but yet the Massett Haidas do not really care to fish on Sunday and would prefer to have one day in the week as a rest day.

MOTOR BOATS.

No motor boats should be allowed to fish for spring salmon or salmon of any kind. This year more than twenty gasoline launches were engaged and some of them had as many as seven poles erected around the mast and the cabin with lines and baited hooks, besides using the hand lines. The Indians and some of the Whites declared that these men in gasoline launches destroyed equally as many fish as they captured, as when the fish bite at the hooks worked on these poles often times their jaws are torn away and they escape only to die, a dead loss to all concerned. This is due to the lines tightening up at once at the rate they are going, and something has to give way, which most frequently are the jaws, &c., of the salmon and not the poles and lines. These motor boats also interfere a good deal with the row-boats and canoes. Being able to go so much faster than a row-boat, they circle around the row-boats and often times get their lines tangled up with those trolling from row-boats and canoes, thereby causing a great deal of unpleasantness. The great majority of the fishermen used row-boats and canoes, one man to the boat, and made very good wages. One Indian in one day by trolling from a canoe made $44; and many of those who used row-boats and canoes made $20 to $25 per diem. This being the case I do not see the necessity of allowing motor boats to take part in these fisheries, and by prohibiting the use of motor boats entirely more men would be able to find employment. My opinion is that only one line should be allowed each boat or canoe as this mode of fishing is adopted by the Indians, and they catch on an average more fish than the Whites. One man to one boat with one line is the way the Indians troll for spring salmon, and they are the successful fishermen.

NUMBERING OF THE BOATS.

Another point that should be considered is the numbering of the boats. All boats and canoes engaged in trolling should be numbered, so that in case of any disturbance or trouble, the number of the boat or boats of the men making the trouble could be taken. Owing to the large number of boats on the fishing grounds it is impossible for strangers to know the names of the owners.

LICENSE FEE.

As the spring salmon fishermen by trolling make as much if not more than the gill-net fishermen for the other kinds of salmon, I would suggest that at the

5 GEORGE V., A. 1915

commencement of each season the fishermen should be compelled to take out a license, the fee to equal that for a gill-net license. As soon as the license is granted the number should be painted on the boat in two conspicuous places.

CLASS OF FISHERMAN.

None but British subjects and pre-emptors who have declared their intention to become British subjects should be allowed to obtain licenses.

The above after careful consideration is my opinion regarding the spring salmon fisheries, and most of these points I discussed with the superintendent when he visited Massett last year.

The gasoline patrol launch *Josephine* was only used barely half the time during the past season as compared with that of 1912, as the humpback salmon being scarce, there was no need of incurring greater expense than was absolutely necessary. This year the Indians only were engaged with seine and gill-net for the humpback and cohoe salmon, and knowing that I was liable at any moment to be on their fishing grounds during the weekly close season, all returned on Saturday morning of each week to the canneries and did not leave until Sunday evening. This year also the Wallace Fisheries did not man the Wallace No. 1 with Dagoes and a Purse Seine, consequently this caused my work to be easier than that of last year. During the past season the *Josephine* was engaged thirty-eight days and travelled 1,556 miles. Knowing that this vessel was chartered for the sole purpose of protecting the fisheries had a deterrent effect, and but for its existence many doubtless would have fished illegally and the weekly close season would probably never have been observed. During my cruises around the rivers I took great care to see that all the fishery signs were in their proper places and distinctly visible.

I have visited also from time to time the various saw-mills on these islands, and the owners are now burning up their saw-dust, so that at the present time no fishing stream is polluted with either saw-dust or any other kind of mill refuse.

NEW BOAT THE 'GANNET.'

The new boat appears to be very satisfactory with the exception of the large mast. It is I think too heavy for the size of the vessel. It should be removed and a smaller one should take its place with a smaller sail, as in a gale of wind the present mast and sail seem to make the boat too top-heavy. She is suitable in my opinion for the purpose for which she was first intended for. *i.c.*, for the purpose of protecting the salmon fisheries from the 1st of April to the end of October between Langara island, Skidegate and Kumshewa Inlet, but not quite suitable for patrol work during the stormy winter months. A larger vessel about 100 or 125 feet in length is necessary to protect the fisheries around these islands during the winter months, and this vessel then could be used to protect the salmon fisheries when the canneries are in operation.

FUTURE DEVELOPMENT.

It is expected that the coming year will eclipse all past records for the fisheries around the Queen Charlotte Islands. The Atlin Construction Company intends to build a saltery, wharf and other buildings at Langara island. It is reported that the Wallace Fisheries will install a cold storage plant at Naden Harbour, and the British Columbia Fisheries are contemplating the expenditure of $150,000 in rehabilitating their plant at Aliford bay. The halibut grounds will be thoroughly tested and all kinds of fish will be handled by the companies interested, giving employment to fishermen the whole year. The fishing for spring salmon which was stopped at an early date this year, will give large returns if carried on during the whole season. It is expected that the humpback run will be a phenomenal one next year. The cod banks and dog-fish grounds will give a good return, which will show as never before, the wealth of the fish in the immediate neighbourhood of Graham and Langara islands. It is also expected that a number of tourists will visit the island next year to engage in the sport of trolling for spring salmon.

I have the honour to be, Sir,
Your obedient servant,

C. HARRISON,
Fishery Overseer.

J. T. WILLIAMS, Esq.,
Inspector of Fisheries,
Vancouver, B.C.

SIR,—I beg to submit the following report in connection with the head waters, and spawning grounds of the Skeena river for the season of 1913.

On Beat No. 1, under the care of Guardian Henry Frank, everything has been quite satisfactory, the fisheries regulations have been well observed, and the Indians have given no trouble. This beat, extends from tide-water to Hazelton, taking in all the tributaries, and at the headwaters of some of these streams, good sockeye spawning grounds exist. The Kitsumkalum spawning grounds were visited three times during the season, and each time sockeye were numerous. Mr. Frank also visited Kitwancool lake twice during the season. This lake which is drained by the Kitwanger river, after a course of twenty-five miles joins the Skeena just below the Indian village of that name. This is the most extensive sockeye spawning area on this No. 1 Beat. The grounds were well stocked with parent sockeye, and the beds were well seeded. He says further in his report, ' The season closes with an abundance of rain, and indications are, that the water will be high in the streams and lakes during the coming winter, thus protecting the spawn from freezing.' I may say, that this presentiment turned out correct for rivers and lakes near the coast remained high until after the ova had hatched.

Beat No. 2 has been under the care of Guardian R. L. D'Egville. This beat, with headquarters at Hazelton, embraces practically the whole of the Skeena watershed above Hazelton, with the exception of the principal, and by far the most extensive sockeye spawning ground namely, the Babine river, lake, and tributaries. The Bulkley river, also is included in, and cared for by the Guardian on this beat. On one section of his charge he follows the Yukon telegraph trail for one hundred and thirty miles which takes him over the divide, and on to the headwaters of the Naas river, and at Blackwater lake ends the most tiresome and arduous trip to be covered on this beat. Mr. D'Egville found large quantities of spring salmon spawning in the river that drains Blackwater lake, and after staying around the locality for three days, recrossed the divide, and visited Schalm Geese lakes, which are drained by a stream running into the Skeena. Considerable salmon have found their way on to these spawning grounds, and though of no great extent, I have always considered them the best in this part of the watershed. This section of the mountains seems to have had more than the average share of rain. The streams were in flood all the time the guardian was there, and he was five weeks visiting the different places of interest along this route, and the Indians were having a hard time catching their supply of salmon for food purposes, on account of the high water. He says in his report, ' the travelling beggars description, the trail being under water for long distances, and I, covered with mud to the hips every day.' At Kuldo, Kispiax, Glen Vowel, Hagwilgat, and Morristown, the Indians respected the fisheries regulations. Kiskagas was not visited this season. The Indians

5 GEORGE V., A. 1915

at this village fish entirely in a narrow swift canyon, which is situated about five miles up the Babine river from its junction with the Skeena, and is about fifty-six miles north of Hazelton. A visit to this place once in every two seasons I think is sufficient, for the Indians only take what they require for food, and there is no place suitable for a barricade.

On Beat No. 3 Guardian MacKendrick, who was assisted by Guardian Collins, sends in a very satisfactory report. This being the year the Babine Indians were to receive their new nets, it devolved upon Mr. MacKendrick (who was at the last distribution) to apportion them out in the usual way, and which was done to the satisfaction of every one concerned. On the Neel-kit-kwah river, a stream which joins the Babine some few miles below Babine lake, the run of sockeye was a little below the average, and the spring salmon were much more plentiful, than for quite a number of years past. On salmon creek, (the creek on which the Babine hatchery is built) the run of sockeye is larger than ever before, and Mr. MacKendrick says in his report, 'At no time have I seen the hatchery able to secure the full complement from this creek alone, yet this season Mr. Gibbs, the superintendent, filled the establishment from this creek in ten days, and would have had no trouble in securing double the amount if necessary, also on this creek, the fish seem larger and of a more uniform size than on most of the other streams.' On Taché creek, the run is greatly improved, and this was the worst stocked stream running into the lake last season. Pierre creek is well supplied, and on fifteen mile the run is larger than it has ever been since Mr. MacKendrick has been in the service, and Mr. Crawford from Stuart lake hatchery collected nearly all the ova necessary to fill that establishment from this stream. Four mile creek is fairly well stocked, Grizzly and Beaver creek are both below the average and on the mile of river (Babine) just below the bridge sockeye were scarce, and this stretch of river last year was the best ground we had. Spring salmon are not plentiful above the bridge, and the run of cohoe which had got well started when Mr. MacKendrick left, was he considered up to the average.

I am Sir,
Your obedient servant,

STEWART NORRIE,
Overseer of Fisheries.

NANAIMO, B.C., June 11, 1914.

F. H. CUNNINGHAM, Esq.,
Chief Inspector of Fisheries,
New Westminster, B.C.

SIR,—I have the honour to submit my annual statistical report of the fisheries of Vancouver Island and the adjacent mainland, District No. 3, in the Province of British Columbia, for the fiscal year ended March 31, 1914.

The various branches of the fishing industry throughout the district show satisfactory development:—

Salmon.—The salmon catch for the past season amounted to 297,450 cwts. showing an increase of 76,040 cwts. over the catch of 1912-1913. This year's pack of sockeye reached 129,925 cases, an increase of 51,887 over that of last year. The run of spring salmon was also above the average. The run of salmon in the Nimpkish river was much larger than in any previous year, and the catch at Barclay Sound and in the traps of the west coast of Vancouver Island was also very satisfactory, but the catch at Quatsino and Clayoquot did not come up to the average. On the mainland coast the packs were not equal to those of last year. After the close of the season the Jervis Inlet Canning Company sustained a great loss on account of the destruction of their cannery by fire. A salmon cannery was operated at Nanaimo for the first time, and put up a pack of 2,500 cases of sockeye.

Four salmon salteries operated this season, three at Nanaimo and one in the Pender harbour district. All of these obtained good catches. The salmon packed at these places were shipped to the Orient.

Cod.—Cod fishing received more attention during the past season than ever before, and the catch therefore was the largest in the history of these fisheries. I consider that, owing to the greater part of the cod fishing being done in the extensive channels lying between Vancouver Island and the mainland, it is necessary that a close season be enforced for the protection of this valuable industry. The spawning season extends from about the middle of January to the end of February, and to save the codfish from depletion, fishing should be prohibited during that time.

Herring.—The herring fisheries in this district continue to increase, the returns showing an increase of 41,340 cwts. over the catch of last year. The greater quantity of herring was taken in the Nanaimo district, where fifteen salteries were operated. Three of these salteries had been newly erected at Cowichan gap on Galiano island, as the fishing areas in this vicinity are among the most prolific in the district. On the west coast of Vancouver Island at Barclay sound and Clayoquot, more attention is being paid to the herring fisheries, and with the improved facilities for shipping, and unlimited markets in the Orient, the outlook for this branch of the fishing industry is bright, and no doubt it will assume much larger proportions in the future. The herring spawning areas were well protected during the spawning season and were well stocked with ova.

Halibut.—The halibut fisheries again show a substantial increase over previous years. This was owing to the fact that more boats were operated on the west coast of Vancouver Island than ever before. The success of the halibut fishing depends to a great extent on the weather conditions, as the halibut banks on the west coast of Vancouver Island are exposed to the full force of the Pacific, and the catch of last season would have shown even better results if more favourable weather had prevailed.

Clams.—The catch of clams this year was 10,000 barrels. Two clams canneries were operated and 7,328 cases were put up.

Whales.—The whaling stations at Sechart and Kyuquot on the west coast of Vancouver Island captured 486 whales; the former station taking 4 sperm whales, 5 sulphurbottoms, 30 finbacks, and 236 humpbacks, a total of 275; and the latter station, 7 sperm, 5 sulphurbottoms, 63 finbacks, and 136 humpbacks, a total of 211. At these stations in the year of 1912-1913, 16 sperm whales were taken, and in the year previous 24.

Fur seals.—The sealing operations carried on by the Indians along the west coast of Vancouver Island resulted in the capture of 119 fur seals valued at $3,570. The Indians hunt the seal under primitive conditions using their canoes and killing the seal with spear. The Fishery Overseers on the west coast have been vigilant to see that the regulations governing the taking of the fur seal are strictly complied with by the Indians.

In concluding this report I am pleased to be able to state that the Fishery Regulations throughout the district were well enforced by the various Overseers. This is largely owing to the facilities provided by the Department for the Overseers in the proper patrol of their districts.

I am Sir,
Your obedient servant,

EDWARD G. TAYLOR,
Inspector of Fisheries.

MARINE AND FISHERIES

5 GEORGE V., A. 1915

SUMMARY

Of the Quantities and Values of all Fish caught and landed in a Green State, and of the Quantities and Values of all Fish and Fish Products Marketed in a fresh, dried, pickled, canned, etc., state, for **District No. 1,** Province of **British Columbia,** during the year 1913-14.

Kinds of Fish.		Caught and Landed in a Green State.		Marketed.		Total Marketed Value.
		Quantity.	Value.	Quantity.	Value.	
			$		$	$
Salmon	cwts.	797,524	4,871,406			
„ used fresh	„			63,881	958,215	
„ canned	cases.			732,059	4,026,324	
„ salted (dry)	cwts.			92,445	574,450	
„ mild cured	„			38	646	
„ smoked	„			1,825	31,025	
						5,590,660
Cod	„	12,690	63,450			
„ used fresh	„			12,094	96,752	
„ green-salted	„			181	1,810	
„ dried	„			78	1,048	
						99,610
Herring	„	29,502	88,506			
„ used fresh	„			14,204	113,632	
„ dry-salted	„			72	720	
„ smoked	„			7,588	94,850	
						209,202
Shad	„	11	56			
„ used fresh	„			11		168
Halibut, used fresh	„	93,677	538,642	93,677		929,160
Flounders	„	580	1,832	580		3,480
Smelts	„	1,835	8,257	1,835		18,350
Oulachans	„	232	928	232		1,856
Octopus	„	211	1,969	211		2,329
Sturgeon	„	1,090	8,720	1,090		16,350
Bass	„	565	2,542	565		3,955
Tom Cod	„	418	2,090	418		3,344
Mixed Fish	„	3,058	10,703	3,058		24,464
Oysters	brls.	2,680	9,380	2,680		9,380
Clams	„	5,567	11,134			
„ used fresh	„			5,567		27,835
Dulse, crabs and other shell fish.	cwts.	2,285	12,869	2,285		20,030
Shrimps	„	43	435	43		870
Guano	tons.			589		18,974
Fish Oil	gallons.			97,650		32,770
Totals			5,632,919			7,012,787

SESSIONAL PAPER No. 39

SUMMARY

Of the Number of Fishermen, &c., and of the Number and Value of Fishing Vessels, Boats, Nets, &c., in **District No. 1**, Province of **British Columbia**, for the year 1913-1914.

—	Number.	Value.
		$
Steam fishing vessels (tonnage 1,060)	5	272,875
Sailing and gasoline vessels	12	30,000
Boats (sail).	385	32,725
" (gasoline)	2,174	869,600
Carrying smacks	34	25,500
Gill nets, seines, trap and smelt nets, etc	3,479	508,415
Trawls	22	1,100
Hand lines	388	970
Oil factory	1	40,000
Salmon canneries	34	2,116,410
Salteries	12	1,200
Freezers and Ice-houses	5	470,000
Smoke and Fish-houses	11	330,000
Fishing piers and wharves	46	1,431,689
Totals		6,130,484

Number of men employed on vessels	75
" " boats	5,142
" " carrying smacks	68
" persons employed in Fish-houses, freezers, canneries, &c.	3,493
Total	8,778

5 GEORGE V., A. 1915

RETURN showing the Number of Fishermen, etc., the Number and Value of Vessels and Boats, and the Quantity and Value of all Fishing Gear, etc., used in the Fishing Industry in **District No. 2**, Province of **British Columbia**, during the year 1913-14.

| | Vessels, Boats and Carrying Smacks. | | | | | | | | | | Fishing Gear. | | | | | | Canneries. | | Other Material. | | | | | | Persons employed in Canneries, Freezers and Fish-Houses. | Number. |
| Fishing Districts. | Steam Vessels. | | | | Sailing and Gasoline Vessels (10 to 20 tons). | | | Boats. | | | Gill Nets, Seines, Trap, etc. | | Skate of Gear 400 f-1 skate. | | Haul Lines. | | Salmon Canneries. | | Freezers and Ice houses. | | Whaling Stations. | | Fishing Piers and Wharves. | | | |
	Number.	Tonnage	Value $	Men.	Number.	Value $	Men.	Sail.	Value $	Men.	Number.	Value $	Number.	Value $	Number.	Value $	Number.	Value $	Number.	Value $	Number.	Value $	Number.	Value $		
1 Skeena and Prince Rupert	34	1536	398250	388	38	105700	85	850	59150	1700	1411	175344	1360	27200			13	735000	5	612000			14	154000	1660	1
2 Rivers Inlet					15	55000	49	700	27500	820	747	95800					7	400000					8	68000	950	2
3 Naas River					10	19000	20	240	23500	477	444	51100					4	165000					6	38000	350	3
4 North Coast					25	101500	65	339	17800	575	415	61425					9	285000	2	120000	2	240000	11	73000	580	4
5 Queen Charlotte Islands	7	375	100000	55	11	46800	33	60	5250	120	66	15200	528	16560	1000	1000	2	80000	1	129000			8	144000	400	5
Totals	41	1911	498250	423	99	328000	252	2189	164200	3692	3083	398869	1888	43760	1000	1000	35	1665000	7	732000	2	240000	47	477000	3950	

THE CATCH.

RETURN showing the Quantities and Values of all Fish caught and landed in a Green State, in the District No. 2, Province of British Columbia, during the year 1913-14.

Number	Fishing Districts	Salmon, cwts.	Salmon, value. $	Cod, cwts.	Cod, value. $	Herring, cwts.	Herring, value. $	Halibut, cwts.	Halibut, value. $	Trout, cwts.	Trout, value. $	Oulachans, cwts.	Oulachans, value. $	Shrimps, cwts.	Shrimps, value. $	Mixed Fish, cwts.	Mixed Fish, value. $	Clams, brls.	Clams, value. $	Dulse, cockles, crabs, and other shell fish.	Dulse, cockles, crabs, and other shell fish, value. $	Whales, number.	Whales, value. $	Number.
1	Skeena and Prince Rupert	172970	370664	1205	6025	42710	42710	89479	447395	45	450	1500	7500	75	750	80	500			125	750			1
2	Rivers Inlet	57650	104250			30	30	35	175	5	50	5000	25000			17	85							2
3	Naas River	54200	98865			1000	1000	7000	35000	6	60	7000	33000			60	300							3
4	North Coast	119220	220557			11000	11000	1000	5000	7	70	450	2250			80	400	1000	2000			219	65700	4
5	Queen Charlotte Islands	10440	12528			7500	7500	9974	49870	8	80					60	2500							5
	Totals	414380	807864	1205	6025	62240	62240	107488	537440	71	710	13950	69750	75	750	757	3785	1000	2000	125	750	219	65700	

Cwt. = 100 lbs.

THE CATCH MARKETED.

RETURN showing the Quantities and Values of all Fish and Fish Products Marketed in a fresh, pickled, canned, &c., state, for District No. 2, Province of British Columbia, during the year 1913-14.

Number.	Fishing Districts.	Salmon, used fresh and frozen, cwts.	Salmon, canned, cases.	Salmon, salted, cwts.	Salmon, mild cured, cwts.	Salmon, smoked, cwts.	Cod, used fresh, cwts.	Herring, used fresh, cwts.	Herring, smoked, cwts.	Herring, pickled, brls.	Herring, used as bait, brls.	Halibut, used fresh, cwts.	Trout, used fresh, cwts.	Oulachans, used, cwts.	Shrimps, used, fresh, cwts.	Mixed fish, used, fresh, cwts.	Clams and quahaugs, used fresh, cwts.	Dulse, crabs, cockles and other shell fish, used fresh.	Fur seal skins, number.	Hair seal skins, number.	Fish oil, gals.	Whale oil, gals.	Number.
1	Skeena and Prince Rupert	9397	164055		16940	290	1205				21350	89479	45	1500	75	100	11 fl./oz.	125		370	900		1
2	Rivers Inlet		68096		850	200		1000		10		35	5	5000						400	700		2
3	Naas River	7200	53423		294	500						7000	6	7000		80				450	15000		3
4	Mass coast	522	129799		1020	5500					5500	1000	7	450		80				900	800		4
5	Queen Charlotte Islands		2080	3000		2000		3000	250	1333		9574	8				1000		285	400	29000	292556	5
	Btals.	17119	417453	3000	19014	8400	1205	4000	250	1343	26855	107488	71	13956	75	757	1000	125	285	2520	46400	292556	
	Rates. $	5	5	5	10	10	5	1	10	2.50	1	5	10	5	10	5	2	6	30	.25	.30	.30	
	Values. $	85595	2087285	15000	190140	84000	6025	4000	2500	3357	26855	587440	710	69750	750	3785	2000	750	8550	630	139920	87766	

Total value $3,220,788

* Cwts = 100 lbs. † Quintal = 112 lbs.

SESSIONAL PAPER No. 39

RECAPITULATION

Of the quantities and Values of all Fish caught and landed in a Green State, and of the Quantities and Values of all Fish and Fish Products Marketed in a fresh, dried, pickled, canned, &c., state, for **District No. 2,** Province of **British Columbia,** during the year 1913-14.

Kinds of fish.		Caught and Landed in a Green State.		Marketed.		Total marketed value.
		Quantity.	Value.	Quantity.	Value.	
			$		$	$
Salmon	Cwts.	414,380	807,864	....	....	
„ used fresh	„	....	....	17,119	85,595	
„ canned	Cases.	....	....	417,453	2,087,265	
„ salted (dry)	Cwts.	....	....	3,000	15,000	
„ mild cured	„	....	....	19,014	190,140	
„ smoked	„	....	....	8,400	84,000	
						2,462,000
Cod	Cwts.	1,205	6,025	....	....	
„ used fresh	„	....	....	1,205	....	6,025
Herring	„	62,240	62,240	....	....	
„ used fresh	„	....	....	4,000	4,000	
„ smoked	„	....	....	250	2,500	
„ pickled	Brls.	....	....	1,343	3,357	
„ used as bait	„	....	....	26,855	26,855	
						36,712
Halibut, used fresh	Cwts.	107,488	537,440	107,488	....	537,440
Trout	„	71	710	71	....	710
Oulachans	„	13,950	69,750	13,950	....	69,750
Shrimps	„	75	750	75	....	750
Mixed Fish	„	757	3,785	757	....	3,785
Clams	Brls,	1,000	2,000	....	....	2,000
„ used fresh	„	....	....	1,000	....	2,000
Crabs, Cockles and other shell fish	Cwts.	125	750	125	....	750
Fur seal skins	No.	....	....	285	....	8,550
Hair seal skins	„	....	....	2,520	.....	630
Wales	„	219	65,700	....	....	
Fish oil	Gals.	....	....	46,400	....	13,920
Whale oil	„	....	....	292,556	....	87,766
Totals	....	....	1,557,014	....	....	3,230,788

39—18

5 GEORGE V., A. 1915

RECAPITULATION

Of the Number of Fishermen, &c., and of the Number and Value of Fishing Vessels, Boats, Nets, &c., in **District No. 2**, Province of **British Columbia**, for the year 1913-14.

	Number.	Value.
		$
Steam Fishing Vessels (tonnage 1911).........	41	408,250
Sailing and Gasoline Vessels......	99	323,000
Boats (sail)	2,189	164,200
Gill Nets, Seines, Trap and Smelt Nets, etc.........................	3,083	396,869
Skates of Gear (400 fath=1 Skate).	1,888	37,760
Hand Lines	1,000	1,000
Salmon Canneries	35	1,695,000
Freezers and Ice-houses	7	732,000
Fishing Piers and Wharves..............................	47	477,000
Whaling Stations......	2	240,000
Totals................ ....!	....!	4,475,079

Number of men employed on Vessels..................................... 675

       "   "   Boats.. 3,692

       "   persons employed in Fish-houses, Freezers, Canneries, &c............. 3,950

            Totals.. 8,317

RETURN showing the Number of Fishermen, &c., the Number and Value of Vessels and Boats, and the Quantity and Value of Fishing Gear, &c., used in the Fishing Industry in District No. 3, Province of British Columbia, during the year 1913-14.

Vessels, Boats and carrying Smacks.

	Fishing Districts.	Steam Vessels.				Sailing and Gasoline Vessels.					Boats.					Carrying Smacks.		
No.		Number.	Tonnage.	Value. $	Men.	Number (40 tons and over).	Number (20 to 40 tons).	Number (10 to 20 tons).	Value. $	Men.	Sail.	Value. $	Gazoline.	Value. $	Men.	Number.	Value. $	Number.
1	Nanaimo	1	40	8000	8		1	20	32300	175			67	40945	190	85	16350	1
2	Cowichan	4	240	140000	24		2		80000	76	65	3250	63	32300	254	20	8000	2
3	Alberni	4	285	137450	57	1		3	17780	24	10	900	24	12290	71			3
4	Clayoquot							3	25500	27			2	1500	4	9	6300	4
5	Quatsino	3	225	105000	30			1	4600	3	10	900	2	1600	22	12	8400	5
6	Alert Bay						2		32000	20	47	4430	37	19000	167	14	1400	6
7	Quathiaski						2	4	11000	16	170	4200	14	7000	195	12	7800	7
8	Comox						1	2	7000	18			1	705	2	3	1200	8
9	Pender Harbour							3	10000	25	200	5000	50	25000	320	1	700	9
	Totals	12	790	390450	119	1	8	36	80	324	502	20680	290	148550	1221	156	50150	

5 GEORGE V., A. 1915

RETURN showing the Number of Fishermen, &c., the Number and Value of Vessels and Boats, and the Quantity and Value of all Fishing Gear, &c., used in the Fishing Industry in of District No. 3, Province of British Columbia, during the year 1913-14.—*Concluded.*

Number	Fishing Districts	Fishing Gear				Canneries				Other Material						
		Gill Nets, Seines, Trap and Smelt Nets		Hand Lines		Salmon Canneries		Clam Canneries		Freezers and Ice-houses		Smoke and Fish-houses		Whaling Stations		Persons employed in Canneries, Freezers and Fish-houses
		Number	Value $	Number	Value $	Number	Value $	Number	Value $	Number	Value $	Number	Value $	Number	Value $	
1	Nanaimo	62	22100	330	330	1	9000					15	50800			276
2	Cowichan	39	61400	500	500	2	75000									345
3	Alberni	37	7000	350	350	1	40000			2	125000	2	6000			397
4	Clayoquot	8	5600	250	250	1	25000	1	3000	1	175000					80
5	Quatsino	16	5000	150	150	1	25000							1	130000	255
6	Alt Bay	74	10000	150	150	4	90000							1	130000	425
7	Quathiaski	15	3640	500	500	1	20000									110
8	Comox	2	800	150	150											
9	Pender Harbour	60	7000	500	500	1	22500					1	1500			150
	Totaux	303	125840	2900	2900	12	304000	1	3000	3	300000	18	58300	2	260000	1948

RETURN showing the Quantities and Values of all Fish caught and landed in a Green State in District No. 3, Province of British Columbia, during the year 1913-14.

Number	Fishing Districts	Salmon, cwts.	Salmon, value $	Cod, cwts.	Cod, value $	Herring, cwts.	Herring, value $	Halibut, cwts.	Halibut, value $	Flounders, cwts.	Flounders, value $	Smelts, cwts.	Smelts, value $	Trout, cwts.	Trout, value $	Oulachans, cwts.	Oulachans, value $	Mixed fish, cwts.	Mixed fish, value $	Clams, brls.	Clams, value $	Dulse, crabs, Cockles, etc., cwts.	Dulse, crabs, Cockles, etc., value $	Whales, Number	Whales, value $	Number
1	Nanaimo	41600	208000	3150	15750	506360	506360	4800	24000	250	1250	30	180	35	350			1150	5750	4310	8620	450	1800			1
2	Cowichan	80503	402515	2950	14750	1200	1200	8200	41000	165	825	60	360	165	1650			1100	5500	5025	10050	250	300			2
3	Alberni	45600	228000	350	1750	41650	41650	3500	17500	35	175	15	90	150	1500			450	2250	55	110	75	260	275	82500	3
4	Clayoquot	7738	38690	250	1250	400	400	5500	27500	25	125	15	90	40	400			350	1750	85	170	65	220			4
5	Quatsino	27398	136990	275	1375	300	300	300	1500	25	125	10	60	25	250			325	1625	60	120	65	60	211	63300	5
6	Alert Bay	56520	282600	450	2250	300	360			75	375	60	360	35	350			975	1875	110	220	160	640			6
7	Quathiaska	74915	749915	850	4250	700	700			225	1125	75	450	60	600	175	875	660	3300	95	190	165	640			7
8	Comox	4000	20000	950	4750	350	350			150	750	50	310	75	750	375	1875	750	3750	185	370	200	800			8
9	Pender-Harbour	19108	95540	6100	30500	6000	6000			650	3250	40	240	65	650			1300	7500	75	150	150	600			9
	Totals	297450	1487250	15325	76625	557320	557320	22300	111500	1600	8000	355	2130	650	6500	550	2750	6660	33300	10000	20000	1560	6240	486	145800	

*Cwt. = 100 lbs

THE CATCH MARKETED.

Return showing the Quantities and Values of all Fish and Fish Products Marketed in a fresh, dried, pickled, canned, etc., state, for District No. 3, Province of British Columbia, during the year 1913-14.

Number	Fishing Districts.	Salmon, used fresh and frozen, *cwts.	Salmon, canned, cases.	Salmon, salted, cwts.	Salmon, mild-cured, cwts.	Salmon, smoked, cwts.	Cod, used fresh, cwts.	Herring, used fresh, cwts.	Herring, smoked, cwts.	Herring, pickled, brls.	Herring, used as bait, brls.	Herring, dry salted, cwts.	Halibut, used fresh, cwts.	Number
1	Nanaimo	4500	2500				3150	1000		300		296741	4800	1
2	Cowichan	7200	60968				2950	1200					8300	2
3	Alberni	3600	50000	28000	6150	3300	350	16600	100		80	14600	3500	3
4	Clayoquot	3600	4926				250	400					5000	4
5	Quatsino	3600	28331				275	300					300	5
6	Al-rt Bay	3600	63000				450	300						6
7	Quathiaska	3600					850	700						7
8	Comox	4000	13652				950	350						8
9	Pender Harbor	3600	18463				6100	3000				1765		9
	Totals	37300	250740	28000	6150	3300	15325	23810	100	300	80	313106	22300	
	Rates	12 00	6 50	2 00	4 00	10 00	10 00	10 00	10 00	2 50	2 00	1 50	12 00	
	Values	$ 447600	1629810	56000	24600	33000	153250	238100	1000	750	160	469659	267600 $	

* Cwts. = 100 lbs.

THE CATCH MARKETED.

RETURN showing the Quantities and Values of all Fish and Fish Products Marketed in a fresh, dried, pickled, canned, etc., state, for District No. 3, Province of British Columbia, during the year 1913-14.—*Concluded.*

Number	Fishing Districts	Flounders, used fresh, *cwts.	Smelts, used fresh, cwts.	Trout, used fresh, cwts.	Oulachans, used fresh, cwts.	Mixed Fish, used fresh, cwts.	Clams and Quahaugs, used fresh, brls.	Clams and Quahaugs, canned, cases.	Dulse, crabs, cockles and other shell fish, used fresh, cwts.	Fur Seal Skins, number.	Whale Bone, cwts.	Whale Bone Meal, cwts.	Fertilizer, tons.	Whale Oil, gals.	Number.
1	Nanaimo	250	30	35		1150	1000	3310	450						1
2	Cowichan	165	60	165		1100	1007	4018	250						2
3	Alberni	35	15	150		450	53		75		24	5500	549	6813	3
4	Clayoquot	25	15	40		350	85		65		80	4250	511	5637	4
5	Quatsino	25	10	25		325	60		55	116					5
6	Alert Bay	75	60	35		375	110		150	3					6
7	Quathiaska	225	75	60	175	660	95		165						7
8	Comox	150	50	75	375	750	185		200						8
9	Pender Harbor	650	40	65		1500	75		150						9
	Totals	1600	355	650	550	6660	2672	7328	1560	119	104	9750	1060	12450	
	Rates	5 00	10 00	10 00	10 00	5 00	2 00	4 00	10 00	30 00					
	Values	8000	3550	6500	5500	33300	5344	29312	15600	3570	478	107125	40280	164135	

Total value..... $3,647,823.

*Cwt. = 100 lbs.

5 GEORGE V., A. 1915

RECAPITULATION.

OF the Quantities and Values of all Fish caught and landed in a Green State, and of the Quantities and Values of all Fish and Fish Products Marketed in a fresh, dried, pickled, canned, &c., state, for **District No. 3**, Province of **British Columbia**, during the year 1913-14.

Kinds of Fish.	Caught and landed in a Green State.		Marketed.		Total marketed value.
	Quantity.	Value.	Quantity.	Value.	
		$		$	$
Salmon ... . cwts.	297,450	1,487,250	....	....	....
" used fresh and frozen..... "	....	....	37,300	447,600	....
" canned............. .... cases.	....	....	250,740	1,629,810	....
" salted.............. .... cwts.	....	....	28,000	56,000	....
" mild cured...... .... "	....	....	6,150	24,600	....
" smoked............ .... "	....	....	3,300	33,000	... .
					2,191,010
Cod....... .... "	15,325	76,625	....	....	....
" used fresh............. "	....	....	15,325	....	....
					153,250
Herring .... "	557,320	557,320	....	....	....
" used fresh.............. "	....	....	23,810	238,100	. . ..
" smoked...... .... "	....	....	100	1,000	. . ..
" pickled ,............. brls.	....	....	300	750	. . ..
" used as bait............. "	....	....	80	160	...:.......
" dry salted......... ...... cwts.	....	....	313,106	469,659	
					709,669
Halibut.... .... cwts.	22,300	111,500	22,300	....	267,600
Flounders..................... "	1,600	8,000	1,600	....	8,000
Smelts..... .... "	355	2,130	355	....	3,550
Trout.... .. .. "	650	6,500	650	... .	6,500
Oolachans .... "	550	2,750	550	....	5,500
Mixed Fish.... .... "	6,660	33,300	6,660	....	33,300
Clams and Quahaugs............ brls.	10,000	20,000	....	....	....
" " used fresh.. "	....	....	2,672	5,344	....
" " canned..... cases.	....	....	7,328	29,312	34,656
Crabs, Cockles and other shell fish cwts.	1,560	6,240	1,560	....	15,600
Whales.... .... "	486	145,800	....	....	....
Fur Seal Skins................. No	....	....	119	....	3,570
Whale bone............... .... cwts.	....	....	104	....	478
Whale bone meal.. .... "	....	....	9,750	....	10,725
Fertilizer....................... tons.	....	....	1,060	....	40,280
Whale Oil....................... gals.	....	....	12,450	....	164,135
Totals.....	....	2,457,415	....	....	3,647,823

SESSIONAL PAPER No. 39

RECAPITULATION.

OF the Number of Fishermen, etc., and of the Number and Value of Fishing Vessels, Boats, Nets, &c., in **District No. 3,** Province of **British Columbia,** for the year 1913-14.

	Number.	Value.
		$
Steam Fishing Vessels (tonnage 790)	12	390,450
Sailing and Gasoline Vessels	45	220,180
Boats (sail)	502	20,680
Boats (gasoline)	260	148,550
Carrying Smacks	156	50,150
Gill Nets, Seines, Trap and Smelt Nets, etc.	303	125,840
Hand Lines	2,900	2,900
Salmon Canneries	12	304,000
Clam Canneries	1	3,000
Freezers and Ice-houses	3	300,000
Smoke and Fish-houses	18	58,300
Whaling Stations	2	260,000
Total		1,884,050

Number of men employed on Vessels	443
" " Boats	1,221
" persons employed in Fish-houses, Freezers, &c	1,948
Total	3,612

5 GEORGE V., A. 1915

RECAPITULATION.

OF the Quantities and Values of all Fish caught and landed in a Green State, and of the Quantities and Values of all Fish and Fish Products Marketed in a fresh, dried, pickled, canned, etc., state, for the **Whole** Province of **British Columbia,** during the year 1913-14.

Kinds of Fish.	Caught and landed in a green state.		Marketed.		Total Marketed Value.
	Quantity.	Value.	Quantity.	Value.	
		$		$	$
Salmon............................ cwts	1,509,354	7,166,520	....	....	
" used fresh................ "	....	....	118,300	1,491,410	
" canned .... cases.	....	....	1,400,252	7,743,399	
" salted (dry).. .... cwts.	....	....	123,445	645.450	
" mild cured..... . .. . "	....	....	25,202	215,386	
" smoked................. "	....	....	13,525	148,025	
					10,243,670
Cod .... "	29,220	146,100	....	....	
" used fresh.................... "	....	....	28,694	256,027	
" green-salted................. "	....	....	181	1,810	
" dried .... "	....	....	78	1,048	
					258,885
Herring...................... "	649,062	708,066	....	....	
" used fresh .... "	....	....	42,014	355,732	
" dry salted .... "	....	....	313,178	470,379	
" smoked..... .... "	....	....	7,938	98,350	
" pickled........ .... brls.	....	....	1,643	4,107	
" used as bait....... . .... "	....	....	26,935	27,015	
					955,583
Shad .. .. cwts.	11	56	11	....	168
Halibut, used fresh.............. "	223,465	1,187,582	223,465	....	1,734,200
Flounders...................... "	2,180	9,832	2,180	... .	11,480
Smelts . . .. "	2,190	10,387	2,190	....	21,900
Trout..... .... "	721	7,210	721	....	7,210
Oulachans... .... "	14,732	73,428	14,732	....	77,106
Octopus .... "	211	1,969	211	....	2,329
Sturgeon.... .... "	1,090	8,720	1,090	....	16,350
Bass........................... "	565	2,512	565	....	3,955
Tom Cod....................... "	418	2,090	418	....	3,344
Mixed fish..................... "	10,475	47,788	10,475	....	61,549
Oysters .... brls.	2,680	9 380	2,680	....	9,380
Clams .... "	16,567	33,134	....	....	
" used fresh........ ...... "	....	....	9,239	35,179	
" canned .... cases.	....	....	7,328	29,312	
					64,491
Dulse, cockles, shrimps, etc..... cwts.	4,088	21,044	4,088	....	38,000
Fur seal skins........... .... No.	....	....	404	....	12,120
Hair seal skins .... "	....	....	2,529	....	630
Whales .... "	705	211,500	....	....	
Fish oil......................... gals.	.. . ..	....	144,950	....	46,690
Whale oil........................ "	....	....	305,006	....	251,901
Whale bone.... .... cwts	....	....	104	....	478
" meal................. "	....	....	9,750	....	10,725
Fertilizer............... .. .. tons.	....	....	1,649	. ...	59,254
Totals	....	9,647,348	....	....	13,891,398
NOTE.—Other fish, not included in the above. caught by Indians and Whites for their own use.. cwts	160,960	1,036,225			

The number of fur seal skins taken during 1912 was 192, not 205 as shown in last year's report.

RECAPITULATION

Of the Number of Fishermen, &c., and of the Number and Value of Fishing Vessels'
Boats, Nets, &c,, in the Whole Province of British Columbia, for the year
1913-14.

	Number	Value.
		$
Steam Fishing Vessels (tonnage 3,761).........................	58	1,071,575
Sailing and Gasoline Vessels	156	573,180
Boats (sail)...............	3,076	217,605
" (gasoline).........................	2,434	1,018,150
Carrying Smacks........................	190	75,650
Gill Nets, Seines, Trap and Smelt Nets, etc....	6,865	1,031,124
Trawls ;.........	22	1,100
Hand Lines..........	4,288	4,870
Skates of Gear..........	1,888	37,760
Salteries......	12	1,200
Salmon Canneries........	81	4,115,410
Clam "	1	3,000
Freezers and Ice-houses.....	15	1,502,000
Smoke and Fish-houses	29	388,300
Fishing Piers and Wharves.....	93	1,908,689
Whaling Stations	4	500,000
Oil Factory..........	1	40,000
Totals.....................	....	12,489,613

Number of men employed on Vessels.....	1,193
" " Boats...........	10,055
" " Carrying Smacks....	68
" persons employed in Fish-houses, Freezers, Canneries, &c	9,391
Totals.................	20,707

APPENDIX No. 10.

Imports and Exports of Fish

IMPORTS.

STATEMENT showing the Quantities of the Chief Commercial Fish and Fish Products Imported into Canada, for Home Consumption, during the fiscal year 1913-14.

(From Report of Customs Department.)

Cod, Haddock, Hake and Pollock (fresh)	cwts.	2,508
" " (dried)	"	86,470
" " (smoked)	"	291
" " (green salted)	"	620
" " (pickled)	"	2
Halibut (fresh)	"	54,524
Herring, (fresh)	"	5,944
" (pickled)	"	75,532
" (smoked)	"	62
Mackerel, (fresh)	"	9
" (pickled)	"	198
Salmon, (fresh)	"	25,974
" (smoked)	"	80
" (canned)	"	21
" (pickled)	"	62,294
Bait fish	"	989
Lobsters, (fresh)	"	455
" (canned)	"	593
Oysters, (fresh, in shell)	brls.	60
" (shelled, in bulk)	gals.	249,518
" (canned, one pint and under)	cans.	451,787
" " (one quart and under)	"	7,146
" " (over one quart)	quarts	2,212
" (preserved)	cwts.	487
Fish oil, cod	gals.	22,326
Seal oil	"	20
Whale oil	"	19,081
Other oil	"	24,763

The value of the imports of Fish and Fish products for the year 1913-14 amounted to $2,542,310.

EXPORTS.

STATEMENT showing the Quantities of the Chief Commercial Fish and Fish Products (the Produce of Canada) Exported during the fiscal year 1913-14.

—From Report of Customs Department.

TO	Cod, including Haddock, Hake and Pollock.				Herring.				Mackerel.		Lobsters.	
	Fresh.	Dry.	Green salted.	Pickled.	Fresh.	Pickled.	Smoked.	Canned.	Fresh.	Pickled.	Fresh.	Canned.
	cwts.	cwts.	cwts.	cwts.	cwts.	brls.	cwts.	cwts.	cwts.	brls.	cwts.	lbs.
United Kingdom		25,073			108,916	1,499	295	325	40,239	22,767	49,439	2,997,990
United States	20,521	198,833	10,711	6,117	33	62,340	23,507		1	4,851		2,273,849
B. W.		37				30,214	9,360					888
Bermuda		3,690			2	51	230	2		102		48
B. Guiana		16,217				1,827	1,064			978		1,088
B.		163										
Newfoundland		5,254	2,018	16	3,750	69	225		31			
Austria-Hungary		1,724										
Brazil		96,527										
Costa		1,560				17	42			116		
Cuba		75,887				25	445			364		
Danish West Indies		560				253	72			17		
Dutch		5,713				3	353			22		24
Siam												
French West Indies												
Guatemala		160					24					
Hawaii		2,592										
Hayti		3,503				3				5		
Italy		61,192					14					
Miquelon and St. Pierre		10			5					1		
Nicaragua		18,993										
Panama		76,092				90	391			1,217		
Porto Rico		7,065				3,061	5			68		
Portugal		3,816										
San Domingo		468					115					
Spain		1,601				10	7			116		
United States of Columbia												

Venezuela												
Australia												222
Hong Kong	14,576											
China							83,925					
Fiji						2	39,149					
British Straits Settlements												
Norway	36,650											
...ka												
Japan							124,179					
Sweden												
Chili												
France	1,996,674											
New Zealand	1,864					49						
British Oceania, other												
Belgium	9,928						6					
Denmark	89,182											
Germany	487,016											
Holland	78,816											
Mexico												
Russia in Europe												
British South Africa	5,160											6
...sh Idia												
Dutch East Indies												
Ecuador												4
French ...												
Sweden	175,053											
Philippines												
Dutch West Indies												179
	8,318,656	49,439	30,628	40,271	327	36,178	346,627	112,706	6,135	12,729	746,482	20,521

EXPORTS—*Concluded.*

STATEMENT showing the Quantities of the Chief Commercial Fish and Fish Products (the Produce of Canada) Exported during the fiscal year 1913-14.

—From Report of Customs Department.

TO	Salmon				Oysters	Bait, including clams.	Oil			Halibut
	Fresh.	Canned.	Smoked.	Pickled.	Fresh.		Cod.	Seal.	Whale.	Fresh.
	cwts.	cwts.	cwts.	brls.	brls.	brls.	gals.	gals.	gals.	cwts.
United Kingdom	10,959	488,626	385	66	36	18	28,734		427,571	3,264
United States	31,556	11,812	60	4,149	297	85,849	335,177	858	234,880	42,783
B. W. Indies	1	212	7	873		1	1,438			
Bermuda		112		22						
B. Guiana	3	4		208						
B. Honduras	10									
Malta					4					
Newfoundland	2	22		13			8,086			54
Austria-Hungary										
Brazil										
Costa Rica				27						
Cuba							82			
Danish West Indies				18						
Dutch Guiana				117						
Siam		576								
French West Indies										
Guatemala										
Hawaii										
Hayti					4					
Italy		7								
Miquelon and St. Pierre										1
Nicaragua				259						
Panama										
Porto Rico										
Portugal										
San Domingo										
Spain				25						
United States of Columbia										
Venezuela										
Australia	181	26,701		359						
Hong Kong	13	2,021								

China							5,681		903	
Fiji										
British Straits Settlements									7,049	
Norway									21,916	
Alaska									132	
Japan	1						28,381		44	
Sweden							180		30	
Chili									816	
France									20,198	180
New Zealand & other									13,544	5
British						1			231	
Belgium									3,113	
Denmark										1,488
Germany							1,274		948	
Holland							6,410		709	
Mexico										
Russia in Europe										
British South Africa									1,164	
British India									5,331	
Dutch East Indies									3,779	
Ecuador										
French Oceania									229	
Sweden										
Philippine									241	
Dutch West Indies										
Totals	46,103	662,451	858	373,517	85,868	342	48,058	452	610,974	44,398

39—19

The value of fish and fish products (the produce of Canada) exported during the year 1913-14, amounted to $20,698,849.

APPENDIX No. 11.

FISHERIES PATROL SERVICE

The following reports on the work of the vessels comprising the Fisheries Patrol fleet during the year 1913-1914 are submitted by the Inspectors whose districts the boats respectively serve:—

REPORT ON THE WORK OF PATROL BOAT *DAVIES,* IN DISTRICT No. 1, NOVA SCOTIA.

SYDNEY, N.S., 1914.

To the Superintendent of Fisheries,
     Ottawa.

SIR,—The patrol boat *Davies,* Captain D. L. Stewart, commenced patrol work on the 25th of July. Unusually blustery weather precluded anything approaching constant or active service for such a small boat assigned to this work. Following the close of the lobster fishing season attention was given to illegal fishing and in all about one hundred traps were destroyed along the coastal shores and at the western entrance of the Bras D'Or lakes.

                         I am, Sir,
                              Your obedient servant,

                                   A. G. McLEOD,
                                   Inspector of Fisheries.

REPORT ON THE WORK OF PATROL BOATS *C* AND *E.*

DISTRICT No. 2, N.S.

PICTOU, N.S., 1914

To the Superintendent of Fisheries,
     Ottawa.

                              ' C.'

SIR,—During the season from June 1 to December 15 this boat was employed patrolling the coast from Lunenburg County line eastward to Canso, also in the Chedabucto bay.

July 8, Overseer George Rowlings on board, patrolled to Ship harbour. July 10 Overseer R. Gaston came on board and patrolled eastward to Guysboro county line. Found a crate with lobsters at Sober island, which were liberated, but were unable to find the owner.

July 11 Overseer R. V. Cooper came on board and patrolled to Goldboro, but found no violations of the lobster regulations. Overseer John A. Dillon joined the boat and patrolled to Canso and Half Island cove; found no illegal fishing.

July 18 found some lobster traps but they were 'leftovers' with no bait in them.

July 28 found some lobster traps in Shad bay which had not been baited recently.

August 7 heard of some illegal fishing at McNab's island; made careful search, but found nothing. From previous date to August 14 patrolled to Canso, found in Dover bay five lobster traps, old gear. Continued patrol to Hubbards cove and returned to Halifax. The officer in charge, Mr. Edward De Young, had to go to hospital under medical advice, and the boat was in charge of Mr. Wm J. Nauss, and patrolled to Whitehead. Leaving that place about 3 a.m., October 1, they, with Overseer Dillon on board, proceeded to Dover, and after waiting some time observed two fishermen going out and fishing lobster traps. The boat was seized and the fishermen summoned before the stipendiary magistrate at Canso and fined one dollar each and costs, and the boat confiscated.

October 7, found some lobster traps set in Dover bay which were confiscated.

November 15, found a few traps at McNab's Island which were confiscated.

December 2, destroyed a crate containing lobsters at Terrance Bay and liberated lobsters.

'E.'

From May 12 to July 12 this boat was engaged collecting spawn for the lobster hatchery at Georgetown, and from the latter date was employed patrolling the coast waters of Cumberland and Colchester, on the Strait of Northumberland; also for two weeks on the coast waters of Westmorland County, New Brunswick.

July 19, found a fishing boat, owned by C. W. Murray, with about 300 pounds of lobsters. Towed the boat to Pugwash, where the owner was tried and convicted, and fined $25.

July 29, found a boat owned by W. E. Spence, having about three hundred pounds of lobsters on board. Towed the boat to Pugwash, where the owner was tried and convicted, and fined $25.

August 22, off the east end of Saddle Island sighted a boat fishing lobster traps and headed for the boat, and when about a mile from her they started for the shore, when the *E* gave chase and endeavoured to head her off. Got quite close and saw a man in her throw overboard a parcel which was supposed to be lobsters. Signalled them to stop but they went on and landed, and ran for the woods. As the water was shoal *E* had to come to anchor, went ashore in the boat, found bait and bait boxes in this boat and took it in tow to Wallace and left it in charge of the Customs Officer.

Having reason to suspect that lobsters were concealed in certain premises, I obtained search warrants and searched a number of places. In one of these a case of canned lobsters was found without a label; this was seized and delivered to the Customs Officer at Pugwash, and the owner summoned, but upon trial the Justice did not convict.

During the season much time was occupied in dragging the coast waters wherever it was believed it was probable that gear was set, and 1,302 lobster traps were found and confiscated, also twenty one anchors and about seven thousand fathoms of rope.

The officer in charge of this boat is Mr. A. E. Seaman of Pugwash.

I am, sir,
Your obedient servant,

ROBERT HOCKIN,
Inspector of Fisheries.

5 GEORGE V., A. 1915

REPORT ON WORK OF PATROL BOATS 'A' AND 'B.' DISTRICT No. 3— NOVA SCOTIA.

To the Superintendent of Fisheries,
Ottawa.

SHELBURNE, N. S., 1914.

'A.'

SIR,—Patrol boat *A*, Captain Hadley Blackford, was placed in commission April 1, 1913, from which date until close of lobster fishing season in Digby and Annapolis Counties, June 30, she was engaged in collecting seed lobsters for the Long Beach lobster pound.

At the conclusion of this work the boat was engaged, in patrolling the waters of Digby and Annapolis until September 11, when she proceeded to Yarmouth County, where she was engaged in patrol work until October 1, after which she returned to Digby, and continued in service until the opening of the lobster season January 6, 1914.

The district in which *A* is employed is not a difficult one, insofar as the enforcement of the fishery regulation is concerned. The lobster fishermen generally are obedient to the close season law. About one hundred and fifty traps were destroyed by the *A* during the season.

'B.'

Patrol boat *B*, Captain John Batemen, was employed in the general patrol of the waters of the south shore of this district during the closed lobster season, beginning May 15, 1913.

Special attention was given to Yarmouth County, where the extensive lobster grounds demand thorough and regular patrol. The work of the boat during the coming year will probably be wholly confined to this County.

Five hundred and fifty traps were destroyed during the season. Three hundred and seventy-five of these were taken in Yarmouth County, and the remainder in Queens and Lunenburg Counties.

Grapling irons are used with much success, as many of the traps are set in trawls, and can only be discovered by grappling.

The boat has given excellent satisfaction, and the officers are fast getting control of illegal fishing.

I am, sir,
Your obedient servant,

WARD FISHER,
Inspector of Fisheries.

REPORT ON WORK OF PATROL BOATS 'SEA GULL' AND 'PHALAROPE.'

To the Superintendent of Fisheries,
Ottawa.

CAMPOBELLO, N.B., 1914.

'SEA GULL.'

SIR,—I have the honour to make the following report on the work of patrol boat *Sea Gull* and *Phalarope*. As you are aware the *Sea Gull* was loaned to us by the Biological Board during July last, taking the place of patrol boat *No. 2*, which was unfit for further service. During the summer and fall she was in charge of Captain

Mitchell, now of the *Phalarope*, operating from Campobello. When the *Phalarope* went into commission the *Sea Gull* was transferred to Grand Manan, and placed in charge of Captain Green, with the waters around Grand Manan Island for her territory. She has done fairly good work there, but as there is a large area to be covered, and as the waters are exposed and generally rough, a much larger and faster boat is needed for that place.

' PHALAROPE.'

This boat was late in getting in commission, as it was December before she was received from the builder, and as it was sometime later before the engine was running satisfactorily. The *Phalarope* gave a good account of herself after that, both in having the lobster size limit carried out during the winter and preventing dynamite violations during the spring as well.

I would like to add that the *Phalarope* is an excellent patrol boat, reasonably fast, staunch and seaworthy, and in every way satisfactory .

I am, sir,
Your obedient servant,

J. F. CALDER,
Inspector of Fisheries.

REPORT ON WORK OF PATROL BOAT 'HUDSON.'

To the Superintendent of Fisheries,
  Ottawa.

NEWCASTLE, N.B., 1914.

SIR,—The following is the report given me by Mr. Chapman on the work of the above named boat.

'Previous to the 8th of July this steamer was engaged in gathering spawn for the Shemogue Hatchery. On that date Captain Goodwin reported to me from Shediac as being ready for patrol work, and I instructed him to proceed to Shippigan. He sailed from Shediac on the 10th and arrived at Shippigan on the 14th. I met him there on the 15th and on the 17th made the circuit of Shippigan and Miscou Islands in the steamer accompanied by the local officers. She remained patrolling in Gloucester County with head-quarters at Shippigan, most of the time around Shippigan and Miscou Islands; also with Overseer Arseneau on board until the 14th of September, when on visiting the Islands I ordered him to Port Elgin, Westmorland County. She arrived at Port Elgin on the 17th, but virtually did no work there on account of trouble with boilers and machinery. Under instructions from the Department I paid off crew on the 10th of October and the steamer was beached and laid up at Baie Verte in charge of Overseer Prescott.

I am, sir,
Your obedient servant,

D. MORRISON,
Inspector of Fisheries.

5 GEORGE V., A. 1915

REPORT ON WORK OF PATROL BOATS 'RICHMOND,' 'D,' AND 'J. L. NELSON.'

CHARLOTTETOWN, P.E.I.

To the Superintendent of Fisheries,
Ottawa, Ont.

SIR,—Patrol boat *Richmond,* Guardian Cameron, went into commission on the 22nd of April. Commenced by taking up some lobster lines run contrary to regulations, then patrolled Richmond Bay, preventing the taking of spawn lobsters. and after the season was over, looked after illegal fishing and destroyed a number of traps and lines, also prevented the taking of small oysters for stocking private beds. looked after quahaug fishermen and collected licenses.

After the first of October, this boat was almost continually on Grand river, seeing that all undersized oysters were returned to the beds. Guardian Cameron did good work although it was a difficult task to look after over one hundred boats on the river.

Patrol *D* left Halifax on the 3rd of May for Tracadie, N.S., arrived on the 13th of June, and commenced distributing lobster fry on the 21st instant, and continued to do so up to the 9th of July, then left for Charlottetown where some repairs were made. On the 15th left for Tignish, arriving there on the 17th instant and commenced patrolling the lobster ground between North Cape and Malpeque up to the 25th of October. During that time a large number of traps, rope, and anchors were taken and destroyed. For some time after was in Richmond assisting in preventing the illegal fishing of oysters. After returning to Tignish continued patrolling until this boat went out of commission, on the 29th of November. I am pleased to say Captain McCarthy did good service.

Patrol *J. L. Nelson,* Captain Wrayton, patrolled from North Cape to West Point and occasionally to Egmont Bay, making head-quarters at Miminegash. This boat has too much draft of water for the harbour, and consequently looses a good deal of time, but succeeded in destroying a large number of traps, besides a quantity of rope.

I am, sir,
Your obedient servant,

J. A. MATHESON.
Inspector of Fisheries.

REPORT ON WORK OF PATROL BOAT 'C. E. TANNER.'

MAGDALEN ISLANDS, P.Q., 1914.

To the Superintendent of Fisheries,
Ottawa.

SIR,—I have the honour to, herewith, submit a brief statement of movements and work of patrol boat *C. E. Tanner* during season of 1913.

Having received instructions on the 2nd of May to proceed to Pictou we accordingly took passage on the ss. *Lady Sybil* on the 6th of May, arriving in Pictou on the morning of the 8th following, and immediately reported to Commander W. Wakeham, aboard ss. crusier *Princess,* who directed us to taken charge of patrol boat *C. E. Tanner,* and make ready immediately to leave for the Magdalen Islands.

We were taken in tow of ss. *Princess,* and left for the Magdalens at 6.30 p.m., arriving on the morning of the 9th at nine o'clock at Grindstone, where we took leave of ss. *Princess,* and proceeded to House Harbour.

From the 10th of May to the 20th, we were occupied in patrolling the different herring trap stations, visiting foreign fishing schooners, and examining their licenses, which caused some to take out bait-licenses. On the 20th, the hatchery being ready, we began the work of carrying spawn from Grand Entry and Amherst; going alternatively, one day to Grand Entry, next day to Amherst, which work we did up to July the 11th, being afterwards engaged in distributing the matured spawn in the different lagoons, ending our work on the 30th of July.

On the 31st of July we resumed our patrolling service of the different lagoons at Grand Entry, House Harbour and Havre Aux Basques, operating up to November, during which time we seized about 1,490 fathoms of rope, two cases of cans, and destroyed 442 traps, all of which were reported to the Department of Marine and Fisheries.

The fishing being now over and further patrolling unnecessary, we are moving our boat into its winter quarters, having ended a fairly successful season.

I am, sir,
Your obedient servant,

CAPTAIN WM. S. ARSENEAU,
In Charge of *C. E. Tanner.*

REPORT ON WORK OF PATROL BOAT 'LADY OF THE LAKE.'

SELKIRK, Man., 1914.

To the Superintendent of Fisheries,
Ottawa.

SIR,—I have the honour to report with respect to the Fisheries patrol boat, *Lady of the Lake,* for the fiscal year, ended March 31, 1914. The *Lady of the Lake* is employed on the waters of Lake Winnipeg and tributaries, comprising all rivers flowing into and out of Lake Winnipeg. The boat is built of wood, a screw steamer, 105 feet in length, 18 feet 5 inches in width, 8 feet 9 inches in depth of hold. It was fitted up with electric light and searchlight, but the dynamo is out of commission. This boat makes about an average of eight miles an hour, and carries a crew of ten men. The necessary repairs, painting, etc., were commenced on the first of April, and she made her first trip to the lake May 12, for Berens river hatchery. Encountered ice, and arrived back to Selkirk May 19. Left Selkirk again May 22 with lighthouse supplies for the lighthouse on lake Winnipeg, we ran into ice after leaving Berens river and arrived back to Selkirk May 19. Left Selkirk again May 22 with lighthouse supplies for the the protection of the fisheries, on Lake Winnipeg, up to August the 1st. After that, she was engaged under Mr. Brunel, delivering wood and supplies from Selkirk to Berens river in connection with hatchery service. She was engaged during the balance of season gathering whitefish eggs for Gull harbour and Selkirk hatcheries. This boat went into her winter quarters on the 12th of November, and was laid up in the harbour at Selkirk. I am pleased to report that the steamer accomplished the work laid out for her in a very satisfactory manner, with one exception; she had the misfortune to break her stern bearing, and had to be dry docked. She was under orders from the writer, who was on board all the season, with the exception of the time she was engaged in the building of the Dauphin river hatchery.

I am, sir,
Your obedient servant,

J. A. HOWELL,
Inspector of Fisheries.

NEW WESTMINSTER, B.C., 1914.

To the Superintendent of Fisheries,
          Ottawa.

SIR,—I beg to submit the following reports on the work of the various patrol boats
that were under my supervision during the year 1913-1914.

I am, sir,
          Your obedient servant,

F. H. CUNNINGHAM,
          Chief Inspector of Fisheries.

NEW WESTMINSTER, B.C., 1914.

To the Chief Inspector of Fisheries,
          New Westminster, B.C.

SIR,—I have the honour of handing you my report covering the work done by the
patrol launch *Fispa* from the date of her commission up to and including March 31,
1914.

On June 14, 1913, the *Fispa* made her initial trip on the Fraser river, under
the supervision of her designer and builders, accompanied by the Chief Inspector of
Fisheries and Captain Crichton. The trial was considered satisfactory and arrangements were made to take the *Fispa* officially over subject to a few necessary adjustments
in detail. These adjustments having been made and everything considered satisfactory
and in accordance with her specifications, the *Fispa* was definitely taken over on
July 2, 1913.

During the early part of July, 1913, the *Fispa* patrolled the waters in the gulf of
Georgia along the boundary line in the gulf and from the Fraser river mouth to
Vancouver and Howe Sound. During this patrol, I had a favourable opportunity of
having the necessary adjustments made on deck and in the engine room to ensure the
safe navigation of the boat.

On the 14th of July, 1913, we had the honour of conveying the Honourable T. W.
Crothers, Minister of Labour, with his official staff, and the Chief Inspector of
Fisheries, from Vancouver to Union bay, and thence to Nanaimo, on official duty.

On the 26th of July, 1913, the *Fispa* sailed on her first long cruise to the north
coast of British Columbia, with the Chief Inspector of Fisheries and Deputy Commissioner of Fisheries for the Provincial Government on board. .

We visited the fishing areas and canneries along the coast, Alert bay, Bella
Coola, Warke Island, Belle Bella East, Digby island fish curing establishment, Tuck's
Inlet, Prince Rupert cold storage. We remained at Prince Rupert while the Chief
Inspector and Deputy Commissioner attended upon the Minister of Marine and
Fisheries, the Honourable J. D. Hazen, and on August 4, we again proceeded south on
an inspection of the various centres and canneries at Rivers Inlet and Namu. We
arrived at Vancouver on August 9, having covered a distance of 1,050 geographical
miles without any discomfort or mishap, which was a severe and satisfactory test on
this, the *Fispa's* maiden voyage.

We then went on patrol in the Gulf of Georgia and down to the boundary line,
and remained at that work until August 21, 1913. Upon this date, she had the honour
of conveying the Honourable Mr. Hazen, Minister of Marine and Fisheries, with his
official party, and the Chief Inspector of Fisheries and Deputy Commissioner for the
Provincial Government, on a cruise of inspection down the Fraser river, through the
Gulf of Georgia to the boundary line, and across to Boundary bay, when the Minister

had an opportunity of observing graphically the sockeye fishing by gill nets on the Fraser river, and also seeing and crossing the boundary line in the height of the sockeye season. We also ran around Boundary bay when the Minister had an opportunity of having demonstrated to him the American and B. C. stake nets in full operation.

The Honourable, the Minister, left us temporarily at a point in Boundary bay on the B. C. line, and we then ran around to Vancouver, when we had the honour of again receiving the Minister and his party and conveying them across the Gulf of Georgia to Nanaimo and again back to Vancouver on the following day.

On September 4, 1913, the *Fispa* sailed on her second northern cruise. We ran up the Straits of Juan de Fuca for the west coast of Vancouver island, and were joined at Alberni by the Chief Inspector of Fisheries, the Deputy Commissioner for the Provincial Government, and Mr. Taylor, Inspector of Fisheries for District No. 3.

We visited all the centres of fishing along the coast of west Vancouver island, which included Barclay, Clayoquot, Nootka and Quatsino Sounds, practically covering all the inland waters on the west coast of Vancouver island.

We proceeded around Cape Scott and touched at the various centres on the north-east side of Vancouver island, working down the east coast, and the mainland, and around the Gulf of Georgia south to Vancouver, making thus a complete inspection of the whole coast line and inland waters of Vancouver island and adjacent mainland waters. We arrived at Vancouver on September 18. This cruise covered a distance of 1,000 geographical miles.

Between September 18 and October 14, 1913, the *Fispa* after undergoing some slight overhaul, was again on patrol in the Gulf of Georgia from Texada island to the boundary line between Fraser river and Roberts Head.

On October 14, I was instructed by the Chief Inspector of Fisheries to take the *Fispa* north and to visit and explore the various inlets on the mainland and north Vancouver island; also the rivers entering these inlets and adjoining lakes, and to report upon in detail the physical condition of the salmon spawning beds within these waters, and to definitely mark down and locate the grounds frequented by the various species of salmon.

I have already placed before you, full reports in detail dealing with the work you entrusted me with, which, as you are aware, covered a very wide field, and took some considerable time.

In abstract: I visited Quatsu river and lake, north Vancouver island, and having explored the river and reached the lake I found it as you see from my report, practically closed with the serious results which you have before you.

I next ran to the mainland and up to Drury Inlet, Actæon Sound and McKenzie Sound. Here I ascertained that the sockeye spawning grounds that contracted the supply of fish for this huge area of water, was confined practically to two lakes, Hauskin and Keagh. I reached both of these lakes and explored the ground and rivers entering and coming out, and gave you my report in detail pointing out the appalling decline of this species of salmon in this area and its undoubted cause, and suggesting to you the remedy for its recovery, in order that something may be done to revive this magnificent area. Cohoe salmon frequent these waters in great numbers, but the blocking up of the lakes has not the same effect upon these fish as they run up the lesser rivers and creeks, which sockeye instinctively avoid.

Having made a complete examination of all the rivers and waters within this area, I proceeded up Knights Inlet, and my first visit was up the Glendall Cove, when I landed and went up the river here, reaching by trail both the lakes, Tom Brown lake, and another lake some 7 to 9 miles further back on the west arm of this river. This is practically the only sockeye spawning ground within Knights Inlet area, and I have placed in detail before you the result of my work here. The spawning grounds were well stocked, and free, with the exception of a very bad obstruction below the upper lake. I then proceeded to the head of Knights Inlet and went up the main river

5 GEORGE V., A. 1915

for a great distance. This river, as I have already reported upon, is frequented by spring salmon which run away up nearly to the Glacier before stopping to spawn. All the lower reaches and creeks were splendidly stocked with cohoe in spawn. I may mention in passing that there is a species of wild duck, the 'Saw Bill,' which does a fearful amount of destruction to the ova on the spawning beds all along the northern waters. I have personally watched them devour the roe on the beds, practically while the salmon are spawning. In my opinion, these birds do more damage than any other salmon enemy, and I would suggest that your officials within these districts destroy as many as possible. These birds are extraordinarily prolific, hatching out and rearing as many as 20 young birds to each brood. I had as many destroyed as possible—from three to four hundred.

Having fully explored the whole of these waters, I ran towards the Kingcomb Inlet waters, running up to Thompson's and Wakeham Sounds, and exploring all the creeks around the shores, and reported upon them in detail; arriving at the main river at the head of Kingcomb Inlet, I had a good opportunity of getting up this river by boat and trail, and gave you my report upon 40 miles of this river. This river, similar as it is to all the glacier rivers are essentially spring and cohoe spawning grounds, and were splendidly stocked and free from all obstructions, but being terribly ravaged by thousands of the Saw Bill duck. I am fully convinced that these birds destroy fully 30 per cent of the ova deposited upon the beds in the lesser creeks. Of course, we are aware that the Mallard and other wild duck do a certain amount of damage, but they are not so persistent and bold as the Saw Bill. Having completed the work on the grounds you entrusted me with, and fully reported, I returned to Westminster. Here the *Fispa* went under her first general overhaul, and was taken up on the Marine Ways, cleaned down, painted, and her propellors adjusted.

On January 13, 1914, we again left Westminster for the north. We conveyed your engineer, Mr. McHugh, as far north as Bella Bella east, and your fishery officer, Mr. Norrie, joined us there. We then proceeded up Ellerslie Channel and ran up to the head, a distance of 30 miles, so as to allow your engineer to reach the rivers and lake, and enable him to survey the obstructions and make the necessary arrangements for a fish pass. While here, I had an opportunity of getting up the lake on the ice, and exploring the feeding creeks and lake shore line, which I duly reported to you.

Mr. McHugh having completed his survey, left us for the south. At Bella Bella east, my instructions from you, sir, were to patrol the inner waters of Queen Charlotte Sound, and at the same time to ascertain as far as possible, what species of fish, and in what numbers, existed along the waters there. I continued at this work (but was handicapped by the exceptionally severe weather) until the 10th of March, 1913, and on that day, I returned to Vancouver and handed you my report, together with my remarks upon the commercial value of these now practically unfished waters. Quite apart from the staple fish, halibut, these waters hold enormous quantities of the smaller, but yet more valuable fish, cod, bass, sole, brit, flounders of eight varieties, and anchovies. With regard to the latter, I would like to ascertain more accurately in what quantities they exist, and of what type. The numerous lagoons appear swarming with them, but later on in the season, I hope to give you a more accurate report upon these lesser, but extremely valuable fish.

The *Fispa* returned north on March 17, 1914, and in accordance with your instructions, I have patrolled the waters of Queen Charlotte Sound, and made further research upon the fishing grounds.

I now have the honour of closing my report for the fiscal year ending March 31, 1914, and append the distance covered. Total geographical miles run from date of commission, July 2, 1913, up to and including March 31. 1914—5,912.

Yours respectively,

J. F. CRICHTON,
Captain.

SESSIONAL PAPER No. 39

NEW WESTMINSTER, B.C.

To the Chief Inspector of Fisheries,
    New Westminster, B.C.

SIR,—I beg herewith to submit a brief report of the services performed by the patrol launches of district No. 1, for the fiscal year ending March 31, 1914.

LAUNCH 'SWAN.'

This boat, under the charge of Patrolman Thomas Hembrough, has been employed in regular patrol work, covering the following waters: Fraser river and tributaries, including Coquitlam, Pitt, Lillooet, Stave, Sumas and Harrison rivers, and Harrison and Sumas-lakes, Boundary bay, and the Gulf of Georgia. This craft travelled, in all. 9,430 miles in the course of her work. The new 27-32 h. p. Eastern Standard engines, installed about a year ago, have done good service and given excellent satisfaction in every way. These engines have increased the value and efficiency of this launch very materially.

This boat has proven to be suitable in every way for the river-work.

LAUNCH 'ELK.'

This boat, under the charge of Patrolman William Dauphinee, has been engaged chiefly on the north arm of the Fraser and Sandheads opposite, enforcing regulations and performing regular patrol work. The efficiency of this boat was greatly increased at the beginning of the year by the installation of the 24 h. p. Lamb engines, which were transferred from the *Swan*. She is of quite shallow draft, and is well adapted in that respect for the service required of her. She travelled, during the year, 6,950 miles.

LAUNCH 'FOAM.'

This is, as you are aware, a new boat, built and put into commission in April, 1913. She is 45 feet long, by 10 feet beam, and is equipped with 27-32 h. p. Eastern Standard engines. She has done good service under the charge of Patrolman Samuel Waddell. The waters covered by this launch were Fraser river, principally between New Westminster and Stefeston, Canoe Pass, Gulf of Georgia, Sandheads and Boundary bay. She also assisted in distributing fry, making a number of trips for this purpose to Nanaimo, Cowichan harbour, Howe Sound, and Lake Buntzen. She logged, in all, 5,434 knots.

LAUNCH 'SEMIAHMO.'

This launch was built in 1901, and was purchased by this Department in 1909. During this time, up till last year, she did good service, but as the hull is old, and has become considerably weakened in parts, her usefulness as a patrol boat is at an end. During the year she has patrolled the waters of Howe Sound, Squamish, English bay, and Burrard Inlet. Patrolman Wm. McC. Moore, who has been in charge, has been compelled to confine his patrol to calm weather. It will be necessary for this boat to be replaced by a larger and more suitably equipped craft, in order to properly protect the interests of the fisheries in this portion of the district. This boat has travelled, during the year, 5,105 miles.

Respectfully submitted,

A. P. HALLADAY,
Assistant Inspector of Fisheries.

5 GEORGE V., A. 1915

REPORT ON WORK OF PATROL BOATS OF DISTRICT No. 2.

VANCOUVER, B.C., 1914.

To the Chief Inspector of Fisheries,
        New Westminster, B.C.

SIR,—I have the honour to submit my annual report in connection with the services performed by the Fishery patrol boats under my control during the fiscal year ended March 31, 1914.

LAUNCH 'MERLIN.'

During the season the gasoline launch *Merlin* was engaged in patroling the waters of Rivers inlet, and vicinity, and making trips of inspection to Smith Inlet. She travelled as near as can be computed, 3,694 miles, of this 1,515 miles were run during the weekly close season. One seizure was made during the close season, the boat having a long net over two hundred fathoms.

Overseer Saugstad reports that the Fishery Regulations were closely observed.

LAUNCH 'KINGFISHER.'

This launch patrolled the waters Burke Channel and Dean Channel which include the Bella Coola and Kimsquit fisheries. She travelled about 2,000 miles during the season, and was in commission from May 1 to September 30. She had a new cabin top built this season, in place of the canvass as heretofore, which was most unsatisfactory in rough weather. The wooden cabin enables her to stand off the rough water better than the canvass top, and Overseer Widsten was much pleased with this improvement rendering her work much more effective.

C. G. S. 'FALCON.'

This steamer was under my control from May 10 until November 1, 1913, and during that time logged 5,730 statute miles, making one hundred and ninety-seven calls, at the different fisheries in the district. She patrols a coast line of some 1,100 miles. Her special work is the protection of the salmon and herring fisheries, and to convey the Inspector to the outlying fisheries that require supervision.

During July we had the Honourable the Minister of Marine and Fisheries aboard as well as Mr. H. S. Clements, M.P., and Chief Inspector Cunningham. We took the Minister around Prince Rupert Harbour, and also as far up the Skeena as Port Essington. During the season we had many other officials aboard, on visits of inspection. Below I give the approximate distances travelled each month.

	Miles.	Calls made.
May	280	20
June	1,436	48
July	1,599	64
August	935	49
September	460	23
October	576	11
Total	5,286	197

LAUNCH 'LINNET.'

The launch *Linnet* was under the control of Overseer Adamson, and patrols the Naas river and Portland Inlet, she travelled some 3,354 miles approximately, and was in commission from April 1 to September 30. She visited the fish traps and seining grounds in the Naas river and Portland Inlet as often as possible, and found everything satisfactory. A few fishing boats were seized for fishing during the weekly close season, and a few more for using gill-nets exceeding 200 fathoms in length. A larger and faster launch would be more effective in enforcing the regulations.

LAUNCH 'HAWK.'

The launch patrols the Upper Skeena and Oxstahl rivers, and during the season travelled some 3,000 miles. She was commissioned from April 1 to October 31. She is under the supervision of Overseer Norrie. Very few seizures were made during the season owing to the effective patrol. The *Hawk* was assisted during the sockeye season by the launch *Pilgrim* which boat was chartered by the Department.

LAUNCH 'KAYEX.'

This launch owing to the defective engine, was unable to perform her duties as patrol boat for the Skeena, satisfactorily, consequently the *Pilgrim* was chartered to help her out. The Department has decided to instal a new thirty horse power engine in the launch, in place of the old one, and she will take up her patrol duties on the Skeena in conjunction with the *Hawk*, on or about 15th April next.

LAUNCH 'GANNET.'

The above boat has been in commission since the 3rd of October, 1913, her log showing a distance travelled of approximately 4,000 miles. During the summer months, she is engaged in patrolling the salmon fisheries in the vicinity of Queen Charlotte Islands, and in the winter months she gives very effective service in guarding the halibut fisheries along the three-mile limit against poachers.

LAUNCH 'EVELYN B.'

This launch was chartered by the Department for a short time during the season on the upper waters of the Naas river, for the protection of the salmon fisheries. The Indians were catching salmon illegally and running them down to the canneries in their gasoline launches, this was entirely stopped by the action of the Department, in placing the launch in commission for a short time.

'ANNIE D.'

This launch was chartered by the Department for the Central division, and was engaged in patrolling the waters between Fitzhugh Sound and Granville Channel, an area of some 1,500 miles. She travelled some 3,000 miles during the five months she was in commission, making four seizures during the season. She was under the control of Overseer Boyd, who states that the fishery regulations were well observed.

I am, Sir,

Your obedient servant,

JOHN T. C. WILLIAMS.

REPORT ON THE WORK OF PATROL BOATS IN DISTRICT No. 3, BRITISH COLUMBIA.

NANAIMO, B.C., 1914.

To the Chief Inspector of Fisheries,
New Westminster, B.C.

PATROL BOAT 'ALCEDO.'

SIR,—During the season the *Alcedo* patrolled the waters between Vancouver Island and the mainland, from the south end of Vancouver Island to Queen Charlotte Sound.

Eight thousand three hundred and eighty-six miles were logged. Twenty seven prosecutions for illegal fishing were made. Under instructions from the Department I placed the *Alcedo* at the disposal of Mr. Fletcher, Inspector of Post Offices. Mr. Haynes, Assistant Inspector, made the inspection of the Post Offices in the northern part of the district during the latter part of October and completed his work the last week in November.

LAUNCH 'COHOE.'

The launch *Cohoe* went into commission on October 24 and since that date has logged 2,176 miles. Her district extends from Howe Sound to Bute Inlet on the mainland coast. She has done good service and is giving satisfaction.

LAUNCH 'RAVEN.'

The launch *Raven* was only in commission for a few weeks during the close of the fishing season, and has been laid up during the winter. Her headquarters are at Alert bay on Cormorant island. Her district comprises the waters lying between Vancouver island and the mainland including Knight, Seymour, and Kingcomb Inlets. This boat is well equipped for her work, and in the coming season will do effective service.

LAUNCH 'GULL.'

The launch *Gull* went into commission on May 30, 1913, and patrolled the waters on the east coast of Vancouver island between Big Qualicum river and Maple bay. The number of miles covered during this time was 7,413. The fishery regulations were well enforced. Only two cases of illegal fishing was reported. This boat is also well equipped for service.

The patrol boats *Egret* and *Heron* now completed and ready for service on the west coast of Vancouver island will prove a valuable addition to the patrol service of the district.

I am, sir,
Your obedient servant,

EDWARD G. TAYLOR,
Inspector of Fisheries.

APPENDIX No. 12.

REPORT ON OYSTER CULTURE BY THE DEPARTMENT'S EXPERT FOR THE SEASON OF 1913.

CHARLOTTETOWN, P. E. ISLAND,

To the Superintendent of Fisheries,
Ottawa.

SIR,—I have the honour to submit to you my annual report on last season's work in connection with oyster culture in the lower provinces.

On the opening of navigation the *Ostrea* was removed from her winter quarters and prepared for sea, and when ready, received instructions from your department to proceed to Pownal bay and assist the provincial survey officers in surveying Pownal and Orwell bays. This work was commenced on the 20th May under the superintendence of Professor H. H. Shaw, Provincial Engineer for Prince Edward Island, and his staff, and continued surveying until the 16th July, when these areas were completed. I then returned to Charlottetown, landed all the surveying instruments, coaled and watered *Ostrea* and sailed the next morning for Caribo.

CARIBO, N.S.

This area was formerly a barren bottom situated on the south side of Caribo harbour about 150 yards off the shore from high water mark, and contains about four acres running parallel with the shore. In 1911 this area was shelled and prepared for planting young oysters, but owing to the lateness of the season only twenty-five barrels of small oysters were obtained from Richmond bay, Prince Edward Island, and planted on one acre on the eastern end of the area. In 1912 arrangements were made by the Department to obtain 75 barrels of seed oysters from Warren, Rhode Island. These oysters arrived in good condition and were planted on the remaining three acres on the 25th of May of last year. It was my intention to have examined this area later in the season if opportunity permitted, but my time was otherwise occupied, and at the end of the season the weather was too wild and stormy to make the attempt, so left it until this season. I sailed from Charlottetown on the 17th July arriving in Caribo the same day, and on Friday the 18th laid out the area and examined same, and found the bed to be clean and free from weed or sediment, the oysters are growing and are in a healthy condition. The oysters which were transplanted from Prince Edward Island the year before last are looking much better than the American oyster, and have grown considerably.

With the American oysters I noticed a small per centage of death, due probably to their long journey, and their growth has not been very rapid, but they are very firm and their shells are hardening, and I look forward to a much larger growth next year. I took three hauls of the dredge of American oysters and their numbers were as follows: 115, 125 and 195 respectively, and two hauls of Prince Edward Island oysters being 251 and 179. I did not detect any signs of spat among the cultch, but the season so

5 GEORGE V., A. 1915

far had been very backward, wild and cold, the temperature of the water at above date was 63 Fahr. The weather too was most disagreeable and wet, I left there on Saturday returning to Charlottetown.

MALPEQUE, P.E.I.

I left Charlottetown on the 21st July arriving in Malpeque on the 23rd to examine Bird Island bed which is an artificially made bed comprising 4 acres of water bottom. It is a barren area situated on the north side of Richmond bay, off the south shore of Bird or Middle island just to the eastward of where the mouth of Bideford river empties into Richmond bay. This area was selected in 1910 when it was cleaned, prepared, shelled and planted with young hardy growing oysters from Ram island. I visited and examined this area in 1911 and found the bottom clean and in perfect condition and also obtained samples of oysters from different parts of the area, they were all found to be in a perfectly healthy condition and had grown considerably, there was no mortality noticed, but could detect several traces of young spat attached to shells, stones, &c., giving every satisfaction. In 1912 I again visited this area and found the bottom clean and free from seaward or eelgrass and the oysters growing in a very satisfactory manner, but it was particularly noticeable, that this bed had been raided on by poachers and the oysters were much scarcer in numbers than formerly.

Upon my arrival at the bed this season I was much surprised to find an enormous growth of eelgrass over the whole area. The growth of eelgrass in Malpeque and Richmond bays this summer has been extraordinary and everyone has remarked that they have never remembered seeing such a quantity before, and before an examination could be made the eelgrass had to be removed so I obtained the services of three men to assist in raking over the area to remove and clean up the ground generally and continued to do so until the 28th August, when I found all the grass had been removed and the area clean. I took a few hauls of the dredge over different parts of the bed with the following results:—The first haul consisted of 124 oysters and small brood, second haul 80 oysters, third, 37 oysters, fourth, 57 oysters, fifth, 18 oysters and sixth haul 25 oysters. The largest proportion of oysters found on the bed consists of small ones which have grown there since the bed was planted, and the original oysters that were placed there are very scarce. Had the bed not been molested by poachers the results would have been very gratifying. I found a few well grown oysters which were planted when very small but the majority of them have disappeared. There are quite a few scattered young oysters to be found along the eastern shore of Bird island, the spat, no doubt orginating from the bed after it had been planted. After completing my work I removed the stakes which marked the area, and sailed from Malpeque on the 2nd September, but owing to bad weather did not arrive in Richibucto until Saturday the 6th of September.

RICHIBUCTO, N.B.

Having received instructions to proceed to Richibucto Village bay for the purpose of examining the bottom to ascertain if it would be possible and advisable to plant quahaugs as an experiment in this locality I thoroughly examined the whole area of the bay, and found the shores to be of a sandy nature, but as the water deepens to two or three feet the bottom is found to be exceedingly soft, the water continues to deepen to a depth of seven or eight feet with the same bottom. In the middle of this bay there is a middle ground or ridge composed of a mixture of sand, mud, clam shells, mussels, and in some places is thinly covered with eelgrass, and is bare at others, with a depth of from five to six feet around the edges, and gradually shoals towards the middle and southeast side to about two feet at low water. I placed stakes around this area and ran a line around it to ascertain the size. The north side had a length of two hundred and forty five fathoms, the south side which was rounded was three hundred and fifty fathoms, the western side was one hundred and five fathoms, and the eastern side was one hundred and eighty fathoms long. The water is pure and a fair current in this locality, and while the soft shell clam is found along the shores, mussels also grow here, and oysters are also found in small quantities in the Narrows, which is only a short distance from the area in question, but no quahaugs are found in this harbour; yet I see no reason why they would not grow if planted here, and also in the Narrows which has a sandy bottom with a depth of ten feet of water, and it is the channel leading to the Village bay. The quahaugs could be planted around this area with a depth of four feet over them at low water.

Should the Department decide to make an experiment of planting any quahaugs here, no labour of preparing the ground is at all necessary, it is only a matter of obtaining them and depositing them on the grounds. They could be obtained at Buctouche at about one dollar per bushel or thereabouts which I believe is the present market price, and would suggest that about one hundred bushels would be sufficient for distribution in the different localities as an experiment, but owing to the lateness of the season the planting of these quahaugs was deferred until the following spring when further arrangements will be made.

I then left Richibucto and proceeded to Bay du Vin arriving there on the 25th September.

BAY DU VIN, N.B.

Last year I made an examination of the water bottoms in Bay du Vin and locality for the purpose of finding an area of barren bottom where an experimental bed could be formed. A suitable piece of ground comprising an area of four acres and having a depth of from seven to ten feet water was laid out, off Horton's creek on the west side of the bay, consisting of a firm smooth bottom with a few scattered stones, and is in close proximity to the Fishery Officer's residence. But before I could prepare this area I received instructions from the Department to take up patrol in Prince Edward Island.

I returned to Bay du Vin this fall in the hope of completing my work which was left undone from last year and endeavoured to make arrangements with the oyster fishermen to save their shells while fishing for oysters, this they promised to do, and on the 3rd October obtained twenty-nine barrels shells, and on the 6th secured seventy-eight barrels from nine boats, afterwards these men would not catch them for the price paid. I then went over to Oak Point to make arrangements with other oyster fishermen to save their shells, but owing to gales of wind and broken weather I was unable to secure more than one hundred and thirty-six barrels up to the 3rd November, making a total of two hundred and forty-three barrels of shells laid to date. The fishermen then stopped fishing for the season, and after preparing for sea I awaited the first favourable opportunity to proceed to Charlottetown. On the 7th November the weather moderated when I sailed from Bay du Vin Island, arriving in Charlottetown on the following day. The season being too far advanced to carry on any further work I dismantled the *Ostrea* placing all her gear in the warehouse and had her hauled into her winter quarters and blocked up snug for the winter. I may here state that during the whole of the past season the weather was very unsettled and disagreeable and much rain fell, which retarded my work to a considerable extent in a boat the size of the *Ostrea*.

PRIVATE OYSTER AREAS IN P.E.I.

In my last year's report I referred to the work which had been taken up by the Prince Edward Island government in surveying the water bottoms and laying out areas to be leased for the purpose of private cultivation of oysters. This work they are still following up, and surveys have been made during the past season of the following

39—20

5 GEORGE V., A. 1915

localities, viz.: St. Peters bay, Pownal bay, Orwell bay, New London bay, Cascumpeque bay, Tryon river, Bentick Cove, Oyster Cove, Indian river, Barbaraweit river, Shipyard river, and Bedeque bay. South West river and New London have been partly surveyed.

There are approximately over five thousand acres already leased and under cultivation, and applications to lease seven thousand acres are also under consideration.

The McNutt Malpeque Oyster Company and the Standard Cup Oyster Company have obtained an up to date plant in the shape of modern gasoline dredging boats from the United States, and have laid large quantities of seed and other oysters. There are at the present date twelve or fourteen oyster companies fairly started in Richmond bay with capital ranging from twenty-five thousand to one hundred and fifty thousand dollars, and new companies are being continually formed, so that it is confidentially expected that under private culture, a new era will begin and the prospects for the future are looking very bright in the oyster industry on Prince Edward Island.

I have the honour to be, sir,
Your obedient servant,

ERNEST KEMP,
Oyster Expert.

APPENDIX No. 13.

FISH BREEDING

OTTAWA, September 30, 1914.

The Deputy Minister,
  Department of Marine and Fisheries,
    Ottawa, Ontario.

SIR,—I have the honour to submit herewith my annual report on the Fish Breeding operations conducted by the Department during the season of 1913-14.

The total distribution of fry and older fish for the season was approximately 1,228,000,000 and while the number of some of the species propagated was not as large as last year, the total distribution was increased to the extent of over 154 millions. A general outline of the work as regards each species propagated is given below.

ATLANTIC SALMON.

Atlantic salmon is propagated principally in the Maritime Provinces and Quebec. With the exception of Tadousac, Que., where they are taken in nets operated by the hatchery staff, the parent fish are purchased from the commercial fishermen. The early run of fish, which enters the river during the legal fishing season, is retained at Tadousac, Quebec, Restigouche and St. John, N.B.; but at Miramichi, N.B., and Margaree, N.S., none are impounded until September, after the beginning of the close season. The total distribution of this species was slightly less than it was last year. This is due to the eggs obtained at St. John and Miramichi not being up to their usual standard. The number of eggs obtained from each retaining pond and the manner in which they were distributed to the different hatcheries is as follows, viz.:—

  Tadousac Pond............................ 2,239,000 Eggs.

These were laid down and brought to the eyed stage in the Tadousac hatchery, when 1,000,000 were sent to the subsidiary hatchery on the Bergeronnes river, and 600,000 were sent to the subsidiary hatchery on the St. Marguerite river. The balance was distributed as fry from Tadousac.

  Restigouche Pond......................... 2,356,500 Eggs.

On the Restigouche 530 salmon were taken in the government net at Tide Head, and 98 were purchased from the commercial fishermen at New Mills. These were retained in the fresh water pond at Tide Head on the Restigouche and yielded 2,356,500 eggs, which were all brought to the eyed stage in the Restigouche hatchery, when 500,000 were transferred to the new subsidiary hatchery on the Nepisiguit river.

  Miramichi Pond.......................... 8,965,000 Eggs.   ..

On the Miramichi river 2,100 salmon were taken, which yielded 8,965,000 eggs, which were distributed as follows, viz.:—

Bedford Hatchery, N.S...................	1,500,000
Kelly's Pond Hatchery, P.E.I............	1,265,000
Windsor Hatchery, N.S..................	1,750,000
Gaspe Hatchery, Quebec................	2,300,000
Miramichi Hatchery, N.B...............	2,150,000

5 GEORGE V., A. 1915

St. John Pond...............................   6,896,500

One thousand two hundred and seventy-five fish stripped at the St. John Pond yielded 6,896,500 eggs. These were all laid down in the hatchery at Grand Falls until the new hatchery at St. John was completed. To accommodate this large number, the trays had to be placed two or three tiers deep in the trough, necessitating more handling and washing, which caused a little heavier loss than would otherwise have been the case. After the eggs reached the eyed stage, the following shipments were made from Grand Falls, viz.:—

St. John Hatchery, N.B..	2,500,000
Newcastle Hatchery, Ontario..	202,900
Cowichan Lake Hatchery, B.C..	100,000
St. Alexis Hatchery, Que..	75,000
Lake Lester Hatchery, Que..	75,000
Margaree Pond..	6,730,000

Six million seven hundred and thirty thousand eggs were obtained from 862 salmon in the retaining pond at Margaree Harbour, N.S. 1,200,000 of these were at once transferred to the Middleton Hatchery, N.S., and the balance, 5,530,000, were laid down in the hatchery at N.E. Margaree, N.S. In February, after they reached the eyed stage, 400,000 were transferred to the Middleton hatchery and 600,000 to the Windsor hatchery, and in March, 1,000,000 were transferred to the subsidiary hatchery at Lindloff creek, near St. Peter's, N.S., leaving a balance of 3,530,000 which were distributed as fry from the Margaree hatchery. Very satisfactory results were obtained from all of these eggs, with the exception of those transferred to Lindloff. For some reason, which up to the present the department has been unable to ascertain, these eggs were almost a total loss.

PACIFIC SALMON.

I am pleased to report the satisfactory increase of over $33\frac{3}{4}$ millions in the different species of Pacific salmon distributed in British Columbia over the numbers distributed in 1913. The parent salmon are intercepted on their way from the ocean to their spawning grounds by fences placed across the streams. These fences in some places were swept away by freshets; but the hatchery officers were able to replace them or to procure the number of fish required by means of nets.

At the Lakelse lake, Skeena river, hatchery the fences were swept out and all the parent fish had to be taken by nets on their natural spawning beds. This necessitated a great deal of difficult work on the part of the staff; but to their credit they succeeded in obtaining the usual supply of eggs.

SPECKLED TROUT.

The distribution of speckled trout fry almost doubled that of last year.

At the St. Alexis hatchery, were the department has in the past obtained practically all the eggs of wild trout that it has handled, heavy rains and consequent freshets during the collecting season made the work difficult in the first place, and a large number of fish were lost by the breaking of a dam, which swept away some of the crates and raised the water so high over other enclosures that the parent fish retained in them escaped. The total collection in the district was 486,000 eggs which were distributed to the different hatcheries as follows, viz.:—

Lac Tremblant Hatchery..	100,000
Lake Lester Hatchery..	75,000
St. Alexis Hatchery..	311,000

The shortage at St. Alexis was made up by the increased collection in the streams of Prince Edward Island, and the purchase of a larger number of eggs than has been customary. The increased collection in Prince Edward Island is very satisfactory. Three years ago, the officer in charge of the Kelly's Pond Hatchery succeeded in collecting 7,000 eggs. In 1912 the collection was increased to 134,000, and last year it was further increased to 459,600. The fry resulting from these eggs were distributed from the Kelly's Pond Hatchery.

The number of speckled trout eggs obtained from the Margaree river was not as large as last year. Unfortunately, by the time that the retaining pond for the parent trout was ready the run had passed up the river and dispersed in the smaller lakes and streams at the headwaters. A small number of trout were taken; but these did not yield as well as could be desired. I am pleased to say, however, that profiting from last year's experience a trap net was set at the proper time to intercept the first run of sea-trout and 700 splendid fish were taken in a short time, a number of which run from 3 to 5 lbs. in weight. These have kept well in the pond throughout the summer and will be liberated after they are stripped. The fry resulting from the eggs obtained last fall were distributed from the Margaree Hatchery.

An attempt was made for the first time to collect trout eggs in the Middleton District, N.S. As this was entirely new work on the part of the men engaged, they only succeded in taking 20,000 eggs. This number was, however, increased by the purchase of 290,000 eggs, making a total of 310,000 trout eggs laid down in the Middleton hatchery.

An effort was also made to collect trout eggs in the Miramichi district. The different streams in the neighbourhood of the hatchery were prospected in September and October. The conditions, however, were not favourable, as the waters were extremely low. The northwest branch, seven miles of the southwest branch, as well as six miles of the Main Miramichi river were gone over without getting any trout. The Tabusintac was also inspected without result; but 180 trout were taken in the Bartibog river. Owing to the injuries these fish received in being transferred to the hatchery pond, the yield of eggs was small. The information, however, obtained indicates that if proper arrangements are made a considerable number of speckled trout eggs can be collected in the river, and, as it is under lease the Provincial government has made provision by Order in Council for the taking of parent fish therein for hatchery purposes. Operations are again being carried on this fall, and next season it is proposed to build a retaining pond at some suitable place and take the parent fish in a pound net adjacent thereto as they ascend the river, much in the same way as was done this year in the Margaree river.

Some 600,000 speckled trout eggs were also collected by the Officer in Charge of the Port Arthur hatchery in Lake Nipigon, Ontario, while engaged on this lake in the collection of whitefish and salmon trout eggs. While not more than 50 per cent of these eggs hatched, this return is all that could be expected under the conditions that prevailed. The eggs had to be kept at the spawning camp for a considerable time and owing to poor transportation facilities were subjected to rough handling in being transferred to the hatchery.

As the Provincial Government has agreed to attend to the propagation of sporting fish in Ontario, no collection of trout eggs in these waters will be made in future. 145,000 fry resulting from the eggs obtained there last fall were, this spring, handed over to the Provincial Government for distribution. The balance of the eggs were distributed to the different hatcheries, viz.:—

Granite Creek Hatchery, B.C.	50,000
Grand Falls Hatchery, N.B.	50,000
Bedford Hatchery, N.S.	75,000

5 GEORGE V, A. 1915

The speckled trout eggs that were hatched in the new establishment at St. John were purchased from the Caledon Mountain Trout Club, of Brantford, Ont.

OUANANICHE.

An attempt was again made by the Officer in Charge of the Bedford Hatchery to collect ouananiche in Grand lake and tributary streams; but without success. Trap and gill-nets were set on October 13, and the first fish were taken on October the 16th. From that date to December the 18th, 75 fish were taken, of which only two yielded eggs, the first on November the 21st and the second on December the 1st. As the remainder did not show any indication of ripening, and from their appearances would not ripen until midwinter and were becoming weak, they were liberated. Only 4,000 eggs in all were obtained, and as the operations were somewhat expensive, the department would not be justified in continuing them.

SALMON TROUT.

The collection of salmon trout eggs and the distribution of fry resulting therefrom was not as large as in 1912-13. The salmon trout eggs are obtained in the Great Lakes from commercially caught fish, and as the weather on Lake Huron and Georgian Bay continued mild right up to the commencement of the close season, a much smaller percentage than usual of the fish taken were ripe and the collection of eggs was in proportion thereto. The collection in Lake Superior was about the same as in former years.

The staff of the Newcastle hatchery collected 7,868,000 eggs in Lake Huron, which were distributed as follows, viz.:—

Southampton Hatchery, Ont..	2,548,000
Lake Tremblant Hatchery, Que.. . . . .(Eyed)	1,000,000
Lake Lester Hatchery, Que.. . . . .(Eyed)	507,600

The balance were hatched and distributed from the Newcastle Hatchery.

The staff of the Wiarton hatchery collected 12,572,000 eggs, which were distributed as follows, viz.:—

Lake Lester Hatchery, Que..	378,000
Grand Falls Hatchery, N.B..	50,000

The balance were hatched and distributed from the Wiarton hatchery.

The staff of the Southampton hatchery collected 8,103,000 which in addition to the shipment from Newcastle were all hatched and distributed from Southampton.

The Port Arthur hatchery staff collected 7,000,000 eggs in Lake Superior and 1,000,000 in Lake Nipigon, of which 1,000,000 were sent in the eyed stage to the Banff hatchery, and the balance were hatched and distributed as fry and fingerlings from the Port Arthur hatchery.

CUT-THROAT TROUT.

I am sorry to say that the collection of cut-throat trout eggs in the Banff district was not a success. Every effort was made to collect these eggs in the Jumping Pond and its tributaries, in Pirmez, Robinson and other creeks; but only a small number were procured, as fish of spawning size could not be found. Unfortunately, most of the eggs that were obtained and placed in the Banff hatchery, were killed by heavy blasting in the immediate vicinity.

SESSIONAL PAPER No. 39

KAMLOOPS TROUT.

Up to last season, Kamloops trout have only been propagated occasionally and in small numbers. The new hatchery at Gerrard is principally for the incubation of this species, and its operations resulted in a distribution of upwards of 770,000 fry. The parent fish were taken in a trap net operated in Trout lake near the hatchery, and the fry were distributed in the various waters of the Kootenay district. This establishment will be of the greatest value in filling many urgent requests for fry from the Kootenays, which the department has been unable to fill from the other hatcheries in the province.

WHITEFISH.

The whitefish operations are conducted in the Great Lakes of Ontario and in Lakes Winnipeg and Winnipegosis, Man. The collection of such eggs was carried on last fall in the Great Lakes on a larger scale than ever before. 57,000,000 were collected in the Bay of Quinte and were hatched and distributed from the Sarnia hatchery. The run of fish in the Bay of Quinte last fall was lighter than usual; but this was made up by the fish yielding better than they have in any previous season since eggs were collected in these waters. As usual, the greatest number was obtained from the Detroit river and Lake Erie. The following quantities were obtained from these waters, viz. :—

Fighting Island	107,560,000
Bois Blanc	18,880,000
Amherstburg	20,600,000
Kingsville	51,640,000
Port Dover	23,320,000
Dunnville	10,000,000

These were all transferred to the Sandwich hatchery and from that establishment were distributed as follows, viz. :—

Sandwich Hatchery. Ont	100,000,000
Collingwood Hatchery, Ont	30,000,000
Port Arthur Hatchery, Ont	33,600,000
Selkirk Hatchery, Man	56,400,000
Detroit River	12,000,000

The following quantities of whitefish eggs were also collected for the Collingwood hatchery, viz. :—

French River	39,000,000
Christian Islands	2,000,000
Naiscotyang River	2,000,000

In addition to the 33,600,000 obtained from Sandwich, 6,000,000 eggs were collected in Lake Nipigon and incubated in the hatchery at Port Arthur.

While the distribution in the Great Lakes of Ontario was materially increased. that in Manitoba waters was not as large as it was last year, and owing to a series of accidents the eggs placed in the Selkirk and Gull harbour hatcheries did not turn out well.

The eggs for the three hatcheries on Lake Winnipeg, namely, Selkirk, Gull harbour and Dauphin river, were collected in the Little Saskatchewan or Dauphin river. As usual, no difficulty was experienced in taking a large number of parent fish. The greater portion of these were enclosed in a small creek; but in some way they escaped before they were stripped. There is some opposition to the closing of this river by the

5 GEORGE V., A. 1915

pound-net when taking parent fish, and the Officer in Charge of operations is of opinion that this objection was the cause of the fish escaping. The greatest difficulties, one of which is usually encountered, are the early snow falls and the freezing up of the river. Last season, the prospects for filling the hatchery were promising up to October 19th, when there was a snow fall of 18 inches, which filled the river and bay full of slush. On October 29th, the slush was heavy in the river and the bay outside was frozen over. Under these conditions, as there was danger of being caught in the ice, the eggs then collected were transferred to Gull harbour and the boat was unable to return. The balance of the eggs taken, some 60,000,000, were placed in the hatchery at Dauphin river.

The eggs for the Winnipegosis hatchery are taken in pound-nets in the Waterhen river. The conditions here are somewhat similar to what they are on Lake Winnipeg. The Waterhen river, where the fish are taken is liable to freeze up earlier than the lake outside, and when this occurs the eggs cannot be taken to the hatchery in good condition, as the tug is unable to come into the river on account of the ice and the lake outside is not safe to cross with teams. To overcome this difficulty a number of the fish that were taken during the early part of the season were conveyed in a pontoon and impounded in a lagoon at the hatchery. This practice ensures a certain supply no matter what conditions may be at the spawning camps. It was also necessary last year to transfer the last eggs taken to the hatchery by dog teams, as the lake began to freeze on October the 28th, and between that time and November 14, when the last eggs were taken, no other means of conveyance was feasible for the reasons above explained. The collection of eggs and the distribution of fry resulting therefrom was, last season, considerably better than during the previous year.

PICKEREL.

The distribution of pickerel was increased to the extent of 21,700,000 over last year. For the first time pickerel fry were distributed in Manitoba waters. The eggs were collected at Swampy Island, Lake Winnipeg, and hatched in the Gull harbour hatchery. The operations were somewhat handicapped by the ice in the outer harbour, which moved with every change of wind and left too small an area of open water to enable the fishermen to operate to advantage. Some 11,000,000 eggs, however, were collected and most of the fry resulting therefrom was distributed in the neighbourhood of Big Island, Lake Winnipeg.

The distribution of pickerel in Lake Huron from the Sarnia hatchery was increased by 14,000,000 over that of last season, notwithstanding the fact that the ice conditions were not favourable and prevented more than half the number of nets, from which eggs are usually obtained, being set in time to take spawn fish. The better quality of eggs, however, more than made up for the smaller number of fish available for spawning purposes.

An attempt was also made to collect pickerel eggs in the Naishcotyang river, Georgian bay, for incubation in the Collingwood hatchery. With the experience gained it is hoped that a fair number of eggs can be obtained at this place next year.

SHAD.

The floating shad hatchery on the St. John river was again operated in Washademoak lake, Queen's county. With a view to meeting the contention raised last year that a larger number of eggs would have been obtained had operations been started earlier in the season, the hatchery was put in readiness and the fish taken at the stands in the immediate vicinity were examined on the night of May the 22nd, but only immature eggs were obtained. These stands were attended nightly with the same result until June the 1st, when the first ripe eggs were procured. From June 1st to 24th, 2,579 shad were handled, of which 6½ per cent were productive females and yielded

2,105,000 eggs. This is a slightly smaller number than was taken last year and is largely due to less favourable weather conditions and to the fact that the resident fishermen, with one or two exceptions, did not appear to appreciate that the hatchery was operating in their interests and did not assist as they could have in bringing their catch to the hatchery collecting boats. An exceptionally large number of small shad was reported in the Kennebecasis river and some of the fishermen believe them to be the result of the fry distributed in 1912. This season, 1,025,000 vigorous fry were liberated in the Washadamoak lake in the immediate vicinity of the hatchery.

CATFISH.

For the first time an effort was made to comply with the numerous applications received to stock the smaller lakes in southern Manitoba, that do not now contain fish. A number of these lakes were examined and it was ascertained that they were not suitable for the better kinds of fish handled in the department's hatcheries. A quantity of young catfish were, therefore, collected in the Red river near Selkirk and were distributed in a number of the lakes in question. From observation, which was made during the summer, these fish appear to be doing well and it is the intention of the department to extend this work next season.

LOBSTERS.

I am pleased to report the satisfactory increase of 53,000,000 in the distribution of lobster fry this season. Although the season was backward and the prevalence of ice prevented the fishermen from setting their traps as early as they usually do, a better quality of eggs was obtained and the percentage of these that hatched was higher than the average.

The number of lobsters retained in the Long Beach Pond was also increased. This year 242 egg bearing lobsters, and 62 umberried lobsters were impounded. With the exception of seven, none of these lobsters were less than eleven inches and some were seventeen inches in length. One hundred and nineteen have been liberated in St. Mary's bay, and the balance are being retained for observation and scientific purposes.

A shipment of 8,000,000 lobster fry was made from the Canso hatchery to Bedford Basin, N.S., in accordance with the recommendation of the Shellfish Fishery Commission that these waters be set apart as a lobster rearing area and young lobsters be distributed therein each season for five years as a test of the efficacy of hatching and planting young lobsters. Bedford Basin was selected for this test as no commercial lobster fishing is carried on there and at one time it was a valuable lobster ground. All arrangements for this transfer were made by the Inspector of lobster hatcheries, and the fry were distributed in the best of condition.

The following tables give the number of the different species of fish distributed during the season of 1914, viz.:—

Atlantic salmon...........................		19,851,830
Pacific Salmon—		
Spring.................	2,251,000	
Cohoe.................	2,274,000	
Landlocked.............	341,000	
Humpback.............	500,000	
Steelhead.............	87,200	
		117,155,900

5 GEORGE V., A. 1915

Salmon trout..	25,707,585
Speckled trout..	1,721,010
Grey trout..	72,000
Kamloops trout..	770,200
Cutthroat trout..	1,260
Ouananiche..	4,000
Whitefish..	285,990,000
Pickerel..	61,700,000
Lobsters..	713,910,304
Shad..	1,025,500
Catfish..	67,000
Total distribution..	1,227,976,589

While the benefits derived from the artificial propagation of fish are evident on all sides, the following results which have come under the direct notice of the hatchery officers might be mentioned, viz. :—

The Officer in Charge of the Windsor hatchery states that quite a number of salmon are now found in the rivers of Hants and King's counties, as a result of the distributions from the Windsor hatchery, and this season a record catch was made in King's county, some of the fish weighing as much as 35 pounds.

The Officer in Charge of the Restigouche hatchery states that the season's catch of fish in the river, bay and coast waters exceeds that of any previous year in the memory of the oldest inhabitants. The fish were larger than usual and very fat. As many as 30 salmon have been taken in one day in the river, with the fly, by two rods. Many scores made with the fly exceed the average weight of any former year. There were no complaints from the netters as to the scarcity of fish. Some stands are reported taking as many as 16,000 to 20,000 pounds, and if the severe storm, which washed a large percentage of the nets on shore just at the height of the season, had not occurred, many more fish would have been taken; but all the dealers, with whom the Officer in Charge conversed, state that they got all they could handle. The anglers, netters and dealers are now all anxious for the future success of the hatchery and are unanimous in the belief that the good results have been brought about by the systematic stocking of the rivers.

The Officer in Charge of the Gaspe hatchery reports that the salmon fishing in the district was unusually good; that the net fishermen took more fish than they have for years; but not apparently at the cost of the anglers, who had an average season.

The Officer in Charge of the Tadousac hatchery reports that the salmon fishing in that district was splendid, and as a result there are now 500 parent salmon in the retaining pond, where last year he succeeded in procuring only 383. The effect of the Bergeronnes subsidiary hatchery was also evidenced in a striking manner by the capture of 1,500 salmon by two fishermen at the mouth of the Bergeronnes river, where in former years none were taken, as it was thought that the results would not justify the expense of setting the nets.

The effect of the Sarnia hatchery is now being seen in an improved catch of legal weight whitefish in that district and the increasing numbers of undersized whitefish. that are of recent years liberated from the pickerel nets.

Conclusive evidence of the way in which the Babine hatchery must be benefiting the commercial fisheries of the Skeena river is shown in the increased numbers of spawning salmon that have been reaching the Salmon river, on which the hatchery is located, during the past few years. Last fall, for the first time the full supply of eggs for this hatchery was obtained in this river, although the run of salmon in other streams flowing into Babine lake was not as large as usual.

Spawning salmon are also appearing in increasing numbers in the creek, on which the Rivers Inlet hatchery is located, and last year over 2,500,000 eggs were obtained from fish taken there. Previous to the erection of the hatchery few spawning salmon resorted to this creek.

The taking of several speckled trout and Atlantic salmon by anglers in the Cowichan river also indicates that these splendid game fish are being established in the rivers of Vancouver Island as the result of the fry that have been distributed from the Cowichan hatchery.

EXAMINATION OF RIVERS.

It appeared from information obtained that salmon were again resorting to certain rivers in the Maritime provinces, to which they have not resorted for years for various reasons consequent upon the clearing away of the forests and the settling of the country. The importance of re-establishing angling in all suitable rivers is fully realized, and with this end in view a complete survey was made of the Kennebecasis river, N.B., last fall to ascertain the extent of the spawning area in this stream and to what extent salmon were resorting to it for spawning purposes. It was found that it contains a considerable number of suitable pools and good spawning areas and arrangements have been made to stock it liberally and systematically for a term of years. The first distribution of salmon fry, 500,000, was made in it this season.

This survey work has been continued and a number of other rivers examined and reported on during the past summer. With the same object in view, that is, re-establishing angling in all suitable streams, the rivers of the Maritime provinces were divided geographically into groups and a certain number allotted to each one of the hatcheries for stocking purposes.

MARKING OF SALMON.

As previously stated, a proportion of the salmon eggs propagated are obtained from the late run of fish. The contention has been raised that the department is not benefiting the fisheries, either commercial or sporting, by propagating these fish, it being claimed that the late run and its progeny are always late run. For the purpose of obtaining reliable information on this point and on the frequency of spawning in the Atlantic salmon, a proportion of all the fish liberated from the different retaining ponds.—750 in all,—were marked last season. '

Two kinds of marks were used, namely, brass and silver tags.

The silver tags are fastened by silver wires to the dorsal fin as near the back of the fish as possible. These tags when properly affixed lie closely against the fin and do not interfere with the movement of the rays.

The brass tags are fastened to the second ray of the dorsal fin by a silk thread. The thread is passed around the ray under the skin on the back of the fish.

Each tag is numbered, and the weight, length, sex and date of liberation of the fish, to which they are attached, is recorded, and to encourage the return a reward of $1.00 is paid for each tag, provided the weight, length, method and date of capture of the fish, from which it is taken, is given. A few of these tags have already been returned. In connection with this work the Officer in Charge of the Restigouche hatchery reports that two fresh run salmon were caught this season at Flatlands with the fly, both of which had punch holes in their tails, the result of marking in October, 1912.

NEW HATCHERIES.

The following hatcheries, which were referred to in my last report as being under construction, were completed and operated during the past season. viz.:

5 GEORGE V., A. 1915

The St. John salmon and trout hatchery, is situated on property leased from the City of St. John at the Little river reservoir, about five miles from the city. The hatchery building is 54 feet 4 inches long by 31 feet wide and 10 feet high from top of sill to bottom of plate. It is fitted up with 30 hatching troughs, which are grouped in clusters of five, each trough being 15 feet 7 inches long by 10½ inches wide and 6½ inches deep. A modern up-to-date dwelling, 32 feet long by 27 feet wide and 20 feet high from top of sill to bottom of plate, is located adjacent to the hatchery. This hatchery will fill a long felt want as regards the rivers and streams of southern New Brunswick. which could not be effectively stocked from the other hatcheries in the province. Two million two hundred and twenty-five thousand seven hundred and fifty salmon and 220,200 speckled trout fry were distributed from it this season.

The fresh water salmon retaining pond at Tide Head on the Restigouche river. has been abandoned and replaced by a salt water pond at New Mills, N.B. The parent fish for the old Tide Head pond were taken in nets operated by the department and at New Mills they are purchased from the commercial fishermen. The New Mills pond is situated on the south side of Bay Chaleur. It is formed of two cribs constructed of logs in courses and saddled into one another. These cribs are 195 feet and 280 feet long, respectively, and form with the bend in the shore line a quarter circle. The cribs are ballasted with stone and sheathed on both sides with planking.

The whitefish hatchery at Dauphin river, Lake Winnipeg, Man., was also completed and operated successfully. The building is situated on the right bank of the Dauphin river, near its junction with Sturgeon bay, Lake Winnipeg. The site comprises 21 acres of land, which has been set apart by the Department of the Interior for hatchery purposes. The building is 76 feet 6 inches x 41 feet 6 inches and is 14 feet high from the top of sill to bottom of plate, and has a capacity of upwards of 75,000,000 eggs. A dwelling for the Officer in Charge is located close to the hatchery. Arrangements have been made to heat this dwelling by steam from the hatchery boilers, and the necessary fittings for this purpose are now being installed.

The Banff hatchery is situated between Glen and River avenues, near the Bow Falls, in the Banff National Park. The hatchery building is 54 feet 4 inches long by 31 feet wide and 10 feet high from the top of the sill to the bottom of the plate. It is fitted up with 30 hatching troughs, grouped in clusters of five, each trough being 15 feet 7 inches long, 10½ inches wide, 6½ inches deep, with passages 2 feet wide between each cluster. The water supply is obtained from the town service, is of excellent quality and varies very little in temperature throughout the year. The dwelling for the Officer in Charge is of the same dimensions as the one at St. John and is fitted complete with modern plumbing, hot-air heating, and a fire-place.

The subsidiary or distributing hatchery at Pirmez creek is located on a tributary of the Elbow river, about 18 miles southwest from Calgary. The building is framed 47 feet 8 inches by 24 feet 6 inches and 8 feet high from top of sill to bottom of plate. It is fitted with 20 hatching troughs of the usual dimensions, namely, 15 feet 7 inches long, 10½ inches wide and 6½ inches deep. Living quarters, 24 feet 6 inches by 12 feet 8 inches, are fitted up in one end of the building. Almost 1,000,000 salmon trout and a small quantity of cut-throat trout fry were distributed from the Banff and Pirmez creek hatcheries this season.

The Gerrard hatchery is situated in the townsite of the same name located at the entrance of Lardeau river into Trout lake. The site was furnished by the Provincial Government and includes lots 6 to 10 in block one; lot 5 in block 4 and lot 8 in block 5 of the above mentioned townsite. The two last mentioned lots provide a right of way for the pipe line from the falls in the creek, from which the water supply is obtained. All of these lots are 30 feet wide by 100 feet deep, with the exception of the last mentioned, which is 100 feet deep on one side by 73 feet 4 inches deep on the other. The hatchery building is 42 feet 4 inches long by 36 feet 4 inches wide and 13 feet from top of sill to bottom of plate. It is fitted up with 40 hatching troughs, each trough being 15 feet 7 inches by 10½ inches by 6½ inches deep. The dwelling for the Officer in Charge is 28 feet long by 25 feet wide. On the first floor there are three rooms and on the second floor three bed-rooms, a store-room and a bath-room.

During the past summer large hatcheries were built at Thurlow, near Belleville, on the Bay of Quinte, Ontario, Kenora, Ontario, and Fort Qu'Appelle, Saskatchewan.

The Thurlow hatchery is a combination whitefish and salmon trout establishment. The hatchery building is 97 feet 8 inches long, 45 feet 6 inches wide and 14 feet from top of sill to bottom of plate. It will readily accommodate 8,000,000 salmon trout and over 60,000,000 whitefish eggs. The whitefish battery extends across one end and along both sides for a short distance. The water from the whitefish jars is utilized for the salmon trout troughs, of which there are 70. Fifty of these troughs are 15 feet 7 inches by 10½ inches by 6½ inches, and 20 of them are of the same width and depth but are only 8 feet long. The floor tank for the whitefish is 17 feet 3 inches long by 9 feet 6 inches wide.

A pier has also been built in front of the hatchery, which consists of 2 cribs, each 20 feet long by 6 feet wide, planked over. The outer crib is in the form of an ell, in which the intake pipe is secured.

A comfortable dwelling for the Officer in Charge has also been built adjacent to the hatchery, 25 feet square and 18 feet 2 inches high. It contains three rooms and a summer kitchen on the first floor and three bed-rooms and a bath-room on the second floor.

The Kenora hatchery is being fitted for the propagation of whitefish only. The site was donated by the town of Kenora and is about one mile from the Kenora post office, on the peninsula between the Lake of the Woods and Kenora bay. The hatchery building is 76 feet 6 inches long by 41 feet 6 inches wide and 14 feet high. It is fitted up with a three tier whitefish battery extending across one end down both sides of the building. The floor tank for the fry is 26 feet 8 inches long by 9 feet wide and 18½ inches deep. This building will readily accommodate 70,000,000 eggs.

A comfortable dwelling for the Officer in Charge is immediately adjoining. It is 25 feet 6 inches square and 18 feet 2 inches high from top of sill to bottom of plate, with an extension one story in height, 10 feet 6 inches long by 10 feet wide and 12 feet high to the point of its roof. There are three rooms and a summer kitchen on the ground floor and three bed-rooms and a bath-room on the second floor. This building is fitted up with all modern conveniences, including hot-air heating, plumbing and electric lighting.

The wharf consists of two cribs, 20 feet long, 6 inches wide, finished 2 feet above water level, which are ballasted and sheathed. The outer crib has an ell, 8 feet by 6 feet, in which the intake pipe is secured. A coal-house, 19 feet 4 inches by 15 feet 4 inches and 10 feet 10 inches high, has been built convenient to the boiler room.

The site for the Qu'Appelle hatchery comprises the reserve in front of lots 12 to 16 in the Qu'Appelle Park, and was granted for the purpose by the Provincial Government. The hatchery is 42 feet 4 inches long by 36 feet 4 inches wide and 14 feet high. The floor tank is 12 feet long, 10 feet wide and 18½ inches deep. The battery extends across one end and along both sides of the hatchery and will carry 500 jars, making the capacity of the hatchery upwards of 50,000,000 eggs.

A pier to secure the intake pipe and a wharf for landing purposes has also been built in the lake in front of the hatchery.

The dwelling is immediately across the road on lots 14 and 15 of the Qu'Appelle park, which were purchased for the purpose. It is 25 feet 6 inches square, 18 feet 2 inches high, with a summer kitchen attached. It is fitted up with hot-air heating and is comfortable and modern in every particular.

These three hatcheries are now nearing completion and will be in operation this season.

The old hatchery at Grand Falls, N.B., which was destroyed by fire in the early part of June, is being replaced by a larger and strictly up-to-date establishment. A

5 GEORGE V., A. 1915

dwelling for the Officer in Charge is also being built adjoining the hatchery. The hatchery building is 54 feet long, 37 feet wide and 10 feet high, from top of sill to bottom of plate. It is fitted up with 40 hatching troughs, grouped in clusters of 5, each trough being of the usual dimensions, namely, 15 feet 7 inches by 10½ inches by 6½ inches. A coal-house and office is provided in one end.

The dwelling for the Officer in Charge is 25 feet 6 inches square, with 18 feet 2 inches wall and 28 feet 6 inches to the ridge of the roof, with an extension 10 feet 6 inches by 10 feet and 12 feet high. A hot-air furnace and up-to-date plumbing are being installed.

A contract has also been entered into for the erection of a dwelling for the Officer in Charge of the Sarnia hatchery. Its dimensions are 32 feet by 27 feet and 20 feet high from top of sill to bottom of plate. On the ground floor are the living-room, dining-room, kitchen and pantry; on the second floor, three bed-rooms and a bath-room. It is fitted complete with electric lighting, hot-air heating and sanitary plumbing.

An extension or annex, 19 feet 5 inches by 10 feet, is being added to the hatchery to house the new boiler that is being installed; and a new 40-foot brick chimney in connection with the same is being built. The old wooden floor, which was in a bad state of repair, is being replaced with concrete, and a concrete well, 12 feet square, from which to obtain the water supply, is being sunk adjacent to the river. This will ensure a filtered water supply for the hatchery and remove the danger that has obtained in the past on account of the heavy storms, which last season washed away the greater part of the dock and a portion of the intake pipe where it enters the river.

SUBSIDIARY OR DISTRIBUTING HATCHERIES.

Owing to the indifferent transportation facilities, a great deal of difficulty has been experienced in distributing salmon fry on the natural spawning beds at the headwaters of some of the most important salmon rivers. To overcome this difficulty and to enable the fry to be planted in the best condition possible, the following subsidiary hatcheries were built during the past year. The eggs are transferred to these subsidiary hatcheries as late in the spring and when they are as near to hatching as conditions permit and as they are located adjacent to the spawning beds and the best distributing grounds, the fry are not subjected to rough handling, and they are planted in a short time in the best of condition in waters most suited for them. Three of these subsidiary hatcheries were built during the past year, viz.:—

The Nepisiguit hatchery is situated on the river of the same name. It is located on Little Church creek, which flows into the Nepisiguit river about one mile below the Grand Falls, and is in close proximity to the main spawning grounds, which extend along the river about two miles. While this hatchery was built principally for the Nepisiguit river, other rivers in that part of the province, which cannot readily be attended to from other hatcheries, can be stocked from it.

The building is framed 30 feet long by 20 feet wide and 8 feet high. It is fitted with 20 hatching troughs, 11 feet 7 inches long by 10½ inches wide and 6½ inches deep. Living quarters for the Officer in Charge in the form of an annex, 12 feet square, are attached to the main building. Five hundred thousand eyed eggs were transferred from the Restigouche to this hatchery last spring, which resulted in the distribution of 469,000 fry.

The Sparkle hatchery was erected on the Southwest Miramichi river, near the Upper Forks, about 18 miles from Glassville, on land leased from the New Brunswick railway company for the purpose. The lease covers the site of the hatchery, 60 feet long by 50 feet wide; right of way for a flume, 300 feet long, running up the bed of the brook from the hatchery site; as well as as right of way therefrom to the river, a distance of about 65 feet. This building will receive its supply of eggs from the hatchery at South Esk, N.B., on the Miramichi river.

The building is framed 26 feet 8 inches by 21 feet 2 inches and 8 feet high from top of sill to bottom of plate. It is fitted with 10 hatching troughs, 15 feet 7 inches by 10½ inches by 6½ inches. Living quarters for the Officer in Charge, 12 feet by 10 feet and 7 feet high, are provided over the hatching room.

The Dartmouth hatchery is situated about 20 miles from Gaspé basin and 2½ miles above the Ladystep Falls, on the Dartmouth river. It is 25 feet long, 20 feet wide and 8 feet high, from the top of sill to the bottom of plate. It is fitted up with 10 hatching troughs, 15 feet 7 inches by 10½ inches by 6½ inches deep. Living quarters for the Officer in Charge are provided in the form of an annex, 12 feet square, attached to the hatchery. The supply of eggs for this hatchery will be obtained from the establishment at Gaspé basin.

A contract has also been let for a subsidiary hatchery on the Tobique river, which will be operated in conjunction with the Grand Falls hatchery.

REPAIRS AND IMPROVEMENTS.

New boilers were, last season, installed in the lobster hatcheries at Bay View; Causo and House harbour, Magdalen Islands; the hatchery grounds at Middleton have been graded, levelled, seeded and planted with trees and a small lot of land has been acquired for the purpose of building rearing tanks and a trout pond. These tanks are now under construction and will be completed this fall. An electric motor and pump to supply water for the Magog hatchery when the river is too low to admit of the water being procured by gravitation, is being installed. The town of Magog is furnishing the power free of charge. The hatchery grounds at Collingwood, Ontario, have been graded, levelled and seeded and arrangements have been made for the installation of a pumping well, into which the water from the hatchery will flow, and from which it can again be pumped into the jars should any trouble be experienced on account of anchor ice clogging the present intake pipe, as happened several times last winter; the Port Arthur hatchery is being lathed and plastered, as a great deal of trouble has been experienced in keeping it warm; the intake pipe at the Gull harbour hatchery, Man., is being moved to a more sheltered location; the dwelling at the Dauphin river hatchery is being fitted with coils and the exhaust steam from the boiler in the hatchery will be used in heating it; the transfer of the hatchery equipment from the old establishment at Bon Accord, on the Fraser river to the Fisheries building in Queen's Park, New Westminster, B.C., and the construction of rearing ponds at the Cowichan hatchery have also been authorized.

At the present time the department has 53 hatcheries, 5 subsidiary hatcheries, 5 salmon retaining ponds and one lobster pond in operation; and three hatcheries, Thurlow, Kenora and Fort Qu'Appelle, and three sub-hatcheries, viz.: Dartmouth, Sparkle and Tobique, under construction.

' I have the honour to be, sir,
Your obedient servant,

J. A. RODD,
Superintendent of Fish Culture.

The following tables give the hatcheries that were operated, their location and date of establishment and the species and number of each species of fish distributed from each one this season, viz.:—

Distribution of Fry. 1914.

Established.	Hatchery.	Location.	Species.	Quantity.	Total Distribution
1876.	Bedford	Halifax Co., N.S.	Atlantic Salmon	1,075,000	
"	"	" " "	Speckled Trout	44,000	
"	"	" " "	Ouananiche	4,000	1,123,000
1902.	Margaree	Inverness " "	Atlantic Salmon	3,200,000	
"	"	" " "	Speckled Trout	2,000	3,202,000
1906.	Windsor	Hants " "	Atlantic Salmon	1,329,410	1,329,410
1912.	Middleton	Annapolis " "	" "	1,369,400	
"	"	" " "	Speckled Trout	307,000	1,676,400
1912.	a Lindloff	Richmond " "	Atlantic Salmon	55,000	55,000
1891.	Bay View	Pictou " "	Lobster	101,0 0,000	101,000,000
1905.	Canso	Guysborough " "	"	66,865,000	66,865,000
1911.	Isaac's Harbour	" " "	"	21,000,000	21,000,000
1911.	Inverness	Inverness " "	"	51,000,000	51,000,000
1911.	Arichat	Richmond " "	"	31,410,000	31,410,000
1911	Antigonish	Antigonish " "	"	64,000,000	64,000,000
1912.	Little Bras D'Or	Cape Breton " "	"	23,785,000	23,785,000
†1913.	Long Brath Pond	Digby " "	"	‡304	304
1874.	Restigouche	estigouche, N.B.	Atlantic Salmon	1,654,700	1,654,700
1871.	Miramichi	Northumber'd Co., N.B	" "	1,944,000	
"	"	" "	Speckled Trout	26,000	1,970,000
1880.	Grand Falls	Victoria Co., N.B.	" "	33,300	
"	"	" " "	Atlantic Salmon	864,700	
"	"	" " "	Salmon Trout	49,000	947,000
1914.	St. John, (Lakewood).	St. John " "	Atlantic Salmon	2,225,750	
"	"	" " "	Speckled Trout	220,200	2,445,950
1914.	a Nepisguit	Gloucester " "	Atlantic Salmon	469,000	469,000
1912.	Shad, St. John River.	River Queen's " "	Shad	1,025,500	1,025,500
1903	Shemogue	Westmoreland "	Lobster	31,000,000	31,000,000
1904.	Shippegan	Gloucester "	"	32,050,000	32,050,000
1912.	Buctouche	Kent "	"	30,000,000	30,000,000
1906.	Kelly's Pond	Queen's Co., P. E. I.	Atlantic Salmon	1,120,000	
"	"	" " "	Speckled Trout	400,000	1,520,000
1904.	Charlottetown	" " "	Lobsters	124,000,000	124,000,000
1909.	Georgetown	King's " "	"	65,000,000	65,000,000
1875.	Tadoussac	Saguenay Co., Que	Atlantic Salmon	649,200	649,200
1875.	Gaspe	Gaspe " "	" "	2,014,400	2,014,400
†1881.	Magog	Stanstead " "	No distribution		
1905.	Lake Tremblant	Terrebonne " "	Salmon Trout	846,800	
"	"	" " "	Speckled Trout	101,800	988,600
1904.	St. Alexis	Maskinonge " "	" "	199,460	
"	"	" " "	Atlantic Salmon	65,270	264,730
1904.	Lake Lester	Stanstead " "	Salmon Trout	763,150	
"	"	" " "	Speckled Trout	158,250	
"	"	" " "	Grey Trout	72,000	993,400
1906	a St. Marguerite	Saguenay " "	Atlantic Salmon	600,000	600,000
1909.	a Bergeronnes	" " "	"	1,000,000	1,000,000
1910.	Port Daniel	Bonaventure "	Lobster	30,000,000	30,000,000
1910.	House Harbour	Magdalen Island "	"	42,800,000	42,800,000
1876.	Sandwich	Essex Co., Ont.	Whitefish	59,000,000	59,000,000
1908	Sarnia	Lambton " "	"	40,000,000	
"	"	" " "	Pickerel	54,000,000	94,000,000
1912.	Collingwood	Simcoe " "	Whitefish	50,000,000	50,000,000
1868.	Newcastle	Durham " "	Salmon Trout	2,077,500	
"	"	" " "	Atlantic Salmon	130,000	2,207,500
1908.	Wiarton	Bruce " "	Salmon Trout	8,556,800	8,556,800
1912.	Port Arthur	Port Arthur City	Whitefish	26,490,000	
"	"	" "	Salmon Trout	5,477,950	
"	"	" "	Speckled Trout	145,000	32,112,950
1912.	Southampton	Bruce Co., Ont.	Salmon Trout	6,909,000	
"	"	" " "	Speckled Trout	48,000	6,957,000
1894.	Selkirk	Selkirk, Man.	Whitefish	12,500,000	12,500,000
"	"	" "	†Catfish	67,000	67,000
†1912	Gull Harbour	Big Island, Lake Winnipeg	Pickerel	7,700,000	
"	"	" "	Whitefish	15,000,000	22,700,000
1914.	Dauphin River	Dauphin River, Lake Winnipeg	Whitefish	41,000,000	41,000,000

Distribution of Fry, 1914—*Continued.*

Established.	Hatchery.	Location.	Species.	Quantity.	Total Distribution
1909.	Winnipegosis	Snake Island, Lake Winnipegosis..	Whitefish....	42,000,000	42,000,000
1914.	Banff	Banff, Alta.	Salmon Trout	987,385	987,385
1914.	a Permiz Creek..	Permiz Creek, "	Cut-throat Trout.......	300	300
1902.	Granite Creek	Shuswap Lake, B.C.	Sockeye Salmon..........	8,662,000	
	" "	" " "	Landlocked Salmon....	341,000	
	" "	" " "	Cohœ Salmon..........	22,000	
	" "	" " "	Speckled Trout....	36,000	9,061,000
1905.	Harrison Lake........	Harrison Lake, "	Sockeye Salmon....	29,923,000	
	" "	· " " "	Spring Salmon..........	1,560,000	
	" "	" " "	Humpback Salmon.... .	500,000	31,923,000
1906.	Pemberton............	Birkenhead River, "	Sockeye Salmon.........	22,950,000	
	"	" " * "	Spring "	50,000	23,000,000
1908.	Stuart Lake	Stuart Lake, "	Sockeye "	5,560,000	5,560,000
1903.	Skeena River........	Lakeles Lake "	" "	4,076,200	4,076,200
1908.	Babine Lake.........	Babine Lake, "	" "	7,767,000	7,767,000
1906.	River's Inlet.... ...	Oweekayno Lake "	" "	12,397,000	12,397,000
1910.	Anderson Lake..... .	Anderson Lake, Vancouver Is. "	" "	3,000	6,717,500
			Spring "		
1910.	Kennedy Lake........	Kennedy Lake, Vancouver Is. "	Sockeye "	8,000,000	8,600,000
1910.	Cowichan Lake......	Cowichan Lake, Vancouver Is. "	Cohœ "	2,752,000	
			Spring "	638,000	
	" - "	" "	Atlantic "	86,000	
	" "	" "	Steelhead "	87,200	
	" "	" "	Lake Trout	32,200	
	" "	" "	Cut-throat Trout.......	960	3,156,360
1914.	Gerrard	Trout Lake, Kootenay District.	Kamloops Trout.....	738,000	738,000
	Nimpkish....	Alert Bay, B.C	Sockeye Salmon.........	5,053,000	5,053,000
		Grand Total.....	....	....	1,227,976,589

‡ Of this number 242 were berried lobster and 62 unberried, (commercial), and all with the exception of 7 were between 11 and 17 inches in length.
* The young catfish were taken from the Red River near Selkirk.
† No distribution was made from the Magog Hatchery as the water supply failed early in the season.
a. Subsidiary or distributing hatcheries.

5 GEORGE V., A. 1915

The total distribution of the various species in each Province in 1914 was as follows,
viz :—

Nova Scotia—

Atlantic Salmon....................	7,028,810	
Speckled Trout	353,000	
Ouananiche	4,000	
Lobsters....	359,060,304	
		366,446,114

New Brunswick—

Atlantic Salmon..........	7,158,150	
Speckled Trout.................	279,500	
Salmon Trout...............•.....	49,000	
Shad	1,025,500	
Lobsters	93,050,000	
		101,562,150

Prince Edward Island—

Atlantic Salmon....................	1,120,000	
Speckled Trout...... ....·	400,000	
Lobsters	189,000,000	
		190,520,000

Quebec—

Atlantic Salmon.	4,328,870	
Salmon Trout.....................	1,649,950	
Speckled Trout.....................	459,510	
Grey Trout...............	72,000	
Lobsters	72,800,000	
		79,310,330

Ontario—

Whitefish	175,490,000	
Salmon Trout.....................	23,021,250	
Pickerel	54,000,000	
Atlantic Salmon. . ...·	130,000	
Speckled Trout	193,000	
		252,834,250

Manitoba—

Whitefish ....'..	110,500,000	
Pickerel..........................	7,700,000	
Catfish....................... ..	67,000	
		118,267,000

Alberta—

Salmon Trout.....................	987,385	
Cut-throat Trout	300	
		· 987,685

British Columbia—

Pacific Salmon....................	117,150,900	
Kamloops Trout...................	770,200	
Speckled Trout......	36,000	
Cut-throat Trout...........	960	
Atlantic Salmon........	86,000	
		118,049,060

Total distribution................. ...	1,227,976,589

The following tables give the names of the waters, the species and the number of each species distributed from the different hatcheries during 1914, viz:—

NOVA SCOTIA.
BEDFORD HATCHERY.

	Ouananiche.	Atlantic Salmon.	Speckled Trout.
Pock-wock Lake, Halifax County	4,000	....	
Indian River, Halifax County........	....	150,000	
Little Salmon River, Halifax County...	....	75,000	
Nine Mile River, Halifax County......	....	150,000	
Musquodoboit River, Halifax County...	....	100,'00	
Sackville River, Halifax County......	....	150,000	
St. Mary's River, Guysboro County....	....	50,000	4,000
South and West River, Antigonish Co.	....	75,000	
Musha Mush River, Lunenburg County.	....	75,000	
Port Joli, Queen's County..	....	50,000	
Roseway River, Shelburne County....	....	50,000	6,000
Shubenacadie River, Halifax County ..	....	150,000	
Cranberry Lake, Halifax County......	.... ..	....	6,000
Nicholson's Lake, Halifax County.....	....	....	6,000
Loon Lake, Halifax County..........	.. ....	....	6,000
Simpson's Lake, Cumberland County...	....	....	6,000
Robertson's Lake, Queen's County.....	....	....	6,000
Robertson's Lake, Halifax County.....	....	....	4,000
	4,000	1,075,000	44,000
Total..............................			1,123,000

MARGAREE HATCHERY.

	Salmon Fry.	Speckled Trout Fry.	Salmon Fingerlings.
N. E. Margaree River—			
Iron Bridge, Big Intervale........	80,000		
McDaniel's...............	50,000		
Iron Bridge, Frizzleton..........	100,000		
McDermid Crossing...............	144,000		
Etheridge Crossing...... .'......	330,000		
Louis Brook.....	100,000		
Hatchery Creek.................	130,000	2,000	20,000
McKenzie Brook................	160,000		
Forest Glen.... ....:	70,000		
Croudis Brook.................	40,000		
Cranton Brook..... ...,.	100,000		
Ross Brook....................	110,000		
Watson Brook.............. ...	110,000		
Big Brook.......	140,000		
Phillip's Brook	160,000		
Stuart Brook......	150,000		
Trout Brook....................	130,000		
S. W. Margaree River...........	520,000		
Little River, Cheticamp..	90,000		
Upper Middle River................	80,000		
Baddeck River......	133,000		
Friar Head Brook.....'...........	53,000		
Mabou Brook..........	90,000		
Indian Brook.........	110,000		
	3,180,000	2,000	20,000
Total..............			3,202,000

5 GEORGE V., A. 1915.

WINDSOR HATCHERY.

	Salmon Fry.	Salmon Fingerlings.
* Kennetcook River, Hants County	100,000	
Meander River, Hants County	240,000	
Avon River, Hants County	552,000	3,300
Hebert River, Hants County	50,000	
Great Village, Colchester County	100,000	
Portapique River, Colchester County	100,000.	
West River, Pictou County	80,000	
Cornwallis River, King's County	100,000	
Gaspereau River, King's County	....	4,110
	1,322,000	7,410
Total	....	1,329,410

* See also Middleton Hatchery.

MIDDLETON HATCHERY.

	Salmon Fry.	Salmon Fingerlings.	Trout Fry.
Annapolis River—			
Morton's Brook	....	4,000	
Walker Brook	20,000	....	10,000
Wisnall Brook	25,000		
Fales Brook	50,000		
Critchell's Brook	25,000		
Willett Brook	50,000	....	20,000
Vroom's Brook	75,000		
Chipman Brook	75,000	....	10,000
Beal's Brook	100,000		
Parker's Brook	75,000		
Nictaux West Brook	....	....	15,000
Nictaux River.—			
Morton Brook	50,000		
Shannon Brook	50,000		
Walker Brook	55,000	....	15,000
Trout Lake Stream	25,000	....	15,000
Critchell's Brook	100,000	3,900	
Stillwater	....	....	10,000
South Fales River	50,000		
Germany Lake	50,000		
Moore's Brook	50,000		
Lequille River	25,000		
Mersey River,	60,000		10,000
Bear River	25,000	†20,000	
Carleton River	25,000		
Pearl Lake	55,000	....	10,000
Ohio Lake	25,000		
H. H. Moore's Private Pond, Anna. Co.	1,500		
LaHave River	50,000		
Sissipoo River	75,000		
Hooper's Lake	50,000		
Salmon River	25,000		
Cameron's River	....	....	10,000
Lake Alma	....	....	10,000
Five Mile River	....	....	10,000

†East Branch.

	Salmon Fry.	Salmon Fingerlings.	Trout Fry.
Eliot Lake....	....	....	10,000
Potter's Lake......................	....	.. ..	10,000
Lake Mt. Hanley...................	.. ....	....	15,000
Upham's · Lake....................	....	....	10,000
Fales River.......................	.. ..	....	10,000
*Kennetcook River.......	....	....	10,000
Cloud Lake........................	......	....	10,000
Pike Brook........................	....	... .	10,000
Habitant River....................	....	......·......	10,000
Moore's Lake, Kedgemakooge Lake....	....	....	10,000 ·
Harris Lake	....	....	10,000
Lake Annis.....	....	....	10,000
Brazil Lake....	....	....	10,000
Lake Skinner.....................	....	....	10,000
Trefry's Lake.....................	....	....	10,000
Hatchery Pond....................	....	....	7,000
	1,370,500	27,900	307,000
Total....:......... 1,676,400			

*See also Windsor Hatchery.

LINDLOFF HATCHERY. ·

Subsidiary to Margaree Hatchery.

	Salmon.
River Denys..	40,000
Tillard River..	15,000
	55,000

BAY VIEW HATCHERY.

	Lobsters.
Cariboo Harbour........................	20,000,000
¬Pictou Island..........................	35,000,000
Cariboo Island.........................	14,000,000
Little Cariboo Island....................	10,000,000
Pictou Harbour	10,000,000
Bay View............................	7,000,000
Gull Rock............................	5,000,000
	101,000,000

CANSO HATCHERY.

	Lobsters.
Canso Harbour..........................	16,140,000
Glasgow...............................	680,000
Bedford Basin	8,000,000
Cranberry Island.......................	3,635,000
Flag Island............................	5,5,455,000
Whitehead..	5,680,000
Canso Islands..........................	5,230,000
Port Felix..	4,545,000·
Fox Island bay........................	4,320,000
Dover..	2,500,000
Cariboo Cove..	3,865,000
St. Andrew's Channel..................	4,770,000
Queensport............................	2,045,000
	36,865,000

5 GEORGE V., A. 1915

ISAAC'S HARBOUR HATCHERY.

	Lobsters
Harbour Island............................	2,000,000
West Shore County Island....................	1,000,000
Black Ledge..............................	1,500,000
Coddles Harbour..........................	1,500,000
Stone Rock..............................	500,000
South east of Big Island....................	500,000
Beckerton...............................	1,000,000
Liscomb................................	1,000,000
New Harbour............................	1,500,000
Goose Island............................	1,500,000
Bear Trap Head..........................	1,500,000
Graham shoal............................	1,000,000
Soo Bay................................	500,000
Charles Cove............................	500,000
Country Harbour.........................	2,500,000
Wine Harbour...........................	1,000,000
Betty Cove point.........................	1,000,000
Scragely Ledge Reef......................	1,000,000
Total..........................	21,000,000

INVERNESS HATCHERY.

	Lobsters.
Chimney Corner..........................	4,000,000
Margaree Harbour........................	1,000,000
Grand Etang............................	4,000,000
Broad cove marsh........................	6,000,000
Point Cross.............................	4,000,000
Eastern harbour.........................	4,000,000
Mabou.................................	3,000,000
Little river.............................	4,000,000
Cheticamp..............................	4,000,000
Cape Rouge.............................	5,000,000
Pleasant bay............................	2,000,000
Pollet's cove............................	2,000,000
Inverness..............................	3,000,000
Belle Cote..............................	2,000,000
Friar's Head............................	3,000,000
Total..........................	51,000,000

ARICHAT HATCHERY.

	Lobsters.
West Arichat...........................	4,600,000
Jersey island...........................	3,150,000
Petit de Grat...........................	6,700,000
Madame island..........................	4,075,000
Little Anse.............................	1,500,000
Bourgeois river..........................	5,869,000
Cape la Ronde..........................	1,000,000
Rockdale...............................	2,500,000
Green island............................	1,000,000
Rocky bay.............................	1,016,000
Total..........................	31,410,000

ANTIGONISH HATCHERY.

	Lobsters.
Tracadie head	9,000,000
Boman head	6,000,000
Monk head	9,000,000
Cape Jock	8,000,000
Little Tracadie head	9,000,000
Harbour au Bouche	10,000,000
Mayett beach	6,000,000
Pourguet island	3,000,000
Breen beach	4,000,000
Total	64,000,000

LITTLE BRAS D'OR.

	Lobsters.
Little Bras D'Or	21,285,000
South bay	2,500,000
Total	23,785,000

NEW BRUNSWICK.

RESTIGOUCHE HATCHERY.

Restigouche River—

	Salmon Fry.	Salmon Fingerlings.
Larry's gulch	325,000	
Trotting ground	300,000	
Red Bank	225,000	
Near hatchery		11,700
Upsalquitch river, Long Lookum	325,000	
Matapedia river	300,000	
Caraquet river	25,000	
Causapscal river	69,000	
Benjamin river	25,000	
Charlo river	25,000	
Jacquet River		24,000
	1,619,000	35,700
Total		1,654,700

NOTE.—The 40,000 Salmon Fingerlings that were in the tanks when last year's report was written were distributed as follows:—

Jacquet river	25,000
Restigouche river, near hatchery	15,000
Total	40,000

MIRAMICHI HATCHERY.

	Salmon Fry.	Trout Fry	Salmon Fingerlings.
North West Miramichi river	450,000		
Hatchery Brook, N. W. River..	....	....	9,000
Main South West Miramichi river..	170,000		
Little South West Miramichi river..	400,000		
Sevogle river..	190,000		
Renous river..	200,000		
Burnt Church river..	75,000		
Tabusintac river..	75,000		
Bay du Vin river..	75,000		
Nashwaak river..	50,000		
Petitcodiac river	75,000		
Buctouche river..	75,000		
Salmon river..	50,000		
Little river, (Coverdale)..	50,000		
Antinory lake..		10,000	
Bartibog river..		16,000	
	1,935,000	26,000	9,000
Total..			1,970,000

GRAND FALLS HATCHERY.

	Salmon.	Salmon Trout.	Speckled Trout.
St. John river, hatchery creek..	864,700	....	33,300
Williamstown lake..		39,000	
Lake Dubé, P.Q..		10,000	
	864,700	49,000	33,300
Total..			947,000

ST. JOHN HATCHERY.

	Salmon.	Speckled Trout.
Jemseg river, Dykeman stream..	100,000	
Kennebecasis river—		
Bushy brook..	100,000	
McLeod brook..	100,000	
Salmon creek..	100,000	
Wards stream..	100,000	
Salmon brook..	100,000	
Washademoak lake —		
Canaan river..	100,000	
North Forks..	100,000	
Washademoak..	100,000	

	Salmon	Speckled Trout.
Musquash river—		
Wetmore brook..	250,000	
Mispec. river..	45,750	
Tynemouth creek..	100,000	
Oromocto river..	50,000	
St. Croix river..	100,000	
Salmon river (St. John Co.)..	100,000	
Little Salmon river (St. John Co.).. ..	100,000	
Quiddy river..	100,000	
Salmon river (Queen's Co.)..	100,000	
Belleisle river..	100,000	
Pocologan river..	100,000	
Shogomoc lake..	75,000	10,000
Skiff lake..	75,000	10,000
Blind Man's lake..	10,000	
Lake Lomond..	20,000	75,000
Crescent lake..		10,000
Fisher lake..		15,000
Fenton pond..		10,000
Magaguadavic lake.. .. .. .. '..		20,000
Walsley lake..		10,000
Alward lake..		10,000
Nashwaaksis river..		20,000
Salt Spring brook..		10,000
McDougal lake..		20,000
Shillington pond..		200
	2,225,750	220,200
Total..		2,445,950

NEPISGUIT HATCHERY.

Subsidiary to Restigouche Hatchery.

	Salmon.
Nepisguit river..	394,000
Tetagouche river..	50,000
Middle river..	25,000
Total..	469,000

SHAD HATCHERY.

	Shad.
Washademoak lake..	1,025,500

SHEMOGUE HATCHERY.

	Lobsters.
Little Cape..	4,000,000
Cape Bald..	5,000,000
Dupuis Corner..	1,500,000
Murray Corner..	5,000,000
Grants..	1,000,000
Cadman Point..	2,500,000
Off hatchery..	4,000,000

5 GEORGE V., A. 1915

	Lobsters.
Ezra..	3,000,000
Leger's brook................................	3,000,000
Jourimain....................................	2,000,000
Total...........................	31,000,000

SHIPPIGAN HATCHERY.

	Lobsters.
Pointe à Peinture........................	8,250,000
Alexander Point..........................	3,550,000
Point Canoe..............................	100,000
Pointe à Marcel..........................	200,000
St. Mary's...............................	225,000
Petit Pokemouche........................	6,225,000
Pointe Brulée...........................	5,000,000
Shippigan gully..........................	8,500,000
Total...........................	32,050,000

BUCTOUCHE HATCHERY.

	Lobsters.
St. Edward's.............................	4,000,000
Buctouche harbour........................	8,000,000
Cormierville.............................	3,000,000
St. Anne's...............................	3,000,000
Cassie cape..............................	3,000,000
Richibucto cape..........................	4,000,000
Cocagne harbour and island................	3,000,000
Cocagne cape.............................	1,000,000
Chockfish................................	1,000,000
Total...........................	30,000,000

PRINCE EDWARD ISLAND.

KELLY'S POND HATCHERY.

	Salmon.	Speckled. Trout.
Winter river....................	80,000	25,000
Morell river....................	240,000	25,000
Dunk river.....................	100,000	25,000
North river....................	160,000	25,000
Belle river.....................	80,000	25,000
Midgell river...................	80,000	25,000
Indian river....................	80,000	25,000
Wheatley river.................	80,000	25,000
East river......................	80,000	25,000
Forbes river....................	60,000	30,000
West river......................	80,000	25,000
Stewart's pond..................		5,000
Marshall's pond.................		5,000
Hardy's pond...................		45,000

	Sabr.on.	Speckled Trout.
Redmond's pond............................		15,000
Black river..		25,000
Hatchery pond..		25,000
	1,120,000	400,000
Total..		1,520,000

CHARLOTTETOWN HATCHERY.

	Lobsters.
Point Prim.. .. ..'	12,000,000
Keppoch reef..	16,000,000
Governor's island..	24,000,000
St. Peter's island..	16,000,000
Holland cove..	10,000,000
Rice point..	18,000,000
Black point.. .. ..•..	8,000,000
Argyle shore..	10,000,000
Seal rock.. .. .. '..	10,000,000
Total..	124,000,000

GEORGETOWN HATCHERY.

	Lobsters.
Between Panmure island and Murray harbour..	10,000,000
St. Mary's bay..	10,000,000
Rollo bay..	10,000,000
Cardigan bay..	5,000,000
Sturgeon bay..	5,000,000
Brudenell river.. .. ..'.	5,000,000
Annandale bay..	5,000,000
Between Broughton island and Souris..	5,000,000
Montague river..	5,000,000
Launching bay..	5,000,000
Total..	65,000,000

QUEBEC.

TADOUSSAC HATCHERY.

	Atlantic Salmon.	
	Fry.	Fingerlings.
Malbaie river..	200,000	
Rivière à Mars..	140,000	
Rivière à St. Jean..	149,000	
Little Saguenay river..	100,200	
Lac de Juge, (flows into Little Saguenay river)..	50,000	
Bergeronnes lake..		5,000
Saguenay. river..		5,000
	639,200	10,000
Total..		649,200

GASPÉ HATCHERY.

	Fry.	Atlantic Salmon. Fingerlings.
St. John river..	590,000	
Dartmouth river..	530,000	
York river..	600,000	9,400
Malbaie river..	35,000	
Bonaventure river..	35,000	
Little Cascapedia river..	60,000	
Port Daniel river..	35,000	
Grand river..	35,000	
Cap Chat river..	35,000	
Magpie river..	50,000	
	2,005,000	9,400
Total..		2,014,400

LAC TREMBLANT HATCHERY.

	Speckled Trout.	Salmon Trout.
Chapleau lake..	5,000	
Carré lake..	10,000	
Cache lake..	5,000	
Des Laurentides lake..	10,000	
Bourdeau lake..	10,000	
Long lake..	10,000	
Violon and Laroche lakes..	10,000	
Morel lake..	10,000	
Wurtele lake..	10,000	
Alarie and Provost lakes..	10,000	
Centre lake..	5,000	
Bleu lake..	5,000	
Vert lake..	1,800	
Walfrid lake..		5,000
Bark lake..		75,000
Duhamel lake..		50,000
D'Argent lake..		25,000
Rond lake..		20,000
Burnet lake..		20,000
Masson lake..		25,000
Charlebois lake..		25,000
Eau Claire lake..		50,000
Equerre lake..		20,000
A la Francaise lake..		10,000
Corbeil lake..		50,000
Noir lake..		50,000
Belanger lake..		50,000
Rochon lake..		50,000
Cook lake..		50,000
Richer lake..		50,000
Sarrasin lake..		50,000
La Truite lake..		10,000
La Grosse lake..		10,000
Des Sables lake..		50,000

	Speckled Trout.	Salmon Trout.
Mont Laurier lake.................		20,000
Lake Kanado....................	.	20,000
Renaud lake...................		20,000
Mercier lake...................		40,000
Tremblant lake..		41,800
	101,800	886,800
Total..		988,600

ST. ALEXIS HATCHERY.

	Fry.	Speckled Trout. Fingerlings.	Salmon.	Fingerlings.
Rat lake..	15,000		10,000	
Mandeville lake....	15,000		10,000	
Simpson lake..			15,000	
Chain of three lakes.	2,000	460	10,000	270
Saccacomie lake.. .			5,000	
Sans Bout river...			5,000	
Cloutier lake..			10,000	
Dickerman brook...	20,000			
Larocque lake.. ...	15,000			
Sans Bout lake.. ...	20,000			
Morin lake..	15,000			
Clair lake..	15,000			
St. Maurice river ..	10,000			
Allaire lake..	5,000			
L'Isle à Pierre lake	10,000			
Noir lake..	10,000			
Loutre lake..	5,000			
McCrea lake..	10,000			
Deer lake..	5,000			
Rouge lake..	5,000			
Francais lake..	5,000			
Lake Murphy..	5,000			
Trois Freres lake..	5,000			
Fouet lake..	7,000			
	199,000	460	65,000	270
Total..				264,730

LAKE LESTER HATCHERY.

	Fry.		Fingerlings.	
	Speckled Trout.	Salmon Trout.	Salmon Trout.	Grey Trout.
Breeches Lake..............	70,000	....	15,000	
Lake Togo................	9,000			
Howard's Pond	5,000			
Lake St. George...........	10,000			
Libby's Lake..............	10,000			
Darker's Pond	8,000			
McIntyre Pond	5,000			
Trout Lake...........	5,000			
Orford Lake..............	....	....	35,000	

5 GEORGE V., A. 1915

	Fry.		Fingerlings.	
	Speckled Trout.	Salmon Trout.	Salmon Trout.	Grey Trout.
Brome Lake......	....	....	30.000	
Brompton Lake.............	.. ..	....	35,000	
Tortue Lake................	9,000			
Nicolet Lake............ ...	9,000			
Fortin Lake	9,000			
Sans Nom Lake.....	9,000			
Massawippi Lake...........	....	50,000	90,000	25,000
Magog Lake..............	....	100,000	75,000	95,000
Muffett Lake......	....	50,000	25,000	
Megantic Lake'	....	150,000	58,000	22,000
Joseph Lake........	....	50.000		
	158,000	400,000	363,000	72,000
Total, Fry, 1914...........			558,000	
" Fingerlings, 1914.................			435,000	

	Two years old. Speckled Trout.	Three years old. Speckled Trout.	Three years old. Salmon Trout.
Lake Lester	150	......	125
Lake Massawippi.	......	......	25
Tomfobia River..............	......	100	
	150	100	150
Grand total, 1914...........			993,400

Note—The Fry and older fish that were in the rearing tanks at the date of last year's report were distributed as follows :—

1913.

FINGERLINGS.

	Grey Trout.	Atlantic Salmon.	Speckled Trout	Salmon Trout.	Red Trout.
Lake Massawippi....	116,000	7,000	....	25,000	. ....
Orford Lake....	13,000	7,000	......	.....	......
Magog Lake........	60,000	10,000	......	.....	.. ...
Brome Lake.......	10,000	10,000	......	....	......
Lake Megantic......	5,000	....	......	15,000	......
Breeches Lake	.....	5,000	......	5,000	......
Lindsay Pond	......	......	......	25,000	..... .
Echo Beech Lake....	......	......	1,000	......	......
Nigar River........	......	. ...	.. ...	......	750
Tomfobia River.....	.....	......	......	......	750
	204,000	39,000	1,000	70,000	1 500

STE. MARGUERITE HATCHERY.

Subsidiary to Tadoussac Hatchery.

	Salmon.
Portage river..	600,000
Total..	600,000

BERGERONNES HATCHERY.

Subsidiary to Tadoussac Hatchery.

	Salmon.
Long lake..	300,000
Gobeil lake..	200,000
Boulanger lake..	200,000
Croche lake..	100,000
Caribou lake..	50,000
Guillaume lake..	50,000
A la Truite lake..	100,000
Total..	1,000,000

PORT DANIEL HATCHERY.

	Lobsters.
Port Daniel west..	1,000,000
Point Macron to Newport..	9,000,000
Hopetown to Port Daniel..	12,000,000
Hatchery to Gascons..	8,000,000
Total..	30,000,000

HOUSE HARBOUR HATCHERY.

	Lobsters.
*Little harbour..	11,000,000
*Cape Vere..	12,000,000
*Narrows..	15,000,000
*Red Cape..	2,400,000
*Harbour Basque..	2,400,000
Total..	42,800,000

ONTARIO.

SANDWICH HATCHERY.

	Whitefish.
Lake Ontario—	
Salmon Point..	1,000,000
Belleville, Bay of Quinté..	1,000,000
Hamilton..	1,000,000
Toronto..	1,000,000
Lake Erie—	
Dunnville..	2,000,000
Port Dover..	2,000,000
Port Stanley..	2,000,000
Kingsville..	1,000,000
Bar Point..	3,000,000
Pigeon bay..	3,000,000
Leamington..	1,000,000

* The distribution was made in the lagoons at these places.

5 GEORGE V., A. 1915

Whitefish.

Detroit river—

Bois Blanc..	9,000,000
Peach island..	3,000,000
Stoney island..	2,000,000
Turkey island..	3,000,000
Fighting island..	10,000,000
Bay below Fighting island..	4,000,000
River at hatchery..	7,000,000

Lake St. Clair—

Mitchell's bay..	3,000,000

Total.. 59,000,000

SARNIA HATCHERY.

Lake Huron—

	Whitefish.	Pickerel.
Along Lake shore from 15 to 25 miles from mouth of St. Clair river.. ..	40,000,000	
Point Edward and Sarnia spawning grounds..		49,000,000
Aux Sable river {Port Frank {Grand Bend		5,000,000
	40,000,000	54,000,000

Total.. 94,000,000

COLLINGWOOD HATCHERY.

Whitefish.

Georgian bay—

Below Christian islands..	10,000,000
Cedar Point..	10,000,000
Tiny Point..	10,000,000
Six Mile Point..	10,000,000
Three Mile Point..	10,000,000

Total.. 50,000,000

NEWCASTLE HATCHERY.

	Salmon Trout. Yearlings.	Salmon Trout. Fry.	Atlantic Salmon. Fry.
Lake Ontario—			
Newcastle..	2,500	100,000	
Port Hope..		400,000	
Cobourg..		400,000	
Whitby..		400,000	
Toronto..		400,000	
Hamilton..		375,000	
Ponds on Coldwater river..			10,000
Lake Simcoe—			
Kempenfelt bay..			15,000
Kawkstone..			15,000
Muskoka lakes			90,000
	2,500	2,075,000	130,000

Total.. 2,207,500

WIARTON HATCHERY.

	Salmon Trout. Fry.	Salmon Trout. Fingerlings.
Lake Huron—		
Duck island	450,000	
Meldrum bay	450,000	
Providence bay	500,000	
South bay	500,000	
Rattlesnake	500,000	
Tobermory	400,000	
Georgian Bay—		
White Cloud island	275,000	
Four Mile Point	275,000	
Hay island	300,000	
Pruder's Landing	300,000	
Griffith island	300,000	
Gravelly Point	300,000	
Cape Croker	400,000	
Port Elgin	400,000	
Jackson Shoal	550,000	
Cape Commodore	400,000	
Vails Point	400,000	
Presqu'Ile	500,000	
Cameron's Point	500,000	
Cape Rich	250,000	
Meaford	200,000	
Squaw island, Killarney	379,000	
Colpoy's bay		27,800
	8,529,000	27,800
Total		8,556,800

PORT ARTHUR HATCHERY.

Lake Superior	Whitefish. Fry.	Salmon Trout. Fry.	Speckled Trout. Fry.	Salmon Trout Fingerlings.
Black bay	2,000,000			
Thunder bay	24,000,000			
Vicinity of hatchery	490,000	580,000		97,950
Rossport		400,000		
St. Ignace island		400,000		
Duck bay		400,000		
Silver island and Tea harbour		400,000		
Thunder cape and Hare island		400,000		
Victoria island		400,000		
Mink island		400,000		
Welcome islands		1,000,000		
Mount McKay		1,000,000		
Ontario Provincial Government			145,000	
	26,490,000	5,380,000	145,000	97,950
Total				32,112,950

5 GEORGE V., A. 1915

In addition to the above distribution the following shipments of eyed eggs were made from this hatchery.

	Brook Trout.	Salmon Trout.
Banff hatchery..		1,000,000
Grand Falls hatchery..	50,000	
Bedford hatchery..	75,000	
Granite creek hatchery..	50,000	
	175,000	1,000,000
Total.. .. ..eyed eggs.		1,175,000

SOUTHAMPTON HATCHERY.

	Salmon Trout.	Speckled Trout.
Lake Huron—		
Big Reef off Kincardine..	750,000	
Nine Mile Point..	1,450,000	
Chief Point..	1,500,000	
Clay banks..	700,000	
Main station..	1,503,000	
Lyal Light and Stokes bay..	1,006,000	
Bowman's lake..		48,000
	6,909,000	48,000
Total..		6,957,000

MANITOBA.

DAUPHIN RIVER HATCHERY.

	Whitefish.
Lake Winnipeg	
Dauphin river..	41,000,000
Total..	41,000,000

GULL HARBOUR HATCHERY.

	Whitefish.	Pickerel.
Lake Winnipeg..	15,000,000	7,300,000
" Louise..		20,000
" Clementi..		60,000
Oak..		60,000
Pelican..		60,000
Max..		70,000
Killarney..		70,000
Rock..		60,000
	15,000,000	7,700,000
Total..		22,700,000

SELKIRK HATCHERY.

	Whitefish.
Lac du Bonnet..	256,000
Shoal lake..	120,000
Lake Winnipeg..	200,000
Red river, near Selkirk.. .. .. ·..	11,924,000
Total..	12,500,000

NOTE.—The following Catfish were collected in the Red river, and distributed in the following lakes:—

	Catfish.
Lake Shoal..	7,000
" Killarney..	10,000
" Rock..	4,000
" Pelican..	6,000
" Overland..	1,000
" Oak..	,9,000
" Clementi..	5,000
" Souris..	7,000
" Swan..	6,000
" Minnedosa..	7,000
" Heatherington..	5,000
Total..	67,000

WINNIPEGOSIS HATCHERY.

	Whitefish.
Lake Winnipegosis, in the neighbourhood of Snake island.	42,000,000
Total..	42,000,000

ALBERTA.

BANFF HATCHERY.

	Salmon Trout.
Lake Minnewanka..	963,000
Retained in tanks at hatchery..	24,385
Total..	987,385

PIRMEZ CREEK HATCHERY.

	Cutthroat Trout.
Pirmez creek..	300
Total..	300

5 GEORGE V., A. 1915

BRITISH COLUMBIA.

Fraser River Watershed.

GRANITE CREEK HATCHERY.

	Cohoe.	Sockeye.	Landlocked. Salmon.	Speckled Trout.
Shuswap lake, Silk-atkwa bay.............	22,000	8,662,000	275,000	
White lake..			6,000	
Turtle lake..			6,000	
Nisconlith lake..			24,000	
Harper lake..			6,000	
Chum lake			24,000	
Tum Water creek.....				35,000
Held in tanks at hatchery				1,000
	22,000	8,662,000	341,000	36,000
Total..			9,061,000	

Note.—In addition to the quantities given in the last report, 1912-13, the following distribution was made from this hatchery during the summer of 1913:—

	Cutthroat Trout.
Kalamalka or Long lake..	23,000
Trout lake..	20,000
Total..	43,000

PEMBERTON HATCHERY.

	Spring.	Sockeye.
Birkenhead river..	50,000	22,950,000
	50,000	22,950,000
Total..		23,000,000

HARRISON LAKE HATCHERY.

	Sockeye.	Spring.	Humpback.
In hatchery ponds..	3,210,000		
Trout creek..	822,000		
Harrison river..	3,610,000		
Morris creek and slough..	2,745,000		
Cascade creek..	604,000		
Harrison lake..	15,182,000	1,500,000	500,000
Seymour slough..	1,500,000		
Bear creek..	750,000		
Silver creek..	750,000		
Chelsales slough..	750,000		
	29,923,000	1,500,000	500,000
Total..			31,923,000

STUART LAKE HATCHERY.

	Sockeye.
Stuart lake, Cunningham creek..	5,560,000

SKEENA RIVER WATERSHED.

SKEENA RIVER HATCHERY.

	Sockeye.
Coldwater creek and Lakelse lake..	4,076,200

BABINE LAKE HATCHERY.

	Sockeye.
Salmon river..	7,767,000

RIVERS INLET HATCHERY.

	Sockeye.
Oweekayno lake, hatchery ponds..	5,690,000
Deer creek..	1,300,000
Quap creek..	1,500,000
Hatchery creeks..	3,907,000
Total..	12,397,000

VANCOUVER ISLAND.

ANDERSON LAKE HATCHERY.

	Sockeye.	Spring.
Anderson lake..	1,790,000	3,000
Ternan creek..	3,565,000	
Clement's creek..	1,359,500	
	6,714,500	3,000
Total..		6,717,500

KENNEDY LAKE HATCHERY.

	Sockeye.
Kennedy lake..	8,600,000

COWICHAN LAKE HATCHERY.

	Spring.	Cohoe.	Steelheads.	Atlantic Salmon.	Lake Trout.	Cutthroat Trout.
Oliver's Creek	111,200	.. ..	.. ..	19,500	8,600	960
Beadnell's Creek..	83,000	10,000	.. ..	.. ..	.. ..	.. .. ..
Beaver Creek..	106,000	316,800	17,000	.. .. ..	9,800	.. ..
Green's "	130,200	162,000	.. ..	.. ..	.. ..	.. ..
Sutton "	146,000	373,200	27,600	39,550	13,800	.. ..
Cowichan River..	.. ..	778,800	.. ..	.. ..	.. .. ..	.. ..
Bear Lake Creek..	.. ..	206,000	.. ..	.. ..	.. ..	.. .. ..
Robinson River..	18,400	313,200	32,000	9,950	.. ..	.. ..
Hatchery Creek	103,200	92,000	.. ..	17,000	.. .. ..	.. .. ..
Foster's "	.. .. ..	.. ..	10,600	.. ..	.. .. ..	.. ..
	698,000	2,252,000	87,200	86,000	32,200	960
Total..						3,156,360

5 GEORGE V., A. 1915

GERRARD HATCHERY.

	Kamloops Trout.
Lardo lake..................................	495,000
Trout lake..................................	200,000
Stobard lake................................	7,500
Christina lake..............................	13,300
North Fork, Kettle river....................	17,000
Retained in tanks at hatchery..............	5,200
Total..................................	738,000

NIMPKISH HATCHERY.

(Operated by British Columbia Packers' Association.)

	Sockeye.
Nimpkish lake..............................	5,053,000

APPENDIX No. 14·

NATURAL HISTORY REPORT

To the Superintendent of Fisheries.

SIR,—I have the honour to submit my report of the Canadian Fisheries Museum for the fiscal year 1913-14.

Since the re-opening of the museum on the 23rd of March last it has been visited by over 12,000 persons.

Mounted examples of rather more than one-fifth of the fishes of the British North American possessions (Canada and Newfoundland), represented by specimens of 116 species, are now on exhibition; and it is anticipated that the museum will contain about two-fifths in the near future.

To facilitate the study of the species and, as much as possible, to make the museum self explanatory, adjoining each kind of fish is a printed label giving its geographical range, and also a short note bearing on some point or points as touching its natural history. These labels, in so far as the range of the fish is concerned, are based upon the subject matter of my book ' Check List of the Fishes of the Dominion of Canada and Newfoundland,' a 1913 publication, and which issued from the King's Printer early in the present year, 1914. The substance of this report is largely based upon those labels, and the numbers which the species bear are those of the Check List.

The names of the species, specimens of which are now contained in the museum, with their geographical range and the short notes to which allusion has been made above, are as follows:—

2. CALIFORNIA HAGFISH.

(*Polistotrema stouti.*)

The hagfishes are the only true vertebrate or back-boned parasites known. They bore their way into the bodies of other fishes, and preying upon them leave nothing but the skin and bones. The only other species of hagfish known to occur in our waters is the American Hagfish of the northeastern coast of the Atlantic.

Ranges from coast of Vancouver island southward to coast of California.

12a. ROUSSETTE.

(*Scylliorhinus profundorum.*)

This uniform coffee-coloured shark is very rare, and as it has been obtained at a depth of over 800 fathoms; no doubt this accounts for its rarity. This individual is from the coast of British Columbia, and it has been obtained also by the United States S.S. *Albatross* in lat. 39° 9 sec. N., long, 72° 3 sec. 15 min. W. It belongs to an oviparous type of shark—the eggs being enclosed in leathery envelopes which are provided with long twining tendrils at the angles for attachment to submarine objects. Were this shark to be had in plenty, it may readily be seen from the specimen that its skin might be turned to account as shagreen for polishing purposes.

5 GEORGE V., A. 1915

14. OIL SHARK OR TOPE.

(*Galeorhinus zyopterus.*)

This fish is also called soup-fin shark on account of the value placed on its **fins**, which are sold in California to the Chinese—the delicate rays of which are dissolved into ·a finely flavoured gelatine, and its liver is manufactured into a coarse oil. It is a viviparous shark—that is, it brings forth its young alive.

Although the oil shark occurs on the coast of British Columbia, it is uncommon in the waters of that Province, and its principal range appears to be from the coast **of** California, from San Francisco to Cerros island, Lower California, Mexico. This specimen was obtained in British Columbia.

18. GREAT WHITE SHARK.

(*Carcharodon carcharias.*)

One of the largest of the sharks attaining a length of thirty feet.

Temperate and tropical seas: occasional on the Atlantic and Pacific coasts of North America—' its distribution evidently girdling the globe.'

20. PICKED DOGFISH.

(*Squalus acanthias.*)

This small shark, and its close ally, the California dogfish of the Pacific coast, **are** very destructive to herrings and other fishes; for they are gregarious and often move in schools. Oil is produced from the livers of the dogfishes, but it is generally used **as** an adulterant to mix with oils of a superior quality; and the whole substance of their bodies may be converted into a fertilizer. The dogfishes are ovoviviparous—that is the eggs are developed, or so to speak hatched, within the mother.

Both coasts of the north Atlantic, extending south to Cuba on the American side: coast of Labrador: Gulf of St. Lawrence: Gaspé bay, and Maritime provinces: common on coasts of the British Isles, including the Orkney islands, and other European coasts.

29. STARRY RAY.

(*Raja radiata.*)

This is one of our smaller skates or rays and is not in any way so common as is **the** barn-door skate. It is conspicuous, as will be seen, by the presence of numerous star-like spines which are arrayed in series upon the head, back, and tail, or indeed upon **the** whole dorsal aspect of the fish, including the pectoral fins.

Both sides of the north Atlantic, on the American side seemingly recorded **at** least as far south as Staten island, State of New York: Maritime Provinces, including the Bay of Fundy: common on the eastern coast of Nova Scotia: on the eastern side of the Atlantic ' inhabits the northern seas only, extending from the British Isles **to** Iceland and Greenland, the coast of Norway, and the Baltic as far as Scania.'

39. RATFISH.

(*Hydrolagus colliei.*)

This is one of the chimæras (so called after a fire-breathing monster of classical mythology) and receives its name from the outward resemblance of its teeth to **the**

65827—Opp. p. 344.

incisor teeth of the rat and of rodents in general. The male is smaller than the female and has a spur-like appendage on the top of its head, and claspers adjoined to the pelvic fins. The eggs of the ratfish are encased in leathery capsules.

British Columbia, ranges from the coast of Alaska southward to Bay of Monterey, California, 'especially plentiful off southeastern Alaska, and about the wharves at Esquimalt.'

42. GREEN STURGEON.

(*Acipenser medirostris.*)

This sturgeon frequents the sea or brackish water, and seldom enters rivers beyond their mouths. Unlike others of its kin it is not esteemed as food, and is even reputed to be poisonous.

British Columbia, ranges from San Francisco northward: 'not common north of the Straits of Fuca.'

43. COMMON STURGEON.

(*Acipenser sturio oxyrhynchus.*)

This is a sub-species of the common sturgeon of Europe, from which it chiefly differs in the number and character of the stellate ossifications—bony scutes which can be readily seen upon the back and along the sides. The flesh of the sturgeons in general is esteemed; caviare made from the roe or ovaries is considered a delicacy, and isinglass is manufactured from the air-bladders.

It is anadromous or ascends rivers from the sea in order to spawn, and perhaps spawns in brackish as well as in fresh water.

Maritime provinces and St. Lawrence river and tributaries: in the United States ranging from Maine to South Carolina.

46. COMMON GARPIKE.

(*Lepidosteus osseus.*)

The garpikes (of which there are several distinct species)—are the nearest extant relatives of the bowfin, but a glance at the respective specimens will at once reveal how great the gap must be between them, and only a study of the fossil remains of extinct intermediate forms can demonstrate their relationship. The flesh of the garpikes is worthless as food.

St. Lawrence river, and Provinces of Ontario and Quebec, westward to Lake Huron, very plentiful near Belleville. Bay of Quinté: in the United States ranges from Vermont westward to the Great Lakes region and southward to the Rio Grande.

48. BOWFIN.

(*Amia calva.*)

Although of little commercial use this species is interesting to the naturalist. It is one of the few survivors of the ganoid fishes which flourished in the waters long before the advent of man, and its congeners heralded the way towards that important group of fishes to which the salmon and the herring belong.

St. Lawrence river, and Provinces of Ontario and Quebec, westward to Lake Huron: very plentiful near Belleville. Bay of Quinté: in the United States, ranges from the Mississippi valley and Great Lakes region southward to the southern States.

5 GEORGE V., A. 1915

50. LADYFISH.

(Albula vulpes.)

The ladyfish is remarkable in that it passes through a metamorphosis, being at first a band-shaped larval form. A small specimen received from the museum of the Natural History Society of New Brunswick for indentification was obtained in 1911 at Black's harbour, Bay of Fundy, N.B.

All warm and tropical seas: ordinarily ranges on the American coasts northward to Long island.

The two mounted specimens were received from the Bahama islands.

53. MOONEYE.

(Hyodon tergisus.)

This, although strictly a fresh water fish, is structurally one of the herrings. It is widely distributed in many of the fresh waters of the eastern portion of Canada.

Provinces of Ontario and Quebec, including the St. Lawrence and Ottawa Rivers and Lake St. Peter: Great Lakes region, including Lake of the Woods: Ohio and Mississippi valleys.

55. COMMON HERRING.

(Clupea harengus.)

Ever since George Benkel a humble Dutch fisherman, who died in 1397, discovered the art of curing the herring this clupeoid has been unsurpassed as a commodity of the sea, and has for centuries afforded one of the chief industries and enterprises of the fisheries.

A marine fish, but according to Boulenger, as first shown by Günther, the fry or 'white-bait' have a predilection for brackish water.

Temperate and colder parts of the northern Atlantic and seas of Europe, including the British Isles: most abundant on the American side north of Cape Cod, extending to the coast of Labrador and embracing Newfoundland: occurs in Gaspé Bay: recorded from south shore of River St. Lawrence: extends in the United States as far south as Cape Hatteras, North Carolina.

56. CALIFORNIA HERRING.

(Clupea pallasii.)

Equally as abundant as its congener the common herring, and distributed, as given below all over the northern Pacific coasts on both the American and Asiatic sides.

British Columbia and Puget Sound: ranging over the entire Pacific coast from San Diego, California, to Alaska and Kamchatka.

62. AMERICAN SHAD.

(Alosa sapidissima.)

An important food-fish, but much rarer than formerly on our Atlantic coast.

Extends, or did extend, from Labrador, Newfoundland, Gulf of St. Lawrence, and Maritime Provinces, to the Gulf of Mexico; but its distribution in our waters more limited and local than formerly: 'occasional in Baie-des-Chaleurs:' still frequents the shores of St. John and Albert counties, New Brunswick; as well as occurring in

SESSIONAL PAPER No. 39

Chignecto, Cobequid and St. Mary's bays, and Bay Verte, Maritime Provinces: mentioned as occurring in Gaspé bay: formerly abundant in the lower Ottawa: has been introduced into Pacific coast waters by the United States Fish Commission, and has ' been established in several of the tributaries of the Mississippi river, notably the Ohio river.'

63. MENHADEN.

(*Brevoortia tyrannus.*)

Rare in Canada. Migratorially erratic in its movements, and not to be depended on. Although not esteemed as food it is in the United States converted into a fertilizer and oil.

Ranges from Nova Scotia, at least from St. Mary's bay, southward to Brazil. Its geographical range varies greatly from year to year, according to Goode as defined for 1877, its wanderings ' bounded by the parallels of north latitude 25° and 45°; on the continental side by the line of brackish water; on the east by the inner boundary of the Gulf stream.'

71. COMMON WHITEFISH.

(*Coregonus clupeiformis.*)

The most important of our fresh-water food-fishes. Extensively propagated artificially.

Distributed from Labrador and New Brunswick westward and northward: abundant in the Great Lakes, especially in Lake Erie; its spawning beds being perhaps more especially on the Canadian side of the lake.

75. CISCO OR LAKE HERRING.

(*Argyrosomus artedi.*)

A fresh-water food-fish of much commercial importance.

Ranges from Province of Quebec and State of Vermont, occurring in Lakes Champlain and Memphremagog and in Thirty-one-mile lake some 60 miles north of Ottawa, westward to Lake Superior: abundant in Lake Erie: extends northward to the Hudson bay region and to Labrador.

84. HUMPBACK SALMON OR PINK SALMON.

(*Oncorhynchus gorbuscha.*)

This is the smallest in size of the typical salmon of the north Pacific. In this species the distorted condition assumed by the males of the salmon of the genus *Oncorhynchus* at the spawning time reaches its maximum. The flesh of the humpback is of good flavour when fresh, but is inferior to that of the quinnat and sockeye as a canned commodity. The six specimens are illustrative of sexual and seasonal features, and the two lowermost of the male and female towards the spawning time.

British Columbia: both coasts of the Pacific and their slopes, ranging from California to Kamchatka and extending northward.

85. DOG SALMON OR CHUM.

(*Oncorhynchus keta.*)

The flesh of this salmon is excellent when fresh and can readily be salted—a condition in which it is largely used in Japan, but does not rank as high as the others as a

5 GEORGE V., A. 1915

canned commodity. The six specimens are illustrative of sexual and seasonal features, and the two lowermost of the male and female towards the spawning time.

British Columbia: both coasts of the Pacific and their slopes, ranging from California to Behring Straits, Kamchatka, and Japan—being ' by far the most abundant species of salmon ' in Japan.

86. QUINNAT, SPRING SALMON OR KING SALMON.

(*Oncorhynchus tschawytscha.*)

The quinnat is the largest of the Pacific coast salmon, and the earliest usually to ascend the rivers. Its flesh which is normally red in colour is subject to turn paler in hue. ·It spawns only in streams of considerable size. The quinnat is second in importance in the canning industry, being surpassed in this respect by the sockeye. The five specimens in this case and the specimen in the basal case are illustrative of sexual and seasonal features, and the two lowermost specimens of the sex of the male and female towards the spawning time.

British Columbia: both coasts of the Pacific and their slopes, ranging from California to Behring Straits and China.

87. COHO OR SILVER SALMON.

(*Oncorhynchus kisutch.*)

The coho bears an outward resemblance to the popular sockeye, from which it is readily distinguishable on account of the scales, which are thin, easily falling off—excepting those on the lateral line. Although not of such value as either the sockeye or the quinnat, its flesh, which is pale in colour, is excellent in flavour. The six specimens are illustrative of sexual and seasonal features, and the two lowermost of the male and female towards the spawning time.

British Columbia: both coasts of the Pacific and their slopes, ranging on the American side from California to Alaska, and on the Asiatic side southward to Japan.

88. SOCKEYE OR BLUE-BACK SALMON.

(*Oncorhynchus nerka.*)

Commercially the sockeye is the most important of the Pacific coast salmon. It is especially desirable for canning on account of the large amount of oil in its flesh. It ascends all the important rivers in British Columbia, and spawns in streams which are tributary to lakes. The six specimens are illustrative of sexual and seasonal features, and the two lowermost of the male and female towards the spawning time.

British Columbia: both coasts of the Pacific and their slopes, ranging on the American side from Oregon to Alaska; and on the Asiatic side southward to Japan, being landlocked in Lake Akan in northern Hokkaido.

90. ATLANTIC SALMON.

(*Salmo salar.*)

Not only is the salmon of the Atlantic coast the treasure of the sportsman, but it is of great importance as a commercial commodity on account of the well known rich and delicious flavour of its flesh. Allusions to the salmon in literature, ever since the days of Pliny until now, are numberless, and the books which have been written upon it, either as to its natural history or as an object of sport, are voluminous.

Both coasts of the Atlantic and its affluents: Maritime provinces, Gaspé bay, St. Lawrence river and gulf with their tributary waters, including La Rivière Jupiter, Anticosti island: formerly Lake Ontario: recently (1905) one specimen found near South bay, Manitoulin island, Lake Huron: Newfoundland and Labrador: northeastern States of North America, and the Delaware river: seas and rivers of Europe, including Iceland, and entering the Baltic: southern limit of distribution in Europe, Galicia, Spain.

91. LANDLOCKED SALMON.

(*Salmo salar sebago.*)

Excepting its smaller size, more plump form, and non-migratory habits, the landlocked salmon differs little from the Atlantic salmon; and there is little to distinguish it from the Ouananich, another land-locked variety, excepting that it attains a larger size.

Certain lakes in New Brunswick, such as Loch Lomond and Sciff and Musquash lakes: and of the States of Maine and New Hampshire: now more widely distributed by having been introduced into lakes of other localities.

93. CUTTHROAT TROUT.

(*Salmo clarkii.*)

This fish receives its name from a deep-red blotch on the membrane connecting the bones of the lower jaw. It has a considerable distribution as mentioned below, and when found in cold streams or seething rapids is considered by anglers to be a very gamy fish.

Southern Alberta and British Columbia: ranging from California perhaps as far north as Alaska.

94. STEELHEAD.

(*Salmo rivularis.*)

The steelhead and certain other species of British Columbia are interesting as instances of the occurrence of salmonoids of the genus *Salmo* in Pacific coast waters —the universally known Atlantic salmon being the type of *Salmo*. These species do not spawn once for all and die as those of the genus *Oncorhynchus* do. The steelhead spends much of its life in the sea, but like its relative, the Atlantic salmon, ascends rivers in order to spawn. It is not a fish favourable for canning purposes owing to the firmness of its bones, otherwise its flesh is excellent as food.

British Columbia to California and eastward to the mountains, extending as far north as Skagway, Alaska: introduced into Lake Superior by the United States Fish Commission, and since found in waters of Ontario.

96. RAINBOW TROUT.

(*Salmo irideus.*)

A choice salmonoid of the Pacific slopes of North America. It has been introduced by the United States Fish Commission with success into certain eastern waters, including Lake Superior.

Ranges, under a number of varieties, from State of Washington to California.

5 GEORGE V., A. 1915

98. SALMON TROUT.

(*Cristivomer namaycush.*)

Commercially one of the most important of our fresh-water fishes. Extensively propagated artificially.

Widely distributed from Labrador, the Maritime provinces, and the State of Maine, to Vancouver island, Alaska, and the Mackenzie river, northward to the Arctic circle.

This species is subject to great variation, and although all the varieties bear the specific name of *namaycush* there is considerable reason for the popular distinctions such as salmon trout, gray trout, and Mackinaw trout. Structurally, however, it has not appeared to ichthyologists that there are sufficient distinctions to warrant the separation of varieties into sub-species, excepting in the instance of the Siscowet.

100. SPECKLED OR BROOK TROUT.

(*Salvelinus fontinalis.*)

A sportsman's favourite, and widely distributed in clear waters in the eastern portion of Canada. Lake Nipigon is noted for its large sized individuals, and there is a sea run variety.

Widely distributed in North America, presumably from the Arctic regions (but 'the northern limits of its range being as yet not well ascertained') southward to Georgia and Alabama, and from Newfoundland to Saskatchewan.

This species varies greatly in size and coloration according to the character of the waters in which it occurs.

106. RED CANADIAN TROUT.

(*Salvelinus marstoni.*)

As yet this beautiful little salmonoid has only been found in certain lakes in the Province of Quebec.

Recorded from the following lakes, among others in the Province of Quebec: Lac de Marbre, near Ottawa; lakes of the Laurentides Club in the Lake St. John region; Lac à Cassette, Rimouski county; and Lake Saccacomi and the Red lakes, Maskinongé county; the above records probably right at the southern limits of its distribution, and that the centre of its distribution is much further north.

113. AMERICAN SMELT.

(*Osmerus mordax.*)

An excellent pan-fish of delicate flavour. Often land-locked as well as marine, being abundant in Lakes Champlain and Memphremagog. It affords a stable and lucrative industry all around our Atlantic coast.

Atlantic coast of North America from Labrador to Virginia: lakes in Maritime provinces, Province of Quebec, and New England states: Lac-des-Isles, Gatineau district, some sixty miles north of Ottawa.

It is known that this species of fish exists land-locked in fresh water lakes in New Brunswick, Nova Scotia, and in the State of Maine, but its occurrence in a lake so far away from the sea as Lac-des-Isles, is perhaps worthy of mention.

124. CHANNEL CATFISH.

(*Ictalurus punctatus.*)

This catfish, as the specimen shows, has the caudal or tail-fin deeply forked. It has, as all the catfishes which occur in Canada have, an adipose fin; but the presence of that fin among catfishes in general is not an universal feature. By some the flesh of the channel-cat is much esteemed.

Rivers of Great Lakes region westward to Manitoba: Mississippi valley, and streams tributary to the Gulf of Mexico.

128. COMMON CATFISH.

(*Ameiurus nebulosus.*)

Well known throughout its range as the bull-pout. A popular boys' fish who catch it with a long stick and a string and even sometimes with a bent pin. Of excellent flavour when properly cooked and served up, and considered by some to taste like spring chicken.

Ranges in Canada from the maritime provinces to Manitoba, including the St. Lawrence river and Great Lakes region: in the United States extending from Maine westward to North Dakota, and southward to the southern states: has been introduced into rivers of California, and into lakes of southern Oregon.

134. LAKE CARP SUCKER.

(*Carpiodes thompsoni.*)

A fish of the carp kind. Conspicuous on account of the elongated or filamentous first rays of the dorsal fin, or fin of the back.

Lake Champlain, upper St. Lawrence river, and Great Lakes region, including Lake of the Woods.

138. NORTHERN SUCKER.

(*Catostomus catostomus.*)

This is a widely-distributed sucker, and is common northward, as its name implies. It abounds in the Great Lakes region, and is nearly cosmopolitan in the Dominion, ranging from Labrador and New Brunswick to British Columbia and from the eastern to the western United States, extending southward at least to latitude 40° N., but has been obtained in West Virginia; and occurs also in Alaska.

140. COMMON WHITE SUCKER.

(*Catostomus commersonii.*)

The best known of our suckers. The suckers as articles of food are much better than they are generally reputed to be, but probably their good flavour is dependent upon the particular waters where they happen to be.

Very cosmopolitan in British North America, from the Maritime provinces, Gaspé district, and Labrador, to Alberta: in the United States extending from the eastern states westward to Montana and Colorado and southward to Georgia.

5 GEORGE V., A. 1915

145. COMMON RED HORSE.

(*Moxostoma aureolum.*)

The red-horses, for there are several kinds of them in our fresh waters, are also like the suckers (see under common white sucker) palatable food-fishes, although their value has been underrated.

St. Lawrence river and Great Lakes.region, including Lake of the Woods; and Manitoba; abundant west of the Alleghany mountains to Nebraska: extending southward to Arkansas and Georgia.

158. SILVER CHUB OR FALLFISH.

(*Semotilus corporalis.*)

This is the giant among the cyprinoids (minnows and carps) of our eastern waters, attaining a length of some eighteen inches.

Widely distributed in the rivers and streams of the Maritime provinces: St. Lawrence River system and streams and ponds of Ontario: abundant also in northern United States east of the Alleghanies.

166. BREAM OR ROACH.

(*Abramis crysoleucas.*)

This cyprinoid is the only representative of the genus to which it belongs in Canada. Its name *crysoleucas* signifies golden white. It is a familiar species of lakes and rivers in the Maritime provinces, including Prince Edward Island—abounding in bayous and weedy ponds.

199. GERMAN CARP.

(*Cyprinus carpio.*)

Originally a native only of Asia, the carp was a few centuries ago introduced into Europe, and in recent years into North America. It seems to thrive wherever it has been introduced, and rapidly makes its way into other waters, and is now common in the Bay of Quinté and waters adjoining or tributary to Lakes Erie and Huron.

201. AMERICAN EEL.

(*Anguilla chrysypa.*)

Although the eel is widely distributed all the way.from the sea coast to far inland, and often occurs in out-of-the-way places and at high altitudes it has never been known to breed in fresh water. It is so constructed that it can make its way through what to any other fish would be insurmountable barriers. A sea condition appears to be essential to the development of its eggs; therefore when it occurs in waters in which access to the sea has been cut off it is supposed to be sterile; and it is safe to say that there never was a true eel which was not once in the sea.

Widely distributed in British North America from Newfoundland and Labrador westward; the falls of Niagara forming a barrier to its further progress in Ontario: in the United States extending from Maine westward to the Rocky Mountains, and southward into Mexico and Central America: occurs also in the West Indies: " caught in considerable numbers in Porto Rico in the small bamboo traps or ' nasas ' set in the small rivers.' '

203. CONGER EEL.

- (*Leptocephalus conger.*)

The only record of the occurrence of the conger eel in Canadian waters is that of a specimen taken in Pokemouche gully, New Brunswick, in October, 1849.

Atlantic ocean: on the American side ordinarily extending from Cape Cod to Brazil: has been also recorded from Porto Rico: and occurs besides at coasts of Europe, Asia, and Africa. This specimen was procured for the museum from the Atlantic coast of the United States.

209. GREEN PIKE.

(*Lucius reticulatus.*)

This, the smallest of our three species of pike, is readily distinguishable, apart from its smaller size for it seldom exceeds 2 feet in length, by having the sides of the head completely scaled all over. It is more common in the United States than it is in Canada.

Ranges from New Brunswick and the St. Lawrence river westward to Ontario: extensively distributed east of the Alleghany mountains to southern United States.

210. COMMON PIKE.

(*Lucius lucius.*)

The common pike is the most widely distributed of all fresh water fishes. It is distinguishable from the green pike (which see) by having only the upper halves of the opercules—or bones covering the gills scaled, and from the maskinonge by having the cheeks completely scaled, the upper parts only of the cheeks of the latter being scaled.

Nearly cosmopolitan in the fresh waters of the northern parts of North America, Europe and Asia: widely distributed in British North America westward to Alberta: extending far north, the limits of its northern distribution not yet well determined, but it occurs in Alaska: extending in the United States from the State of New York westward to the Mississippi valley, perhaps further west, and southward to the Ohio river.

211. MASKINONGE.

(*Lucius maskinongy.*)

The maskinonge is the largest fish of the pike family, and one of the most popular fishes of the sportsman. It is readily distinguishable from the common pike and green pike by the absence of scales on the lower parts of the cheeks and opercules. Although, under one or two varieties, it occurs south of our borders in northern waters of the United States it is practically a fish of our own.

St. Lawrence river and Great Lakes region, embracing the Provinces of Quebec and Ontario: common among the Thousand Islands: said also to occur in Manitoba: ranging from Lake Champlain to the upper Mississippi valley.

264. PESCADO DEL REY OR SO CALLED CALIFORNIA SMELT.

(*Atherinopsis californiensis.*)

The sub-name of this species is a misnomer, as it is in no wise related to the smelt. It is claimed that these specimens were obtained in British Columbia, but by rights it is a fish of the coast of California, where it occurs in schools near the shore. It is important as a food-fish, the flesh being white and of a fine texture.

5 GEORGE V., A. 1915

268. CALIFORNIA POMPANO OT POPPY FISH.

(*Palometa simillimus.*)

This species is one of the butterfishes. Its flesh is rich and delicate, and is highly prized as food.

British Columbia and Puget sound, southward to California.

273. SILVER HAKE OR WHITING.

(*Merluccius bilinearis.*)

The silver hake is said to be roving in its habits, and that it follows the schools of herring, devouring multitudes of that valuable fish. Its flesh is not highly esteemed, and is considered coarse.

Ranges from the coast of Labrador, embracing Newfoundland, Gulf of St. Lawrence, Maritime provinces, and the New England states, southward to the Bahama Islands.

276. POLLACK OR COALFISH.

(*Pollachius virens.*)

Although the pollack, like the codfish and the haddock, often feeds at the bottom of the water it is largely a surface feeding species. It is a well known fish at our maritime coasts, and is very common in the Bay of Fundy. It is an important commercial food-fish. Oil is yielded from its liver.

Both sides of north Atlantic: Maritime provinces southward to State of New York: ' on the shores of Spitzbergen;' ' in all the northern seas and in the Baltic;' Orkney and Shetland islands; coasts of England; ' on the Irish coast from Waterford along the eastern shore to Belfast;' and ' very abundant on the western and northern coasts of Scotland:' ' occurs about Iceland:' on the European side at least as far south as the coast of France.

279. TOMCOD OR FROSTFISH.

(*Microgadus tomcod.*)

A diminutive codfish, seldom exceeding a foot in length. Its flesh is esteemed a delicacy. It loves brackish water, and enters fresh water to spawn. Besides this Atlantic coast species there is a Pacific coast species of tomcod.

Ranges from coast of Labrador, and embracing the Gulf of St. Lawrence, Gaspé bay, and Maritime Provinces, southward to the coast of Virginia.

280. COMMON CODFISH.

(*Gadus callarias.*)

From a commercial standpoint the value of the codfish cannot be over-estimated. As is well known its flesh is used not only in a fresh but in a salted condition; but in order to fully appreciate the flavour of its flesh it should be cooked and eaten just after it has been taken out of the sea. Cod-liver oil is a well known product of commerce. The cod-fish is usually a pelagic spawner—that is it spawns in the open sea, and its eggs are buoyant and float near the surface. The cod-fish is generally taken by baited-line.

Both sides of north Atlantic: ranging on the American side at least from coast of Labrador, and embracing the Gulf of St. Lawrence, Gaspé bay, Maritime Provinces, and Newfoundland, southward to coast of Virginia; and on the European side ' found universally from Iceland very nearly as far south as Gibraltar.'

281. PACIFIC CODFISH.

(*Gadus macrocephalus.*)

This codfish is very similar to the common codfish of the Atlantic, but has relatively a larger head; and it is said to have a smaller air-bladder, an organ popularly termed 'sounds.' As yet it has not been turned to the same commercial account as the common cod, but no doubt will be.

British Columbia: both coasts of northern Pacific, ranging from Behring sea southward to the offshore banks of Oregon, and to Japan. Said to be very abundant in the sea of Okhotsk.

283.. HADDOCK.

(*Melanogrammus æglefinus.*)

Second only in importance to the codfish among fishes of that kind ranks the well known haddock. Everybody knows of the 'haddie,' a Scottish name which has become a byword. In inland places, far from the sea, the flesh of the haddock is probably preferred to that of the cod, and when properly dried in the sun it is palatable without being cooked at all.

Both coasts of north Atlantic: on the American side ranging from the coast of Labrador southward, and embracing Gulf of St. Lawrence. Maritime Provinces, and Newfoundland, to coast of North Carolina; and on the European side from Iceland and 'the Scandinavian coast to East Finmark and Varanger Fjord' southward to the coast of France: 'coast of Great Britain, from extreme north to the Land's End' and 'all round the shores of Ireland.'

286. FRESH WATER LING-OT BURBOT.

(*Lota maculosa.*)

The American Burbot is very close to the common species of the fresh-water ling of northern Europe and Asia, and may prove wholly identical with the latter. If the two are to be regarded as forms of one and the same species then the geographical range of the fresh water ling is in all probability nearly co-extensive with the fresh waters of the northern part of the northern hemisphere. It is a gadoid, or in other words is a relation of such fish as the codfish, haddock, and pollack, and is the only strictly fresh water gadoid in North America known.

Well nigh cosmopolitan in northern North America: recorded from Labrador, and ranging from New Brunswick westward to British Columbia; and from the Arctic regions southward to northern States of the Union.

289. CODLING OR WHITE HAKE.

(*Urophycis tenuis.*)

The codlings, of which there are several species, are better known at our coasts as hake. The flesh is dried and salted, and the sounds or air-bladders, are manufactured into glue.

Ranges from coast of Labrador southward to coast of North Carolina; Gulf of St. Lawrence, Gaspé bay, Maritime provinces, and Newfoundland.

293. CUSK.

(*Brosme brosme.*)

A fish of the cod kind, not very gainly in appearance, but now of considerable commercial account.

39—23½

5 GEORGE V., A. 1915

Both sides of north Atlantic; ranging on the American side from Greenland, and embracing Labrador, Newfoundland, Maritime Provinces, and New England States, southward to coast of Massachusetts: ' occurs in Iceland and Spitzbergen, and along the entire length of the Scandinavian peninsula,' and ' occasionally taken in the Firth of Forth:' frequently found ' in the Orkney Islands, and swarms among those of Zetland:' ' among the Faroe Islands;' ' coasts of Norway as far as Finmark;' and ' just touches the most northern part of Denmark, at Skagen in Jutland.'

303. CALICO BASS OR STRAWBERRY BASS.

(*Pomoxis sparoides.*)

' The silvery olive ground colour mottled all over with olive green renders the calico bass one of our most beautiful fishes. Its congener the crappie strongly resembles it, but the two do not appear to intergrade, so that the genus *Pomoxis* has two well marked species.

Provinces of Quebec and Ontario, through the Great Lakes region, including Lake of the Woods, westward to Manitoba: in the United States ranging from eastern States westward to the Mississippi valley and southward to the southern States.

304. ROCK BASS.

(*Ambloplites rupestris.*)

A well known and widely distributed fresh-water fish—hardy and gamey, and a good pan-fish.

Provinces of Quebec and Ontario, embracing the St. Lawrence river and Great Lakes region, westward to Manitoba: in the United States ranging from Vermont westward to the Mississippi valley and southward to Louisiana and Texas.

312. COMMON SUNFISH.

(*Eupomotis gibbosus.*)

The sunfishes of which there are a goodly number of species in our fresh waters will when first taken out of the water vie' with many a fish of the tropics in beauty of coloration. They belong to the bass kind, and like them make nests in which they deposit their eggs.

Ranging in Canada from the Maritime provinces to Lake Huron; and in the United States from Maine westward to the Mississippi valley, and southward to Florida.

313. SMALL-MOUTH BLACK BASS.

(*Micropterus dolomieu.*)

This bass is the most important of our fresh-water percoids, or the fishes of the perch kind. Not only is it highly esteemed for its flesh, but holds its own as a rival of the salmon as a sport fish, which has led to the construction of ponds where it may propagate. In those ponds, as in a state of nature, the black bass make their bowl-shaped excavations, or so-called nests, in the gravel beds, and with pertinacity guard their young.

Widely distributed in the Provinces of Ontario and Quebec, extending through the St. Lawrence river and Great Lakes region westward to Manitoba and the Mississippi valley, and southward to South Carolina, Mississippi, and Arkansas: introduced into waters of other provinces of the Dominion, and of various States of the Union, and also into waters of various European countries.

314. LARGE-MOUTH BLACK BASS.

(*Micropterus salmoides.*)

Second only in importance among our fresh water percoids. It is readily distinguishable from its ally the small-mouth black bass in the great size of the maxillaries —that is the pair of triangular bones which form the side borders of the mouth, and which pass the posterior borders of the orbits of the eyes.

Provinces of Ontario and Quebec, extending through the St. Lawrence river and Great Lakes region westward to Manitoba: ranges in the United States from the Great Lakes westward to the Mississippi valley, and southward to southern States: also in waters of Mexico.

315. PIKE PERCH.

(*Stizostedion vitreum.*)

Called by the French Canadians the doré. One of the most important of our fresh water .fishes. Widely distributed, and extending far westward to. the lakes of the prairie provinces.

Provinces of Ontario and Quebec, extending through the St. Lawrence river and Great Lakes region westward to Saskatchewan: Hudson bay region and Labrador: ranges in the United States from Vermont westward to the upper Mississippi valley and southward to Alabama and Georgia.

316. SAUGER.

(*Stizostedion canadense.*)

Not considered of such value as food as its ally the pike perch. Plentiful locally, and distinguishable from the pike perch chiefly in its coloration, and in the lack of a black blotch at the end of the spinous dorsal fin, or first fin of the back.

Provinces of Ontario and Quebec: distributed under one or two varieties from the St. Lawrence river and its tributaries westward, and embracing Manitoba, and perhaps Saskatchewan, to Montana, and southward to Arkansas,

317. YELLOW PERCH OR AMERICAN PERCH.

(*Perca flavescens.*)

One of our best known eastern fresh water fishes. The type of its family. Structurally it differs very slightly from the European perch.·

Widely distributed in British North America from the Atlantic sea-board to Saskatchewan, and in the United States from Maine to the upper Missouri valley, and extending southward to North Carolina.

334. WHITE BASS.

(*Roccus chrysops.*)

The white bass is very similar to the striped bass, but, unlike the latter, it is a strictly fresh water fish. It appears to have been introduced with success into France by M. Carbonnier during the years from 1877 to 1879.

St. Lawrence river and Great Lakes region, westward to Manitoba and Mississippi valley, and southward to Arkansas.

335. STRIPED BASS.

(*Roccus lineatus.*)

Essentially a brackish water species—ascending fresh waters to spawn. Very plentiful in Miramichi region. A fish of great commercial importance.

5 GEORGE V., A. 1915

Atlantic coast of North America, from the Maritime provinces to the Gulf of Mexico: ascends the Miramichi and St. Lawrence rivers and tributaries among others: introduced into waters of the Pacific coast by the United States Fish Commission.

336. WHITE PERCH.

(*Morone americana.*)

An ally of the striped bass, and like it ascending rivers to spawn. It often occurs land-locked. It is a very excellent pan-fish.

Atlantic coast of North America from the Gulf of St. Lawrence and Maritime provinces to South Carolina: recorded as abounding 'in the numerous lakes of Nova Scotia.'

337. COMMON WEAKFISH.

(*Cynoscion regalis.*)

The weakfish is regarded to be a casual visitor to the coasts of Nova Scotia. It frequents sandy shores. Its flesh is very tender and easily torn, hence its name, but it is a food-fish of high value. Its range ordinarily is the Atlantic and Gulf coast of the United States from Cape Cod southward to Mobile, and this specimen was procured for the museum from the Atlantic coast of the United States.

338. WHITE SEA BASS.

(*Cynoscion nobilis.*)

This is one of the weakfishes, and among our fishes has hardly a rival for beauty. It is a most valuable food-fish.

Vancouver island, southward to coast of California.

339. SHEEPSHEAD OR FRESH-WATER DRUM.

(*Aplodinotus grunniens.*)

The sheepshead receives its sub-name from the drumming sound it makes under the water, and which is heard above the water. Its flesh is somewhat tough and coarse in fibre, but is coming more into repute as an article of food than it was.

Extends from Ontario, through the Great Lakes region, westward to Manitoba; and ranging through the Ohio and Mississippi valleys southward to Louisiana and Texas: also recorded from the Rio Usumacinta, Tabasco, southern Mexico.

340. STRIPED SURF-FISH.

(*Tæniotoca lateralis.*)

The surf-fishes are in the strictest term of the word viviparous—that is they bring forth their young alive. This fact was first discovered by accident half a century or more ago in cutting slices from the side of a surf-fish which had been caught to use as bait to catch more. There were 19 fully developed young ones, facsimiles of each other, and each a perfect miniature of the mother fish: and they swam about freely in a pail of water. According to the species, which are somewhat numerous. the surf-fishes differ greatly in coloration.

This species ranges from British Columbia to coast of California.

351. PORGEE.

(*Damalichthys argyrosomus.*)

One of the surf-fishes. (See under 349 Striped Surf-fish.)

Pacific coast from British Columbia and Puget sound to lower California: 'entering the inlets in thousands.'

352. CUNNER.

(*Tautogolabrus adspersus.*)

This is one of the most common of the fishes of the Maritime provinces, where it is often called the perch, and is abundant at wharves near the shore. It is a good pan-fish, and useful as a scavenger, but is a pest to fishermen as it nibbles the bait off their hooks.

Atlantic coasts of North America extending from Labrador and Newfoundland to Sandy Hook, and embracing the Gulf of St. Lawrence, Gaspé Bay, Maritime provinces, and New England states.

353. TAUTOG.

(*Tautoga onitis.*)

An ally of the cunner, and a valuable food-fish with flesh of a superior quality. The name Tautog is of Indian origin.

Atlantic coasts of North America from the Maritime provinces to South Carolina: among rocks and kelp.

354. COMMON MACKEREL.

(*Scomber scombrus.*)

The mackerel stands high in the scale of fish life, and is so organized as to move through the water with great rapidity. As a food fish it excels. and affords one of the chief fishing industries.

Both sides of North Atlantic: on the American side from Labrador to Cape Hatteras, North Carolina; and on the European side from Norway to the Mediterranean and Adriatic.

360. CALIFORNIA BONITO.

(*Sarda chilensis.*)

The scales of the pectoral region of the Bonitos form a protective corselet. The few species are of a bright metallic lustre, but are little esteemed as food, as they are said to be coarse and very oily.

The California Bonito occurs on the coast of British Columbia, and ranges in the Pacific ocean from Patagonia to Japan.

5 GEORGE V., A. 1915

363. SWORDFISH.

(*Xiphias gladius.*)

The swordfish is well named, as the bones of the upper jaw are converted into a veritable weapon of attack—horizontally shaped like a sword. Hardly any other fish has greater muscular power for moving with rapidity through the water. To quote the saying of an old fisherman: "Where you see swordfish, you may know that mackerel are about."

Atlantic ocean, and on both its sides: Maritime provinces and Newfoundland banks: occurs also in the Pacific ocean, and in the Mediterranean; and is said to enter the Baltic: otherwise distributed in many seas, being of nearly world-wide distribution, and occasionally occurring on the coasts of Great Britain and Ireland, and also at the South Sea Islands; but is said to be rare off the coast of California, and scarcely known in Japan.

364. PILOTFISH.

(*Naucrates ductor.*)

The pilotfish has long been credited with guiding the shark—hence its name. It is true that it accompanies sharks, and also follows vessels; and as this habit doubtless enables it readily to procure its food, this fable may be thus accounted for.

Occurs in tropical and temperate seas, and appears occasionally on the British coasts: occasional also on the Atlantic coasts from Cape Cod to the West Indies; and has been recorded, at least once, from the coast of Nova Scotia.

The two mounted specimens are from the coast of Massachusetts.

365. RUDDERFISH OR BANDED SERIOLE.

(*Seriola zonata.*)

This is one of the amber-fishes, most of which are valued as food. A specimen was once caught on the banks south of Devil's Island off the coast of Nova Scotia, but its ordinary range is from Cape Cod to Cape Hatteras.

The mounted specimen is from the coast of Massachusetts.

374. HALIBUT.

(*Hippoglossus hippoglossus.*)

This is the giant among the flat-fishes, and by far the most important commercially. It is found in all, or at least most of, the northern seas, which accounts for its occurrence at both our Atlantic and Pacific coasts.

Ranges from the Arctic regions southward: Atlantic and Pacific coasts of British North America and United States; as far south at least as Montauk Point and the Farallone Islands: occurs plentifully in Behring sea northward to Behring straits: along the entire west coast of Greenland, also Iceland, and north to Spitzbergen in latitude 80°: numerous seas of northern Europe southward at least to the coast of France.

377. SAND DAB OR ROUGH DAB.

(*Hippoglossoides platessoides.*)

This flat-fish, as given below, has a very extensive geographical range, and is a rather common food-fish of northward waters.

65827—Opp. p. 360.

Both sides of North Atlantic: ranging on the American side from Greenland southward to coast of Massachusetts, and embracing Labrador, doubtless Newfoundland. Gulf of St. Lawrence, Gaspé Bay, Maritime provinces, La Have bank, and New England states; and on the European side from the Scandinavian coast southward to the coast of England.

381. SHARP-NOSED FLOUNDER.

(*Parophrys vetulus*)

This small flounder occurs in water of moderate depth. It is one of the flounders which possess an accessory dorsal branch to the lateral line.

British Columbia: ranges from Santa Barbara, California, to coast of Alaska.

384. TWO-LINED FLOUNDER.

(*Lepidopsetta bilineata.*)

This flat-fish receives its name owing to the way in which an accessory branch of the lateral line winds upwards upon the back over the head, a feature which can be seen in the specimens.

British Columbia: ranges from coast of California to Behring Straits: ' in Behring Sea it far outnumbers all other flounders.'

385. RUSTY DAB.

(*Limanda ferruginea*)

One of the flat-fishes which are peculiar in having both eyes on the same side of the head. The rusty dab is one of the flounders.

Maritime provinces and Gaspé bay: ranges from coast to Labrador southward to state of New York.

387. WINTER FLOUNDER OR COMMON FLATFISH.

(*Pseudopleuronectes americanus*)

This species is one of the most abundant of the small flounders. It reaches a length of about fifteen inches, and is a very good food-fish.

Ranges from the coast of Labrador southward to South Carolina, and embracing Gulf of St. Lawrence, Maritime provinces, and New England states.

389. STARRY FLOUNDER.

(*Platichthys stellatus*)

This flounder may be distinguished by having the scales substituted by scattered star-like tubercles. It is a Pacific coast species, living in shallow water, and it sometimes ascends rivers. It is an excellent food-fish.

British Columbia: widely distributed from the Arctic ocean southward to the Amur river, and Pacific coasts of Asia: abounds in Behring sea.

5 GEORGE V., A. 1915

393. WINDOW PANE.

(*Lophopsetta maculata*)

Probably this is the flat-fish occasionally mistaken for the European turbot. In fact it is not distantly related to the latter, and agrees with it in having the eyes normally on the left side of the head.

Maritime Provinces: ranges in the United States from Maine to South Carolina.

400. SNAPPER OR ROSEFISH.

(*Sebastes marinus*)

The most remarkable thing about this gaudily adorned fish is that it is viviparous, or brings forth its young alive. It is good food-fish.

Both sides of the Atlantic ocean: on the American side ranging from Greenland and Labrador southward to off the coast of New Jersey, and embracing the Maritime provinces and Newfoundland: coast of Europe northward to Iceland and Spitzbergen and southward to the British channel.

402a. GOODE'S ROCKFISH.

(*Sebastodes goodei*)

This brightly coloured species lives in deep water. This individual is from British Columbia, but ordinarily the fish occurs off the coast of California; and is now taken in abundance about the Coronados Islands, Santa Catalina, and the Cortez banks.

403. JACKFISH OR BOCACCIO.

(*Sebastodes paucispinis*)

One of the rockfishes. It is claimed that these specimens were obtained in British Columbia, but the ordinary range of this fish appears to be the coast of California.

404. BLACK SEA BASS.

(*Sebastodes melanops*)

One of the rockfishes. The rockfishes are a vast assemblage of fishes, all so far as yet known, belonging to the Pacific ocean, and numbering as many as seventy species or more.

Vancouver Island: ranges from Monterey to Kadiak Island, Alaska.

405. BLACK ROCK FISH OR PRIESTFISH.

(*Sebastodes mystinus*)

This is the pêche prêtre of the Californian coast. It abounds in rather shallow water from about San Francisco northward to British Columbia.

408. ORANGE ROCKFISH.

(*Sebastodes pinniger*)

This gaudily coloured rockfish has, as will be seen in the specimens, the lateral line running as a continuous palish coloured streak, which is not crossed by the reddish marking of the body.

Recorded from British Columbia: ranges southward, including Puget Sound, to coast of California.

410. RED ROCKFISH OR TAMBOR.

(*Sebastodes ruberrimus*)

Well named red rockfish from its reddish colour. The tambor is an important food-fish and attains a length of about two and a half feet, being one of the largest of the rockfishes.

Ranges from British Columbia and Puget Sound southward to coast of California.

411. BROWN ROCKFISH.

(*Sebastodes auriculatus dallii.*)

This northern form of the Brown Rockfish which ranges from Vancouver island southward differs from the typical form in a few details of structure, and is said to be darker in colour. It may be caught by hook and line from wharves.

411a. GRASS ROCKFISH.

(*Sebastodes rastrelliger.*)

These two specimens are from British Columbia, but the chief habitat of the Grass Rockfish appears to be the coast of California.

414. BLACK AND YELLOW ROCKFISH.

(*Sebastodes chrysomelas.*)

This rockfish ranges from British Columbia to the coast of California, and frequents rather deep water.

416. BLACK BANDED ROCKFISH.

(*Sebastodes nigrocinctus.*)

One of the most striking of the rockfishes, and known at once by its deep red colour and jet black stripes. It frequents deep water.

Ranges from Vancouver island to coast of California.

417. SKILL OR COAL FISH.

(*Anoplopoma fimbria.*)

This is a fish of the north Pacific, common about the Straits of Fuca, and valued as food.

Pacific coast of Canada from Straits of Juan de Fuca to Queen Charlotte islands: entire range from coast of California to Aleutian islands.

5 GEORGE V., A. 1915

418. BOREGAT OR STARRY ROCK TROUT.

(*Hexagrammos decagrammus.*)

Whilst the males of this species are very uniform the females vary much in coloration. It is remarkable in the possession of five lateral lines on either side. It attains a length of some eighteen inches, and is a good food-fish.

British Columbia: ranges from Point Conception to Kadiak island.

421. CULTUS COD.

(*Ophiodon elongatus.*)

The application of the name cod to this fish is a misnomer, as it is in nowise related to the codfishes. Its flesh, which is vivid blue or green in colour, is used as food; and it is one of the important fishes of the Pacific coast.

British Columbia: ranges from Santa Barbara to coast of Alaska.

425. CABEZON.

(*Scorpænichthys marmoratus.*)

This is one of the sculpins. Used as food, and common in the markets, but its flesh is coarse and tough. The small specimen is a juvenile.

Ranges from British Columbia and Puget Sound to coast of California.

460. DADDY SCULPIN.

(*Myoxocephalus grœnlandicus.*)

This is one of the largest of the sculpins, attaining a length of about 2 feet. It is very voracious, preying incessantly upon smaller fish, and even devours the young of its own species.

Ranges from Greenland, and embracing Labrador, doubtless Newfoundland, and the Maritime provinces, southward of the State of New York.

461. COMMON SCULPIN OR LONG-SPINED SCULPIN.

(*Myoxocephalus octodecimspinosus.*)

The common sculpin is easily distinguished from its allies by the long spine extending along the opercular cover, or the bones which cover the gills.

Atlantic coast of North America, ranging from Labrador to Virginia, and embracing the Gulf of St. Lawrence. Maritime provinces, and New England states.

474. SEA RAVEN.

(*Hemitripterus americanus.*)

One of the sculpins. A remarkable looking fish of our Atlantic coast.

Atlantic coast of North America: Maritime provinces, Gaspé bay, Gulf of St. Lawrence, Labrador, and Newfoundland: extending southward to the coast of the State of New York.

496. LUMPFISH.

(*Cyclopterus lumpus.*)

This dumpy formed fish, of singular appearance, has the pelvic fins (which in it are situated below the pectoral fins or those behind the gills) converted into a suctorial disk by which it is enabled to adhere firmly to rocks or other objects. According to age, sex, and individuality, the lumpfish varies much in colour. It is rarely used as food.

Both coasts of North Atlantic ocean: on the American side ranging from Davis straits, and embracing Labrador, Newfoundland, Gulf of St. Lawrence, Gaspé bay, Maritime provinces, and New England states, southward to Cape Cod: on the European side occurs at the British islands, including the Orkney islands and ' all around the Irish coasts'; and coasts of Scandinavia and Baltic sea southward to coast of France.

514. FLYING GURNARD OR FLYING ROBIN.

(*Cephalacanthus volitans.*)

Like the true flying fishes, to which it is not directly related, the flying gurnard can rise out of the water and for a period move in the air.

Occasionally occurs off the coast of the southern part of the Bay of Fundy: found along the entire coast of the United States south of Cape Cod to the West Indies and coast of Brazil: occurs also in the Mediterranean and in the neighbouring parts of the eastern Atlantic.

The two mounted specimens are from the Bahama Islands.

517. KELPFISH.

(*Heterostichus rostratus.*)

This is the largest of the clinoid blennies, and its pattern and hue agree in coloration with the kelp among which it abounds: and for which reason it receives its name. Its ordinary habitat is the coast of California; but this is a specimen from the coast of British Columbia.

541. WOLF-FISH.

(*Anarhichas lupus.*)

A very voracious fish, as the teeth of the specimens may evidence. Its flesh is not valued.

Both sides of north Atlantic southward to Cape Cod and France: Maritime provinces, Gulf of St. Lawrence, Gaspé bay, Labrador, and without doubt Newfoundland: ' off the coast of Norfolk and Yorkshire, in Berwick bay, in the Firth of Forth, and among the Orkneys, occasionally also on the eastern coast of Ireland, and it is well-known on the northern shores of Europe, and in Greenland and Iceland.'

543. WOLF EEL.

(*Anarrhichthys ocellatus*)

This is one of the wolf-fishes, and one of the most remarkable of our fishes. As will be seen from the specimen it is exceedingly elongate, and the tail tapers to a tip.

5 GEORGE V., A. 1915

In typical wolf-fishes the caudal fin is well developed, whereas in this species there is the merest vestige of that fin, confluent with the dorsal and anal fins, so that it appears as if it had no caudal fin at all.

British Columbia and Puget Sound southward to coast of California.

545. EEL POUT.

(*Zoarces anguillaris.*)

This is a viviparous fish, that is it brings forth its young alive. It is rather a common species north of Cape Cod, and ranges from the coast of Labrador, embracing the Gulf of St. Lawrence and Maritime Provinces, and doubtless Newfoundland, southward to the coast of Delaware.

561. ANGLER, FISHING FROG OR MONKFISH.

(*Lophius piscatorius*)

This singular looking fish has the carpal bones noticeably elongate forming a sort of arm, and the foremost dorsal spine which overhangs the cavernous mouth tipped with a lappet—spine and lappet together presenting the appearance of a baited fishing rod by which it allures its prey. It is a fish of enormous voracity, greedily devouring multitudes of small fishes.

Both sides of north Atlantic: Maritime Provinces and Gaspé bay: extending southward on the American side to the Barbados Islands: in the eastern hemisphere ranges from Norway to the Cape of Good Hope: 'not rare on any part of the coasts of Great Britain and Ireland, and is particularly common in the Solent and in the harbours of Portsmouth and Southampton.'

563. TRIGGER FISH.

(*Balistes carolinensis.*)

This is one of the fishes known as the plectognaths, singular forms, which although differing among themselves, agree in the following points among others. The bones of the upper jaw (the maxillaries and premaxillaries) are united; the gill-openings are greatly reduced; and they nearly all of them develop poisonous alkaloids in the flesh.

This trigger fish occurs occasionally, but very rarely, at the coast of Nova Scotia. It also occasionally occurs in the gulf stream. is common on the Atlantic coast of the United States. in the Mediterranean, very rarely on the coast of England. one specimen obtained in the Bay of Galway is recorded in Thompson's Natural History of Ireland. and is common in tropical parts of the Atlantic.

The mounted specimen is from the Bahama islands.

566. HEADFISH OR SUNFISH.

(*Mola mola.*)

This singular plectognath, very semi-circular in form. outwardly appears to be all head and no body, and looks much as if its body had been bitten off by some sea monster; but this is a mere dissemblance. for although highly aberrant the body with its organs is there.

Temperate and tropical seas: occurring at the West Indies, in the Mediterranean and Adriatic sea, northward to the British islands. occasionally to coasts of the Maritime

65827—Opp. p. 366.

provinces, and to San Francisco; and has been recorded from the coast of Labrador.

Besides the above mentioned there are on exhibition in the museum mounted skeletons of various fishes indigenous to the waters of Canada, whilst others are being prepared; and a skeleton of a fin-back whale, some 50 feet in length, is ready to be mounted upon a steel rod which has been manufactured under contract for that purpose. Specimens of two octopi have also recently been acquired; and will shortly be shown in glass vessels specially made for their reception; there is also a fine collection of Canadian water birds, some of which were acquired by exchange from the museum of the Geological Survey; and the following objects in the museum placed on standards form an attractive feature.

LEATHER TURTLE.

(*Dermochelys Coriacea*.)

Generally distributed between the tropics—a casual visitor to temperate coasts. This specimen was taken off the coast of Nova Scotia.

MODEL OF STEAM TRAWLER SHOWING OTTER TRAWL GEAR.

Description of Trawl Net.

Headline.. .. ..Yellow and Green	Cod-end.. .. ..Yellow
Square.. .. ..Green	Flapper in cod-end.. .. ..Blue
Short upper wings.. .. ..Blue	Pockets laced in.. .. ..Yellow
Lower wings.. .. ..Blue	Ground rope.. .. ..Yellow
Baitings of top.. .. ..Red	Small rope Bolsh line.. .. ..Yellow
Belly bottom of net.. .. ..Red	Cod-line.. .. ..Yellow

Model of steam herring drifter showing section of drift nets.

Model of a fishing schooner—Off for the banks.

Model of a dory used in fishing from vessels on the banks.

Haida Indian dugout, Queen Charlotte Islands.

Model of a herring weir.

Model of a British Columbia salmon trap-net.

There is also contained in the museum a vast amount of natural history material collected from all parts of the Dominion, many of the objects of which are small, and too numerous to mention in detail in this report.

Besides the specimens of the one or two species of the fishes from the Bahama Islands mentioned above as occasionally or casually occurring in the waters of Canada, the museum also contains specimens of the following which were procured from the coral reefs of the Bahamas.

The trumpet-fish, the butterfly fish, the blue-tang, the rock-beauty, the spade-fish, the angel-fish, a file-fish, a trigger-fish (another kind from that mentioned above). several different kinds of coffer-fishes, and the porcupine-fish. There are also from the Bahamas specimens of a few beautiful gastropod shels, viz.: the conch, the king-cassis, and the queen-cassis; and also a specimen of the tortoise shell turtle.

The doors of the museum are open to visitors on week days from 9.30 a.m. to 5.30 p.m.; and on Sundays, during the winter months, from 2 to 5 p.m.

ANDREW HALKETT.
Naturalist of the Department.

5 GEORGE V., A. 1915

APPENDIX No. 15.

EXPENDITURE AND REVENUE.

The total expenditure for all fisheries services, except civil government, for the fiscal year ended March 31, 1914, amounted to $1,070,857.94.

The total net fisheries revenue derived from rents, fines, sales and license fees (including *modus vivendi* licenses to United States vessels) for the same period amounted to $110,994.63.

The following is a summary of the sums appropriated and those expended for the various services during 1913-14.

FISHERIES EXPENDITURE, 1913-14.

Service.	Appropriation	Expenditure.
	$ cts.	$ cts.
Salaries and disbursments of fishery officers..........	230,000 00	229,547 16
Fish-breeding establishments......	400,000 00	354,675 13
Fisheries Patrol Service.............	137,500 00	135,330 87
Fishery patrol boats for British Columbia.....................	75,000 00	66,542 10
Ten fishery patrol boats for Atlantic Coast..........................	50,000 00	15,994 08
Oyster culture	6,000 00	4,434 60
Cold storage and transportation of fresh fish..	100,000 00	90,868 51
Dogfish reduction works...	60,000 00	41,188 37
Fisheries Intelligence Bureau...	10,000 00	8,956 76
Exhibit of fresh fish (Toronto Exhibition)...	10,000 00	9,700 48
International Fisheries Commission	5,000 00	441 59
Building fishways and clearing rivers....	20,000 00	12,341 93
Legal and incidental expenses......	4,000 00	1,100 87
Canadian Fisheries Museum	16,000 00	9,100 54
Services of customs officers in connection with issuing of *modus vivendi* licenses........................	900 00	537 90
Fisheries patrol steamer for Lake Winnipeg	145,000 00	40,146 03
Marine Biological stations and investigations......	17,000 00	17,000 00
Expenses of investigating claims for compensation under the Pelagic Sealing Treaty......	17,000 00	16,713 02
Allowance to Department of Public Works for the loss of the ice-breaking Tug *Sir Hector*...............................	16,238 00	16,238 00
Total	1,319,638 00	1,070,857 94
Fishing bounty.........	160,000 00	158,661 25

The following summaries show the salaries and disbursments of the fishery officers in the several provinces, the expense for maintenance of fish-breeding establishments throughout Canada ; also the expense for that part of the Fisheries Protection Service called the Fisheries Patrol Service.

Details will be found in the Auditor General's Report under the proper headings.

SESSIONAL PAPER No. 39

SALARIES AND DISBURSEMENTS OF FISHERIES OFFICERS, 1913.14.

Provinces.	Officers.		Guardians.		Miscella-neous.	Total.
	Salaries.	Disburse-ments.	Wages.	Expenses.		
	$ cts.	$ cts.	$ cts.	$ cts.	$ cts.	$ cts.
Nova Scotia.	4,811 51	21,636 53	27,497 33	197 72	773 86	54,919 95
Prince Edward Island	2,177 89	2,301 43	4,250 98	34 74	65 20	8,830 24
New Brunswick	4,848 17	13,028 40	30,871 40	1,299 67	163 04	50,210 68
Quebec..................	5,015 85	4,063 17	873 50	134 46	....	10,086 98
Ontario..................	3,300 00	327 05	....	....	305 50	3,932 55
Manitoba ....' ...	2,291 63	1,774 53	5,013 45	1,937 80	58 35	11,075 76
Alberta and Saskatchewan.	10,524 83	6,199 15	4,946 48	3,133 08	127 60	24,931 14
British Columbia.........	33,383 81	7,488 88	7,875 57	577 90	2,064 57	52,390 23
Yukon..................	1,306 25	208 75	. ...	....	5 00	1,520 00
General account.....	....	2,511 83	....	....	9,007 80	11,549 63
Total.....	....	....	....	....	....	229,447 16
Outstanding advances...............................						100 00
Total expenditure........						229,547 16

5 GEORGE V., A. 1915

FISH BREEDING, 1913-14.

Hatcheries.	Salaries.	Maintenance	Total Expenditure of Hatchery.	Total Expenditure of Provinces.
	$ cts.	$ cts.	$ cts.	$ cts.
Nova Scotia.				
Antigonish	111 00	2,257 10	2,368 10	
Arichat.	177 00	2,704 36	2,881 36	
Bayview	120 00	4,656 70	4,776 70	
Bedford	2,112 50	1,226 22	3,336 72	
Canso		3,142 78	3,142 78	
Digby Pond		107 50	107 50	
Inverness	78 00	3,476 55	3,554 55	
Isaac Harbour	75 00	1,644 27	1,719 27	
Petit Bras d'Or	90 00	2,506 41	2,596 41	
Lindloff		635 35	635 35	
Long Beach Pond	75 00	6,454 00	6,529 00	
Margaree	1,133 35	3,334 02	4,467 37	
Margaree Pond		3,682 72	4,682 72	
Middleton	1,595 85	2,198 15	3,794 00	
Windsor	1,325 00	814 05	2,139 05	
				45,732 88
Prince Edward Island.				
Charlottetown		2,523 73	2,523 73	
Georgetown	168 00	2,558 15	2,726 15	
Kelly's Pond	1,525 00	608 57	2,133 57	
				7,383 45
New Brunswick.				
Buctouche	129 00	3,850 89	3,979 89	
Little River	233 34	8,106 25	8,339 59	
Miramichi	1,300 00	4,020 17	5,320 17	
Nepisiguit		862 48	862 48	
New Mills Pond		3,853 38	3,853 38	
Restigouche	2,025 00	3,345 42	5,370 42	
Shad		906 50	906 50	
St. John's Pond	375 00	8,260 44	8,635 44	
St. John's River	1,737 78	5,352 92	7,090 70	
Shemoque	138 00	3,266 33	3,404 33	
Shippegan	180 00	3,346 49	3,526 49	
Sparkle		178 06	178 06	
Tobique		173 67	173 67	
				51,641 12
Québec.				
Dartmouth River		173 43	173 43	
Gaspé	1,325 00	1,131 33	2,456 33	
Lac Lester	1,275 00	1,820 76	3,095 76	
Lac Tremblant	512 50	1,442 02	1,954 52	
Magdalen Islands		5,276 89	5,276 89	
Magog	1,497 93	1,155 47	2,653 40	
Port Daniel	171 00	2,231 11	2,402 11	
St. Alexis	512 50	1,023 09	1,535 59	
Tadousac	1,375 00	2,119 79	3,494 79	
				23,042 82
Ontario.				
Belleville		865 92	865 92	
Collingwood	1,962 50	13,869 41	15,831 91	
Newcastle	1,312 50	2,330 51	3,643 01	
Ottawa	1,933 30	219 44	2,152 74	
Port Arthur	2,987 50	6,818 01	9,805 51	
Quinte Pond		105 00	105 00	
Sandwich	3,135 00	14,843 08	17,978 08	
Sarnia	2,062 50	5,665 77	7,728 27	
Southampton	1,237 50	3,725 28	4,962 78	
Wiarton	1,812 51	3,992 08	5,804 59	
				68,877 81

SESSIONAL PAPER No. 39

FISH BREEDING, 1913-14—*Concluded.*

Hatcheries.	Salaries.	Maintenance	Total Expenditure of Hatchery.	Total Expenditure of Provinces.
	$ cts.	$ cts.	$ cts.	$ cts.
Manitoba.				
BerensRiver..	720 84	595 75	1,316 59	
Dauphin River..	141 66	20,633 47	20,775 13	
Gull Harbour:		7,777 93	7,777 93	
Selkirk	1,777 98	5,675 34	7,453 32	
Winnipegosis	1,845 84	8,601 16	10,447 00	
				47,769 97
Saskatchewan and Alberta.				
Banff	508 58	10,661 08	11,169 66	
Permiz Creek		1,904 04	1,904 04	
Qu'Appelle		896 14	896 14	
				13,969 84
British Columbia.				
Anderson Lake	1,000 00	4,944 47	5,944 47	
Babine	1,000 00	5,374 22	6,374 22	
Cowichan	1,000 00	3,461 34	4,461 13	
Fraser River		2,072 88	2,072 88	
Gerrard		7,319 14	7,319 14	
Granite Creek	1,100 40	7,303 05	8,403 45	
Harrison Lake	1,100 00	8,432 05	9,532 05	
Kennedy Lake	1,000 00	4,702 11	5,702 11	
Pemberton	1,000 00	8,039 10	9,039 10	
Rivers Inlet	1,024 99	6,062 83	7,087 82	
Skeena River	999 96	8,468 22	9,468 18	
Stuart Lake	1,000 00	6,718 55	7,718 55	
				83,123 10
General account	2,800 00	9,334 14	12,134 14	12,134 14
				353,675 13
Outstanding advances				1,000 00
Total expenditure				354,675 13

FISHERIES REVENUE FOR FISCAL YEAR ENDED MARCH 31, 1914.

Provinces.	Amount Collected.	Refunds.	Net Amount.
	$ cts.	$ cts.	$ cts.
Ontario	806 69		806 69
Québec	5,286 89		5,286 89
New Brunswick	17,526 48	19 30	17,507 18
Nova Scotia	7,732 50	50 00	7,682 50
Prince Edward Island	2,245 60		2,245 60
Manitoba	4,846 50		4,846 50
Saskatchewan and Alberta	8,253 05		8,253 05
British Colnmbia	53,035 50	200 00	52,835 50
Yukon	226 00		226 00
Total	99,959 21	269 30	99,689 91
Transfer of licenses issued by D. Morrison in 1914 and to come to account in 1914-15			423 78
			99,266 23
Modus Virendi licenses			11,728 50
Grand total			110,994 63

FISHERIES PATROL SERVICE 1913-14.

Name of Vessels	Pay lists	Fuel	Provisions	Repairs Hull	Repairs Engine	Supplies Engine	Supplies Deck	Charter	Clothing	Sundry	Totals
	$ cts.	$ cts.	$ cts.	$ cts.	$ cts.	$ cts.	$ cts.	$ cts.	$ cts.	$ cts.	$ cts.
'Ada'	5,040 00	916 80	2,118 45	340 69	224 00	146 17	486 17		43 36	1 10	3,316 74
'Annie D'								1,125 00	1 00		1,126 00
'Bee'	449 68	168 98		20 00		31 10	184 68		2 40	38 30	895 04
'Davies'	1,585 65	149 84	114 00	180 46	197 00	39 50	159 27		63 86	111 57	2,601 15
'Elk'	1,020 00	13 08		3 04	237 58	4 00	29 10			11 93	1,305 65
'E. G. Mildred'	776 00					126 71	18 24	487 50		9 37	1,425 90
'Evelyn B'								948 00			948 00
'Falcon'		1,117 75									1,117 75
'Fispa'	862 15	2,502 25	1,516 40	268 98	492 92	972 66	1,324 50		272 13	1,289 30	1,208 29
'Foam'	2,040 00	6 70		26 50	114 55	6 75	348 21			24 71	2,367 42
'Get'	1,445 12	740 36				238 34	785 39		25	123 95	3,335 22
'Gull'	1,734 18	615 74		85 00	26 70	56 41	280 82		2 40		2,716 25
'Hawk'	1,175 00	334 10			23 35	797 22	8 70		85	56 90	2,483 12
'Hudson'	1,720 99	228 62		202 10	298 15	50 57	175 53		111 12	356 97	3,143 75
'Shine'								760 00			760 00
'Kayex'	475 00	151 86		117 50	90 97	280 49	135 36		3 75	1 25	1,256 18
'Kingfisher'	389 35	180 90		62 00	6 60	5 15	30 55		2 85		677 40
'Lady of the Lake'	4,078 39	1,699 57	1,688 30	667 33	81 50	14 24	684 65			1 715 10	10,633 06
'Linnet'	721 13	286 25			52 35	17 06	62 65		2 85	18 50	1,160 19
'Merlin'	846 23	839 75			13 76	9 22	62 02		2 85		1,274 42
'Mary G'	31 50	9 60									50 10
'Dan'	2,118 48	203 61		138 21	92 58	133 68	386 59		19 25	9 00	8,508 76
Patrol Boat 'A'	2,387 00	186 98		77 23	6 44	692 36	12 38		84 57	446 36	3,876 94
" 'B'	2,128 34	19 56		36 50	69 45	368 81	74 32		94 82	79 58	2,848 98
" 'C'	1,623 88	27 59		630 14	81 98	270 65	51 70		18 01	57 18	2,726 29
" 'D'	1,914 62		35 00	473 73	4 00	465 15	191 15		74 81	22 34	3,256 78
" 'E'	2,560 05	83 41	50 60	437 61	33 21	760 33	191 85		123 61	168 32	4,317 39
" No. 1	1,532 46	182 40		142 90	250 00	49 81	75 29		84 77	76 92	2,414 27
" 2	695 82	126 92	22 00		36 02	16 60	36 11		84 77	96 65	1,103 37
" 3						72 76			49	60 41	72 76
" 4						16 52					16 52
'Phalarope'	1,100 51	440 25			112 21	171 73	178 30		73 14		2,081 14
'Pioneer'							10 28				10 28
'Princess'	10,252 29	5,566 36	3,656 80	2,251 11	2,459 83	435 47	2,397 72	450 00	602 54	2,142 17	29,767 29
'Pilgrim'	650 99									5 00	1,100 99
'Raven'	211 93	210 83		196 10		75 16	150 34		2 40	175 15	825 81
'Restless'		412 75									412 75
'Richmond'		89 19		57 15	88 10	59 02	38 90			22 96	210 07
'Somiahmus'	970 00	367 42			13 89	78 26	89 83			88 20	1,877 91
'Sea Gull'	1,722 50					68 17	241 44		53 41	10 50	2,147 06

	1	2	3	4	5	6	7	8	9	10	11
'Sparker'	405 00					190 95	130 19	112 50		1 12	99 57
'Swan'	1,860 60			150 76	117 01	2,138 00	688 11			20 54	4,416 50
'Tanner'	2,885 63	131 55		149 25	418 32	18 60	79 33		225 44	219 21	4,736 11
General account	2,721 00	2,098 97				582 30		992 20	351 41	12,261 33	19,086 54
Totals	66,727 87	19,609 84	9,201 55	6,697 29	5,642 46	9,459 62	9,848 45	4,875 20	2,431 34	19,722 29	154,215 91

COMPARATIVE STATEMENT of Expenditure and Revenue of the

Number.		1893-94.		1894-95.		1895-96.	
		Expenditure	Revenue.	Expenditure	Revenue.	Expenditure	Revenue.
		$ cts.	$ cts.	$ cts.	$ cts.	$ cts.	$ cts.
1	General Account Fisheries....						
2	Ontario.................	22,634 37	23,632 82	21,938 56	33,211 60	24,917 48	35,681 68
3	Quebec................	11,692 82	7,211 82	12,459 34	8,836 18	11,880 43	8,160 98
4	New Brunswick.	18,522 94	8,333 24	21,370 94	11,170 36	20,526 56	10,696 87
5	Nova Scotia............	20,420 81	5,296 27	23,555 38	7,075 07	23,049 41	6,180 93
6	Prince Edward Island........	3,078 55	980 15	3,796 58	3,312 30	3,555 87	2,161 85
7	Manitoba and N.W. Territory	5,331 29	926 99	6,178 71	2,458 80	6,935 20	2,256 69
8	Alberta.................						
9	Saskatchewan...........						
10	British Columbia...........	5,283 21	25,337 90	6,218 74	23,517 25	6,226 77	26,410 75
11	Yukon......						
12	Hudson Bay Territory.....						
13	Fish-breeding and fishways ..	45,024 67		39,730 93		38,050 41	
14	Fisheries Protection Service..	115,147 59		100,207 29		102,021 72	
15	Miscellaneous............	34,892 19		24,619 86		20,203 25	
	Totals ......:..	282,028 44	76,719 19	260,076 33	89,581 56	257,237 10	91,549 76
	Fishing bounties.......	158,794 54		160,089 42		163,567 89	

		1900-01.		1901-02.		1902-03.	
16	General Account Fisheries...	1,117 49		765 78		402 97	
17	Ontario.....	3,819 57	717 35	4,445 93	373 42	4,650 53	1,818 83
18	Quebec.............	7,934 03	4,738 92	6,242 58	2,498 85	6,785 86	4,379 15
19	New Brunswick.......... ...	28,452 51	10,150 40	23,813 62	11,658 34	27,132 84	11,188 02
20	Nova Scotia.............	35,760 39	6,595 94	32,618 00	6,084 65	39,118 79	3,962 45
21	Prince Edward Island........	7,934 03	1,525 30	7,814 02	1,843 45	7,081 60	2,007 35
22	Manitoba..................	2,669 74	1,103 00	2,624 87	2,279 00	3,129 70	1,784 00
23	N. W. Territory...... ..	6,251 39	1,222 55	5,928 22	950 07	7,076 26	1,350 50
24	Alberta..						
25	Saskatchewan.						
26	British Columbia......	17,886 36	52,960 35	18,560 73	41,178 65	17,808 45	43,015 02
27	Yukon..................			2,066 66	1,130 00	1,522 00	320 00
28	Hudson Bay Territory.......						
29	Fish-breeding	68,961 40		79,891 85		77,330 86	
30	Fisheries Protection Service..	124,211 21		152,723 69		145,137 49	
31	Miscellaneous..... ......|	27,833 79	9,178 50	56,131 26	11,223 65	30,903 27	8,925 40
	Totals.......	332,767 07	88,145 11	393,627 21	79,169 58	368,091 12	78,635 82
	Fish bounties............	158,802 50		155,942 00		159,853 50	

		1907-08.		1908-09.		1909-10.	
32	General Account Fisheries....	1,437 28		4,751 36		3,910 03	
33	Ontario..	3,188 34	349 10	4,784 23	770 78	4,836 86	1,620 75
34	Quebec	5,590 94	8,145 97	7,895 53	6,797 91	7,886 85	4,947 46
35	New Brunswick	24,987 70	9,153 08	38,904 12	12,385 14	41,188 19	13,044 88
36	Nova Scotia.............	24,989 09	3,118 73	44,601 04	5,369 70	46,590 66	3,821 81
37	Prince Edward Island.'.....	5,792 32	1,300 94	8,410 25	2,393 66	9,396 08	2,359 93
38	Manitoba.......	2,173 33	2,285 98	3,945 73	3,704 22	5,323 82	6,962 88
39	†Alberta........			5,713 80	915 00	7,938 22	703 00
40	Saskatchewan			6,591 20	1,085 50	6,474 57	1,209 44
41	N. W. Territories....		969 50				
42	British Columbia.....		29,903 95	35,139 58	39,251 65	37,509 61	41,864 80
43	Yukon..............	6,359 22	173 00	1,019 50	228 00	2,316 63	457 00
44	Hudson Bay Territory.	20,381 97	10 00		20 00		501 83
45	Fish-breeding...	1,030 35		190,563 19		180,345 65	
46	*Fisheries Protection Service			242,601 14		295,443 47	
47	Miscellaneous................	118,681 62		196,808 02	9,794 00	345,249 58	10,876 78
		204,837 82					
	Totals..............	115,219 92	4,134 00	791,728 69		994,355 22	
	Fishing bounties.......			159,999 90		155,221 85	
		534,669 90					
	Grand Totals......	159,015 75	....	951,728 59	82,715 56	1,149,577 07	85,070 56

* The Fisheries Protection Service being now under the control of the Naval Department, this expen
† Since 1912-13 Saskatchewan is included with Alberta.

SESSIONAL PAPER No. 39

Fisheries Department, July 1, 1893 to March 31, 1913.

	1896-97.		1897-98.		1898-99.		1899-00.		Number.
	Expenditure	Revenue.	Expenditure	Revenue.	Expenditure	Revenue.	Expenditure	Revenue.	
	$ cts.	$ cts.	$ cts.	$ cts.	$ cts.	$ cts.	$ cts.	$ cts.	
	2,198 47	....	2,389 66		2,632 12	....	652 41	....	1
	21,592 40	32,814 66	19,239 34	30,574 57	11,784 22	5,830 85	3,804 94	794 12	2
	12,910 80	7,876 13	11,440 16	7,571 15	11,350 27	6,287 71	5,452 41	2,543 04	3
	21,671 92	10,110 77	17,063 58	5,317 08	22,922 50	10,430 08	21,659 94	12,015 27	4
	23,682 33	5,239 55	21,683 91	11,511 85	25,348 11	6,668 22	27,461 91	5,494 49	5
	3,744 36	2,932 25	6,775 78	2,707 57	6,832 85	2,242 24	7,364 30	2,207 12	
{	1,908 14	1,719 00	1,206 26	1,515 00	1,883 37	1,537 35	1,723 59	2,028 00	6
{	2,181 58	344 13	2,324 66	393 87	4,005 68	150 50	3,848 25	1,522 50	7
									8
									9
	8,841 64	39,388 82	8,508 79	47,864 75	8,459 47	45,801 75	13,662 17	53,195 35	10
	....	....	....	....	....	....	....	....	11
									12
	27,330 73	....	28,002 32	....	34,522 57	....	38,070 12	....	13
	99,357 01	....	101,807 96	....	105,133 27	....	97,370 11	....	14
	62,777 30	....	59,919 56	....	23,207 73	....	31,125 67	....	15
	289,197 01	100,025 30	280,061 98	107,455 84	427,599 16	75,949 20	411,717 35	79,799 89	
	154,389 77	....	157,504 00	....	159,459 00	....	160,000 00	....	

	1903-04.		1904-05.		1905-06.		1906-07.		
	1,362 11	....	1,314 75	....	3,135 91	....	2,261 66	....	16
	4,500 43	2,578 48	4,294 60	1,471 51	4,857 23	458 00	4,949 67	499 15	17
	7,619 67	4,670 64	6,769 16	4,648 86	8,200 02	6,185 63	8,123 04	7,564 39	18
	27,664 34	10,494 20	25,253 16	11,887 19	36,445 88	11,541 20	35,856 38	11,395 84	19
	30,003 04	3,685 75	32,619 86	6,448 88	45,241 50	4,470 45	49,351 10	4,934 43	20
	7,320 96	1,983 42	6,879 05	2,046 50	9,455 8.	3,013 85	9,351 81	2,206 25	21
	2,786 74	4,002 70	2,800 64	4,875 70	4,638 51	3,527 05	3,687 07	4,148 00	22
	7,317 49	922 50	7,003 55	1,151 50	12,718 15	1,151 10	....	....	23
					31,964 83	48,737 55			24
					1,226 30	274 00	11,124 22	868 97	25
	15,133 65	56,904 34	16,631 37	47,436 00	....	360 00	30,141 33	51,532 50	26
	1,100 00	240 00	1,400 00	340 00	235,660 26	....	1,083 31	282 00	27
		10 00		10 00	225,279 96			10 00	28
	109,286 07	....	149,419 24	....	181,267 38	395 15	209,279 78	....	29
	204,654 66	....	462,082 12	....	956,196 23	....	249,876 37	....	30
	56,858 18	10,166 50	105,892 97	10,472 00	156,114 50	....	194,993 61	14,568 16	31
	475,880 31	95,756 53	822,360 46	90,988 14	1,118,310 79	80,113 98	968,626 00	....	
	158,943 70	....	157,228 24	....	....	....	158,540 65	....	

	1910-11.		1911-12		1912-13.		1913-14.		
	4,540 84	....	9,392 19	....	11,563 48	....	11,549 63	....	32
	7,125 37	280 25	20,255 96	658 45	4,332 25	548 74	3,932 55	806 69	33
	7,695 49	5,336 61	10,558 70	6,044 75	9,784 38	8,095 79	10,086 98	5,286 89	34
	41,593 46	12,996 84	42,708 01	13,902 15	45,136 31	15,152 52	50,210 68	17,930 96	35
	45,800 42	7,749 60	49,540 37	5,912 65	45,828 11	6,780 00	54,919 95	7,682 50	36
	9,415,09	2,499 63	9,116 56	2,477 50	8,890 15	2,927 96	8,830 24	2,245 60	37
	7,163 36	8,137 75	7,152 24	6,334 00	6,862 15	6,039 00	11,075 76	4,846 50	38
	7,867 27	698 50	8,537 07	709 00	17,413 00	4,268 50	24,931 14	8,253 05	39
	7,597 87	1,246 00	8,587 31	1,304 75	....	....	....	....	40
									41
	40,314 16	45,846 70	37,028 05	44,898 51	45,824 40	48,824 50	52,390 23	52,835 50	42
	1,964 95	907 50	2,094 75	203 25	1,909 83	342 00	1,520 00	226 00	43
		100 00							44
	220,727 66	....	235,699 52	....	283,793 43	....	354,675 13	....	45
	....	....	92,666 65	....	36,843 18	....	135,330 87	....	46
	199,762 00	15,076 50	150,519 90	13,785 00	193,764 07	13,500 00	351,404 78	11,728 50	47
	601,567 94	....	683,857 28	....	761,956 74	....	1,070,857 94	....	
	159,166 75	....	159,999 70	....	159,996 40	....	158,661 25	....	
	760,734 69	100,875 88	843,856 98	96,230 01	921,953 14	106,469 01	1,229,519 19	110,994 63	

fliture, from the year 1911-12, is for the Fisheries Patrol Service.

APPENDIX No. 16.

FISHING BOUNTIES.

The payments made for this service are under the authority of the Revised Statutes, 1906, chap. 46, intituled: "An Act to encourage the development of the Sea Fisheries, and the building of fishing vessels," which provides for the payment of the sum of $160,000 annually, under regulations to be made from time to time by the Governor General in Council.

REGULATIONS.

The regulations governing the payment of fishing bounties were established by the following Orders in Council:—

AT THE GOVERNMENT HOUSE AT OTTAWA,
TUESDAY, the 30th day of June, 1908.

PRESENT:

HIS EXCELLENCY THE GOVERNOR GENERAL IN COUNCIL.

Whereas, in view of the revision of the Statutes of Canada in 1906, it is necessary that the regulations governing the payment of fishing bounties which were adopted by order in council on the 10th December 1887, be readopted under chapter 46 of the Revised Statutes of Canada, 1906, "The Deep Sea Fisheries Act":

And whereas new conditions require certain changes in the existing regulations in order to establish a better interpretation of the bounty system.

Therefore, His Excellency the Governor General in Council is pleased to order that the regulations established by the order in council of the 10th December, 1897, under the provisions of the Bounty Act, 1891, 54-55 Victoria, chapter 42, shall be and the same are hereby rescinded and the following substituted therefor:—

1. Resident Canadian fishermen who have been engaged in deep-sea fishing in Canadian vessels or boats for fish other than shell-fish, salmon and shad, or fish taken in rivers or mouths of rivers, for at least three months, and have caught not less than 2,500 pounds of sea fish, shall be entitled to a bounty; provided always that no bounty shall be paid to men fishing in boats measuring less than 13 feet keel, and not more than 3 men (the owner included) will be allowed as claimants in boats under 20 feet.

2. No bounty shall be paid upon fish caught in trap-nets, pound-nets and weirs, nor upon the fish caught in gill-nets fished by persons who are pursuing other occupations than fishing, and who devote merely an hour or two daily to fishing these nets but are not, as fishermen, steadily engaged in fishing.

3. Only one claim will be allowed in each season, even though the claimant may have fished in two vessels, or in a vessel and a boat or in two boats.

4. The owners of boats measuring not less than 13 feet keel, whether propelled by oars, sails or other motive power, which have been engaged during a period of not less than three months in deep-sea fishing for fish other than shell-fish, salmon or shad,

or fish taken in rivers, or mouths of rivers, shall be entitled to a bounty on each such boat.

5. Canadian registered vessels owned and fitted out in Canada, of 10 tons and upwards (up to 80 tons), by whatever means propelled, contained within themselves, which have been exclusively engaged during a period of not less than three months in the catch of sea-fish other than shell-fish, salmon or shad, or fish taken in rivers, or mouths of rivers, shall be entitled to a bounty to be calculated on the registered tonnage which shall be paid to the owner or owners.

6. Owners or masters of vessels intending to fish and claim bounty on their vessels must, before proceeding on a fishing voyage, procure a license from the nearest collector of customs or fishery overseer, said license to be attached to the claim when sent in for payment.

7. The date when a vessel's fishing operations shall be considered as having begun shall be the day upon which she sails from port on her fishing voyage, after the license has been procured, and the date upon which her fishing season shall end shall be the day upon which she arrives in port from her last fishing voyage prior to the 1st December. The three months during which a vessel must have been engaged in fishing to be entitled to the bounty, shall not include such periods as she may have been lying in port, provided that not more than three days may be permitted for the sale, transfer or discharge of her cargo of fish and refitting.

8. Dates and localities of fishing must be stated in the claim, as well as the quantity and kinds of sea-fish caught.

9. Ages of men must be given. Boys under 14 years of age are not eligible as claimants.

10. Claims must be sworn to as true and correct in all their particulars.

11. Claims must be filed on or before the 30th November in each year.

12. Officers authorized to receive claims will supply the requisite blanks free of charge, and after certifying the same will transmit them to the Department of Marine and Fisheries.

13. No claim in which an error has been made by the claimant or claimants shall be amended after it has been signed and sworn to as correct.

14. Any person or persons detected making returns that are false or fraudulent in any particular may be debarred from any further participation in the bounty, and be liable to be prosecuted according to the utmost rigour of the law.

15. The amount of the bounty to be paid to fishermen and owners of boats and vessels will be fixed from time to time by the Governor in Council.

16. All vessels fishing under bounty license, are required to carry a distinguishing flag, which must be shown at all times during the fishing voyage at the main topmast head. The flag must be four feet square in equal parts of red and white, joined diagonally from corner to corner. Any case of neglect to carry out this regulation reported to the Department of Marine and Fisheries will entail the loss of the bounty, unless satisfactory reasons are given for its non-compliance.

RODOLPHE BOUDREAU,
Clerk of the Privy Council.

AT THE GOVERNMENT HOUSE AT OTTAWA,
WEDNESDAY. the 22nd day of February, 1911.

PRESENT:

HIS EXCELLENCY THE GOVERNOR GENERAL IN COUNCIL.

His Excellency in Council, in virtue of the provisions of section 7 of chapter 46 of the Revised Statutes of Canada,—An Act to encourage the development of the Sea

Fisheries and the building of Fishing Vessels,—is pleased to order and it is hereby ordered that section 5 of the regulations governing the payment of claims for Fishing Bounty be rescinded and the following substituted in lieu thereof:—

5. Canadian registered vessels. owned and fitted out in Canada, of ten tons and upwards (up to eighty tons), by whatever means propelled, contained within themselves, which have been exclusively engaged during a period of not less than three months in the catching of sea-fish, other than shell-fish, salmon or shad. or fish taken in rivers, or mouths of rivers, shall be entitled to a bounty to be calculated on the registered tonnage, which shall be paid to the owner or owners: Provided that vessels known as 'Steam Trawlers,' operating 'Beam.' 'Otter,' or other such trawls, shall not be eligible for any such bounty.

RODOLPHE BOUDREAU,
Clerk of the Privy Council.

The bounty for the year 1913 was distributed on the basis authorized by the following order in council, approved by his Royal Highness the Governor General on the 22nd January, 1914.

His Royal Highness the Governor General in Council is pleased to order, and it is hereby ordered that the sum of one hundred and sixty thousand dollars, payable under the provisions of chapter 46 of the Revised Statutes of Canada, 1906, intituled: 'An Act to encourage the development of the Sea Fisheries and the building of fishing vessels,' be distributed for the year 1913-1914, upon the following basis:—

Vessels: The owners of the vessels entitled to receive bounty, shall be paid one dollar ($1) per registered ton, provided, however, that the payment to the owner of any one vessel shall not exceed the sum of eighty dollars (80), and all vessel fishermen entitled to receive bounty, shall be paid the sum of six dollars and seventy cents ($6.70) each.

Boats: Fishermen engaged in fishing in boats who shall also have complied with the regulations entitling them to receive bounty. shall be paid the sum of three dollars and ninety-five cents ($3.95) each. and the owners of fishing boats shall be paid one dollar ($1) per boat.

RODOLPHE BOUDREAU,
Clerk of the Privy Council.

During the year 1913, 13,412 claims were received. an increase of 441 over 1912, while the number paid was 13,533, which includes a number held over from 1912. being 569 more than in the previous year.

The amount of bounty paid to vessels and their crews is $60.887.10, and to boats and boat fishermen $97,774.15, making the total payments during the year. $158,661.25.

Bounty was paid to 910 vessels, a decrease of 55 as compared with 1912. the aggregate tonnage being 22,833 tons, 2,067 tons less than in 1912. The number of vessel fishermen to whom bounty was paid is 5,679, a decrease of 468.

Bounty was also paid to 12,623 boats. and 21,557 boat fishermen. an increase of 625 boats and 1,146 men over 1912.

SESSIONAL PAPER No. 39

DETAILED STATEMENT of Fishing Bounty Claims received and paid during the year 1913.

Provinces.	Counties.	Number of claims.			
		Received.	Rejected.	Held in abeyance.	Paid.
Nova Scotia.............	Annapolis................	159	2	....	157
	Antigonish...............	172	....	....	172
	Cap Breton..............	458	....	....	458
	Cumberland.............	6	....	....	6
	Digby....................	429	2	....	427
	Guysborough.............	918	....	....	918
	Halifax.	1,228	3	....	1,225
	Inverness................	402	1	....	401
	Kings. :....	60	....	....	60
	Lunenburg...............	870	....	....	870
	Pictou...................	103	....	....	103
	Queens..................	185	....	....	185
	Richmond................	593	7	4	582
	Shelburne...............	751	1	....	750
	Victoria.................	341	....	....	341
	Yarmouth	287	....	....	287
	Totals.............	6,962	16	4	6,942
New Brunswick	Charlotte................	488	....	....	488
	Gloucester...............	388	3	....	385
	Kent....................	36	....	....	36
	Northumberland..........	14	....	....	14
	Restigouche	1	....	....	1
	St. John...............	33	....	....	33
	Totals.............	960	3	....	957
Prince Edward Island	Kings....................	558	1	....	* 564
	Prince...................	494	....	....	* 514
	Queens..................	117	....	....	117
	Totals.	1,169	1	....	1,195
Quebec..	Bonaventure..............	932	10	....	* 967
	Gaspé...................	2,761	3	....	* 2,844
	Rimouski	44	....	....	44
	Saguenay	584	....	....	584
	Totals	4,321	13	....	4,439
	Grand totals......	13,412	33	4	13,533

* Claims paid includes a number held over from previous year.

5 GEORGE V., A. 1915

DETAILED STATEMENT of Fishing Bounties paid to Vessels and Boats during the year 1913.

Provinces	Counties	Number of Vessels	Tonnage	Average Tonnage	Number of Men	Amount Paid	Number of Boats	Number of Men	Amount Paid	Total Bounty Paid to Vessels and Boats in 1913
Nova Scotia	Annapolis	6	253	42·16	48	574 60	151	251	1,142 45	1,717 05
	Antigonish	1	14	14·00	3	34 10	171	230	1,079 50	1,113 60
	Cape Breton	16	247	15·44	63	669 10	442	769	3,479 55	4,148 65
	Cumberland	1	20	20·00	1	26 70	5	11	48 45	75 15
	Digby	15	540	36·00	118	1,330 60	412	717	3,244 15	4,574 75
	Guysborough	49	739	15·08	223	2,233 10	869	1,261	6,244 95	8,478 05
	Halifax	88	1,782	20·25	425	4,629 50	1,137	1,550	7,259 50	11,889 00
	Inverness	36	520	14·44	170	1,659 00	365	670	3,011 50	4,670 50
	Kings						60	85	389 70	399 70
	Lunenburg	193	10,373	53·75	2,452	26,801 60	677	834	3,971 80	30,772 90
	Pictou						103	132	624 40	624 40
	Queens	13	178	13·69	36	410 20	172	291	1,321 45	1,740 65
	Richmond	32	824	25·76	194	2,152 40	550	914	4,160 30	6,257 70
	Shelburne	90	1,586	17·62	424	4,427 00	660	1,110	5,044 50	9,471 00
	Victoria	12	164	13·64	62	512 40	329	473	2,197 35	2,709 75
	Yarmouth	42	1,048	24·95	275	2,890 50	245	417	1,892 15	4,782 65
	Totals	594	18,288	30·78	4,464	48,334 80	6,348	9,816	45,121 20	93,456 00
New Brunswick	Charlotte	27	484	17·92	83	1,040 10	461	711	3,269 45	4,309 55
	Gloucester	234	3,209	13·71	939	9,561 10	151	357	1,561 15	11,062 25
	Kent	13	138	10·61	27	318 90	23	38	173 10	492 00
	Northumberland	9	96	10·67	30	230 00	5	11	48 45	278 45
	Restigouche	1	12	12·00	2	25 45				25 45
	St. John	1	31	31·00	3	51 10	32	34	166 30	217 40
	Totals	285	3,970	13·93	1,074	11,166 60	672	1,151	5,218 45	16,385 05
Prince Edward Island	Kings	12	297	24·75	47	611 90	552	744	3,490 80	4,102 70
	Prince	7	94	13·43	22	241 40	507	1,226	5,349 70	5,591 10
	Queens	6	75	12·50	22	222 40	111	267	1,165 65	1,388 05
	Totals	25	466	18·64	91	1,075 70	1,170	2,237	10,006 15	11,081 85

Quebec.....									
Bonaventure.........	6	......	......	......	......	967	1,685	7,622 75	7,622 75
Gaspé.............	......	109	18·16	30	310 00	2,838	5,542	24,729 90	25,039 90
Rimouski..........	......	......	......	......	......	44	66	304 70	394 70
Saguenay.........	......	......	......	......	......	584	1,000	4,771 00	4,771 00
Totals........	6	109	18·16	30	310 00	4,433	8,333	37,428 35	37,738 35
Grand totals......	910	22,883	25·09	5,679	60,887 10	12,623	21,557	97,774 15	158,661 25

5 GEORGE V., A. 1915

GENERAL STATISTICS.

The fishing bounty was first paid in 1882.

The payments were made each year on the following basis:

1882, vessels $2 per ton, one-half to the owner and the other half to the crew; boats at the rate of $5 per man, one-fifth to the owner and four-fifths to the men.

1883, vessels $2 per ton, and boats $2.50 per man, distributed as in 1882.

1884, vessels $2 per ton as in 1882 and 1883.

Boats from 14 to 18 feet keel, $1; from 18 to 25 feet keel, $1.50; from 25 feet upwards, $2. Boat fishermen, $3.

1885, 1886 and 1887, vessels, $2 per ton paid as formerly. Boats the same as in 1884, with the admission of boats measuring 13 feet keel, and fishermen, $3.

1888, vessels $1.50 per ton, paid as formerly. Boats, the same as 1885, 1886 and 1887.

1889, 1890 and 1891, vessels $1.50 per ton as in 1888. Boats $1 each. Boat fishermen, $3.

1892, vessels $3 per ton, paid as formerly. Boats $1 each. Boat fisherman $3.

1893, vessels $2.90 per ton, paid as formerly. Boats $1 each. Boat fisherman $3.

1894, vessels $2.70 per ton, paid as formerly. Boats $1 each. Boat fishermen $3.

1895, vessels $2.60 per ton, paid as formerly. Boats $1 each. Boat fishermen $3.

1896, vessels $1 per ton, which was paid to the owners, and vessel fishermen $5 each, clause No. 5 of the regulation having been amended accordingly. Boats $1 each, and boat fishermen $3.50 each.

	Vessels.		Men.		Boats.		Men.	
1897.	$1 00	per ton.	$6 00	each	$1 00	each.	$3 50	each
1898.	1 00	"	6 50	"	1 00	"	3 50	"
1899.	1 00	"	7 00	"	1 00	"	3 50	"
1900.	1 00	"	6 50	"	1 00	"	3 50	"
1901.	1 00	"	7 00	"	1 00	"	3 50	"
1902.	1 00		7 25	"	1 00	"	3 80	"
1903.	1 00	"	7 30	"	1 00	"	3 90	"
1904.	1 00	"	7 15	"	1 00	"	3 75	"
1905.	1 00	"	7 10	"	1 00	"	3 65	"
1906.	1 00	"	7 10	"	1 00	"	3 75	"
1907.	1 00		7 40	"	1 00	"	4 00	"
1908.	1 00	"	7 25	"	1 00	"	3 90	"
1909.	1 00	"	7 50	"	1 00	"	4 25	"
1910.	1 00		7 60	"	1 00	"	4 30	"
1911.	1 00	"	7 15	"	1 00	"	4 10	"
1912.	1 00	"	6 90	"	1 00	"	3 95	"
1913.	1 00		6 70	"	1 00	"	3 95	"

Since 1882, 27,052 vessels, totalling 867,105 tons, have received the bounty. The total number of vessel fishermen who received bounty is 195,324, being an average of 7.5960 per vessel.

The total number of boats to which bounty was paid since 1882 is 423,001, and the number of fishermen 756,690. Average number of men per boat 1.333,689.

The highest bounty paid per head to vessel fishermen was $21.75 in 1893; the lowest 83 cents, while the highest to boat fishermen was $4.30, the lowest $2.

SESSIONAL PAPER No. 39

COMPARATIVE STATEMENT by Provinces for the Year 1882 to 1913, inclusive, showing :
(1) Total number of fishing Bounty Claims received and paid by the Department
of Marine and Fisheries,

Year.	Nova Scotia.		New Brunswick.		P. E Island.		Quebec.		Totals.	
	Received.	Paid.	Received.	Paid.	Received.	Paid,	Received.	Paid.	Received.	Paid.
1882...	6,730	6,613	1,257	1,142	1,169	1,100	3,162	3,117	12,318	11,972
1883...	7,171	7,076	1,693	1,579	1,138	1,106	3,602	3,325	13,604	13,086
1884...	7,007	6,930	1,252	1,224	923	885	3,470	3,429	12,652	12,468
1885...	7,646	7,599	1,609	1,588	1,117	1,025	3,943	3,912	14,315	14,124
1886...	7,639	*7,702	1,767	1,763	1,131	1,080	4,275	*4,355	14,812	14,900
1887...	8,262	8,227	1,975	1,958	1,201	1,126	4,138	4,105	15,576	15,416
1888...	8,481	8,429	2,065	2,026	1,153	834	4,328	4,310	16,027	15,599
1889...	8,816	8,523	2,428	2,392	1,211	*1,511	4,664	4,652	17,119	17,078
1890...	9,337	*9,429	2,522	2,469	1,352	1,257	4,860	4,804	18.071	17,959
1891...	10,242	10,063	2,831	2,084	1,482	1,446	5,108	4,913	19,663	18,506
1892...	8,272	8,186	1,067	1,001	1,065	1,051	4,425	4,204	14,829	14,442
1893...	7,926	7,844	967	881	1,027	1,012	4,059	3,898	13,979	13,635
1894...	8,640	8,600	925	911	983	963	3,948	3,876	14,496	14,350
1895...	8,835	8,825	979	975	1,009	*1,025	3,904	*3,955	14,727	14,780
1896...	8,597	8,562	1,137	1,064	1,111	*1,120	4,366	4,229	15,211	14,975
1897...	8,450	8,418	1,042	991	1,175	1,171	4,180	4,149	14,847	14,729
1898...	8,446	8,347	934	917	1,143	*1,145	4,156	4,092	14,679	14,501
1899...	7,894	7,754	849	825	1,016	947	4,134	4,102	13,893	13,628
1900...	7,484	7,452	904	904	1,119	*1,169	4,264	4,251	13,771	13,776
1901...	7,346	7,344	829	826	941	937	4,277	4,267	13,393	13,374
1902...	6,710	6,671	802	794	913	912	4,371	4,346	12,796	12,723
1903...	6,297	6,284	832	830	978	974	4,110	4,090	12,217	12,178
1904...	6,750	6,732	879	866	1,027	994	4,095	4,079	12,751	12,671
1905...	7,034	7,018	881	873	921	921	4,350	4,329	13,186	13,141
1906...	7,434	7,415	930	923	918	916	4,251	4,249	13,533	13,503
1907...	7,124	7,087	904	895	1,000	984	4,239	4,227	13,267	13,193
1908...	7,690	7,648	1,002	988	1,030	993	4,250	4,212	13,972	13,841
1909..	7,276	7,250	834	830	877	872	4,024	4,004	13,011	12,956
1910...	6,670	6,659	915	903	900	898	4,159	4,150	12,644	12,610
1911..	6,735	6,722	923	905	1,001	877	4,220	4,141	12,879	12,645
1912..	6.717	6,709	904	890	1,052	*1,142	4,299	4,223	12,972	12,964
1913...	6,962	6,942	960	957	1,169	*1,195	4,321	*4,439	13,412	13,533
Totals..	246,620	245,060	39,798	38,174	34,252	35,588	133,952	132,434	454,622	449,256

* Includes a number of claims held over from previous year.

MARINE AND FISHERIES

(2) NUMBER of vessels, tonnage and number of men who received Bounty in each year.

Year.	Nova Scotia.			New Brunswick.			P. E. Island.			Quebec.			Totals.		
	No. of Vessels.	Tonnage.	No. of Men.	No. of Vessels.	Tonnage.	No. of Men.	No. of Vessels.	Tonnage.	No. of Men.	No. of Vessels.	Tonnage.	No. of Men.	No. of Vessels.	Tonnage.	No. of Men.
1882....	588	22,841	5,343	120	2,171	531	15	389	74	63	2,210	538	786	27,611	6,486
1883....	700	29,788	6,238	126	2,102	496	16	450	66	62	2,236	443	904	34,576	7,243
1884....	700	29,828	6,327	139	2,289	560	16	582	92	56	1,965	382	911	34,664	7,361
1885....	629	27,709	5,897	128	2,120	496	19	597	113	55	1,791	317	831	32,217	6,823
1886....	562	25,375	5,022	145	2.628	520	32	1,071	215	52	1,730	320	791	30,804	6,077
1887....	566	24,520	4,900	154	2,889	563	38	1,677	338	54	1,883	334	812	30,969	6,135
1888....	589	26,008	5,450	150	2,545	544	37	1,245	249	51	1,842	388	827	31,640	6,631
1889....	597	27,123	5,684	153	2,590	565	35	1,274	239	48	1,729	330	833	32,716	6,818
1890....	540	23,955	4,935	133	2,129	447	32	1,002	203	34	1,182	220	739	28,268	5,805
1891. ..	527	22,780	4,618	124	2,051	411	27	778	155	27	924	168	705	26,533	5,352
1892....	507	22,279	4,611	108	1,683	343	30	983	139	23	803	159	668	25,748	5,252
1893....	536	23,195	4,780	210	2,922	634	27	910	151	32	952	179	805	27,979	5,744
1894....	602	24,735	5,077	238	3,189	721	21	594	114	38	1,066	178	899	29,584	6,090
1895....	603	25,018	5,184	238	3,107	764	27	769	129	39	1,262	173	907	30,156	6,250
1896....	553	23,415	4,607	250	3,337	800	23	656	114	36	1,143	144	862	28,551	5,665
1897....	507	21,323	4,829	239	3,079	816	20	490	109	24	833	116	790	25,725	5,870
1898....	505	20,868	4,840	239	3,155	859	24	561	125	16	524	77	784	25,108	5,901
1899....	519	22,538	5,323	238	3,131	885	15	373	76	17	497	78	789	26,539	6,362
1900....	525	22,474	5,352	234	2,969	890	29	737	153	14	459	76	802	26,639	6,471
1901....	508	21,469	5,158	242	3,229	872	23	541	115	13	366	69	786	25,605	6,214
1902 ..	505	21,248	5,126	249	3,293	972	28	630	135	13	350	51	795	25,521	6,284
1903....	546	21,992	5,173	259	3,454	971	36	765	169	10	290	48	851	26,501	6,361
1904....	552	21,285	5,040	257	3,429	981	30	594	126	15	382	73	854	25,690	6,220
1905....	620	21,240	5,238	264	3,600	1,035	28	587	125	10	259	56	922	25,686	6,454
1906 ...	644	20,008	4,891	273	3,753	1,066	32	732	147	8	139	33	957	24,632	6,137
1907 ...	612	17,041	4,178	265	3,720	1,010	41	916	178	9	154	34	927	21,831	5,400
1908....	616	17,804	4,364	269	3,672	1,034	34	643	140	6	87	25	925	22,206	5,563
1909....	591	16,180	3,919	247	3,344	935	30	572	113	6	99	26	874	20,195	4,993
1910 ..	588	17,567	4,294	249	3,321	976	31	612	117	8	178	37	876	21,678	5,424
1911....	664	19,555	4,931	266	3,528	1,025	27	540	115	8	177	41	965	23,800	6,112
1912...	668	20,649	4,983	255	3.336	987	33	648	131	9	267	46	965	24,900	6,147
1913....	594	18.288	4,484	285	3,970	1,074	25	466	91	6	109	30	910	22,833	5,679
Totals..	18,563	720,098	160,796	6,746	95,735	34,783	881	23,384	4,556	862	27,888	5,189	27,052	867,105	195,324

(3) NUMBER of Boats and Boat Fishermen who received Bounty in each year.

Year.	Nova Scotia.		New Brunswick.		Prince Edward Island.		Quebec.		Totals.	
	No. of Boats.	No. of Men.	No. of Boats.	No. of Men.	No. of Boats.	No. of Men.	No. of Boats.	No. of Men.	No. of Boats.	No. of Men.
1882	6,043	12,130	1,024	2,530	1,087	3,070	3,071	5,716	11,225	23,446
1883........	6,458	13,553	1,453	3,309	1,098	3,106	3,266	6,188	12,275	26,156
1884.	6,257	12,669	1,086	2,505	869	2,346	3,344	6,416	11,556	23,936
1885..........	6,970	13,396	1,460	3,254	1,006	2,606	3,857	7,485	13,293	26,741
1886..........	7,140	13,351	1,618	3,567	1,048	2,547	4,303	7,981	14,109	27,446
1887........	7,662	13,997	1,804	3,994	1,088	2,711	4,051	7,550	14,605	28,252
1888........	7,840	14,115	1,876	4,148	797	2,141	4,259	7,852	14,772	28,256
1889..........	7,926	14,118	2,237	5,032	1,475	3,568	4,602	8,807	16,240	31,525
1890.........	8,886	15,738	2,324	5,242	1,192	3,024	4,766	9,241	17,168	33,245
1891	9,525	16,552	1,928	4,126	1,383	3,427	4,865	9,402	17,701	33,507
1892	7,679	12,307	893	1,765	1,021	2,047	4,181	7,693	13,774	23,812
1893..........	7,308	11,748	671	1,314	985	1,962	3,866	7,245	12,830	22,269
1894..,....	7,956	12,899	661	1,281	913	1,813	3,821	7,139	13,351	23,132
1895.....	8,222	13,106	737	1,434	998	2,141	3,916	7,877	13,873	24,558
1896.	8,008	12,454	814	1,553	1,095	2,126	4,189	7,688	14,106	23,821
1897..........	7,911	12,542	752	1,351	1,151	2,147	4,125	7,572	13,939	23,612
1898........	7,872	12,438	678	1,237	1,121	2,199	4,076	7,627	13,747	23,501
1899......... .	7,235	11,305	587	1,027	932	1,710	4,085	7,696	12,839	21,738
1900..........	6,927	10,645	670	1,184	1,140	2,198	4,237	8,004	12,974	22,031
1901....... ...	6,836	10,464	584	1,001	914	1,735	4,254	8,017	12,588	21,217
1902....... ...	6,166	9,442	545	966	884	1,638	4,333	8,180	11,928	20,226
1903.....	5,738	8,775	571	964	938	1,722	4,080	7,688	11,327	19,149
1904.....	6,180	9,556	609	1,082	964	1,792	4,064	7,648	11,817	20,078
1905..........	6,398	9,822	609	1,047	893	1,630	4,319	8,002	12,219	20,501
1906.	6,771	10,138	650	1,139	884	1,648	4,241	7,946	12,546	20,871
1907....... ...	6,475	9,739	630	1 158	943	1,750	4,218	7,873	12,266	20,520
1908..........	7,032	10,685	719	1,365	959	1,810	4,206	7,809	12,916	21,669
1909..........	6,659	10,163	583	1,069	842	1,583	3,998	7,314	12,082	20,129
1910..........	6,071	9,353	654	1,195	867	1,672	4,142	7,451	11,734	19,671
1911...	6,058	9,403	639	1,048	850	1,574	4,133	7,682	11,680	19,707
1912..........	6,040	9,324	635	1,096	1,109	2,131	4,214	7,860	11,998	20,411
1913.....	6,348	9,816	672	1,151	1,170	2,237	4,433	8,353	12,623	21,557
Totals.....	226,597	375,743	31,373	64,134	32,616	69,801	131,515	247,002	422,101	756,690

5 GEORGE V., A. 1915

(4) TOTAL Number of Men who received Bounty in each year.

Year.	Nova Scotia. No. of Men.	New Brunswick. No. of Men.	P. E. Island. No. of Men.	Quebec. No. of Men.	Totals.
1882.	17,473	3,061	3,144	6,254	29,932
1883.	19,791	3,805	3,172	6,631	33,399
1884	18,996	3,065	2,438	6,798	31,297
1885.	19,293	3,750	2,719	7,802	33,564
1886	18,373	4,087	2,762	8,301	33,523
1887.	18,897	4,557	3,049	7,884	34,387
1888	19,565	4,692	2,390	8,240	34,887
1889	19,802	5,597	3,807	9,137	38,343
1890	20,673	5,689	3,227	9,461	39,050
1891	21,170	4,537	3,582	9,570	38,859
1892.	16,918	2,108	2,186	7,852	29,064
1893.	16,528	1,948	2,113	7,424	28,013
1894	17,976	2,002	1,927	7,317	29,222
1895	18,290	2,198	2,270	8,050	30,808
1896.	17,061	2,353	2,240	7,832	29,486
1897.	17,371	2,167	2,256	7,688	29,482
1898	17,278	2,096	2,324	7,704	29,402
1899.	16,628	1,912	1,786	7,774	28,100
1900	15,997	2,074	2,351	8,080	28,502
1901	15,622	1,873	1,850	8,086	27,431
1902	14,568	1,938	1,773	8,231	26,510
1903	13,948	1,935	1,891	7,736	25,510
1904.	14,596	2,063	1,918	7,721	26,298
1905.	15,060	2,082	1,755	8,058	26,955
1906	15,029	2,205	1,795	7,979	27,008
1907.	13,917	2,168	1,928	7,907	25,920
1908.	15,049	2,399	1,950	7,834	27,232
1909.	14,082	2,004	1,696	7,340	25,122
1910	13,547	2,171	1,789	7,488	25,095
1911.	14,331	2,073	1,689	7.723	25,819
1912.	14,307	2,083	2,262	7,906	26,558
1913.	14,300	2,225	2,328	8,383	27,236
Totals.	536,439	88,917	74,367	252,191	952,014

SESSIONAL PAPER No. 39

(5) TOTAL annual payments of fishing Bounty.

Year.	Nova Scotia.	New Brunswick.	P. E. Island.	Quebec.	Totals.
	$ cts.	$ cts.	$ cts.	$ cts.	$ cts.
1882.............	106,098 72	16,997 00	16,137 00	33,052 75	172,285 47
1883.........	89,432 50	12,395 20	8,577 14	19,940 01	130,344 85
1884.......... ...	104,934 09	13,576 0C	9,203 96	28,004 93	155,718 98
1885.............	103,999 73	15,908 25	10,166 65	31,464 76	161,539 39
1886	98,789 54	17,894 57	10,935 87	33,283 61	160,903 59
1887	99,622 03	19,699 65	12,528 51	31,907 73	163,757 92
1888	89,778 90	18,454 92	9,092 96	32,858 75	150,185 53
1889............	90,142 51	21,026 79	13,994 53	33,362 71	158,526 54
1890	91,235 64	21,108 33	11,686 32	34,210 72	158,241 01
1891.............	92,377 42	17,235 96	12,771 30	34,507 17	156,891 85
1892.............	100,410 39	10,864 61	9,782 79	29,694 35	159,752 14
1893	108,060 67	12,524 09	9,328 62	28,320 72	158,234 10
1894	111,460 03	12,690 80	7,875 79	28,040 18	160,066 80
1895.............	110,765 27	12,919 32	9,285 13	30,598 27	163,567 99
1896.........	98,048 95	13,602 88	9,745 50	32,992 44	154,389 77
1897.............	102,083 50	13,454 50	9,809 00	32,157 00	157,504 00
1898	103,730 00	13,746 00	10,188 00	31,795 00	159,459 00
1899	106,598 50	13,514 50	7,822 00	32,065 00	160,000 00
1900.............	101,448 00	13,562 50	10,589 00	33,203 00	158,802 50
1901.............	101,024 50	13,420 50	8,335 50	33,161 50	155,942 00
1902............	100,455 70	14,555 80	8,716 55	36,125 45	159,853 50
1903.....:........	99,714 15	14,872 75	9,652 50	34,704 30	158,943 70
1904.........	99,286 44	15,110 80	9,179 35	33,651 65	157,228 24
1905..	100,664 35	15,379 50	8,317 20	34,185 60	158,546 65
1906	99,518 80	16,247 55	8,839 40	34,410 00	159,015 75
1907	93,381 70	16,454 50	10,175 95	36,102 35	156,114 50
1908.............	98,156 20	17,203 75	9,708 90	34,931 05	159,999 90
1909	95,413 60	15,480 15	8,973 85	35,354 25	155,221 85
1910.............	96,468 20	16,531 05	9,557 80	36,609 70	159,166 75
1911..............	99,424 90	15,795 00	8,669 85	36,109 95	159,999 70
1912.............	97,904 25	15,109 75	11,119 00	35,863 40	159,996 40
1913...........	93,456 00	16,385 05	11,081 85	37,738 35	158,661 25
Totals......	3,102,885 18	493,722 02	321,847 77	1,050,406 65	5,058,861 62

5 GEORGE V., A. 1915

LIST of Vessels which received Fishing Bounty during the year 1913-14.
Province of Nova Scotia.

ANNAPOLIS COUNTY.

Official Number.	Name of Vessel.	Port of Registry.	Tonnage.	Name of Owner or Managing Owner.	Residence.	No. of Crew paid.	Amount of Bounty paid.
							$ cts.
121818	Albert J. Lutz....	Digby........	95	John D. Apt.........	Port Wade......	19	207 30
77740	Elmer	"	15	David Hayden......	"		15 00
80803	Exenia..........	Windsor......	18	Fred Longmire......	Hillsburn......	7	64 90
126873	Myrtle L......	Digby........	47	B. Longmire.......	"	13	134 10
94832	Venus..........	Weymouth.....	42	Jno. W. Snow	Port Wade	3	62 10
121812	Wilfred L. Snow..	Digby.	51	Abraham Holmes....	"	1	57 70

ANTIGONISH COUNTY.

111794	Volunteer	Pt. Hawkesbury.	14	John Brow..	Hr. au Bouche...	3	34 10

CAPE BRETON COUNTY.

112376	Agnes.....	Arichat........	15	William Martell.....	Main-à-Dieu	3	35 10
103858	B. & B. Holland..	Halifax.........	26	John Stacey.	Glace Bay.......	5	59 50
126561	Caberfeidgh....	Sydney........	12	Alex. McDonald.....	Alder Point	5	45 50
122188	Charles A. H.....	Arichat.......	10	Samuel Chislet......	Nth Sydney	3	30 10
116348	Florence M.	"	16	Robert Fudge..... .	"	3	36 10
116883	Grayling...	"	25	T. & W. Moulton. ...	"	4	51 80
126569	Madona May. ...	Sydney.....	16	James Bonar........	Glace Bay........	4	42 80
116915	Maggie and Esther	Pictou........	11	C. L. Miller........	"	2	24 40
117144	Mary E. Faulkner.	Halifax........	14	Angus Nicholson.....	Nth Sydney	4	40 80
107999	Maud S	Canso	12	Jacob Rogers........	"	3	32 10
83104	Minnie Long.	Richibucto......	19	Samuel Gilmot......	Glace-Bay........	4	45 80
115392	Nyanza..........	Sydney.......	15	Geo. Herridge.......	Nth Sydney.....	5	48 50
111799	Rosie G	Pt Hawkesbury.	16	John Gallant.......	Little Lorraine. ..	5	49 50
111902	St. Thomas......	Arichat........	10	Henry Kelly.........	Lingan........	4	36 80
112386	Shamrock.......	Sydney......	11	Andrew Cann.......	Nth Sydney	4	37 80
122184	Two Brothers.....	Arichat.......	19	Patrick Campbell....	Main-à-Dieu......	5	52 50

CUMBERLAND COUNTY.

116687	Myrtle Mac......	Charlottetown. ..	20	John D. McLeod	Tidnish	1	26 70

DIGBY COUNTY.

107603	Augusta Evelyn..	St. John......	31	Horace Thurber......	Freeport........	10	98 00
122145	Cerita	Yarmouth......	10	Luke C. Deveau....	Salmon River....	2	23 40
116236	Cora May	Digby.... ..	64	Chas. F. Finigan.....	Freeport....	16	171 20
126879	Dorothy G. Snow..	"	98	Jos E. Snow........	Digby........ ..	4	106 80
126874	Dorothy M. Smart	"	94	Howard Anderson....	"	16	187 20
121883	Fanny Rose	Yarmouth.. ...;	15	F. J. Doucette.......	Cape St. Mary's ..	3	35 10
122097	George L..........	"	13	Peter LeBlanc.......	Salmon River....	1	19 70
126889	Gyno...........	Digby.......	11	Edward Thomas......	Westport.......	3	31 10
111838	Lavinia D........	"	21	James Doucette	Cape St. Mary's ..	5	54 50
121816	Loren B. Snow...	"	85	Jos. E. Snow........	Digby.........	23	234 10
116660	Nora..........	Yarmouth......	11	P. Doucette........	Mavillette........	4	37 80
111835	Roxana	Digby.........	11	F. B. Comeau........	Meteghan River ..	2	24 40
100609	Swan	"	56	Edwin Hains........	Freeport........	14	149 80
121659	Viola	Yarmouth... ..!	10.	Alex. Frontain. ...	Cape St. Mary's..	3	30 10
122049	Waldo R........	St. Andrews....	47	Jos. A. Robichaud....	Meteghan Centre.	1	53 70

SESSIONAL PAPER No. 39

LIST of Vessels which received Fishing Bounty, etc.—Nova Scotia—*Continued.*

GUYSBORO COUNTY.

Official Number.	Name of Vessel.	Port of Registry.	Tonnage.	Name of Owner. or Managing Owner.	Residence.	No. of Crew paid.	Amount of Bounty paid.
							$ cts.
121700	Agnes E.........	Yarmouth......	10	Simon Horne Jr......	Dover	5	43 50
122302	Albata..........	Lunenburg.....	20	F. H. Hawes........	Canso......... ..	5	53 50
116344	Annie B. M......	Arichat...	18	Thomas Fanning.....	" ....;	5	51 50
122185	Beatrice.........	"	11	Geo. Hendsbee.......	Half Isld Cove....	3	31 10
112016	Blanche...	Canso........	13	Mark Richard...	Charlos Cove.....	5	46 50
112375	C. G. Munroe....	"	14	Vincent Richard.....	"	5	47 50
117060	Dorothy Aleta....	"	11	Daniel Pitts........	. "	6	51 20
126112	Dorothy G.......	Lunenburg.	17	Claude Rhynold:	Canso............	5	50 50
117054	Emma Jane.....	Canso.........	16	Jno. George	White Head.....	5	49 50
122010	Ena T..........	Lunenburg. ...	16	Robert Mosher.	Canso...........	6	56 20
117093	Florence D......	Arichat.......	11	Robt. Creamer......	Philips Harbour...	3	31 10
107993	Florence May....	Canso	10	Jno. Kennedy........	Canso...........	4	36 80
112373	Flying Cloud....	Arichat.......	13	Jas. Mannett........	Larry's River....	4	39 80
107996	Green Linnet....	Canso.	12	Felix Sampson.	Dover........	7	58 90
126297	H. C. R.........	"	18	Harry Kavanagh....	Canso...........	5	51 50
122430	Hattie Maud.....	Halifax........	10	John J. Berrigan....	"	5	49 50
126294	Horman Lee......	Canso.........	17	Edward Kavanagh...	"	5	50 50
108470	Ida M. Burke. ..	Arichat...	16	Jos. Fougere.......	Larry's River....	2	29 40
126292	Irbessa.........	Canso........	17	Edward Hearn......	Canso	4	43 80
112374	J. B. Saint	Arichat.......	18	Samuel Snow......	White Head .	3	38 10
116747	Jessie W........	Halifax.......	12	Jacob Manuel.....	Canso·.........	5	45 50
111910	Lizzie J. Greenleaf	Arichat.......	11	Jos. H. Richard	Charlo's Cove....	5	44 50
117097	Lizzie May......	"	12	W. C. Richard......	"	7	58 90
117100	Louisa Ellen ...	"	11	Angus Feltmate.....	White Head.....	3	31 10
126291	Marg. Katheleen..	Canso..........	16	Jno. Boudroit.......	Dover	5	49 50
111909	Margaret May....	Arichat.......	12	Stephen C. Richard ..	Charlo's Cove....	5	45 50
112379	Mary S.......	"	18	A. D. Feltmate......	Canso	6	58 20
126295	Mary W.Catherine	Canso........	13	Wm. Pelrine.......	Port Félix........	5	46 50
107757	Mayflower........	Charlottetown..	17	Jas. Lumsden..	Canso...........	4	43 80
100450	Minto	Canso........	18	Henry A. Richard ...	Charlo's Cove....	7	64 90
126296	Murray R. Munroe	"	21	Thurlo Munroe	White Head	4	47 80
116500	Oreda..........	Lunenburg.	16	Abner Munroe.....	Cole Harbour....	3	36 10
126298	Petawawa........	Canso........	32	Frank C. Lohnes....	Canso..........	6	72 20
112024	Reta S........	"	13	Wm. Shrader.......	"	4	39 80
108000	St. Patrick.......	"	18	Geo. L. Avery......	Larry's River....	4	44 80
126472	Shiloh....	Halifax..... .	22	Chas. A. Mosher.....	Canso.........	5	55 50
112023	Silver Bell	Canso........	14	Simon J. Pelrine....	Larry's River....	6	54 20
116884	Silver Swan.....	Arichat........	20	Chas. Richard......	Charlo's Cove....	6	60 20
112025	Squanto........	Halifax........	13	Freeman Casey	"	4	39 80
122317	Stanley Hubley...	Lunenburg.	17	Jas. J. Lukeman.....	Hazel Hill.......	6	57 20
116885	T. Lilly........	Arichat.......	10	Levi W. Ehler.......	Queensport.......	5	43 50
117055	Thelma..........	Canso..... ...	15	Geo. Ryan.........	Canso........	3	35 10
116532	Togo	Lunenburg.....	14	Wm. J. Peitzsch.....	White Head	3	34 10
107994	True Love........	Canso........	10	David Walsh	Canso...........	2	23 40
130351	Vennie May.. ...	Arichat.......	17	Thos. L. Richard....	Charlo's Cove....	3	37 10
116887	Wenona...	"	10	Wesley Munroe	White Head.....	5	43 50
126293	Winnie May ...	Canso........	10	Geo. C. Jamieson....	Cole Harbour....	2	23 40
130721	Winnifred Marr .	Lunenburg....	17	Martin Meagher.	Canso..........	4	43 80
122000	Zoraya.....	"	16	Louden Munroe......	White Head....	3	36 10

HALIFAX COUNTY.

94632	A. C. Greenwood..	Shelburne	15	John Beaver........	Spry Bay.........	3	35 10
130596	A. Hubley........	Halifax	69	Ainsley Hubley....	Hackett's Cove....	18	189 60
126812	Adana C..........	"	17	Wm. Hubley, Sr....	Spry Bay........	3	37 10
116526	Adelaide.	Lunenburg.....	13	J. Francis Gray......	Pennant	3	33 10
130952	Adamantine.... .	"	10	Harris Levy.........	Hackett's Cove ...	2	23 40
133802	Adonia S	"	18	David Slaunwhite....	Terence Bay.....	3	38 10

5 GEORGE V., A. 1915

List of Vessels which received Fishing Bounty, etc.—Nova Scotia—*Continued.*

HALIFAX COUNTY—*Continued.*

Official Number	Name of Vessel.	Port of Registry.	Tonnage.	Name of Owner. or Managing Owner.	Residence.	No. of Crew. paid.	Amount of Bounty paid.
							$ cts.
130591	Aileen Gladys	Halifax	16	Geo. E. Siteman et al.	Ship Harbour	4	42 80
130578	Alice M. C.	Lunenburg	12	Creighton Covey	Indian Harbour	1	18 70
130960	Alvin S.	"	27	J. Foster Rood	Halifax	8	80 60
122422	Annie G. W.	Halifax	17	Edward Markie	Sober Island	4	43 80
126380	Annie Hilton	"	10	John May	Owls Head	5	43 50
133665	Arena	"	12	Edwd. S. Marryatt	Pennant	3	32 10
126196	Bonnie B.	Lunenburg	19	Francis B. Martin	Ketch Harbour	4	45 80
130571	Brenda C.	"	11	Harold Harrie	Terence, Bay	4	37 80
130074	C. L. Miller.	"	10	A. Zinck	West Dover	3	30 10
130954	Comet G.	"	11	Herbert Little	Terence Bay	3	31 10
126033	D. C. Mullhall	Halifax	42	Geo. Pelham	Herring Cove	14	135 80
111428	Duchess	"	12	James Morash	West Dover	3	32 10
130585	Edith Adele	"	33	John C. Martin	Ketch Harbour	4	59 80
130568	Ella M. Young	Lunenburg	12	Maynard Young	West Dover	3	32 10
122424	Ella May	Halifax	57	Leander Hubley	Halifax	10	124 00
96726	Ellen Maud	"	16	Richard Drew	Terence Bay	3	36 10
111434	Ermynthrude	"	36	Fred J. Durrach	Herring Cove	3	56 10
117141	Etha May	"	11	Geo. Johnson	West Dover	3	31 10
130565	Ethel M. G.	Lunenburg	11	Arthur Johnson	Indian Harbour	2	24 40
130687	Eva E. L	"	11	Manuel Morash	West Dover	3	31 10
133668	F. C. Twohig	Halifax	10	Andrew Twohig	Pennant	3	30 10
100247	Fairy Queen	"	11	Geo. H. Nickerson	"	4	37 80
116290	Flora M. J.	"	78	Jas. Julien et al.	Grand Desert	15	178 50
106259	Florence G.	"	15	Caleb Gray	Sambro	3	35 10
130738	Frances Lenore	"	12	Russell Garrison	Indian Harbour	3	32 10
122282	G. M. Stephens	Shelburne	12	Lindsay Zwicker	Indian Harbour	3	32 10
130584	Gladys E. B	Halifax	24	Walter Brown	Herring Cove	5	57 50
111432	Gladys Elena	"	16	Chas. Twohig	Pennant	4	42 80
126817	Gladys G. Hart	"	27	Jas. L. Hart	Sambro	7	73 90
130945	Gladys Irena	Lunenburg	16	Wm. L. Smith	Terence Bay	5	49 50
116731	Grand Desert	Halifax	65	Martin Julien et al.	Grand Desert	16	172 20
116738	Gretta	"	14	Edward Drake	Clam Harbour	4	40 80
116287	Handy Andy	"	15	Jno. P. Westhaver	Sheet Har. Passage	3	35 10
112129	Hattie	Lunenburg	12	Raymond Beck	East Dover	4	38 80
130472	Hattie M. J	"	12	Richard Coolen	"	3	32 10
126374	Hazel Levy	Halifax	14	Cyrus Levy	Owls Head	4	40 80
100544	Helen Maud	"	26	Howard Jennex	East Jeadore	9	86 30
116740	Hilda M. Horton	"	29	Jas. Westhaver	Sober Island	4	55 80
131072	Howker	"	12	Chas. H. Thomas	Herring Cove	3	32 10
130594	I Wonder Y	"	16	Wm. S. Henneberry.	Sambro	5	49 50
126373	Ideal	"	16	Chas. W. Schnare	Pennant	5	49 50
130577	Irene L	Lunenburg	11	Wm. C. Slaunwhite	Terence Bay	2	24 40
126825	Joseph Earle	"	29	Alex. Slaunwhite	"	7	75 90
126136	Kathleen W	Halifax	22	Robt. J. Slaunwhite	"	12	102 40
111404	Kimberley	Lunenburg	92	Seybert Coplen	Hubbard's Cove	14	173 80
126915	Lola B.	Halifax	10	C. W. Boutillier	Spry Bay	3	30 10
131078	Lola R	Lunenburg	13	Jas. V. Reno	Herring Cove	2	26 40
126132	Lottie V. M.	Halifax	10	Isaac Morash	West Dover	3	30 10
131075	Margaret E.	Lunenburg	11	Neil Flemming	Ketch Harbour	1	17 10
130592	Margaret M. Gray	Halifax	23	Angus Gray	Pennant	5	56 50
133667	Marjory N	"	11	Harry W. Nickerson.	"	3	31 10
130595	Marona	"	25	Edwd. Parker et al.	Owls Head	4	51 80
85964	Mary E	"	14	H. Zinck	West Dover	3	34 10
133669	Mary K	"	12	Wm. H. Henneberry.	Eastern Passage	3	32 10
131071	Mary Maude	Lunenburg	10	Geo. Johnson	West Dover	2	23 40
131064	Mattapex	"	12	Chas. Scott	Indian Harbour	3	32 10
130821	Mianus	"	15	D. M. Duggan	East Dover	3	35 10
103539	Neva	Halifax	11	Hiram Marryatt	Pennant	3	31 10
131167	Ovila	Lunenburg	23	G. Henneberry	Sambro	2	36 40
107317	Pearl	Halifax	30	Lewis Murphy	East Ship Harb.	3	50 10
130727	Pearl Beatrice H.	Lunenburg	32	Wm. Hubley	Indian Harbour	7	78 90

SESSIONAL PAPER No. 39

List of Vessels which received Fishing Bounty, etc.—Nova Scotia—*Continued.*

HALIFAX COUNTY—*Concluded.*

Official Number.	Name of Vessel.	Port of Registry.	Tonnage.	Name of Owner or Managing Owner.	Residence.	Number of Crew paid.	Amount of Bounty paid.
							8 cts.
116745	Perseverance......	Halifax.........	12	Alfred Boutilier......	Indian Harbour...	4	38 80
130563	Phoebe M.......	Lunenburg......	12	David Morash.......	West Dover.......	2	25 40
131076	Plymouth Rock...	"	24	Otis Scott.	East Dover.......	5	57 50
116749	Reliance.........	Halifax.........	14	Jas. Howard.......	Terence Bay......	7	60 90
126823	Rosie L	Lunenburg.....	20	Geo. Little.........	"	5	53 50
122307	Sadie H....... ..	"	17	Chas. Beaver.......	Harrigan Cove....	3	37 10
130958	Shianne	"	21	Lubin Duggan......	East Dover.......	4	47 80
130722	Tacoma....	Halifax.......	11	A. J. Wambolt..	Indian Harbour..	2	24 40
130949	Titus McLeod....	Lunenburg....	11	Jas. Berringer.......	West Dover......	3	31 10
133661	Una E. Hart....	Halifax........	21	Jas. L. Hart....... ..	Sambro	6	61 20
122429	Uncas..	"	11	Mark Nickerson.....	"	5	44 50
131171	Valerie S.........	Lunenburg.....	17	J. Slaunwhite........	Terence Bay.....	5	50 50
117142	Valkyria...	Halifax........	13	David Levy........	Sober Island.....	4	39 80
130686	Vera May........	Lunenburg.....	22	Joel Zinck........	Halifax........	5	55 50
133666	Village Leaf... ..	Halifax........	78	John E Wolfe et al..	Grand Desert	18	198 60
126912	Viola G. Hartlin..	"	25	Peter Hartlin........	East Jeddore.....	9	85 30
126917	Violet C..	"	14	Jas. H. Smith.......	Sambro........	4	40 80
130566	Violet F.........	Lunenburg.....	12	W. Frederick..	Indian Harbour...	3	32 10
116283	Vixen	Halifax........	15	Henry MacKenzie...	Gerrard's Island .	2	28 40
126478	Willetta....	"	15	Jos. Gray...........	Sambro...... ...	6	55 20
130600	Willie Roy........	"	13	Andrew Sullivan.....	Herring Cove.....	3	33 10

INVERNESS COUNTY.

Official Number.	Name of Vessel.	Port of Registry.	Tonnage.	Name of Owner or Managing Owner.	Residence.	Number of Crew paid.	Amount of Bounty paid.
96778	Campania	Pt. Hawkesbury	11	Robin, Jones & Whitman.............	Eastern Harbour..	4	37 80
126575	Cheticamp.......	" ..	10	Leonie Chiasson	"	4	36 80
103325	Elizabeth Ann....	"	11	David Bourgeois.....	"	4	37 80
130781	Flora Matthews..	"	16	Anselme Cormier....	Point Cross.... ..	5	49 50
122004	Florence B.... ...	Lunenburg	46	Robin, Jones & Whitman...	Eastern Harbour..	7	92 90
103317	Flying Star.	Pt. Hawkesbury	11	Simon Bellefontaine..	"	4	37 80
126873	Great Dipper. ...	"	10	David R. Doucett....	Grand Etang......	4	36 80
126577	Gros Ours.......	"	14	Emillien LeBlanc....	"	5	47 50
126579	Hattie L. B	"	12	Wm. Desveau......	Eastern Harbour..	5	45 50
126578	Hennepin........	"	12	Jos. M. Cormier......	Grand Etang......	5	45 50
130785	J. S. M	Pt. Hawkesbury	16	John S. Muise	Cape Rouge	6	56 20
130782	Karina II........	" ..	21	Lubin S. Chiasson....	Little River	8	74 60
126101	Lantana	Lunenburg......	17	Robin, Jones & Whitman.............	Eastern Harbour..	6	57 20
103316	Laura	Pt. Hawkesbury	10	" "	"	4	36 80
126574	Laurent Aucoin...	"	10	Louis L. Aucoin....	Point Cross......	3	30 10
103315	Lillie	"	12	Matthews & Scott...	Eastern Harbour .	4	38 80
96775	Louise...........	"	11	Simon Bellefontaine..	"	4	37 80
103330	Lucy.	" ..	11	Robin, Jones & Whitman.............	"	4	37 80
126576	M.C.G. Boudreau.	"	22	Simon Bellefontaine..	"	7	68 90
126104	M. Unity.	Lunenburg.....	26	Robin, Jones & Whitman	"	5	59 50
117056	Margaret.........	Canso....	16	Matthews & Scott....	"	5	49 50
96771	Marie	Pt Hawkesbury.	10	Robin, Jones & Whitman.............	"	5	43 50
96777	Marie Joseph.....	" ..	11	" "	"	3	31 10
103314	Mary........ ..	"	10	Wm. R. Doucet......	Grand Etang......	4	36 80
111797	Mermaid..........	"	13	Thomas Harris.....	Plateau	4	39 80
103326	Mizpah...........	"	10	Thos. LeBrun.....	Grand Etang......	4	36 80
126580	Paul V....	"	14	Robin, Jones & Whitman.............	Eastern Harbour..	4	40 80

5 GEORGE V., A. 1915

LIST of Vessels which received Fishing Bounty, etc.—Nova Scotia—*Continued.*

INVERNESS COUNTY—*Concluded.*

Official number.	Name of Vessel.	Port of Registry.	Tonnage.	Name of Owner or Managing Owner.	Residence.	No. of Crew paid.	Amount of Bounty paid.
							$ c.
122128	Reliance.........	Halifax.......	18	Robin, Jones & Whitman..............	Eastern Harbour..	6	58 20
130786	St. Clements.. ...	Pt Hawkesbury.	12	"	"	5	45 50
111792	Saint Aubin	"	15	"	"	5	48 50
116889	Saint Dominique..	Arichat	21	Chas. A. Smith	Port Hood Island.	6	61 20
103329	Saint Helier.	Pt Hawkesbury.	12	Robin, Jones & Whitman..............	Eastern Harbour..	4	38 80
122238	Violet & Annie...	Halifax..... ...	12	"	"	5	45 50
96773	Virgin	Pt Hawkesbury.	10	"	"	4	36 80
126571	Warbler.........	"	10	"	"	4	36 80
130783	Zambuck........	"	17	"	"	4	43 80

LUNENBURG COUNTY.

130466	A. G. Eisnor	Lunenburg	96	Willis A. Ernst......	Mahone Bay......	18	200 60
130675	A. L. Conrad. ...	"	11	Albert Conrad. . .	Rose Bay........	3	31 10
130947	Abacena..........	"	88	J. W. Sarty...	Pleasantville......	21	220 70
131173	Accrescent..	"	11	Chas. Mason.	Eastern Points...	3	31 10
130739	Ada M. Westhaver	"	100	E. F. Zwicker........	Lunenburg.......	19	207 30
111641	Aguadilla.......	"	100	William Arenburg....	"	22	227 40
130790	Albert A. Young..	"	92	Jacob Hiltz..........	Indian Point.....	17	193 90
112107	Alexandra...	"	93	Freeman Anderson.	Lunenburg.......	19	207 30
130956	Alfarata.........	"	92	Willis A. Ernst . ..	Mahone Bay......	19	207 30
130475	Alma M..........	"	15	Henry Miller........	Eastern Points ...	3	35 10
130942	Amy B. Silver....	"	99	Kenneth Silver...	La Have........	21	220 70
116522	Anita...........	"	16	Wm. Cleversey	West La Have....	3	36 10
133816	Anita P..........	"	12	Clarence Publicover	Blandford...	2	25 40
126585	Annie L. Spindler.	"	95	E. F. Zwicker........	Lunenburg.......	19	207 30
131165	Araminta........	"	95	Theophilus Creaser..	Riverport........	19	207 30
130818	Araucania.......	"	92	J. M. Rhodenizer ...	Lunenburg......	17	193 90
131176	Areola...	"	97	H. W. Adams..		19	207 30
130465	Artisan	"	98	Wm. Arenburg....	"	20	214 00
130737	Asaph F.........	"	14	Albert Fleet........	Blandford	3	34 10
131163	Associate........	"	90	J. E. Backman . .	Riverport... ...	19	207 30
126857	Assurance.......	"	99	Wm. C. Smith.......	Lunenburg......	20	214 00
133814	Austin B	"	10	Albert Bush	West Dublin....	2	23 40
126830	Benevolence......	"	99	Wm. C. Smith.......	Lunenburg......	19	207 30
131061	Bernice........	"	10	James Langille......	Tancook	2	23 40
130679	Bessie A. P.	"	11	Manuel Publicover...	Blandford.......	2	24 40
130726	Beulah W........	"	11	Herbert Young	Taucook........	3	31 10
111734	Blake..	"	99	J. N. Rafuse........	Conquerall	18	200 60
131070	Blanche L. G.....	"	11	Henry Gates........	Blandford.......	3	31 10
131080	Blanche S.......	"	10	Noah Baker	East River.......	3	30 10
126393	Burnett C........	"	105	A. V. Conrad.......	Parks Creek	19	207 30
131066	C. W. Mason.....	"	10	Solomon Richard...	La Have........	2	23 40
111732	Calavera........	"	90	Willis A. Ernst.....	Mahone Bay.....	13	167 10
112128	Campania.	"	90	S. W. Oxner........	Lunenburg......	17	193 90
130953	Cantow.........	"	13	Elias Publicover.....	Blandford	2	26 40
126119	Carrie L. Hirtle...	"	99	Wm. C. Smith.......	Lunenburg......	19	207 30
131164	CarrieM, Wamback	"	109	Wm. Duff	"	19	207 30
121999	Cavalier........	"	93	Kenneth Cleveland.	Blandford.......	3	33 10
126586	Cecil L. Beck.....	"	93	Wm. C. Smith.	Lunenburg......	17	193 90
130944	Cento...........	"	90	Dean Fralick........	Pleasantville....	18	200 60
122315	Clintonia........	"	96	Wm. C. Smith.......	Lunenburg	19	207 30
111743	Corean...........	"	70	Jas. Fralick, sr	Pleasantville.....	19	197 30
117736	Coronation..... ..	"	98	H. W. Adams.......	Lunenburg......	18	200 60
130823	Dagon	"	12	Ernest Covey........	Tancook..........	3	32 10
130731	Daisy Z.........	"	11	Solomon Zinck......	Blandford.......	2	24 40
126824	Dan Patch..	"	12	Robert Levy........	Lunenburg	4	38 80
131177	Delawana........	"	95	Wm. C. Smith.......	"	19	207 30

LIST of Vessels which received Fishing Bounty, etc.—Nova Scotia—*Continued.*

LUNENBURG COUNTY—*Continued.*

Official Number.	Name of Vessel.	Port of Registry.	Tonnage.	Name of Owner or Managing Owner.	Residence.	No. of Crew paid.	Amount of Bounty paid.
							$ cts.
111711	Defender..........	Lunenburg.....	98	Alex. Knickle......	Lunenburg.......	17	193 90
130948	Delia H..........	"	11	Joseph Hirtle......	Tancook........	3	31 10
131069	De Witt........	"	11	Francis Mason......	Eastern Points....	2	24 40
130562	Donald L. Silver..	"	94	Wm. Arenburg.....	Lunenburg........	20	214 00
130729	Dora C..........	"	12	Hugh Cleveland.....	Blandford........	2	25 40
130463	Doris V. Myra....	"	99	Clarence Myra.. ..	Riverport........	19	207 30
116540	Douglas Adams...	"	99	H. W. Adams,.....	Lunenburg........	17	193 90
133805	E. B. Walters	"	98	Cyrus Walters......	Parks-Creek......	23	234 10
116506	E. M. Zellars...	"	84	Fraser Gray........	La Have........	18	200 60
122009	Earl Grey........	"	96	E. F. Zwicker......	Lunenburg........	17	193 90
126391	Edith Marguerite.	"	95	F. Himmelman.....	Riverport........	19	207 30
112099	Electro..........	"	88	W. N. Reinhardt..	La Have	19	207 30
83308	Ella	"	10	Jennis C. Hanson ...	Mahone Bay......	1	16 70
121944	Ella Mason.... ..	"	74	J. W. Publicover ..	Getson's Cove.....	16	181 20
133815	Elma M..........	"	10	Steadman McDonald.	Black Rocks... .	3	30 10
130690	Elsie C..........	"	10	Wm. Cross........	Tancook........	4	36 80
130827	Elsie L. Corkum..	"	97	Auriel Corkum.....	La Have........	19	207 30
122318	Elsie M. Walters.	"	97	W. N. Reinhardt ...	"	17	193 90
130819	Elsie Porter	"	100	"	"	19	207 30
131079	Elsie S..........	"	10	Robert Schnare ..	Blandford..	3	35 10
131073	Estey	"	10	Nathan Silver. ...	Lunenburg	4	36 80
116518	Eva June....... ..	"	93	Wm. C. Smith......	"	17	193 90
126814	Evelyn V. Miller..	"	89	H. W. Adams.......	"	20	214 00
130728	F. M. Toro.......	"	100	E. F. Zwicker......	"	20	214 00
122304	Falcon..........	"	85	Edmen Walters.....	Parks-Creek......	18	200 60
130734	Falka	"	100	E. F. Zwicker......	Lunenburg........	17	193 90
130575	Forman F........	"	14	Obed Fleet........	Blandford........	2	27 40
126581	Frank H. Adams..	"	93	Freeman Anderson...	Lunenburg........	20	214 00
130825	Frank J. Brinton..	"	92	William Gillfoy.....	"	20	214 00
116525	Gatherer...... ...	"	15	Henry Pub.cover....	Blandford........	3	35 10
130464	Gigantic....	"	99	A. V. Conrad......	Parks Creek.	19	207 30
130812	Gladys and Lilian.	"	84	H. W. Adams	Lunenburg........	17	193 90
121851	Gladys B. Smith..	"	100	Wm. C. Smith	"	21	220 70
133810	Granite	"	92	Wm. Richard......	La Have........	22	227 40
116527	Guide	"	73	W. N. Reinhardt ..	"	17	186 90
131068	H. Mason..........	"	10	Casper Mason.	Eastern Points....	3	30 10
133807	H. H. McIntosh..	"	99	Wm. C. Smith......	Lunenburg........	19	207 30
130678	Harper..........	"	10	Harris Publicover...	Blandford.. ...	2	23 40
130461	Harry W. Adams.	"	99	H. W. Adams......	Lunenburg......	20	214 00
126392	Hawanee........	"	99	Wm. C. Smith......	"	19	207 30
126102	Hazel L. Ritcey...	"	92	Lemuel Ritcey.	Riverport........	18	200 60
122005	Hy. L. Montague.	"	96	Wm. C. Smith......	Lunenburg........	20	214 00
121857	Hiawatha.	"	99	"	"	17	193 90
130684	Hollo............	"	11	Ozen Hubley	Bayswater.......	4	37 80
131077	Hosie........ ..	"	10	Steadman Wilneff...	Tancook..........	3	30 10
132813	Howard Stanley ..	"	15	Stanley Langille.....	"	4	41 80
130950	Hurrah....	"	13	Otis Stevens........	"	5	46 50
130673	Hughie V. L	"	11	Rogers Levy.......	Little Tancook....	2	24 40
112089	Iona W...........	"	78	Willis A. Ernst......	Mahone Bay......	13	165 10
126813	Itaska..........	"	100	E. F. Zwicker......	Lunenburg........	19	207 30
107116	Ivy.............	"	12	John Backman.	"	3	32 10
126584	J. B. Young......	"	100	John B. Young......	"	21	220 70
130943	J. D. Hazen	"	99	Wm. C. Smith......	"	20	214 00
126822	Jennie E. Ritcey..	"	97	"	"	20	214 00
133804	John Parker . ..	"	99	W. N. Reinhardt ...	La Have..........	23	234 10
111726	Juanita.	"	100	Wm. C. Smith......	Lunenburg........	17	193 90
126819	Laura M. Levy...	"	11	Maynard Levy......	"	3	31 10
130473	Lavinia B	"	11	Amos Boutilier	Mill-Cove.......	3	31 10
131170	Leone G.......	"	12	Alex. Greek........	Blue Rocks......	3	32 10
130959	Leta J. Schwartz..	"	95	E. F. Zwicker.......	Lunenburg.......	20	214 00
130462	Lewis H. Smith...	"	98	Wm. C. Smith......	"	19	207 30

5 GEORGE V., A. 1915

List of Vessels which received Fishing Bounty, etc.—Nova Scotia—*Continued.*

LUNENBURG COUNTY—*Continued.*

Official Number.	Name of Vessel.	Port of Registry.	Tonnage.	Name of Owner or Managing Owner.	Residence.	No. of Crew paid.	Amount of Bounty paid.
							$ cts.
130815	Lilian B. Corkum.	Lunenburg	97	E. F. Zwicker	Lunenburg	20	214 00
133817	Lillian G	"	11	David Graves	Chester	2	24 40
130811	Lillian M. Richard	"	98	Elias Richard Jr	La Have	19	207 30
126821	Lloyd George	"	99	G . Himmelman	Riverport	17	193 90
131065	Lois M. C	"	12	Alvin Cross	Tancook	4	38 80
130820	Lottie A. Silver	"	96	Russell Silver	Lunenburg	20	214 10
130570	Lottie B. L	"	11	Albert Levy	Lit. Tancook	3	31 10
130688	Lottie M. Blanche.	"	12	David Moland	East Chester	3	32 10
130730	Lowell F. Parks.	"	99	Perry Parks	Parks Creek	17	193 90
130814	Lucille B. Creaser.	"	99	Arthur Creaser.	Riverport	19	207 30
131074	Lunenberg	"	10	George Baker	Lunenburg	2	23 40
130732	M. M. Gardner	"	100	Wm C. Smith	"	20	214 00
130477	Madge A. P	"	10	Chauncey Publicover.	Blandford	2	23 40
131180	Malada	"	21	Harris Fleet	"	3	41 10
116523	Mankato.	"	76	Edmen Walters	Parks-Creek	18	196 60
121862	Marina.	"	78	A. V. Conrad.	"	18	198 60
111709	Mariner	"	100	E. F. Zwicher.	Lunenburg	17	193 90
130816	Marion Adams.	"	99	H. W. Adams.	"	20	214 00
130829	Marion A. Silver.	"	99	Robert Silver	Riverport	21	220 70
126820	Marion Mosher	"	93	J. M. Rhodenizer.	Lunenburg	19	207 30
126829	Mark Twain	"	12	William Wight	Eastern-Points.	3	32 10
130941	Mary & Mildred	"	100	Christian Iversen.	Lunenburg	20	214 00
131169	Mary D. Young	"	99	John B. Young	"	18	260 00
133803	Mary F. Fleming.	"	94	Christian Iversen.	"	20	214 00
130822	Matanzas	"	96	Wu. C. Smith	"	20	214 00
130736	Matapedia	"	98	J. E. Backman	Riverport	20	214 00
130676	Mathilda H.	"	11	Collins Heisler	Tancook	4	37 80
121854	Mattawa	"	96	E F. Zwicker.	Lunenburg	17	193 90
121861	Medina A	"	74	Amiel Corkum.	La Have.	17	187 90
133818	Mildred Baker	"	10	Howard Baker	Lunenburg	3	30 10
121865	Millie Louise.	"	80	Willis A. Ernst	Mahone Bay	17	193 90
126107	Minnie M. Mosher	"	73	William Duff	Lunenburg	15	173 50
126113	Muriel B. Walters	"	98	Angus Walters.	"	20	214 00
130733	Muriel E. Winters	"	100	Freeman Anderson	"	19	207 30
130573	Muriel L	"	15	Peter Lowe	Mahone Bay	3	35 10
122007	Muriel M. Young.	"	100	John B. Young	Lunenburg	20	214 00
126463	Nellie J. Banks.	Shelburne	35	Whildon Bowers	Vogler's Cove.	9	95 30
94833	News Boy	Port Medway.	16	James Bell	Dublin Shore.	3	36 10
126827	Nobility	Lunenburg	99	Hiram Ritcey	Riverport	19	207 30
131178	Nordica	"	98	J. E. Backman	Vogler's Cove.	19	207 30
130955	No Tow	"	15	Harry Publicover.	Blandford	3	35 10
133806	Orante.	"	96	Joseph Conrad.'	Dayspring	22	227 40
130826	Original.	"	98	Wm C. Smith	Lunenburg	20	214 00
130683	Oriole L.	"	10	William Levy	Little-Tancook.	3	30 10
130572	Otokia.	"	89	Willis A. Ernst.	Mahone Bay	17	193 90
131067	P. C. Mason	"	11	Phineas Mason	Eastern Points	2	24 40
133801	Pasadena	"	91	Willis A. Ernst.	Mahone Bay	19	207 30
126589	Percival S. Parks.	"	109	Simon Parks	Parks Creek	19	207 30
130828	Phyllis L. West- haver	"	99	J. M. Rhodenizer.	Lunenburg	19	207 30
130817	R. L. Borden.	"	99	A. Himmelman.	Riverport	19	207 30
130951	Rakwana	"	11	Albert Meisner	Lunenburg	3	31 10
130569	Rebecca M. L.	"	11	Nathaniel Levy	Little Tancook	3	31 10
130674	Reggie P.P.	"	11	Norman Publicover.	Blandford	2	24 40
126114	Revenue	"	99	Wm C. Smith	Lunenburg	19	207 30
130478	Review	"	74	J. N. Rafuse.	Conquerall	16	181 20
130480	Roland A. T.	"	11	Abraham Knickle.	Blue Rocks	4	37 80
130946	Ronald C.	"	14	Clarence Tanner.	Black Rocks	4	40 80
121856	Ronald G. Smith	"	100	Wm C. Smith.	Lunenburg.	20	214 00
130689	Rosanna T	"	11	Israel Tanner	Black Rocks.	2	24 40
126034	Russel H. Pentz.	"	99	A. V. Conrad.	Parks-Creek	17	193 90

LIST of Vessels which received Fishing Bounty, etc.—Nova Scotia—*Continued.*

LUNENBURG COUNTY—*Concluded.*

Official Number.	Name of Vessel.	Port of Registry.	Tonnage.	Name of Owner or Managing Owner.	Residence.	No. of Crew paid.	Amount of Bounty paid.
							$ cts.
130685	S. F. Levy	Lunenburg	12	Hezekiah Levy	Little Tancook	3	32 10
130580	Sadie Evelyn	"	11	Marcus Publicover	Blandford	2	24 40
130724	Sealer	"	11	Amos Levy	Cross Island	4	37 80
133808	Selma M	"	12	Albert Mason	Tancook	3	32 10
126582	Sesame	"	15	Joshua Ernst	Pleasantville	2	28 40
130474	Shant Alec	"	11	Robert Wight	Eastern Points	3	31 10
130471	Skip	"	11	Arthur Mason	" "	4	37 80
131161	Thelma C	"	13	F. Cleveland	N. W. Cove	1	19 70
126590	Uda A. Saunders	"	95	E. F. Zwicker	Lunenburg	19	207 30
122306	Undaunted	"	15	James Oxner	Dublin Shore	4	41 80
131179	Vera E. Himmelman	"	99	Wm. Duff	Lunenburg	21	220 70
130681	Verna L	"	12	E. Corkum	La Have	3	32 10
131063	Vernie S	"	10	Albert Stevens	Tancook	2	23 40
131166	W. Cortada	"	108	E. F. Zwicker	Lunenburg	19	207 30
131174	W. C. McKay	"	99	William Deal	Riverport	20	214 00
131172	W. G. Robertson	"	90	J. W. Publicover	Getson's Cove	21	220 70
130824	W. H. Smith	"	94	Wm. C. Smith	Lunenburg	17	193 90
131175	W. T. White	"	99	Wm. C. Smith	"	19	207 30
130682	Warren G. C	"	12	Daniel Gilfoy	Feltzen South	2	25 40
126120	Warren G. Winters	"	95	Freeman Anderson	Lunenburg	19	207 30
133809	Warren M. Colp	"	92	Wm. Duff	"	22	227 40
126115	Watauga	"	99	H. W. Adams	"	17	193 90
126818	William C. Smith	"	99	Wm. C. Smith	"	20	214 00
121852	Winnifred	"	99	Willis A. Ernst	Mahone Bay	15	180 50
111419	Yukon	"	97	Wm. C. Smith	Lunenburg	18	200 60
130813	Zelma T. Young	"	15	Victor Zinck	Blandford	2	28 40

QUEENS COUNTY.

121685	Augusta	Yarmouth	11	Ambrose Verge	Port Medway	3	31 10
130677	Cunner	Lunenburg	10	Thomas Smith	S. W. Port Mouton	3	30 10
122235	Ena E	Barrington	12	Walter Leaman	Port Medway	2	25 40
121877	Florence	Liverpool	15	Nathan Boutilier	Black Point	3	35 10
116352	G. B. Zwicker	Port Medway	13	Chas. Zwicker	Port Medway	3	33 10
130247	Gaetta	"	16	Stanley E. Parke	"	3	36 10
122239	Hilda Brennan	Liverpool	10	Merril F. Pentz	Black Meadows	3	30 10
121887	Lena	Yarmouth	11	Robert Fisher	S.W. Port Mouton	3	31 10
122105	Lottie G	"	10	Andrew Leaman	Port Medway	3	30 10
131201	Lydia May	Liverpool	39	D. C. Mulhall	Liverpool		39 00
126184	Marion C	"	11	Bert Payzant	Port Medway	3	31 10
122103	Muriel S	Yarmouth	10	Albert McLeod	S.W. Port Mouton	4	36 80
130725	W. Baker	Lunenburg	10	William Baker	Liverpool	3	30 10

RICHMOND COUNTY.

116657	Alice M	Yarmouth	26	Thos. R. Boudrot	Petit de Grat	5	59 50
111472	Annie May	Arichat	17	Peter Landry	"	4	43 80
103463	Annie May	"	11	Henry LeLacheur	Martinique	4	37 80
74100	andid	"	23	Chas. LeBlanc	River Bourgeois	3	43 10
130355	E. L. Comeau	"	14	Alex. A. Boudrot	Petit de Grat	6	54 20
121866	Eldora	Lunenburg	79	Adelina Poirier	Descousse	18	190 60
116343	Eva May	Arichat	11	Henry Fougere	Poulamond	2	24 40
80829	Florence B	"	32	Chas. Boudrot	River Bourgeois	9	92 30
117091	Hazel Maud	"	10	Alcide Goyetche	Cape Auguet	5	43 50

5 GEORGE V., A. 1915

List of Vessels which received Fishing Bounty, etc.—Nova Scotia—*Continued.*

RICHMOND COUNTY—*Concluded.*

Official Number.	Name of Vessel.	Port of Registry.	Tonnage.	Name of Owner or Managing Owner.	Residence.	No. of Crew paid.	Amount of Bounty paid.
							$ cts.
122183	Justina............	Arichat......	10	Isiah Boudreau	River Bourgeois..	2	23 40
103469	Katie B..........	"	16	John Burke	"	5	49 50
111480	Lady Laurier....	"	12	Henry LeBlanc	Poulamond.....	2	25 40
117092	Lass of Gowrie ...	"	14	Jos. Petitpas	Arichat.........	5	47 50
107374	Leah Hardy	Sydney....	20	Peter Landry........	St. Peters	4	46 80
111905	Lena Jane.......	Arichat.........	11	Leo Miller...... ...	Poulamond.......	2	24 40
116350	Maggie F........	"	15	Alexis Baccardax....	Gully............	5	48 50
111798	Marie C........ ..	Pt. Hawkesbury	18	Alex. R. Boudrot....	Petit de Grat...	6	58 20
116345	Mary Alice..	Arichat........	10	P. E. Sampson....	Lr. L'Ardoise..	3	30 10
11479	Mary Atalanta...	"	15	Isaiah Burke	River Bourgeois ..	4	41 80
122182	Mary Elizabeth...	"	11	Placide Burke......	"	2	24 40
103462	Maud.	"	20	Henry Duon.......	Arichat..........	2	33 40
72067	Minnie	Pt. Hawkesbury	26	John Pelham	Janvrin Island...	5	59 50
121869	Petite...........	Lunenburg.....	61	Alex. P. Poirier....	Poirierville....	16	168 20
117095	Rodrid Grace. ...	Arichat	17	Hubert Birette......	Lr. L'Ardoise ...	2	30 40
116272	Rosie M. B......	Halifax.........	75	Anselm Sampson....	River Bourgeois ..	15	175 50
122189	Rostand *.......	Arichat........	95	D. Y. Stewart.....	St. Peters	16	190 40
122189	Rostand	"	95	D. Y. Stewart......	"	16	187 20
96962	Sunrise	Yarmouth.......	18	Chas. Fougere.......	River Bourgeois ..	5	51 50
103460	Two Brothers....	Arichat	18	Jos. Fougere	Poulamond.......	5	51 50
122190	Virginie S.......	"	16	Elias V. Landry....	Petit de Grat	7	62 90
116292	Wilona Fraser....	Charlottetown..,	13	Isaac Dugas.........	West Arichat....	3	33 10
100812	Wyvern	Barrington......	25	Jas. D. Walker ...	Walkerville	4	51 80

SHELBURNE COUNTY.

Official Number.	Name of Vessel.	Port of Registry.	Tonnage.	Name of Owner or Managing Owner.	Residence.	No. of Crew paid.	Amount of Bounty paid.
121802	Abbie May.......	Barrington	10	Chas. E. Rapp......	McNutt's Island..	2	23 40
121801	Alice M. Atwood..	Yarmouth.......	10	Geo. L. Nickerson...	Woods Harbour..:	3	30 10
116235	Alcyone	Digby..	52	Lockeport Cold Storage Co., Ltd....	Lockeport	14	145 80
122133	Alter C.........	Yarmouth.......	10	John Y Smith.......	Baccaro	3	30 10
122149	Alva	"	11	Lewis Cunningham ..	Stoney Island.....	4	37 80
122093	Anita	"	11	Willard Mathews....	E. Ragged Island.	3	31 10
117134	Annie Lue.......	"	10	John A. Smith. ...	Port La Tour.....	5	43 50
121890	Annie Smith.....	"	13	E. P. Crowell........	"	1	19 70
100612	Ardella.........	Shelburne	10	Eleazar Crowe......	Sandy Point... ..	4	36 80
122453	Bertha A........	Yarmouth.....	12	David H. Flemming..	Cape Negro......	3	32 10
130508	Blanchard C.....	Shelburne	11	Austin Swansburg ...	Little Harbour..	4	37 80
121906	Blanche	Yarmouth.......	10	P. W. Stoddart......	Woods Harbour..	3	30 10
103186	Brittania... !....	Shelburne	11	Jas. Enslow, Jr...	West Green H'br..	4	37 80
122288	Buema............	"	36	Daniel Ryder.......	Central Argyle..	7	82 90
121681	Claymore........	Yarmouth	10	J. R. Shand	Bear Point......	4	36 80
121683	D. E. Nickerson..	"	10	Jno. W. Hemeon.....	Sandy Point.....	4	36 80
122462	Daniel S.........	"	10	Albert P. Ross......	Stoney Island....	3	30 10
122002	Dolly Gray......	Lunenburg.....	13	Ross Enslow........	West Green Hbr..	3	33 10
121791	Eddie C	Yarmouth	10	N. E. Smith........	Smithville	4	36 80
116830	Edith Pauline....	Barrington..	10	J. L. Nickerson.....	Woods Harbour ..	3	30 10
122570	Edna M.........	Yarmouth..	11	Wilbur Halliday....	Bear Point.......	2	24 40
130504	Ella M. Rudolph-.	Shelburne	54	Wm. McMillan.....	Lockeport	14	147 80
122470	Elva Belle.......	Yarmouth.....	11	Burns McKenzie....	East Green Hbr..	3	31 10
122467	Enterprise.......	"	10	Oscar Gardner.....	Port La Tour....	4	36 80
121796	Etta N..........	"	10	Chas. B. Locke....	Lockeport.......	2	23 40
121901	Eva M.........	Barrington	11	Edwd. Goodick.....	Sandy Point.....	4	37 80
126345	Eva S...........	"	10	Louis Crowell......	Port La Tour.....	2	23 40
117048	Evangeline.	"	11	Foster Crowell·....	Clark's Harbour..	3	31 10
122146	Flirt	Yarmouth.....	16	H. D. Smith........	Port La Tour	5	49 50
122106	Florence M......	"	10	Percy Ross........	Stoney Island....	4	36 80
117045	Fred. C..........	Barrington.	12	C. E. Nickerson.....	Clam Point......	4	38 86
122142	Gertrude.........	Yarmouth	10	Mitchell Smith	Doctor's Cove.....	3	30 10

* For 1912.

SESSIONAL PAPER No. 39

List of Vessels which received Fishing Bounty, etc.—Nova Scotia—*Continued.*

SHELBURNE COUNTY—*Concluded.*

Official Number.	Name of Vessel.	Port of Registry.	Tonnage.	Name of Owner or Managing Owner.	Residence.	No. of Crew paid	Amount of Bounty paid.
							$ cts.
112138	Gladiator.........	Shelburne	11	Hugh McAlpine.....	Lockeport	3	31 10
122468	Gladys............	Yarmouth..	11	Clayton Shand... ..	Shag Harbour.....	3	31 10
122463	Gladys M..... ...	"	10	Ransom Chetwynd...	Up. Port La Tour.	3	30 10
130507	Gladys Thorburn..	Shelburne	39	Jno. H. Thorburn....	Sandy Point	9	99 30
121797	Hattie and Ina ..	"	10	E. W. A. Doane, ...	Carleton Village..	2	23 40
122139	Hazel	Yarmouth......	10	Geo. Crowell........	Atwoods Brook...	3	30 10
122100	Helen C	"	10	N. Crowell.........	Woods Harbour...	3	30 10
131094	Helen G. McLean	Shelburne	33	Kenneth B. Backman.	Shelburne........	7	79 90
122232	Helen Doris......	Barrington.....	12	Floyd Ross.........	Stoney Island.....	4	38 80
126185	Helen Glenn.....	Shelburne......	10	Edwd. Hammond....	Lr. Jordan Bay...	4	36 80
122237	Helena Maud.....	Barrington.....	11	A. B. Smith	Newellton........	3	31 10
122141	Hillside	Yarmouth	10	Jno. C. William s...	West Green Hbr..	2	23 40
126347	IdaM.Cunningham	Barrington. ...	16	Br F. Cunningham..	South Side........	2	29 40
117131	Ilona & Ida.	Yarmouth......	13	H. A. Brannen.....	Stoney Island.....	4	39 80
121904	Ilena & Maggie...	Barrington	11	Whitman Ross... ..	"	4	37 80
116822	Jennet..	"	11	Kenney & Gardner..	McNutt's Island.	3	31 10
122138	Jennie L........	Yarmouth	10	Jas. A. Smith.....	Port La Tour....	4	36 80
121795	John L...........	"	11	Bert. Hipson	Sandy Point	2	24 40
126670	Julie Opp........	Shelburne	38	H. R. Swim........	Lockeport	12	118 40
122131	Katie M..........	Yarmouth....	10	Geo. A. Acker.....	Birch Town.....	2	23 40
122290	Kernwood	"	84	Lockeport Cold Storage Co., Ltd....	Lockeport........	14	173 80
117136	Laura B.. ..,,....	"	10	C. D. Atkinson....	Stoney Island ..	4	36 80
122458	Lila A...........	Barrington... .	10	Howard Atkinson...	"	3	30 10
130627	Lily M. Hodge...	Yarmouth......	28	Lockeport Cold Storage Co., Ltd.	Lockeport..-.....	7	74 90
121693	Little Charley....	"	10	Howard Newell.	West-Head.......	3	30 10
126188	Lulu S...........	Shelburne	23	H. R. Swim........	Lockeport.......	5	56 50
121880	Mabel C..	Barrington.....	10	Wm. R. Reed...	Stoney Island ...	5	43 50
121888	Margaret........	Yarmouth.....	10	Albert Adams.......	Barrington.......	3	30 10
83434	Mary May.......	Shelburne	20	Adam J. Firth.....	Shelburne........	5	53 50
117043	Mattie & Charlie..	Barrington.	10	Frank Francis......	Brass Hill.......	2	23 40
121905	Mira L. Smith....	"	14	E. P. Crowell........	Port La Tour.....	4	40 80
103800	Nellie I. King....	Shelburne	99	Geo. H. King	Sandy Point.....	21	220 70
131091	Nellie Viola	"	40	Jno. T. McKenzie..	Lockport.....	12	120 40
122457	Nema & Millie....	Yarmouth.....	11	Sanford Slate.......	Slateville.........	3	31 10
117132	Nema D..........	"	10	Wm. Hipson,......	Shelburne.......	....	16 00
131096	Ohio..........	Shelburne	42	Jas. R. Bower......	"	14	135 80
117050	Olive R	Barrington.....	14	H. R. Swim.......	Lockeport	1	20 70
122233	R. H. Milford ...	Barrington.....	13	Isaiah S. Newell.....	West Head.....	2	26 40
130506	R. L. McKenzie..	Shelburne	33	Ralph McKenzie...	Lockeport........	8	86 60
131095	Ronald B........	"	40	McKenzie Bower...	Jordan Ferry.....	5	73 50
130509	Roseway........	"	37	Jas. R. Bower......	Shelburne.......	12	117 40
126342	Sakotis..........	Barrington.	11	B. J. Newell	West Head.....	4	37 80
121878	Selma	Yarmouth.....	14	H. Crowell	Charlesville.....	4	40 80
122108	Seretha....	"	10	N. C. Nickerson...	Clark's Harbour ..	2	23 40
103783	Springwood......	Shelburne	98	Wm. McMillan	Lockeport.......	19	207 30
90648	Stranger........	Barrington,....	20	Lovitt Banks.......	Barrington Passage	5	53 50
122236	Thelma B........	"	12	H. R. Swim	Lockeport	3	32 10
117046	Three Brothers..	"	13	Wilfred Atkinson....	Stoney Island.. .	4	39 80
116825	Three Sisters	"	11	Wallace Penny	N. E. Point......	4	37 80
116448	Togo	Shelburne	18	Edmund C. Locke...	Lockeport........	5	51 50
121792	Twin Sisters......	Yarmouth.....	10	Osborne D. Smith...	Hawk............	4	36 80
117143	Valmore.........	Halifax........	11	Clayton Collupy....	Lockeport.......	3	21 10
121873	Viola S...	Yarmouth .. .	16	C. E. Van Amburg...	"	4	42 80
77744	Whip-poor-Will...	Shelburne .. .	17	Isaac Ringer......	Sandy Point	4	43 80
122150	Wilford H......	Yarmouth.....	11	David T. Horton ...	Port La Tour.....	3	31 10
122464	Willie M	"	14	Durkee Chetwynd...	Up. Port La Tour.	4	40 80
121690	Winnifred........	"	10	Allan Nicker 2n...	Clark's Harbour ..	3	30 10
121656	Zilpha...........	"	10	Alamander Atwood..	Hawk.....	5	43 50

MARINE AND FISHERIES

List of Vessels which received Fishing Bounty, etc.—Nova Scotia—*Continued.*

VICTORIA COUNTY

Official Number.	Name of Vessel.	Port of Registry.	Tonnage.	Name of Owner or Managing Owner.	Residence.	No. of Crew. paid.	Amount of Bounty paid.
							$ cts.
126028	Beatrice Donovan.	Sydney..	18	Wm. Donovan......	South Ingonish...	4	44 80
130368	Bridget Dunphy .	"	11	J. W. Dunphy......	"	3	31 10
130369	Edna R. Hines...	"	18	Angus J. Hines......	Ingonish Ferry..	6	58 20
131213	Elizabeth Donovan	"	11	Wm. T. Donovan....	South Ingonish...	5	44 50
126569	Hawley Brothers..	"	11	Jas. Hawley	Ingonish Ferry...	5	44 50
122120	Julia F. C.......	"	12	Thos. A. Young.....	South Ingonish...	7	58 90
130362	M. A. McDouald..	"	17	Angus McDonald....	"	4	43 80
107355	Mary E.........	"	10	Allen McIntyre......	Ingonish Ferry...	3	30 10
117026	Mary E. Daisley..	"	16	Avery Daisley.......	Dingwall........	3	36 10
131214	Phœbe Jordan....	"	15	Chas. J. Williams...	South Ingonish...	3	35 10
100444	Stella May......	Canso....	12	Simon P. Hawley....	Ingonish Ferry...	6	52 20
130363	V. F. Williams. ..	Sydney.....	13	Vincent Williams....	South Ingonish...	3	33 10

YARMOUTH COUNTY.

Official Number.	Name of Vessel.	Port of Registry.	Tonnage.	Name of Owner or Managing Owner.	Residence.	No. of Crew. paid.	Amount of Bounty paid.
121876	Adoriam..........	Yarmouth....	15	Oscar Van Amburg ..	Pubnico Head....	5	48 50
122132	Aerolite..	"	16	Jas. J. Duncan	Deep Cove Island.	4	42 80
116898	Agnes M.........	"	11	Geo. Doucett........	Tusket	2	24 40
126808	Agnes Pauline....	"	71	R. N. D'Entremont..	West Pubnico	15	171 50
111879	Annie B..........	"	20	Theo. D'Eutremont..	M. W. Pubnico ...	8	73 60
121698	Argo	"	10	Theo Jacquard	Comeau Hill......	3	30 10
122586	Aspinet...........	"	14	Arthur McComiskey.	Lr. E. Pubnico...	4	40 80
122109	Bella.............	"	17	Ulysse J. Amiro.....	West Pubnico	2	30 40
121694	Columbia ..,.....	"	10	Fred Murphy........	Pubnico Head	4	36 80
116205	Eddie James......	"	79	Yarmouth TradingCo.	Yarmouth........	18	199 60
116528	Edith F. S.	"	67	"	"	17	180 90
126807	Elizabeth D	"	79	S. D. D'Entremont..	Lr. W. Pubnico...	20	213 00
121809	Estella..........	"	11	Albt. E. Carland.....	Pubnico Head	3	31 10
122461	Eva E............	"	10	Aaron Allen.........	Yarmouth	5	43 50
121872	Francis A....	"	93	Yarmouth TradingCo.	"	20	214 00
122092	Georgie M. Smith.	"	13	Thos. E. Smith	Yarmouth Bar....	5	46 50
122574	Gladys Olia......	"	10	Wm. McNair........	Argyle Sound.....	3	30 10
117137	Glorianna	"	10	Henry LeBlanc	Abram's River...	2	23 40
116894	Harry M. Johnson	"	14	M. A. Nickerson.....	Deep Cove Island.	4	40 80
122099	Hilda	"	17	Henry Boudreau.....	Wedgeport	5	50 50
122454	Industry	Barrington ...	11	Nathaniel Sears.....	Port Maitland ...	3	31 10
130626	Joseph Lester. ...	Yarmouth......	15	Raymond Amiro.....	West Pubnico ...	5	48 50
116204	Laurie J........ .	"	65	E. J. D'Entremont..	"	17	178 90
103709	Lizzie E..........	"	19	E. Juston Ellis.....	Port Maitland ...	3	39 10
130625	Louis P....	"	60	L. P. D'Entremont..	West Pubnico	18	186 60
116899	Lydia L..........	"	14	Adolphe LeBlanc ...	Wedgeport. .. .	2	27 40
121903	M. F. Atwood...	Barrington	15	John Surette........	Morris Island....	5	48 50
122240	M. L. Nickerson..	"	10	Wm. H. Nickerson...	Argyle Sound.....	4	36 80
116658	Mabel A........	Yarmouth....	15	Yarmouth TradingCo.	Yarmouth........	4	41 80
121879	Matilda...........	"	10	Wm. C. Hatfield.....	"	4	36 80
122231	Minola...........	Barrington	13	Stillman Smith	Lr. Argyle.......	2	26 40
121687	Monitor...........	Yarmouth......	10	Wm. H. Adams	Port Maitland	4	36 80
126187	Nathalie..........	Yarmouth......	28	Yarmouth Trading Co	Yarmouth........	5	61 50
111875	Nelson A.........	"	72	"	"	18	192 60
103706	Regine............	"	10	T. A. D'Entremont ..	West Pubnico ...	5	43 50
117044	S. B. Millard....	Barrington	20	Louis A. Amiro......	"	6	60 20
121875	Toronto	Yarmouth......	13	Howard Atkins......	Port Maitland ...	3	33 10
103711	Venite............	"	24	Jas. E. Crosby.......	Yarmouth........	4	50 80
122134	Venus...........	"	10	L. A. D'Entremont...	West Pubnico....	3	30 10
121894	Vice Reine.	"	12	Hugh McManus......	Yarmouth........	6	52 20
122452	Virginia	"	17	Jas. L. Purdy.......	Rockville........	2	30 40
122465	White Wings. ...	"	11	Joseph Harris.......	Yarmouth........	1	17 70

SESSIONAL PAPER No. 39

List of Vessels which received Fishing Bounty during the year 1913-14   Province of New Brunswick.

CHARLOTTE COUNTY.

Official Number.	Name of Vessel.	Port of Registry.	Tonnage.	Name of Owner or Managing Owner.	Residence.	No. of Crew paid.	Amount of Bounty paid.
							$ cts.
92517	Ada..........	St. Andrews....	10	Wm. Matthews.....	Letete	2	23 40
107903	Ava M...........	"	17	Geo. A. Johnson	Woodward's Cove.	3	37 10
122573	Bohemia....... ..	"	10	T. M. Dakin........	North Head......	5	43 50
122250	Bonita..........	"	15	Benj. Carter	Seeley's Cove....	2	28 40
103114	Edward Morse...	"	32	Alex. Calder	Campobello.....	4	58 80
80832	Ella Mabel.	"	14	Eldorado G. Lee	Beaver Harbour..	2	27 40
111527	Etta H...........	Digby.	10	Geo. Justason.......	Black's Harbour..	3	30 10
130426	Fannie May.....	St. Andrews...	25	Wm. McLellan ...	Campobello... ...	1	31 70
111552	Flora B...........	"	13	Nelson Ingersoll. ...	Woodward's Cove.	3	33 10
112282	Florence H..	Digby.	20	John Malloch... ...	Wilson's Beach...	2	33 40
107910	Grace & Ethel....	St. Andrews...	16	A. Ingersoll........	Woodward's Cove.	5	49 50
111839	Harry C..........	Digby..........	16	Lewis Matthews.....	Letete.	4	42 80
83463	Havelock....,....	St' Andrews...	33	William James.......	Wilson's Beach...	2	46 40
112590	Helen & Beatrice..	"	29	Gordon C. Calder....	Campobello......		29 00
103121	Island Girl......	"	17	Birdell Lambert.....	Woodward's Cove.	3	37 10
122591	Junnie T	"	31	Jas. Nesbitt........	North Head......	5	64 50
103997	Jessie James. ...	"	11	Josephine Frankland.	White Head......	3	31 10
122242	Mary M. Lord...	"	21	Leonard Bros.	St. John........	1	27 70
130427	Mollie G. Gaskill.	"	23	Jos. E. Calder......	North Head......	4	49 80
103993	Pythian Knight..	"	19	Frank Ingeroll......	Grand Manan....	3	39 10
107904	Quoddy Queen. ..	"	13	Chas. H. Matthews .	Letete...........	1	19 70
107806	Rena F..........	"	12	Jno. Ingersoll......	Woodward's Cove.	4	38 80
121660	Squanto	Yarmouth....	11	Howard Calder.....	Campobello.....	3	31 10
85390	Susan C..........	Barrington....	21	Sewall Newman.....	Wilson's Beach...	6	61 20
59387	Telephone........	St. Andrews....	19	Alfred Stanley.......	North Head	6	59 20
107440	Three Links.. ...	"	12	Robt. A. Main.......	Woodward's Cove.	2	25 40
103111	Volunteer........	"	14	Geo. Ingersoll........	"	3	34 10

GLOUCESTER COUNTY.

130658	Abutilon ...	Chatham..,....	19	Jos. Lacroix......'..	Caraquet..........	4	45 80
72099	Adelina.	"	12	Patk. Blanchard	"	4	38 80
103081	Albatross	"	13	Wm. Fruing & Co. .	"	5	46 50
112156	Albert W.......	"	10	Philorome Chiasson ..	"	5	43 50
130985	Alexisna..........	"	17	Romain A. Noel.....	Little Lamèque...	5	50 50
122037	Alice..........	"	15	Sevère Duguay.......	"	5	48 50
130332	Alika P......... ...	"	15	Zoël G. Paulin.	Lamèque........	4	41 80
112162	Alma.......... ...	"	12	Agapit Duguay......	"	5	45 50
92419	Anna	"	12	Jérémie S. Aché.....	"	4	38 80
100960	Annie M.........	"	11	W. S. Loggie Co....	Chatham........	5	44 50
96739	Argeline..........	"	14	F. T. B. Young....	"	5	47 50
130988	Aviator..........	"	17	Pierre S. Lanteigne .	Caraquet	4	43 80
103072	Ben-Hur..........	"	12	Adolphe Leclerc.....	"	5	45 50
100975	Big Bear...... ...	"	10	Gervais Plourde.....	"	3	30 10
100299	Blanchard........	"	12	Robin, Jones & Whit-man	"	4	38 80
103589	Blenheim........	"	13	"	"	4	39 80
130657	Bolina..........	"	20	"	"	4	46 80
103780	Britannia........	"	13	Wm. Fruing & Co...	"	4	39 80
100780	Britannic........	"	12	W. S. Loggie Co.....	Chatham........	4	38 80
111465	C. R. C..........	"	13	Robin, Jones & Whit-man............	Caraquet.	4	39 80
100988	Caesar..........	"	10	G. P. Chiasson......	"	4	36 80
100774	Calliope..........	"	12	Raphaël Hébert......	"	4	38 80
130339	Caraquet..........	"	19	Philias Doiron.	"	5	52 50
130996	Castaleno....... .	"	28	Robin, Jones & Whit-man......	"	6	68 20
103271	Celia	"	11	D. D. Landry........	"	4	37 80
103585	Cerdric...... .. .	"	14	Henri X. Chenard....	"	4	40 80

5 GEORGE V., A. 1915

List of Vessels which received Fishing Bounty, etc.—New Brunswick—*Continued.*

GLOUCESTER COUNTY—*Continued.*

Official Number.	Name of Vessel.	Port of Registry	Tonnage	Name of Owner or Managing Owner.	Residence.	Number of Crew paid.	Amount of Bounty paid.
							$ cts.
100784	Charlotte	Chatham	13	F. T. B. Young	Caraquet	3	33 10
138911	Contribution	"	11	Guillaume Chenard	"	3	31 10
103083	Corsair	"	10	Wm. Fruing & Co	"	3	30 10
133920	Cute	"	12	Abraham Chiasson	Island River	4	38 80
100913	Daffodil	"	10	Wm. Fruing & Co	Caraquet	4	36 80
130998	De Grace	"	10	Jas. De Grace	Shippegan	3	30 10
103076	Dipper	"	12	W. S. Loggie & Co	Chatham	4	33 80
130982	Dit-on	"	12	John Poirier	Caraquet	4	38 80
103948	Dora	"	12	Robin, Jones & Whitman	"	4	38 80
112155	Dora	"	10	Séraphin Doiron	Miscou Harbour	4	36 80
122053	Dorie	"	10	Peter P. Chiasson	Island River	4	36 80
100999	Dove	"	11	Wm. Fruing & Co	Caraquet	4	37 80
100998	Eagle	"	10	Alfred Gauvin	Mizonette	5	43 50
116070	Elie Anne	"	17	Jos. J. Doiron	Caraquet	4	43 80
100293	Eliza	"	15	F. T. B. Young	"	5	48 50
103590	Eliza	"	13	Robin, Jones & Whitman	"	4	39 80
130986	Emerencienne	"	17	Théophile Noël	Lamèque	5	50 50
92585	Emma	Gaspé	19	Sydney Des Brisay	Petit Rocher	1	25 70
100911	Emperor	Chatham	10	Wm. Frning & Co	Caraquet	4	36 80
133925	En Avant	"	11	André Aché	Lamèque	3	31 10
100772	Estelle	"	13	Harry Rive	Caraquet	3	33 10
100787	Ethel	"	11	F. T. B. Young	"	4	37 80
133916	Etoile d'un Marin	"	20	Octave Noël	Lamèque	5	53 50
122058	Evangeline	"	10	Vilas Frigot	Mizonette	3	30 10
92417	Evangeline	"	11	Xavier B. Noël	Little Lamèque	5	44 50
103901	Falcon	"	10	Jos. X. Chiasson	Caraquet	4	36 80
103077	Fame	"	10	Geo. D. Mallet	Shippegan	4	36 80
133926	Fidelis	"	11	Amédée L. Duguay	Little Lamèque	2	24 40
122621	Fillera	"	18	Harry Rive	Caraquet	4	44 80
130654	Fish Seeker	"	20	Gust. J. Gallien	"	3	40 10
100298	Fisher	"	12	Louis Guignard	Lamèque	4	38 80
61445	Flavie	"	13	Alex. Frigault	Caraquet	4	39 80
111468	Fleetwing	"	14	Wm. Frning & Co	"	4	40 80
112165	Flying Cloud	"	13	W. J. Robichaud	Shippegan	4	39 80
100782	Flying Foam	"	12	F. T. B. Young	Caraquet	4	38 80
112151	Flying Foam	"	15	Robin, Jones & Whitman	"	4	44 80
116479	Fortuna	"	10	Prosper Boudreau	Mizonette	3	30 10
111497	Four Brothers	"	13	Henri Albert	Caraquet	5	46 50
100778	Gambetta	"	13	W. S. Loggie Co	Chatham	4	39 80
111464	Gazelle	"	13	Robin, Jones & Whitman	Caraquet	4	39 80
100954	Gazelle	"	10	W. S. Loggie & Co	Chatham	4	36 80
100968	Gem	"	12	G. G. Doiron	Blue Cove	5	44 50
96733	Gem	"	12	Wm. Fruing & Co	Caraquet	5	45 50
103766	Genesta	"	12	Jos. G. Chiasson	Island River	4	38 80
116980	Georgina	"	15	W. S. Loggie Co	Chatham	5	48 50
103282	Gilknockie	"	11	Harry Rive	Caraquet	5	44 50
131336	Ginger	"	20	Luc L. Friolet	"	4	46 60
103086	Gipsy	"	20	W. S. Loggie Co	Chatham	4	46 80
111848	Gipsy	"	15	Wm. Fruing & Co	Caraquet	3	35 10
107775	Gold Seeker	"	13	Robin, Jones & Whitman	"	4	39 80
122491	Good Intent	"	10	André D. Chiasson	Lamèque	2	23 40
112157	Grasshopper	"	16	Harry Rive	Caraquet	4	42 80
92418	Grip	"	12	Gustave Chenard	"	4	38 80
111849	Happy Home	"	16	Harry Rive	"	4	42 80
100994	Hercules	"	10	Léandre Paulin	"	4	36 80
107771	Heron	"	13	Wm. Fruing & Co	"	4	39 80

SESSIONAL PAPER No. 39

LIST of Vessels which received Fishing Bounty, etc., New Brunswick—*Continued.*

GLOUCESTER COUNTY—*Continued.*

Official Number.	Name of Vessel.	Port of Registry.	Tonnage.	Name of Owner or Managing Owner.	Résidence.	No. of Crew paid.	Amount of Bounty paid.
							$ c.
103939	Hope	Chatham	11	John Michon	Caraquet	5	44 50
103765	Hirondelle	"	11	Agapit Leclerc	"	5	44 50
100906	Hotspur	"	10	Isaïe Lanteigne	"	3	30 10
130992	Hoy	"	11	Clément Lanteigne	Lamèque	3	31 10
117181	Ida	"	16	Jos. Savoy	"	4	42 80
103931	Irene	"	12	Wm. Fruing & Co	Caraquet	4	38 80
96724	Isabel	"	11	J. Bte. Hébert	"	5	44 50
131000	J. L. B	"	13	J. N. LeBouthillier	"	4	39 80
103289	Jersey Lily	"	12	Wm. Fruing & Co	"	3	32 10
100958	John B	"	11	W. S. Loggie Co	Chatham	4	37 80
130991	Joseph Marie G	"	22	Charles Gauvin	Little Lamèque	4	48 80
100965	Josephine	"	11	Harry Rive	Caraquet	3	31 10
112169	Kathleen	"	15	Wm. Frning & Co	Caraquet	4	41 80
111466	King Edward	"	14	Robin, Jones & Whitman	"	4	40 80
103949	Kingfisher	"	13	Wm. Frning & Co	"	5	46 50
103288	Kite	"	11	P. E. Lanteigue	"	2	24 40
107774	Klondyke	"	14	Robin, Jones & Whitman	"	4	40 80
103283	Koh-i-noor	"	13	Joseph A. Doiron	"	3	33 10
130984	L'Acadie	"	17	Lange Aché	Lamèque	5	50 50
130337	L'Acadienne	"	18	Jno. S. Noël	"	4	44 80
111461	Ladysmith	"	17	Hypolite Chiasson	Little-Lamèque	5	50 50
130983	Lamecca	"	19	Camille Aché	Lamèque	5	52 50
103003	Lark	"	10	Wm. Fruing & Co	Caraquet	4	36 80
130987	L'Assomption	"	18	J. J. Z. Chiasson	"	5	51 50
133927	Lefebvre	"	11	Sebastien Savoy	Shippegan Isld	3	31 10
107773	L'Etoile	"	15	Prudent Gallien	Caraquet	5	48 50
122059	Letty Jane	"	15	Wm. Frning & Co	"	5	48 50
112152	Lillian	"	15	Robin, Jones & Whitman	"	4	41 80
100972	Lizzie D	"	11	F. T. B. Young	"	4	37 80
130981	Lobelia	"	21	Théotime Gallien	"	4	47 80
116977	Mabel	"	16	W. S. Loggie Co	Chatham	4	42 80
130999	Mabel Luce	"	11	Philip Luce	Little Shippegan	3	31 10
102154	Mac	"	11	Wm. J. Ward	Miscou-Harbour	3	31 10
110955	Majestic	"	10	W. S. Loggie Co	Chatham	5	43 50
112158	Maple Leaf	"	13	Wm. Fruing & Co	Caraquet	5	46 50
116978	Margaret	"	16	W. S. Loggie Co	Chatham	5	49 50
112163	Margaret Ann	"	13	John Jones	Little Lameque	4	39 80
72100	Marie	"	11	Pierre A. Doiron	Caraquet	4	37 80
107779	Marie	"	15	Gaspard Savoie	Shippegan	3	35 10
103278	Marie Celia	"	13	Frank Baudin	Miscou Harbour	5	46 50
133919	Marie Delphine	"	16	Jos. H. Savoie	Lamèque	5	49 50
117182	Marie Etoile	"	20	Jos. A. Doiron	Caraquet	4	46 80
100292	Marie Joseph	"	12	Pierre P. Noël	Little Lamèque	5	45 50
133994	Marie Justine	"	24	Jos. A. Doiron	Caraquet	3	44 10
100295	Marie Louisa	"	18	Jos. A. Paulin	"	4	44 80
116471	Marie Louise	"	10	Gustave Chiasson	"	3	30 10
111847	Mary	"	14	David Albert	"	4	40 80
130655	Mary E. Rive	"	21	Harry Rive	"	5	54 10
103084	Mary Emma	"	11	Wm. Fruing & Co	"	4	37 80
130995	Mary J. Margaret	"	25	Harry Rive	"	4	51 80
92413	Mary Jane	"	14	Harry Rive	"	3	34 10
130994	Mary M. Florence	"	32	Harry Rive	"	5	65 50
116478	Mary O	"	11	Jos. O. Cormier	Mizonette	3	31 10
100957	Mary R	"	12	W. S. Loggie Co	Chatham	4	38 80
116475	Mary Rose	"	17	Robin, Jones & Whitman	Caraquet	3	37 10
112161	Mary Star	"	15	H. LeBouthillier	"	5	48 50
112150	Mary Star of the Sea	"	15	Luke Friolet	"	5	48 50

5 GEORGE V., A. 1915

List of Vessels which received Fishing Bounty, etc., New Brunswick—*Continued.*

GLOUCESTER COUNTY—*Continued.*

Official Number.	Name of Vessel.	Port of Registry.	Tonnage.	Name of Owner or Managing Owner.	Residence.	No. of Crew paid.	Amount of Bounty paid.
							$ cts.
111844	Mary Star of the Sea..........	Chatham	14	Robin, Jones & Whitman....	Caraquet........	1	40 80
116477	Mary Star of the Sea..............	"	20	Ferdinand Savoy.....	Shippegan........	4	46 80
107777	May Flower	"	11	Fred. Lanteigne......	Little Shippegan..	4	44 50
103768	Mayflower.......	"	13	Robin, Jones & Whitman.............	Caraquet.........	5	46 50
130997	Médaille d'Or	"	24	Huguet Lanteigne...|	"	4	50 80
100779	Mermaid...... ...	"	11	W. S. Loggie Co.....	Chatham	4	37 80
112164	Merry Christmas..	"	13	Celestin Jean.......	Little Lamèque..	4	39 80
133924	Merveille.........	"	12	Arthur J. Aché......	Lamèque.........	2	25 40
100300	Mikado...........	"	13	Robin, Jones & Whitman.............	Caraquet......	3	33 10
130659	Mildred Elaine...	"	20	Wm. Fruing & Co..	"	5	53 50
133922	Morning Dew...	"	10	Edmond E. Robichaud	Shippegan Island.	3	30 10
117188	Morning Star....	"	14	Alexis Noël.........	Lamèque........	4	40 80
122055	Olive...... . .	"	14	Thos. A. Lanteigne .	Caraquet.........	4	40 80
103004	Oriole...........	"	11	Wm. Frning & Co...	"	3	31 10
103005	Osprey..........	"	10	Thos. J. Mallet	Shippegan........	3	30 10
133917	Overseer	"	20	F. F. Chiasson......	Island River.....	4	46 80
130656	P. A. L	"	17	P. A. Lanteigne.. ...	Caraquet	5	50 50
100904	P. T. S........	"	11	E. O. LeBouthillier......	"	4	37 80
100297	Palma...........	"	14	Amédée Aché........	Lamèque.......	4	40 80
100776	Patrick.	"	11	W. S. Loggie Co....	Chatham........	5	44 50
112125	Pearl...........	"	14	Luc Lanteigne......	Caraquet........	4	40 80
103778	Pelican	"	13	Wm. Fruing & Co..	"	3	33 10
133923	Pembina.........	"	17	Jean Aché.........	Lamèque	5	50 50
103674	Petrel..........	"	12	Philorome Ross	Caraquet	4	38 80
122623	Pride of the Fleet.	"	24	Robin, Jones & Whitman.............	"	5	57 50
116974	Providence......	"	18	M. L. Lanteigne....	"	5	51 50
96740	Providence......	"	13	Prospere Leger.....	Caraquet..	4	39 80
130335	R. J. W.........	"	26	Robin, Jones & Whitman.............	"	4	52 80
100775	Red Gauntlet.....	"	11	J. H. LeBouthillier...	"	3	31 10
103586	Remus	"	17	W. S. Loggie Co....	Chatham	4	43 80
103078	Reward.........	"	13	L. B. Albert........	Caraquet.	4	39 80
130661	Richibucto Pearl..	"	12	A. T. Mallet.......	Shippegan........	4	38 80
97191	Rita	"	12	Robin, Jones & Whitman.	Caraquet.	4	38 80
111470	River Branch.....	"	11	Wm. Fruing & Co...	"	4	37 80
133992	Robichaud........	"	10	P. G. Robichaud....	Shippegan........	3	30 10
103946	Robin...	"	12	Robin, Jones & Whitman	Caraquet....	4	38 80
103587	Romulus.........	"	19	W. S. Loggie Co.....	Chatham.	4	45 80
92404	Rosa.............	"	17	Fredk. Lanteigne....	Caraquet........	4	43 80
100908	Rosalie..........	"	10	P. G. Lanteigne.....	"	4	36 80
100773	Rupert..........	"	12	Eustazade L. Albert..	"	4	38 80
116972	St. André.......	"	15	André A. Aché......	Lameque........	4	41 80
116473	St. Anne.......	"	14	Onésime Chiasson ...	"	5	47 50
111469	St. John.......	"	13	John Aché.........	"	4	39 80
112167	St. Joseph.......	"	16	Raphael Gionet......	Caraquet.....	4	36 80
103008	St. Joseph.......	"	12	Eugene H Gauvin...	Lameque	5	45 50
130660	St. Sauveur......	"	18	Isaie Chiasson.	"	5	51 50
107776	St. Peter.......	"	12	Jno. G. Chiasson....	Caraquet........	4	38 80
117187	Ste. Anne.......	"	13	Jean P. Noel.......	Lameque.........	3	33 10
117189	Ste. Cecelia......	"	13	Gelas Aché....	Little Lameque..	4	39 80
122051	Ste. Julie........	"	12	Marcelin Noel.......	Lameque	4	38 80
133915	Samuel LeGrand..	"	14	Alex. J. Robichaud..	Shippegan.......	3	34 10
74401	Sara........	"	11	Francis S. Doiron....	Caraquet.........	4	37 80

SESSIONAL PAPER No. 39

List of Vessels which received Fishing Bounty, etc., New Brunswick—*Continued.*

GLOUCESTER COUNTY - *Concluded.*

Official Number.	Name of Vessel.	Port of Registry.	Tonnage.	Name of Owner or Managing Owner.	Residence.	No. of Crew paid	Amount of Bounty paid.
							$ cts.
100907	Sarah	Chatham	10	F. T. B. Young	Caraquet	4	36 80
117190	Saturn	''	10	Dominick Blanchard	Mizonette	4	36 80
103584	Saxon	''	13	Jos. Baudin	Caraquet	4	39 80
100959	Sea Bird	''	10	W. S. Loggie Co	Chatham	4	36 80
126254	Sea Duck *	''	16	Edward P. Roy	Bathurst	4	43 60
126254	Sea Duck	''	16	''	''	3	36 10
100914	Sea Flower	''	11	Ernest Marks	Miscon Harbour	3	31 10
96926	Sea Foam	''	15	Jno. M. Ward	Miscou Centre	4	41 80
100901	Sea Flower	''	12	J. P. Lanteigne	Caraquet	4	38 80
96731	Sea Star	''	13	Patrick Albert	''	4	39 80
133913	Selonia	''	11	A. T. Chiasson	Shippegan	3	31 10
133914	Shippegan Pearl	''	10	Jos. Brideau	''	2	23 40
130993	Shippegan's Best	''	10	W. S. Loggie Co	Chatham	4	36 80
133928	Sillery	''	12	Jos. Aché	Lameque	2	25 40
100961	Silver Moon	''	14	W. S. Loggie Co	Chatham	4	40 80
100788	Sir Charles	''	11	Napoleon E. Gionet	Caraquet	4	37 80
122060	Spark	''	10	Wm. Frning & Co	''	3	30 10
100963	Stanley	''	10	André D. Gionet	''	3	30 10
103087	Stanley	''	10	Jos. Chiasson Jr	Island-River	4	36 80
133912	Star of Shippegan	''	11	M. D. Chiasson	Shippegan	3	31 10
103767	Stella Maris	''	19	Robin, Jones & Whitman	Caraquet	4	45 80
122056	Sunbeam	''	14	Wm. Frning & Co	''	4	40 80
111845	Superior	''	14	Robin, Jones & Whitman	''	4	40 80
103947	Swallow	''	13	Marcin Doiron	''	4	39 80
103006	Swallow	''	11	Wm. Fruing & Co	''	2	24 40
103762	Swan	''	14	''	''	4	40 80
100777	Teutonic	''	11	W. S. Loggie Co	Chatham	4	37 80
96738	Three Brothers	''	13	J. N. E. Lanteigne	Caraquet	4	38 80
117184	Three Brothers	''	16	D. F. Chiasson	Shippegan Island	5	49 50
100918	Tickler	''	12	Robin, Jones & Whitman	Caraquet	5	45 50
112159	United Empire	''	17	T. O. LeBouthillier	''	5	50 50
103285	Valkyrie	''	12	Jos. F. Hébert	''	4	38 80
103775	Victoria	''	16	W. S. Loggie Co	Chatham	4	42 80
133921	Vika	''	29	Maximin Paulin	Little Lameque	5	62 50
117183	Vina	''	14	Jacques Noel	Lameque	5	47 50
100995	Voltaire	''	10	Luc Mailloux	Caraquet	4	36 80
100966	Von Moltke	''	11	Pierre J. Frigot	''	3	31 10
103588	Vulture	''	13	W. S. Loggie Co	Chatham	4	39 80
122054	White Fish	''	13	Eurrope Chiasson	Lameque	5	46 50
100953	White Wings	''	10	F. T. B. Young	Caraquet	4	36 80
100973	World's Fair	''	11	''	''	4	37 80
103079	Wren	''	11	Jos. B. Paulin	''	4	37 80
100920	Zephyr	''	12	Robin, Jones & Whitman	''	4	38 80

KENT COUNTY.

126771	Dorothy F	Richibucto	12	W. E. Forbes	Richibucto	2	25 40
130665	Fulta	''	14	Geo. H. Long	''	1	20 70
116688	Harry Dickson	''	10	W. E. Forbes	''	2	23 40
130662	Jardineville	''	10	A. J. Arseneau	Jardineville	2	23 40
116689	Joseph Doucett	''	10	Albert Daigle	Lit. North West	2	23 40
130664	Lapewalem	''	10	Mrs. Jos. Doucett	ReXton	2	23 40
116684	Ocelot	''	11	W. E. Forbes	Richibucto	2	24 40

* For 1912.

39—26½

LIST of Vessels which received Fishing Bounty, etc., New Brunswick—*Continued.*

KENT COUNTY.—*Concluded.*

Official Number.	Name of Vessel.	Port of Registry.	Tonnage.	Name of Owner or Managing Owner.	Residence.	No. of Crew paid.	Amount of Bounty paid.
							$ c.
126773	S. and G....,...	Richibucto.	10	Sylvester Gray.......	St. Charles.......	2	23 40
126777	Samuel G........	"	10	A. & R. Loggie......	Richibucto	2	23 40
116685	Sea Alder.... ...	"	10	W. E. Forbes........	"	2	23 40
126772	Sylvalee.........	"	10	James Legoof.......	"	4	36 80
126778	3 0 3	"	10	W. E. Forbes.......	"	2	23 40
126774	Wawota........	"	11	Wm, H. Long......	"	2	24 40

NORTHUMBERLAND COUNTY.

Official Number.	Name of Vessel.	Port of Registry.	Tonnage.	Name of Owner or Managing Owner.	Residence.	No. of Crew paid.	Amount of Bounty paid.
122499	Beat the Wind....	Chatham..	10	T. B. Williston......	Baie du Vin.......	1	16 70
96725	Bessie T..........	"	10	Donald Loggie......	Burnt-Church.....	4	36 80
130338	Financier...	"	10	Bernard Williston....	Baie du Vin......	1	16 70
130333	Maggie Swift.....	"	11	Gordon Murdoch. ...	Hardwick........	1	17 70
92420	Mary Louise......	"	13	Donald Loggie......	Burnt-Church.....	4	39 80
116683	Plum...........	"	10	Michael Jimmo......	Escuminac........	2	23 40
100952	Replevin	"	10	Henry Albert........	Neguac	3	30 10
130340	Skidoo....	"	11	Harrison Murdoch...	Hardwick.	1	17 70
126252	White Cap	"	11	Wm. Jimmo.........	Escuminac........	3	31 10

RESTIGOUCHE COUNTY.

Official Number.	Name of Vessel.	Port of Registry.	Tonnage.	Name of Owner or Managing Owner.	Residence.	No. of Crew paid.	Amount of Bounty paid.
103826	Superbe	Paspebiac	12	Geo. A. Jarvis.......	Fredericton.	2	25 40

ST. JOHN COUNTY.

Official Number.	Name of Vessel.	Port of Registry.	Tonnage.	Name of Owner or Managing Owner.	Residence.	No. of Crew paid.	Amount of Bounty paid.
103704	Whisper..........	Yarmouth	31	Chas. Harkins.......	Dipper Harbour ..	3	51 10

PROVINCE OF PRINCE EDWARD ISLAND.

KINGS COUNTY.

Official Number.	Name of Vessel.	Port of Registry.	Tonnage.	Name of Owner or Managing Owner.	Residence.	No. of Crew paid.	Amount of Bounty paid.
112021	Annie M.	Canso..........	29	Thomas Poole........	Souris..........	4	55 80
94643	Carrie M. C	Lunenburg	39	Allan McLeod.......	Murray Harbour..	8	92 60
100696	Emerson....... ...	Pictou..........	30	Jno. McKenzie	Beach Point......	4	56 80
116308	Francis D. Cook .	Charlottetown...	47	Herbert Cahoon......	Murray Harbour..	6	87 20
122081	Frank..........	"	10	Jos. M. Cheverie. ...	Souris....	5	43 50
122086	Florence.... . .	"	14	Philip Rillard..	Beach Point......	2	27 40
126063	JohnG.Scrimgeour	"	14	Herbert Williams....	Murray Harbour..	2	27 40
107751	Minnie Laura.....	"	31	Reuben Penny.	"	4	57 80
107985	Muriel	Shelburne	25	M. Sencabaugh.......	"	4	51 80
112378	Olive S.	Charlottetown...	26	Albert Gosbee......	"	1	32 70
116296	Outlook	"	21	Hugh Jackson....	"	4	47 80
96727	Ryse.............	Chatham	11	Wm. R. Chennel.....	Souris....	3	31 10

SESSIONAL PAPER No. 39

LIST of Vessels which received Fishing Bounty, etc., Prince Edward Island—*Con.*

PRINCE COUNTY.

Official Number.	Name of Vessel.	Port of Registry	Tonnage.	Name of Owner or Managing Owner.	Residence.	No. of Crew. paid.	Amount of Bounty paid.
							$ cts.
117096	Alaska	Arichat	10	G. N. Matthews	Alberton	3	30 10
103279	Alice Maud	Chatham	10	Jos. Gallant	Ebbsfleet	4	36 80
121860	Aurora	Lunenburg	10	Jno. T. Stewart	West Point	3	30 10
116513	Laurie H	"	16	Wm. C. Leavitt	Alberton	4	42 80
100580	Maggie E. C	"	20	Jas. Mountain	Malpèque	4	46 80
94793	May English	Richibucto	10	Daniel English	Miminegash	1	16 70
103592	Rosamond	Charlottetown	18	Geo. A. Champion	Darnley	3	38 10

QUEENS COUNTY.

100445	Carrie O	Canso	12	Thos. Hiscott, Sr	Stanley Bridge	4	38 80
117059	Fortuna	"	14	J. Delaney	French River	2	27 40
107763	Guinea	Charlottetown	10	Boyce Harding	"	4	36 80
130343	Libby P	"	11	Jos. Pineau	North Rustico	6	51 20
126068	Mary E. Spears	"	10	David Spears et al	French River	3	30 10
92745	Surprise	"	18	Frank Pidgeon	"	3	38 10

PROVINCE OF QUEBEC.

GASPÉ COUNTY.

116294	Charlotte S	Charlottetown	14	J. Cassidy	Amherst, M.I	3	34 10
85400	Minnie M	Magdalen Isl'ds.	13	Honoré Cormier	"	5	46 50
85399	Minnie May	"	10	Wm. Boudreau	"	5	43 59
85408	Onato	"	35	Vital Boudreau	Grindstone	7	81 90
92571	Primrose	Halifax	14	Fortune Cormier	Amherst, M.I	5	47 50
111430	Shamrock	"	23	A. V. Vigneau	"	5	56 50

5 GEORGE V., A. 1915

APPENDIX No. 17.

The following are lists of United States Fishing Vessels which have entered Canadian Ports on the Atlantic and Pacific Coasts, and of United States Fishing Vessels to which *Modus Vivendi* Licenses were issued during the year ended March 31, 1914.

ATLANTIC COAST PORTS.

Number	Name of Vessel	Tonnage	Number of men	Magdalen Islands	Souris	North Sydney	Louisburg	Arichat	Port Hawkesbury	Canso	Liscomb	Halifax	Lunenburg	Liverpool	Lockeport	Barrington	Clark's Harbour	Shag Harbour	Wood's Harbour	Port La Tour	Yarmouth	Digby	Shelburne
1	Ada	62	16			1				2	4			2									1
2	Arcadia	90	18			3						1		1									
3	Alert	74	18		2	1	1							1	3								
4	Arkona	97	21							6	1		1	2						19		4
5	Avalon	85	18							2	1	2		2						2	1	1
6	Agie	36	16				1		1	1	1	1		1								
7	Acton	16	10							4				1								
8	Asp	83	18			2	1			3	1	2		3						3		3
9	Atlanta	77	18			6	1			3	1			1	1					1	4	3
10	Arthur James	95	19				2			2	1			2		2						
11	Arthusia	107	24		1	1				3	1			1		1						
12	Athlete	96	21				2			1	2	3		2	1							1
13	Annie M. Parker	92	21			2				2	1											
14	A. P. Andrew	100	22				1							1								
15	Aloha	23	23			1								1								
16	Agnes	86	18																			
17	Arabia	75	19																			
18	Ida M.	57	19			1																
19	Appomatox	49	16																			
20	A. D. Willard	39	8											1								
21	Admiral	19	4			1																
22	Annie Louise	7	2																			
23	B. P.	71	16			1				1		1		1		2						1
24	Boyd & Leeds	37	17																			
25	Bohemia	86	18			1				1		1		1		1		3				
26	Bernice & Bessie	29	7																		1	
27	Blanche F. Irving	25	7								1									4		2
28	Benj. A. Smith	91	20				1			2	3	1		5						2	1	2

SESSIONAL PAPER No. 39

	No.	Vessel		Total
29	Bessie			6	21
30	Bay State			22	102
31	Blanche			8	78
32	Catherine Burke			22	92
33	Cavalier			20	96
34	Corona			10	82
35	Cy ...			19	89
36	Cy ...			9	95
37	Clintonia			17	105
38	Chester A. Kennedy			18	13
39	Ezra A. Marston			2	20
40	Conqueror			7	104
41	Commonwealth			22	93
42	Corsair			9	78
43	C. A. Dollivar			17	19
44	Carrie C.			3	75
45	Dorcas			13	28
46	Diana			7	89
47	Ella			20	98
48	Eglantine			16	66
49	Eliza Bernier			6	14
50	F. C. Husey			18	41
51	Edmund F. Black			12	35
52	Eva & Mildred			12	43
53	Elk			18	83
54	Elmer E. Gray			17	84
55	Eugenia			16	66
56	Esperanto			18	91
57	Elizabeth N.			19	102
58	Essex			8	85
59	Evelyn N. Thompson			19	57
60	E. M. Morrisey			18	85
61	Etta Mildred			16	45
62	Ella G. King			17	52
63	Ethel B. Penny			17	57
64	Eva M. Martin			3	11
65	E. ...			3	21
66	Elizabeth			17	45
67	Eliza A. Bannet			6	14
68	Elphranto			19	101
69	Fannie E.			18	87
70	Francis P. ...			17	71
71	Flora L. Oliver			19	71
72	Francis S. Grueby			22	94
73	Fannie B. ...			18	82
74	Fannie A. Smith			18	87
75	Francis J. O'Hara			22	83
76	Flirt			8	82
77	Johny Reed			20

ATLANTIC COAST PORTS—Continued

Number	Name of Vessel	Tonnage	Number of crew	Totaux
78	Fannie Bell	16	3	4
79	Georgiana	87	22	9
80	Gala	65	18	5
81	Governor Foss	88	23	8
82	Georgie Bell	78	8	2
83	G. W. Turner	26	18	3
84	Gladys & Nellie	64	17	9
85	Gertrude de Costa	61	20	5
86	Good Luck	66	18	1
87	Hortense	52	18	1
88	Harriet	58	18	1
89	Harvard	76	18	6
90	Harmony	80	21	11
91	Hazel R. Hines	79	9	1
92	Hookomack	22	9	1
93	Hookonock	46	9	1
94	Hattie A. Hickman	72	16	1
95	Helen G. Wells	66	20	2
96	Helen B. Bliss	48	16	2
97	Harold	76	18	1
98	Hiram Lowell	95	20	1
99	Higeo	12	3	3
100	Imperator	99	22	10
101	Independence II	109	20	3
102	Ingomar	103	22	20
103	Judique	89	9	1
104	John J. Fallon	77	18	3
105	Jeannette	66	7	1
106	Jennie H. Gilbert	25	23	8
107	James W. Parker	96	9	21
108	James & Fadner	47	10	6
109	Josephine de Costa	84	17	12
110	Juno	85	20	3

SESSIONAL PAPER No. 39

No.	Name		
111	Jennie B. Hodgson	8	85
112	John Hays Hammond	23	92
113	J. J. Flaherty	25	124
114	Jubilee	19	61
115	J ...	19	89
116	J. R. Atwood	11	41
117	...	18	83
118	Katie L. Palmer	7	30
119	Lucania	18	107
120	Lillian	18	95
121	... H. Merchant	18	76
122	Layfayette	6	13
123	La Verina	25	95
124	Lucinda P. Lowell	15	77
125	Lochinvar	9	34
126	Lizzie Giffen	18	71
127	Lsie Elsie	3	11
128	Moonam	18	82
129	Marguerite Haskins	19	70
130	Marion E. Turner	14	45
131	... E. Silviera	16	63
132	Margaret	18	79
133	Muriel	22	83
134	Mary E. Curtis	20	85
135	... Turner	11	44
136	Monitor	18	100
137	Manhasset	18	79
138	Mary E. Sennett	8	11
139	Matthew L. Greer	18	75
140	Maxime Elliott	8	75
141	Mattie Winship	12	73
142	...	13	64
143	...	10	96
144	Mildred Robinson	22	86
145	Morning Star	23	85
146	Mary	23	93
147	Margaret Dillon	14	48
148	Massachusetts	22	97
149	Mertie H. Perry	14	36
150	Mertie H. Hartley	18	78
151	Mary E. Fallon	17	48
152	Mystery	18	79
153	...	10	40
154	Massasoit	8	32
155	Motor	8	35
156	Muriel Milliard	17	45
157	Norua	18	78
158	Nellie Dixon	19	68
159	Nellie G. Davis	11	36

ATLANTIC COAST PORTS—*Concluded.*

Number.	Name of Vessel.	Tonnage.	Number of men.	Magdalen Islands.	Souris.	North Sydney.	Louisburg.	Arichat.	Port Hawkesbury.	Canso.	Liscomb.	Halifax.	Lunenburg.	Liverpool.	Lockeport.	Barrington.	Clark's Harbour.	Shag Harbour.	Wood's Harbour.	Port LaTour.	Yarmouth.	Digby.	Shelburne.	Totals.
160	Ada..	50	16			4		1	1	1	2	1		1							2		3	4
161	Onata..	106	23			1		1		2		1								1		1	19
162	Iva..	77	18			2				1				2								5
163	Oriole.	104	27			1				3										6		4
164	Pinta..	66	18							3		3		1						2		16
165	Premier.	97	22											1							10
166	Paragon..	80	18							1				1					1		13
167	Pontiac.	75	19					2						1							5
168	Patriot.	77	17									1		1						12		4
169	Preceptor.	89	18			2		1		5		1		1							7
170	Priscilla.	27	8							4									3		1
171	Pythian.	46	15			2	1	1		1		1		1					3		17
172	Dewitt	111	20					2		2		1		2					1		8
173	Regina.	66	23									1		1					1		8
174	Ruth.	81	18					1		1				2							10
175	Rhodora.	77	19											1							11
176	Bob Roy.	93	16							1		1							6		13
177	Rex..	49	18									1							2		5
178	Rebecca.	90	22			1				6	1	2		1		1					4
179	Richard.	90	14			2				2				2							3
180	Romance.	54	10					1													1
181	Richard J. Newman.	55	15																		5
182	Ralph E. Hall.	91	17							1											8
183	Ralph Russell.	48	17					2		2		1		2			1				17
184	Rockin..	77	18																		7
185	Senator.	74	17							1				2							9
186	Selma.	87	22							1	1	1							3		11
187	Stilletto.	99	23			1				6		1		1							4
188	Sylvania.	99	21			2				2				1				1			7
189	Squanto.	95	16									1									
190	Sadie M. Nunan.	36	9																		
191	Senator Gardiner.	97	20			2		1		2	1			2					1		4
192	Smuggler	91	18			2															7
193	Saladin.	89	19			2		1		1		1		1					2		

194	Stranger	8	26
195	Spartel	5	126
196	Thalia	12	47
197	Tacoma	16	71
198	Miss S. Gorton	23	92
199	Terra Nova	22	94
200		18	59
201	Thomas B. Cromwell	18	89
202	Tatler	25	135
203	T. M. Nickerson	9	90
204	Thelma	17	28
205	Theo. Roosevelt	17	90
206	Valerie	23	97
207	Vaness	21	84
208	Virginia	16	73
209	Var	17	75
210	Viking	9	40
211	Veda McKeweon	18	83
212	W. M. Goodspeed	18	64
213	Watagua	7	18
214	Mo L.	18	81
215	Wm. H. Ryder	12	45
216	Washaka	14	47
217	William Matheson	18	72
218	Yakima	18	71
219	Yankee	9	18
	Totals	3507	15071

MARINE AND FISHERIES

5 GEORGE V., A. 1915

PACIFIC COAST PORTS.

Name of Vessel.	Tonnage.	Number of crew.	Vancouver.	Nanaimo.	Victoria.	Prince Rupert.	Totals.
Kingfisher......	141	37	11	3	.....	4	18
Manhattan......	134	37	10	3	.....	1	14
New England...................................	70	34	10	3	....	1	14
Totals............	345	108	31	9	....	6	46

SESSIONAL PAPER No. 39

UNITED STATES Fishing Vessels to which Licenses were issued under the Act entitled "An Act~to protect the Customs and Fisheries," during the Fiscal Year ended 31st March, 1914.

Name of Vessel.	Port of Registry.	Ton-nage.	Port of Issue.	Amount.
				$ cts.
Maxime Elliott................ ...	Gloucester........	75	House Harbour........	112 50
Alice....	Boston	62	"	93 00
Tacoma....	Tacoma	71	"	106 50
Atlanta......	Gloucester........	74	Arichat................	111 00
Alaha	"	100	"	150 00
Fannie Prescott......	Boston............	87	Woods Harbour........	130 50
Rex	Gloucester........	94	Canso.....	141 00
Premier	"	97	"	145 50
Monitor	"	100	"	150 00
Fannie A. Smith........	"	87	. "	138 00
A. Pratt Andrew................	"	92	"	126 00
Vanessa ...., ...	Boston	84	"	147 00
Elsie	"	98	"	132 00
Govoner Foss...........	Gloucester...........	88	"	148 50
Stilleto.................... .. .	"	99	"	144 00
Cavalier....................	"	96	"	138 00
Thos. S. Gorton................	"	92	"	106 50
Zakima	"	71	"	124 50
Elk	Boston	83	"	115 50
Lucinda I. Lowell........	Gloucester	77	" ....:	106 50
Flora L. Oliver	"	71	"	139 50
John Hays Hammond.............	"	93	"	106 50
Frances P. Mesquito ... .:.....	"	71	"	75 00
Olympia....................	"	50	Halifax................	135 00
Richard	"	90	"	145 50
Arkana....................	"	97	Liverpool...............	126 00
Elma E. Gray................ ...	Boston	84	"	111 00
Senator	Gloucester....	74	"	121 50
Rhodosa	"	81	"	148 50
Sylvania.................... ..	"	99	North Sydney	160 50
Arithusa..	" ....:..	107	Port Hawkesbury......	124 50
Keneo	"	83	"	117 00
Mystery	"	78	"	133 50
Preceptor....................	"	89	"	127 50
Avalon....	"	85	"	144 00
Athlete......	"	96	Shelburne..............	202 50
Tattler...	"	135	"	142 50
Laverna	"	95	Sand Point....	118 50
Margaret.................	"	79	"	156 00
Conqueror....................	"	104	"	154 50
Ingomar	"	182	Shelburne	156 00
Oriole.	"	104	"	144 00
Jas. W. Parker......	Boston	96	Sand Point	157 50
Oneta.....................	"	105	"	127 50
Mary F. Curtiss	Gloucester............	85	"	138 00
Catherine Burke................	"	92	Louisburg	115 50
Olga	"	77	Wedgeport	186 00
J. J. Flaherty....................	"	124	"	141 00
Senator Gardner...............	"	94	Yarmouth	54 00
Byron H. Mayo	South West Harbour....	36	"	135 00
T. M Nicholson.................	Bucksport, Me.........	90	St. Peters.............	55 50
Boyd & Leeds	Salem, Mass............	37	Woods Harbour........	67 50
Wm. H. Rider.................	Gloucester	45	Canso	70 50
Thalia	"	47	Liverpool....	163 50
Independence 2..................	"	109	Shelburne	130 50
Selma.......................	Boston	87	Sand Point............	73 50
Rebecca	"	49	"	87 00
Romona	Gloucester.....	58	North Sydney	142 50
Hiram Lowell..	Bucksport, Me.........	95	Louisburg	106 50
Lizzie Griffin....	Bangor, Me............	72	"	136 50
Smuggler......................	Gloucester............	91	Wedgeport	123 00
Flirt.....	"	82	"	72 00
Ralph Russell................. ..	"	48	Pubnico	67 50
Etta Mildred.................	"	45	"	

5 GEORGE V., A. 1915

UNITED STATES Fishing Vessels to which Licenses were issued under the Act entitled "An Act to protect the Customs and Fisheries," during the Fiscal Year ended 31st March, 1914—*Concluded.*

Name of Vessel.	Port of Registry.	Ton-nage.	Port of Issue.	Amount.
				$ cts.
Eugenia	"	66	Yarmouth	99 00
Patrol	"	58	Pubnico	87 00
Ella G. King	"	52	Souris, P.E.I.	78 00
Agnes	"	75	North Sydney	112 50
Morning Star	Boston	85	Yarmouth	127 50
Thos. A. Cromwell	"	85	Sand Point	133 50
Arthusa	Gloucester	107	"	160 50
Jessie Costa	Boston	89	Yarmouth	133 50
Morning Star	"	55	"	127 50
Mertis H. Perry	S. W. Harbour, Me	54	"	54 00
Georgia	Gloucester	65	Sand Point	97 50
Esperanto	Not known	91	"	136 50
Francis P. Mosquito	Gloucester	71	"	106 50
Jas. W. Parker	Boston	96	"	144 00
Ingomar	Gloucester	103	Shelburne	154 50
John Hays Hammond	"	93	Halifax	139 50
Harmony	Boston	81	"	121 50
Sylvania	Gloucester	99	"	148 50
Yakima	"	71	Liverpool	106 50
Lillian	Boston	95	Yarmouth	142 50
Mystery	Gloucester	79	Halifax	118 50
Georgiana	Boston	87	"	130 50
Independence 2	Gloucester	109	Liverpool	163 50
Athlete	"	96	Lockeport	144 00
Hazel R. Hines	"	79	Pubnico	118 50
Bohemia	"	86	Tusket	129 00
J. J. Flaherty	"	124	Wedgeport	186 00
Annie M. Parker	"	100	Tusket	150 00
Senator Gardiner	"	94	Wedgeport	141 00
				11,729 50

APPENDIX No. 18

THE OUTSIDE STAFF OF THE FISHERIES BRANCH

LIST OF INSPECTORS OF FISHERIES IN THE DIFFERENT PROVINCES OF THE DOMINION OF CANADA, 1913–14.

Names.	P.O. Address	Extent of Jurisdiction.
McLeod, A. G.........	Whitney Pier, Sydney, N.S.........	District No. 1.—Cape Breton Island.
Hockin, Robt...........	Pictou, N.S........	District No. 2.—Cumberland, Colchester, Pictou, Antigonish, Guysboro', Halifax and Hants Counties.
Fisher, Ward..........	Shelburne, N.S. ...	District No. 3.—Lunenburg, Queens, Shelburne, Yarmouth, Digby, Annapolis and Kings counties.
Calder, John F.........	Campobello, N.B....	District No. 1.—The counties of Charlotte and St. John.
Morrison, Donald......	Newcastle, N.B.....	District No- 2.—Restigouche, Gloucester, Northumberland, Kent, Westmorland and Albert counties.
Harrison, H. E.	Fredericton, N.B....	District No. 3 —Kings, Queens, Sunbury, York, Carleton and Victoria counties.
Matheson, J. A.........	Charlottetown.......	Prince Edward Island.
Wakeham, Wm., M.D..	Gaspé Basin, Que...	Lower St. Lawrence river and gulf.
Bernard, C. A.........	St. Césaire..........	Eastern Townships.
Riendeau, Jos..........	Longueuil, Que.	The counties of the province of Quebec bordering on the St. Lawrence from Huntingdon to Three Rivers.
Foster, T. J............	Sault-Ste.Marie,Ont.	The districts of Rainy River, Thunder Bay, Algoma, Nipissing, Parry Sound, Muskoka ; and the counties of Simcoe, Grey and the Georgian bay side of Bruce county to Cape Hurd, including the waters of and around Manitoulin island and islands in its vicinity, as well as the waters of Georgian bay, North channel, and the Canadian waters of Lake Superior.
Sheppard, O. B........	Toronto, Ont.......	That portion of the county of Bruce bordering on Lake Huron from Cape Hurd south, and the waters within the said county, as well as the counties of Huron, Lambton, Essex, Kent, Elgin, Norfolk, Haldimand, Welland, Middlesex, Oxford, Perth, Brant, Waterloo, Wellington and Dufferin, and the Canadian waters of Lakes Huron, St. Clair and Erie, and connecting waters and Niagara river down to Niagara falls.
Hurst, J. S	Belleville, Ont	The remainder of the province of Ontario, embracing the Canadian waters of Niagara river from Niagara falls, as well as the Canadian portion of Lake Ontario and the St. Lawrence river, and the Ontario half of the Ottawa river up to, and including, the portion thereof in the county of Renfrew, as well as the whole of Lake Simcoe.
Howell, Capt. J. A.....	Selkirk, Man........	Lake Winnipeg & Northern Waters. Chief Inspector.
Reid, D. F..	509 Boyd Bldg, Winnipeg, Man..	Southern and Western parts of the Province.
Davidson, Geo. S.......	Fort Qu'Appelle..	Province of Saskatchewan. Alberta and district of McKenzie } Chief Inspector.
MacDonald, G.C.	Prince Albert, Sask.	Province of Saskatchewan.
Wilson, Justus...	Noyes Crossing, Alta	Northern Alberta.
Payson, C. C...	Dawson City..	Yukon District.
Cunningham, F. H.. ...	New Westminster...	Province of British Columbia—Chief Inspector for the Province.
Halladay, A. P.	"	Province of British Columbia—Assistant Inspector—No. 1, Southern district.
Williams, J. T	Port Essington......	Province of British Columbia—No. 2, Northern district.
Taylor, E. G.....	Nanaimo.......	" " " No. 3, Vancouver Island.

5 GEORGE V., A. 1915

LIST OF FISHERY OFFICERS IN THE DOMINION OF CANADA, 1913-14.*

NOVA SCOTIA.

Annapolis County.

Name of Officer.	P. O. Address.	Extent of Jurisdiction.
Purdy, Walter..........	Deep Brook	Annapolis County.

Antigonish County.

McDougall, Hugh	Cross Roads, Ohio..	Antigonish County.

Cape Breton County.

King, H. A............	Little Bras d'Or. ...	Cape Breton County.
Gillis, D. M	Grande Mira........	" "
McCuish, John..........	Bateston............	" "
Hall, Edward	Main-à-Dieu........	" "
McDonald, Allan........	Gabarouse Lake.....	" "
McLean, Murdock......	Jacksonville	" "
Ferguson, N	Port Morien........	" "
Sullivan, Timothy......	Florence, Sydney M.	" "
Burke, Wm......	Mira Ferry	" "
Gillis, J. A.	Grand Mira.... ..	" "

Colchester County.

Marsh, Lowell..........	Central Economy....	Colchester County.
Langille, B. S..........	Tatamagouche..	"
McCleave, J. H....	Lower Stewiacke....	"

Cumberland County.

Angevine, Frank......	Middleboro.........	Cumberland County.
Hunter, Clark T........	Linden.............	"
Holmes, Capt. D. W....	Parrsboro...........	"
Kirwan, Frank.	Wallace	"
Smith, R. S	Pugwash.	"
Embree, Jas. E.........	Oxford...	"

Digby County.

Torrie, G .E...........	Digby.......	Municipality of Digby, Digby County.
Aymar, Wm........ ...	Meteghan	Municipality of Clare. "

Guysboro County.

Dillon, John A	Guysboro....	Guysboro County.
Cooper, R. V.....	Wine Harbour......	"

Halifax County.

Gaston, Robt...........	Tangier...........	Halifax County.
Kennedy, Thos........	Hubbards..........	"
Rowlings, George.......	Musquodoboit Harb.	"

* Revised up to October 1914.

LIST of Fishery Officers in the Dominion of Canada—*Continued.*

NOVA SCOTIA—*Continued.*

Hants County.

Name of Officer.	P.O. Address.	Extent of Jurisdiction.
Salter, R. J. U.........	Brooklyn.	Hants county.
Rose, Thos.............	Urbanian	"

Inverness County.

LeBlanc, Lazare........	Eastern Harbour....	From Big Pond lobster factory north, including Cheti-camp, Eastern Harbour, Little River, Pleasant Bay and Follets Cove.
Cody, M. J.............	S. W. Margaree	Inverness coast from Broad Cove Chapel to Delany's Cove, also East Lake Ainslie and streams, Loch Ban, S.W. Margaree river and tributaries, and Margaree river from forks of Margaree harbour.
Ross, Jas. J............	N. E. Margaree......	Coast of Inverness Co., from Delany's Cove northward, including Big Pond, Eastern Harbour, etc., also N. E. Margaree river from Margaree forks to source, and all other streams to Victoria County line.
McLellan, D. N	Dunvegan	Inverness county.
McIntosh, Geo. P.......	Pleasant Bay...... ..	Coast of Inverness County extending from Pleasant Bay to Meat Cove (inclusive).
McLennan, Jno. B......	Kingsville	Inverness County.
McDonald, A. J........	Seaside, Port Hood .	W. division coast south of Mabou Harbour, including S. W. Mabou river, Port Hood, Judique, Long Point, Pt. Hastings and Hawkesbury, to N.W. Arm River Inhabitants in interior; and north side Victoria Co., from Js. McKinnon's to Whycocomagh bay; and through Glencoe and S. W. ridge of Mabou to Mabou bridge.

Kings County.

Chute, Capt. Edward....	Canada Creek.	Kings county.
Reid, Reuben F.........	Wolfville......	"
Rathbone, C. F. A.....	Hortonville..... ...	"

Lunenburg County.

Hebb, L. J.	Lunenburg..........	Lunenburg county.
Evans, Austin..........	Chester.	"

Pictou County.

Sutherland, Robert	River John........	Western division Pictou Co., comprising coast water from Colchester county line to Cole's reef, Pictou Harbour and streams flowing into it, viz., River John and tributaries, Toney river, and Big and Little Cariboo rivers.
Germain, Wm.........	Reidway...........	Pictou county.
McDonald, D. L.......	Bailey's Brook.....	"
Pritchard, A. O........	New Glasgow.......	Pictou harbour, Pictou island, East, West and Middle rivers, Pictou county.

39—27

'5 GEORGE V., A. 1915

LIST of Fishery Officers in the Dominion of Canada—*Continued.*

NOVA SCOTIA—*Concluded.*

Queens County.

Name of Officer.	P. O. Address.	Extent of Jurisdiction.
Fraser, W. E.........	Liverpool........	Queens county.
Young, Chas.........	Mill Village........	"

Richmond County.

Sampson, Anthony......	Lower L'Ardoise....	That portion of sea coast, lakes and inland waters lying east of St. Peter canal.
Boudrot, Capt. Sylvester.	Petit de Grat.....	Coast and inland waters of Isle Madame, including southerly half of waters of Lennox passage.
Thibeau, P. J..........	Thibeauville........	Richmond county.

Shelburne County.

Stoddard, Henry........	Shag Harbour	From and including Clyde river to Yarmouth county line.
Walls, George..........	Shelburne....	Shelburne county.

Victoria County.

Campbell, Jno. M......	Care Marine Agent at Halifax.	St. Paul's Island.
McAulay, Allan	Big Baddeck........	Victoria county.
Hellen, Wm........ ..	Cape North	Northern part of Victoria county.
McDonald, A. M.......	Plaster, North Shore.	Englishtown north to Smoky cape at South Ingonish.
Grant, Dan. J........	Boulardarie East....	Big Bras d'Or north to Englishtown.
Donovan, J. T..........	Ingonish....,....	North and South Ingonish, including Ingonish island.
McDonald, Wm. A	Brook, Middle River	Victoria Island.

Yarmouth County.

D'Entremont, J . G.....	Middle, W. Pubnico.	Yarmouth county.

NEW BRUNSWICK.

Albert County.

Conner, N. D	Alma........	Albert county.

Charlotte County.

Worrell, Robert.........	St. Andrews.......	Waters in vicinity of St. Andrews, extending from Owen head to Oak bay.
Fraser, W. A..........	Woodward's Cove, Grand Manan.....	Island of Grand Manan, and waters surrounding the same.
Brown, Burden.........	Wilson's Beach.. ..	Campobello, and the West Isles. Charlotte county.
Lord, C. H....	Lord's Cove, Deer Is.	West Isles.
Justason, E. C..........	Pennfield	Charlotte county.
McNichol, Elgin........	Letete............	"
Daley, Patrick..........	Lepreaux	"

SESSIONAL PAPER No. 39

LIST of Fishery Officers in the Dominion of Canada—*Continued.*

NEW BRUNSWICK—*Continued.*

Gloucester County.

Name of Officer.	P. O. Address.	Extent of Jurisdiction.
Canty, Thomas	Bathurst	Gloucester county.
Doucet, Jas. P.	Elm Tree	"
Arseneau, Edmond	Inkerman	"
Sewell, Edmund	Pokemouche	"
Mourant, John A.	Caraquette	"
Ache, Adolphe	Shippegan	"

Kent County.

Hannah, Wm F.	Richibucto	Kent county.
Allain, P. A	Buctouche	Coast line and inland waters of the parishes of Wellington and St. Marie.
Després, E. T.	Cocagne Bridge	Kent county.

Madawaska County.

Gagnon, L. A	Edmundston	Madawaska county.

Northumberland County.

Abbott, Lemuel H	Chatham	Both shores of Miramichi river from Point au Quart on south and Oak point on north to junction of N.W. and S.W. Miramichi rivers, with all islands therein and tributary streams.
McDonald, Ronald	Bayside	Northumberland County.
Williston, Wathan	Baie du Vin	"
Parker, L. P	Derby	"
Sutherland, M	Red-Bank	"

Queens County.

Holmes, Wm.	Gagetown	Queens County.
Worden, A. C	Cody's	"

Restigouche County.

Hamilton, Wm C.	Black Lands	Baie des Chaleurs, and tributaries, from Belledune to Dalhousie.
Ferguson, Ebenezer	Pointe La Nim.	Restigouche river and its tributaries in the counties of Restigouche and Victoria.
Mowat, Max.	Campbellton	Restigouche County.

Sunbury County.

Babitt, Fred	Swan Creek	St. John River from Indiantown, Sunbury county, to the county line of York.

St. John County.

Brittain, B. B.	55 Middle street, St. John West	St. John county.

5 GEORGE V., A. 1915

LIST of Fishery Officers in the Dominion of Canada—*Continued.*

NEW BRUNSWICK—*Concluded.*

Victoria County.

Name of Officer.	P. O. Address.	Extent of Jurisdiction.
Watson, Chas. F........	Drummond.........	Victoria county.

Westmorland County.

Name of Officer.	P. O. Address.	Extent of Jurisdiction.
Vienneau, Siffroid	Barachois	Coastal and inland waters of parish of Shediac, and portion of Botsford parish, north of Big Shemogue Hr., and road from same to near Bristol corners, past Bristol corners and Lowthers to parish of Sackville, with jurisdiction in parishes of Moncton and Salisbury.
Belliveau, Philip.......	Pré-d'en-haut......	Parish of Dorchester, including Petitcodiac river.
Prescott, Robert........	Baie-Verte.	Part of Botsford parish, county of Westmorland.
Prescott, Joseph........	"	Parishes of Westmorland and Sackville.

York County.

Name of Officer.	P. O. Address.	Extent of Jurisdiction.
Niles, Thos.............	Fredericton.........	York county.
McNally, Alex.	Lr. French Village..	"

PRINCE EDWARD ISLAND.

Kings County.

Name of Officer.	P. O. Address.	Extent of Jurisdiction.
Keays, John............	Souris	Kings county.

Prince County.

Name of Officer.	P. O. Address.	Extent of Jurisdiction.
McFarlane, John.......	Cape Traverse ...	Prince county.
Quinn, Geo.....	Leoville ....·..	"

Queens County.

Name of Officer.	P. O. Address.	Extent of Jurisdiction.
McAulay, A. C.	Tracadie Cross......	Queens county.

PROVINCE OF QUEBEC.

Gaspé and Bonaventure Counties.

Name of Officer.	P. O. Address.	Extent of Jurisdiction.
Kennedy, Frederick	Douglastown.	That portion of the province south of the St. Lawrence, to and including county of Bellechasse, but especially the counties of Bonaventure and Gaspé.

Quebec County.

Name of Officer.	P. O. Address.	Extent of Jurisdiction.
Migneault, T...........	140 St. François St., Quebec.	From Quebec to the Saguenay river on the north shore and from Quebec to Rimouski on the south shore.

LIST of Fishery Officers in the Dominion of Canada—*Continued.*

PROVINCE OF QUEBEC—*Concluded.*

Magdalen Islands.

Name of Officer.	P. O. Adress.	Extent of Jurisdiction.
Chiasson, Cirice	House Harbour	Magdalen islands.
Chevrier, J. A	Havre Aubert	That part of Magdalen islands comprising Entry, Amherst and Grindstone islands, also Harbour Basque lagoons.

Saguenay County.

Comeau, N. A	Quebec	Saguenay county.
Levesque, Elzear.	Seven Islands	"
Le Blanc, E	Esquimalt Point	"
Landry, Wilfrid	Natashquan	"
Cormier, A	Esquimalt Point	"
Evans, T. W.	St. Augustine'	"
Kennedy, Jas.	Baie des Rochers	"

PROVINCE OF MANITOBA.

White, C. L.	Winnipegosis	Manitoba.
Stevenson, E. H	Le Pas	Keewatin district.
Daly, Daniel S	Selkirk	Manitoba.

SASKATCHEWAN.

McNicol, Duncan	Wadena	Wadena district.
Hunter, G. S	Dilke Lake	District of Long Lake, Qu'Appelle river, bounded on south by base line Tp. No. 16, on north by Tp. No. 30, on east by east side of Range 19, and on West by West side of Range 27, all west of 2nd meridian.
Fitzgerald, Ira	Meota	Jackfish lake district.
Clarke, Thos	Montreal' Lake P.O.	Red Deer Lake and Lac la Rouge District.
Beatty, Edward	Green Lake P.O	Isle la Crosse district.

ALBERTA.

Hoad, Nelson J	639 6th Ave. west, Calgary.	Southern Alberta.
Wood, Ingram	Wetaskiwin	Pigeon lake, etc.
Travers, Oliver	Grouard	Lesser Slave Lake and vicinity.

BRITISH COLUMBIA.

District No. 1.

John McLeod	Nelson	Kootenay district.
Charles J. Godwin	Vernon	Yale district.
Horatio Shotton	Kamloops	Kamloops district.
J. L. Hill	Quesnel	Lillooet district, north of Clinton, Cariboo and Cassias.
D. J. M. Perkins	Fort George	

District No. 2.

Gunner Saugstad	Rivers Inlet	Rivers Inlet District.
Stewart Norrie	Prince Rupert	Prince Rupert district.
W. T. Adamson	Naas	Naas River District.
John Widsten	Bella Coola	Bella Coola and Kinsquit districts.
Chas. Harrison	Massett, Q.C.I.	Queen Charlotte Islands.
James Boyd	Vancouver	Central Division, district No. 2.

5 GEORGE V., A. 1915

List of Fishery Officers in the Dominion of Canada—*Concluded.*

BRITISH COLUMBIA—*Concluded.*

District No. 3.

Name of Officer.	P. O. Address.	Extent of Jurisdiction.
W. M. Galbraith	14 Ridge Road, Victoria	Cowichan River district.
John Grice	Clayoquot	Clayoquot Sound district.
J. B. Wood	Alberni	Alberni district.
R. M. Colvin	Cowichan Station	Cowichan district.
Harry McIndoo	Nanaimo	Nanaimo district.
Capt. Harry Beadnall	Courtnay	Comox district.
A. F. Lloyd	Quathiaski	Campbell River district.
Alex. Lucas	Alert Bay	Alert Bay district.
Arthur Newland	Welcome Pass, Pender Har	Pender Harbour district.

LIST OF OFFICERS IN CHARGE OF GOVERNMENT FISH HATCHERIES, ETC., 1913-14.

Name.	P. O. Address.	Province.	Rank.
McLeod, A. W	Belleville	Ontario	Officer in charge Government Fish Hatchery.
Parker, Wm	Sandwich	"	"
Parker, Ray	Kenora	"	"
McNab, A. J	Port Arthur	"	"
Eldridge, W. J	Wiarton	"	"
Laschinger, A. G	Sarnia	"	"
McDougall, A	Southampton	"	"
Clark, Matthew	Collingwood	"	"
Lindsay, R. C	Gaspé	Quebec	"
Meilleur, Jos	Mont Tremblant	"	"
Audet, L. A	Magog	"	"
Elliot, Jos	St. Alexis des Monts	"	"
Catellier, J. N	Tadousac	"	"
Belknap, W. G	Baldwin's Mills	"	"
Mowat, Alex	Campbellton	New Brunswick.	"
McCluskey, F. J	Grand Falls	"	"
Sheasgreen, Wm	South Esk	"	Acting Officer in charge.
Brittain, B. B	St. John West	"	Officer in charge.
McAfee, Geo.	R. F. D. No. 4, Lakewood	"	"
Ogden, Alfred	Bedford	Nova Scotia	"
McDiarmid, Donald	N.E. Margaree	"	"
Burgess, Frank	Windsor	"	"
Burton, L. J	Middleton	"	"
Holroyd, A. W	Winsloe Station	P. E. Island	"
Paulson, C. P	Selkirk	Manitoba	"
Grenon, Jos. O	Winnipegosis	"	"
Craig, Samuel	Fort Qu'Appelle	Saskatchewan.	Acting Officer in charge.
Rodd, R. T	Banff	Alberta	Officer in charge.
Robertson, Alex	Harrison Springs	British Columbia	"
Mitchell, D. S	Tappen	"	"
Graham, T. W	Lillooet	"	"
Gibbs, H. L	Hazelton	"	"
Martin, J. E	Tofino	"	"
Bothwell, David	Kildonan	"	"
Castley, J. H	Duncan	"	"
Crawford, H. C	Fort St. James	"	"
Hamer, J. N	Rivers Inlet	"	"
Catt, James	Lakelse	"	"
Ogilvie, L	Gerrard	"	"

APPENDIX No. 19.

ANNUAL REPORT ON THE BIOLOGICAL STATIONS OF CANADA FOR THE YEAR 1914.

The three stations were in operation as usual during the season, the British Columbia station being of course open all the year and the work was under the superintendence of Dr. McLean Fraser at the Departure Bay station, Nanaimo, B.C., Dr. J. W. Mayor (of the University of Wisconsin) at Go-Home Bay, Ont., and Dr. A. T. Huntsman at St. Andrews, New Brunswick.

As is to be expected, there are fluctuations each year in the number of workers who form the staff at each station, and both at Nanaimo and at Georgian Bay, the researches carried on were mainly conducted by the curators in charge, no other regular workers resorting to these laboratories. At St. Andrews, however, the accommodation was taxed to its utmost, and some applications had to be refused. Faunistic work was carried on actively and the collections of specimens, which are being preserved for future important uses, were considerably increased at Nanaimo and at St. Andrews. The work at the former station has been aided by the securing of a fine gasoline launch, at present called the *Ordoness* which is unusually well fitted and equipped for marine researches. The station at St. Andrews has a similar advantage in now possessing a fine gasoline boat, the *Prince,* which enables several most profitable trips to be made to Grand Manan, St. Mary's Bay, Nova Scotia, and many other localities important for fishery investigations, but hitherto not very accessible to the staff.

At the Departure Bay station Dr. McLean Fraser who had expressed a wish to be released of the office of curator, and officer in charge of researches, again remained in responsible control and carried on a very important research upon the herring, obtaining the ova in March and hatching out the young fry towards the end of the month. He aided Dr. A. T. Cameron, of the University of Manitoba, in an investigation of certain marine algæ, valuable as yielding iodine. Dr. Cameron's iodine investigations are of extreme economic value and it is hoped that they will be continued and completed during the season of 1914. Dr. E. M. Walker, and some scientific assistants, also carried on marine researches, and the usual programme of fishery and technical investigations was carried on.

The library of the former curator, the late Rev. George W. Taylor, was purchased by authority of the Biological Board, and is now the property of the station.

At Georgian Bay the staff was small but J. W. Mayor, now of the University of Wisconsin, carried on important researches and Dr. Klugh of Queen's University, and some other workers spent brief periods at the Go-Home Bay station. Some extremely valuable reports from this station, including Dr. B. A. Bensley's beautifully illustrated report on the 'Fishes of Georgian Bay,' will be printed in the forthcoming volume of 'Contribution to Canadian Biology,' now in the hands of the King's Printer.

At the St. Andrews station, New Brunswick, a most successful season was experienced, Dr. A. T. Huntsman again acting most efficiently as curator. Professor Knight (Queen's), Cox (Fredericton), Perry (Acadia, N.S.), and Prince, chairman of the Biological Board, made a more or less lengthy stay at the station, and valuable work was carried on. Mr. A. R. Cooper, Toronto, Mr. Millar (Queen's), Mr. Detweiler (Queen's), Mr. Wallace (Toronto) and others were amongst those who

5 GEORGE V, A. 1915

conducted special researches, upon some of which, reports are already completed or in an advanced state. Several most fruitful trips were made to Grand Manan, and some interesting grounds near St. Andrews. The launch *Prince* gave great facility in making these trips when dredging, townetting, physical, chemical and other work was carried on. One trip was to Sandy Bay, St. Mary's Bay, Nova Scotia and the visits to fishing centres such as Tiverton, Westport, etc. proved of great value to the staff. A large collection of valuable material was secured including interesting fish eggs and newly hatched larvæ. Mr. Martin (Toronto University) left the party at Long Beach in order to spend some time in chemical, physical and biological investigations at the Government lobster pound, Digby Neck. The officer in charge of the lobster pound, aided in the work and Mr. Martin has completed a very interesting report, which has been submitted to the Deputy Minister of Marine and Fisheries. Official consent was given, at the earnest request of Mr. Hartt, M.P., for a fisheries exhibit, by the biological station at the St. Stephen exhibition in September. It was one of the great features of the exhibition and proved most attractive, not only to fishermen, but to the general visitors who crowded the building all the time. The cases and glass-vessels with their wonderful contents (fishes, crustaceans, echinoderms, etc.) excited very great interest. One special piece of work was carried on by Professor Knight at a small waterfall near St. Andrews, viz.: the testing of an elevator fish-pass on a new principle suggested by Professor Prince. The Department of Marine and Fisheries have given instructions that a pass on this model is to be erected at the impassable falls on the Magaguadavic River, St. George, N.B. The fish-pass has been built at St. John and will be in operation in the season of 1914. . If completely successful, this fish pass perfected by the experiments at the Biological Station, will be a vast utility on rivers and streams obstructed by dams and not surmountable by fish.

It only remains to add that the Government are so convinced of the possibilities and the value of the work of the biological stations, that the parliamentary vote is to be increased by $4,000 for 1914-15 to enable halibut, herring, and other special researches to be carried on.

Mention may be made of the fact that notable fishery investigations and technical researches carried on under the auspices, and in some cases at heavy cost by the biological board, are published by other boards and organizations.

It was pointed out that Mr. F. A. Potts, Trinity Hall, Cambridge, England, published most important scientific results of work done at the B.C. biological station in German and English journals. Miss Pixell and Miss Haddon also published remarkable scientific papers, the materials for which were obtained at the Canadian biological stations. The Conservation Commission published a book on the oyster by Dr. Stafford, which embodied many years of work done at the expense of the board, and illustrated by drawing, the artist of which was paid by the board. Reference may be made to a paper on Pacific salmon by Professor McMurrich, a paper on B.C. hydroids by Dr. McLean Fraser, Royal Society of Canada, in the 'Provincial Museum Journal,' Victoria, B.C., and a paper on Tunicates by Dr. Huntsman, in the 'Canadian Institute Transactions' and other important papers published under other auspices than those of the board. Doubtless when full credit is given by authors to the biological stations, such outside publication is of value to the stations, but it is to be hoped that all results of valuable work carried on at the Canadian stations may, in some form, be published in the volumes issued by the board.

A. B. MACALLUM, F.R.S., *Secy.-Treasurer,*
Biological Board of Canada.

REPORTS

ON

FISHERIES INVESTIGATIONS

IN

HUDSON AND JAMES BAYS AND TRIBUTARY WATERS

IN

·1914

BY

C. D. MELVILL

A. R. M. LOWER

AND

NAP. A. COMEAU

DEPARTMENT OF THE NAVAL SERVICE

APPENDIX

TO THE ANNUAL REPORT OF THE DEPARTMENT OF THE NAVAL SERVICE
FOR THE FISCAL YEAR ENDING MARCH 31, 1914.

OTTAWA.
PRINTED BY J. DE L. TACHÉ, PRINTER TO THE KING'S MOST
EXCELLENT MAJESTY
1915

[No. 39a—1915]

REPORT

ON THE

EAST-COASTAL FISHERIES OF JAMES BAY.

BY

C. D. MELVILL, F. R. G. S.

OTTAWA, October 20, 1914.

To the Deputy Minister,
    of the Naval Service,
        Ottawa, Ont.

SIR,—I beg to submit my report on investigations undertaken during the summer and fall of 1914 into the Fisheries of the South and East Coasts of James Bay.

In submitting this report I desire to acknowledge the hospitality and kind consideration shown me by the officers of both the Hudson Bay Company and Messrs. Revillon Frères Trading Company. The help and information given also by these companies greatly assisted the expedition.

I have the honour to remain, sir,
                Your obedient servant,

C. D. MELVILL.

INSTRUCTIONS RECEIVED FROM THE DEPARTMENT.

This report is the result of investigations undertaken during the summer of 1914 into the value of the commercial fisheries of the South and East Coastal Waters of James bay, and as far as possible the tributary waters.

The full instructions received from the Department of Marine and Fisheries being as follow:—

To proceed by canoe by the most feasible route to Moose Factory (a Hudson Bay Company trading post on James bay) thence around the coast as far north as Cape Jones (54° Lat.) the northeastern limit of James bay, obtaining as far as possible information on the following points:—

(1) To ascertain the different kinds of food fish to be found in the bay and its tributary waters; and the extent of each kind of fishery;

(2) To ascertain the period of time of the runs of anadromous fish;

(3) Keeping in view local conditions, to report on the most feasible methods of catching the various kinds of fish, and what regulations should be adopted for their conservation, should commercial fishing be undertaken;

(4) As far as possible, to obtain all information regarding the spawning areas available in the upper reaches of the rivers for the different kinds of fish;

(5) To investigate the conditions of climate and all local influences relative to their affecting the value of the fisheries;

ITINERARY.

In accordance with the instructions received from the department, in company with Mr. A. M. Lower, I left Ottawa for Cochrane on the evening of June 4. The canoemen engaged for the expedition (Messrs. Duncan McNab and Angus Chevrier) had previously been instructed to meet me at Haileybury.

Three days were spent at Cochrane, outfitting and waiting for the National Transcontinental Railway construction train on which we were able to obtain passage to Missanaibie.

Leaving Cochrane on June 9, we reached the Missanaibie river that evening and at once loaded up the canoe and went a few miles down the river. The river being very low for the time of year, great care had to be exercised at the numerous rapids, which with higher water would have been drowned out. In all six portages (four short and two long) were made between Missanaibie and tide water.

On June 18 the junction of the Mattagami and Missanaibie rivers was passed, the river hereafter being known as the Moose. On June 20 we reached Moose Factory, the headquarter post of both the Hudson Bay Company and Messrs. Revillon Frères Trading Company in James bay.

We left Moose Factory for Charleton island on June 25 in the small steamer *Inninu*, being indebted to Mr. F. D. Wilson, District Manager of the Hudson Bay Company, for his kindness in giving us passage. Charleton was reached on the afternoon of June 26, after a very cold miserable passage. Three days were spent at Charleton island, thence we proceeded to Rupert's House (45 miles south). Leaving Rupert's House on July 2, we arrived at Sherrick mountain* on July 3, and East

---

* Elevation about 700 feet, a conspicuous landmark, being the highest hill or mountain around James bay.

5 GEORGE V., A. 1915

Main Fort on July 6. Stormy weather delaying us two days, it was not until July 9 we could start for Fort George, accompanied by three Indian families in four canoes, and by a large canoe belonging to Mr. W. G. Todd, of Pittsburg, U.S.A., who was making a collection of the birds of James bay.

We reached Fort George (Big river) on July 15. Bad weather again delaying us, we were unable to leave before July 19. On July 23 we arrived at Cape Jones, after three very cold and wet days' travel. Three days were spent at Cape Jones near the Indian and Eskimo camps and some valuable information was obtained regarding the so-called Arctic salmon. We were fortunate in catching a few of these fish, although the natives stated it was yet too early for them in any quantity.

Leaving Cape Jones (the northern limit for the expedition according to my instructions from the department) on July 27, we arrived at Kakashewan point on July 28, and Brandy bay the following day. Bad weather delayed us here one day, but this delay enabled us to make an excellent whitefish fishery. Fort George was reached on July 31.

Arrangements were here made with Messrs. Revillon Frères for the hire of the small schooner *Violet* for a trip to the North Twin island.

Leaving Fort George on August 5, stormy weather repeatedly prevented us beating out into the bay and making the 65-mile crossing, so that it was not until August 12 that we reached the island, but meanwhile a few days had been spent at Long Point, Eskimo Duck islands and other places, which proved excellent fishing grounds.

We returned to Fort George on August 15; since my instructions were to return to Ottawa about the beginning of October, I considered it nearly time to commence our long return journey south. On August 17 we left Fort George, arriving at East Main on August 27, after many days of northwest gales and foggy weather in which we were unable to travel.

Rupert's House was reached on September 1, and here bad weather again stopped us until the 6. On September 8 we camped near the mouth of the Nottaway river and Sawayan point on September 10.

After experiencing some difficulty in Hannah bay through our ignorance of the tides, we reached the Harricanaw river on September 13 and West river September 15. Thence travelling day and night we eventually arrived at Moose Factory on September 17 and here heard the first news of the European war.

A week was spent at Moose fishing and obtaining such information as was possible regarding the fisheries in the neighbourhood.

Leaving Moose on September 24, and travelling by the Mattagami and Ground Hog rivers we reached the railroad on October 7, and Cochrane on the 8. The water in the rivers was extremely low, causing us to wade and drag the canoe in many places for over a mile, thus travelling was very slow.

At Cochrane I at once paid off the men, and after settling up all other accounts returned to Ottawa as soon as possible.

The expedition travelled (measured from the railroad back to the railroad) about 1,400 miles. The weather on the whole was very cold and wet; the last two weeks, however, were fairly fine, the few days spent on the Moose river seeming very hot after the cold winds of the bay.

The canoe, fishing nets and other gear provided by the department were satisfactory. I would also like to add that the two canoemen, Duncan McNab and Angus Chevrier, performed all necessary work most efficiently.

DESCRIPTION OF THE SOUTHERN AND EASTERN SHORES OF JAMES BAY.

James bay is that portion of Hudson bay lying south of a line drawn from Cape Henrietta Maria on the west to Cape Jones on the east coast. From the most southerly point in Hannah bay the distance due north to a line drawn between the two capes,

SESSIONAL PAPER No. 39a

is, roughly, 300 miles, while the average breadth of the bay is 145 miles. The area. therefore, of the whole bay is very considerably greater than that of lake Superior.

From the mouth of Moose river to Rupert's bay, the general coast line is very l'''' and flat, with extremely shallow water, deepening slowly from the shore seawards. On the southern shore at low water only mud flats covered with large and small boulders can be seen looking seawards. The shore is in most places marshy, covered with grasses, alders and willows, with numerous brackish pools for a considerable distance from high water mark; in fact, in many places it is difficult to say where the land begins and the sea ends, or vice versa. Beyond, on higher ground, is the usual forest growth of spruce, tamarack and poplar.

Situated between the Moose and Rupert's rivers, Hannah bay is so shallow th''' with the exception of the channels of the Harricanaw and West rivers, the whole bay is practically dry at low water. When a boat or canoe is left by the tide, as very oft'' cccurs, the thoughts and language of the crew can be better imagined than exprbsse'l. as they wait, perhaps out of sight of the low-lying shore line, for the return of the water to float them off.

On the east side of the bay (north of Sherrick mountain) the character of the coast changes considerably, the low marshy shores giving place to a rocky, sandy coast line fringed with innumerable islands of all sizes from a mere pile of boulders to islands some thousands of acres in area.

The water becomes very much deeper and the landing from a small boat at lo''' tide, impossible on the south coast except at the expense of a walk through two or three miles of mud and clay, becomes easy.

Navigation is comparatively easy, although many shoals and hidden boulders are present: The waters can be safely navigated by small craft, the islands and bays affording excellent shelter, the only danger perhaps for canoes when running from island to island is being caught in a heavy squall. This danger can, of course. be considerably reduced by a proper knowledge of the local weather conditions.

The country inland from the east coast appears to consist mostly of swan''' although along the rivers the soil is good. Further inland the country gradually changes to a rough plateau gradually rising to over 2,000 feet above the sea level. I cannot describe the country better than by quoting Mr. A. P. Low, of the Geological Survey, who explored this country in the summer of 1887. Mr. Low says, " The edge of the tableland leaves the coast to the north of Cape Jones and runs in a SSE. direction so that to the southward there is an interval varying from 10 to 30 miles between it and the coast. In this portion the general level is not much over 100 feet above the sea, and the soil is of post-pliocene clays and sands with alluvium, affording good land for cultivation, but as the climate is colder than on the west side it is doubtful if it will allow the successful growth of any but the hardiest cereals; good crops of potatoes, however, and other roots could be and are grown as far north as the mouth of Big river (Lat. 53° 50′)."

Rivers.

Eleven large rivers and numerous smaller ones flow into James bay on the south and east coasts, the principal being the Moose (which is composed of the Abitibi, Mattagami, Missanaibie rivers and other smaller although important tributaries), West, Harricanaw, Broadback, Rupert's, East Main, Old Factory, Big, Bishop Roggan and Seal rivers. Bishop Roggan is not, as might be supposed, the name of some enterprising missionary, but is the more interesting, from the point of view of this expedition, as this word is the English corruption of the Cree Indian word " Peshipwaytok ", meaning Fish Weir. It was on this river that the Indians in former days made basket weirs from willows for catching fish as they descended the stream.

All the rivers flowing into the south and east coasts of James bay are swift and very much broken by falls, rapids and shallows and are without exception only fit for canoes or boats of very shallow draught; the chief characteristic of them all being the great width of their beds in comparison to the amount of water to be carried. After the ice leaves the rivers and during the fall rains high water covers for a short period all or a great many of these obstructions so that navigation with larger boats might be feasible. During this summer (1914) so shallow was the Abitibi river that the Hudson Bay Company was unable to send a loaded canoe from Moose Factory to their trading post (situated some 100 miles from the junction of the Abitibi and Moose rivers).

Lakes.

The principal lakes in this district under review are as follow: Mesakami lake situated at the head of West river, Nemiskau lake on Rupert's river, Sherrick Mountain lakes. Wabstaka and Opinaka lakes on tributaries of the East Main river, and White Fish lakes on Salmon river (near Cape Jones).

None of these lakes are of a large size, Nemiskau, probably the largest, being a narrow, irregular shaped body of water about 30 miles long, but only 3 or 4 in breadth. This lake is spoken of by the Indians as being by far the best fishing lake in that part of the country; this report is borne out by its name ("Nemis"—Cree Indian for "fish"). It is situated 100 miles from Rupert's House and, roughly, 180 miles north of the National Transcontinental railroad. Should a railroad ever be built north from the National Transcontinental railroad to Rupert's bay, this lake, and others further south—such at lake Evans on the Broadback and lake Mattagami on the Nottaway river—would prove of value for commercial fishing provided the railway passed sufficiently near. The expedition had no time to visit these lakes; indeed, under present conditions a full summer would be required to reach them and properly investigate their fisheries.

Harbours.

The question of harbours relative to the fishing grounds is an important one. At the present time the Hydrographic Survey, under Mr. Jobin, are doing very considerable work in James bay sounding and surveying the natural harbours and river mouths. The current of all the rivers brings down such a great quantity of sediment that shoals and bars almost completely block the estuaries except for narrow channels in each.

At Moose River Roads the Hudson Bay Company's ship, drawing about 16 feet of water, used to anchor in what is called Ship-Hole, some 8 miles from Moose Factory, and there discharge her cargo into barges sent from the factory for this purpose. The 8 miles of estuary (from Ship-Hole to the Factory) is very shallow, so much so that it is only with the greatest care the company's small steamer *Inninu* can approach the latter place. The last few years, however, the annual ship has discharged her entire cargo at Charleton island, and this is now her only port of call in James bay. Strutton island, some 7 miles north of Charleton, is the distributing centre for Messrs. Revillon Frères, the only other fur traders in the bay. The goods for the various posts are distributed by small steamers and by schooners (sent from the various out-posts).

Although Moose Factory is no longer the principal port of the bay, it must undoubtedly be considered the capital. The inhabitants around the coast and in the interior looking on a trip to Moose in much the same light as the country people in civilized life consider a visit to their largest city. It may be one of the events of a life time.

Moose Factory is built on an island and has a beautiful site overlooking Moose river. The mission church, school, the large and numerous warehouses and dwelling

SESSIONAL PAPER No. 39a

houses of the Hudson Bay Company all whitewashed and arranged in symmetrical order, and the field cannon underneath the flag-staff all combine to make the place in some small degree resemble a government or even a military institution, at any rate from a distance. A closer inspection, however, of the average inhabitant will very quickly dispel any illusion of this kind, he is anything but martial in appearance or manner.

The Ontario Government have recently been exploring the mouth of Moose river with a view to the further extension of the Temiskaming and North Ontario railway to James bay. Whether this project has been abandoned or merely temporarily postponed, I am unable to say.

It would seem that the cost of a railroad to the mouth of Moose river, the last few miles of which would have to be built on an embankment in practically the open sea and exposed to all northerly storms, would be enormously expensive out of all proportion to any possible source of revenue. This remark applies also to Rupert's river; but Sawayan point, the peninsula which separates Hannah and Rupert bay, having deep water at low water might be made into an excellent harbour by the building of a breakwater, which would be considerably less expensive than the construction of an embankment at Moose or Rupert rivers. The large bay immediately north of Sherrick mountain (called Boatswain bay on the map) would also appear to be a natural harbour which with a comparatively small expenditure could be made into a very fair anchorage. The mouth of Big river is probably the best harbour in the bay and with a little dredging would provide good anchorage for large ships. Four miles north of Big river is Stromness harbour formed by two or three islands. This being sheltered on all sides, and having plenty of water makes an excellent anchorage. Sabaskunika and Old Factory bays will also probably prove good harbours for fishing vessels, if not for larger ships.

Islands.

The largest island of the south and east coast, but the second largest in the bay (Agumiski island on the west being the largest) is Charleton. This island is about 18 to 20 miles long and 9 in breadth, and is situated some 125 miles northeast of Moose Factory, and 45 miles north of Rupert's House. The formation of this island, the North and South Twins, the Struttons, and other islands lying between Charleton and the Twins, is what is geologically known as "Drift", being composed wholly of sand, clay and boulders with no rock "in place." The forest growth of the island is mostly small spruce with a few birch and poplar. Numerous lakes full of speckled trout are found in the interior. The Hudson Bay Company, in 1846, introduced some beaver and a few are still to be found. This probably was the first fur farm established in Canada. This year, I believe, a fox farm is to be started.

The narrow straits between Danby and Charleton islands do not freeze in winter owing to the strong current (about 5 knots at the first of the ebb or flood). Owing to this fact numerous vessels have wintered here. Early in the history of the Hudson Bay Company this island was used as a depot for the distribution of their goods, but was abandoned for nearly 250 years. A year or so ago the annual ship made this again the only port of call. The Hudson Bay Company's buildings consist of a large warehouse, two dwelling houses and a small wharf; a wreck of a Norwegian three-masted schooner completes the rather lonely scene.

Besides Charleton, the only other of the outer group of islands I visited was the North Twin. This is the largest of four islands situated some 60 or 70 miles west of East Main river, the other three being the South Twin, Walter and Spencer islands. For the trip to this island (being too far to make in a canoe) I hired at Fort George a small 10-ton schooner belonging to Messrs. Revillon Frères. Only one or two of the oldest natives at the Fort had ever been to the island and their trip had ended in disaster, the sailing boat being wrecked on a reef close to their destination; they them-

5 GEORGE V., A. 1915

selves having the unpleasant experience of having to return to the mainland in a birch bark canoe. As pilot, I had an old Indian named Matthew, who had been one of the wrecked crew already mentioned; together with three others and my two men we had what Matthew considered sufficient crew for this small boat; in reality two men ought to have been able to take her anywhere.

Delayed by bad weather we were seven days reaching the island. Seen some distance away the North Twin looms very high, though in reality the whole island is only about 100 feet above sea level; the cliffs, which from the sea look most precipitous, being only 30 or 40 feet high.

The harbour in which we dropped anchor is a deep crescent-shaped bay exposed to all 'north or east winds and is an unsafe anchorage with poor holding ground, the bottom being sand; a reef of rocks extending from the southern point of the bay gives some protection from southeasterly winds.

The island near the shore line is very marshy, with small shallow lakes filling all the depressions. Inland, at a higher elevation, the ground is covered with arctic plants; no trees growing on the island except a few small stunted spruce close to the harbour. This group of islands is a favourite breeding ground of the Canada Goose and there were hundreds of these birds to be seen in a moulting state at the time of our visit. The island is sometimes visited by Polar bears after a heavy storm, according to the report of an Eskimo family who have wintered there. Seals abound in the waters between the Walter and North Twin islands, and from the presence in such quantities of these animals, hopes were not unreasonably entertained that fish would prove to be also plentiful. Except for some tullibee we caught nothing. Perhaps if time could have been spared for a lengthy stay we might have been more successful.

The second group of islands can best be described as a maze of islands extending from Sherrick mountain to Cape Jones. Those in this group are composed of rock or boulders, the more southerly situated being heavily timbered while those north of Cape Hope, generally barren of trees, are covered with mosses and arctic plants. They are of no great elevation above the sea; Cape Hope island and Wastikun, two well-known landmarks, which loom high seen from a distance, are in reality only 200 feet or so above sea level.

It would be difficult to find the boat channel through this labyrinth of islands without a pilot, although the Indians have, it is true, set up tree logs and cairns at frequent intervals along the route. But the turns and twists in the channel are crooked and many, in addition to the fact that on these barren islands the natives have a habit of erecting upright poles on which traps are placed for the capture of the Snowy Owl, a bird which they consider a highly esteemed delicacy.

Climate.

The climate of the south and east coasts of James bay may be divided into two zones; the first, which may be said to extend from the south shore as far north as **Big** river, can be described as temperate. The second, from Big river northwards, as sub-arctic, or certainly cold temperate.

The climate with regard to fishery conditions alone matters only in so far as the freezing up of the waters may stop or impede fishing. Information on this was obtained from the white residents and natives.

Generally speaking, it would seem that the southern rivers are free of ice about the beginning of May, and about two weeks later in the bay itself there is a channel between the main body of the ice and the shore.

On the east coast the rivers open a little later than on the south, and the islands are free enough of ice to permit of fishing about the middle of June.

SESSIONAL PAPER No. 39a

Navigation on the bay commences about June 20, the date depending largely on the wind. The prevailing wind being northwesterly it is apt to drive large quantities of ice into the southern extremities of the bay.

As a matter of fact it is difficult to state with any degree of accuracy at what date the bay as a whole can be navigated. It is really entirely a question of the wind. The northern part is generally full of ice until the end of July and, I believe, the Hudson Bay Company's steamer *Inninu* rarely attempts going to the Whale River Post until August. This summer the Whale River schooner was unable to leave the river mouth until August owing to the ice completely blocking the channel.

The southern rivers freeze up about November 20, those on the east coast a few days earlier; James bay itself is not frozen sufficiently to bear sled-travel until about Christmas. The ice is said to reach a thickness of over four feet. It is probable, however, that the centre of the bay never freezes at all.

The temperature of the sea water taken at various places is given below with the date and place of observation:—

Cape Jones	July 27	40°F.
Twin islands	August 12...	45°F.
Sabaskunika bay.......................	" 24.................	50°F.
Factory bay	" 26.................	50°F.
Cape Hope..	" 27................	52°F.
Boatswain bay... ,..........	" 30....	54°F.
Mouth of Moose river..........	September 16..............	62°F.

In all cases the temperature was taken some 2 to 4 miles from the mainland (except at the Twin islands). The main body of water, undoubtedly, has a low temperature, possibly below 45° F. The comparatively higher temperature found around the coast being on account of the numerous rivers and the general shallowness of the water. I believe the greatest depth of the bay is only about 65 fathoms.

This large body of cold water exercises a very unfavourable influence for agriculture, although excellent potatoes and other vegetables are grown at Moose Factory, Rupert's House and as far north as Fort George on Big river. Oats and barley can be grown at Rupert's House and Moose Factory, and there is an abundance of wild hay in the neighbourhood of these places and also at East Main and Fort George. The cattle kept at all the posts appear to be in excellent condition. There is little doubt that further inland, away from the cold winds off the bay, surer crops could be raised. It is probable that this adverse climatic influence extends some 30 miles inland.

The soil appears to be mostly sandy loam, but very extensive draining would be necessary before farming operations on a large-scale could be undertaken.

An instance showing the lateness of the arrival of spring, wild strawberries and other berries were found on the east coast to be ripe in the middle of August and the leaves of the poplars and willows were only just out at the beginning of July at Charleton island. This would compare unfavourably with the Mackenzie River country where at Lat. 65°, or nearly 900 miles further north, the leaves are all out about the middle of June, and wild strawberries and other berries are ripe at the end of July. On the other hand, winter sets in earlier in the Mackenzie basin than in James bay, but the drier and hotter summer (although shorter) of the former is far more favourable for the ripening of crops and the growing of garden produce than the latter. The same remark applies with more force still to the Peace River country, Northern Alberta (Lat. 58°).

Some people, too eager to ' boom'" and praise, have the hardihood to liken Hudson and James bays to the Mediterranean sea. Even on a brilliant summer's day a very vivid imagination is needed to compare these stormy northern waters to that genial southern sea; there is no point of resemblance anywhere.

5 GEORGE V., A. 1915

But there is an European sea, the Baltic, the conditions of which are superficially at any rate very much akin to James bay. The rocky islands, low sandy coasts, shallow depth, comparative low range of tide, and the general climatic conditions common to both, all make points of close resemblance.

However, this statement is not meant to imply that warm and beautiful summer days do not occur; on the contrary, a glance at the Meteorological records at Moose Factory will show that 90° F. is no uncommon temperature in July or August.

Nevertheless, from our observations this summer, the conclusion arrived at was that the cold water and prevailing winds from the north retarded all growth to such an extent that the heat of a few fine summer days comes too late. Cold, fog and mist are common, and rain seems to fall unceasingly for days. Out of 83 days spent this summer on James bay, rain fell on 44, and 16 were foggy—the majority of which were foggy enough to stop travelling.

A warm sultry day generally brings up a sharp thunder-storm from the southward; the winds afterwards " backing " round to the north with great violence, and turning bitterly cold.

Heavy clouds in the south should always be a warning to canoes to avoid a long crossing, as a heavy squall very quickly raises a dangerous, choppy sea in the shallow waters of the bay.

In winter time the cold from December to the end of February is very severe (the thermometer not uncommonly recording 40° to 45° F. below zero).

The treeless coast and inlands of the northeast are uninhabited at this season of the year except by a few Eskimo families. The Indians who live there during the summer retire to the more sheltered rivers and only venture back to set a few fox-traps, or in the early spring to catch cod.

Tides.

The rise and fall of the tide in James bay is about 5 feet, causing a current of about 3 knots per hour at the ebb and a little less at the flood. Considerable advantage is taken of the tide by the natives in making crossings from the mainland to the island, indeed they seldom start from camp except with a fair wind and tide in their favour.

In many places amongst the islands the current flowing through the numerous channels over a rocky uneven bottom makes with any wind a race which a canoe and small boat should be careful to avoid.

During the course of our stay in the bay it was impossible to make many observations on the influence of the tide on the movements of the fish. References are made elsewhere to such information as was obtainable.

Timber suitable for boat building.

The timber of James bay consists of spruce, tamarack, poplar and some birch, none of it of first-rate quality or of large size.

Messrs. Revillon Frères and the Hudson Bay Company import nearly all the timber which they at present use in the construction of their sail-boats, although in former days, I am informed, local timber was used.

On the construction of railroads to the bay and the development of the fisheries, it would seem that the boat and ship-building industries (which are closely allied to that of fishing) cannot depend on local timber for their needs.

Boats.

The Indian boat of James bay is the canoe, not now made of birch bark, but of wood (cedar or basswood) or more commonly of a cedar frame covered with a heavy canvas. Their dimensions are generally about 16 feet long and 18 to 20 inches in

SESSIONAL PAPER No. 39a

depth, and built with an extremely rockered keel, giving them when out of the water a very curious appearance. It is claimed that this design is superior to any other in rough, choppy water. One does not usually associate a canoe with a stormy northern sea, but owing to the shoal water of the bay and the scarcity of harbours for even shallow draught boats a large canoe is really a suitable craft for a coasting trip, and if fitted with a motor and centre-board would be very hard to excel.

The Indians in their canoes generally rig up a sprit-sail made of a blanket and with a fair breeze do not hesitate to make long crossings from island to island or point to point.

The Eskimo use the well-known "Kayak." In this little boat about 16 feet long by 30 inches wide, made of seal-skin or heavy canvas stretched on a wooden frame, these people make long sea passages (60 or 70 miles) to the outer islands.

The Hudson Bay Company have a number of half-decked, Ketch-rigged boats from 30 to 45 feet long and about 3 feet draught. They also have two or three schooners of about 20 tons for carrying freight from Charleton to the out-posts. These boats are well enough in a fair wind, but can make little way against a head wind and the short heavy sea in the bay.

Natives.

The Indians inhabiting the eastern and southern coasts of James bay are mostly Crees, and locally they are divided into two classes—the inlanders and the coasters. The inlanders seem to be the favourites with the trading companies, as they apparently travel far inland for their trapping, while the coasters spend most of their time fishing amongst the islands, hunting ducks and rabbits or begging from the trading companies.

These Indians have been in contact with British people for nearly 250 years, and for the last 50 years have had missionaries amongst them. They all, or nearly all, profess the Christian religion, and without exception all dress in cheap European clothes obtained from the traders. For a long time past they have obtained a high price for their furs and, in fact, so great has the competition been between the two trading companies, that they have been able to obtain large advances on the future prospects of their hunt, with the result that they are almost without exception well off. The present war will, no doubt, be temporarily destroying the markets for fur, cause the trading companies to entirely curtail this credit system.

With all these advantages it would be thought that these Indians would have the appearance of enjoying some prosperity, but the reverse is the case, a more hang-dog, miserable looking lot of people in the aggregate it would be impossible to conceive.

The Indians have practically undisputed ownership of the coast line as far north as Fort George, but beyond this point Eskimo are to be met, although they are not very numerous south of Cape Jones. One or two Eskimo families live on an island near Cape Hope, these being probably the most southerly representatives of their race.

The Eskimo can be described as a littoral people, inhabiting the bays and islands of the Arctic and sub-Arctic coasts. Of the two races (Indian and Eskimo) there can be no question of which is the more desirable from the point of view of an employer of labour; the Indian being, if not lazy, absolutely indifferent to time and quickly tiring of any work. They are also timid sailors in any large boat, that is in venturing far from land, and will only consent to go provided the total crew is double the number really necessary; a curious fact, since they would appear from the manner of handling their small canoes to be quite capable sailors.

The Eskimo, on the other hand, are a manly race, excellent seamen and will prove invaluable as fishermen. The Eskimo living as they do all the year round on the coast or islands are unable to make as large a fur catch as the Indian, with the result that they do not enjoy the same credit with the trading companies. They are, therefore, considerably poorer, but infinitely harder working, and in every way a more deserving people.

5 GEORGE V., A. 1915

Nearly all travellers in the Arctic speak of the Eskimo in a kindly fashion. Captain Coates (elsewhere mentioned as the author of the book "Remarks on the Geography of Hudson's Bay") has many generous thoughts regarding these people. Although written over 150 years ago, the sentiments therein expressed so coincide with others of the present day that they seem worth while quoting. Captain Coates says, " It will be necessary before I quit these parts to set down my own sentiments and that of others in regard to the Usquemous, the natural inhabitants of all the northern borders of Hudson's bay and the streights which swarms with robust, hardy fellows fit for the severest exercise and, indeed, with such dispositions as if God's providence in fulness of time had prepared them to receive the yoke of civility. And I do assert of my own knowledge that these people are nothing near so savage as is represented by our early voyagers, and that their confidence is in their innocence, not in their numbers, which I have often experienced, when one or two has put themselves into my hand without reserve or caution." Elsewhere he describes them as, " bold, robust, hardy people, undaunted masculine men, no token of poverty or want, with great fat, flat, greasy faces, little black piercing eyes, good teeth, etc.", and he propounds a pious scheme whereby these tribes " may be made useful to us and acquire salvation to themselves."

The question of food supply is an important one for all natives living as the Indian and Eskimo do, by hunting and fishing. Fish there is no difficulty in getting, but meat is harder. Rabbits, ducks and geese are after fish their principal food. Caribou, which a few years ago were plentiful on the east coast on the barren islands and mainland, are now very scarce, while moose are unknown north of East Main river; the last named animal is probably migrating northwards, being driven back by the building of the National Transcontinental railroad. Judging by the numbers of moose seen on the Missanaibie and Mattagami rivers it would appear that about 100 miles north of the National Transcontinental railway is as good a moose country as there is anywhere in Canada. In the fall of the year the Indians kill large quantities of geese and ducks. The southern end of Hannah bay is notorious for its wild fowl; Snow Geese, Canada Geese' Blue Geese (chen Coerulescens) and many varieties of ducks gathering on the marshy plains in immense flocks and fattening on berries and grass seeds before the final flight south at the first touch of winter. To the natives from Rupert's House and Moose Factory the annual goose hunt in Hannah bay is an event of much importance.

With the comparative nearness of James bay to the outside world (220 miles from Moose Factory to the National Transcontinental railway) it would be supposed that some white men (prospectors and trappers) would have by this time penetrated to this by no means remote region; but this is not the case and there does not appear to be a single white inhabitant on the south and east coasts, except the officials of the two trading companies and the missionaries. This is remarkable, as in northern British Columbia and in the Canadian Arctic (in actual mileage far further from civilization and with greater difficulties of transport and, therefore, more expensive supplies) it is not uncommon to meet white trappers and prospectors.

All freight for the two trading companies is brought by ship to the bay, and although there are risks of navigation, goods and food supplies appear reasonably cheap.

Money is practically unknown among the natives, the companies pricing fur and goods on the basis of a value in what is locally called a " Made Beaver "; an arbitrary value having absolutely nothing to do with the skin of that animal. Thus a skin of a marten or fox is said to be worth so many " Made Beaver," against this a cotton shirt or one pound of tobacco is also valued at so many " Made Beaver." At Rupert's House there is still used the old brass coins or tokens representing one, a half, and a quarter " Made Beaver."

SESSIONAL PAPER No. 39a

Historical.

While tradition has assigned to French fishermen the honour of first reaching Hudson bay about the year 1590, James bay was, undoubtedly, first discovered in 1610 by Henry Hudson on his third unsuccessful attempt to discover a route to China and the East Indies through the northwest passage.

Sailing through Hudson straits and bay, he, late in the year 1610, explored the southeastern shores of James bay, and eventually wintered in a small bay full of islands about Lat. 53°. (Probably Old Factory Bay.)

After spending a winter of great hardships, due principally to scurvy, he started to return, but his crew mutinying while off the mouth of Little Whale river (Hudson bay), cast him and his son and the few faithful sailors adrift in a small boat.

Hudson's ultimate fate and that of his companions is unknown, but it is probable that he survived for some time after reaching shore. Miserable though his end may have been, his name given to Hudson bay and straits and the Hudson river (New York) will live until the end of time. The mutineers eventually reached England with about half their number gone, the rest having been murdered by Eskimo on an island in Hudson straits.

While two or three northern expeditions left England during the succeeding years, it was not until 1631 that James bay was visited again. Captain James, outfitted by English merchants, sailed through Hudson straits and thence southward to Cape Henrietta Maria and eventually wintered at Charleton island. According to his story, he suffered great hardships from the extreme cold. He returned to England in the autumn of 1632, after having explored the southern and western shores of James bay.

The next expedition to the bay was for the purposes of trading with the natives. Two Frenchmen, Radisson and Groisselier, who had been trading with the Indians in the western interior, engaged some of them to act as guides to James bay. On their return in 1666 they endeavoured to induce some of the French fur traders of Quebec to outfit a trading expedition to the bay. Being unsuccessful they proceeded to Paris, but with no more success than they had met with in Canada. However, eventually they obtained an introduction to the English court, and armed with this they were successful in having a favourable hearing granted to them by Prince Rupert and a group of wealthy and influential men of London.

In 1668 the ship *Nonsuch* was outfitted and despatched to James bay under command of one, Zachariah Gilham,—Radisson and Groisselier accompanying the expedition. They passed safely through Hudson straits and sailed southward, eventually reaching Rupert river, which was then called the Nemiskau (Lat. 51° 30″).

Here they built a trading post or fort, naming it Fort Charles, and after numerous friendly meetings with the natives returned to England the following summer.

In 1670 Prince Rupert, and others associated in this trading venture, obtained a charter from Charles II, styling themselves "The Governor and Company of Adventurers of England Trading into Hudson Bay."

In 1670 the Hudson Bay Company sent out Charles Bayley to establish a post at Rupert river. This post, known as Rupert's House, is the oldest post of the Hudson Bay Company, and is also therefore undoubtedly one of the very earliest British settlements in Canada.

In 1674, and succeeding years, the company gradually extended their trading operations, establishing posts at Moose, Albany and East Main rivers.

In 1693 war broke out between France and England. The French in Canada, the following year, sent a force overland (probably by way of Michicopoten and the Missanaibie river) and took Albany, Moose and Rupert's posts.

In 1695 the company with the help of two ships of the English navy re-took these forts.

39a—2

5 GEORGE V., A. 1915

In 1697 the Treaty of Ryswick assigned only Fort Albany to the Hudson Bay Company. This was the condition of affairs until the Treaty of Utrecht in 1713. By this treaty, France ceded all her rights in the bay to England.

From 1713, until the present time, little change has happened to James bay.

Early in the 19th century the company established forts on the east coast at Big river (Fort George), and at other points in Hudson bay, and about this time several exploration parties were sent out both to the southern district (Nottaway river) and the Labrador peninsula. Ten years or so ago Messrs. Revillon Frères, of Paris, established posts in close proximity to the Hudson Bay Company's establishments; this firm being the first competitor the company have had in this region.

Since the year 1871 numerous expeditions of the Geological Survey Department have been despatched to James bay. The principal of these to the east coast was sent in the year 1877-8 under Dr. R. Bell, F.R.S., and to the south and east coasts in 1887-8 under Mr. A. P. Low. In 1898-9 Dr. G. A. Young made a micrometer survey of the south and east coasts from Cape Jones to the Harricanaw river. This resulted in the excellent map published by the Geological Survey Department.

The Hydrographic Survey (already mentioned) are doing considerable and much needed work in charting the principal river estuaries and mapping the larger islands. The only chart of practical use is largely compiled from notes and memoranda made by a Captain Coates, who was in command of one of the Hudson Bay Company's ships during the years 1727 to 1751. These notes have been published in book form called, "Remarks on the Geography of Hudson's Bay."

When we left Moose Factory late in September, 1914, news of the great war was just beginning to trickle in to all the outlying camps and posts of the bay. The natives seemed far more concerned at the thought of the possible rise in the price of their sugar and a corresponding fall in the price of fur than the all-important outcome of the struggle. To them Germany means absolutely nothing and the British Empire not much more; their minds cannot grasp the fact that their future destinies are being settled on the battle-fields of Europe.

In the Anglo-French wars of the 17th and 18th centuries there was a very good reason for the capture of a Hudson bay fort. The fur trade at that time was the only trade of Canada and a Hudson Bay Company's fort was a point of great strategical value. In common with other posts, Rupert's House was for those times very strongly fortified and armed. The fortifications are now gone, but the cannon can still be seen doing duty as bollards for mooring vessels to the wharf.

LIST OF FOOD FISHES.

The following is, I think, a comprehensive list of the food fishes to be found in the south and east coast waters and tributaries of James bay:—

Name of Fish.	Description of Habitation.
Sturgeon...........................	Anadromous, lake and river.
(Acipenser Rubicundus.)	
Whitefish..........................	Anadromous and lake.
(Probably two species, Coregonus Clupeiformis and Labradoricus.)	
Tullibee...........................	Anadromous and lake.
(Tullibee Argyrosomus.)	
Speckled Trout.....................	Anadromous, lake and river.
(Salvelinus Fontinalis.)	
Lake-Trout or Salmon-Trout.........	Lake and river to a certain extent.
(Cristivomer Namayush.)	
Land-Locked Salmon.................	Lake.
(Ouananiche.)	
(Salmo Salar Ouananiche.)	

SESSIONAL PAPER No. 39a

Name of Fish.	Description of Habitation.
Long-Finned Charr..	Anadromous and lake.
(Salvelinus Alpinus Alipes.)	
Jackfish or Pike..	Lake and river.
(Lucius Lucius.)	
Pickerel, Doré, Wall-eyed Pike..	Lake and river, and to some extent found in
or Pike Perch (Stizostedion Vitreum.)	tide water.
Sucker (two species)..	Lake and river and also in tide water.
(Catostomus Commersonil.)	
(Catostomus Catostomus.)	
Ling, or Maria..	Lake and river, and also in tide water.
(Lota Maculosa.)	
Codfish..	Marine.
(Gadus Ogac.)	
Moon Eye, or Toothed Herring..	Lake and river.
(Hyodon Tergisus.)	
Silver Chubb, or Fall Fish..	Lake and river.
(Semotilus Corporalis.)	

DESCRIPTION OF FISH.

A description of each of these fish is given, but at present, with the comparative small amount of information available, only the whitefish, speckled trout, tullibee, and possibly the sturgeon, can be considered commercially valuable.

Whitefish.

Commercially the two species can be considered identical. There can be little doubt that the whitefish fishery of James bay will prove to be one of the most prolific in Canada, equalling, if not surpassing, the fisheries of the Great Lakes.

The whitefish of the bay, that is the sea-run fish, are small, averaging 2½ to 3 pounds (the largest caught by us weighed 4½ pounds). The fish of the interior lakes on the other hand are larger, averaging 4½ to 5 pounds, or possibly more. They are both of excellent quality, but more especially those taken in salt water. The very noticeable difference in weight, between the fish living in the sea and those in the lakes, leads to the belief that there is no connection between the two. Those inhabiting the sea are apparently distinctly marine or, at least, coastwise in their habits except at the spawning time when they ascend the rivers; while the lake fish are believed never to descend to the sea at any time.

Little need be said regarding the lake fish, their habits not differing from those in other parts of Canada.

Range of Whitefish.

The sea-run fish are said to occur in large quantities in the estuaries of the rivers, and along the coast with the first open water in the spring. They apparently go back to deep water amongst the numerous islands as the season progresses. About the middle of August another movement towards the shore takes place, and this increases as the spawning season (beginning of October) draws near.

Their range from the time of open water in the spring until the spawning time in the fall is from Sherrick mountain as far north as cape Jones (the most northerly boundary of the bay and the limit of our investigations).

Practically all the larger rivers, including those on the south coast, are ascended for spawning, but since the majority are considerably broken by rapids and falls comparatively close to their mouths, few fish ascend much higher than 50 or 60 miles. So far as is known they stay in the rivers until the middle of December, when they return to the sea, probably remaining in deep water until the following spring.

Unquestionably the best fisheries are around Big river and among the islands north and south of the mouth of this river.

5 GEORGE V., A. 1915

The best fishery made by the expedition was in Brandy bay, some 12 miles north of Big river. At this place at the beginning of August some 60 fathoms of 4½-inch mesh net, 60 fathoms of 4-inch mesh net and 40 fathoms of 2¾-inch mesh net caught about 600 pounds of fish. This, I think, will compare most favourably with any of the fresh water lakes. Undoubtedly, we should have done considerably better by using a 3½-inch or 3¾-inch mesh net, as the 4½-inch net caught only a very few.

Eskimo Duck islands, a group of islands, some 10 miles from the mainland south-west of Big river, and the islands around Long point, we also proved to be excellent fishing grounds. Cape Hope, and the islands in the vicinity, can also be favourably mentioned.

Gill-netting is the only method of catching whitefish adopted in the bay, 3 or 3¾-inch mesh nets being generally used for the sea fishing, but in the lakes of the interior owing to the larger fish, 5-inch and 5½-inch nets are necessary.

The nets are generally set only a few yards from shore; the best location being on a gravel bottom. The Indians on the whole are very poor fishermen, being extremely conservative and never condescending to experiment or try new methods.

The fishing for the winter food supply is done in the general Indian fashion during the spawning season; the fish being "put-up" frozen in barrels. On the east coast seal-blubber is generally fed to the sled-dogs, and is preferred to fish owing to its being a far stronger food, so I did not hear of any large fisheries being established by the trading companies as is generally done in the northwest provinces of the Dominion.

Sturgeon.

I regret not being able to give more information regarding this, the most valuable, individually, of all Canadian fish.

I find that sturgeon frequent in more or less degree Moose river and its tributaries, the Nottaway, the Broadback, and possibly the Harricanaw. Rupert river and others as far north as Big river. This last appears to be their most northerly limit.

There is only one species, I think,—the lake sturgeon; and the same fish occurs in many lakes, such as Nemiskau, Opinaka (East Main), and Wabstaka (East Main).

As a coastwise fish, it inhabits the estuaries and travels up the rivers early in June for spawning. The majority ascend only the first few rapids, but some undoubt-edly go higher. They stay in the rivers about three weeks, some (the smallest), perhaps, staying all summer in the deep pools and eddies, and only returning to the estuaries at the freeze-up; it is possible, indeed, that they do not return even then, but winter in the rivers.

The lakes and smaller tributaries at the head of such rivers as the Rupert, the Nottaway, and the Harricanaw all contain sturgeon; this information comes from Indian report.

The largest authentic catch that I heard of was 200 fish taken in one night about four years ago by an Indian on the East Main river. They were all small, probably only averaging about 10 pounds in weight.

We saw no big sturgeon, 35 pounds being the largest, and I should judge that 70 pounds would be a large fish for these waters; although stories are certainly told of fish that by the measurements recounted would weigh well over 100 pounds.

The Indians take these fish by gill-nets (about 7 or 8 inch mesh) or spearing them on the way up the rivers or by very occasionally setting lines.

It is probable that James bay is the last virgin fishing ground for sturgeon in the world, virgin that is to say only as regards commercial fishing, for the trading com-panies and the natives have for centuries taken their toll for food.

The statistics regarding the Canadian sturgeon fisheries published in the blue-book of the Department of the Naval Service show that for the year 1912-13 there was a slight increase in the amount of sturgeon caught over the preceding year. While this is certainly satisfactory, unfortunately there can be little doubt that the sturgeon, unless protected, is in time doomed to become as extinct as the American buffalo.

In his last annual report, the United States Commissioner of Fisheries says, "The story of the sturgeon is one of the most distressing in the whole history of the American fisheries." *The Scientific American* of April, commenting on this, makes the following interesting remark on the report, which, coming from such an authoritative paper, should do much to draw attention to the danger: "For years these large, inoffensive fish were supposed to be of no value, and when, as often happened, they became entangled in fishermen's nets, they were knocked on the head and thrown back into the water. When it was discovered that the sturgeon's eggs were valuable as caviar and the flesh as food a period of reckless fishing began, and in a few years the best and most productive waters were depleted, and what should have been made a permanent fishery of great profit was destroyed. On the Atlantic 7,000,000 pounds to less than 1,000,000 in fifteen years, and an even more rapid decline occurred on the Pacific coast and in the Great Lakes. At present the total annual yield for the whole country is less than 1,000,000 pounds and is decreasing. Meanwhile the demand for the eggs and flesh has steadily increased, with the natural result on prices. A mature female sturgeon now often brings more than $150.

"The worst of the situation is the fact that all attempts at artificial propagation have failed; so that unless prompt steps are taken to protect the sturgeon by law this fish will be practically extinct in American waters in a very few years. The commissioner recommends that the legislatures of all states, in which this fish exists, or has existed, should absolutely prohibit its capture or sale for a period of at least ten years. Meanwhile the Bureau of Fisheries proposes to transplant into our waters young sturgeon from foreign countries; especially a species from the Danube and the Caspian sea, specimens of which have been offered by the Roumanian government."

Experiments have been made in the United States with a view to the artificial propagation of the sturgeon, but as yet have met with very little success, the chief reason being the great difficulty of obtaining the two sexes "ripe" at the same time.

Sturgeon meat marketed is worth about 12 cents per pound, and the roe prepared as caviar $1 per pound. The Dominion fisheries blue-book for 1912-13 records that over one million pounds of the fish, and 96 hundredweights of caviar were marketed during the period under review.

The dried air-bladders, commonly called "Isinglass," are also of commercial value; and, I believe, are considered an article of trade with the fur companies in James bay;—but whether a sufficient quantity is obtained from the Indians to export to Europe, I am unable to say.

Speckled Trout.

These fish occur in great quantities, both in the sea as a coastwise fish and in all the suitable streams and lakes of the interior. The small lakes and creeks on Charleton island also contain these fish in immense quantities. The sea-run variety attain a large size, $4\frac{1}{2}$ to 5 pounds in weight being frequently caught.

A net set at random among the islands on the east coast would always catch trout; 40 of these fish averaging $2\frac{1}{2}$ pounds in weight being our best catch for a 40-fathom $2\frac{3}{4}$-inch mesh net.

It may possibly be thought that this species of trout would never occur in such numbers as to make them commercially valuable, but with prices at 10 cents per pound (which is the present price paid by any wholesale dealer in Quebec) it will be seen that even if they were in comparative small quantities, they are fish well worth the catching.

5 GEORGE V., A. 1915

The movements of the coastwise fish are practically the same as the whitefish, that is—with the first open water at the mouths of the rivers, they appear in great numbers. As summer advances they are to be found everywhere amongst the islands, entering the rivers again as the spawning time (about September 15) draws near. During the winter, from information given me, they appear to stay in large quantities in the estuaries of the rivers, the Indians catching them through the ice by angling and to some extent with gill-nets. The record catch that I heard of was about 140 pounds of trout in two hours made by an Indian woman on a small stream near East Main.

Tullibee.

Tullibee occur in vast quantities as a coastwise fish. A 3-inch or 2¾-inch net* set haphazard off the shore or amongst the islands and left for a tide will generally be full. As with the whitefish and trout, the best fishing grounds are, undoubtedly, amongst the islands of the east coast, but they also occur in the estuaries of all the rivers and around the larger islands of the bay, such as Charleton, the Struttons and the North and South Twins. Their movements are almost identical with the whitefish, coming into the rivers as early as September and leaving again in December. They do not appear to go above the first rapids in any of the rivers, but the Indians state they catch them in many lakes of the interior.

These fish should prove to be of great commercial value, if placed on the market absolutely fresh. But the fact is, they deteriorate very quickly, and also are very inferior when caught in the rivers, but the freshly caught sea-fish are delicious and will prove to be most valuable. Owing to their vast quantities, a canning establishment would be a paying industry, at any rate the scheme would be well worth looking into.

The average size caught was about 1 pound in weight, the largest caught weighed 2 pounds.

Lake Trout, Pickerel, or Doré, Pike and Ling.

With the general development of the James bay fisheries quantities of these fish will be caught which alone would scarcely make the business worth while pursuing.

The pickerel, doré, or wall-eyed pike of the district are especially a fine fish, specimens being frequently caught over 8¼ pounds in weight. They occur in every stream and lake, and while not entering the sea are frequently caught in nets set in the estuaries in extremely brackish water.

The above remarks apply also to pike or jackfish. Stories were told us by the natives of the great size of the fish inhabiting the rivers and lakes of the east coast. We were, however, unsuccessful in catching any monsters, 12 pounds being our largest, but there is no doubt considerably heavier pike than this are to be found in that district.

Ling are found in most of the lakes and rivers, particularly in the estuaries during the winter time. They grow to a large size—up to 25 pounds, or even more,—and are considered an excellent food fish by all the inhabitants, Europeans and natives alike; although this is contrary to the general opinion held in the western provinces of Canada.

Ling spawn in February or March, but very little is known regarding their habits.

They do considerable damage to the whitefish fisheries, following these fish up the rivers to their spawning grounds and eating vast quantities of eggs and later, no doubt, fry.

* During the spawning season, in the rivers or other suitable places the Indians sometimes use " Seine nets " for catching this fish. These nets are also very often placed below some rapids or falls (generally the first rapids near the sea) and then dragged swiftly ashore. I am informed incredible numbers of Tullibee and other fish are caught in this manner.

Lake Trout.

Lake trout are found in nearly all the larger lakes and, to some extent, the rivers. The Indians, though, report they never catch them in the estuaries, in this respect differing from the speckled trout.

Their habits are the same as lake trout in other parts of Canada.

They spawn in September, frequenting the shallow gravel bars of the lakes; in places where they are plentiful, great numbers at this time are caught in gill-nets and "smoked" by the Indians.

Land-locked Salmon.

(Called So-a-sa-so by the Indians at Rupert's House, the same name as given the long-finned charr.)

The fish is known in Nemiskau lake, and probably will be found in other lakes on the Rupert and Nottaway rivers. Commercially its numbers are too few to make it valuable, but as a fish for the angler it is considered to excel even its near relative the Atlantic salmon and, therefore, ranks high in the estimation of the world and is entitled to important recognition.

No specimens were caught by the expedition, but information of its occurrence was obtained from a trustworthy source.

Long-finned Charr.

(Commonly called "salmon" in James bay.)

This fish occurs only in the extreme northern limits of the bay; Kapsewis river being practically their most southerly boundary. Beyond Cape Jones, northward into Hudson bay, the Eskimo and Indians report catching these fish in large quantities during the months of August and September.

The movements of this little-known fish are as follow: Towards the middle of August the run begins into the rivers (in James bay the only rivers which they frequent being the Seal, Salmon and Kapsewis). They proceed up these streams as far as the lakes at the head of each river. In these lakes, according to report, they spawn and stay all winter, coming back to the sea at the break-up in the spring.

The natives catch them in gill-nets set at right angles close to the shore, and by spearing them in the rivers, making, as they term it, "a house" of rocks into which by means of wing-dams the fish must enter; there they are speared.

I only caught a few specimens of this fish, but can testify that if they occur in large quantities a fishery would certainly prove to be a paying proposition, as the fish are first-rate in every way.

While I was unable, owing to the limited time at my disposal, and to the instructions received from the department, to proceed further north than Cape Jones, the Eskimo and Indians gave me such information as to lead one to suppose that in Hudson bay proper these fish must be very plentiful, frequenting every stream with a sand or gravel bottom, eschewing the very rocky.

8½ pounds was the weight of the largest "salmon" caught, but they run considerably heavier. The Rev. W. G. Walton, of Fort George, told me that Eskimo had brought him three "salmon" weighing altogether 90 pounds.

It is probable that in James and Hudson bays there are two specimens of charr, very closely allied to one another, one is the species already described, the other being the Greenland charr or Hearnes salmon (*Salvelius Alpinus Stagnalis*).

The old records of residents of Hudson bay are interesting regarding this fish; one writes as follows: "Salmon are in some seasons very numerous.. .. ... I once found them so plentiful that had we been provided with a sufficient number of nets and salt, we might soon have loaded the vessel with them. But this is seldom the case, for

5 GEORGE V., A. 1915

in some years they are so scarce that it is with difficulty a few meals of them can be procured during our stay at these harbours. They are in some years so plentiful near Churchill river that I have known upwards of 200 fine fish taken out of four small nets in one tide within a quarter of a mile of the fort; but in other years they are so scarce that barely that number have been taken in upward of twenty nets during the whole season."

Codfish.

(Greenland codfish.)

The above are found on the east coast from a few miles south of East Main river northwards.

The expedition was unable to prove its existence in great quantities, but Indian report tends to the belief that large catches are made in February and March by hook and line.

The largest fish caught by my party was 5 pounds only in weight, but the Indians have told me that they catch them up to a size which, as they express it, " it takes only four to fill a flour sack." This would certainly mean a weight of about 20 pounds per fish.

The extreme north of James bay (Cape Jones) is spoken of as being the best winter fishing grounds for large fish, but many Indians fish in the early spring around Paint Hill islands. Old Moar bay, and, in fact, I believe anywhere around East Main river. I believe a catch of about 200 fish is considered a good morning's work, but I do not suppose the average would weigh much more than 5 or 6 pounds.

Cod are known to occur in Hudson bay. A few schooners from St. John's fish in Hudson straits and Ungava bay every year. There would apparently be, therefore, no reason why these fish should not exist in larger quantities than have yet been found.

Sucker, Moon-eye and Chubb.

These fish while edible must be considered commercially worthless, so long as better flavoured fish can be caught in the same vicinity and with as much ease.

Suckers (of which there are two or three varieties) occur in prodigious quantities, weighing from a few ounces up to 4 pounds. A small mesh net set in any river will be full if left only for a few hours.

Moon-eye appear to be very common in the Moose river and its tributaries and, doubtless, also in other rivers. This remark applies also to chubb.

A Species of Flounder or Flatfish.

An Eskimo gave me the information that he had in July (1914) caught a flatfish in a net set for whitefish near the Cape Hope islands. He described this fish, the first he had ever seen, as being about a foot in length, nearly as broad, and sand-coloured with a few red spots, but white or whiteish underneath.

This was the only flatfish I heard of in the bay, although repeated inquiries were made.

Another fish I heard of as inhabiting Rupert bay bears, from the description, a close resemblance to the " shad." I was informed that this fish came into the Nottaway river (so far as my informant knew this was the only river these fish entered) in late June or early July to spawn. They are never seen except at that time, possibly after the spawning season returning to the deep waters of the bay.

Shellfish.

Mussels, scallops and clams are found in great quantities everywhere on the sea-shore and among the rocks at low tide. In the report of the Dominion Shell Fish

SESSIONAL PAPER No. 39a

Commission of 1912 and 1913, mention is made of the great value of mussels as bait for cod, and the report further urges that the mussel resources of the country be given more attention by fishermen. The attention is also drawn to the great decline generally in the shellfish fisheries.

The expedition, owing to lack of equipment, were unable to dredge, so that the only evidences of the occurrence of shell fish in large quantities were the empty shells found on the beach. This, however, is sufficiently conclusive.

Crabs.

The expedition was unsuccessful in catching any live crabs and the natives generally did not seem to know of them. I saw only shells of two on an island near Sabaskunika bay. An Eskimo told me that in certain parts of Hudson bay large crabs were to be found, but they were not regarded as edible by the natives, who, it must be admitted, have generally managed to find out what is good in the provision line long before Europeans arrived in the country.

Oysters.

While in James bay I was requested by a resident to give some particulars regarding the possible successful culture of oysters in the southern part of the bay.

As is well known, oysters only spawn in water varying from 60° to 70° F. temperature, and also require a certain salinity of the water. The highest water temperature I obtained in the bay was 62° F. in the estuary of Moose river, but this water taken from the estuary was nearly fresh. The highest temperature of salt water obtained was only 54° F.

In Puget Sound, on the Pacific coast, about Lat. 48°, the temperature of the water is found too low for the oyster to spawn, and the beds are kept up by annual plantings of seed oysters.

It may, therefore, be conclusively stated that James bay is not a suitable place for oyster-culture, owing to the low temperature of the water. Apart from this fact, the hard clay bottom of the southern portion of the bay might in many places make a suitable oyster ground.

Seals.

Two species of seals are common in the bay, the "Harbour" (*Phoca Vitulina*) and the "Bearded" (*Erignathus Barbatus*). The skins of both animals are of great value to the Eskimo, who from them make boats and other articles of clothing and also the covering for their "Kayak." They also highly esteem the meat and blubber as food.

The Indians hunt and kill seals, but the meat and blubber is given to their dogs and the skins traded to the Eskimo who make the long sea boots indispensable along the coast, which in return the Indians buy. It would seem that neither race encroaches on the work that by tradition and custom is done by the other.

Whether seals occur in the northern part of Hudson bay or straits in the great herds such as are found in the early spring on the ice floes off the coast of Newfoundland or Jan Mayen island is as yet unknown, but it would seem not unlikely.

Even should this be the case, it would yet be very problematical whether the ice conditions in James or Hudson bays would permit of vessels proceeding from Port Nelson or other ports of the bay to hunt them.

The Newfoundland sealers leave St. John's not later than March for the sealing grounds, and at that time of the year all Hudson bay is still in the grip of winter. It is claimed (and, no doubt, is the truth) that the bay only freezes around the coast, and that a powerful ice-breaker could very quickly make passage to the open sea. The advent of the railroad will, no doubt, promote much enterprise, and in a few years it may be that Hudson bay sealing vessels will meet with as much success as those of Newfoundland.

39a—3

5 GEORGE V., A. 1915

Walrus.

Judging by the name walrus or "sea-horse" (the old English name) given to as many islands and points in James bay, this animal must at one time have been fairly plentiful. In northern Hudson bay it is common still, and the old records of the Hudson Bay Company have stories of their sloops and barges being attacked by herds of these animals. I think on the east side of Whale river may be said to be its southern boundary, so it scarcely comes within the purview of this expedition.

White Whale.

(Toothed whale—*Delphinapterus Catadon.*)

The white whale occurs in varying degree over the whole of James and Hudson bays, being probably more plentiful on the west than on the east side.

Fisheries for this animal were carried on by the Hudson Bay Company over 150 years ago, both at Fort Churchill on the west and Whale river on the east, the latter being discontinued according to the old records in the year 1758.

Of late years an effort has been made to re-establish this fishery, but without success.

The method of killing these animals is by the hunter waiting in his canoe and harpooning one when he is fortunate enough to get within range. At Whale river nets were spread across the river and arranged in such a manner that they lay well below the surface. On a whale being sighted in the river the nets were "sprung" and the animal, surrounded by people armed with rifles and harpoon guns, and unable owing to the nets to return to the sea, quickly succumbed.

GENERAL CONDITIONS.

The present methods of fishing in the bay are primitive, but since the fishermen can with ease catch all the fish they want, there is no need for better methods.

In the spring and summer seasons the sea-fishing begins in the middle of June and closes towards the middle of October, a short four months. The river and lake season is, of course, considerably longer, being only stopped a week or so at the break-up and a little longer during the freeze-up.

No fishing takes place off the coast in winter, but the Indians angle for trout and ling in the estuaries and, perhaps, have a net set for whitefish up to Christmas, and in the early spring they angle for codfish off the islands.

The expedition, therefore, is unable to report on the value of the winter fisheries of the bay as no information is available, but there would appear to be no good reason why winter fishing for white fish and tullibee should not be carried on in the bay under possibly only a little severer conditions than fishermen are now experiencing in the more northerly lakes of the western provinces.

The east coast is, undoubtedly, the best fishing ground, of which Fort George may be considered the centre.

As far north as this point it is possible, as I have already written, to find land which when cleared would be suitable for a limited amount of agriculture. Granted this, there is nothing to hinder a fisherman making a very good living from his business, and having as comfortable a home as the average homesteader in the West. But, first, must be assured railroad connection, of which at present in James bay there is none, nor the very immediate likelihood of any. Further away, 800 miles from Fort

SESSIONAL PAPER No. 39a

George, the centre of the James bay fishing grounds, is Port Nelson,[1] the terminus of the Hudson Bay railway now building. The questions which must be asked and which require very careful investigation, are:—

(1) Would it be feasible and profitable commercially to run refrigerating boats or vessels solely for carrying fish from Fort George to Port Nelson?

(2) The best method or methods of preserving fish for shipment to a market some hundreds of miles away?

Whitefish (of all kinds) unless well frozen deteriorate very quickly. Trout and cod can be salt cured, but this is not very suitable for whitefish, although in the Great Lakes in former years considerable quantities were thus treated.

Smoking would be feasible and the Indian smoke-cured whitefish is delicious. In Alberta I have eaten these fish cured by a German patent process and found them excellent, equal to the best finnan-haddie. For the tullibee, occurring as they do in such great quantities, I believe a cannery would be a paying venture; the fish should be put up in the same style as the canned herrings, now largely sold.

Regarding the codfish fishery, I am not sure whether climatic conditions are very favourable to drying fish, but if not, the catch can, of course, be salted.

The fisheries of the interior lakes must depend for their development on the railroad, which is, in the future, expected to reach the southern shore of James bay from some point on the National Transcontinental railway.

Nemiskau lake and others of the same group near the Nottaway, Broadback, and Harricanaw rivers, are excellent fish lakes, but at present being considerably over 150 miles from the National Transcontinental railway, are absolutely worthless for commercial fishing. I should judge 100 miles to be the farthest distance it is profitable to freight fish by "sled haul," and to accomplish this successfully, it would be necessary to have a fairly good road, and in summer this country would be absolutely impossible to travel over, unless roads were built at great cost over country which is to a large extent swamp.

From information I received, it seems that some of the smaller rivers flowing into lakes Mattagami, Evans, and others of that region, are in the spring and summer very prolific of sturgeon.

The lakes of the east main coast are, from all reports, excellent fish lakes, but far too distant to be worth considering commercially either at present or probably for many years to come.

The finding and exploitation of minerals on the east coast, always a great possibility, may lead to railroad development in the near future, which is at present undreamed of. Apart from this, however, it would seem that the terminus of the James Bay railway will be on the southern shore.

From Fort George to Moose river (the possible terminus of a future railroad) the distance is about 200 miles in a straight line; and this distance must be taken into account and reckoned with in considering the value of the James bay fisheries.

SUMMARY.

The result of the investigation of the fisheries of James bay may be summarized as follows:—

(1) The question of railroads is of paramount importance; without them the fisheries are worthless; unless the ice conditions of northern James bay and southern

---

[1] From Fort George to Winnipeg (via Port Nelson and the Hudson Bay railway now under construction) the distance is approximately 1,920 miles. From Prince Rupert (the headquarters of the northern pacific fisheries) the distance to Winnipeg via the Grand Trunk Pacific railroad is 1,745 miles. If it has proved profitable to send fish to Winnipeg, Chicago and other points in Central Canada and the United States from Prince Rupert, the comparatively small extra mileage from Fort George should be no impediment to James bay fish being marketed in the same places under practically equal terms.

5 GEORGE V., A. 1915

Hudson bay are such that a "fish carrying vessel" can make continuous journeys in summer between Fort George and Port Nelson. At present these conditions are unknown except in August and September.

(2) Provided that the conditions mentioned above are satisfactory, it would be necessary to investigate very carefully the cost of running such a boat and its general feasibility.

(3) It has been mentioned in the general report that the conditions of climate on the James bay coast, while perhaps severe in winter, are sufficiently favourable for growing some garden produce and the hardier crops, so that there is nothing to prevent Europeans living in health and, let us hope, in comparative happiness, as the fur traders and missionaries, their wives and families are doing now and have done for generations past.

(4) With the first two questions favourably settled, there can be little doubt the whitefish fishery will prove one of the greatest in Canada, and with its development the other fisheries will become of immense value.

The following specimens were collected by the expedition on the east coast of James bay during the summer of 1914, and determined by Mr. A. Halkett, of the Dominion Fisheries Museum, Ottawa:—

> Trout Perch (*Percopsis Guttatus*).
> Long-Finned Charr (*Salvelinus Alpinus Alipes*).
> Sand Launce (*Ammodytes Americanus*), (Possibly the Form "A"—*Dubius*.)
> Cottoid.
> Daddy Sculpin (*Myoxocephalus Groenlandicus*).
> Sculpin.
> Whitefish (*Coregonus*).
> Tullibee (*Argyrosomus*).
> Greenland Codfish (*Gadus Ogac*).

BIRDS.

Mr. W. G. Todd (Curator of Ornithology, Carnegie Museum, Pittsburg, U.S.A.), whom I met this summer in James bay, kindly gave me a list of birds found in the bay destructive to fisheries. They are as follows:—

> Ring-billed Gull.
> Herring Gull.
> Bonaparte Gull.
> Arctic Tern.
> Common Tern.
> Double Crested Cormorant.
> Two Species of Eider Duck.
> Surf Scoter.
> Merganser.
> Mandt's Guillemot.
> Loon (Great Northern Diver).
> Red Throated Loon.

Ruperts House.

Rupert's House.

East-Main Wharf.

Hudson's Bay Co.'s steamer " Inenew " at Charleton Island.

The Canoe and Outfit, Department of Fisheries.

Rapids on the Mizsauaibi River.

Indian Canoes at Wastikum.

Whitefish taken at Brandy Bay.

Indian Canoe at Hannah Bay.

Salt Marshes, Hannah Bay.

Hannah Bay—Left by the Tide.

The schooner "Violet" hired for the trip to the North Twin Island.

Setting a Whitefish Net.

Indians fixing the fishing nets on board the "Violet."

Indian Fishing Camp.

Dinner at Cape Hope.

Indian Canoe at Sabaskunika.

Eskimo in Kayak at Cape Jones.

Cape Jones.

Eskimo at Cape Jones.

Indian Women at Cape Jones.

Eskimo Women at Cape Jones.

Net full of Tullibee.

Trout taken at Long Point.

A REPORT ON

THE FISH AND FISHERIES

OF THE

WEST COAST OF JAMES BAY

BY

A. R. M. LOWER, B.A.

DEPARTMENT OF THE NAVAL SERVICE,

OTTAWA, December 1, 1914.

To the Deputy Minister of the Naval Service.

SIR,—I have the honour to present to you the following report, being an account of an expedition sent out during the past summer to collect information in regard to, and to investigate, the fisheries of the west coast of James bay and of the rivers flowing into it.

I have the honour to be, sir,

Your obedient servant,

A. R. M. LOWER,

INTRODUCTION AND SUMMARY.

It has been deemed that the usefulness of this report would be much increased if a summary, containing the salient points and essential facts, were appended. Accordingly the following short synopsis of my investigations is appended :—

The region about James bay is underlaid by a series of sedimentary rocks, mostly bedded limestone. These rocks not only underlie the land but extend for many miles under the water; as a result the land is very flat and the water, having such a slight and gradual slope, for many miles out from such coast as there is, is very shallow. The rivers, discharging over these limestone flats, and bringing down vast quantities of sediment from the soft clay country through which they run, have naturally large bars at the mouths. These bars spoil the entrances to the river mouths and there is thus not one harbour for large ships on the coast. About eight feet of water at high tide is the best that can be obtained in the biggest of these, the Albany. There is only one island on the coast—Agumiski; it is seventy miles long and lies eight miles off the coast at its north end and about sixty at its south. Between this island and the shore the rapid tide that races up and down the strait has worn out a channel of considerable depth. This channel is about three miles wide and outside of it the water is usually very shallow; it is not unusual to find only six feet of water at a distance of four or five miles from land. The tides average about five feet in height but are very much affected by the winds. Continued south winds almost destroy the tides, while continuous north winds pile the water up to great heights at the south end of the bay. In the rivers the tide runs up from four to twelve miles but the salt water does not penetrate much beyond the bars.

There are sixteen different species of fish found on the west coast or in the rivers flowing into it. The peculiarity about these fish is that with exceptions of no economic value, they are nearly all fresh-water species. The most valuable from a commercial standpoint is the whitefish and the river in which it is found to the greatest extent is the Albany. From the estuary of the Albany, there is known to be taken year after year, the amount of 13,000 pounds; besides this amount, a population of some four hundred finds abundant sustenance. Unfortunately the presence of the fish in these rivers is not continuous; from the middle of June to the middle of August nearly all the estuaries are devoid of fish of any sort; it is probable that the anadromous fishes are out in the deep water of the middle and eastern part of the bay. All the above amount is taken in the last two weeks of October, though the supply is just as great from the last of August as it is at that period.

It should be noted that the above figures include the fish called tullibee which differs from the whitefish merely in the shape of its jaw, and in a slight inferiority in food value.

Sturgeon are not numerous on the west coast but they are caught regularly every spring and fall; sometimes they are obtained up to a length of seven feet but that is very rare, the more usual size being about three feet. There are vast numbers of suckers to be obtained in every river. As the country has agricultural possibilities, these may possibly be useful some day for fertilizer; at present their value is nil—as, for food, it always will be.

There are a great many speckled trout caught each year in the rivers of the northern part of the coast; these fish average a pound and a half in weight and form about the best food fish that can be obtained. They are anadromous in their habits, their movements coinciding with those of the whitefish. They are much larger than the ordinary brook trout and altogether one of the best fishes of the coast. In this con-

33

5 GEORGE V., A. 1915

nection it should be stated that the whitefish ordinarily taken in the fall and making up the bulk of the catch, is not a mature fish that has entered the rivers for the purpose of spawning but an immature fish of one or two years growth that has merely come in in obedience to the habit that will later cause it to return to spawn.

The rivers of this coast contain fair quantities of both pike and pickerel but it is likely that if anything like fishing on a wholesale scale were introduced they would soon be stripped of these two species. The pickerel are probably numerous enough in the Albany to withstand the inroads that one or two fishermen in a small way would make on them but if systematic exploitation of the waters were allowed they would soon disappear.

The other species of the west coast are not of a great deal of economic importance; rock-cod occur in the bay and it is said that the true cod does also, but there is no record of a single specimen of either of them ever having been found on the west coast.

Besides the strictly fish wealth, there are other forms of marine life possessed of considerable value; one such is the white whale, another is the seal. The former occurs in great numbers and as he is quite valuable for his oil—of which he yields 100 gallons each, it may be expected that an industry founded on his products will develop as soon as the market is brought near enough to the place of production. The seal is not the fur seal but its hide makes extraordinary waterproof bags, boots, gun-covers, and so on.

There are two large rivers on the coast, three of fair size and numerous smaller ones besides unlimited creeks. There are some fish to be found in all of these but the most valuable are the Albany, the Kapiskau, the Lowashy and the Attawapiskat.

The best way at present to get to James bay is to go down the Kenogami-Albany system; there is no need to take very much food along, as trading posts are abundant and their prices not very much higher than at Cochrane. Travel along the coast, owing to the peculiar tidal conditions is slow and tedious and seldom is undertaken without the help of an Indian guide, who knows the landing and camping spots. The fishing gear used is made up entirely of gill and seine nets; the latter are used by the fur companies, the former by the Indians. Nets of a small mesh are the only useful ones; the Indians' nets are never more than two inches. It is often difficult to set nets, owing to the strong currents and the rubbish they carry, but the Indians have devised a method of staking nets which overcomes that obstacle to a certain extent. Deep water fishing on the west coast has never been undertaken so that it is hard to say what are the conditions attached to it. So far as the party with the limited apparatus at its command could determine, the tidal currents are as dirty as are the river ones and quickly fill up a net with sea weed.

The greatest difficulty a fisherman on the west coast would have to contend with would be the lack of harbours or of any shelter to which he might run in case of bad weather; only the very smallest of fishing craft would be able to get into the rivers at low tide and if caught by an offshore gale and an ebbing tide they would not be able to get in at all unless they possessed some mechanical power. This disadvantage is to a certain extent balanced by the infrequency of really serious gales. To sum up in a general way, it may be stated that the west coast of James bay offers at present great opportunities for fishing at certain limited times of the year but that at others, its value for this industry—apart from the unknown contents of the water several miles offshore—is nil.

I should like to express my thanks to the officers of Revillon Frères and of the Hudson's Bay Company at all the posts on the bay at which I was present for their unfailing courtesy and willing assistance.

A. R. M. LOWER.

I.—A REPORT ON THE FISH AND FISHERIES OF THE WEST COAST OF JAMES BAY.

A.—INSTRUCTIONS.

The instructions I received from the Superintendent of Fisheries directed that I should proceed from Ottawa to Cochrane, Ont., where I was to find men and equipment for an overland journey to James bay. I was to go down the most suitable river leading to the bay and from its mouth, travel along the west coast until I should come to the last river of importance on that side of the bay. From thence I was to turn south and work my way back to Cochrane. It was to be my business to gather all possible information about the fish and the fisheries of James bay—and also any other facts that would be useful in connection therewith—more particularly about the fish and fisheries of the rivers and river estuaries, but also so far as circumstances would permit, about the conditions in the main body of bay itself.

ITINERARY.

I left Ottawa on June 4 and went at once to Cochrane, in the company of Mr. C. D. Melvill, who was undertaking similar work on the east side of the bay. Leaving him at Haileybury, to arrange for the transport of the four men he had engaged for us at that point, I went to Cochrane to arrange for that part of our journey which was to be performed over the National Transcontinental railway. Both parties spent the next two days in Cochrane, purchasing supplies and making other preparations for the trip. A journey of 200 miles which owing to the unsatisfactory state of the train service was stretched out over several days, brought us to the Nagogami river, down which it had been decided to go.

Our second day's paddling brought us to the junction point of several rivers— locally known as the Mattawa. At this point is situated the small fur trading post of "English river." Having passed a day or two for inquiries at this point, we went on down the Kenogami, as the river is termed after its numerous branches come together, and in two days reached the Albany. The Albany is a very large river with a very rapid current and as a consequence we did not stop along its course to set nets or to perform other fishing operations. Eight days after our departure from the railway we came to Fort Albany, situated about seven miles from the mouth of the river. As my instructions called mainly for an investigation of the river estuaries, I judged it wise to spend considerable time at Albany, both for the purpose of fishing, myself, and for gathering information from Indians and others whom I met there. The results of these investigations as well as others made during the summer are noted in the main body of this report.

Leaving Albany, I secured the services of an Indian guide and set out for the next river of importance—the Kapiskau. This was reached after a trip of a few days along the open sea-coast. After a brief stay there we left and on the same day got to Lowashy river—really the most southerly mouth of the Attawapiskat. This latter river is the site of the only permanent settlement north of Albany. Here both the Hudson's Bay Company and Revillon Frères have establishments and here all the Indians from the north and the west over a region of several hundred miles, congregate. The party remained at Attawapiskat for several days, adopting much the same methods as had been employed at Albany. A small schooner was obtained in which we crossed to the large island of Agumiski about fifteen miles distant from the mainland. After a few days there, we again came to Attawapiskat and continued our inquiries. Then taking advantage of the sailing of Revillon's schooner with supplies for the small winter post they maintain on the Opinegau river, we proceeded

5 GEORGE V., A. 1915

northward once more. making no stops until the Opinegau was reached. This is only a small river but it proved to have several interesting features. Only one river of any size flows into James Bay to the north of the Opinegau. the Nagedowzaky, and accordingly when we had reached that point we turned back. The route pursued on the return journey to Albany was of course the same as that followed while going north. From Albany I crossed in the French company's steamer to Stratton island and from thence to Moose Factory. At that point I met Mr. Melvill and we arranged to make the up-river journey together: we followed the Moose river up to the forks of the Missinabie, at which point we branched off onto the Metagami. Two days journey from the railway, we came to the Ground-Hog river which was the route followed until October 9 on which date we reached the railway.

HISTORICAL.

The great land-locked seas of our northland have been the scene of trade and adventure for three centuries. Despite this long stretch of time, their resources are still very much matters of speculation. It was in the year 1610 that Henry Hudson perished there miserably, set adrift by his mutinous sailors. Nineteen years later Captains James and Fox passed through the straits and explored the bodies of water which bear their names to-day. James sailed along the west coast from Nelson southward, landing at the long flat point which marks the entrance to the smaller bay. This he named Cape Henrietta Maria, from the Queen consort of the day. A few miles further south, on one of the low gravel ridges that stand out as the only breaks in a shoreline of incredible monotony, he buried one of his men. "Mourning Point" a distance south of the river Opinegau bears testimony to the event. James wintered that year at Charlton island, thus marking out the spot that was to serve as rendezvous to trader and explorer till the present day. In 1662. came Radisson. French fur trader and wood-runner. Meeting with naught but rebuffs on his return to Quebec, he was driven to offer his services to the English. As a result of the voyages that he and his brother-in-law Groseilliers—naively referred to by his employers as ' Gooseberry "—undertook, the first post of what was to be the Hudson's Bay Company was established at the mouth of the Rupert river. From that time on the history of James bay becomes the history of the Hudson's Bay Company. In another half dozen years the " Gentlemen Adventurers " had established themselves more securely, were possessed of outposts at Moose and Albany and a depot on Charlton island for their annual ships. When a few more years had passed. these places had grown into substantial establishments and others had been begun.

Meanwhile friction with the French went on unceasingly. Cargoes were seized, crews were massacred and forts were taken and retaken with commendable regularity. A decline in the trade on the east of the bay was found to be due to the appearance of the ubiquitous French wood-runner on the head waters of those rivers down which the Indians were in the habit of travelling to the English posts. Such a situation caused the more attention to be directed to the tribes of the west coast about the Albany and the Severn. But it was not long that the company was allowed to remain undisturbed in the possession of the enormously profitable trade of that region. Trouble was brewing at Quebec, and in 1685, after several mysterious visits of individual Frenchmen, who one and all declared that they came merely "to see the country " an expedition was organized by permission of Denonville. Governor of New France. It made a successful journey overland and speedily reduced all the English possessions on James bay. The following extract from the writings of Miss Agnes Laut gives a vivid picture of its proceedings. more especially of the taking of Albany. the chief fort on the west side.

SESSIONAL PAPER No. 39a

"THE OVERLAND RAID ON THE POSTS OF THE HUDSON'S BAY COMPANY."

" Sixty-six swarthy Indians and thirty-three French wood-runners, led by the Chevalier de Troyes, the Le Moyne brothers and La Chesnaye, the fur trader, were threading the deeply forested, wild hinterland between Quebec and Hudson's bay. After taking Moose Factory and Rupert's House, with prisoners, ship, cannon and ammunition, the French set sail westward across the bay for Albany. The wind proved perverse. Ice floes, drifting toward the south end of the bay, delayed the sloops. Pierre Le Moyne D'Iberville could not constrain patience to await the favour of wind and weather. With crews of voyageurs he pushed off from the ship in two canoes. Fog fell. The ice proved brashy, soft to each step and the men slithered through the water up to the arm-pits as they carried the canoes. D'Iberville could keep his men together only by firing guns through the fog and holding hands in a chain as the two crews portaged across the soft ice.

" By August 1, the French voyageurs were in camp before Albany, and a few days later de Troyes arrived with the prisoners and the big sloop. Before Albany, Captain Outlaw's ship, the *Success,* stood anchored; but the ship seemed deserted and the fort was fast sealed like an oyster in a shell. Indians had evidently carried warning of the raid to Sargeant (the factor) and Captain Outlaw had withdrawn his crew inside the fort. The Le Moynes, acting as scouts, soon discovered that Albany boasted forty-three guns." But " if the French had but known it, bedlam reigned inside the fort. While the English had guns, they had very little ammunition. Gunners threw down their fuses and refused to stand up behind their cannon till old Sargeant drove them back with his sword hilt. Men on the walls declared that while they had signed to serve they had not signed to fight, and if any of us lost a leg, the Company could not make it good." The Chevalier de Troyes with banner flying and fifes shrilling, marched forward. . . . . .Bombs began to sing overhead. Bridgar came under a flag of truce to Sargeant and told him that the French were desperate. It was a matter of life and death. They must take the fort to obtain provisions to return to Quebec. If it were surrendered, mercy would be exercised. If taken forcibly, no power could restrain the Indians from massacre. And Sargeant. . . . . had his family in the fort. Just at this moment one of the gunners committed suicide from shear terror and Captain Outlaw came from the powder magazine with the report that there was not another ball to fire. Before Sargeant could prevent it, an underling had waved a white sheet from one of the upper windows in surrender. The old trader took two bottles of port, opened the fort gates, walked out and sat on a French cannon while he parleyed with de Troyes for the best terms obtainable. The English officers and their families were allowed to retire to Charlton island to await the coming of the company's yearly boats.

The Chevalier de Troyes bade his men disband and find their way as best they could to Quebec. Only enough English prisoners were retained to carry the loot of furs back overland. The rest were turned adrift in the woods. Of fifty prisoners only twenty survived the winter of 1686-87. Some perished while trying to tramp northward to Nelson, and some died in the woods after a vain effort to save their miserable lives by cannibalism."

Within the next decade the fur posts changed hands frequently. At the treaty of Ryswick in 1697, it was provided that each nation should retain what possessions they had at the time the treaty was made. This left the Hudson's Bay Company owners of but one fort and that was Albany. But the fortunes of war varied again in succeeding years and at the Peace of Utrecht (1713), England was able to force the French to give up all their claims to territory in the Hudson bay region. Since that date all interference by force of arms has ceased, but the company has had to meet the competition of the Coureurs de Bois, who, by 1733, had succeeded in penetrating into the very interior of Labrador, and of the North-West Company which from its

5 GEORGE V., A. 1915

inception in 1770 to its amalgamation in 1821 proved a very troublesome opponent. For the last century James bay may be said to have had no history, other than what is included in the unceasing round of trading, and hunting, of fierce struggles with the cold of the arctic winters and of long trips of exploration in the pleasant summers.

B. GEOLOGICAL.

To give a general idea of the bay and the regions surrounding it, it will be necessary to set down a brief résumé of the chief geological features that are met with.

The North American continent is built about two axes or backbones—the Rocky mountain system in the west and the Appalachian and Laurentian areas in the east. The latter of these, variously known as the Laurentian, the Archean or the region of Igneous rocks, extends in one direction from the northern part of the North Atlantic states through Quebec, Labrador, on to the shores of Hudson straits and Baffin's land. Another great wing is flung off from the main branch in the west of northern Quebec and passing through Ontario—where it forms the rock masses of Muskoka and the northern districts—extends in a northwesterly direction to the shore of the Arctic, some distance to the east of the Mackenzie river. The whole mass is thus roughly triangular in shape with the apex to the south and the base to the north. Hudson bay may be considered as a huge ' V ' cut out of the base of this triangle. At the close of the Archean period. with the exception of some geological "islands" in the Appalachians, this was the only portion of North America that had risen above the water. It thus supplied the shores, both to the sea that has since become the Atlantic ocean and to that great inland body of water whose disappearance has given us the wide prairies of the west. On these shores was laid down in layers of varying width and thickness the sediment that was erided from the mass of the Archean "backbone." At the same time the swarming marine life of the time, contributed a vast amount of lime to the ooze that was constanly sinking to the bottom and being hardened by the pressure of the accumulations above it. Sooner or later the sea bottoms began to rise and when that process had gone on long enough, dry land-began to appear at the edges of the Archean shore lines and the second great series of rocks made their appearance. These were the rocks of the Palæozoic period and they differed from the Archean type in that they were formed out of sediment and lime, in layers under water and gradually hardened by pressure from above, as has been stated. They consisted mainly of limestones. With those on the southern border of the Archean mass, we have nothing to do but with those on the northern, we must deal as being those rocks which underly most of the district around the south and south-west coasts of James bay and also a large part of the bay itself. The new land thus formed extended outward from the shores of the old rock mass a distance of two or three hundred miles to the east and along those shores from north to. south, considerably farther. Thus a region as large as old Ontario had arisen above the sea. At the close of that first movement. James and Hudson bays had almost received their present outlines. All that was lacking was the area lying between a line drawn from about the Ekwan river north-west to the Severn and the present Cape Henrietta Maria. When the same process of deposition of sediment and calcareous shells had gone on for some time longer and when another raise in the sea bottom had taken place, a new space of dry land which occupied the above mentioned area, made its appearance. This land consisted of limestone rocks differing but little from the last. The geological centre of all this palæozoic area is supposed to be located under the waters of James bay off the mouth of the Albany river; the bed rock thus extends unbroken from far inland to a long distance out in to the sea. In forming an idea of the appearance of the country on the west side of James bay, then, it will be important to remember that it is all underlaid by level floor-like areas of limestone and that the only elevations. or depressions that can occur must necessarily be made out of the material overlying these areas—that is, there can be no rocky hills or ridges.

The great ice-age came and went and the palæozoic limestones became covered with a thick layer of glacial " drift ";—that is, the glaciers in retreating left behind them the debris that they had carried; this debris consisted of layers of boulder clay which were deposited almost uniformly over the whole territory. Dr. Bell, of the Geological Survey, in describing this feature of the country says: "The drift is a continuous sheet varying in thickness between 30 and 90 feet.... it becomes thinner as we rise higher and get further inland. It is of a looser and less clayey nature in the higher grounds and consists largely of washed gravel and shingle."

The ice sheet was of enormous thickness and it is supposed that its weight was sufficient to depress the level of the country to far below the surface of the sea; in some places this submergence is supposed to have been as much as five hundred feet. The result was that all the area so depressed became silted up with pretty much the same material as had been deposited to form the limestones of the earlier period. But still another elevation of the land taking place before the pressure had become strong enough to harden these materials into solid rock, the new deposit rose from the water in the form of a marine clay and it is this marine clay of which most of the land around the bay consists to-day.

These two deposits—the glacial drift or boulder clay and the later sedimentary or marine clay make up all that country which commonly is called the clay belt and which extends from about the line of the National Transcontinental railway northward almost beyond the limits of the district of Patricia.

That such is the case is born out by the present appearance of the country; it is one vast wooded plain with a gradual and uniform slope to the north and east; in the whole course of the Kenogami and Albany from English river post to the sea there is not a single elevation of any one point above the surrounding country. As might thus be expected the sea coast is singularly flat and low; it is so flat that land is lost sight of when but a few miles out . And in the same way the shore presents no variation in appearance; its features are absolutely the same throughout its length; the whole vast plain slopes down to and under the water at a very slight angle; if one can imagine a board of a few feet in length, part of which is under water and part of which is not and which at its one end is immersed but an inch or two and at its other is elevated but an inch or two, he will have a good idea of the nature of the country. The water on the west coast is exceedingly—unbelievably—shallow just as the land is exceedingly flat. A typical piece of coast line is that at the mouth of Chikeney creek. At that point. the woods are about three miles back from the average high tide mark. Between the forest and the tide mark is an open, level plain, the first mile of which is covered with scrubby willows. The other two miles support a growth of luxuriant grass. This grass gradually gets thinner as it approaches the water until at last only scattered bunches of it remain. Between the extremes of high and low tide, a space of about three miles of soft clay mud intervenes; this is also perfectly flat and covered with small boulders. On the mud when the tide is out lie shallow pools of salt water. From the last bunches of grass it is hardly possible at low tide to see the open water. Under the surface at low tide the same level stretch continues so that even a small sail boat of very light draft has to keep several miles out in order to obtain sufficient water. The slope becomes a little more rapid about ten miles off shore and when the middle of the bay is arrived at a fair depth is obtained.

The only variation that is met with consists in the mounds of pebbles which line the shore from Neakwow point northward. These mounds may reach a height of several feet—when they do so, they are locally referred to as " bluffs ". They sometimes form

5 GEORGE V., A. 1915

long sweeping points and are undoubtedly the result of ice action; the winter ice which in this part lingers about until the middle of July, tossed and retossed upon the coast, has scraped them up from the mud flats between and beyond the tide marks and deposited them in heaps upon the beach. At a few points the action has been vigourous enough to grind the pebbles into sand; with the exception of the creeks and rivers, these are the only places at which it is possible to land directly onto dry land; at all others, the canoe must be left lying on the mud and the camp outfit portaged through the mud to a dry spot further back.

Throughout the wooded plain the rivers of the country run. They are nearly all quite similar in general characteristics. In the first place there are few portages; on the Albany and Attawapiskat, for two or three hundred miles, there are none at all; and the Albany is navigable for fair sized craft for all this distance when the water conditions are good. They are all swift and carry down vast quantities of sediment to the sea. Few of them afford quiet places or backwaters; in all of them the current usually sweeps straight along, wearing down the points and straightening out the channels between the islands. Islands are formed quickly: first a sand bar, then a little grass appears; next a few willows begin to grow and if the ice is not too devastating in its effects trees such as small poplars take root. More soil is added by the ice every spring until at last an island is formed. But no sooner does it attain a level of a few feet above the water than it begins to disappear again; the incalculable force of the spring breakup, works on the up-stream end, tearing away bank, trees and soil, carrying all down stream to be deposited in another place. Thus a constant process of island formation and island destruction is going on; we could almost imagine the same island beginning hundreds of miles up stream and gradually travelling downward until it reached the river mouth and was carried out to sea there to add its contribution to the enormous bar that stretches across the river mouth.

The sides of all the rivers are concave in shape and vary from four or five feet in height at the sea to fifty or more up country. They are all quite free from undergrowth and afford excellent walking; their openness makes them peculiarly suitable for such a process as seining. The winter ice as it rushes down in the spring sweeps them clean and presses the boulders that it carries deep into the clay; thus are formed the well-known "pavements"—stretches along which the bank is literally and uniformly paved after the manner of a cobblestone roadway.

While the above remarks are true of the rivers of the west coast they will not apply in their entirety to the Moose system of the south west. This system travels a much shorter distance from the Archean highlands and in consequence has worn down its bed further below the level of the country; in fact for the greater part of its course it has worn away all the surface clay and travels over bedded limestone. As a result small rapids occur constantly, the river bed is very wide and the water very shallow; only at moderately high water is it even easily navigable for canoes. The limestone exposures however form numerous coves and quiet pools where nets can be easily set and where fish congregate in large numbers. When this river enters the Archean area its character, of course, changes again; rapids occur at intervals only and when they do so are of considerable size and length; between them the water is held up in long quiet lake like stretches of little current. The banks too become rocky and abrupt, usually steep and when not so, covered with a dense growth of shrubs.

WATER CONDITIONS.

Under this heading I propose to set down those observations I made which I have reason to think would be helpful to anyone wishing to gain a knowledge of the local conditions of that region, either for practical purposes such as those of the sailor and fisherman or for more academic ones.

SESSIONAL PAPER No. 39a

A general idea of the coast line is given under the section dealing with the geological structure of the country; it may be added here that the coast is almost straight and runs slightly to the west of north; there are no natural harbours except the river mouths—which will be dealt with later. The only variations from the straight line are wide shallow bays which afford no protection from the weather. A typical bay might be ten miles across and a mile or so deep. Twelve miles off shore, the land is lost sight of and at two miles the entire visible portion of coast to be seen would probably be not more than fifteen miles. At that distance the shore presents the appearance of an even black line, subtending an angle of about 120 degrees; outside of this the black gradually fades out against the skyline.

Owing to the slight depth and the muddy bottom, the water is seldom clear; its usual colour is brownish-yellow; after a long period of calm, however, it may get to have not very large traces of this colour. Of course, the deeper it gets the clearer it also becomes and at four miles or so from the high tide mark—the nearest course to land that small sail-boats can follow—it is clear as often as not. James bay water has been described as "slightly brackish" but it is unmistakeably salt; it is only in the neighbourhood of great rivers such as the Albany that it can be called "slightly brackish".

Over the slightly submerged plain that forms the bottom and shores of James bay, the rivers discharge. The channels they have worn out for themselves are in comparison to the amount of water carried, surprisingly shallow. Some of the rivers indeed which are of a fair size almost lose themselves at low tide, spreading out over the flat expanse of mud to such an extent that they may be said to have no channel beyond the grassy plain; even a river as large as is the Bowashy spreads its waters out over such a wide area on the mud banks at low tide that a canoe can hardly enter it. Thus all these rivers must be entered at high tide, at which time one has beneath him the depth of the tide plus whatever water is naturally in the river.

About two miles from low water mark or about an average of five from high, one gets six feet of water. Another two miles gives a depth of twelve feet, or more. At fifteen miles out from Albany, the lead registers seven fathoms, at twenty miles fifteen fathoms, at sixty miles from Albany on the course to Stratton island the maximum depth of the southern portion of the bay 35 fathoms is reached; this depth decreases slightly between that point and Stratton Island. The maximum depth to the west of a line drawn from the mouth of the Moose river through the Gasket shoal to Cape Henrietta Maria is according to the soundings of the master of Révillon Frères' steamer *Emilia*, between 20 and 28 fathoms. The Gasket shoal lying about east by north sixty miles out from the Albany is a low heap of clay and boulders about three miles long, with bad approaches of smaller shoals and boulders; quite near it on the south side a depth of twelve fathoms is obtained. The gentleman referred to above believes it to be a continuation of Agumiski island whose southern extremity is surrounded by vast stretches of shoals leading in the direction of the Gasket.

Agumiski island ("Agoomiskik"—"the land across") is about 70 miles long and roughly triangular in shape with the base at the northern end and the main axis running in south-east by easterly direction. Its most northern part lies a little to the north of Neakwow point and its southern is between 50 and 80 miles north-east by east of Albany. The island is not shown correctly on any of the maps of that region. It is similar to the mainland in appearance but its western shores are heaped high with banks of pebbles and the forest comes within a few feet of the water's edge. In winter the strait between it and the main shore is frozen over at its northern end; it is seldom, however, that a space wider than 15 or 20 miles freezes; this means that the only portion connected with the mainland by ice is the projecting westerly point lying off the mouth of the Attawapiskat. Here in one place the strait is only about 8 miles wide and is broken up by a few small, flat islands—the Manowinau—the only

5 GEORGE V., A. 1915

ones on the coast. The ice, however, for a few miles north and south along that part of the coast and island becomes quite solid and is regularly crossed both on foot and with loaded dog sleighs.

The main shore north off the Attawapiskat river sweeps round to the northeast in a large curve terminated by Neakwow, that is "Sand" point. Into this curve the western projection of Agumiski fits, the Manowinan islands being given off at the western extremity of the island and the narrowest point of the curved strait. North of Neakwow point the land slopes rapidly back to the west. This conformation, as will be seen, has a great deal to do with the action of the tides. The only other point at all similar to Neakwow is that locally known as the "Cock" situated midway between the Moose and Albany rivers.

TIDES.

In height the tides are very uniform throughout the bay. A high tide is six feet or over, an average tide is about four feet. As entrance or exit to or from the rivers is so absolutely dependent upon the tide and its freaks, it is essential to have a clear idea of the factors affecting these movements. Owing to its shape—a huge, almost landlocked body of water lying from north to south—and its shallowness, the James bay tides are very much affected by the wind. As a general rule, it may be stated that a north wind makes a good tide and a south wind a poor one. The reason of this will appear later.

The tide enters the bay from the north, travelling from the straits in a south-westerly direction. It spreads uniformly over the entire body until it reaches Agumiski island and Neakwow point. At these places it splits; that portion of the water that comes to Neakwow point divides, the main stream turning north and flowing along the coast toward the cape. The rest penetrates between Agumiski and Neakwow, flowing on down the strait. At the same time, that portion of the main tide that had gone down the east shore of Agumiski, travels southward until it reaches the "Cock," and there divides in its turn, part of the water going on south to Moose river and part turning north, penetrating the Albany, flowing along the coast, becoming pursed up in the narrowing strait between Agumiski and the mainland, and finally meeting the northern half of the tide in the neighbourhood of the Manowinan islands. The results are: (a) Four high tides a day around the Manowinan islands; these come in pairs and the crest of each member of the pair is not far apart. That is, shortly after tide A has begun to ebb, tide B becomes full. (b) A tide race of considerable violence in the strait. The currents are so strong here that a sound of considerable depth has been hollowed out. This sound or deep channel lies close to Agumiski—about one mile off shore—and is about three miles in width. Between Lowashy river and the fur posts on the island a depth of 18 fathoms has been found and the average depth is said to be in the neighbourhood of 10 fathoms. The banks are quite abrupt and the water that flows through this channel is filled with various kinds of floating seaweed in great quantities. From the west bank of the sound towards the mainland the deepest water would probably be twenty feet, but that depth cannot be relied upon as it becomes rapidly shallower as the shore is approached.

The tidal currents here as elsewhere in the bay are too swift for a sailboat to make headway against, unless it has a wind from aft. In fact, a steamer with a speed of seven or eight miles an hour makes very slow progress.

From the diagram given may be seen the manner in which the wind affects the tides. A north wind blows the water into the bay and by thus aiding the tide, raises it and holds it up for a greater length of time. A south wind does exactly the reverse of this. A west wind too, delays and lowers the tide. During the past summer the steamer *Emilia* was fast on the Albany bar for over a week owing to gales of heavy west winds. At no time in that period did the tide, which usually averages $5\frac{1}{4}$ feet at that point, exceed three.

THE FISH OF JAMES BAY.

The various species of fish found on the west coast are as follows:—

1. Acipenser Sturio (Lin.)... ... ... Common Sturgeon.
   Catostomidæ, Suckers:—
2. 　　Catostomus Catostomus.. ... ...Northern or Long-nosed Sucker.
3. 　　Moxostoma Aureoleum... ... ...Common Red Horse.
4. Erimyzon Sucetta.. ... ...Chub Sucker.
5. Hiodon Alosoides.. ... ...Mooneye or Goldeye.
   Salmonidæ :—
6. 　　Coregonus Clupeiformis.. ... ...Common Whitefish.
7. 　　Coregonus... ... ...Species uncertain.
8. 　　Argyrosomus Tullibee.. ... ...Tullibee.
9. 　　Salvelinus Fontinalis... ... ...Speckled Trout.
10. Mallotus Villosus.. ... ...The Capelin.
11. Esox Lucius... ... ...Common Pike.
12. Stizostedion Vitreum.. ... ...Pickerel (Doré).
13. Perca Flavascens... ... ...Yellow Perch.
14. Lota Maculost... ... ...Fresh-water Ling or Burbot (Marl).
15. Cottidæ Icelus Hamatus... ... ...Northern form of common Sculpin.
16. Cottus Ictalops.. ... ...Miller's Thumb (Blob).

On the Hudson's Bay watershed occurs;—

Cristivomer Namaycuch.. ... ...Lake Trout.

COMMON STURGEON.

It seems a contradiction in terms to state that the common sturgeon is not at all common, but such is nevertheless the case. It is caught, of course, regularly but nowhere in very great abundance; it is never caught in the sea but a few are taken every year in the rivers and river estuaries. The usual practice is to bait large hooks and to suspend a great number of these from a horizontal cord so that they are lying on the bottom. Sometimes, however, the sturgeon are taken in ordinary nets. In no case do the inhabitants use a special sturgeon net for these fish. The largest size reported was seven feet (a specimen taken in the North Albany) but the average is very much smaller, probably two feet, certainly not over three. It is generally considered useless to try for sturgeon during July and the first part of August but by the middle of the latter month, fishing is supposed to be good. It continues so until the late fall and begins again during the spring. The habits of the sturgeon on this coast do not differ from the habits of the same fish elsewhere.

THE SUCKERS.

The habits and life history of all these fish are too well known to need further comment at this point. They are not observed to enter salt water. They spawn in the spring, running up the rivers and small creeks for this purpose. Strangely enough, they seem to be scarce in the middle of the summer as if they had moved away from their usual haunts. In August, however, they return in great numbers and may be seen at any time moving along the shore close to the bank. They form a very important item of the food of the Indian and his dog since they are easily obtained. They are probably present in greater numbers than any other fish and doubtless the waters of the north would be more productive of valuable fish if some way could be found of removing them.

The third species mentioned whose identification is doubtful is a rather handsome fish of much less common occurrence than the other two. All the specimens I examined were under 12 inches in length. Their bodies are much compressed and

5 GEORGE V., A. 1915

deep in proportion to their thickness. Their colour, which of course like that of all fish would be very variable, tends to a light metallic green above, with the fins reddish. Their scales are as large as those of the common sucker. Their lateral line is not straight but curved.

GOLDEYE OR MOONEYE.

But a few specimens of this fish, taken in the upper waters of the Moose river, were met with. It is valuable as a food fish but very local in its distribution, being unknown in the Albany or Attawapiskat systems.

COMMON WHITEFISH.

The whitefish is found almost without exception in all the waters of the north. It abounds in James bay, although its movements are such as to cause its absence from large portions of that body for considerable periods of time. It averages not more than a pound and a quarter in weight and sixteen inches long. The largest individual taken this past summer weighed about four pounds and measured about 21 inches by 6 inches. It is said that in the head waters of the Ekwan river and also in the Trout river, there are places where they may be obtained two feet in length. These, however, are the largest of which even the oldest Indians have ever heard so that it is safe to say that the whitefish of James bay do not grow to the size of that of the Great Lakes.

The movements of the whitefish are as follows:—When the ice leaves the rivers in the spring, the fish are found in great quantities; fishing continues good for about a month or until the first part of June. As the sun gets hotter and the water warmer, the fish disappear until in July there is scarcely a fish to be obtained in any estuary along the west coast. This absolute dearth continues until about the middle of August, at which time the whitefish come back. This return takes place quite quickly, as a few days will suffice to fill the tidal estuaries with fish. The time of the return, of course, depends on the season; a fine summer prolongs the period during which there are no fish and a cold one shortens it. The fish seem to come back to the whole west coast at the same time. Thus when the fishing becomes good at Opinegau river, two hundred miles north of Albany, it also becomes good at Albany. One might expect that since the northern water gets colder before that further to the south the fish would return to it sooner, but such does not seem to be the case. Evidently conditions in their summer home—which may be in the depths of Hudson bay or may be merely out in the deep water of James bay—determine their return and not the local conditions of the west coast.

When they come back they are all very fat and many carry eggs or milt. These latter are, however, in the minority and are always the larger fish. Although I examined a great number of specimens I was unable to find one less than 16 inches in length or 1½ pounds in weight which was prepared to spawn. The fish under this size evidently come back to the rivers merely in accordance with the same migratory instinct that leads them back as adults to spawn. I found numerous specimens, too, quite unprepared to spawn that were as large as a good many of those that were ready to spawn. These whitefish or " Atikameg " as the Indians call them, are taken every fall in vast numbers from about four inches in length upward to the sizes named before.* The immature fish congregate in vast schools in the river estuaries and are commonly taken with the seine nets. As in the case of herring, they come suddenly and make their presence known by " skipping " on the surface of the water. Usually fish of about the same size keep together; thus in September the seine will capture individuals averaging three-quarters of a pound in weight and about 12 inches in length; later on the size

---

* See section of report dealing with individual fisheries.

most commonly taken is 7 inches. The fishing continues extraordinarily good until about the last of October when as a rule the rivers freeze. During the winter but little fishing through the ice is done; at least as there are no Indians around the posts and as the companies have already secured their supply of fish but little is done around the posts and in the river estuaries. It is thus impossible to state whether the catch would prove as abundant during that season as at the other periods mentioned. The fact however that there are plenty of fish in the spring just after the ice leaves goes to indicate that the fish remain in the rivers all winter.

Spawning takes place in October or late September; the spawning grounds are usually but a few miles up the rivers; that is, the spawning grounds of most of the fish; very probably a good many penetrate further, this apparently being the case on one river, at least—the Ekwan. The depth of water does not seem to be uniform, but is never more than a couple of fathoms.

The most interesting problems about the James bay whitefish are thus: (1) The annual return to fresh water of vast numbers of immature fish; (2) the disproportion between these and the mature fish; (3) the whereabouts of the fish in the summer. To solve this last problem a deep sea expedition is necessary. They are not present near the shores of the bay as none can be found at Agumiski island until the return to the rivers takes place. At that time they are not only abundant in the estuaries but are found along the coast in fair quantities also. As whitefish are taken pretty regularly at Stratton island all summer, the probability seems to be that they frequent during the summer the deeper and colder waters of the east coast.

SECOND SPECIES OF WHITEFISH.

With the exception of one or two well differentiated species, the existence of different kinds of whitefish in bodies of water even so well known as the Great Lakes is still more or less a matter of controversy: scientists are unable to decide whether certain forms are only varieties of the common kind or whether they are distinct species. As a second species of whitefish has been from time to time reported from James bay, it is mentioned here but it must be stated that if this second species does exist, it differs in its habits in no observable manner from its better known relative. None of the Indians recognize a second species. The only observable difference in the whitefish in the bay lies in their shape. In some there is a pronounced "hump" on the back as in the whitefish of the lakes; these fish are inclined to be short and deep. In others there is no "hump" and they are inclined to be longer, thinner and less deep than the first kind. Of five males examined, two were plainly the common whitefish, three lacked the latter's characteristic "hump." Of eight females examined, three were common whitefish, five lacked the "hump." In addition these latter had rather sharper jaws than the former. As no specimens could be brought back it is impossible fully to decide the matter.

TULLIBEE.

This fish is distinguished from the whitefish by its projecting lower jaw and by its softer flesh. It grows to about 18 or 20 inches in length and reaches a weight of three pounds. In every feature except its size it is similar to the herring of the Great Lakes. Its movements correspond exactly with those of the whitefish and the two species are always found closely associated. It is said not to resemble the tullibee of Manitoba very closely. The smallest specimen examined was 8 inches in length; this fish, in the beginning of September was filled with eggs which were in a condition to be shortly deposited. All others examined were in the same condition. From the first few days of August until the first of September a distinct development was noted in the egg masses. The eggs themselves became larger and harder, the ovaries more richly

5 GEORGE V., A. 1915

supplied with blood. All during this period the fish were very fat (as were also all the whitefish taken). Tullibee and whitefish were caught in about equal numbers both in the seine net and in gill nets. Tullibee are usually reported as spawning in the latter part of October but on September 10 in the North Albany river I took a couple of specimens which had apparently already spawned; they had lost all their fat and contained no milt. Still it is hardly to be expected that the spawning season should be so much earlier in James bay than elsewhere even taking into consideration the latitude and the very considerable difference in season.

The females outnumber the males in a proportion of about three to one. The average size would be about twelve inches long and the average weight about three quarters of a pound. Large specimens are however by no means uncommon. The flesh is excellent if used very shortly after the fish is caught but if it is allowed to remain for any time—even overnight—it deteriorates and becomes soft.

SPECKLED TROUT.

The range of this fish extends over the whole western James bay watershed from Albany north; it is also found in the lower reaches of the Moose river. It is not, however, very common in the Albany or in any of the more southern rivers, but abounds in all the rivers and creeks from Mourning point northward. The largest river in which it is found abundantly is the Opinegau. It is reported that it increases in numbers as one goes northward and that some of the rivers of the Hudson bay slope, notably the Trout and the Winisk. are filled with it.

Its movements coincide almost exactly with those of the whitefish and tullibee; it disappearing from the fresh water and the coast when the water becomes warm, returning later on to spawn when the temperature is falling. Unlike the whitefish, the autumn migration does not comprise immature individuals who merely accompany the adult fish; all the fish that return to the estuaries come back laden with eggs and milt and ready to spawn. The average female carries about 2,500 eggs.

The Indians usually calculate on the return of the fish occurring about August 10. August is known locally as the "Trout Month." All the rivers and creeks along the northern portion of the coast are of the same character; at the mouth they are mere beds of stones and mud when the tide is out, and usually very wide. When the tide comes in, it fills them up for several miles from the mouth and makes them look like rivers of a very fair size. Higher up they narrow down, the banks become steeper and they have a fair depth of water. As they approach their head waters they run over bedded limestone and at those places considerable rapids occur. The speckled trout enter these streams and for a considerable time remain in the estuaries; gradually they advance farther up until by the latter part of September when spawning takes place they have reached the rapids. In the swift water there, they deposit their eggs. After spawning they distribute themselves throughout the stream or river and in the winter may be caught at almost any point through the ice. When spring comes, the time of open water finds them again on their way to the sea and by the middle of June they have disappeared into the deep water once more. A few stragglers, however. remain in the rivers throughout the summer and these fish may be caught in certain places at any time. Some people assert that the fish of certain rivers—notably the Opinegau—acquire a swampy taste late in the winter; if this be true it will detract considerably from their food value.

The average size of the "Maskmaygus." as the Crees call the fish. is about 16 inches in length and a pound and a half in weight. The largest specimen was taken in one of the southern rivers where they are not usually very common, and weighed 5 pounds. No fish could be better eating; their flesh is a salmon pink or sometimes yellowish. With their brilliant colours of red and bluish, their shining bodies fresh from the salt water. they are most attractive in appearance. Although the speckled trout of the north is a most valuable and interesting fish.

THE CAPELIN.

This little fish is found along the shores of the west coast and especially at Agumiski island. Its life history is well known and its habits do not differ in the bay from those of its kind in other bodies of water. It is often found in considerable quantities. It spawns along the shore preferably in the surf and during rough weather. A good description of this process as well as other interesting facts in regard to the capelin is found in Goode's "*American Food and Game Fishes.*"

THE PIKE.

This fierce, submarine pirate dominates the waters of the north as he dominates all other localities in which he is found. Fortunately, he is confined to the fresh water so that his ravages must for the most part be directed on the less valuable kinds of fish. However, during the autumn when every creek is filled with toothsome morsels the destruction he works must be terrific. He does not seem to grow to the enormous size that he sometimes attains elsewhere, neither is he present in as great numbers as in other bodies but he is always hungry and always combative. The largest specimen taken weighed about 7 pounds, the average was about 4. The pike is a food fish of considerable value if cooked properly, especially the fish of larger size. He is about the easiest of all fish to catch as he may be taken by almost anything that glitters, whether on a troll or an ordinary "hook-and-line." The pike spawns in the spring. He retreats from the tidal estuaries during the summer months, preferring the upper reaches of the rivers; one reason for his course of action is that there he probably finds more food; the estuaries are devoid of fish at this time, most having gone out to sea, but a few up river There the pike goes after them. It is a pity that some plan could not be worked out whereby our waters could be cleared of such fish, for the destruction they entail among food fish must, every year, be enormous.

THE PICKEREL (DORÉ).

The pickerel is a member of the perch family and as such has an important position in the list of food fishes. It is locally known by the servants of the Hudson's Bay Company as the "Perch," this name doubtless being due to the resemblance it bears to the perch of the British isles from whence in bygone days the name has been carried. The pickerel is met with abundantly in the waters of the Albany system, in those of the Attawapiskat, the Kapiskau and the Ekwan. It doubtless occurs in the rivers of the north half of the coast also but no specimens were taken in them during the past summer. The largest obtained were two taken in concert with Mr. Melvill on the Metagami river. These weighed 8 and 9 pounds each. The average would be about three pounds and a half or a little more.

The pickerel does not enter the salt water, but seems at home in the tidal estuaries which often become a bit brackish. It, like the pike, is most abundant during the cold months and its flesh is then in better condition; of course it is always good eating but in the autumn and spring it may be kept longer before cooking. It spawns in the spring. It not only is a good food fish but provides good sport when taken on a troll. Its spiny dorsal fin when erected makes an efficient weapon of defense and unless its captor exercises care, he will pay for his prize with a lacerated hand. The pickerel is almost as voracious as the pike, disdaining very little in the shape of food that comes near it.

YELLOW PERCH.

This well-known little fish is found in the upper waters of the Albany system, in limited quantities. I have not heard of its being found elsewhere, though it is quite possible that its range extends to the other near-by river systems. The perch spawns in the early spring, depositing its eggs in a long, semi-transparent strip of sticky mucus. It never exceeds a pound and a few ounces in weight and twelve inches in length. As a food fish it ranks high. some people considering it of as good quality as the pickerel and both of them as superior even to whitefish or trout.

FRESH WATER LING (" MARI ").

This fish is well distributed over Ontario and is found in all the waters investigated last summer. It has a hundred different local names; thus in Michigan it is known as the "Lawyer"; in southern Ontario, the people of the inland lakes call it the "Dog-fish"; in the north, it is almost universally called the "Mari". The Indian name is "Malaskachoosh". It is the only fresh-water representative of the cod family and shows its affinity to that valuable stock by possessing an enormous liver which is of considerable food value. The flesh of this fish is sometimes eaten but he who has once tried it will not readily do so again; the fish is not only disagreeable to the taste but also repulsive to the sight. The flesh, especially the liver, is said to improve in winter. It is of importance to the Indians as it may be taken at almost any time.

It is a bottom feeder and as such possesses the wide mouth equipped with feelers that most fish of this type exhibit. Its head is flat and its body tapers rapidly to the tail. It has no rays in its fins and is scaleless, the skin being covered with a coat of slime. A large one is two and a half feet in length, an average one about twenty-two inches. They are caught quite commonly on lines set for sturgeon. Spawning takes place under the ice in January. So far as known, this fish is exclusively restricted to the fresh water. .

SCULPIN.

This fish is known as the "Anotinamek" by the Indians—a word probably meaning "wind-fish", perhaps because of its curious habit of puffing out its cheeks as it breathes. It never exceeds 14 inches in length. It is covered with spines about the head and has two rows of small horny plates down each side of its back; these plates or scales number about twenty-five. Its pectoral fins are very large and are mottled yellow and black. Its ventrals consist of three soft rays. The sculpin is usually considered a scavenger but the stomachs of the specimens examined were for the most parts filled with small slugs. Some specimens were badly infected with worm-like parasites. The sculpin is said to make its home almost entirely in the sea but those found were taken in the mouth of a river and one or two specimens a good way above salt water though not beyond the tide. They are not very numerous and beyond the fact that their liver is eatable and that their fleshy tail is sometimes eaten—especially by the Esquimo—they are of little economic importance.

MILLER'S THUMB.

But one specimen was taken of this fish and that was a dead one picked up in a pool on a rock in the Metagami river. It is very small and not important. Its range may likely extend farther to the north as it is very likely to be overlooked or taken for the young fry of some other species.

LAKE TROUT.

Reports of huge fish inhabiting the waters of the Trout river and Sutton Mills (or "Trout") lake are very common. All the Indians who have been in that region assert that these fish are half as long as a man and that in nets of the largest mesh they are only caught by the teeth. While accepting such tales with a considerable degree of doubt we may be reasonably certain that large fish exist in those waters and as all agree in saying that they are trout of some sort it is fairly safe to put them down as lake trout. The waters of the above named lake which are very deep and clear and cold would be well fitted for this fish. It is said to occur in exceedingly large numbers.

ROCK COD.

This fish is not included in the list given of those specimens found on the west coast because there is no record of it ever having occurred there. While the writer was at Stratton island, however, which lies in the deep, clear-water part of the bay, he ascertained that rock-cod were taken there in abundance. The report of the expedition up the east side of the bay will doubtless contain information in regard to this fish.

OTHER MARINE LIFE.

WHITE WHALES.

The white whale (*belunga catadon*, Gray) is very common. Its range extends through the Bay and it often enters the rivers. It grows to a very considerable size and is fairly approachable. It is useful for its blubber, hide and flesh. The Indians make great use of the flesh for dog food but they do not eat it themselves "except in case of necessity". An average whale yields 100 gallons of oil and is worth, all told, about $15. As the supply of them seems to be unlimited they are a very valuable resource. It is by no means an uncommon sight to see fifty or a hundred of them from the deck of one of the small schooners used in that country. As a general rule it may be said that they are more common in the northern part of the bay than in the southern.

SEALS.

Seals are not very common on the west coast but occur with enough frequency to furnish the natives a fairly constant supply of hide for bags, gun covers and so on. They often come into the river mouths and it is here that they are usually secured. The only means taken of obtaining them is by shooting them; as they are heavier than water, they often sink before the canoe containing the hunters can get to where they are. How wasteful this process is may be judged when it is known that but only one out of four or five killed is ever secured. I did not have an opportunity to see any at close range but from what I could learn there are two species that frequent the west side—the grey seal (*halichoerus grypus*) and the common seal (*phoca vitulina*). The walrus is also taken very occasionally but only in the extreme northern parts. It is not known whether the seals produce their young on the west side or whether they are mere visitors from the east coast.

CRAYFISH.

In all the river estuaries a single specimen of crayfish is found. It is about 5 inches in length and of a blueish colour. One of the gentlemen of Révillon Frères makes use of them for food purposes and says he finds them very palatable; he catches them by means of a small net stretched on a hoop on which are spread bits of meat or fish. They are most commonly taken in about eight feet of water.

IN GENERAL.

The most interesting condition in regard to the fish of the west coast is that although all the conditions of the sea are present, practically all the fish found are fresh-water species. The capelin and the sculpin are the only exceptions to this rule, and neither of these fish are of very much economic importance. One finds all the Salmonidæ taking on migratory movements of the most distinct character and gradually transferring the main phases of their existence from the fresh water which is their

5 GEORGE V., A. 1915

natural home to the salt. If it were not geologically certain that James and Hudson bay always have been salt and connected with the ocean as at present, one would be inclined to think from the fish life present that they had originally been bodies of fresh water which had become salt and that in the process the fresh water fish had adapted themselves to the salt water conditions. Even were we to examine the fish life of the whole Bay, but few conditions would present themselves in contradiction to such a theory. The only salt water fish of much note throughout appears to be the rock cod; all the most important ones are really fresh water fish. And yet there is unlimited access to all the species that frequent the north Atlantic. One wonders how it is that they do not come in and take up their abode in the Bay. Gunther in his *Introduction to the Study of Fishes,* remarks in this connection: "The sturgeons and salmonids evidently belonged originally to the fresh water series, and it was only in the course of their existence that they acquired the habit of descending to the sea, perhaps because their fresh water home did not offer a sufficient supply of food. These migrations of fresh water fishes have been compared to the migrations of birds, but they are much more limited in extent and do not impart an additional element to the fauna of the place to which they migrate as is the case with birds." . . . "There is a constant exchange of species in progress between fresh water and marine fauna, yet certain groups have apparently been, during the whole course of their existence inhabitants of the one or the other. . . . A genus of fresh water fish is regularly dispersed and most developed within a certain district, the species and individuals becoming scarcer as the type recedes more from its central home." At that time then, when the sturgeons and salmonids of the north had not adapted themselves to the salt water the whole vast stretch of the west coast must have been without fish life.

The manner in which those fish have distributed themselves is problematic. The whitefish, of course, is found almost throughout Canada; I do not know of another locality, however, in which it enters the sea so freely and its movements also have a regularity which is unusual. Something parallel to them occurs in Lake Erie where it moves from the deeper water in the eastern end of the lake in spring up on tb the "platform" at the western end; during the summer, it retires to the deep water again returning in the fall once more to the shallows, this time to spawn. A large body of fish is always to be found in the deep water even during the spawning time, but there is no evidence that these fish spawn there. This large body of non-spawning fish may correspond with the immense number of immature fish that enter the tidal estuaries of James bay in the fall. In lake Simcoe, so far as is known, the whitefish frequent the deep parts of the main body of the lake in summer and during the late fall or winter move up into the bays near shore where they are caught through the ice; they move out again in the spring. One would hardly expect river fish to have such migratory movements and, as the James bay whitefish would hardly acquire such movements if it had merely adapted itself to the salt water after having lived a river existence, it is reasonable to suppose that these fish found their way into the bay from some other point. Gunther says on this subject: "Since salt water often proves no barrier to fresh water fish, their distribution has probably been in some cases from river mouth to river mouth through the sea."

Speckled trout are known to enter the sea in other localities so that it is not a matter of surprise to see them doing so in this case. Frank Forester, an author of a work on American fishes refers to their anadromous habits as follows: "The brook trout run down and remain permanently in the sea, more or less, along the whole south side of Long island and probably at many other points along the eastern coast." The only eccentric feature in regard to them is the manner in which they are distributed; at Moose Factory, there is only one stream of the entire system that contains them (Doctor's creek); there are only a few in the rivers and streams south of Mourning point but north of that spot, although there is absolutely no change in the character of the country, they are more abundant than any other kind of fish. There seems no good reason for this.

At the time of the return in the fall, the fish, especially the whitefish, appear to eat next to nothing. Many stomachs were examined but very few contained more than a little gravel. Trout, during the summer are fond of the larvæ of the dragon fly; stomachs examined at that time contained large numbers of these insects.

While by far the most of the fish make their summer home in the sea, there seem to be others which live permanently in the rivers. Whether any distinct line could be drawn between sea-going fish and river fish of the same species is not known but the individuals of the two classes are not hard to distinguish. For instance at the time of their return, the whitefish present a shining, silvery appearance, brown or greenish or bluish on the back and splendidly clean and white; the whitefish that have stayed in the rivers all summer, on the other hand, are duller, not as attractive in appearance, their backs tinged with yellowish, they dry up more quickly and lack the silvery glitter of the sea-run fish. Whether, however, these fish never enter the sea or whether they are mere stragglers whom chance has detained for a summer is yet to be determined. It is quite certain that fish of all the sea-going species do stay in the rivers all summer and in some cases and some favorite localities in considerable quantities.

FISHERIES INVESTIGATED.

1. NAGEDOWZAKY RIVER.

This is a small stream that flows in about fifty miles south of Cape Henrietta Maria; it is very shallow at the mouth and fishing boats could only enter at high tide. On August 3, when I was there, the Indians were catching fair numbers of trout in their small nets, which they had set in pools, a couple of miles from the mouth. Whitefish were also being taken but the prevailing opinion was that the season was too early for the best fishing, as the water had not yet got cold.

2. OPINEGAU RIVER.

On the banks of this river, about five miles from its mouth are situated the last outposts of the fur-companies. About seven miles up is a deep pool in which the fish congregate in the winter, at which time the trout may be taken, in unlimited numbers, on the hook. Farther up still, are other such places. The river at the fur-posts is about fifty yards wide, rather sluggish and perhaps, six feet in depth. It broadens out so much as it gets near the sea that sailboats are unable to come up it more than a mile. The trout that are caught in this and the other small streams nearby furnish the staple article of food for the thirty families of Indians that make the region their hunting ground.

The expedition arrived there at the end of July, at which time a few trout were being taken every day. We caught a few ourselves in nets that we had placed well out to sea. When we returned on August 4, more trout still were being taken besides quite a few whitefish and the nets were not being placed quite so far out as before. All fish taken were ready to spawn that fall. It was the opinion of all the Indians with whom I conversed that the real autumn run had not commenced at that time and would not commence until the tenth or twelfth of the month. We took a few more fish ourselves, the number being about equally divided between trout and whitefish.

Other northern streams, Chickeney, Lowashy, the Kenopwenik and the Black Duck, besides a few smaller ones of less importance, yield trout in about the same quantity as the Opinegau.

5 GEORGE V., A. 1915

3. SWAN RIVER.

This is an unimportant stream about midway between Opinegau and the Ekwan. It contains a few pike and suckers but no other fish of value resort to it.

4. THE EKWAN.

The Ekwan enters the sea about 30 miles north of the Attawapiskat; the mouth is surrounded by shoals and low grassy islands.. Boats of 3 or 4 feet draught can get up a short distance. The peculiarity about the Ekwan is that the fishing is never very good at the mouth while at points higher up, the largest whitefish in the region are said to be taken. These points are chiefly two: one 100 miles up stream, in a deep pool, the other 200 miles up in another pool. As the Ekwan has only one small portage in all this distance it is not difficult for fish to make their way up it. The whitefish said to be taken that far inland are commonly reported to be 2 feet long and 7 inches deep; that is, they would weigh probably six or seven pounds.

5. ATTAWAPISKAT RIVER.

The information that I collected about Attawapiskat fish from others' reports and from my own observations is as follows:—

Sturgeon: This fish is not abundant. There are favourite spots for it such as at the so-called rapids two miles below the settlement. At these places sometimes two or three of a night are taken by one fisherman. They in no case exceed 3 feet in length.

Suckers: There is no limit to the numbers of suckers that may be obtained, both the common, or northern sucker, and the red-horses. They are usually caught at all times of the year but in much greater quantities during the spring and fall than at other periods. They form the chief summer food of the innumerable husky dogs about the place.

Common Whitefish: We arrived at Attawapiskat on July 9 and immediately put out our nets. We were rewarded with, among others, one whitefish. That sufficiently indicates the state of things during the summer. When we came back, we fished from August 17 to August 20 and had little better luck. This is not to state that no fish are to be found at Attawapiskat but rather that we were not there at the right time.

The water off the Attawapiskat is very shallow and thus very warm, also the season was particularly fine, and most likely the whitefish stayed out in the sea longer than they do most years. There were signs that they were beginning to come in when we left; our own catch had increased slightly and the number of Indian nets being put out was very much in excess of what it had been earlier in the summer. Then, too, the French company officials at this place depend on the whitefish catch for the winter food supply of their dogs. The usual practice is to seine in certain well-known localities late in the season—as short a time before the freeze-up comes as possible. This is done because the fish are kept in a frozen state all winter and of course are ruined if they are not so kept from the first. It often happens thus that seining, in waiting for the steady cold weather, is left too late and the ice catches the fishermen unawares.

Seining Dates: The seining dates for 1912 were from October 9 to October 25. For 1913 they were from October 20 to October 25. The freeze-up in these years was October 26 and October 28 respectively. The best catch reported is a canoe load in three hauls of a 100-yard seine. A canoe would probably hold about 600 pounds of fish. Annually the company's officers aim to put down about 100 tubs of fish, a tub containing 100 pounds. This amount of course includes suckers, but not very many in the average year. The whitefish obtained in this way average about 15 inches in length and a pound in weight. They are not as small as those taken at Albany and known at that place as "seine fish." They do not seem to congregate in such large schools as do the Albany fish.

Tullibee are not differentiated from the whitefish in all the reports I have received. My own observations show that at this river they are about as numerous as the whitefish.

Trout: Speckled trout are rare in this shallow, dirty river; a few are caught every season but not enough to make their capture a commercial proposition. Reports of "salmon trout" occurring in this river may be due to the capture of occasional lake trout; these fish I never saw myself.

Pike and Pickerel: These fish, while not abounding are caught regularly, except in the middle of the summer when they seem to seek haunts that are as yet unknown to their would-be captors. Any net put down at other times is sure to contain a few of them; sometimes the pickerel are of good size; they are always good eating.

Mari: The above remarks will apply to this fish also, with the exception that it is to be taken at any time of the year. It is not valued when anything else can be obtained.

6. THE COAST.

A few whitefish are taken on the coasts all summer in nets set as described above. When the water increases in cold the fish increase in number. A fisherman could be sure of getting about 5 pounds of fish to 10 fathoms of net all the time and considerably more than this in the fall. The deeper the water he fished in the larger would probably be the supply of fish. At Agumiski island however, where the writer was in the middle of July, there were no fish to be obtained, though the huge number of netsticks along the shore, bore testimony to the fishing activities of the autumn. The presence of a fair number of seals, too, a few miles off the coast and of hundreds of white whales, besides constituting a valuable resource in itself, indicates considerable quantities of fish. At Neakwow point, where the dividing tides have worn out a deep channel close to the bank, much fishing is done in the fall and whitefish can be obtained most of the summer. From descriptions of the Gasket shoal, I should fancy cod might be found there, but the skipper of the *Emilia* tells me he tried for them there on one occasion, without success.

7. LOWASHY RIVER.

Lowashy has the distinction of being considered the one river along the coast where plenty of fish may be obtained at any time. Indians, who ordinarily live at Attawapiskat during the summer, visit this river for a few days at a time in order to load up their canoes with fish, smoke them and take them back to Attawapiskat. And as there are about 400 Indians living at the latter place, each of whom can consume incredible quantities of fish, the demand made upon Lowashy is not small. The party visited Lowashy on July 7 and though it was impossible to set our nets well owing to the swift tide, we got many more fish than up to that time we had at any other point along the coast. On our return in August (22), we discovered a quiet spot about three miles from the river's mouth which was reputed good for fish. At this place we obtained a very good catch, consisting mainly of whitefish but with a few good pickerel; the Indians who were camped near us, were at this time also getting good catches regularly. As there are innumerable pools and backwaters among the islands that lie in the mouth of this river, I should fancy that a constant and fairly large supply of fish could be obtained here. I should add that the largest trout we took during the summer and about the largest that is ever taken—5 pounds—came out of this river. The south bank projects about two miles further out to sea than does the north bank—this may act as does the leader of a pound net and thus account for the constant supply of fish.

8. KAPISKAU RIVER:

This river lies about 60 miles north of Albany. It has been described under "Harbours." Owing to our ignorance of the good places, when there in July, we were forced to set our nets in the open current. As the river carries down a great deal of

5 GEORGE V., A. 1915

debris after we had got them put out with a good deal of difficulty, they became very
dirty. We did not get many fish at that time. When we returned in August, we were
informed of a good place about 7 miles up-stream. Putting our nets down there, we
soon had them filled. Four or five whitefish would be in the net in the time that
elapsed from setting it to going back over it and straightening it out—a matter of a
few minutes. Most of these fish had come up to spawn but there were a good many
of the whitefish that had no eggs in them—roughly, all under 15 inches long. Tullibee
were about equal in number to the whitefish. We also caught more pickerel here than
we had at any other point up to that date. Suckers and pike were also common. In
two days from the same pool, out of which we mainly fished, an Indian in the fall of
1913 got 400 whitefish in two of the little nets used by these people. We tried our seine
at various points on the river bank but met with little success. Seining has never
been done in this river but it is altogether likely that if the right places were found,
it would yield just as well as do the other good 'seine-fish' rivers.

9. ALBANY RIVER.

To deal with the upper waters of this huge system first, let me set down the
information I acquired at English River post. At this place, four large rivers come
together; the banks of all of them deepen very quickly from the shore and they are
all very swift. In the Nagogami, which is one of them, just below the rapids, in the
spring the Indians are accustomed to get two or three sturgeon of a night. The
longest on record is 5 feet. Around the "Mattawa" or confluence, an occasional
sturgeon is obtained all summer long. At this point also they get large quantities of
suckers and pickerel in the fall, but during the summer the fishing is very poor. I
saw the results of a gill net set opposite the post for two days; the catch was one
trout, one whitefish and several suckers; this was in the third week of June. But
very few whitefish are found here at any time of the year. Trout are captured more
often though never in large quantities; the maximum size is seven or eight pounds
(speckled trout).

At Martin's Falls post which is located at the first portage on the Albany, about
three hundred miles above Fort Albany, it is reported that they take very large
quantities of tullibee and whitefish in the fall; it is possible that these are sea-run
fish as up to that point the Albany offers no impediment in the way of rapids for
fish that wish to ascend it.

Albany Estuary.

The mouth of this splendid river is the scene of the greatest fishery on the whole
of the bay, but like all the other waters of the west side, the time of that fishery is
limited to the fall and spring. All the kinds of fish caught elsewhere are also
obtained here, although the trout and the sturgeon are not abundant. The best
sturgeon catch is three or four in a night and the largest one on record measured 7
feet and was taken in the North River. The fish wealth of Albany consists almost
wholly of whitefish and tullibee. Pickerel and pike are caught in probably greater
numbers here than elsewhere on the coast and the former are usually above the
average in size. As in the other rivers, there are certain favourite fishing places and
it is in these that practically all the fishing is done. The most usual place for nets
is directly opposite the settlement, on the south side of the long, low island lying
opposite to it. Another good place is on the south channel of the north river, out
beyond the tree line. Seining is usually done in "Fishing Creek" which enters the
main river about opposite the posts of Revillon Frères: this creek of which about
one mile may be ascended in the canoe, is also usually pretty well filled up with

SESSIONAL PAPER No. 39a

Indian nets. It is at the limit of tide water on this creek that the seining is done. The place where our party had its best luck was not Fishing creek but a small bay one mile above the village at that point, we several times obtained seventy-five pounds in one haul of a 90-foot seine. Large hauls have also been made at other points; as for instance last fall (1913), Revillon Frères obtained one of the largest single hauls that have ever been made (3,000 pounds) immediately in front of the warehouse, where every " old inhabitant " of Albany predicted that no fish could be caught. If it has been a rainy season or if the water in the river is unusually high a poor catch generally results. Sometimes too, the frost comes and catches the fishermen unawares, it being the aim here as at Attawapiskat to leave the seining as late as possible so that the fish will remain frozen from the time they are taken out of the water.

August 15 is given as the date of the commencement of the fall fishery. The fish are usually first procurable out towards the mouth of the river and gradually work in. When we arrived on September 2, every Indian was taking many pounds of fish every time he lifted his nets. The catch at this time is about evenly divided between whitefish and tullibee. All the fish without exception are very fat and in splendid condition.

The average size taken in our seine was 12 inches in length and three quarters of a pound in weight; the average taken in the nets would be larger than these. The tullibee were all ready to spawn but only the very largest whitefish (those over 16 inches) were. Every body assured me that the characteristic " seine-fish," as it is locally termed, had not yet arrived in any number. These fish come about the 1st of October and make their presence known by the flipping of their fins, of a bright day, on the surface of the water. They are much smaller than the ordinary whitefish caught, averaging not more than 6 inches in length. They travel in vast schools so that if the fishermen once locate the school, they are a very short time in getting as many fish as they want. The greatest cloudiness of ideas prevails in regard to this fish; many Indians will say that it is a different kind of fish from the others, being although small, mature and coming to spawn; others maintain that it is just an immature whitefish and returns, following the adults which come to spawn. I caught many small whitefish, ranging in size from 3 inches up, all of which all the Indians who saw them, declared to be the regular " seine-fish." It is hardly probable that, if the " seine-fish " be another species, either some individuals would not have straggled in by the time I left Albany, or those people who saw the small whitefish I was getting would have named the latter "seine-fish." The only possibility of another species occurring is that the so called seine-fish may be a species of lake herring; but as the Indians all recognize the slight distinction between the whitefish and tullibee, they would be almost sure to recognize the same difference between the immature whitefish and another fish. It seems highly improbable that the " seine-fish " is anything but an immature whitefish of one or two years growth.

Spawning takes place in the ends of the creeks and shallows about Albany. One of the spawning places is in the creek that enters just below the " rapids," about three miles above the post. At Chickeney, where many whitefish resort to spawn, the operation takes place about two miles from the sea. Besides all the fisheries mentioned almost any one of the numerous small creeks along the coast is resorted to in fall for spawning purposes.

5 GEORGE V., A. 1915

DETAILS of the fall fishery at Fort Albany. 1 tub—100 pounds of fish; seine used; 100 yards in length. From the records of Messrs Revillon Frères.

Post established 1903.

Year.	Tubs.	Year.	Tubs.	Year.	Tubs.	Year.	Tubs.	Year.	Tubs.	Year.	Tubs.	Year.	Tubs.
1907.		1908.		1909.		1910.		1911.		1912.		1913.	
Oct. 15	7½	Oct. 17	2	Oct. 15	21	Oct. 18	1	Oct. 16	33	Oct. 10	24	Oct. 20	0
„ 16	0	„ 19	8	„ 16	15	„ 19	9	„ 17	46	„ 11	40	„ 21	1
„ 17	24	„ 20	10	„ 18	0	„ 20	½	„ 18	42	„ 12	19	„ 22	2
„ 18	0	„ 21	26	„ 19	0	„ 21	42	„ 19	21½	„ 16	14	„ 24	0
		„ 22	28	„ 20	8	„ 22	14	„ 20	3	„ 17	35	„ 27	1
		„ 23	30	„ 21	22	„ 23	40½	„ 21	18	„ 18	9	„ 28	30
				„ 22	14	„ 26	51	„ 23	8½	„ 19	10		
				„ 25	39	„ 29	0			„ 21	20		
				„ 26	20					„ 26	24		
				„ 27	28					„ 31	6		
				„ 28	73								
4	31½	6	104	11	232	8	158	7	172	10	201	6	34

Totals—Days and Tubs.

Results in Pounds.

Year.	Total Pounds.	Best Catch.	Date.	No. of Days.	Average Catch per day.
1907	3,150	2,400	Oct. 17	4	787½
1908	10,400	3,000	„ 23	6	1,733
1909	23,200	7,300	„ 28	11	2,200
1910	15,800	5,100	„ 26	8	1,975
1811	17,200	4,600	„ 17	7	2,457
1912	20,100	4,000	„ 11	10	2,010
1913	3,400	3,000	„ 28	6	566
	93,250			52	1,793

Earliest date, October 10, 1912.   Latest date, October 31, 1912.

From the records of the English Mission :——

Mission Established, 1858.

Year.	Tubs.	Year.	Tubs.	Year.	Tubs.	Year.	Tubs.
1900.		1901.		1902.		1903.	
Oct. 15	20	Oct. 23	2	Oct. 20	22	Nov. 4	51
„ 22		„ 24	7	„ 21	11		
„ 29	25	„ 25	32	„ 22	11		
„ 30	23	„ 26	1	„ 23	21		
		„ 28	40	„ 24	9		
		„ 29	19				
7 days	100	6	101	5	74	1	51

RECORD IN POUNDS.

Year.	Total.	Best Catch.	Date.	No. of Days.	Average Catch.
1900........................	10,000	2,500	Oct. 29. ...	7	1,428
1901........................	10,100	4,000	„ 28......	6	1,683
1902...................	7,400	2,200	„ 20......	5	1,480
1903.....	5,100	5,100	Nov. 4.....	1	5,100
	32,600	....	....	19	1,716

Earliest date, October 15, 1900. Latest date, November 10, 1900. (Record incomplete for that year).

Hudson Bay Company Post Established, 1675: I am not at liberty to publish in detail the records of the Hudson Bay Company, but the general information I acquired about their operation is as follows:—

They have been seining ever since they have had a post there and the catch shows no appreciable falling off from year to year. They usually put down about two hundred tubs, invariably all of which are the small whitefish described above. During the last four years, there have only been four days on which they have seined without result. The catches have varied all the way up from 50 pounds a day to 6,800. The following figures, while not official, are reliable:—

Year.	Total Pounds.	Best Catch.	Date.	No. of Days.	Average Catch.	Dates.
1910.........	16,000	2,700	Oct. 22.....	10	1,600	Oct. 15-28
1911........	24,400	5,500	„ 26.....	11	2,218	„ 13-25
1912.....	23,300	6,800	„ 25.....	12	1,942	„ 14-30
1913..........	7,650	2,100	„ 27.....	8	956¼	„ 17-29
	71,350	....	....	....	1,740	... .

*The catch by Revillon Frères in 1914 was 12 tons (240 tubs). This was all taken in a few days. The largest catch in one day was 86 tubs, taken in three sweeps of the seine. All that prevented a much greater haul was the size of the boat used. In one sweep 42 tubs were taken. All this fish was whitefish. The Hudson Bay Company obtained about the same amount. The River Albany froze on November 5th.

Totals for all three:—Total pounds, 197,200; total years, 15; average per year, 13,147 pounds. Greatest catch recorded, 1912, 43,400 pounds. Poorest catch recorded, 1913, 11,050 pounds.

The old figures of the mission reduce the average a good deal.

Catholic Mission, Established about 1904: Besides all the above, the Catholic mission also seines every year; its catch is reputed to be about 15,000 pounds, but a great many of these are suckers caught in the channel known as the " Gutway." Still the whitefish there taken would probably easily make the average of whitefish for all Albany, annually, 18,000 pounds.

GENERAL INFORMATION FOR FISHERMEN, SAILORS, ETC.

WEATHER CONDITIONS.

Spring may be said to begin in earnest at the south end of James bay (Albany) about the last week in April. By the middle of May, the river is usually free from ice and the snow has gone. Snow storms, however, occur in an

5· GEORGE V., A. 1915

irregular fashion much later than that date and it is no unusual·thing to see snow falling in small quantities late in June. The rivers all break up suddenly and, in the course of a day or two, the whole sweep of ice, which probably has extended for several hundred miles almost intact, rushes down and out to sea. If it should pile up on the bars or meet with other obstacles at the mouths of the rivers, a flood is the result and all the people living near the mouths (where the posts are situated) are forced to retreat to platforms previously prepared in the woods or take to the second story of their dwellings—if they possess one. Long piles of ice are also deposited on the banks of the rivers and, as these are pretty well covered with mud and gravel, it is only before the best efforts of the July sun that they disappear. The havoc wrought in the beds and banks of the rivers is enormous; huge caverns are gouged out of the banks and hundreds of trees are carried away; the river bottom becomes a series of deep holes and shallow bars.

The shore ice is said to linger about, dashed back and forth on the shores, till the middle of June. James bay freezes for a few miles out and this ice after it is loosened up, is detained for a good while by the action of the tides and by the prevailing winds. As, however, the tide flows south (comes in) for only five hours while it flows north (ebbs) for seven, the shore ice gradually works off to the north and finally loses itself in the wide expanse of Hudson bay. This does not occur until the end of July and ice is said to hang around Cape Henrietta Maria even longer than that; this last summer huge fields of shore ice were visible off Neakwow point on July 24. The small coasting steamers of the Hudson Bay Co. and Revillon Frères never enter any of the rivers on which their posts are situated much before July, though it would probably be neither very difficult or very dangerous for them to do so by June 15.

Frost is apt to occur almost at any time. On June 15, when camped on the Kenogami a few miles above its junction with the Albany, we experienced a severe frost—severe enough to form ice of considerable thickness on the water in the camp utensils. I am inclined to think that frosts occur late in the season more frequently inland than on the sea coast as we had no noticeable ones in this latter locality until the first week in August. On August 3, we had a heavy frost while at a little stream a few miles north of Opinegau river. This was the most noticeable one of the entire month for although there were others, they were not severe. During September, too, there was scarcely any frost while we remained on the coast. When we began our journey up the Moose river, we had not been out many days before we experienced low temperatures at night. When one considers the distance north, the coast makes a very creditable showing in this regard; it is by no means unusual for frosts to occur in the early part of August in the country between Sudbury and Porcupine, hundreds of miles to the south—a country demonstrated to be suitable for agriculture.

Whenever the wind blows from the north, cold weather results instantaneously; this is due to the above-mentioned fact of the presence of the shore ice to the north. The bay is not notorious for winds or for bad weather conditions. If the past summer be an average one, it will compare favourably with any large body of water in existence. There were only one or two winds during the whole four months spent there which would make dangerous weather for steamers; there were perhaps a dozen storms which would have made it rather hazardous for small sailing schooners or fishing smacks. Compared with a large inland sheet of water—lake Nipigon—on which the writer spent the previous summer, James bay stands out as safe and dependable for navigation. This year south winds were very prevalent and almost without exception, they were very hot. It is not known whether this is invariably the case. During the first two weeks of August there were about nine days on which

SESSIONAL PAPER No. 39a

south winds blew. Without exception they were light and balmy. During the middle weeks of September there was an eight days' gale of violent south winds. This was the longest blow from any one direction without intervening change within the memory of any of the white sailors.

Rainstorms were infrequent and there was not very much thunder. There were numerous days on which a little rain fell. There was a very large proportion of days on which the sun shone. The finest weather of the season was in September; the first three weeks of that month would compare favourably with the average weather of the same month anywhere in Canada. There was no sign of snow up to the date when the party left the bay (September 25). The temperature would of course average less than during any corresponding period of time in southern Canada, but there were plenty of days on which the sun made it uncomfortably warm. The hours of sunshine in that high latitude are unusually long; at Albany, it was possible to read in June by the twilight at 10 p.m. On July 8, at Lowashy river we had 16½ hours of sunlight and on August 3, at Nagedowzaky river (Lat. 54.30) we had 16 hours of sunlight.

A continued blow from the north brings rain, sooner or later; the wind then usually changes to the south and after a heavy blow from this quarter, fine weather comes again. Fogs were practically non-existent, though the Indians informed me that they occur more frequently, late in the fall. Owing to that and to other climatic conditions the coasting steamers and schooners usually try to get done their work by the end of September. This past summer, the Hudson Bay steamer *Inninu* was delayed in her work, and by September 25, had still several cargoes to take out from the depot on Charlton island to the various posts around the bay. She was considered by other men accustomed to the bay to be in rather an unfortunate condition, though every one was willing to concede that she could perform her trips without a great deal of danger.

By October 20, heavy frost and low temperature has become almost constant. By the end of the third week of that month the rivers are in a freezing condition and the more northern ones are frozen. By the end of the first week of November the Albany has frozen and winter has set in. From that date on until the end of April, winter is continuous; there are no thaws and no soft weather. The thermometer does not register any lower minimums than many places in Ontario and the West but the low temperatures are continuous, and for days at a time the thermometer will stand at thirty or forty or even forty-five below zero. Travelling is then performed by means of dogs, the broad band of ice along the coast making excellent going. This ice is quite smooth and glare with no snow upon it, that being all turned into ice by the tide rising over it. The country is beyond the line of greatest snowfall and the snow is not extraordinarily deep at any time. In fall, the freeze-up comes before much snow has fallen and in spring the snow has gone from the clearings before the rivers melt. For every 75 miles north, the difference in season is about five days.

Altogether it may be said that while James bay has a long and severe winter, it also has a summer equally unbroken and of very fair lenght. All the year is divided between winter and summer—the between-season is very short. During the summer, conditions there are not different from conditions elsewhere in the country and there is no obstacle that would hinder the carrying on of all the activities customarily associated with summer.

5 GEORGE V., A. 1915

WEATHER RECORDS FOR POINTS ON JAMES BAY.

Opinegau river (Lat. 54·15).

River open.	River frozen.
1914, May 5.	
Average, May 28...Oct. 25	

Attawapiskat river.

1912..Oct. 26
. 1913, May 21...Oct. 28
·1914, May 12.

Albany river.

(Compiled chiefly from the Journals of the Anglican Mission).

WEATHER RECORDS FOR THE ALBANY RIVER.

Year.	River Open.	Events of Interest.	River Frozen.	Events, etc.
1883 ..	....	... ...	Nov. 11..	
1884....	May 17..	First spring goose, May 1.........	" 2..	
1885....	" 15..	April 26, first goose..............	......	
1887....	" 18..	Bad flood......	Oct. 30..	
1888....	" 12..	Small flood. April 21, first goose..	....	November 18, 5 degrees F. November 25, 20° F.
1889....	" 6..	April 12, first goose..............	Nov. 5..	October 2, potatoes raised.
1890 ...	" 28..	May 3, 15° F....	" 5..	
1891....	" 14..	April 15, first geese	" 15..	
1892....	" 21..	....	" 6 ..	
1893....	" 14..		....	December 3, 30° F.
1894....	Apr. 30..	Bad flood	....	
1895....	May 5..	....	Oct. 27..	
1896....	" 10..	April 15, first geese	Nov. 15..	
1897....	" 7..	April 12, first geese	" 10..	
1898 ...	Apr. 27..	April 14, first geese	" 23..	
1899 ...	" 29..		" 12..	
1900....	May 2..	April 6, first goose..............	" 13..	November 6, first snow to stay.
1901....	" 1..	....	" 7..	October 11-20, 50 bags of potatoes raised
1902..	" 15..	April 22, largest snowstorm of the year.	" 11..	
1903....	" 20..	... ...	" 13..	
1904....	" 8 ..	May 15, North Albany still frozen.	....	
1905....		....	Oct. 23..	Exceptionally early fall.
1906....	May 4..	....	Nov. 11..	
1907....	" 26..	June 1, North Albany breaks up...	Oct. 22..	
1908 ...	....		Nov. 17..	
1909....	May 21..	April 27, snow going..............	" 17..	
1910..	Apr. 22..	May 2, 21° of frost..............	" 8..	October 1, potatoes raised.
1911 ..	May 3..	Apr. 27, river began to break ...	Oct. 31..	
1912....	" 14..	....	Nov. 17..	October 30, heavy frosts.
1913....	Apr. 28..	....	" 5..	October 30, 0° F. at night.
1914....	May 6..	May 7, bad flood..................	.. ..	

1898 Apr. 27, Earliest .... Oct. 22, 1907.
1890 May 28, Latest...... .... Nov. 23, 1898.

NOTE.—The coming of the first goose indicates about the same facts as does the first appearance of the robin further south.

By the end of April as a general rule, the river banks are cleared of snow, and by the middle of May all snow is gone except that in the depths of the woods.

ROUTES.

Undoubtedly the best way to get to and from the bay outside of the sea route is by means of the Albany river. The traveller may take advantage of the National Trans-continental westward from Cochrane. After he has made a run of 175 miles, he will get off at the crossing of the Nagogami river. From this point a portage of a mile and a half leads around the rapids that occur just at the railway. The Nagogami is very rapid for the first few miles of its route but all the portages are passed in the first 15 miles. Exclusive of the first there are only three, each of which is very short. All other rapids are run when the water is high and the canoe is waded down them when it is low. Once past these places, the traveller has clear sailing until he reaches the bay; there are no more rapids in the Nagogami and none in the Kenogami; this latter is a very large river with plenty of water in it and having a very rapid current. The flow of the Albany is just as swift and its water deeper. In size it is fully as large as the Ottawa. About one week will suffice to make the 300-mile trip down stream and about fifteen days will be occupied in the return journey.

FOOD SUPPLIES, ETC.

Both the Hudson Bay Company and Revillon Frères maintain numerous posts on the bay at which can be obtained all the staple articles and at the larger posts a good many luxuries as well. Considering the number of times these goods must be transhipped and the risks of the trade, prices are very reasonable; at one place, indeed, I found them just as low as at Cochrane. The personal equation, of course, always is a big factor in such matters. Plenty of fresh meat may be obtained along the coast. Ducks and other smaller wild fowl are present in incredible quantities and are not hard to obtain. Caribou meat may sometimes be purchased from the Indians. The moose has not yet penetrated north of Albany. The greatest obstacle one has to contend against in the matter of food is provided by the "husky" dogs. These brutes are everywhere and have an appetite that is surpassed nowhere on earth; the only means of securing anything from them, be it food, boots or camp supplies is to raise it all up on platforms well out of their reach.

COASTING.

The aboriginal mode of travelling along the coast is by canoe; the fur companies however make use of small schooners of from 30 to 40 feet of keel. Both these methods give rise to extremely vexatious delays, occasioned for the most part owing to the tide. In using a canoe, it is next to impossible to get an Indian to venture out on the water when there is a bit of a head wind and he positively refuses to travel when the tide is out. Coasting resolves itself into paddling for about three or four hours every day from half tide through the full tide to half tide again. There is justification for this process when the tide is high, say at noon and again during the middle of the night. In this case, if one were not to land on the grass-plain at the edge of the high tide mark, when the tide began to retreat, he would have to keep on following the water out until he was almost out of sight of land and then continue paddling all night until the tide came again; or he would have to sleep in his canoe when evening came as it would be next thing to impossible to transport the camp outfit over the miles of mud that would intervene between him and dry land. When the course is from river to river, however, there is no need to lay up when the tide is out as all the rivers except those in the north can be entered by a canoe at low water; it is merely a case of going far enough out to pass the sand and mud that the river has piled up for miles from its mouth. This native usually is afraid to do. Sail-boats nowhere on the west coast can beat against the tide except they have a beam or fair wind. They thus have to anchor when the tide turns on them. As most of the posts are situated well

5 GEORGE V., A. 1915

up the rivers—inside the tree line as a usual thing—it is often a matter of two or three days before even the river can be cleared. These little boats are used constantly to make the run out to Stratton or Charlton island, which entails a journey of 50 or 60 miles from one point of land to another. Open boats are also used for coasting work, but numerous wrecks occur among these, especially on the badly exposed shore north of Agumiski island; no one, however, has ever been known to be drowned in these mishaps.

FUEL.

One must depend largely on driftwood ·for his firewood; if that is lacking, he must break dead boughs off the scrub willows that grow within a mile or two of the sea; these are usually damp and very small; the fire they make is sufficient to boil tea, but will do little else. At all the larger rivers, of course, the canoe may be run up to the tree-line where plenty of wood is obtainable.

NATIVE HELP.

If at any time commercial fishing should be undertaken, people would possibly look to Indians as the source of the labour required. That source at the present day is abundant and contrary to the general opinion not decreasing, but it is not of high quality. Nature never intended the Indian to be a hum-drum working man and civilization cannot make him so. He will accept work but the monetary features of it have very little attraction for him and he does not hestitate to quit and do nothing if he is not pleased with it. The fur companies have developed a sort of patriarchial arrangement whereby they keep the Indian employed at nominal tasks all summer in order to induce him to give his employers his fur in the winter. He has thus never been schooled to real work outside of the hardships he experiences in his own method of life. His dependence on the Government has made him lose whatever ambition he ever possessed and he is now utterly improvident. Some things he does well, as work that involves the use of tools, but the most of the peculiarly white man's tasks he does very ill. He fishes, and fishes very successfully after his own fashion, but is too conservative to change for a better one. His extreme dislike for the terrors of the sea would make it hard to make a deep-sea fisherman out of him. Yet, here and there are individuals to be found who are thoroughly reliable and courageous men. More than that, the Indian has been employed in certain places and although he is not as satisfactory as white labour, he has filled the gap when no one else was obtainable. The fur companies pay even their bonded servants a very small wage but it is impossible for a stranger to get a guide or helper for much less than $2 a day and board. Most of the Indians do not return to the posts until June and they begin to leave for their winter hunting grounds towards the end of August. Those who have not far to go, may stay as late as the end of September. Although they do not excel at steady labour they are unbeaten as guides and never fail to pilot the white man through to this destination safely if they are allowed to take their own time and go about it in their own way.

NETS AND CONDITIONS OF FISHING.

All the rivers of the west coast are, owing to the absence of dams in the shape of high ridges of rock (which form rapids and falls on other systems) very swift and with few quiet backwaters, deep pools or calm reaches. It is thus difficult to set nets in them. What usually happens when this is attempted is that the net is dragged from its natural position at right angles to the shore and cast up on the beach further down or even torn completely away and lost. To add to the fisherman's difficulties, these streams are for the most part very dirty and carry along with them large quantities of sticks, stumps and other debris. In the upper reaches of the Albany so hopeless a proposition is it that the Indians do not depend greatly on fish at all though without doubt at certain periods of the year they are present in plenty. Probably such a state of affairs could be solved by using drift nets. The same remarks apply to the tidal waters; the tide is equally as difficult to reckon with as is the river current. Nets set off the coast of Agumiski were filled with sea-weed and dragged along by the tidal currents as if they were not anchored at all. In such cases, too, the remedy would be drift nets. The Indians have adopted the following plan to enable them to set their nets in water where the tide runs strong; they select the bank of a river and, beginning on land which is uncovered at low tide, they plant a row of strong stakes at right angles to the tide. This row they continue out as far as it is practicable to drive them into the bottom; they seldom reach thus a depth of more than 7 feet. On the side of the stakes against which the strongest current comes—that is, river current or tidal current—they place their net which is thus held in position by the stakes. That portion of the net placed on the tide mark is covered by the incoming tide. The same method is also employed on the coast, stakes being placed at right angles to the shore. As no west-side Cree ever yet attempted to fish in water that was not in immediate proximity to the shore, the problem of how to meet the tidal currents of the deep water has not yet been solved. So, too, but little use of the stake-plan is made in the upper waters of the Albany because the river gets deep too quickly from the shore to permit of the stakes being securely fastened.

Most of the other rivers afford more or less room for nets. Thus in the Kapiskau there is a place about 7 miles from the mouth where the river makes a sharp turn and has gradually cut off an elbow in making that turn. That particular place is splendid for putting nets down and also, incidentally for getting them filled with the best of fish. In the estuaries of all the rivers there are abundant backwaters in which it is easy to set nets. That statement is especially true of the Attawapiskat which by reason of its large number of mouths has numerous islands and back channels in which the water is still. It is impossible to seine without the greatest of difficulty in the whole stretch of the Kenogami and Albany until one gets among the islands near the mouth of the latter; the banks are too steep and the current is too strong. Nets should at all times be very securely anchored as the wind and current have great power in washing them away, especially when there is added the force of an ebbing tide. Two nets were lost during the summer owing to the wind and tide carrying them out to sea. The backing of the nets should also be well seen to, not only because of the strength of the current but because of the frequency with which white-whales and seals come into contact with them; if the backing be secure they will merely go right through the nets; if it be weak, it will break and the whole net will be carried away.

Repeated observations show that nets of the type that the Indians use are best suited to the work. These nets are narrow—about 20 meshes wide—and quite short, never more than 15 fathoms in length. Nets of the enormous length of those used in the Great Lakes would be quite out of the question. The mesh too, must be small— from 1½ to 2½ inches is the best size. I will quote a few figures to show the greater utility of the small meshed-net. Out of three short nets set parallel to each other a short distance apart, 40 fish were obtained. The middle net was about 3½-inch mesh and considerably longer than the other two. The other two were the regular Indian size—about 2 inches. Out of the large meshed net came 4 fish, all the others were taken in the small-meshed nets. One of the small nets was raised and in another hour out of the other one were taken half a dozen whitefish, out of the large meshed net, none at all. And the small nets took the largest fish that we were in the habit of obtaining just as readily as did the large meshed nets. Again, on one occasion out of a total of 18 pounds of fish from five short nets, two of which were small meshed, all but one or two individuals were in these two small-meshed nets. It is abundantly apparent that nets with a small mesh are the most suitable for the fish of the west side of James bay.

5 GEORGE V., A. 1915

HARBOURS.

Outside of the rivers, there are no harbours. Of the rivers only one—the Albany—will admit vessels drawing up to 8 feet; the Kapiskau, it is said, will admit a craft drawing about 7½; the Attawapiskat takes a sail boat of 4½ and the Ekwan and Lowashy are even shallower. All these depths are at high tide and the 8-foot craft that comes into the Albany must come in on a more than average high tide. The depth of water on the bars of these rivers at low water is about 3 feet or less. The Albany outer bar is just about out of sight of land—10 or 12 miles away at the least. A ship approaching the mouth of that river is forced to drop anchor in the open sea and lie there in no matter what kind of weather until the tide becomes high enough to allow it to cross the bar. One feature in the navigation of the west coast is that the slope of the bottom is so regular and so gradual that mariners by sounding can tell at any moment just how far off the coast they are.

THE ALBANY.

The Albany enters the sea by three mouths—the North and South rivers and Chickeney creek. Between the South or so-called main river and the North, lies Albany island, about three or four miles in breadth. Between the North and Chickeney, there is a stretch of 12 or 15 miles. The South river is the river on which all the posts are located and the one that has been used for generations. From the "rapids" three miles above the settlement, which are the head of tide water, to the outer bar, is about 15 miles; an arc of a circle with the north and south shores of the river and the bar as points upon it, would have a length of 10 miles; within that arc the water is fresh at all times. The channel in front of the settlement is about a mile wide and three or four fathoms deep in places.

A good channel for ships of almost any depth is obtained once the half-mile long outer and inner bars are crossed. The estuary contains numerous small islands. shoals, bays and backwaters in all of which, at the times of the year indicated above, fish abound; there are, too. also several small rivers and creeks which enter the main stream at this point, in which the number of fish caught is prodigious. The North river is even deeper though not quite so wide; its current is much swifter and it perhaps carries the greater volume of water. It does not afford as much scope for the setting of nets as does the south river, although islands and backwaters are numerous. If properly investigated it would probably be found to have the best ship channel of the two. Altogether, in the estuary of the Albany. there are probably 30 or 40 square miles of good fishing grounds.

The next river to the north is the Kapiskau. All the sailing craft on the west coast enter it and it was this year proposed to send the steamer *Inninu* there. Once over the bar, there is plenty of water for a small steamer of say a couple of hundred tons. The bar is situated far out to sea and is probably travelling farther away from land all the time. This river has piled the mud up on either side of its narrow channel for miles beyond the grass limit; on this account it gets its name, which means "Shut-up river." The tide extends for a dozen miles above the grass and nowhere does it flow with greater velocity than at this river.

The Attawapiskat river, the second largest on the coast, flows into the sea by five mouths, of which the Lowashy river is the most southern.

Lowashy divides from the main channel 40 miles above its mouth and enters the sea 10 miles to the south of it. At its mouth the shoals are extraordinarily wide. When the tide is out the sea literally cannot be seen from the shore, impassable stretches of mud and boulders intervening between. The river has a very shallow channel at low tide but a small fishing smack could get in at high water or even half tide.

The Attawapiskat's delta is about five miles across not including the portion between Lowashy and the next mouth north. The only branches of importance are the Boat river and the Main river. Boat river is no longer used by the schooners on account of its narrowness, but it is said to have a better channel than has the Main river. The other two branches both leave the Main river within a few miles of the post. Boats have fair anchorage opposite the settlement though the river is filling in on the north side so quickly that it is making the main channel narrower all the time and thus more difficult for a sailboat to ascend from the sea—a distance of 9 miles. The channel out from the grass banks is usually indicated by beacons of latticework erected on poles. These extend for 3 or 4 miles out from the grass and when they are left behind the only safe course for the sailor is to head straight out till all danger of grounding has, as revealed by the lead, passed.

The Ekwan is the last river on the coast into which boats may go. It enters by two main mouths but soon after one leaves the sea it becomes very shallow. Numerous shoals mark the approaches to it and it would be impossible to lie behind these in case it was impossible to get into the river. The Manowinan islands also, a few miles off shore and to the south would give shelter from some winds.

If a boat enters the Opinegau river, it must be prepared to lie high and dry on boulders and mud when the tide goes out and even at that it can barely get into the mouth of this little river.

POSSIBILITY OF OYSTER CULTURE.

I take the following extract from the work of Dr. Jos. Stafford on the *Canadian Oyster* in the report of the Commission of Conservation for 1913:—" The physical conditions of natural-oyster producing, as compared with non-oyster producing, areas will determine the prime essentials, not only for the life of the oyster, but for the successful production of eggs, larvae and spat. Along our coasts the oyster lives and breeds in comparatively shallow bays, coves and estuaries of rivers that are sheltered from the deep, cold, stormy waters of the gulf and ocean by islands or projecting, long sand-bars; that have areas of less than three fathoms depth, a tidal fluctuation of only three to five feet, and some admixture of river water; with rather hard bottom of rocks, stones, clay or sand, often overlaid with a dark-coloured, light, loose, fluffy ooze of organic origin, but no deep heavy, sticky mud or shifting sand. The salinity generally lies between 1·012 and 1·020 (distilled water being 1·000) but varies a few degrees with the ebb and flow of the tide and with the amount of river water. In the early part of July the temperature approximates to 20 degrees Cent. (68 deg. Fahr.) and, owing to the small exchange of tidal water and the great amount of heated sand, there is no great and sudden variation. Such physical conditions are also favorable to the presence and multiplication of numerous diatoms and other minute food-supplying organisms......

"A river.........may discharge over or in proximity to oyster beds.....".

"Lime is required by the oyster for the construction of its shell which forms the greater part of the weight of the oyster. The amount of this existing in.... oyster shells is enormous, all of which, or the constituents of which, must be contained in the water. It comes from the disintegration of old shells, from rocks in the ocean and along the shores, but especially from the river water that has drained through the land and over the rocks of river basins.

"The temperature of the water where oysters abound, varies with the year, the month, the physiography of the contiguous land, prevailing winds, the size, shape and depth of the body of water, the nature of its entrance, the presence of islands, reefs, sand bars, shoals, flats, the extent of the shore, the amount of river-water, evaporation, sunshine, fog and such-like conditions. The oyster itself can withstand considerable changes of temperature—it is the developing young which suffer. Accordingly there

5 GEORGE V., A. 1915

has risen a periodicity in the spawning, which falls in the warmest parts of the season. As soon as the snow and ice have disappeared and the spring freshets subsided, the water gradually rises in temperature and becomes inhabited by increasing numbers of microscopic plants and animals. In May and June, oysters like other large animals that live on such minute plankton organisms, begin to ripen their eggs and spawn in time to give their offspring the advantage of the long spell of comparatively calm and warm water." Dr. Stafford then notes that on July 7, 1909, at Shediac, the water was at 63¼° F. and the warmest water he records is 72½° F. on August 2.

When applied to James bay, these facts mean that the only possible places for oyster culture would be in the river estuaries. In these locations, the tidal rise, the depth, the salinity of the water, the supply of lime would all be most satisfactory; the factors likely to prove unfavourable would be the temperature and the nature of the bottom. There is no doubt but that spots could be found where the bottom was hard and where the mud would not be soft enough to allow the oyster to sink in it and thus be suffocated. There might be danger from shifting sand but if sheltered places were chosen, this would be very immanent. The temperature is at the south end of the bay high enough but it is probable that the season is too short; by the end of June the water has almost reached its maximum temperature but this begins to fall rapidly about the middle of August. The natural northern limit of oysters is supposed to be around the St. Lawrence but, so far as is known, no really serious effort to acclimatize them further north, has ever been attempted.

AGRICULTURAL POSSIBILITIES.

It is important to note that there are very good prospects of some of the region near James bay becoming fit for agriculture when drainage and deforestation have taken place. The fact that garden produce can already be grown is most satisfactory in view of the influence this would have in inducing fishermen to take up their residence in the country. I believe it would be quite possible for a man to live comfortably from the products he had himself raised. The soil is exactly the same as all through the much-talked of Ontario 'clay-belt' and the climate is, though rigorous in winter, one of long sunshine in the summer. Already at Moose Factory, everything in the way of roots has been grown, oats have been raised regularly for years and even wheat was ripened last year. At Albany potatoes are a good annual crop while such things as lettuce, radishes and turnips also do well. At Attawapiskat, so far attempts to ripen potatoes have not been a success, but I have reason to think that a fair trial has not been made. The country is one of the finest imaginable for cattle as untold quantities of hay grow along the coast—which already sustain a few head at each of the posts. All the different kinds of domestic animals have at one time or another been brought to the bay and all without exception have done well. It is quite possible that the future will see this country a well settled farming community.

CONCLUSION.

In conclusion it is only necessary to draw attention to one or two of the most salient features of the conditions relating to fish in James bay. By far the most valuable fish is the whitefish; this fish has been taken in great quantities for very many years and so far as information can be obtained shows no signs of decreasing; nature has provided that few of the spawning fish should be destroyed and man has confined his operations to fish which are not ready to reproduce. In view then of the peculiar situation existing, it hardly seems necessary to enact the usual rigid close-season laws in this case. The speckled trout will no doubt in time provide good sport for the angler and the tourist; at present of course, the entire lack of communication or transportation facilities prevents—except by people actually resident in the country—any use being made of any of the fisheries. When the Hudson Bay railroad is opened, it will be possible to get the products of the smaller bay out to the cities of the west in the course of a few days and then we shall expect to see greater use made of them. Such a traffic will require substantial fishing tugs, quite independent of wind and tide, for these latter are the great bugbear of all traffic carried on with the old-fashioned craft. If a line should ever be built from Ontario to the bay, the Ontario north will be supplied in the same way. The great rivers of the west coast of Hudson bay are as yet unknown quantities. Winisk, Severn, the huge Nelson, the Churchill and all those of the Chesterfield inlet, remain to be investigated; when these have added their stores of wealth to the sum already obtained it will be found that in the great seas of the north we have a food-resource of the first magnitude. It is not too much to predict that some time in the future the supply of fish that comes from the salt water of the north will be as constant and as plentiful as that which at the present is yielded by our great inland lakes.

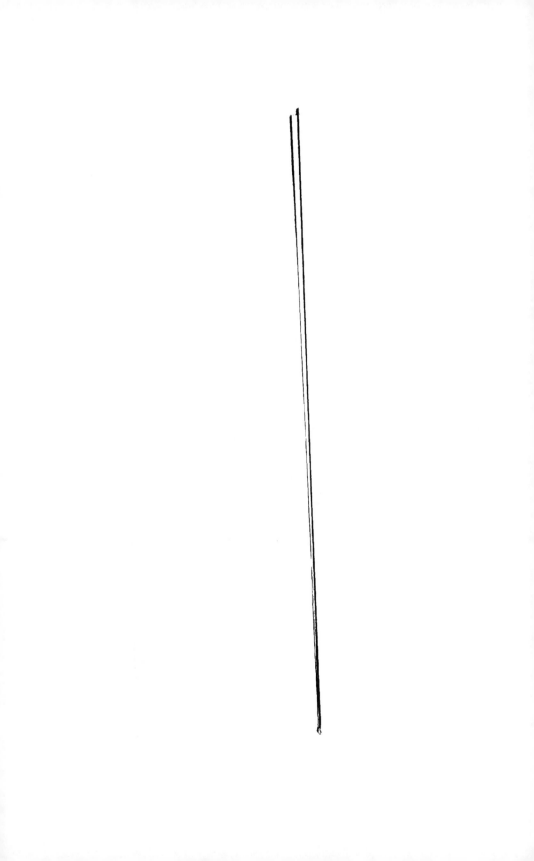

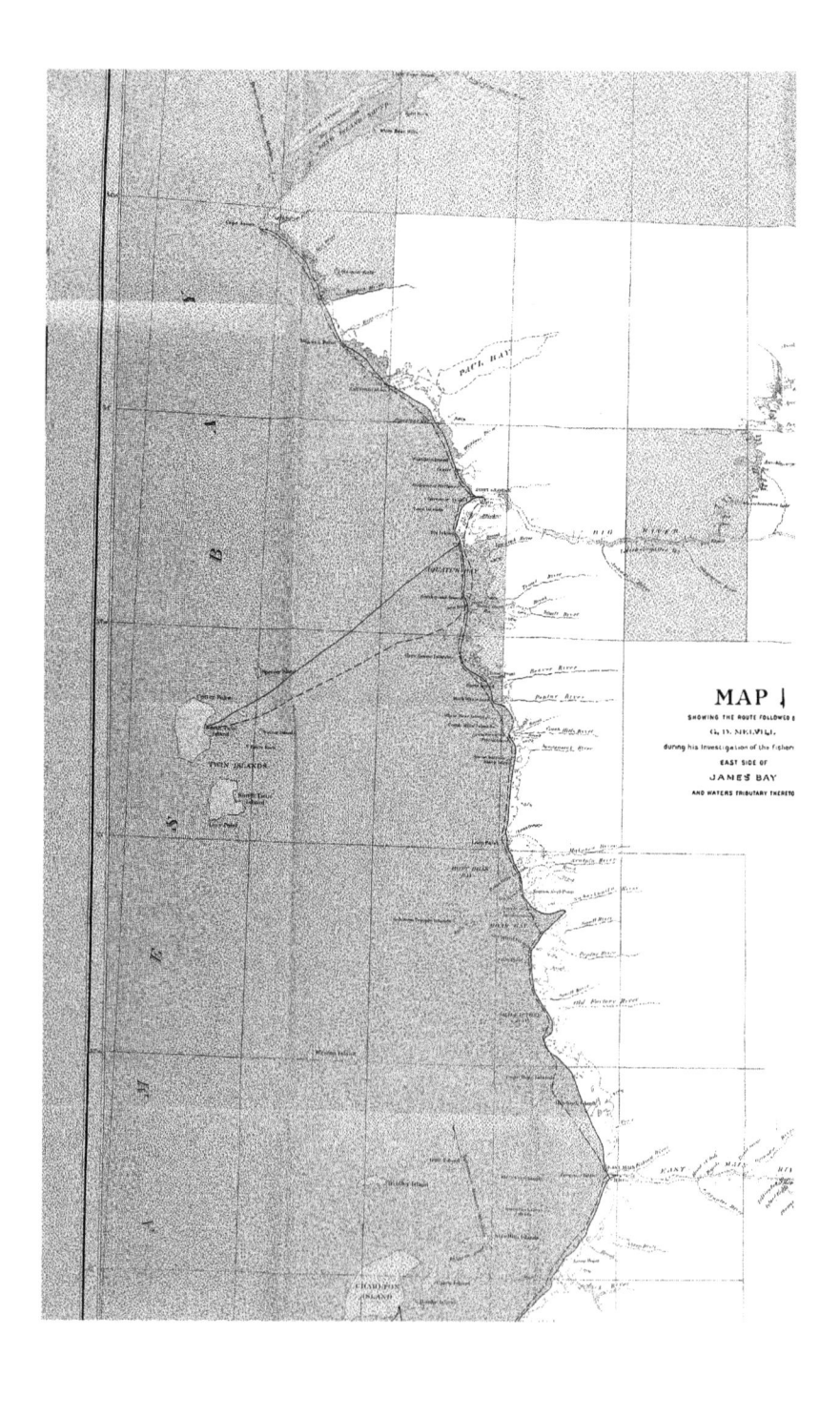

MAP I

SHOWING THE ROUTE FOLLOWED :

G. D. MELVILL.

during his investigation of the fishery

EAST SIDE OF

JAMES BAY

AND WATERS TRIBUTARY THERETO

Insert

Foldout

Here

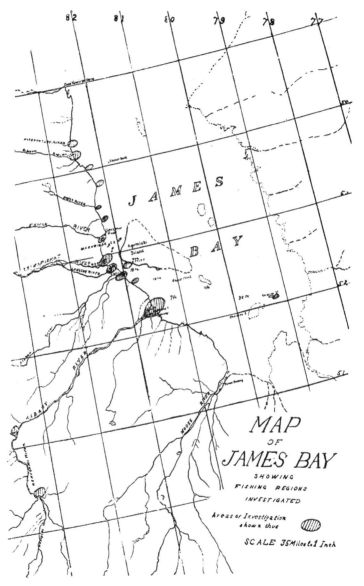

MAP
OF
JAMES BAY

SHOWING
FISHING REGIONS
INVESTIGATED

Areas of Investigation
shown thus

SCALE 35 Miles to 1 Inch

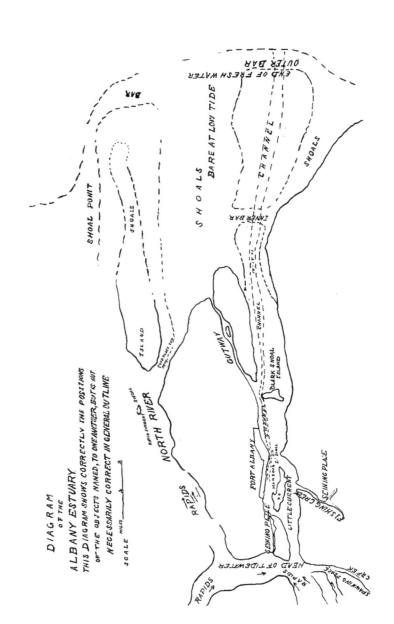

DIAGRAM
OF THE
ALBANY ESTUARY.
THIS DIAGRAM SHOWS CORRECTLY THE POSITIONS
OF THE OBJECTS NAMED, TO ONE ANOTHER, BUT IS NOT
NECESSARILY CORRECT IN GENERAL OUTLINE

SCALE MILES 1   2

NORTH RIVER

RAPIDS

RAPID ROBERT SHOAL

SHOAL POINT

SHOALS

ISLAND

BAR

END OF FRESH WATER

OUTER BAR

SHOALS

BARE AT LOW TIDE

CHANNEL

SHOALS

INNER BAR

CUTWAY

CLARK SHOAL ISLAND

FORT ALBANY

SEINING PLACE

LITTLE CURRENT

FISHING CREEK

SEINING PLACE

HEAD OF TIDEWATER

RAPIDS

RAPIDS

SPAWNING PLACE

CREEK

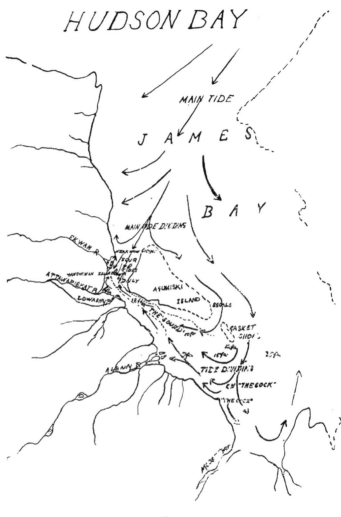

HUDSON BAY

MAIN TIDE

J A M E S

B A Y

DIAGRAM TO ACCOMPANY

DESCRIPTION OF TIDES

Speckled Trout from the Opinegau.

Father Boissean's residence at Atawapiskat
Post.

Spring Ice on the Albany, June 22.

Fish caught at Lowashy River, July 8. The largest one is a Speckled Trout.

Appearance of the Coast at Low Tide—Two miles from Land.

A Cree Encampment.

Preparing the Nets to set—Kapiskau River.

Fish caught at The Kapiskan in August.

" Farthest North "—Tide entering Mouth of Nagedowzaky River, August 4.

REPORT ON THE FISHERIES EXPEDITION TO HUDSON BAY IN THE AUXILIARY SCHOONER "BURLEIGH"

1914

BY

NAP. A. COMEAU

The Deputy Minister of the Naval Service,
Ottawa.

SIR,—I have the honour to enclose herewith my report on the *Burleigh* Expedition to Hudson Bay. As will be seen by it, the season was an abnormally late one, which left me a very short time for work there. I regret this very much as I am sure that the southeastern shore would have proved to be of greater value in fisheries than the western portion, though the latter is very promising. It would take one full season to investigate these properly and I trust the department will continue this interesting work.

I have the honour to be, sir,

Your obedient servant,

NAP. A. COMEAU,
Officer in charge of " Burleigh " Expedition.

REPORT OF TRIP TO HUDSON BAY—*BURLEIGH* EXPEDITION,

N. A. COMEAU, OFFICER IN CHARGE.

As per instructions received, I left Godbout on the 22nd of June, for Halifax, there to join the schooner *Burleigh* and proceed to Hudson Bay, as soon as possible thereafter.

Arrived at Halifax on the 26th June, and next day went to the Department of Marine and Fisheries agency and reported. I found that the vessel was far from ready. However, work was proceeding as quickly as practicable. Supplies were ordered, crew engaged by the captain, sails bent on, etc., and we were finally ready to sail on the 8th of July. In the end this delay did not make much difference, as we could not have entered Hudson strait before the date we arrived there, July 31. The ss. *Bonaventure* which left Halifax about July 3 arrived off the Button islands on the 14th and was 17 days in the ice, before she could force an entrance on the same day that we reached there. I had expected this all along, owing to the late season all over the rest of the province; so it could not differ much so far north, in fact we might expect worse. The weather was very cold, from the time we passed Belle Isle till we reached Port Nelson, the thermometer was only a couple of times above 40° Farenheit, and as low as 25°. (See table of temperature for the trip.) During the passage of 37 days, we had 25 of more or less fog and rain. We had intended going to Fort Churchill in the first place, but owing to some trouble with our compass, bringing us further south than expected, we headed for Port Nelson, so that I could report our arrival to the department. We reached there on the 14th of August. Next day prepared our camp outfit and provisions, and left on Monday the 17th to investigate the fishing on the Nelson river. This work was later continued along the coast line and in other rivers until the 24th of September. As the *Burleigh* could not be kept there so late, I had ordered her back on the 8th of the month, thus giving me two weeks more time there for my work. I returned on the *Bonaventure*, leaving Port Nelson on the 27th September and arrived at Sydney on the 6th of October, p.m., and the *Burleigh* reached Halifax on the 19th of October. In this report each subject will be treated under separate headings.

WHALES.

Many whales were seen on the trip, both going and returning between the straits of Belle Isle and Hudson strait. All in that section were apparently of the kind known as "finners" *Balaenoptera sibbaldii.* In Hudson bay I only saw one, a "bow head" *Balaena mysticetus,* it was seen about one hundred miles from Cape Tatnam. The beluga or white whale was seen in great abundance in the estuaries of the Nelson and Hayes rivers and also in those of all the rivers and creeks of any size in that vicinity. In the Nelson I saw them up near Seal Island just a mile or so below the tide limit. In the smaller rivers and creeks they only enter the mouth at high water and move out as soon as the tide begins to fall. From examination of the contents of four specimens, (one shot at Partridge river), I found that their chief food consists of whitefish and capelin; in one there seemed to be remains which looked like that of a carp or sucker.

They enter the streams shortly after the ice is gone out, and when the fish are on their way to sea, disappear for a short while, and then return with the migrating fish on their way back to spawn. The quantity of fish consumed by the beluga must

5 GEORGE V., A. 1915

be enormous. When the locality can be more easily reached, and labour and cost of living cheaper, I believe they can be taken in paying quantities, by the use of stake fisheries as used in the St. Lawrence. The water being discoloured with clayed substances would be a great advantage. I noticed they were much less shy than in the St. Lawrence. Their destruction should be encouraged.

SEALS.

At certain periods, notably in September, when the fish enter the rivers, seals follow them up. In the western portion of the bay, visited, they all appear to be of one species, which I take to be the "barbed seal" *Erignatus barbatus*. I shot three of them, but unfortunately secured only one, which floated some days later and was partly eaten by bears. I have, however, secured the best part of the skull which will be sent to the Department. In the straits I observed some harps or Greenland seals, *Phoes groenlandica*, floe rats, *P. hoetida P. vitulina* and one hood seal, *Cystophora christata*. Around the McDonald islands and Charles island they were especially numerous. I had no chance of securing any. Four narwhals, *Monodon monoceros*, were also seen in that vicinity. No walrus were seen at all. The stomach of the seal killed contained remains of suckers and whitefish, but I presume that all kinds of fish found there fall a prey of them. The destruction of fish by the seals in that western section from Churchill to James bay, cannot be very considerable, because they are not numerous anywhere in the vicinity and at times are not seen at all. Very few are killed, the Indians not appearing to care much for them either for food or otherwise, quite in contrast with the Esquimaux, who pursue them constantly, further north.

SALMON.

No appearance of any kind of salmon was seen in any of the several rivers that I visited, although I tried many times with drift and fixed nets and also the seine. The locality is certainly not very favourable for that kind of fish, owing to the clayey bottoms of all these streams and the extensive mud flats in their entrances. Many of these rivers cannot be entered even by small fish before the tide is more or less high. The whole of this western coast line is very low, swampy ground; near the sea coast there is a grassy beach two or three feet above the summer river level. The streams cut their way through these low banks for a short distance from the mouth and then the water spreads itself out like a fan over the extensive flats and loses all semblance of a river. One can often be only half a mile from the entrance and find no indication of any river, unless it is a good sized one.

Further north beyond Churchill and in the south eastern portion of the bay where the rivers have sandy and gravelly bottoms, the natives tell me they have taken a species of fish, which, according to the description, belongs to the salmon kind.

I have a report that one salmon was found dead on the ice, in Stuparts bay on the 5th of August, by one of the crew of the s.s. *Bonaventure*. From the description given to me, it was apparently a specimen of our Atlantic salmon *Salmo Salar*.

TROUT.

From reports gathered from residents, Hudson Bay Company's agent, and natives, trout migrate to the sea immediately after the opening of the rivers in spring. This usually takes place in that section early in June in ordinary years. The fish return about the middle to end of July, and probably in August. Sergeant Walker, an ex-member of the R.N.W.M. Police, showed me a diary that he kept of his catches of trout, on Sam creek where he resides, and the heaviest runs appear to be about the end of June, this was with a net, but he also had some good scores by angling. When I arrived at Port Nelson the run of the fish was considered to be over and no one fished for them, either with net or fly. I was told that I was too late for the fishing. However when I went up the river a few miles, I found there were still plenty to be had, both by net and angling. In the main stream of the Nelson I could not get any by angling either with bait or fly, but I got plenty in the net. In the small tributaries and creeks wherever the water was bright and clear they took the fly readily. They ranged in size from half a pound to five and a half pounds, this last being the heaviest fish I caught on the fly. This fish was taken at the entrance of North Seal creek, a tributary of the Nelson. The probable cause of their not frequenting the main river is due I believe to the discoloration of the water by clay and the immense quantity of sediment in suspension, especially after rain, which washes down its steep clayey banks. .The residents claim there are two kinds of trout. One they call salmon trout and the other river trout, basing their distinction on the colour of the skin and also of the flesh, one being brightly coloured with red spots and the other of a darker appearance and the spots more obscure. I examined carefully many of those we got and could find no difference at all in them. They are all of the same species, *Salvelinus fontinalis*. The different colour of the flesh is no distinction. It is due to other causes not satisfactorily explained yet. It is not a sexual difference either. All the trout that migrate to the sea lose the bright colours that one sees on the fish that remain in fresh water. The fish that migrate to the sea lose their colour very quickly in the salt water but take it on again after their return to fresh water—exactly the same thing as takes place with our St. Lawrence fish, and I could see no difference between the two. I did not get or see a single specimen of *Salvelinus oquassa* in that region.

.Judging from the size of the spawn in the ovaries, the date of spawning of these fish is about the same as ours, about October. It seems as if in the smaller streams and creeks many of the fish never migrate to the sea, and not having the same abundance of food never attain the same size. These are what the people call the river trout, or with us, brook trout. If, perchance there really is another variety, I did not see any specimens during my trip.

STURGEON.

This species of fish is reported to be abundant in the upper waters of both the Nelson and Hayes rivers. An odd one is sometimes taken in the whitefish nets set by the Indians in the estuaries, but the proper fishing grounds are said to be at and above the Limestone rapids. We tried both fixed and drift nets in the entrance and tideway, but did not succeed in getting any. I also tried in the same way the pools below Deer island without success. I believe the water was rather too deep in that section, in one place I found thirty-eight feet of water. The bottom of the river certainly appears very favourable and well suited to the habits of these fish. In all the eddies and sheltered places long grasses and other water plants are abundant, and must afford good feeding grounds. Dr. Sinclair, of Le Pas, who came down the Nelson in September, told me he had seen one killed by the Indians on the way. It was about four feet long. This was during the present season. Possibly earlier in the season, in June or July, the fish may be found near the estuaries.

WHITEFISH.

This is certainly the most abundant and valuable fish of that region. Early in the spring, after the ice has gone out of the rivers, they are caught in short nets set along the shores at random, in any place, not only in the rivers and estuaries but along the coast line as well. This would indicate a migration to the sea, and possibly they may have been seen or taken some distance off shore, which has given grounds for the report that large herrings had been seen in the bay. To the ordinary observer, they could certainly be easily mistaken for one, as they look very much like herrings. Another good proof that some such migration does really take place is the fact that in September we got them in nets everywhere along the coast that we visited and in the

5 GEORGE V., A. 1915

entrance of all the rivers, big or small, and the run was a'l one way, *heading up stream.* The settlers, what few there are, and natives, fish for them in the most primitive way. A short net of seldom more than five or six fathoms long, frequently less, and four to five feet in depth, is attached to'a few poles driven into the muddy bottom. This net is run out in a straight line from the shore, has no trap or pound of any kind, no floats or sinkers. Very often no canoe or boat is used in setting or visiting the nets, they simply wade out as far as they can go comfortably, it serves as a bath at the same time, not an unnecessary thing. In these short nets an average catch of about fifteen fish is taken each day, just about enough to supply their needs, which is all they care for. This fishing lasts from the opening of the rivers until about the end of July when the fish disappear for a time, and commence running again about the end of August and from that on till the fall. These were the dates given to me by the natives. I found however, by actual experience, that while at the time they were getting no fish at all (August 17) in their shore nets, I could get as much as two barrels in one haul of our small seine. We also set fixed floating nets with a short winger at the end and seldom got less than thirty or forty fish, each day, or night. Later in September the quantity increased considerably, at Seal creek on the Nelson, we got close on to a ton of fish in three hauls of our seine. This seine of ours was only twenty fathoms long and only intended for experimental fishing. Going over so much ground in a few days the places selected for fishing may not have been the best that there is. It takes a considerable time to locate suitable places, having first to ascertain depth of water, kind of bottom, snags, if any, etc. If, as seems to be an accepted fact, these fish move out to sea, the use of drift nets outside, as soon as the ice disappears in July, might prove successful. Of all the fishes of that locality the whitefish will certainly be the one that will prove most valuable and easiest and cheapest to fish.

OTHER SPECIES OF FISH.

There appears to be a considerable quantity of pickerel in the Nelson river, some of very fair size, up to nine pounds weight, taken in our seine. I did not see this fish in any other river, but it will likely be found in the Hayes river also. This fish, although locally called pickerel or jackfish, is really a pike, *Esox lucius,* of our southern waters.

A species of sucker, *Moxostoma,* is also found in the Nelson and Hayes rivers, but it is not much esteemed as a food fish. They do not seem very abundant, as we seldom got more than three or four in a haul of the seine.

Little or nothing seems to have been known so far of the existence of capelin, *Mallotus villosus,* in Hudson bay. I was therefore exceedingly surprised when I found out that the beluga or white whale subsisted partly on them, as proven by examination of contents af the stomach. Dr. Marcellus, at present in charge of the medical department of Port Nelson, but formerly of Port Churchill, states that the beach was some years covered with a small fish, which from his description was capelin. Sergeant Jones of the R. N. W. M. Police also gave me the same information. He said the natives around Fort Churchill collected them for food. This was during the month of June when the fish came near shore to spawn, later on they move out to deeper water and could not be so easily observed. This is what affords subsistence to these immense quantities of white whales that frequent these shores early in the season, before any fish could migrate to the sea from the rivers. One specimen of *goldeye,* weighing two and a half pounds, was taken in a net on the Nelson river, unfortunately the specimen was lost. I had placed it near our tent in some cool moss to preserve it, and it was carried away by a mink during the night. It appeared to be a subspecies of the genus *Coregonus,* and evidently not very numerous, as we did not see it anywhere else on that coast.

Some of the Indians from York Factory spoke to me of a species of fish resembling cod, which is sometimes seen late in the fall along that shore. Mr. Macdonald, one of the agents of the Hudson's Bay Company, whom I met at Port Nelson, said there was an abundance at times of a species of rock-cod in the south eastern part of the bay. They ranged in weight from two to seven pounds, and were frequently taken in shallow water and some times by fishing from the rocky points. I tried fishing in September, in depths of six to seven fathoms, but got nothing. I don't believe the bottom is suitable for such fish in that section, being too muddy. Mr. Macdonald said he had caught them himself in quantities. The stomach of a polar bear, which I examined. contained, amongst other things, remains of some species of sculpin, shreds of skin. which were mottled grey and black, probably *Cottus Axillaris*. No grayling were seen or taken in any of the streams visited, although they are reported as being common around Churchill and further north.

VALUE OF FISHERIES.

Little or no benefit can be derived for the moment, from the fisheries on the western side of Hudson bay, beyond supplying the local demand. The distance is too great and the difficulties of reaching it too numerous, to enable any sailing vessel from the Maritime Provinces or Quebec making profitable trips, no matter how considerable these fisheries might be. They must be developed by local fishermen and this will only be done profitably when the Hudson Bay Railway is completed. Then I have no doubt they will prove of great value to the northwestern provinces. From what I could ascertain during my short stay there, the best fishing season would be from the middle of June to August, when the fish are on the coast shores after the migration, and then from September until about October, when the ice sets in. From Churchill south to James bay, most of that coast is fringed with long and low muddy flats, strewn over more or less with small boulders, these extend several miles from shore in most places. There are practically no harbours even for light draught vessels, if we except the Nelson river. Fishing will therefor have to be prosecuted in light draught boats, that could enter most of the small creeks or rivers at high tide. Drift nets of two or three inch mesh in extension will be found, I think, the most suitable for coast fishing. In the estuaries and rivers fixed floating nets of moderate length, two and a half inch mesh, are the most suitable. Seines are very quick and handy engines for catching fish of all kinds, but they cannot be used except on clear bottoms and where the currents are weak.

During the time I was on the Nelson, especially on my last trip in September, we could easily, had we devoted all our time to it, have taken with our short seine, from eight hundred to a thousand pounds of fish daily. In one single haul we caught 689 whitefish. On the Hayes river, near Fishing island, a haul of 100 fish is frequently made. Even at a moderate price, this would be quite a lucrative business, provided it could be shipped fresh to market, as will be the case in a year or two. Some experiments that we made in salting a few proved very satisfactory. They were equal to the best quality of herring in richness and more delicate in taste. Valuable as these fisheries may be to the Western Provinces, they will be totally eclipsed by another source of revenue and that is in,—

GAME PRESERVES.

I have visited and shot over most of the celebrated game resorts of this continent, the Northwestern States with its famed duck ponds, Lake Champlain in its palmy days, the famous Longue Point and Sorel marshes, seal reef in the St. Lawrence and the Labrador shore with its myriads of birds; but never have I seen anything that could compare to this Hudson bay shore. Geese of various kinds, black and pintail ducks, many species of plovers and the smaller members of this family, are to be found there in countless thousands. All that low marshy belt of land extending from Churchill to James bay, several hundred miles in length and eight to ten wide, on an average, is nothing but an immense breeding ground. Resting in our canoe at night on the mud flats, waiting for a rising tide, we actually could not sleep owing to the continuous honking of the geese around us. Flocks of several hundreds were constantly rising as we sailed or paddled along. Closer to the shores, in the ponds and marshes in the early mornings, or at night, masses of smaller birds were continuously on the wing. At high water, the grassy ridges near the creeks were lined with immense bunches of pintails, *Dafila acuta* and green winged teal. One single shot in a bunch on the wing would generally be sufficient to keep three hungry men satisfied for two days. A list of all the different birds found in this section is appended to this report. It is to be hoped that proper protection will be given to this immense breeding ground of these birds, and that the government will make a "Game Preserve" of it. In a few years, there is sure to be an influx of sportsmen in that region, and certain points at *proper seasons* could be rented and yield considerable revenue.

Larger game is also to be found in that same region. We saw some caribou and their tracks frequently.

Polar bears are common. We saw as many as five in one day, all going in the same direction. They apparently get on the ice in the northern portion of Hudson bay and drift south with it. Then in July and August, as the case may be, the ice melts and breaks up and they are forced to make the coast line by swimming, and work back north again by following the shore. Along the rivers, or inland, eight or ten miles from the sea shore there are some wooded sections, where black bears appear to be common. We saw some of their tracks occasionally. Not being in want of meat we made no attempt to secure any large game. Our canoe was too much cumbered with our fishing gear and baggage to burden ourselves with such a weight, and there was no sport in wanton killing. In October and later on in the fall and winter, this fringe of wood, a considerable part of which is willow, swarms with ptarmigan. Around Port Nelson several thousand were killed last winter. Like most other good things, this "sportsman's paradise" has its drawbacks. It is difficult of access, walking soft and bad, a nasty coast to travel along either in boats or canoes. Low and marshy camping grounds, with no fuel except drift wood and you can go miles sometimes without finding any. We used to carry a small supply in our canoe as we went along and found it very useful. It is also a wise precaution to select the highest ridges, as being somewhat drier, for a camping place, and also to avoid occasional high tides. These may come without any warning, you may wake up in the night with water all around. They are due to strong northerly winds in the bay, driving and piling up the water on these long mud flats, with no chance for an undertow, consequently the waters rise several feet above their ordinary level, varying with the strength of the gale and state of the moon. If one happens to be obliged to ground on these shoals with a falling tide, it is very risky and unwise to leave the canoe and walk ashore to build or cook a meal. The tide comes in so quickly that it may be impossible to reach the canoe in time. We saw two accidents of this kind, while we were on that coast, and it happened to trappers, and all their kit was lost.

HARBOURS.

As already referred to, there are no harbours or shelter to be had, except for very small boats, between Fort Churchill and the Nelson river, and from there south to the Severn, where a light draught vessel may enter at rising tide, while further south to James bay it appears to be the same low lands and shallow water for miles out. This will practically prevent any fishing being ever done on that coast line with sailing vessels, supposing the fish were there for it. I noticed that the government were

SESSIONAL PAPER No. 39a

thoroughly alive to the necessity of having the approaches to that low and dangerous coast rendered safer and that steps had been taken to light the entrance of the Nelson. Although this does not exactly come under the object of my trip to Hudson bay, I would beg to offer the following suggestions and trust they may be of some use to the department.

While I was there I was informed that one pole light was to be placed on the highest portion of the Nelson shoal just off the Stoney river, another on Cape Tatnam. These lights will undoubtedly be of some use, but I do not believe adequate. During the summer season these flats and low marshy shores, are always more or less covered with mist or haze towards night, or early in the morning, which disappears with the rise and heat of the sun, so that many times they would be invisible even at a short distance. It must also be borne in mind, that there still remains outside of these two points several miles of shallow water. In my opinion the only safe way for lighting that route when it will be open for commerce, will be to have two good lightships, with suitable fog alarms placed outside of the above two points. A good position for the Cape Tatnam one, would be about twenty-five miles off that shore. These lightships could be safely wintered at Port Nelson and put in commission early in the season, fully two weeks before any vessel could enter the bay. Under existing conditions the wireless apparatus placed on vessels, is not of much use except to communicate with each other, but on the supposition of their not being able to do that in a case of need, the vessel would be perfectly helpless. It would therefore be important to have two or three stations along the straits, say one at, or near, Port Burwell, one at Erik cove, and possibly one on Coutts island. They would be invaluable aids to the vessels coming in early or going out late in the fall, by giving information of the ice conditions, and under this head I will now give our own observations.

ICE CONDITIONS.

The first ice we observed was near Point Amour and gradually increased in quantity with a few icebergs here and there until we got off Battle harbour, when we met packed ice. We were then about sixty miles off shore. From this last point to Hudson strait, it was apparently one continuous belt of ice, with small patches and lanes of open water here and there on its outer edge. We made an attempt to keep outside of it as much as possible and estimated that we were, at certain points, fully one hundred and seventy-five miles from land. Off Cape Mugford we got within thirty miles of land, and finding the same conditions headed out again for more open water, and kept fifty or sixty miles off until we neared the straits. About half way across the straits we came to open water and from there to Resolution island and further northeast there was apparently no ice. This was on July 31. On August 1 we went over the same course and found no ice again so that we had probably reached the end of that long strip of drift ice coming from the coast north of Hudson strait. In the straits the ice seemed to drift out all the time on the south side, while on the north it was affected more regularly by the tides and moved in and out, but the main direction was in. We observed this by our own drift and that of many icebergs, which moved up the straits as far as Big island. Beyond this point we did not see any bergs, but close in towards the land the influence of the flood was still felt. Near Charles island there was a strong ebb tide when we passed there at 6 p.m., August 8. We judged it was running at about 2½ knots an hour. From Big island inwards, up to near Mansel island, we passed some patches of very heavy ice, much of it dirty, discoloured with mud and stones. At some time during the past winter it had been subject to great pressure near the land, because it was piled up in layers, showing the same discoloration. It was evidently old ice that had been drifting around since the previous year or longer. In Hudson bay itself we met very little ice until we reached about sixty miles from Cape Tatnam. Here there was a patch of small

5 GEORGE V., A. 1915

broken-up ice of the past winter's formation, with occasionally here and there some heavy pans twenty to thirty feet thick. We were informed later that this patch was over one hundred and fifty miles long. It was not heavy enough to impede any good steamer. From the time we left Resolution island until we reached the western end of the straits, we were mostly on the north side of it, and owing to the prevailing winds being from that direction we escaped much of the heavy ice that was encountered by other vessels that were on the south side. We were never at any time subject to any pressure that would have damaged any ordinary vessel. All the hard knocks and the little damage that we received was had through reckless sailing and mistakes. We were held up several times, for more or less long spells in the ice, until a change of wind or tide would release us. A large quantity of the ice that we saw along the north side of the straits, was last winter's formation, moving out of the innumerable bays and passages among the islands of that coast. It had never been rafted or subject to pressure because the edges were intact and the snow had not even been disturbed on it. Probably had we been there a week or two earlier it would have been found fast to the land. We were told that sledging with dog teams had been kept up on the bays of the south side until the end of July. All this goes to prove that the season of 1914 was an unusually late one as far as navigation was concerned in Hudson straits. Such conditions I believe will always be found to follow mild and open falls, which release large quantities of heavy ice in the Fox channel and elsewhere. This drifts down and is caught with the fresh formed ice of the winter and finds its way out in the following spring. We saw no field ice at all, perhaps because we were there late, or else it may not have had a chance of forming owing to the strong gales on this large expanse of water, which breaks it up. Strongly built ice patrol boats, fitted with wireless apparatus, would render immense service to shipping, when that route will be opened to commerce. To guard against loss of life and property as much as possible, no vessel should be sent to Hudson bay without a wireless installation, otherwise if wrecked they might be weeks or months before obtaining any assistance, especially if it happened to be in some out of the way place.

All the time I was on the *Burleigh* we did not experience any very heavy gales, much less than I expected in that region. I presume this was due to the late season at which we were there. We had much worse weather in September on our canoe trips, and several frosty nights, but no ice worth mentioning had yet formed up to the time we left (Sept. 27th), and we had only one light snow fall. Going out October 1st we observed considerable patches of fresh snow on the mountains, but none on the low lands.

THE "BURLEIGH."

Although this vessel is probably good, staunch and suitable for navigating these waters, she was far too slow with her auxiliary engine to be adapted for an expedition of the kind we were on. At her best she could do no more than two and a half knots an hour, and more frequently it was only two knots. This was not sufficient to stem the least wind or head sea and consequently it could not be relied upon to any extent. We were thirty-eight days out and it took forty-one days back, nearly three months, or practically the whole season consumed in sailing alone. A steamer of moderate size, capable of steaming eight or nine knots, on low consumption of coal and drawing about nine or ten feet of water, would be the right kind of vessel for such work, if the Department intends to continue these investigations. I am inclined to believe that the most valuable fisheries will be found in the southeastern portion of the bay, which we could not visit for want of time.

I have the honour to be
Your obedient servant,

NAP. A. COMEAU,
Officer in charge "Burleigh" Expedition.

SESSIONAL PAPER No. 39a

WESTERN COAST LINE.

The whole of the coast line which we visited on the western side of Hudson bay, from the Owl river to near Cape Tatnam, is very low, swampy ground, perfectly level. It is composed of clayey deposits, which have been leveled by the action of the water. The outer ridge near the tide-way is covered with grasses and wild flowers of various kinds, common to that latitude. A little further in towards the land this changes to low scrub, about knee high, mixed here and there with small stunted patches of willow—until about eight to ten miles from the beach, when the timber line is reached. These woods consist for the most part of black spruce, poplar, larch and pussy willow, at first of low stature, but increasing in size as one goes further into the interior, and afford shelter to game and fur bearing animals. Outside of the grassy ridges are extensive mud flats, sprinkled with stones and boulders, that extend from ten to twelve miles out in some places. The larger boulders on these shoals are constantly shifted about by the ice in the spring. During heavy gales large quantities of seaweed, especially of the long leaved round stem variety, *fucus,* are detached in the shallow waters and thrown up in ridges along the beach, where we find them in various stages of decomposition. There is also a great abundance of the common blue mussel, *Mytilus edulis,* which get washed ashore and are to be found all over the inside flats miles from the present tide mark. We have also observed old pieces of driftwood a long way inland which the tides have not been anywhere near for years. The above would indicate that this western shore is gradually extending out each season, from accumulations of deposit carried out by the rivers and thrown up on the beaches by the sea. On the Stoney river we visited the site of what is supposed to have been an old whaling station. This must have been for the white whale fishery only and possibly fur hunting in winter. Traces of the foundation of the two huts, about 15 x 20 feet, are still plainly visible, and also a rendering place where we found old iron hoops and pieces of fire bricks. This site is fully a mile inside of the present estuary of the river. We believe when it was originally built, it must certainly have been placed in a handy and convenient spot for handling their products and was probably close to the estuary of that period. To-day it would be in an extremely awkward position for conducting anything of that kind. Stones show the location of a pier where likely they came alongside with boats but which was too shallow to float my canoe. Marsh Point, at the entrance of the Hayes river, shows the same indications, and has probably lengthened considerably within the past hundred years or so. When the beacon was erected by the Hudson Bay Company we must presume it was placed in the most prominent position on the point, now it is quite a distance back. It shows signs of age and is pretty shaky. The south east corner post is rotten and cut right through about the middle. It is a wonder how it stands the severe gales in that condition. From Marsh Point westward going up along the south shore of the Nelson river for a distance of about ten miles, the shore is covered with a dense growth of rich grasses and hay, sufficient to maintain a large herd of cattle, with plenty of good water at hand.

LIST OF BIRDS OBSERVED OR SHOT BY NAP. A. COMEAU, ON "BUR-LEIGH" EXPEDITION TO HUDSON BAY, 1914.

1. Pied billed Grebe. *Podilymbus podiceps.* Two seen on the Partridge river, apparently breed there (August 30.)
2. Loon. *Urinator imber.* Seen several times along the west coast and in Hudson strait common.
3. Red throated loon. *Urinator lumme.* Common, seen frequently in straits and bay. I was particularly watchful as regards loons, hoping to see a specimen of the Arctic loon, *U. articus,* but am sorry to say did not come across any. This bird must be rare as I never saw a live one.

5 GEORGE V., A. 1915

4. Guillemot, Sea pigeon. *Cepphus grylle.* Seen all the way up the Labrador coast, many miles from land and in Hudson straits, but saw none on the west coast.

5. Murre. *Uria troile.* An exceedingly abundant bird everywhere from Belle Isle north to Cape Chidleigh and in the Hudson straits, less common as we enter the bay. Have seen hundreds sitting on icebergs or flying near them a hundred miles from land. In a fog, the presence of these birds in any number is a pretty sure indication of the proximity of some ice or bergs. Saw some immense flocks in Hudson straits, and some female birds with small fish in their bills, carrying it to their young when land was fully thirty miles distant.

6. Razor-billed Auk. *Alca torda.* Not very common, a few in Hudson strait, none in bay.

7. Dovekie. *Alle alle.* Seen frequently, especially numerous under the lee of icebergs, they are fond of sitting on them like the murres. Common in the straits, none seen on west shore.

8. Skua or Jaeger. *Stercorarius pomarinus.* Fairly common, from Belle Isle north.

9. Parasitic Jaeger. *S. parasiticus.* A commoner bird than the above. We were aware that these birds were regular pirates and had often watched them robbing the poor Kittywake gull and other larger ones of their hard earned fish, but we never thought they were murderers. This season, while we were on the west coast of Hudson bay, between Churchill and the Nelson, we saw them hunt regularly in pairs, and kill small birds. It was most interesting to see the intelligence they displayed in chasing the birds. As soon as one of them started after its game, the mate would sweep along low, and get under the bird to prevent it from diving into the brush or grass and thus evade pursuit. They would thus continue in company and worry the poor thing, until it was exhausted and fall a prey to one of them, because both did their best to get hold of it. It was not struck by swooping like some hawks do, but caught with the beak and swallowed while on the wing. On one occasion we saw the jaeger go out some distance and sit on the water. We could not see on account of the distance if the bird was disgorged and then torn, but one thing we noticed was that immediately after the bird was captured by one, its mate would leave it alone. We saw no quarrelling for the spoils. Most of the birds chased were of the smaller species of the sand piper family and sparrows.

10. Kittywake. *Rissa tridactyla.* A well known and familiar bird all over the St. Lawrence and Labrador right into Hudson bay. Especially abundant in the entrance of rivers.

11. Ice gull. *Larus glaucus.* A few were seen in the straits, common in the entrance of the Nelson river, and many observed on the way up.

12. Saddle back. *Larus marinus.* Observed in same places, but not so common as glaucus.

14. Herring gull. *Larus argentatus.* A common bird seen during the whole trip in one place or other, but more numerous near the land and mouth of rivers.

15. Ring-billed gull. *Larus delarawensis.* Fairly common along the west coast of bay. Not observed in the strait.

16. Fulmar. *Fulmarus glacialis.* Fairly common from Belle Isle to Resolution island. Not seen in bay.

SESSIONAL PAPER No. 39a

17. Common Tern. *Sterna hirundo.* A few seen going up, but only two or three observed in Hudson strait. Other terns were observed but could not be identified.

18. *Shearwater.* Puffinus. These birds were frequently seen, but none were secured and so could not be identified as to species. None in straits or bay.

19. Stormy Petrel. *Procellaria pelagica.* Seen often sometimes quite numerous from Halifax to Hudson strait. None in bay.

20. Merganser. *Merganser americanus.* One female and brood seen on the Nelson river.

21. Common Sheldrake. *Merganser serrator.* Very common near all the rivers, of the western part of Hudson bay. None seen in straits as we were probably too far from land most of the time.

22. Black duck. *Amas obscura.* Common in the marshes and ponds of the west coast, breeds.

23. Green winged teal. *Anas carolinensis.* One of the most abundant birds of the species. Found all over the west shore, wherever there is a small patch of fresh water. Clouds of them near the Partridge and Stoney rivers.

24. Pintail. *Dafila acuta.* The commonest of all the ducks of that region.

25. Golden eye. *Glaucionnetta islandica.* One small bunch seen near Macdonald group of islands, in the straits and a few on the west shore of bay.

26. Old Squaw. *Clangula hyemalis.* We were disappointed seeing so few specimens of this duck. Some were seen in the straits, but very few in the bay, and these were in the estuary of the Nelson.

27. Eider ducks. *Somateria.* One small bunch probably King eider, were seen near Charles island, Hudson straits. Most likely abundant along the land. None observed in bay.

28. American scoter. *Oidemia americana, O. fusca and O. deglandi.* The three varieties were seen in straits and bay; a sight of these birds in any number is a pretty sure sign of being near land.

29. Snow goose. *Chen hyperborea nivalis.* Two of these birds were seen near the Owl river, (Aug. 28) none seen elsewhere on trip.

30. Canada goose. *Branta canadensis.* Extremely abundant on the shore line and on the shoals. We observed their tracks quite a distance inland, but saw no nests. We happened to spend some hours on the flats at night, on account of the falling tide and their honking in our vicinity was so continuous, as actually to prevent us from sleeping.

31. Brant. *Branta bernicla.* Quite numerous olong this western shore and breeds, because we observed some young birds. They are not often seen on land, preferring to feed on the flats, or while swimming some distance out, on floating grasses and roots carried out by the current.

32. Swan. *Olor.* One swan was seen on the wing during a heavy gale, on August 31, near the Partridge river, probably, buccinator.

33. Blue heron. *Ardea herodias.* Seen on the Hayes peninsula and on South Seal creek, not common. 19th August.

34. Sora Rail. *Porzana carolina.* One young bird seen at Marsh point, Hayes river, September 9.

5 GEORGE V., A. 1915

35. Phalaropes. *Phalaropus hyperboreus and P. tricolor.* (Were quite common in flocks, in the open water outside the ice belt, from Belle Isle to Hudson strait. In strong winds they were frequently seen under the lee of icebergs, in big flocks. They have a peculiar habit of hovering around the bow, or under the shelter of the sails of a vessel at night, especially on dark nights. They look like bats on the wing and utter a most plaintive and monotonous cry. None seen in Hudson bay.

36. Gray snipe. *Macrorhamphus griseus.* Very abundant in flocks of about twenty or so. All over the marshes, between Churchill and Nelson. Very tame bird there.

37. Sand piper. *Tringa minutilla.* Seen in immense flocks on all the west coast of the bay.

38. Yellow legs. *Totanus melanoleucus* and *T. flavipes.* Both very common birds, in pairs and flocks, easily decoyed, seen only on west shore of bay.

39. Solitary Sand piper. *T. solitarius.* Found along the small creeks and rivers.

40. Bartramian plover. *Bartramia longicauda.* Fairly abundant.

41. Spotted sand piper. *Actitis macularia.* Common along the rivers.

42. Curlew. *Numenius hudsonicus* and *N. borealis.* Both very abundant at one time along the Labrador coast, now getting, for some unknown reason very rare. Some people are inclined to believe that they are exterminated by excessive shooting. If that is the case, it must be during their migration or in their winter haunts, as very few are killed on their breeding grounds. We saw two small flocks and a few stragglers, north of the Nelson.

43. Ox-eye plover. *Charadrius squatarola.* Common, in flocks.

44. Golden plover. *C. apricarius.* Common, but never seen in large numbers.

45. Killdeer plover. *Aegialitis vocifera.* A few seen in small bunches.

46. Ring plover. *A. hiaticula.* Seen here and there, but not abundant.

47. Ptarmigan. *Lagopus lagopus.* Saw abundant traces of their passage in the way of droppings and feathers, in the willow patches but saw none of the birds. They were killed in hundreds at Port Nelson last winter 1913 and 1914.

48. Marsh hawk. *Circus hudsonius.* A common bird along the marshy beaches.

49. Coopers hawk. *Accipiter cooperii.* Shot one that had lit on the ridge pole of our tent. It had been attracted by a small bunch of plover and teal that was hanging on the end of it. South side of Nelson.

50. Rough legged hawk. *Archibuteo lagopus.* Seen occasionally.

51. Golden eagle. *Aquila chrysaetos.* Some of these birds were seen several times around the west coast of bay.

52. Gyr falcon. *Falco islandicus.* Found a dead specimen of this bird along the banks of North Seal creek, it had been shot and wounded and afterwards perished, it had lain there for some months.

53. Sparrow hawk. *Falco sparverius.* Frequently seen.

54. Osprey. *Pandion haliaetus.* Several seen on the rivers and coast Hudson bay.

55. Barred owl. *Syrnium.* A very large specimen of this family, probably, *cenercum,* was flushed in the woods near Deer island, Nelson.

56. Horned owl. *Bubo virginianus.* Was heard several times along the Nelson river.

SESSIONAL PAPER No. 39a

57. Black backed wood pecker. *Picoides arcticus.* Observed several times.

58. Night hawk. *Chordeilus virginianus.* Common on the barren heights along the Nelson river and around Port Nelson and Hayes river.

59. Horned lark. *Otocoris alpestris.* Seen around Nelson and along the grassy beaches, not numerous.

60. Canada jay. *Perisoreus canadensis.* Common along the shores of the rivers and coast wherever there are any trees.

61. Northen raven. *Corvus corax, principalis.* A very common bird and considered a great pest by the trappers, who lose no chance of shooting them whenever possible.

62. Common crow. *Corvus americanus.* Seen often but not abundant.

63. Black bird. *Scolecophagus carolinus.* Quite common in flocks about the west coast of bay.

64. Red poll. *Acanthis linaria.* Common in flocks, frequently chased by the jaegers.

65. Savanna sparrow. *Ammodramus savanna.* One of the commonest sparrows seen.

66. Swamp sparrow. *Melospiza georgiana.* Observed in the low brushes and swampy regions near the rivers.

67. Chickadee. *Parus hudsonicus.* Observed only in the wooded portions up the Nelson river.

68. Hermit thrush. *Turdus pallasii.* Heard and seen along the rivers, in wooded sections.

Various smaller species of owls were seen, some hawks, many small birds of which we only got a glimpse or saw at too great a distance to identify them. The special work I was on, did not warrant my losing any time in their pursuit. These notes were taken simply because we take an interest in bird life and it may interest others.

NAP. A. COMEAU,

Officer in charge, "Burleigh" Expedition to Hudson Bay, 1914.

Beset in the Ice, Hudson Strait, August 7, 1914.

York Factory.

Seining on the Nelson.

A Cree Camp, Hayes River.

Some of our Catch.

Five and a half pound Trout, N. Sacol Creek.

5 GEORGE V.   SESSIONAL PAPER No. 39b   A. 1915

SUPPLEMENT

TO THE

47th ANNUAL REPORT OF THE DEPARTMENT OF MARINE AND FISHERIES, FISHERIES BRANCH

CONTRIBUTIONS

TO

CANADIAN BIOLOGY

BEING STUDIES FROM THE

BIOLOGICAL STATIONS OF CANADA

1911-1914

FASCICULUS I.—MARINE BIOLOGY

OTTAWA
PRINTED BY J. DE L. TACHE, PRINTER TO THE KING'S MOST
EXCELLENT MAJESTY
1915

[No. 39b—1915]

PREFACE.

By PROFESSOR EDWARD E. PRINCE, Commissioner of Fisheries, Chairman of the
Biological Board of Canada, Canadian Representative on the International
Fisheries Commission, and Member of the Advisory Fishery Board of the
Dominion.

When the last series of Biological papers was published two years ago, I
stated, in my introductory note appearing as the preface to the publication, that
some memoirs were nearly in shape for publication, but could not be included in
the volume issued in 1912.

These papers were subsequently placed in my hands, and others have been
completed, so that no less than twenty-two important original contributions
to the Biology of Canadian waters, marine and fresh-water, are now ready for
publication.

This series is indeed more voluminous than had been anticipated, and it has
been found desirable to issue them in two parts,—One, Fasciculus I. composed of
papers dealing with sea-fisheries and marine subjects, and Fasciculus II. issued
separately, including papers which refer to the interior fresh-water fisheries, and
to subjects relating to the Great Lakes.

The researches, embodied in the first series of papers, were conducted chiefly
at the St. Andrews Biological Station, on the Atlantic Coast, while the second
series of papers embraces work done by the members of the staff at the Georgian
Bay Station on the Great Lakes. Many papers representing work done at the three
Biological Stations, and authorized by the Biological Board, and indeed carried
on under the direction and auspices of the Board, have been published elsewhere
or the present series would have been much more extended. Credit should be given
to the Biological Board, and to the Biological Stations, for such investigations
published in reports issued elsewhere or appearing in journals or magazines in
Canada or abroad.

Thus it may be mentioned that Dr. Stafford, who has practically carried on
all his marine biological studies under the Board, and who commenced his fishery
investigations when the Atlantic Station was opened at St. Andrews in 1899, and
has continued until recently a member of the staff of workers, has published
two papers on the Canadian oyster, its life-history, conservation etc., in the reports
of the Commission of Conservation,* while Mr. F. A. Potts of Cambridge, Eng.,
Professor McMurrich of Toronto, Miss Katherine Haddon and others, have published their results in various scientific journals on this continent and in Europe.†

The present series includes two important papers on the minute floating life
in the sea, a source of food for fishes, especially in the early stages of their life,
and an important part of the food of the oyster and other shell-fish.

* See Fisheries of Eastern Canada, Comm. of Cons. Report, 1912, pp. 26 to 49, and the
Canadian oyster, Comm. of Cons. Report 1913, pp. 1 to 158.

† Spengel's Zoologisches Jahrb. 1912, pp. 575 to 594; Roy. Soc. of Canada 1913,etc.

5 GEORGE V., A. 1915

Professor Willey, of McGill University, deals with St. Andrews' Plankton, and Professor Bailey, of the University of New Brunswick, treats of the Diatoms in the Bay of Fundy waters. The paper on certain diseases of fish, completed by Dr. J. W. Mavor, is of special scientific and practical value. Comparatively little has been done in this difficult field of research, although our sea-fishes and fresh-water fishes often perish in vast numbers, no doubt owing to some epidemic of disease about which little is accurately known. The study of fish diseases is the readiest method of coping with this serious loss. Last season, 1913, it may be mentioned that the herring fisheries of the Gulf of St. Lawrence suffered serious loss by the death of vast numbers of fish from some such cause. Dr. Mavor's fame as a specialist, and the unique character of his paper on the Sporozoa of New Brunswick fishes, gives it unusual importance and it will be welcomed by all interested in our fish and fisheries, and by scientific men generally.

Dr. Huntsman's paper on a new Crustacean, a Caprellid, not previously described or determined, is of special value. Much remains to be done in the field of Crustacean research in Canada. Mr. J. D. Detweiler gives a list of New Brunswick Mollusca, this being another of those contributions, published by the Board, which will aid in the preparation of a complete marine faunistic list for our Atlantic Coast.

The paper on the fungi collected at St. Andrews by the late Miss Van Horne, aided by Miss Adaline Van Horne, has a melancholy interest for the MS. was handed to the late Professor Penhallow for publication. Neither Professor Penhallow nor Miss Mary Van Horne survived to see the paper printed.

The relation between the fisheries and the land fungi may not appear to be very intimate, though it is well known that insects abound, and, indeed, feed upon decayed fungi, and insect food is important from a fishery point of view.

The report by Professor A. T. Cameron, University of Manitoba, calls for special mention on account of its important commercial bearing. It has long been known that a valuable chemical product is present in certain seaweeds, and Dr. Cameron has completed an original research, in which he has studied no less than twenty species of marine plants, including the giant Pacific Kelps. He studied six species of sponges; five species of jelly fish and fourteen higher forms in order to determine the amount of Iodine present in them, and at the conclusion of his paper, he adds an Appendix on the commercial aspect of the Kelp beds on the Pacific Coast as a source of Iodine production. Mr. A. B. Klugh, (Queen's University) rendered Dr. Cameron assistance in this work.

Two papers by Mr. Stock and Mr. Martin of the University of Toronto, treat of some Parasites (Copepods) of certain Bay of Fundy fishes, and on the effect of freezing upon living fish. Both are of the nature of preliminary reports and they are of very special interest.

Since the last issue of the Biological Contributions, the Board has been deprived by death of two esteemed colleagues, Professor Penhallow and Rev. George W. Taylor, both of whom devoted much time and labour to the work of the Biological Stations and contributed substantially to Biological Science in Canada.

OTTAWA, Jan. 1914.

CONTENTS.

THE PLANKTON IN ST. ANDREW'S BAY.

BY A. WILLEY, D.Sc., F.R.S.

Professor of Zoology, McGill University, Montreal.

Few imagine, when crossing the ocean, that the prow of the ship is cleaving its way through teeming myriads of foam-like creatures and that every turn of the screw is a marine catastrophe, bringing sudden death to multitudes of sensitive beings.

That this is a fact is frequently demonstrated in the darkness of the night when the swarm of life approaches nearer the surface which it illuminates by phosphorescent scintillations.

An ingenious method of testing the vitality of the sea from the seemingly unfavourable situation of a passenger on an ocean liner, has been adopted in recent years by Professor Herdman of Liverpool, the founder of the successful Marine Biological Station at Port Erin (Isle of Man). The method simply consists in straining the sea-water as it flows from a tap through a silk bag, at intervals during a voyage.

Even in the daytime, in calm weather, the presence of living matter may be made manifest by the occurrence of smooth oily-looking streaks and patches in the midst of the rippling water. The remarkable character of these so-called animal currents was first recognized by Carl Vogt so long ago as 1848. A graphic description of their appearance around Lanzarote, one of the Canary Islands, was published by Professor Richard Greeff in 1868. Similar streaks may be observed in the bay of St. Andrews; they are due in part to the tidal currents and in part to the organisms which are contained in them.

The floating fauna and flora of the oceanic and coastal waters constitute what is known as the Plankton or drifting life of the sea. This technical term, which is now universally employed at Biological Stations, was introduced by Professor Victor Hensen of Kiel in 1887. The only single vernacular term, previously in use, which conveyed the same meaning, was the German word 'Auftrieb', this was commonly borrowed by other tongues, and the custom of using it continued for several years after the more international expression 'Plankton' had been happily suggested; but now it is seldom heard.

The originator of the special study of the marine Plankton and, therefore, the father of planktology, was the greatest naturalist of the nineteenth century in Europe during the period which intervened between the death of Cuvier (1832) and the rise of Darwin (1858), namely, Johannes Müller of Berlin. It was he who introduced the method of towing through the water a very fine-meshed gauze-net of muslin or silk, which he used in furtherance of his researches on the free-swim-

5 GEORGE V., A. 1915

ming larvæ and the metamorphosis of starfishes and sea-urchins at Heligoland between 1845 and 1855.

An enormous advance in the qualitative description of the Plankton of the five oceans resulted from the collections and observations accumulated during the voyage of H.M.S. Challenger (1873-1876).

The intensive quantitative determination of the Plankton was inaugurated by Professor Hensen, who led the well-known Plankton Expedition in the Atlantic Ocean in 1889. The finely illustrated reports which have been issued from that time to this, sufficiently attest the value of the results obtained; but the actual significance of the countings and calculations can only be appreciated fully by professional statisticians.

The principal object aimed at by the promoters of the Plankton Expedition was a physiological one: the discovery of the factors which control the metabolism of the sea, *i.e.*, the assimilation and interchange of nutritive materials under the influence of light, heat, and oxygen, on the part of pelagic organisms which have no place in popular esteem, but which nevertheless are the prime sustenance of all the marketable food-fishes.

The scientific interpretation of the Plankton is thus a physiological problem and its bearing upon human welfare lies in opening the way to a rational conception of the fertility of the sea. The prodigality of marine life in its less conspicuous aspects is a natural phenomenon which must be investigated by methods as rigorous as those that are applied to the elucidation of other natural phenomena, in order that progress may be reported all along the line. It is impossible to avoid the problem; and the multiplication of biological stations in all the progressive countries of the world, proves that it is impossible to rest contented with temporary achievements, however brilliant they may appear to be.

After the quantitative method has been adequately tested, the next way of dealing with the great question of the metabolism of the sea is the experimental method. Perhaps unnecessary emphasis has been laid upon the distinction between observation and experiment, although it is by no means easy at all times to draw the line of demarcation. When Pasteur in 1860 drove the last nail into the coffin of the doctrine of the spontaneous generation of micro-organisms, the contrast between the methods of observation and experiment was indeed brought into high relief by the futile opposition of an otherwise excellent zoologist, Georges Pouchet, whose name is perpetuated by its having been applied to a peculiar member of the micro-plankton, *Pouchetia*.

This is one of the Flagellata, distantly related to a very common species at St. Andrew's named *Peridinium divergens*, shaped like a miniature chafing-dish with a conical cover, which is probably responsible, at least in part, for the display of phosphorescence to be witnessed there, according to the testimony of the staff at the Biological Station. Of course Pouchet's opposition to Pasteur was the one sad mistake of his life, but he did much good work besides. Amongst many other things, he reported upon the Sardine Industry of France. On one occasion, in company with a colleague, he found the stomachs of the sardines which they were examining, filled with *Peridinium divergens* and an allied species of the same genus. They calculated that there would be, at a minimum, twenty million Peridinia in a single fish.

The truth seems to be that all methodical observation has an experimental basis, and the merit of advancing biology to the rank of an experimental science does not rest entirely with the mechanists of the present decade, nor even with the hybridists, great as have been the results of their respective labours.

With reference to the constitution of the Plankton, Haeckel (Plankton-Studien, 1890, p. 66) insisted upon the fact, known to every experienced planktologist, that the first and most striking peculiarity is the variable combination of its component units. The differences of composition are both qualitative and quantitative and are as noteworthy when comparing different localities at the same time, as when comparing different seasons at one and the same station.* Under these circumstances, in order to secure complete and reliable data respecting the periodical fluctuations of the Plankton, it is necessary to institute continuous series of observations at a given locality throughout at least one entire year, and better still through several successive years, after the manner adopted in recording meteorological conditions, with which the various planktological conditions are directly and intimately correlated.

In illustration of the kind of data concerning the circulation of Plankton in coastwise currents, which may be obtained by the co-ordination of observations made at different stations at the same season or at different seasons, I may mention that a certain small Crustacean species, named *Acartia clausi*, was the most abundant representative of its order (the Copepoda) at St. Andrew's in July and August 1912. It was not found at Woods Hole, Mass. during the same two months in 1899 (W. M. Wheeler); but it occurred abundantly in Naragansett Bay in January and February 1906 (L. W. Williams).

This species belongs to a section of Copepoda termed Calanoida by G. O. Sars, the veteran author of "An Account of the Crustacea of Norway," one of the standard works of reference on this subject. The Copepoda of this group afford nutriment to several common food-fishes. *Calanus finmarchicus*, a relatively large species attaining a length of four millimetres and a leading type of the North Atlantic zooplankton,† is known to be the food of the herring along the Norwegian coasts. Very few examples appeared in my tow at St. Andrew's, and these were immature, not exceeding three millimetres in length. If it should ever be found in quantity within the bay of St. Andrew's, it would make a notable record. Arctic specimens of *C. finmarchicus* attain a maximum length of five millimetres (G. O. Sars).

Associated in the tow with the *Calanus*, and not so rare as the latter, was a transparent, fragile being belonging to the group called Tunicata-Appendiculariæ, named *Fritillaria borealis*. This little creature is shaped like a miniature hammer,

---

* On this point attention should be drawn to Professor Herdman's Plankton Investigations in the Irish Sea. *Vide* 26th Ann. Rep. Liverpool Mar. Biol. Committee. December 1912, p. 36. Also Professor McIntosh's Plankton Reports, Scot. Fish. Bd. Rep., 1890, etc., and Dr. Williamson, Plankton Reports, Scot. Fish. Bd. Rep., 1898, etc.

† Animal plankton as distinguished from Phytoplankton comprising the pelagic Algæ.

5 GEORGE V., A. 1915

a relatively long body and a still longer muscular tail; but the latter, instead of being continuous with the hinder end of the body, is inserted at the centre of the body, at a right angle, like a handle of a hammer or pick-axe. It is more at home in the open sea than in confined, inshore waters, though the specimens were perfectly healthy. This form is found both in the Arctic and in the Antarctic Oceans, and is therefore described as being bipolar.

Another characteristic Arctic Appendicularian, *Oikopleura labradorensis*, was not observed at St. Andrew's. Both of these species extend their range in the spring and summer, when the polar water spreads southwards; and at that season they have been taken in the North Sea (H. Lohmann).

The principal factor governing the distribution of the organisms of the Plankton is the temperature of the sea; this is even more effectual than the salinity of the water.* From the open sea where the salts are dissolved at a concentration of 35 per cent, *Fritillaria borealis* is periodically transported to the brackish water of the Baltic Sea with a salinity of 15 per cent. It is therefore not so surprising as it appeared at first sight, to find this delicate form near the mouth of the St. Croix river at St. Andrew's, more especially since the Appendicularians are known to feed largely upon the Peridinians.

In what has been said it is implied that the physiological aspect of the Plankton is that which concerns the practical questions of nutrition and distribution. What is known as its morphological aspect cannot be regarded as having any bearing upon the fisheries, except in respect of the fundamental distinction between the zooplankton and the phytoplankton. The true relationship of any planktonic species have nothing to do with their food-value to other species. In this connection the contrast between morphology and physiology is exactly analogous to that which, as we have seen, can sometimes be drawn between observation and experiment. In a complete scientific presentation of the subject it is impossible to divorce the one from the other, especially if we desire to penetrate into the obscure origins of the Plankton.

It may therefore be of interest to recall that Haeckel looked upon the Appendicularians as representing the common stem-form or ancestral stock both of sea-squirts (fixed Tunicata) and of fishes (Vertebrata). It is worth while examining this opinion from the standpoint of the marine Plankton as a whole, of which the Appendicularians are one of the most constant constituents. Haeckel's view involves the assumption that they are primarily pelagic; and as this assumption is the crux of the entire question, it is certainly one which should be scrutinised with the utmost circumspection.

In dealing with this matter it should be borne in mind that adaptation is the first consideration and that it is not necessary, at the outset of the discussion, to dwell upon details of structure or life-history. In very many instances (*e.g.* the pelagic Mollusca) it is usually taken for granted that the pelagic habit of the organisms of the Plankton is a special adaptation from a bottom dwelling or benthonic life to a surface-frequenting or planktonic life.

---

* Carl Chun. Die Beziehungen zwischen dem arktischen und antarktischen Plankton. Stuttgart, 1897.

Indeed so far as the zooplankton is concerned, it is an open question whether the planktonic habit is not in every case the consequence of secondary adaptation. In any case it is obvious that it must be an arbitrary proceeding to select one of the leading planktonic types as representing a primarily pelagic, ancestral stock.

Two kinds of plankton are to be distinguished by their situation, namely, the oceanic and the coastal or neritic* plankton. These associations naturally merge into one another, but the latter is much the richer. It seems natural to suppose that the oceanic plankton is but an expansion of the neritic plankton, just as southern forms are carried northwards by the Gulf Stream, while northern forms are borne southwards by the Labrador current.

The next suggestion which might occur to the mind is one that cannot be advanced definitely without a prolonged analysis; and it is only too likely that even then it would fail to carry conviction. It may nevertheless be proposed as a thesis that the neritic zooplankton is to be derived ultimately from the littoral fauna.

There are two kinds of large and well-known jelly-fishes or umbrella-shaped medusae, several inches in diameter, which are commonly seen floating near the surface in St. Andrew's bay or left stranded on the beach by the receding tide. One of them is the common American *Aurelia flavidula* with its four horse-shoe-shaped rosettes; the other is called *Staurostoma laciniatum*, with a simple St. Andrew's cross showing through the transparent disc. The first recorded specimen of the *Staurostoma* was brought to L. Agassiz in a jar containing *Aurelia* taken in Boston harbour in 1849; he says he had scarcely ever valued any discovery more highly.

Besides these true jelly-fishes there is another class of pelagic animals which bear some resemblance to medusae from which, however, they differ in shape as well as in many more fundamental characters. They are usually barrel-shaped and, running lengthwise from one end of the barrel towards the other, there are eight equidistant rows of vibratile, comb-shaped flappers, whence the class was named Ctenophora by Eschscholtz in 1829.

The Ctenophora are the most exquisite creatures imaginable and always excite the unbounded admiration and astonishment of those who see them alive for the first time. The body is generally as clear as glass, of filmy consistency, and sometimes it will undergo complete liquefaction so that nothing visible is left behind. They were represented in St. Andrew's bay, at the time of my visit,† by a form which was described in 1849 by Louis Agassiz from examples collected off the coast of Massachusetts, under the name *Bolina alata*.

In this species the fluids of the body are so exactly adjusted to its conditions of life, being separated from the surrounding water only by a cellular membrane

* This is one of Professor Haeckel's useful terms; from Nerites, the son of Nereus and grandson of Pontus and Gæa. It differs from littoral in that the latter term refers to the inshore bottom-dwelling forms. The entire plankton of St. Andrew's bay, considered as a unit, belongs to the neritic group.

† The specimens were placed at my disposal by Dr. A. G. Huntsman to whom they were familiar and who found a shoal of them at about 7 a.m. in shallow water, at the foot of the wharf belonging to the Biological Station, during a very low tide on August 14th.

5 GEORGE V., A. 1915

of extreme tenuity, that any alteration in the density of the water, as for example when a preservative liquid is added to it, causes speedy disruption.

An interesting analogy of distribution is presented by *Staurostoma* and *Bolina*: *S. laciniatum* from the north Atlantic coast of America is as nearly related to *S. mertensi* from the coast of Alaska, as *B. alata* is to *B. septentrionalis* from Behring's Straits. All of these species are doubtless descended from circumpolar forms which have streamed down along the different coast lines from the Arctic ocean.

The neritic plankton is enriched at certain seasons by free-swimming larval forms belonging to the littoral fauna. One of the most bizarre of these was first described by Johannes Müller as *Actinotrocha branchiata*, and was subsequently shown by A. Kowalevsky to be the larva of a worm called *Phoronis* which lives in sand-tubes. Without entering into details, it may be stated that the chief peculiarity of this form is that in effecting the transformation from the larval to the adult condition, the body becomes, up to a certain point, turned inside out. One example of *Actinotrocha*, identified with a species previously described from

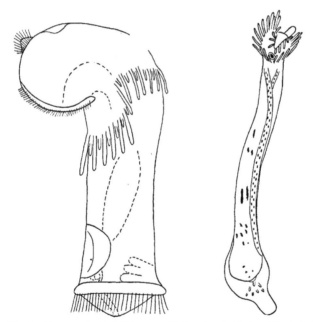

Fig. 1. The figure to the left is a magnified outline sketch from life of *Actinotrocha Brownei* [de Selys-Longchamps]; that to the right is a similar sketch of *Phoronis Brownei* immediately after the metamorphosis. Observed at St. Andrew's, New Brunswick, August 19th, 1912.

SESSIONAL PAPER No. 39b

Plymouth, England, appeared in the tow at St. Andrew's on August 19th. Whilst it was under examination in a glass vessel, the eversion took place, and the previously free (planktonic) larva was converted into the sedentary (benthonic) worm.

Almost equally rococo are the free-swimming larvæ of the common starfishes at St. Andrew's. Whilst not very abundant, yet they were detected on most days in the plankton between August 10th and 20th. These larvæ possess many long, trailing arms. There are in all fourteen arms arranged in two sets of eleven and three respectively. The eleven arms of the first category are simple, elongated, tentacle-like processes, slightly clubbed at their orange-coloured extremities. Along their borders, up one side and down the other, is a narrow refringent zone clothed with vibratile cilia. The ciliated zone or band is continuous at the bases of the tentacles from one to another, excepting that the two tentacles immediately in front of the mouth have their own band continuous with the upper lip of the mouth; while the median anterior tentacle together with the eight posterior tentacles have their band continuous with the lower lip of the mouth. Thus there is a pair of pre-oral tentacles with a pre-oral ciliated band; and a series of nine tentacles (one median and four pairs) with a post-oral ciliated band skirting them from end to end. It is called post-oral because most of it lies behind the mouth, although as mentioned, it is continued over the median anterior tentacle. ·

Occupying the area of the pre-oral lobe in front of the two pre-oral tentacles, there are three arm-like processes crowned with adhesive papillæ, and in the middle of the pre-oral lobe, between the bases of the arms, there is a somewhat oval thickening with a few small papillæ irregularly scattered around it; this is a median adhesive disc or suctorial plate which can be retracted, *i.e.*, the area which contains it can be pulled back. Of the three adhesive arms, two are ventral, occurring as a pair, and actually arising in the axils of the pre-oral tentacles; the third is median and dorsal. The pre-oral ciliated band is not continued upon the adhesive arms but ceases on each side at the base of the paired arms. This interruption of the pre-oral ciliated band was observed in a young larva which possessed neither arms nor tentacles.

The cilia are the means of locomotion which consists in an even gliding through the water. The tentacles themselves, although mobile, are not organs of progression, but are sensitive balancers, assisting to suspend the larva in the water. They would represent, therefore, a temporary adaptation to the pelagic habit. When the time of metamorphosis approaches, the tentacles become flaccid and wrinkled, the ciliated rim begins to lose its continuity, and the larva sinks to the bottom where it adheres by means of its adhesive processes and the median sucker.

Meanwhile the body of the young starfish has been developing in the hinder region of the larva. There is still a certain amount of obscurity surrounding the disappearance of the provisional larval structures and the definite assumption of the starfish form. Soon after the fixation of the larva, the young starfish once more becomes free, but this time as a denizen of the littoral zone of the sea-bottom.

5 GEORGE V., A. 1915

Here again, as with *Actinotrocha*, the free-swimming larva gives place to the shore-dwelling adult.

The starfish is known as a serious enemy of oysters, but there can be no question that the larvæ are a valuable commodity of the neritic plankton.

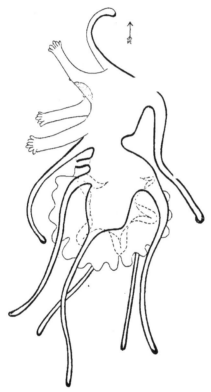

Fig. 2. *Brachiolaria* larva identified with that of the common starfish (*Asterias vulgaris*); represented in the attitude of swimming, in the direction indicated by the arrow. The outline of the developing starfish occupies the hinder end of the larval body; the position of the mouth is seen behind the tentacle (one of a pair, the other not shown) which follows the three adhesive arms.

The fixed Tunicata or sea-squirts, to which reference has been made above, produce tailed larvæ known as Ascidian tadpoles. At the front end of the body these tadpoles are provided with three adhesive processes which in some cases are borne upon relatively long stalks. Of these processes one is median and

SESSIONAL PAPER No. 39b

dorsal, the other two forming a ventral pair. Here, therefore, we have an apparatus of fixation precisely comparable to that of the starfish larva, which, by the way, is known as *Brachiolaria,* on account of its adhesive arms. The three adhesive arms of *Brachiolaria* and the three adhesive processes of the Ascidian tadpole are only comparable as physiological mechanisms of like nature, though of independent origin.

The few remarks offered, bearing upon certain aspects of the zooplankton of St. Andrew's bay, as it appeared in July and August, although without any pretence of completeness, may serve as an indication of the results that would follow from an extended and organised survey embracing the whole of Pasamaquoddy bay and continued from year to year. To make such a survey effective, what may be called the resident or benthonic (bottom-dwelling) fauna and flora should be known with some degree of thoroughness; and in fact that is in course of being worked out by the temporary staff of biologists at the station.

Special attention would naturally be given to the leading planktonic types; and an attempt would be made to bring the records into line with the existing data concerning the northern plankton. This is really an ambitious scheme requiring much preparation in matters of detail; but it offers a programme worthy of consideration.

The microscopic plants or algæ which make up the phytoplankton are enormously abundant in our region, and as these constitute the fount of all life in the sea, their importance for the fisheries is clear enough. The northern diatoms have been observed to accumulate on the under surface of the ice where they form a vast brownish incrustation [E. Vanhöffen]. If the biological station could be kept open the year round, there is no doubt that much could be found out as to what goes on under the ice-sheet.

In conclusion it may be stated with confidence that the seasonal, diurnal, and tidal fluctuations of the plankton in St. Andrew's bay, would well repay a more intensive investigation than has hitherto been accorded.

II.

THE PLANKTON DIATOMS OF THE BAY OF FUNDY.

By L. W. Bailey, LL.D., F.R.S.C., Etc.

Emeritus Professor of Natural History and Geology, University of New Brunswick,
Fredericton, N. B.

(Plates I, II, and III.)

The Plankton Diatoms constitute a group of peculiar interest in a division of miscroscopic plants which, in all its branches, afford to the naturalist a field of pleasurable and instructive study.

The term "Plankton" is one which is applied to the entire assemblage of minute, mostly microscopic organisms, including both plants and animals, which are found, often in vast numbers, swimming or floating freely, in the waters of ponds, lakes or in the open ocean, having no connection with the solid earth, but deriving their food supply from the medium in which they live. So far as the animal kingdom is concerned this floating population embraces members of several groups, such as Infusoria, Foraminifera, and Radiolaria, together with larval forms of Echinoderms, Annelids, Polyzoa, Crustacea and Mollusca, but, as regards plant life, this is confined, with the exception of the small group known as the *Peridineae*, to the family of the *Diatomaceae*. These are minute unicellular algæ of which the most notable peculiarity is the secreting of a siliceous shell or lorica, determining their form and strength, and which is practically indestructible.

Existing as they do in such enormous numbers in the purer oceanic waters, the plankton Diatoms constitute a very large part of the food of higher oceanic organisms, as is proved by the fact that they are found in such large numbers in the stomachs of marine animals such as echinoderms, crustacea, molluscs and even fishes. Even where these animals are not themselves direct plankton feeders, like the members of the herring and mackerel families, they nevertheless rely for their nourishment upon smaller animals, Copepods and the like, which are thus supported, so that the Diatoms may very properly be regarded as affording the basic food supply for marine life, even in its highest forms.

The features which especially characterize the so-called Plankton Diatoms are those of their adaptation to a life of *flotation*. This is partly effected by a relative reduction in the amount of silica contained in their cell walls, reducing their specific gravity, but mainly in other ways, such as by the nature of their forms or the development of expedients which favor buoyancy. Thus in certain genera (*Coscinodiscus, Actinocyclus, Actinoptychus* &c.) the form is that of a nearly flat or slightly convex disc, exposing a large surface in proportion to the thickness of the

- 5 GEORGE V., A. 1915

cell; in others (such as *Biddulphia, Melosira, Skeletonema, Rhabdonema, Tabellaria* &c.) the frustules, though individually small, are attached to each other to form filaments or chains; while in still others, and these the most characteristic, the desired object is attained by the development of processes, arms or horns, projecting from the cells, and which are often of extravagant length—(*Chaetoceros, Bacteriastrum, Nitschia,* &c.)—the presence of oil globules in the cells probably also assists in certain cases.

The Biology of the Plankton-Diatoms.has, until within a few years, been the subject of comparatively little study; but now that their importance in connection with their relations to the support of other forms of life has been generally recognized, observers at the principal Biological Stations, both in America and Europe, have been giving them much attention. Both their classification and nomenclature are, however, still in a very unsatisfactory state, and the literature of the subject is comparatively small. So far as New Brunswick is concerned absolutely nothing has previously been published, though references to some of the characteristic species have been made by the writer in earlier articles on the Diatoms of the New Brunswick seaboard. Dr. Ramsay Wright has also described and figured some of the species found by him in the Plankton of Canso, N.S. (Contributions to Canadian Biology, 1902-1905).

The materials upon which this paper is based were obtained mainly from Passamaquoddy Bay and the adjoining waters of the Bay of Fundy, and in connection with the work of the Marine Biological Station at St. Andrews. In making the collections very fine silken tow-nets were employed, and their contents examined while still fresh and in their proper element, chemical treatment being apt to disintegrate the concatenate forms, while mounting in balsam will often cause delicate forms, though readily seen in water or dry, to become nearly or quite 'invisible. It is, however, often necessary to treat the material, after removal of salt by washing and decantation, with Nitric acid, in order to remove the vast number of Copepods and other organisms with which they are associated as well as foreign matters adherent to the Diatoms themselves.

The most frequent accompaniments of the plankton-diatoms, in addition to the Crustacea, are silico-flagellate Infusoria of the genera *Amphorella* and *Tintinopsis*.

The literature available to the author in his study of the Plankton of the New Brunswick waters includes the following:—

Smith's Synopsis of British Diatomaceae.

Van Heurck's Diatomées de Belgium.

Wolle's Diatoms of North America.

Nordisches Plankton—Brandt and Apstein—Kiel.

Brightwell—On the Filamentous, Long Horned Diatomaceae. (Quarterly Microscopical Journal, London, Vol. IV.)

Of these the first three are of a general nature. Only the last two relate especially to the Plankton. In an article by Prof. Ramsay Wright in "Contributions to Canadian Biology," published in 39th "Annual Report of the Department of Marine and Fisheries—Canada" some descriptions and figures of the forms occurring at Canso, N.S. are given, and these have been of service in the study of the New Brunswick forms.

In the following account of the species entering into the composition of the Plankton those which may be regarded as especially characteristic of the latter, exhibiting the most marked adaptations to a life of flotation, will be first considered, to be followed by those which, though less marked in this respect, are nevertheless of general or frequent occurrence.

Prof. W. A. Herdman, F.R.S. of Liverpool University, who has been in charge of special plankton investigations around the Isle of Man, gives six genera as those which are especially characteristic of the plankton flora of that region, and it is interesting to note that, with one possible exception (*Lauderia*) all of these occur and in most instances are abundant in the Bay of Fundy and adjacent waters. These genera are *Chaetoceros, Rhizosolenia, Biddulphia, Coscinodiscus, Thalassiosira* and *Lauderia*, to which may be added *Skeletonema, Bacteriastrum* and *Asterionella*.

DESCRIPTIONS OF GENERA.

CHAETOCEROS. This genus is probably the most remarkable one among the Plankton Diatoms, and exhibits the widest divergence from the ordinary type of these plants, leading some authors to doubt whether they should really be considered as Diatoms at all. Their most noticeable feature is that of their being provided with spines, awns, or bristles, which, though usually very thin, greatly exceed in length the diameter of the frustule to which they are attached, and sometimes exceed the latter fifty times or more. The frustules are usually arranged in chains, embracing a considerable number of individuals, which may be united either by a band or cingulum, or by the interlocking of the horns. These latter vary in number from two to four, and most of them are arranged laterally or at right angles to the chain, being sometimes attached to or proceeding from the usually convex valves of the frustules above and below, so as to interlock and thus add strength to the chain, or in some instances from the cingulum, or from both. In addition to the lateral bristles there are often terminal ones as well, usually two in number which are either longer or shorter than the others, and may also differ from them in other respects as well.

Though usually single, the spines may sometimes bifurcate near their point of origin, and while commonly smooth throughout, are often spinous or serrated or present the appearance of bearing imbricated scales. Occasionally they seem to have a spiral twist, like a screw. In the case of the terminal awns, though usually bristle-like, they are also sometimes clavate or somewhat spatulate, suggesting comparison with the antennae of lepidopterous insects. Some awns are stout and rigid, others fine or hair-like and flexible. Their length seems to be connected with their age, the terminal awns being also often much longer than the lateral ones. The angle of divergence of the awns and the disposition of the chromatophores have both been regarded as of diagnostic importance, but the observations of the writer hardly accord with this view, different frustules of the same chain exhibiting considerable diversity in both these respects. The shape of the cells and therefore

5 GEORGE V., A. 1915

of the intervening spaces also differ at different seasons of the year. Finally sporangial frustules also differ considerably from the ordinary ones, the valves being provided with short branched processes, and forms of this character have, as in the case of *Dicladia*, been constituted into different genera, though they are now believed to be auxospores of the genus *Chaetoceros*.

The genus *Chaetoceros* embraces a considerable number of species, but these have as yet been very imperfectly differentiated, and much confusion exists as to their identity and synonymy. In the descriptions which follow, and in Plates to which these refer, only such forms are included as the writer has himself observed in the coastal waters of New Brunswick and mainly from the Bay of Fundy, with suggestions as to their probable indentity with those found elsewhere.

PLATE I. FIG. I. *Chaetoceros decipiens.—Cleve.*

This is perhaps the most common of the species found about Passamaquoddy Bay, as Prof. Ramsay Wright reports it to be at Canso, Nova Scotia. The frustules are quadrangular, with concave faces, producing between adjoining cells a vacant space which is elliptical or approximately hexagonal in outline, while the lateral bristles arise from the points of contact between the frustules, and for short distances may be confluent. The bristles are four in number at each point, but of these only two belong to each frustule. They are filiform and of only moderate length, perhaps three or four times that of the diameter of the frustule. The terminal bristles are shorter, bearing transverse striae, and, though divergent at a considerable angle, are more nearly parallel than the lateral ones with the axis of the chain.

PLATE I. FIG 2. *Chaetoceros decipiens.—Cleve, (Var).*

This form differs from the preceding in the much closer approximation of the frustules, together with the very slight concavity of their opposed surfaces, the intervening space being narrowly linear. Two smooth and filiform lateral awns arise on each side of the junction lines, diverging at an angle of about 30°, and, by intersection with their fellows, produce the appearance of lattice-work. The terminal awns have not been observed. The form is believed to be a variety of *Ch. decipiens*, Cleve, the shape of whose cells, and therefore of the interval separating the latter, are known to vary with the seasons and other conditions.

PLATE I. FIG. 3. *Chaetoceros.*

This form resembles that of *Ch. decipiens, Cleve*, in the general form of the frustules and in the arrangement of the horns or bristles, but the terminal awns are clavate and symmetrically curved to enclose a space forming about one half of a broad ellipse. The chromatophores are condensed in the centre of each frustule. In the clavate form of its terminal awns it resembles what some authors have described and figured under the name of *Ch. dicladia*, but these are now usually regarded as varieties of *Ch. decipiens*.

SESSIONAL PAPER No. 39b

PLATE I. FIG. 4. *Chaetoceros* species?

This form also resembles *Ch. decipiens* Cleve in the cup-like form of its frustules and in the number and attachment of the lateral awns, but the terminal curved awns are not clavate, and the lateral bristles, which are spinous, after slight diver-gence at the base become nearly parallel.

PLATE I. FIG. 5. *Chaetoceros.*

This specimen has the general form and structure of *Ch. decipiens*, Cleve, but in certain of the cells (primary) are inner transverse partitions which project in the form of two high cone-shaped processes, each of which at the apex bears a conspicuous dichotomously divided spine, which is very characteristic, while the other (or secondary) cells are almost flat. It is to forms like these that the name of *Dicladia mitra* has been given, but they are now thought to be resting spores of *Ch. decipiens.*

PLATE I. FIG. 6. *Chaetoceros.*

This form is probably related to the last, but between the two cingula the lateral surfaces are conspicuously undulate, with a prominent median cone on either side, separating two equally marked depressions, while upon the ends of the frustule two diverging filiform spines arise from the centre of each cingulum.

PLATE I. FIG. 7. *Chaetoceros.*

This is probably also a series of resting spores of *Ch. decipiens* but the branching processes are more numerous.

PLATE I. FIG. 8. *Chaetoceros.*

In this case there are also numerous processes, arising from a single convex enlargement or dome, but these are alternately long and short and unbranched.

PLATE I. FIG. 9. *Chaetoceros chriophyllum—Cast.*

This form differs from the preceding in the fact that the quadrangular valves of the frustules, instead of being flat or concave, are convex, while the setae or horns, which are of great length, arise from towards the middle of the valves and not from the corners, being at first turned downwards and then, somewhat abruptly, curving upwards, the single awns on either side making with those of the opposite side nearly a right angle, while the terminal awns are much shorter, and diverge at an angle of about 38°. Except in this latter character the species bears much resemblance to *Chaetoceros volans* of Cleve. It is probably a variety of *C. chriophyllum—Castracane.*

5 GEORGE V., A. 1915

PLATE I. FIG. 10. *Chaetoceros.*

In this form the frustules are in lateral view elliptical in outline, the lateral bristles, which are smooth, arising without curvature from between the convex apposed surfaces of the valves. The terminal awns are straight and filiform, diverging at an angle of about 45°. It is probably another variety of *C. chriophyllum—Cast.*

PLATE I. FIG. 11. *Chaetoceros Peruvianum—Bright?*

This form is remarkable in the fact that the awns, which arise in pairs from each joint of the chain, are noticeable for their length and stoutness, as well as for their spinous character. The portions of the spines nearest to the chain are small, numerous and thick-set, but, like the spines themselves, become larger as the distance increases, as well as more widely separated, The terminal appendages are much shorter, smooth (?) and sigmoid, resembling a pair of horns. The form would seem to be nearly related to *Chaetoceros Peruvianum, Bright*, of the North Atlantic.

PLATE II. FIGS. 1-7. *Rhizosolenia.*

This genus differs from the ordinary type of Diatoms even more widely than does the genus *Chaetoceros*, the most noticeable features being the great elongation of the cylindrical frustules, the crossing of the latter by distinct transverse lines or annuli, and the frequent presence of a calyptriform base, terminating in one or more short but conspicuous spines.

At least three different species of the genus have been observed in the waters of the Bay of Fundy and the Gulf of St. Lawrence.

PLATE II. FIGS. 1-2. *Rhizosolenia setigera—Bright.*

What is believed to be this species has been observed in Passamaquoddy Bay, St. John Harbor and Bathurst, as well about the shores of Prince Edward Island. The figures here given are taken from those of Prof. Ramsay Wright, who refers especially to the peculiar spear-blade-like enlargement about the middle of the length of the terminal spines (Fig. 2). I have also observed this feature in some instances, but it does not seem to be a constant characteristic, and is given in only one of Dr. Ramsay Wright's figures.

PLATE II. FIGS. 3-4. *Rhizosolenia styliformis—Bright.*

Prof. Ramsay Wright refers to this species as being the most abundant at Canso, N.S., but on the New Brunswick coasts it seems to be less common than the preceding species. It has been found as yet only in St. Andrews Harbor. Figs. 3 and 4 show its general appearance, as well as the peculiar character of the cell junctions. (Fig 4a).

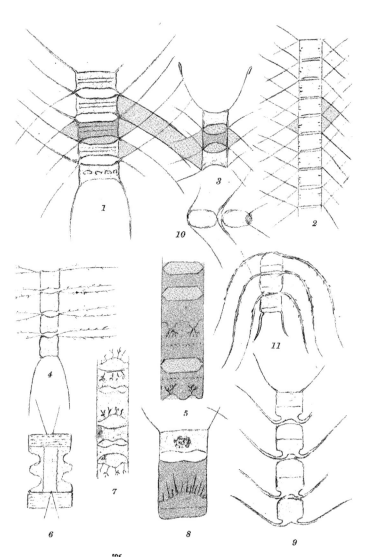

Plate II. Fig. 5. *Rhizosolenia.*

In its narrowly linear form and in the absence of lateral bristles the form here figured resembles *Rh. setigera*—but the entire frustule is divided into oblique segments of which the terminal ones are attenuated in opposite directions to be produced into fine spinous processes.

Plate II. Fig. 6. *Rhizosolenia.*

This form is considerably broader than the last, but lacks the oblique transverse lines and terminates in a more pronounced calyptra, of which the spine, as in the last form described, is turned to one side. The frustule is also distinctly punctate.

Plate II. Fig. 7. *Rhizosolenia.*

The form here figured is noticeable for its wide diameter and for the fact that the annuli curve from either side to a central or axial line, while the terminal setae, the only ones present, are quite short and spine-like, recalling Ehrenberg's first description of the genus, which is described as "attenuate and multifid, as if terminating in little roots." It may be a variety of *R. imbricata, Bright.*

Plate II. Fig. 8. *Rhizosolenia?*

In its general aspect the form here figured is that of a *Rhizosolenia,* and I have little doubt that it belongs to that genus. Its most marked feature is the apparent contraction of the ends of the valves, suggesting the idea of puckering. At their ends, in addition to a central sharp spine of considerable length, are two little teeth or processes, projecting laterally. In the general form and terminal spines, the species bears a close resemblance to that figured in Gran's Nordische Plankton as *Ditylium Brightwellii,* and, as as indicated below, may have some relationship with the latter. In Fig. 8b, two frustules are shown as connected and with the terminal awns oblique and decussating. The endochrome in these was gathered in circular masses at the points where the frustules approach.

Plate II. Figs. 9-10-11. *Triceratium.*

In connection with the forms last described, those represented in Figs. 9-11 are of very great interest. Thus the resemblance between Fig. 9 and Fig. 8 will be at once apparent, so far as the general outline is concerned, but at one extremity of the delicate gelatinous (?) cylinder in Fig. 9 is a triangular and at the other a quadrangular enclosed form, both of which recall variant forms of the polymorphic genus *Triceratium.* Mr. Brightwell, in his paper illustrating this genus, gives somewhat similar figures in the case of species *Triceratium undulatum,* the valves as in this case being enclosed in a cylinder and, again as in this case, bearing a prominent central spine. This gives strong confirmation to the view that

5 GEORGE V., A. 1915

Amphitetras, Amphipentas, etc., are but varying forms of *Triceratium* and that this is closely connected with *Rhizosolenia*. Whether *Ditylium*, as first described by the writer should also be regarded as a related form, seems to him more doubtful— Dr. Mann regards them as quite distinct.

PLATE II. FIG. 12. *Skeletonema costatum—Grev.*

In the general form and structure of its frustules the species of this genus closely resemble those of the genus *Melosira*—especially *M. nummuloides* or *M. varians*; but differ in the much wider separation of the frustules and the presence in the intervening space of numerous fine hair like processes, connecting the valves. The chains thus formed are of considerable length and well adapted to flotation. The specimens found were from Chamcook Bay and Deadman's Harbor, as well as elsewhere about the Western Isles, and are not uncommon. The species found by Dr. Ramsay Wright at Canso, are referred by him to *S. costatum* in reference to the ribbed sides of the slender cylinders, and it is altogether probable that the New Brunswick forms are of the same species.

PLATE II. FIGS 13-14. *Thalassiosira—Cleve.*

The genus *Thalassiosira* is very abundantly represented in the waters adjacent to the New Brunswick coast about the entrance of the Bay of Fundy, especially in early summer, when they often form a considerable portion of the plankton. Their generic identity is easily recognized by the somewhat wide separation of the frustules and the fact that these are connected into chains solely by the agency of a usually single fine thread, often of considerable length; and they probably include several species, but, from the want of sufficient literature, the writer has not in all cases been able to identify these with certainty.

FIG. 13. *Thalassiosira Nordenskioldii—Cleve.*

A form which is believed to be this species and which corresponds quite closely to the latter as described and figured by Prof. H. H.. Gran (Nordisches Plankton 1905) is very abundant in early June in the waters of the Western Isles, being usually accompanied by *Chaetoceros decipiens* and *Rhizosolenia*. The frustules are noticeable for their distinctly octagonal outline, from which at the four external angles project minute spines, while the connecting filaments do not usually exceed and are often less than the smaller diameter of the cell. The chromatophores are somewhat variously disposed, but usually along the interior of the cell wall, those of one side being connected with the other by a slight isthmus.

Fig. 14. *Thalassiosira.* The distinctive feature of the form here represented is the shape of the frustules, these being in the form of lengthened cylinders, which are connected into chains by threads arising from the centres of their opposite circular ends. The chromatophores, which are minute and granular, are condensed at the same points. No external processes were observed.

PLATE II

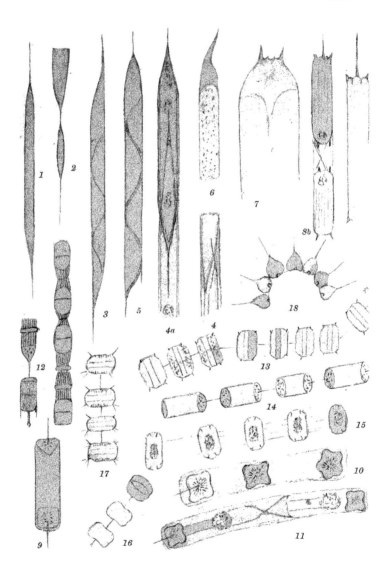

Fig. 15. A group resembling *Th. Nordenskioldii*, but having the cells connected not by one but by many threads. It may be *Coscinosira polychorda*.

Fig. 16. *Thalassiosira*. A series of biconcave discs, connected by a single fine thread or filament. It may be *Th. hyalina* of Grun.

Fig. 18. *Asterionella*. Forms of this beautiful genus are not very uncommon in the plankton of Passamaquoddy Bay and adjacent waters. In the specimen represented six frustules were observed as grouped in a semi-circle by the attachment of their bases, each frustule being cuneate or triangular with the apex of each prolonged into a rigid spine. The species may be *As. Japonica*, *Cleve*, which occurs in the North Sea, but the spines are more clearly differentiated than in that species as figured by Gran.

PLATE III. FIG. 1-2. *Chaetoceros boreale—Bail.*

I have not myself observed this species with certainty, but it is common in the North Atlantic and is doubtless to be found in the waters of the Bay of Fundy. It is mentioned by Dr. Ramsay Wright as seen by him at Canso, N.S. The figure here given is taken from that of Dr. Gran in the Nordisches Plankton.

FIG. 3. A chain of auxospores probably of *Chaetoceros decipiens*.

FIG. 4. This form, like many species of *Thalassiosira*, bears much resemblance to a *Melosira*, but, unlike the forms referred to this genus, has the cells connected not by a single thread but by several. In this respect it resembles the species described and figured by Gran as *Coscinosira polychorda*. Only one specimen was seen, gathered in early June from Deer Island. Without closer examination of the cell-structure its identity could not be determined with certainty.

FIG. 5. This species may be a variety of *Thalassiosira gravida*, Cleve.

FIG. 6. A chain of frustules of *Thalassiosira Nordenskioldii*.

FIG. 7. This is apparently a *Thalassiosira*, but has not been determined.

FIG. 8. This would appear to be *Th. gravida*, Cleve, the slightly separated quadrangular frustules bearing bristles at the slightly truncated angles.

FIG. 9. This form has been figured and described under the name of *Dicladia capreolus*, but is probably only an auxospore of some species of *Chaetoceros*.

FIG. 10. *Sydendrium diadema* Gr. This form is occasionally, but rarely met with. It belongs to the *Chaetoceros* family.

PLATE III. FIG 11. *Actinoptychus undulatus—Kutz.*

This beautiful form is too well known to require description here. It is one of the most common forms in the coastal waters of New Brunswick and Prince Edward Island, and is to be found in nearly all gatherings therefrom.

5 GEORGE V., A. 1915

Fig. 12. *Hyalodiscus subtilis—Bail.*

This species is not uncommon in plankton gatherings, both from the Bay of Fundy and Gulf of St. Lawrence, but its representatives are usually much smaller than those of the same species found at the more southern points on the Atlantic sea-board.

Fig. 13-14. *Coscinodiscus—Ehr.*

This genus is more abundantly represented than any other, except perhaps *Chaetoceros*, in the plankton flora of the Bay of Fundy as elsewhere. The species most commonly met with are *C. asteromphalus* (Fig 13) of which *C. oculus-iridis* is a variety, *C. eccentricus*-Ehr. (Fig. 14) and *C. radiatus* Grun, though quite a number of others have been observed.

Fig. 15-16. *Grammatophora*. This can hardly be regarded as a true planktonic genus, being usually, perhaps always, attached, and having a somewhat littoral habitat. Yet scattered frustules and sometimes chains are not uncommon in plankton gatherings. The species most commonly met with are *G. marina* (Fig. 15) and *G. serpentina* (Fig 16)

PLATE III. Fig 17. *Synedra.*

The genus Synedra is not uncommon in planktonic gatherings, being well adapted by its lengthened form to a life of flotation. This feature is most pronounced in *Synedra undulata, Bail*, a species which, while rare in the waters of the Bay of Fundy, is not very uncommon in those of the Gulf of St. Lawrence and Prince Edward Island. In addition to its almost extravagant length it has the further peculiarity, to which its name refers, of being corrugated or undulatory through the larger part of that length, thus adding materially to its strength.

PLATE 3. Fig 18. *Nitschia.*

This genus exhibits the same adaptation to flotation as the preceding genus, the length being greatly disproportionate to its breadth. This is seen more or less conspicuously in all the Nitschias, but is especially marked in *N. longissima* (Fig 18) of which all but the central part is extremely narrow and spinous, the total length, as in the figure, being often nearly twenty times its widest diameter.

PLATE III. Fig. 19. *Biddulphia.*

This is eminently a planktonic genus, its representatives being found in most tow-net gatherings. The individual frustules are provided with more or less prominent horns, aiding flotation, but this is probably much more effectually brought about by the adherence of the frustules in long chains, sometimes containing twenty or more individuals. The four species represented are *B. aurita*,

PLATE III

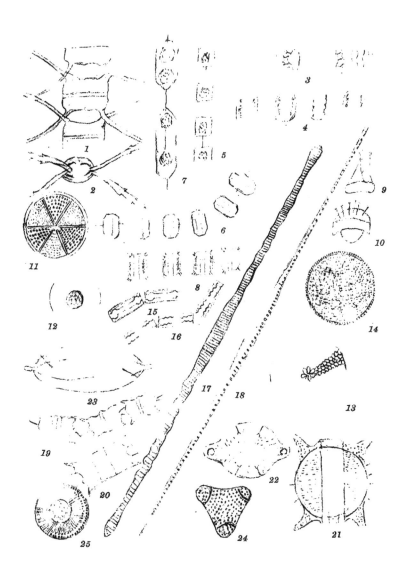

B. laevis, B. rhombus and *B. Mobiliensis* (= *B. Baileyi*) Figs. 20-23 of which the latter, at some points along the Bay of Fundy coast, makes up nearly the whole of the plankton, occurring in vast numbers. *B. aurita* (Fig. 19) is also common, while *B. laevis* and *B. rhombus* Figs. 21-22 are comparatively rare.

The following genera and species of Diatoms, though less distinctly planktonic than the preceding, are met with in more or less frequency in tow net gatherings.

Fragillaria capucina—Des.	Pleurosigma angulatum—W.S.
Acnanthes longipes—Ag.	Pleurosigma attenuatum—W. S.
Acnanthes subsessilis—Kutz.	Pleurosigma Balticum—W. S.
Amphiprora alata—Kutz.	Pleurosigma fasciola—W. S.
Bacillaria paradoxa—Gmel.	Pleurosigma strigilis—W. S.
Campylodiscus	Pleurosigma strigosum—W. S.
Cocconeis scutellum—Ehr.	Pleurosigma acuminatum.
Cyclotella compta—Kg.	
Epithemia musculus—Kutz.	
Grammatophora marina—Kutz.	Rhabdonema arcuatum—K.
Grammatophora serpentina—Ehr.	Rhabdonema Adriaticum—K.
Isthmia nervosa	
Licmophora Lyngbei—G.	Schizonema crucigerum—W. S.
Melosira nummuloides—Kutz.	Stauroneis anceps—Ehr.
Melosira Borerii—Grev.	Stauroneis obliqua.
Navicula Smithii	Striatella unipunctata—Ag.
Navicula didyma—Kutz.	Surirella gemma—Ehr.
Navicula viridis—Kutz.	Surirella ovalis—Breb.
Nitschia bilobata—W. S.	Synedra ulna—Ehr.
Nitschia closterium—W. S.	Synedra undulata—Bail.
Nitschia sigmoidea—W. S.	Synedra longissima.
Nitschia vermicularis—Hanty.	Synedra radians—W. S.
Nitschia sigma—W. S.	Tabellaria.
Nitschia longissima—Ralfs.	

No quantitative measurements have as yet been made to determine the relative abundance of plankton Diatoms at different localities in New Brunswick or at different seasons. It is, however, interesting to note in this connection the results of observations made by Prof. W. A. Herdman, F.R.S., and others in the waters about the Isle of Man. Dr. Herdman states that in a single haul made in the latter part of April, 49 millions of the genus *Chaetoceros* were found. The maximum however, was in August, while in late September the number had fallen to 3 millions, and in October was only one million. *Rhizosolenia* was feeble in April, reached its maximum (13 millions) in June, was absent in August, and had a second maximum (470,000) in late September. *Lauderia* (*L. borealis*) was rare until April, was absent in August, reached a maximum (20 millions) on April 22, and was rare throughout the summer. *Biddulphia*, chiefly *Bid. Mobiliensis* had its maximum in April.

Of Diatoms in general there was a marked minimum in August while the maxima were in August and June, the former consisting chiefly of *Chaetoceros* and

5 GEORGE V., A. 1915

the latter of *Rhizosolenia.*   In September there is a second small rise of *Chaetoceros,*
but *Rhizosolenia* was nearly wanting.   The April rise is supposed by Sir John
Murray to be due to the increasing amount of sunlight at that time, but may
be due to variations in food supply.

As the Biological Station at St. Andrews was open only during the months
of July, August and part of September, the opportunity for comparisons of this
kind was wanting.   It will be obvious, however, that if similar variations in the
phyto-plankton of New Brunswick waters exist, as is probable, they must have
some effect upon the relative abundance, at different seasons, of the higher forms
of life, of which they are the food supply.

EXPLANATION OF PLATES.

NOTE—The figures in these plates have been drawn, with few exceptions, with the eye, as
seen under a ¼″ in objective, but not to scale.

PLATE I.

Fig. 1.   Chaetoceros decipiens—Cleve.
  2.   "   "   "   "
  3.   Chaetoceros decipiens—Cleve ?
  4.   "   "   "   ?
  5.   "   "   "   Resting spores.
  6.   Chaetoceros sp?
  7.   "   "   Resting spores ?
  8.   "   "   "   "
  9.   Chaetoceros chriophyllum.—Cast.
  10.   "   "   "   var. ?
  11.   Chaetoceros Peruvianum—Bright ?

PLATE II.

Figs 1–2.   Rhizosolenia setigera—Bright.
  3–4.   "   styliformis—Bright.
  5.   "   setigera ?
  6.   sp ?
  7.   "   imbricata ?
  8.   "   ?   (Ditylum Brightwellii) ?
  9–11.   Triceratium undulatum—Bright ?
  12.   Skeletonema costatum—Grev.
  13.   Thalassiosira Nordenskioldii—Cleve
  14–15.   "   sp ?
  18.   Asterionella.

PLATE III.

Fig. 1–2.   Chaetoceros boreale—Bail.
  3.   Chain of auxospores of Chaetoceros decipiens ?
  4.   Thalassiosira?
  5.   "   gravida—Cleve?

SESSIONAL PAPER No. 39b

6. Thalassiosira Nordenskioldii. Chain of frustules.
7.   "   sp ?
8.   "   gravida—Cleve ?
9. Dicladia capreolus—Probably a Chaetoceros.
10. Syndendrium diadema—Gy.
11. Actinoptychus undulatus—Kutz.
12. Hyalodiscus subtilis.
13. Coscinodiscus asteromphalus Ehr. var. oculus-iridis.
14.   "   eccentricus—Ehr.
15. Grammatophora marina—Kutz.
16.   "   serpentina—Ehr.
17. Synedra undulata—Bail.
18. Nitschia longissima.
19. Biddulphia aurita—Brel.
20.   "   Baileyi—B. Mobiliensis.
21.   "   rhombus—W.S. zonal view.   ·
22.   "   "   lateral view.
23.   "   Baileyi—Valvular view.
24. Triceratium alternans—Bail.
25. Cyclotella compta—Kg.

III.

STUDIES ON THE SPOROZOA OF THE FISHES OF THE ST. ANDREW'S REGION.

By J. W. Mavor, B.A., Ph.D., etc.

Instructor in Zoology, University of Wisconsin, Madison, U.S.A.

(Curator of the Biological Station of the Canadian Government on the Georgian Bay, Canada.)

(Plate IV.)

INTRODUCTION.

The only papers published on the Myxosporidia of American fishes are two by Gurley ('93 and '94) and a short one by Tyzzer ('00). During the twenty years since Gurley's papers our knowledge of the Sporozoa has greatly increased. Only comparatively recently however has special attention been directed to the Myxosporidia. The researches of Doflein, Mercier, Schroeder, Awerinzew, and others have shown this group to be one of great interest, and to-day there is perhaps no group of the protozoa which offers so many interesting features and about the life-cycle of which there is so much doubt.

The writer was of opinion that a study of the Myxosporidia living in the gall bladders of fishes from the Eastern coast of America would lead to interesting results, not only with regard to the distribution of these parasites, but also, it was hoped, with regard to some of the disputed points of their life-history. The present paper deals with the first of these subjects. Another paper to be published later deals with the life-history of one of the parasites found, Ceratomyxa acadiensis n. sp.

While searching for myxosporidian parasites two other parasites were found, a Coccidian and a Haemosporidian, which seem of sufficient interest to be included in this list.

MATERIAL AND METHODS.

The material for the present investigation was collected in Passamaquoddy Bay at or near the mouth of the St. Croix river while the author was at the Marine Biological Station at St. Andrews, New Brunswick, Canada. The fish were brought

(1) The writer wishes to acknowledge his indebtedness to the Board of Directors of the Marine Biological Station at St. Andrews, New Brunswick, Canada for the privilege of working at the station during the summer of 1912.

5 GEORGE V., A. 1915

in a "car" to the laboratory, where they were kept alive either in the car or in tanks supplied with running water. The study of the living parasites was made during the months of July, August and September, 1912, and all the preserved material was collected during the same period.

In searching for parasites of the gall bladder, the bile duct of the fish was ligatured and the gall bladder removed to a carefully cleaned watch glass where it was cut open. Into a pipette freshly made from new glass tubing a small quantity of the bile was drawn. If a fresh preparation was desired this was dropped on a slide and covered with a coverglass. Both slides and coverglasses were prepared as follows: After being cleaned in a mixture of one part bichromate of potash and one part concentrated sulphuric acid to ten parts water they were washed first in tap water and then in distilled water and stored in 95% alcohol. When required for use, the alcohol was burned from them by passing them through the flame of an alcohol lamp. If fixed and stained smear preparations were desired the bile was dropped from the pipette on a coverglass and then sucked back again so that only a very thin film of bile remained on the coverglass. The coverglass was then inverted and allowed to drop on the fixing fluid in such a way that it was supported by the surface tension of the liquid. In this manner the preparations were given no opportunity to dry. This is practically the method of Doflein ('98), with the exception that in all cases no blood was added to the gall. The fixing fluids were Schandinn's fluid, consisting of two parts saturated aqueous solution of corrosive sublimate to one part absolute alcohol used either hot or cold and Hermann's fluid consisting of 75 cc. of 1% platinic chloride, 4 cc. of 2% osmic acid and 1 cc. of glacial acetic acid. These fluids were allowed to act for from five to ten minutes and the coverglasses were then transferred (after Schandinn's fluid) to 60% alcohol containing iodine, or (after Hermann's fluid) to distilled water. The stains used were Giemsa's azar-eosin or Dalafield's haematoxylin. Both were diluted before use to one or two per cent and allowed to act for from twenty-four to forty-eight hours. After staining in Giemsa's mixture the smears were washed in tap water and destained in a mixture containing 95% acetone and 5% xylol. When sufficiently destained they were passed in succession through the following mixtures: (1) acetone 70% and xylol 30%; (2) acetone 50% and xylol 50%; (3) pure xylol, and were finally mounted in Canada balsam. For the details of this method of using Giemsa's stain, Kisskalt and Hartmann ('10, p. 14) may be consulted. After staining in Dalafield's haematoxylin, smears were either first destained in acid alcohol or mounted directly in Canada balsam.

For the study of attached stages, the wall of the gall bladder was sectioned. Pieces of the bladder, opened in a watch glass as described above, were fixed in Schaudinn's fluid, imbedded in paraffine, and cut into sections from four to seven microns in thickness. The sections were stained in Giemsa's mixture or in Dalafield's haematoxylin, diluted as for the smear preparations, or in Heidenhain's iron haematoxylin. In the case of Giemsa's stain the best results were obtained by washing in water rapidly, for twenty seconds or so, and then destaining in a mixture of acteone 95 cc. and xylol 5 cc. for eight to ten minutes.

SESSIONAL PAPER No. 39b

TABLE OF FISHES SEARCHED WITH THE SPOROZOAN PARASITES FOUND IN THEM.

Host and Organ	Parasite	Number examined	Number Infected
Clupea hareugus			
Testis..........................	None	12	0
Gall Bladder....................	None	1	0
Cryptacanthodes maculatus.			
Gall bladder....................	None	1	0
Hemitripterus americanus.			
Gall Bladder....................	Ceratomyxa sp?	1	1
Myxocephalus octodeoemspinous.			
Gall bladder....................	None	1	0
Myxocephalus groenlandicus.			
Gall bladder....................	None	4	0
Melanogrammus aeglefinus.			
Gall bladder....................	Myxidium beregense	1	1
Air bladder.....................	Gaussia gadi	1	1
Osmerus mordax.			
Viscera..........................	No cysts	22	0
Pseudopleuronectes americanus.			
Gall bladder....................	Ceratomyxa acadiensis	25	25
Gall bladder....................	Myxidium sp?	25	few
Viscera.........................	No cysts	82	0
Raja ocellatus.			
Gall bladder....................	None	1	0
Urophycis chuss.			
Gall bladder....................	Ceratomyxa acadiensis	10	9
Gall bladder....................	Myxosporidian sp?		
Blood...........................	Haemogregarina sp.	1	1
Zoarces angularis			
Gall bladder....................	Ceratomyxa acadiensis	8	8

LIST OF SPOROZOAN SPECIES.

I. *Ceratomyxa acadiensis* n. sp.

The *Myxosporidium* (Pl. IV Figs. 1-5, 10-13) is typically club-shaped with a long tail, often many times the length of the thicker part of the body (Pl. IV, Fig. 10). Large individuals may be irregularly stellate (Pl. IV, Fig. 12). The pseudododia often show a rigidity as if possessed of a rigid endoplasmic axis. The protoplasm of certain of the pseudopodia may be collected into clumps, the clumps being connected together by thin hyaline filaments of ectoplasm. A division into ectoplasm and endoplasm though not always clear is often to be seen in the anterior rigion. In the parasite of Urophycis chuss the mxyosporidia were very often found attached to the myxospqridium of an undetermined species (Pl. IV, Fig. 7 and 8) described in the fourth part of this section. An examin-

5 GEORGE V., A. 1915

ation of freed individuals showed the attachment to be brought about by short pseudopodia at the anterior end.   In the parasite of Zoarces angularis the attachment is probably to the epithelium of the gall bladdeer, since the fine pseudopodia are found and the myxosporidium found in U. chuss seems to be absent.   In the parasite of Pseudopleuronectes americanus no attachment has been seen.   The dimensions of a typical myxosporidium are:—

Length, excluding tail.....12–25 μ
Width..................10–20 μ
Tail..................up to 60 μ

In studying the structure of the *spores* of the Myxosporidia it is convenient to use the method of orientation employed by Thélohan ('95, p. 250-251) and generally adopted by subsequent writers.   Where there is a single polar capsule or two (cps. pol. Fig. 1) or more close together the part of the spore in which the capsules lie is called anterior (a Fig. 1).   The plane (pa Fig. 1) passing through the suture separating the two valves is called the sutural plane.   The spore is

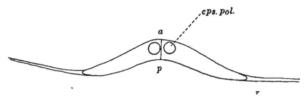

Fig. 1

Fig. 1.   Spore of Ceratomyxa acadiensis n. sp. drawn to show method of orientation and nomenclature.   Explanation in text.   × 2000.

orientated by placing it with the polar capsules in front and the sutural plane vertical (Fig. 1).   Then the front is anterior and the part behind is posterior (p Fig. 1), the upper surface dorsal and the lower surface ventral, the right side the right and the left side the left.   The sutural diameter (Thélohan '95, p. 251) is the greatest diameter in the sutural plane.   The bivalve axis (l r, Fig. 1) is the line which measures the greatest distance between the two valves perpendicular to the sutural plane.

The general shape of the spore of Ceratomyxa acadiensis n. sp. (Fig. 1) may be described as that of a spindle, of which the longitudinal axis has been bent into an arc of a circle.   The chord of this arc is the bivalve axis, and may be called the width of the spore.   The convex side of the arc is anterior, the concave side posterior and the opposite ends right and left.   The sutural axis extends in the anteroposterior direction and is equivalent to the length of the spore.   The two valves are cone-shaped, the pointed ends being directed one to the right and the other to the left and the bases meeting in the plane of suture.   The spore is slightly compressed dorso-ventrally.   A slight variation in the form and dimensions of opposite valves of the same spore was often noticed.   The lateral filaments, extending outward from the tips of the valves on either side, are very long and thin.

Their exact length in the spore of the parasite from Urophcxis chuss was not measured. Their extreme fineness and great length make this very difficult except in very favorable preparations. This was, however, done in the case of the parasite of Zoarces angularis (Pl. IV, Fig. 9) where they were found to measure 250-300 μ or about six times the width of the spore exclusive of the filaments. The cavity of the valves does not appear to extend into the filaments. The length of these filaments is greater both relatively to the width of the spore and absolutely, than the length recorded for the lateral filaments of any other species of Ceratomyxa. Long filaments are most common in the two genera Ceratomyxa and Henneguya. It is generally believed that the filamentous appendages of Myxosporidian spores function in aiding the distribution of the spores by retarding the rate at which they sink and by rendering them more easily carried by currents.

The polar capsules (Fig. 1, cps. pol.) are almost spherical and lie close together at the anterior end of the spore. They are so oriented that the polar filaments when extruded cross each other (Pl. IV, Fig. 14). The extrusion of the polar filaments was effected by concentrated sulphuric acid but was not brought about by a solution of iodine in potassic iodide or by ammonia water. The failure of these two reagents may have been due to the spores not having been ripe. When extruded the filaments appear as very fine threads of uniform thickness.

The sporoplasm as seen in fixed and stained preparations is eccentrically placed, being in one valve, and contains, in all the spores examined from the gall bladder two compact darkly staining nuclei.

The dimensions of a typical spore are:

Length = sutural axis	7-8 μ
Width = bivalve axis	40-50 μ
Diameter of polar capsule	3-4 μ
Length of lateral filaments	205-300 μ
Length of extruded polar filaments	70 μ

Triradiate spores are of frequent occurrence. These spores may show a fairly regular radial symmetry, both as regards the valves and the polar capsules (Pl. IV, Fig. 16) or one of the valves may be smaller than the other two while the three polar capsules are of equal size and symmetrically arranged (Pl. IV, Fig. 15). Cases where a triradiate spore and a normal spore were developing in the same myxosporidium were found (Pl. IV, Fig. 12) as were also cases where two triradiate spores were developing together.

Ceratomyxa acadiensis has been found in three hosts and perhaps in a fourth from the coast of New Brunswick, Canada. In the gall bladder of Urophycis chuss, the hake, it is usually found attached to an undetermined parasite, probably a species of Myxidium or Chloromyxum which is itself attached to the gall bladder. Nine out of ten U. chuss examined for the parasite were found to be infected. In the gall bladder of Zoarces angularis, the eel pout, C. acadiensis was not found attached although the modification of the anterior end for attachment was found. Each of the eight Zoarces angularis examined for the parasite was found to be infected. In the gall bladder of Pseudopleuronectes americanus, the winter flounder, no evidence of attachment was seen, vegetative forms were found relatively abun-

5 GEORGE V., A. 1915

dantly, spores only rarely. Twenty-five flounders examined all contained the parasite. In Hemitripterus americanus myxosporidia resembling closely the myxosporidia of Ceratomyxa acadiensis were found. As no spores were found it was not possible to make a complete identification of this parasite.

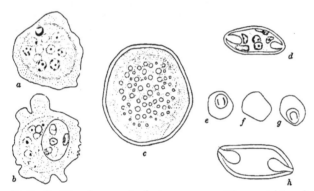

Fig. 2. Myxidium bergense Auerbach. *a*, myxosporidium containing eleven nuclei in the endoplasm and showing the intermediate zone and the ectoplasm; from a preparation stained with Delafield's haematoxylin. *b*, a similar myxosporidium containing a sporoblast with six nuclei and ten other nuclei in the endoplasm; from a preparation stained with Grenacher's borax carmine. *c*, myxosporidium showing outer resistant membrane (indicated by the clear area between the two contour lines) and numerous green granules; from a fresh preparation. *d*, spore showing the two polar capsules and the six nuclei; the two germ-nuclei lie one over the other near the centre, the two polar nuclei lie against the polar capsules and the valve-nuclei are more faintly stained and lie against the valves of the spore; from a preparation stained with Delafield's haematoxylin. *e, f, g*, optical cross sections of a spore; *e* and *g*, at either end and *f* at about the middle. *h*, spore showing shell and polar capsules and placed so as to correspond in position to the sections *e, f, g*. Figures *e-h* from fresh preparations. × 1900.

The spores of Ceratomyxa acadiensis resemble in size most closely those of C. appendiculata Thél. (Thélohan '95). As Thélohan does not give a figure of the spore and the only measurements given are those of the length and width it is impossible to carry the comparison further. The myxosporidium differs from that of C. appendiculata in being found attached. The spore resembles in form that of C. drepanopsettae Awer. (Awerinzew, '09) but differs from it in size.

Some interesting stages in the life history of this parasite have been worked out and will form the subject of a separate paper.

2. *Myxidium bergense* Auerbach.

The *Myxosporidium* is spheroidal, 25-35 μ in diameter or elongated up to 50 μ in length. There is a clear differentiation into ectoplasm, an intermediate zone resembling that described in M. lieberkuhni, Butschli, by Cohn ('96) and

SESSIONAL PAPER No. 39b

endoplasm. In the living parasite the ectoplasm is hyaline, the intermediate zone very finely granular and slightly less transparent than the ectoplasm while the endoplasm is filled with yellowish green granules (Fig. 2, c). In stained preparations this differentiation of the protoplasm becomes more apparent, the intermediate zone being more deeply stained than either the ectoplasm or endoplasm (Fig. 2, a and b). The nuclei are confined to the endoplasm. The pseudopodia may be of two forms:—lobose, relatively large and rounded (upper and left side of Figure 2, b) or fine and short in which case they are usually numerous and arranged so as to give the part of the surface where they occur a villate appearance (right of Fig. 2, b). The latter attach the myxosporidium to the epithelium of the gall bladder. Under certain conditions the myxosporidium may become surrounded by a distinct doubly contoured membrane (Fig. 2, c) giving the whole the appearance of a cyst. At times the protoplasm may be seen in fresh preparations to be shrunken within this membrane leaving a clear space between the membrane and the ectoplasm. The sporoblasts are formed without the previous formation of pansporoblasts. One to six sporoblasts may be found in a myxosporidium. The sporoblasts are usually not arranged in pairs but are scattered in the myxosporidium. Figure 2, b, shows a myxosporidium with one sporoblast. The sporoblast shows the usual six nuclei:—the two nuclei of the valve cells, the two of the capsulogenous cells, and the two germ nuclei. The two nuclei of the valve cells will be seen each to have adherent to the periphery at one point a dark body. This dark body seems to be of frequent or constant appearance at this point. Its significance is not clear to the writer. A later stage where the polar capsules are forming is shown in Figure 2, d. Here also there are two germ nuclei. In every spore examined from the gall bladder there were two germ nuclei.

The spores are spindle shaped with the axis of the spindle slightly bent in the form of an enlongated S, the two ends of which have been bent at right angles to the plane of the letter and in opposite directions. Corresponding to this curving of the axis of the spindle, the polar capsules are placed with their axes approximately tangent to the curve described, i.e., their axes make angles (of about 20°) on opposite sides of the line joining their points of contact with the spore shell. The polar filaments are visible within the capsules in the fresh state but the number of coils of the spiral in one capsule could not be counted. The filaments were not extruded when treated with a solution of iodine in potassic iodide. The dimensions of a typical spore are:

Length..16–18 μ.
Width... 6–7 μ.
Length of polar capsule.......................... 4 μ.
Width of polar capsule...........................2.5–3 μ.

This description will be found to agree with that of Auerbach ('09, '09[a] p. 61, and '12, pl. 2), in all particulars with the exception of the cyst-like condition described in the present paper. The presence of this cyst may however be due to some exceptional condition of the parasite.

5 GEORGE V., A. 1915

3. *Myxidium* Sp.?

The myxosporidium of this rare parasite was not seen in fresh preparations of the bile. In stained smears there occurred a large spheroidal myxosporidium containing twenty-two nuclei, and having numerous long lobose pseudopodia on one side. The general arrangement of the pseudopodia suggested that they served for the attachment of the myxosporidium to the gall bladder. It contained no spores.

The pansporoblasts are spherical 15-16 μ. in diameter.

a   b

Fig. 3

Fig. 3. Spores of Myxidium sp. from Pseudopleuronectes americanus. *a*, with polar filament extruded by ammonia water. $\times$ 660   *b* $\times$ 1320.

The spores (Fig. 3) are spindle shaped with the long axis slightly bent in the form of an S. The polar capsules are pear-shaped and situated at either end of the spindle. The polar filaments were visible in the fresh state within the capsule. The polar filaments were extruded in ammonia water (Fig. 3, a).

The dimensions of a typical spore are:

Length..	14–15 μ.
Width...	6–7.5 μ.
Length of polar capsule.........................	4 μ.
Width of Polar capsule....'.....................	2.5 μ.
Length of extruded polar filament................	90–95 μ.

This species of Myxidium was found in the gall bladder of Pseudopleuronectes americanus on the coast of New Brunswick, Canada.

The spores found resemble most closely those of M. bergense Auerbach (:09, p. 74 and '09ª. p. 61) but differ from these by their small size and longer polar filaments. They resemble also the spores of M. sphericum Thél. but differ in the relatively smaller polar capsules (Thél. '95, Pl. 7, Fig. 28) and the longer polar filaments.

4. *Myxosporidium* of an undetermined species.

Attached, usually in large numbers, to the epithelium of the gall bladder in Urophycis chuss, occurs a spherical or ellipsoidal myxosporidium which in stained preparations is found to contain numerous nuclei (Pl. IV, Figs. 6-8). The examination of a large number of these myxosporidia has not revealed the presence of any developing spores in them. Very often clusters of C. acadiensis are found

SESSIONAL PAPER No. 39b

adhering to the free surface of the myxosporidium (Pl. IV, Figs. 7 and 8) i.e. the surface not in contact with the epithelium. In fresh preparations the appearance is that of budding from a parent organism (Pl. IV, Figs 7-8). For a time this was thought possibly to be the case for some of the adherent individuals. An examination of sections has shown a sharp division between the myxosporidium and C. acadiensis. No other spores than those of C. acadiensis were found in the gall bladder of U. chuss.

5. *Goussia gadi* Fiebiger.

The haddock in which this parasite was found was caught on the sixth of August. The abdominal organs were cut out and the fish was put on ice. Next day when the fish was being prepared for the table it was proclaimed unfit for cooking on account of a creamy exudation in the dorsal part of the body cavity. It was at this time that the fish was brought to the notice of the writer. On examination a creamy mass, yellowish white in color was found adherent to the inner surface of the air bladder. This had the appearance of being due to the breaking down of the lining membrane. The kidneys and surrounding muscular tissue appeared quite normal. A microscopic examination revealed the presence of numerous ellipsoidal spores arranged in groups of four in the creamy mass. "Wet" smears were fixed in Schaudinn's sublimate-alcohol mixture and in Hermann's platinic chloride-osmium-acetic mixture. They were subsequently stained in Grenacher's borax carmin and in Delafield's haematoxylin. The preservation proved to be not all that could have been desired but seems sufficient to determine the systematic position of the parasite.

The macerated condition of the cells of the air bladder both when examined fresh and in preserved preparations has made it impossible to determine any of the schizogonic or syngamic stages. There can however be no doubt that the form is tetrasporous from the almost constant occurrence of the spores in groups of four usually surrounded by a structure which appears membranous in the preparations.

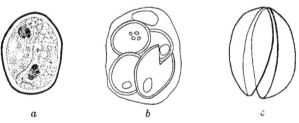

a   b   c

Fig. 4. Goussia gadi Fiebiger. *a*, spore stained with Delafield's haematoxylin showing the two sporozoites with their nuclei, × 1900. *b*, tetrad of spores inclosed in mass which is probably remains of host cell; drawn from fresh preparation, × 970. *c*, two valves of spore cell drawn from preparation fixed in Hermann's fluid, × 1900.

5 GEORGE V., A. 1915

Figure 4, *b* drawn from a fresh preparation of the creamy mass in the air bladder shows the arrangement of the oval spores in tetrads. In this the tetrad is inclosed in what may have been one of the cells of the air bladder.

In fresh preparations the spores measure 16 μ in length by 12 μ in width. A spore stained with Delafield's haematoxylin is drawn in Figure 4, *a*. The two sporozoites are seen filling the spore. Each has a nucleus situated near one end. The nucleus of a sporozoite is usually, though not always, situated at one end and the nuclei of the two sporozoites in a spore are usually at opposite ends of the spore. There is no residual protoplasm in the spore.

The shell of the spore is ellipsoidal. The line of suture of the two valves does not lie in a focal plane of the ellipsoid but is shaped so as to give each valve somewhat the form of a spoon. In fresh preparations the spore shell could be seen to consist of two layers, an outer yellowish layer and an inner dark green layer. Figure 4, *c* drawn from a preparation preserved in Hermann's fluid shows the shape of the valves of the spore shell.

From the above description there can be no doubt that the organism we are concerned with belongs to the order Coccidiida. Following the classification of Labbé ('99) since the number of archispores (sporoblasts) is limited to four we have:—

> Order Coccidiida
> Sub-order Oligoplastina
> Tribe Tetrasporea,

and since the spores are oval and bivalve the parasite is to be placed in the genus Goussia, Labbé ('96). Fiebiger ('08) has described under the name of Goussia gadi a species of Goussia infecting the air-bladder of Gadus morrhu and Gadus virens and has identified it with the parasite found by J. Müller in the air-bladder of Gadus callarias. Auerbach ('09, p. 74, 81) has also described briefly a parasite from the air-bladder of Gadus aeglefinus which he identifies as a species of Goussia. The writer is of opinion that in the present stage of our knowledge these parasites are to be regarded as all belonging to the same species and that the parasite found by him is probably also of this species.

The microscopic appearance of the diseased air bladder as described by these authors is the same as that found by the writer. The chief difference between the parasites described by Fiebiger and he are in the size of the spores and the form of the sporozoites. The spores of the parasite described by Fiebiger measure only 11 μ × 7·5 μ as against the 16 μ × 12 μ of those found by the writer. In describing the sporozoites Fiebiger ('08) says "Es sind dies schlanke Gebilde mit einem vorderen zugespizten und einem hinteren abgerundeten Ende von 10μ Lange und 4 μ Breite." Those found by the writer are proportionately shorter and wider. As these characters are usually considered to be of great systematic importance considerable doubt may be expressed as to the two parasites being of the same species. However, the writer considers that other similiarities make it possible that the variations in size may be due to the different environments of the hosts and the difference in the form of the sporozoites, to his not having seen the final stage in their development or to defective preservation. It is worthy of note that Fiebiger found also such sporozoites in his preparations ('08, Fig. s).

6. *Hæmogregarina* sp?

In order to insure against the confusion of elements of the blood with stages in the life-history of the parasites of the gall-bladder of Urophycis chuss, smears of the blood were made. In these smears a hæmogregarine (Fig. 5)-was found. The infection was a rather abundant one, some hundred or so individuals being found in a single smear and at times two in one field of the oil-immersion objective. All the individuals found had the characteristic sausage shape of the merozoite of hæmogregarines. Usually one side of a red corpuscle was completely filled by the parasite and often the nucleus of the corpuscle was forced to one side (Fig. 5).

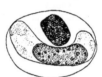

Fig. 5.   Hæmogregarina sp.? from the blood of Urophycis chuss.   × 3000.

The nucleus of the haemogregarine was usually about half as long as the individual and filled its complete thickness; it was usually situated nearer one side. In the nucleus could usually be distinguished a number of deeply staining granules. Sometimes the merozoits were bent upon themselves. In such cases, however, the corpuscles were shorter than usual and the curling of the parasite was probably due to the drying of the smear.

The host of the hæmogregarine, Urophycis chuss, occurs on the coast of North America from the banks of Newfoundland to Cape Hatteras (Jordan and Evermann 1898; III, p. 2555). The writer is not aware of the description of any hæmogregarines from the fishes of these waters.

ON THE GEOGRAPHICAL DISTRIBUTION OF THE PARASITES FOUND.

Certain of the parasites found in the fishes of Passamaquoddy Bay are believed by the writer to be of the same species as parasites found in the same fishes occurring on the coast of Europe.

Myxidium bergense has been found by Auerbach ('12) in Sebastes viviparus, Anarrhichas lupus, Gadus callarias, Gadus aeglefinus, Gadus merlangus and Pleuronectes platessa, caught at points on the coast of Norway extending from Christiania in the South to Vardö in the North, and by the writer in Gadus aeglefinus from the eastern coast of Canada.

Goussia gadi has been found by Fiebiger ('08) in Gadus morrhua and Gadus virens from the coast of Iceland but not in Gadus aeglefinus from the same region

5 GEORGE V., A. 1915

which he also searched for the parasite. Fiebiger attributes his failure to find the parasite in the latter species to his not having examined a sufficient number of fish. Assuming that the parasite described by Auerbach ('09, p. 74, 81) is Goussia gadi, as seems probable, it has been found in Gadus aeglefinus on the coast of Norway at Bergen. The coccidian described by J. Müller ('42) from Gadus callarias is identified by Fiebiger ('08) as Goussia gadi. The parasite found by the writer is also identified as Goussia gadi. The distribution of Goussia gadi is therefore from the Cattegat to the North of Norway, Iceland and Eastern Canada.

There can be no doubt that the parasites in question, Myxidium bergense and Goussia gadi complete their life cycle in the host fish, in other words there is no intermediate host. Hence their spread occurs only from fish to fish, and a fish becomes infected only by coming into such relations to an infected fish that the spores of the parasite are carried to it from the latter by water currents. This probably means the fairly close proximity of the two fish. The investigation of infectious diseases, where the method of infection is contaminative, has shown that their spread over large areas is almost invariably due to the migration of diseased animals. It is possible that the spread of Myxidium bergense and Goussia gadi over the North Atlantic is due to the migrations of the host fishes in these waters.

The places mentioned in the discussion of the distribution of Myxidium bergense and Goussia gadi are shown on the map (Fig. 6).

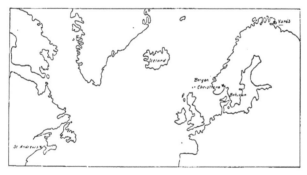

Fig. 6. Map on Mercator's projection showing places mentioned in the section upon geographical distribution.

The fact that no cysts of Sporozoa were found in the 82 specimens of Pseudopleuronectes americanus is interesting. The writer found fifty per cent of the fish of this species caught in the Wood's Hole region in the summer and winter of 1910 infected with Glugea stephani Hagenmüller. At this time he also found Osmerus mordax from Wood's Hole frequently infected with a microsporidian, apparently Glugea stephani. The twenty-two examples of Smelt Osmerus mordax examined from the St. Andrews region contained no microsporidian cysts.

SESSIONAL PAPER No. 39b

BIBLIOGRAPHY.

Auerbach, M.: '09· Bemerkungen über Myxosporidien. Zool. Anz. Bd. 34, p. 65-82.

'09a. Biologische und Morphologische Bemerkungen über Myxosporidien. Zool. Anz. Bd. 35, p. 57-63.

'12· Studien über die Myxosporidien der norwegischen Seefische und ihre Verbreitung. Zool. Jahr. Abt f. Systematik. Bd. 34, p. 1-50, pl. 1-5.

Awerinzew, S. '09· Studien über parasitische Protozœn. I. Die Sporenbildung bei Ceratomyxa drepanopsettae mihi. Arch. f. Protist Bd. 14, p. 74-112.

Cohn, L. '96· Uber die Myxosporidien von Esox lucius und Perca fluviatilis. Zool. Jahrb., Abt. f. Morph. Bd. 9, p. 227-272.

Fiebiger, J. '08· Uber Coccidien in der Schwimmblase von Gadus-Arten, Vorläufige Mitteilung. Annalen des K. K. Naturhistorischen Hofmuseums Wien. Bd. XXII, Nr. 2-3. 1907-08., p. 124-128.

Gurley, R. '93· On the Classification of the Myxosporidia. Bull. U. S. Fish. Comm. for 1891, Vol. II., p. 407-420.

'94· The Myxosporidia or Psorosperms of Fishes and the Epidemics produced by them. Report U. S. Comm. Fish and Fisheries. Pt. 18, p. 65-304.

Jordan, D. S. and Evermann, B. W. '96-'00. The Fishes of North and Middle America. U. S. National Museum, Bull. No. 47, 4 pts., Washington.

Kiskalt, K. and Hartmann, M. '10· Praktikum der Bakteriologie und Protozoologie. Teil II., Protozoologie, Jena.

Labbé, A. '96· Recherches zoologiques, cytologiques et biologiques sur les Coccidies. Arch. Zool. Exp. Ser. 3, vol. 4, p. 517-654., pl. 12-18.

'99· Sporozoa in Tierreich Das. Deutsch. zool. ges. 5. Liefg., Berlin.

Müller, J. und Retzius, A. '42· Uber parasitische Bildungen, Müller's Archiv. f. Anat., Physiol. u. wiss. Medizen, p. 193-198.

Thélohan, P. '95· Recherches sur les Myxosporidies. Bull. Scient., France et Belgique, Vol. 26, p. 100-394.

Tyzzer, E. E., '00· Tumors and Sporozoa in Fishes. Journ. Boston Soc. Vol. 5, p. 63-68, Pl. 6.

5 GEORGE V., A. 1915

EXPLANATION OF PLATE.

PLATE IV.

Ceratomyxa acadiensis, n. sp.; myxosporidia and spores drawn from fresh preparations of the bile of the host.

Fig. 1. Young myxosporidium of C. acadiensis from the gall bladder of Urophycis chuss. × 390.

Figs. 2-5. Young myxosporidia of C. acadiensis from the gall bladder of U. chuss. × 830.

Fig. 6. Undetermined myxosporidium from gall bladder of U. chuss. × 600.

Fig. 7. Undetermined myxosporidium from gall bladder of U. chuss with attached C. acadiensis. × 830.

Fig. 8. Same subject as figure 7, drawn three hours later. × 830.

Fig. 9. Spore of C. acadiensis from gall bladder of Zoarces angularis. × 270.

Fig. 10. Myxosporidium of C. acadiensis from gall bladder of Pseudopleuronectes americanus × 830.

Fig. 11. Myxosporidium of C. acadiensis from gall bladder of Pseudopleuronectes americanus. × 830.

Fig. 12. Myxosporidium of C. acadiensis containing two sporoblasts, one forming a normal spore, the other forming a triradiate spore with three polar capsules. From the gall bladder of P. americanus. × 390.

Fig. 13. Myxosporidium of C. acadiensis from the gall bladder of Zoarces angularis. × 830.

Fig. 14. Spore of C. acadiensis from the gall bladder of U. chuss. × 390.

Fig. 15-16. Triradiate spores from the gall bladder of U. chuss. × 390.

All drawings were made with an Abbe camera lucida.

PLATE IV.

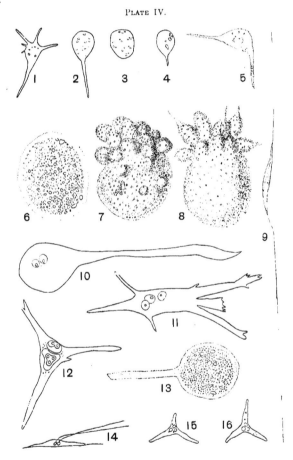

IV.

A NEW CAPRELLID FROM THE BAY OF FUNDY.

By A. G. Huntsman, B.A., M.B., Biological Department,
University of Toronto.

(*Plates V and VI.*)

In the summer of 1912 at the Biological Station, St. Andrews, New Brunswick, an attempt was made to collect large numbers of the smaller Crustacea by attaching to the dredge a bag of sacking in place of the ordinary net bag. Among other things two specimens of an interesting new species of Caprellid were obtained, one a male and the other a female. Both were obtained on muddy bottom in shallow water, the one in Oak Bay and the other near Niger Reef. A habitat on muddy bottom has been given by Sars (1895, p. 656) for an European Caprellid, *Pariambus typicus*, which has also been found upon starfish.

The rudimentary condition of the legs on the fifth peraeal segment attracted my attention. As in the genus *Pariambus* the legs on that segment are rudimentary, I at first thought that I had a species of that genus. Investigation has shown that it does not belong to that genus and in fact it will not fit into any of the current genera. Mayer's admirable monographs have made possible a ready comparison with the known genera.

Almost every character possessed by the new species is to be found in one or other of the known genera, but the combination it shows has not been observed up to the present. The most striking features are,—the presence of two joints in each of the 1st and 2nd pairs of pereiopods, three joints in the 3rd pair, mandibular palp three-jointed, its terminal joint with a single bristle, abdomen of female with two pairs of spines (representing legs?) and abdomen of male with a pair of rudimentary legs and a pair of large spines behind these, representing another pair.

In determining the affinities of this form, there are many possible choices and I cannot see that one is more probable than another.

The third pair of pereiopods are remarkably similar to those figured by Mayer (1903, t. VII, f. 45) for *Piperella grata*. The maxillipeds are almost identical with those of *Triantella solitaria* (Mayer, 1903, t. IX, f. 36). The mandibular palp is in all essentials identical with that of *Protomima denticulata* (Mayer, 1903, t. IX, f. 6). The condition of the first and second pereiopods is similar to that in most of the species where the number of the joints is reduced to one, two or three, that is, the terminal joint has three bristles, the middle one being feathered.

The condition in the abdomen of the male may be peculiar, not with respect to the amount of reduction of the appendages, for similar conditions are known, but

5 GEORGE V., A. 1915

with respect to the armature of the legs. I have not been able to find a similar condition of armature figured, although it may occur in many of the well-known species. The small size of the abdomen renders examination difficult in most cases. From the foregoing facts it has seemed necessary to form a new genus for the reception of this species. The number of genera in the group of the Caprellids is large and the majority are monotypic. It seems impossible, however to avoid creating a new one without doing violence to the principles laid down by Dr. Mayer for classification in this group. The classification that he has built up is doubtless as sound as any that could be devised.

I should like to call attention to the way in which the various genera in this group result from a ringing of the changes on a comparatively small number of characters. Nearly all possible combinations of these characters are to be found. This is analogous with the way in which among chemical compounds a large proportion of the possible combinations of certain radicles or elements may be obtained. I believe that the analogy is due to the fact that in each case the basis is a chemical one.

I propose to name the genus in honour of Dr. P. Mayer, to whom we owe the major part of our knowledge of the Caprellidae. His monographs will long form the foundation of any work in this group.

Mayerella, gen. nov.

Inferior antennae.—Flagellum two-jointed.

Mandible.—Palp three-jointed, terminal joint with a single bristle, which is terminal in position.

Maxilliped.—Inner plate half as long as outer and with three bristles.

Branchiae.—On third and fourth segments of peraeon.

First and second pereiopods.—Two-jointed, terminal joint short and with three bristles.

Third pereiopod.—Three-jointed, terminal joint with four bristles.

Abdomen or pleon.—In female, with two pairs of bristles but without legs. In male, with one pair of unjointed legs and behind these a series of bristles on each side, representing another pair of legs; each leg bearing from five to seven bristles and terminating in a series of hooked teeth.

M. limicola, sp. nov. (Pls. V and VI, figs. 1-12).

Surface of body smooth, with scattered minute bristles. Length (exclusive of appendages), of male 5¼ mm., of female 4¾ mm.

The proportionate lengths of the segments of peraeon are roughly,

	Head + 1st	2nd	3rd	4th	5th	6th + abdomen
Male	2.5	2.5	3.5	4	5	3
Female	2	2	3	3	4	2.5

Superior antennae one-third length of body in male, somewhat less in female. First joint of peduncle slightly shorter than second, third about half as long as second. Flagellum eight-jointed in male and four-jointed in female.

Inferior antennae about four-fifths the length of superior. First two joints subequal, together somewhat less than third joint. Fourth joint slightly longer than third.

Merus and carpus of anterior gnathopoda scarcely produced, setigerous. Propodus narrowly ovate, three groups of bristles on dorsal margin, minutely and evenly denticulate on palmar margin. Dactyl curved, several long sharp teeth along inner edge, denticulate between the teeth.

Posterior gnathopoda of male,—palm of propodus notched beyond middle, with a strong tooth just behind notch, and a bristle on each side of tooth, without serrations except near proximal end and with a prominent bifid spine at proximal end; dactyl long, sickle-shaped, scarcely serrate. In female, the palm of propodus has a smooth sinuate margin and at the proximal end a prominent process bearing a bifid spine; dactyl as in male.

Anterior branchiae about twice as long as posterior branchiae.

First, second and third pereiopods as described above for the genus, similar in the two sexes.

Fourth and fifth pereiopods very slender. Propodus slightly exceeding the carpus in length. Dactyl very long and slender.

Habitat. In from 5 to 10 fathoms on muddy bottom. St. Croix River, New Brunswick.

Literature.

1903. Mayer, P. Die Caprellidae der Siboga-Expedition. Siboga Expeditie, Monographie XXXIV.

1895. Sars, G. O. The Crustacea of Norway. Vol. I. Amphipoda. Christiania.

EXPLANATION OF PLATES.

All the figures are of *Mayerella limicola*.

PLATE V.

Fig. 1.—Female. × 27.
Fig. 2.—Left first maxilla of male, anterior view. × 250.
Fig. 3.—Left mandible of male somewhat crushed, medial view. × 200.
Fig. 4.—Right maxilliped of male, posterior view. × 375.
Fig. 5.—Head of male. Some of the appendages of the mouth have been removed and the remainder are displaced. × 45.

* In the summer of 1913 numerous specimens have been found at several localities in the Bay of Fundy, in depths ranging up to 50 fathoms and on muddy sand bottom.

5 GEORGE V., A. 1915

PLATE VI.

Fig. 6.—Right second pereiopod of female, lateral view. × 290.
Fig. 7.—Right first.pereiopod of female, lateral view. × 290.
Fig. 8.—Left anterior gnathopod of male, lateral view. × 80.
Fig. 9.—Right third pereiopod of female, lateral view. × 320.
Fig. 10.—Abdomen of female, right lateral view. × 320.
Fig. 11.—Abdomen of male, left lateral view. × 200.
Fig. 12.—Abdomen of male, oblique ventral view. × 200.

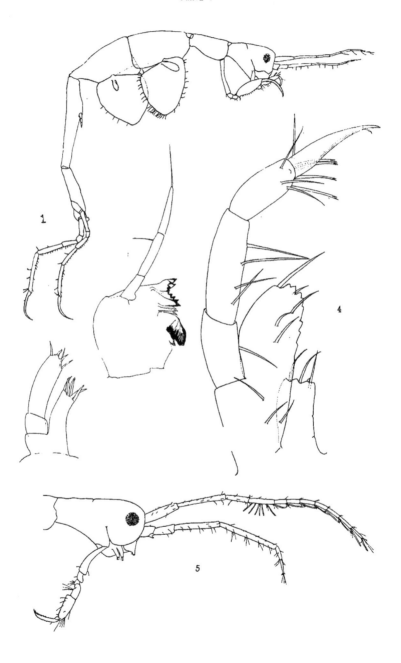

PLATE VI.

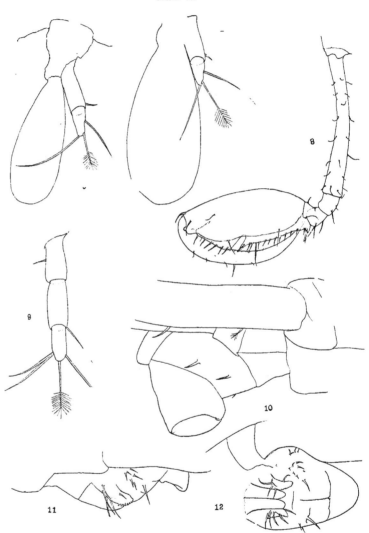

V.

PRELIMINARY NOTES ON THE MOLLUSCA OF ST. ANDREWS AND VICINITY, NEW BRUNSWICK.

By John D. Detweiler, B.A. (of Queens University).

St. Andrews College, Toronto.

From the middle of August to the middle of Sept. 1912, I spent at the Biological Station, St. Andrew's, N.B., working on the distribution of the Mollusca. Collecting was done by dredging and by collecting on the shore at low tide. On account of the limited time spent in the work and the want of a complete supply of literature for reference the number of species identified was not very large. The writer hopes to complete the work in a later season.

PELECYPODA.

Anomia oculeata. Mueller.
    Dredged in 20–30 fathoms at the Wolves, Aug. 17.
Pecten magellanicus (Gmelin).
    Dredged at the Wolves, Aug. 17.
Mytilus edulis. L.
    Common throughout this region in the littoral zone.
Modiolaria discors. (L).
    Dredged south of the Wolves in 50 fathoms, Sept. 10.
Modiolaria nigra (Gray).
    In 40 fathoms, Pendleton's Island, Aug. 29.
    In 20 fathoms Sand Reef Light, Sept 5.
    In 1 fathom Grand Harbour, Sept. 9.
Modiolaria corrugata (Stimpson).
    Dredged in 17 fathoms off Robbinston, Sept. 4.
Crenella glandula (Totten).
    Dredged in 20–30 fathoms at the Wolves, Aug. 17.
Nucula delphinodonta Mighels.
    In 15 fathoms off Minister's Island, Sept. 6.
Yoldia sapotilla (Gould).
    In 40 fathoms off Pendleton's Island, Aug. 29.
Megayoldia thraciaeformis. Storer.
    Dredged in muddy bottom off Pendleton's Island, Aug. 29.
Cardium pinnulatum, Conrad.
    Dredged in 20–30 fathoms off the Wolves, Aug. 17.

Cardium ciliatum Fabricius.

In 17 fathoms off Robbinston.

Cyprina islandica (L).

Dredged in muddy bottom off Pendleton's Island in 40 fathoms, Aug. 29.

Cytherea convexa Say.

Off Robbinston in 15 fathoms Sept. 5.

Astarte subaequilatera Sowerby.

Dredged in 50 fathoms off the Wolves, Sept. 10.

Astarte undata Gould.

Dredged on sandy bottom in 50 fathoms off the Wolves, Sept. 10.

Astarte castanea Say.

In 7 fathoms off Robbinston, Aug. 14.

Venericardia borealis (Conrad)

Dredged in 20–30 fathoms off the Wolves, Aug. 17.

Macoma balthica (L).

On beach at Biological Station, Sept. 2.

Pandora gouldiana Dall.

In 5 fathoms off Joe's Point, Aug. 20.

Lyonsia hyalina Conrad.

Small specimens dredged off Gleason's Cove in 14 fathoms, Aug. 29. Large specimens dredged off Robbinston in 15 fathoms, Sept. 5.

Thracia truncata Mighels and Adams.

Dredged in 20 fathoms off Sand Reef Light, Sept. 5.

Mya arenaria L.

Common throughout the region in the littoral zone.

Saxicava rugosa (L).

In littoral zone at St. Andrew's Point, Sept. 12.

SCAPHOPODA.

Dentalium entalis L.

Dredged in 20-30 fathoms at the Wolves, Aug. 17.

AMPHINEURA.

Tonicella marmorea (Fabricius).

Off Gleasons Cove in 14 fathoms, Aug. 29.

GASTEROPODA.

Lepeta caeca, (Mueller).

In 17 fathoms, off Robbinston, Sept 14.

Acmaea testudinalis, (Mueller).

Common on rocks in littoral zone.

Puncturella noachina (L).

Dredged off the Wolves in 20–30 fathoms, Aug. 17.

Margarita cinerea (Couthany).

    Dredged in 20–30 fathoms off the Wolves, Aug. 17.

    In 17 fathoms off Robbinston Sept. 4.

Margarita unduluta, Sowerby.

    In 5–10 fathoms off the Wolves, Aug. 17.

    Off Robbinston in 15 fathoms, Sept. 5.

Margarita helicina (Fabricius)

    In littoral zone at St. Andrews' Point, Sept. 11.

Scalaria groenlandica Perry.

    In 10–15 fathoms on gravel bottom, off Robbinston, Aug. 2.

Lunatia heros (Say).

    Common in littoral zone.

Lunatic heros triseriata (Say).

    Off Robbinston in 5–10 fathoms, Sept. 11.

Natica clausa, Broderip and Sowerby.

    Dredged off the Wolves in 50 fathoms on a Sandy bottom, Sept. 10.

    Off the Wolves in 20–30 fathoms, Aug. 17.

Crucibulum striatum (Say).

    Dredged 17 fathoms off Robbinston, Sept. 4.

Littorina palliata (Say).

    In littoral zone at Biological Station, Sept. 10.

Littorina litorea (L).

    Common in littoral zone.

Littorina rudis (Maton).

    Littoral zone at Biological Station and Woodward's Cove.

Velutina undata (Brown).

    In 15 fathoms off Robbinston, Sept. 5.

Trichotropis borealis Broderip and Sowerby.

    Off Robbinston, Sept. 4.   Dredged in 20–30 fathoms off the Wolves, Aug. 17.

Aporrhais occidentalis Beck.

    Dredged in 20–30 fathoms off Wolves, Aug. 17.

Purpura lapillus (L).

    Common in littoral zone.

Tritia trivittata (Say).

    Off Joe's Point in 5 fathoms.

    Off Robbinston on gravel bottom in 10–15 fathoms, Aug. 20.

Buccinum undatum L.

    Common in littoral zone.

Neptunea decemcostata, Say.

    Common in the sublittoral zone and at the lowest limits of the littoral zone.

Sipho stimpsoni (Mörch).

    Near Green Island in 5–10 fathoms, Sept. 10.

Sipho pygmaeus (Gould).

    Dredged off Robbinston in 10–15 fathoms, Aug. 20.

5 GEORGE V., A. 1915

Bela scalaris (Moeller).
  Dredged off Wolves in 20–30 fathoms, Aug. 17.
  Off Robbinston on gravel bottom in 10–15 fathoms, Aug. 20.
Bela decussata (Couthouy).
  Dredged in 20–30 fathoms off Wolves, Aug. 17.
Bela harpularia (Couthouy).
  Dredged in 20–30 fathoms, off the Wolves, Aug. 17.
Bela cancellata (Mighels).
  Dredged in 20–30 fathoms off the Wolves, Aug. 17.
Bela bicarinata (Couthouy).
  Dredged in 20–30 fathoms, off the Wolves, Aug. 17.
Bela pleurotomaria (Couthouy).
  Dredged in 20–30 fathoms, off the Wolves, Aug. 17.
Retusa pertenuis (Mighels).
  In 1 fathom at Grand Harbour, Grand Mannan, Sept. 2.

CEPHALOPODA.

Illex illecibrosus (Lesueur).
  Common throughout the region.

VI.

A LIST OF FLESHY FUNGI COLLECTED AT ST. ANDREWS, NEW BRUNSWICK.

By Miss Adaline Van Horne and the Late Miss Mary Van Horne.

The following 108 species of Fungi have been found in the vinicity of St. Andrews, New Brunswick, from 1895–1908 by the late Miss Mary Van Horne, and Miss Adaline Van Horne.

Critical species, it may be stated, have been submitted to Professor Charles Peck, State Botanist of New York for identification or verification.

1.—Amanita muscaria, Linn. Minister's Island—August 1899.

2.—   "   phalloides, Fr.   "   "   September 1904.

3.—   "   verna, Bull.   "   "   September 1904.

4.—Amanitopsis vaginata var fulva, Schaeff   Minister's Island, July 1901.

5.—   "   "   var livida, Pers.   Minister's Island, August 1902.

6.—   "   var alba,   Minister's Island, August 1901.

7.—Lepiota naucinoides, Pk. Fort Tipperary, St. Andrews, Sept. 1905 an grounds of Risford, near St. Andrews, September 1901.

8.—Armillaria imperialis, Fr. Minister's Island, September 1905.

This was sent to Professor Peck for identification. It was the first specimen he had seen, and it was kept for the N.Y. State Museum Herbarium. He says regarding it, "It is a magnificent species, and I am very glad you sent me this specimen."

9.—Armillaria mellea, Vahl.—Minister's Island,   October 1901.

10.—Tricholoma personatum, Fr. Minister's Island September 1907.

11.—   "   subacutum, Pk.   "   "   September 1904.

12.—   "   rutilans, Schaeff   "   "   July 1900.

13.—   "   equestre, Linn.   "   "   September 1904.

14.—   "   vaccinum, Pers. (very abundant),
Minister's Island,   September 1904.

15.—Clitocybe nebularis, Batsch, Minister's Island, October 1901.

16.—   "   laccata, Scop. var pallidifolia Pk., Minister's Island, October 1901.

17.—   "   odora, Bull., Chamcook Mt., September 1907.

18.—Pleurotus ostreatus, Pk. Minister's Island, June 1900.

19.—Hygrophorus pudorinus. Fr., Minister's Island October 1904.

Also in great quantity in woods about Chamcook Mt. October 1907.

5 GEORGE V., A. 1915

20.—Hygrophorus chrysodon Fr., Minister's Island, September 1907.
21.—   "   puniceus Fr.,   "   "   September 1907 and
                              Chamcook Mt.,   October 1904.
22.—   "   virgineus Fr., Sheep Pasture, Minister's Island,
                              August and September 1897.
23.—Lactarius affinis, Pk. Minister's Island   October 1901.
24.—   "   theiogalus, Fr., "   "   October 1901.
25.—   "   aquifluus, Pk.—var. brevissimus, Pk. Minister's Id., Sept. 1904.
26.—   "   deliciosus, Fr. Minister's Id., July 1895, in great quantity on
                              MacMaster's Island, Aug. 1896.
27.—   "   exsuccus, Sm. Minister's Island, July 1897.
28.—   "   lignyotus Fr. Minister's Island,   August 1900.
29.—   "   torminosus, Schaeff var. necator, Minister's Island, October 1901
30.—   "   piperatus (Scop), Fr. Minister's Island, September 1897.
31.—   "   glyciosmus, Fr. Minister's Island, September 1904.
32.—   "   rufus, Scop.   "   "   September 1904.
33.—Russula alutacea Fr. Minister's Island,   July and August 1895.
34.—   "   emetica Fr.,   "   "   July and August 1895.
35.—   "   virescens Fr., Ghost Road, Chamcook, N.B., August 1897.
                              and also Minister's Island, August 1900.
36.—   "   heterophylla Fr., Minister's Island, August 1895.
37.—   "   aurea Fr., Minister's Island.   August 1901.
38.—   "   brevipes Pk.,   "   "   July 1900.
39.—   "   albella Pk.,   "   "   July 1900.
40.— Cantharellus cibarius. Fr. August to October 1895.
41.—   "   aurantiacus Fr. var. pallidus Pk. Minister's Island,
                              October 1901.
42.—   "   floccosus. Schw. Minister's Island, August and September
                              1900.
43.—Marasmius oreades Fr., Minister's Island, and Golf Links, August 1902,
                              Mr. Maxwell's lawn, August 1907.
44.—   "   urens, Fr., Minister's Island, August 1900.
45.—   "   cohaerens, (Fr.) Bres., Minister's Island, October 1904.
46.—Lentinus lepideus, Fr., Minister's Island,   July 1897.
47.—Entoloma lividum, Bull., Minister's Island,   October 1900.
48.—Clitopilus prunulus, Scop.   "   "   August 1900.
49.—   "   orcellus, Bull.,   "   "   Aug. and Sept. 1900.
50.—   "   subvilis, Pk.;   "   "   October 1901.
51.—Pholiota caperata, Pers. (rare), Minister's Island, September 1904.
                              Rather abundant in August 1908.
52.—   "   squarrosa, Müll., Minister's Island, August 1908.
53.—   "   lutea, Pk., growing on birch tree, Minister's Island, Aug. 1908.
54.—Inocybe fastigiata, Schaeff, Minister's Island,   September 1899.
55.—Flammula alnicola, var. marginalis Pk., Minister's Island, Sept 1904.

SESSIONAL PAPER No. 39b

56.—Cortinarius ochroleucus (Schaeff) Fr. Minister's Island,
September 1904.

57.—Cortinarius violaceus, Fr. Minister's Island   September 1897.

58.—   "   armillatus, Fr.   "   "   September 1897.

59.—   "   turmalis, Fr.   "   "   October 1905.

60.—   "   coerulescens, Fr.   "   "   September 1904.

61.—   "   collinitus, Fr.   "   "   July 1900 and
October 1905.

62.—   "   albo-violaceus, Pers. Minister's Island, September 1904.

63.—Cortinarius cinnamoneus, Fr. var. semi-sanguineus, Minister's Island,
September 1904 and October 1901.

64.—   "   evernius Fr., Minister's Island,   October 1904.

65.—Paxillus involutus (Batsch) Fr., Minister's Island, September and Oct. 1910, and September 1904.

66.—Agaricus campestris, Linn. Minister's Island   September 1900.

67.—   "   silvicola, Vitt. Minister's Island,   July and Sept. 1907,
and Ghost Road, Chamcook, N.B. July 1899.

68.—   "   semi-orbicularis, Bull., St. Andrews, July 1900.

69.—Hypholoma perplexum, Pk. Minister's Island, October 1901,
also near St. Andrews, same date.

70.—   "   incertum, Pk. Covenhoven Garden, Minister's Island,
October 1901.

71.—   "   sublateritium, Schaeff, Minister's Island, September 1904.

72.—Coprinus atramentarius (Bull.) Fr., Minister's Island, July 1899
and September 1901, and September 1907.

73.—Panaeolus retirugis, Fr., Minister's Island,   September 1904.

74.—Boletus edulis, Bull., Minister's Island,   July and August 1899,
and in great quantity Senator MacKay's place, September 1905.

75.—Boletus edulis clavipes, Pk., Minister's Island,   October 1901 and
September 1907.

76.—   "   felleus, Bull., Minister's Island,   September 1900.

77.—   "   scaber, Fr.   "   "   July and August 1897.

78.—   "   chromapes, Frost "   "   September 1899.

79.—   "   clintonianus, Pk. "   "   September 1899.

80.—   "   piperatus, Bull.   "   "   August 1899.

81.—   "   luridus, Schaeff.   "   "   July and August 1899,
and Golf Links October 1901.

82.—   "   versipellis, Fr. Minister's Island,   August 1899.

83.—   "   cyanescens, Bull.   "   "   August and Sept. 1897.

84.—   "   chrysenteron Tr.   "   "   August 1899, and Bar
Road same month.

85.—Polyporus perennis, Fr. Ghost Road, Chamcook, July 1897.

86.—   "   betulinus, Fr., Minister's Island, Sept. 1899, and St. John
Road near Chamcook, September 1900.

5 GEORGE V., A. 1915

87.—Hydnum　imbricatum, L. Golf Links, St. Andrews, August 1899.
88.—　　"　　repandum, L. Chamcook Lake, August 1899 and Minister's
　　　　　　　　　　Island, September 1900.
89.—　　"　　rufescens, Pers. Golf Links, St. Andrews, August 1899.
90.—Clavaria purpurea, Fr. (rare), Minister's Island, August 20th, 1908.
　　　　　　New to N. Y. State Herbarium. Name confirmed by Professor
　　　　　　Peck.

91.—Clavaria	formosa, Pers. Minister's Island	August and September 1899
92.—　"	amethystina, Bull. "　"	July 1900.
93.—　"	fastigiata, D. C.,　"　"	September 1899.
94.—　"	coralloides, L.,　"　"	September 1899.
95.—　"	cristata. Holmsk.　"　"	September 1899.
96.—　"	aurea, Schaeff　"　"	September 1897.
97.—　"	botrytes Pers.　"　"	October 1904.
98.—Helvella lacunosa Afzel.　"　"		September 1902.
99.—Leotia lubrica, Pers.　"　"		September 1907.

100.—Gyromitra esculenta, Fr. Chamcook, August 1901, and Minister's Island,
　　　　　　　　　　　　　October, 1904.*
101.—Mitrula vitellina, Sacc. var. irregularis, Pk., Minister's Island Sept. 1904.
102.—Spathularia velutipes, Cooke and Farlow, Minister's Island, September 1900.
103.—Peziza aurantia, Pers., Minister's Island, October 1901.
104.—Hypomyces lactufluorum, Schw., Minister's Island, and in woods near
　　　　road approaching Chamcook Lake; in great quantity during the summer
　　　　of 1896.
105.—Phallus impudicus, Linn., Minister's Island, July 1897. Not found since.
106.—Lycoperdon pyriforme, (Schaeff), Minister's Island, September 1899.
107.—　　"　　gemmatum, Batsch, Minister's Island, August 1899.
108.—Scleroderma vulgare Fr., Minister's Island, September 1897. Not seen
　　　　in recent years.

———

*Suggested by Professor J. H. Faull, Toronto, as possibly *Helvella enfula*, Schaeff., *G. esculenta*
being a spring form.—(ED.)

/II.'

THE IODINE CONTENT OF THE MARINE FLORA AND FAUNA IN THE NEIGHBOURHOOD OF NANAIMO, VANCOUVER ISLAND, B.C.

(With an Appendix on the Economic Value of the Pacific Kelps)

By A. T. CAMERON, M.A., B.Sc.,
Assistant Professor of Physiology and Physiological Chemistry, University of Manitoba.

The two outstanding facts of biological importance in the history of the element iodine are the discovery of the element by Courtois in sea-weeds, in 1811, and the discovery of its presence in the thyroid gland by Baumann in 1885. Subsequent to the latter discovery, most of the biological investigation of the element was directed to discover its function in the thyroid gland. In spite of a very large number of papers which have appeared during the last twenty years, this function is still obscure. It has been shown with fair certainty that iodine is a constituent of all normal thyroid tissue,[1] and that the amount present is dependent on the amount present in the diet. I showed recently that iodine is present in the thyroid of the dog-fish *Scyllium canicula* in amount greater than any previously recorded,[2] and this fact suggested the desirability of making comparative determinations of the iodine content of the various forms of sea-life, since the element is known to be present in sea-water, and since here the relative effect of a constant iodine diet should show itself distinctly.

Iodine is known to be present in most Sea-weeds, and has further been discovered in Sponges and Corals. In these two kinds of animals it has been established beyond doubt[3] that it is present in organic combination, and at least in part in a protein complex in a radical derived from di-iodo-tyrosine. Definite proof has yet to be adduced of the presence of a similar complex in the thyroid, for though evidence supporting such a hypothesis has been put forward by Oswald and others,[4] the exact nature of the combination of iodine in Oswald's "thyreoglobulin"[5] has yet to be determined. I considered that further data as to the amount of iodine present in different kinds of marine organisms, and especially as to the kind of

---

[1] I have summarised the evidence in favour of this view in a paper on "The presence of Iodine in the Thyroid Gland," J. *Biol. Chem.*, 1914, **16**, 465.

[2] *Biochemical J.*, 1913, **7**, 466.

[3] See for example, Wheeler and Mendel, J. *Biol. Chem.*, 1909, **7**, 1; Drechsel, *Zeitschr. f. Biol.*, 1896, **33**, 85; Mörner, *Zeitschr. f. physiol. Chem.*, 1907, **51**, 33; 1908, **55**, 77, 223.

[4] Cp. for example, Oswald, *Arch. f. exp. Path. u. Pharm.* 1908, **60**, 115; Nürnberg, *Biochem Zeitschr*, 1909, **16**, 87.

[5] Oswald, *ibid.*, 1901, **32**, 121.

5 GEORGE V., A. 1915

tissue found to contain it, might throw fresh light upon the problem of its presence in the thyroid gland.

Further, should kelp be utilized extensively as a source of potash for fertilizing purposes, as seems not unlikely from recent investigations,[1] the iodine present in the kelp would become the chief bye-product of such an industry; hence additional information as to its distribution and variation in different algae seemed also likely to lead to results of value.

With the permission of the Biological Board of Canada, I collected material during August and part of September, 1913, at and near the Biological Station at Departure Bay, B.C. This material I have subsequently analysed in the Physiological Chemical Laboratory of the University of Manitoba.

A large number of specimens of different species of algae were obtained, and also specimens of representatives of most of the animal phyla. The selection of the latter was made more or less at random, and analysis of different tissues of the species examined was also not systematic; the investigation is to be regarded as preliminary, with the purpose of indicating the direction for further work. Complete examination of the tissues of the dog-fish *Squalus sucklii* was carried out.

The various specimens were collected at the following points: At the Biological Station, or at points within half a mile of it (including Jesse Island); north-west of the Station, in the neighbourhood of Hammond Bay and the Lagoon; near Snake Island, two miles east of the Station; near Protection Island, two miles south-east; in False Narrows, about eight miles south-east of the Station; north of Breakwater Island, two miles east of False Narrows; on Mudge Island, two miles south of False Narrows;

Methods of Preservation and Analysis of Material: The algae were air-dried, further dried over sulphuric acid, and finally heated at 100° C. to constant weight. The remaining material (except in the case of a few shells and tests which were air-dried) was preserved in absolute alcohol, or in a few cases in dilute formaldehyde. In all cases before analysis the alcohol (or formaldehyde) was evaporated and the material heated to constant weight in the water oven at 100°, so that the results are all expressed for dry tissue.

Hunter's method of analysis has been adopted.[2] It has been shown by Seidell[3] and others that this is a very accurate method for analysis of small quantities of organically combined iodine. The results given by it are slightly higher than those obtained by the Baumann method or its various modifications, one or other of which have hitherto usually been employed.

---

[1] See "Fertilizer Resources of the United States," Senate Document, No. 190, 62nd Congress 2nd Session, 1912.

[2] Hunter, J. *Biol. Chem.*, 1910, **7**, 321.

[3] Cp. Seidell, *ibid.*, 1911, **10**, 95.

The results obtained are shown in the following tables:

(A). PLANTS.

I. Algae.

(1) Sub-class *Chlorophyceae*, family *Ulvaceae*.
A large number of complete plants were taken in each case, so that the results can be regarded as a fair average.

Species	Where obtained	Amount taken	Iodine found	Per cent. Iodine
Monostroma fuscum	Station; at low tide.	0.500 g.	0.000024 g.	0.005%
	Breakwater I., at low tide	0.500	0.000021	0.004
Ulva lactuca var. latissima (?)	Dredged in Departure Bay.	0.500	0.000103	0.021
Enteromorpha compressa	Station; at low tide.	0.500	0.000043	0.009
		0.500	0.000045	0.009
				Mean 0.009%
	Breakwater I.	0.200	0.000006	0.003
		0.197	0.000006	0.003
				Mean 0.003%

(2) Sub-class *Phaeophyceae*
i. Family *Desmarestiaceae*, species *Desmarestia ligulata*.
A single specimen, dredged near the north end of Breakwater Island.

Species	Amount taken	Iodine found	Per cent. Iodine
Desmarestia ligulata	0.500 g.	0.000171 g.	0.034%

ii. Family *Laminariaceae*
The following analyses were carried out on single plants, and on parts of the same plant.

Species	Where obtained	Part examined	Amount taken	Iodine found	Per cent. Iodine
Agarum fimbriatum	Dredged; Breakwater I.	Frond	0.500 g.	0.000112 g.	0.022%
Laminaria saccharina	Dredged; Breakwater I.	Frond	0.500	0.000770	0.154
			0.500	0.000790	0.158
					Mean 0.156%
		Stipe and holdfast	0.500	0.001045	0.209

5 GEORGE V., A. 1915

Species	*Where obtained*	*Part examined*	*Amount taken*	*Iodine found*	*Per cent. Iodine*
	Jesse I., just below low water	Frond (small)	0.250	0.000370	0.148
			0.250	0.000411	0.164
					Mean 0.156%
		Frond (average)	0.2002	0.000354	0.177
		Frond (sample of 3	0.500	0.000895	0.179
Laminaria bullata	Breakwater I. Dredged.	Frond	0.500 g.	0.000300 g.	0.060%
Nereocystis lütkeana	Near Station (small specimen)	Frond	0.500	0.000920	0.184
		Float	0.500	0.000602	0.120
		Stipe	0.0825	0.000121	0.147
	Near Station (Average specimen)	Frond	0.500	0.000855	0.171
		Float	0.500	0.000449	0.090
		Stipe	0.500	0.000804	0.161
		Holdfast	0.500	0.000419	0.084
	Protection I. (small specimen)	Frond	0.500	0.000321	0.064
			0.500	0.000318	0.064
					Mean 0.064%
		Float	0.250	0.000543	0.217
		Stipe	0.498	0.000427	0.085
		Holdfast	0.500	0.000528	.0.105
			0.399	0.000413	0.103
					Mean 0.104%
	Protection I. (Average specimen)	Frond	0.500	0.000649	0.130
		Float	0.200	0.000216	0.108
		Stipe	0.500	0.000229	0.046
		Holdfast	0.500	0.000855	0.171
	Breakwater I. (Small specimen)	Frond	0.500	0.000801	0.160
		Float	0.500	0.000058	0.011

iii. Family *Fucaceae*

The whole plant was taken, and each sample analysed was taken from a number of plants.

Species	Where obtained	Amount taken	Iodine found	Per cent. Iodine	Average Iodine
Fucus evanescens	Near Station; above low tide mark.	0.500 g. 0.500	0.000093 g. 0.000094	0.019% 0.019	0.019%
	Jesse I.	0.500	0.000063	0.013	0.013
	Breakwater I.	0.500 0.500	0.000040 0.000042	0.008 0.008	0.008
Fucus furcatus	Near Station; above low tide mark.	0.500 g.	0.000087 g.	0.017%	0.017%
	Jesse I.	0.500 0.500	0.000071 0.000063	0.014 0.013	0.013
	Protection I.	0.500 0.500	0.000129 0.000130	0.026 0.026	0.026

(3) Sub-class *Rhodophyceae*

i. Family *Nemalionaceae*

A number of specimens of *Gelidium amansii* were sampled, the whole plant being taken.

Species	Where obtained	Amount taken	Iodine found	Per cent Iodine
Gelidium amansii	Dredged; Departure Bay.	0.400 g.	0.000369 g.	0.092%

ii. Family *Gigartinaceae*

A single plant of *Gigartina radula* was examined; a number of specimens supposed to be *Gigartina mamillosa* were sampled.

Species	Where obtained	Part examined	Amount taken	Iodine found	Per cent iodine
Gigartina radula	Breakwater I.	Frond	0.500 g.	0.000037 g.	0.007%
		Frond without papillae	0.500	0.000032	0.006
		Papillae	0.250	0.000016	0.006
Gigartina mamillosa (?)	Breakwater I.	Whole plant	0.499 0.250	0.000082 0.000038	0.016 0.015
				Mean,	0.016%

5 GEORGE V., A. 1915

iii. Family *Rhodomeliaceae*

Samples of a number of specimens of *Rhodomela larix* were examined, the whole plant being taken.

Species	Where obtained	Amount taken	Iodine found	Per cent Iodine
Rhodomela larix	Breakwater I.	0.500 g.	0.000073 g.	0.014%

iv. Order *Delesseriaceae*

Samples from several plants in each case.

Species	Where obtained	Amount taken	Iodine found	Per cent Iodine
Nitophyllum ruprechteanum	False Narrows	0.1000 g.	0.000155 g.	0.155%
		0.1500	0.000241	0.161
				Mean 0.158%
Nitophyllum violaceum	Breakwater I.	1.500	0.000636	0.127

v. Family *Cryptonemiaceae*

Samples from a number of plants in each case.

Species	Where obtained	Amount taken	Iodine found	Per cent Iodine	Average
Prionitis lyallii	Departure Bay	0.500 g.	0.000216 g.	0.043%	0.043%
Corallina officinalis	Breakwater I., above low water	0.500	0.000028	0.006	
		0.500	0.000024	0.005	0.005

vi. Family *Bangiaceae*

The fronds of single plants of *Porphyra vulgaris* were examined.

Species	Where obtained	Amount taken	Iodine found	Pec cent Iodine
Porphyra vulgaris	Jesse I., just below low water mark.	0.500 g.	0.000057	0.011%
	"	0.500	0.000026	0.005
		0.500	0.000030	0.006
				Mean 0.005%
	Breakwater I., dredged	0.500	0.000056	0.011
	Protection I.	0.500	0.000047	0.009

II. Flowering Plant.

Species	Where obtained	Part examined	Amount taken	Iodine found	Per cent Iodine	Average
Zostera	Near Station	Blades	0.500 g.	0.000015 g.	0.003%	
marina			0.500	0.000007	0.001	0.002%
		Stalk	0.300	0.000010	0.003	
			0.300	0.000005	0.002	0.002
		Roots	0.1500	0.000019	0.013	
			0.1000	0.000014	0.014	0.013

(B) ANIMALS.

(1) Phylum *Porifera*

Six species of sponges have been examined, one calcareous, *Aphrocallistes whiteavesianus*, and five non-calcareous. Single specimens were examined in each case.

Species	Where obtained	Amount taken	Iodine found	Per cent Iodine
Aphrocallistes whiteavesianus	Dredged off Snake I.	0.500 g.	0.000097	0.019%
Rhabdocalyptus dowlingii	" " "	0.548	0.000075	0.014
Bathydorus dawsonii	" " "	0.499	0.000045	0.009
Myxilla parasitica	" " " (adhering to scallop shells)	0.500	0.000049	0.010
Esperella adhaerens	" " " "	0.501	0.000073	0.015
		0.501	0.000074	0.015
			Mean	0.015%
Reniera rufescens	Found near Station at very low tide.	0.500	0.000058	0.012

(2) Phylum *Coelenterata*

The specimens of *Obelia* were attached to the wharf at the Station. They were washed free from dirt, and preserved in alcohol. Foreign organisms present (diatoms, ostracods, caprellae) certainly did not amount to one per cent. of the total weight. A number of *Aequorea* were obtained in False Narrows. The *Aurelia* were obtained in the same region. The sea-anemones were obtained on rocks at Jesse I. The complete organism was not obtainable, but the larger part was removed by cutting as the organisms hung above low water. The comb-jellies, probably a species of *Pleurobrachia*, were obtained near the Station. These four species were preserved in dilute formaldehyde. Their weights, after hardening

5 GEORGE V., A. 1915

by the formaldehyde, were determined, and then the whole evaporated to dryness. The "formaldehyde" and dry weights are quoted, although I do not know in how far the original weight was altered by the addition of the formaldehyde. The dried material appeared to consist chiefly of crystalline salts. Some of the iodine if present, may have been lost by the evaporation of what was initially a slightly acid solution.

Class	Species	(Fresh) Weight	Dry Weight	Amount taken	Iodine found	Per cent Iodine
Hydrozoa:	Obelia longissima			0.500 g.	0.000067 g.	0.013%
				0.500	0.000064	0.013
				0.500	0.000066	0.013
					Mean	0.013%
	Aequorea forskalea	317 g.	17.20 g.	0.500	0	0
Scyphozoa:	Aurelia flavidula	158	9.96	0.500	0	0
Actinozoa:	Metridium marginatum	83	7.74	0.500	0	0
Ctenophora:	Pleurobrachia (?)			0.500	0	0

(3) Phylum *Vermes*, sub-phylum *Annulata*, order *Polychaeta*

Unfortunately, especially in view of the strikingly high figures obtained for some species, I have been unable so far to have all of the species examined definitely identified. The worms were preserved in alcohol, the tubes air-dried.

Species	Where obtained	Part examined	Amount taken	Iodine found	Per cent Iodine	Average
A *Nereis* worm	Mudge I.	Whole worm	0.500 g.	0.000043 g.	0.009%	
			0.500	0.000035	0.007	0.008%
	Lagoon	" "	0.500	0.000094	0.019	
			0.500	0.000082	0.016	0.017
A *Nepthys* worm	Mudge I.	" "	0.400	0.000035	0.009	0.009
			0.500	0.000124	0.025	
Diopatra (Sp.?)	Mudge I.	Worm	0.500	0.000109	0.022	
•			0.500	0.000115	0.023	0.023
		Horny tube {0.300	0.001247	0.416	
		Inner layer {0.1000	0.000411	0.411	0.414
		Horny tube {0.500	0.001358	0.272	
		Outer layer {0.300	0.000741	0.247	0.262
Serpula columbiana	Jesse I.	Worm	0.500	0.000192	0.038	
			0.500	0.000198	0.040	
			0.500	0.000189	0.038	0.039
		Calcareous tube	0.500	0.000159	0.032	
			0.500	0.000156	0.031	
			0.500	0.000137	0.027	0.030

SESSIONAL PAPER No. 39b

(4) Phylum *Molluscoïda* class *Polyzoa*, family *Cellularina*, species *Bugula flabellata*.

The specimens examined were obtained on a plant of *Laminaria bullata* dredged in Departure Bay. They were washed free from adhesive material (examination under the microscope revealing the presence of only a few foreign forms) and were preserved in absolute alcohol.

Species	Amount taken	Iodine found	Per cent Iodine	Average
Bugula flabellata	0.2500 g.	0.000039 g.	0.016%	
	0.1000	0.000017	0.017	0.016%

(5) Phylum *Echinodermata*
 i. Class *Echinoidea*

Species	Where obtained	Part examined	Amount taken	Iodine found	Per cent Iodine
Strongylocentrotus drobrachiensis	False Narrows	Aristotle's lantern	0.500 g.	0	0 %
		Internal organs	0.0697	0.000014 g.	0.02
		Gonads and contents	0.500	0.000018	0.004
			0.500	0.000015	0.003
				Mean	0.003%
Strongylocentrotus franciscanus var. purple	False Narrows	Test	0.500	0	0
		Spines	0.500	0	0
		Internal organs	0.250	0.000125	0.050
			0.300	0.000139	0.046
			0.1000	0.000058	0.058
				Mean	0.049
		Gonads and contents	0.500	0.000004	trace
Strongylocentrotus franciscanus var. red	False Narrows	Aristotle's lantern	0.500	0.000010	0.002
				0.000007	0.001
				Mean	0.001

 ii. Class *Holothuroidea*

A specimen of *Stichopus californensis* (dredged up north of Hammond Bay) was examined. I am not satisfied with the results, but they indicate that if iodine is present, it is present in relatively very small quantity.

5 GEORGE V., A. 1915

Species	Part examined	Amount taken	Iodine found	Per cent Iodine
Stichopus californensis	Integument (preserved in alcohol)	0.500 g.	0.000003	0.001%
	Integument (air-dried)	0.503	0.000018	0.004
	Internal organs	0.250	0.000005	0.002
		0.250	0.000005	0.002
			Mean	0.002%
	Muscle	0.1000	0	0

iii. Class *Asteroidea*

One complete ray of the whole animal was preserved in alcohol, and a sample of the whole ray examined.

Species	Where obtained	Amount taken	Iodine found	Per cent. Iodine	Average
Pyknapodia helianthoides	Jesse I.	0.500 g.	0.	g. 0.	%

(6) Phylum *Arthropoda*, class *Crustacea*

The barnacles, *Balanus*, were attached to posts at the station Pier; the specimen of *Cancer* was obtained in shallow water at the same point.

Species	Part examined	Amount taken	Iodine found	Per cent Iodine
Balanus balanoides	Shell	0.500 g.	0. g.	0 %
	Soft part	0.200	0.000010	0.005
Cancer productus	Carapace	0.500	0.000016	0.003
		0.500	0.000015	0.003
			Mean	0.003%
	Muscle	0.2000	0	0

(7) Phylum *Mollusca*, class *Pelecypoda*

Species	Where obtained	Part examined	Amount taken	Iodine found	Per cent. Iodine
Mya arenaria	Station	Shell	0.500 g.	0.	0 %
		Soft part	0.400	0.000035	0.009
			0.400	0.000035	0.009
				Mean	0.009%

Species	*Where obtained*	*Part examined*	*Amount taken*	*Iodine found*	*Per cent. Iodine*
Schizothoerus nuttalli	Mudge I.	Shell	0.501	0	0
		Outside cuticle of foot	0.300	0.000893	0.298
		Inside muscle of foot	0.1995	0	0
		Heart and Kidney	0.0350	(0.000009)	(0.02)
		Gonads and contents	0.·500	0	0
		Gills	0.2000	0	0

A second analysis of the outer cuticle of the foot of Schizothoerus indicated a result of the same order but was spoilt before completion. The figure given for heart and kidney requires confirmation.

(8) Phylum *Chordata*.

i. Sub-phylum *Tunicata*.

Only a few specimens of one form were obtained (at Mudge Island, at low tide), and these did not yield sufficient material for definite results except in the case of the test.

Species	*Part examined*	*Amount taken*	*Iodine found*	*Per cent. Iodine*
Pyura haustor	Test	0.300 g.	0.000605 g.	0.202%
		0.300	0.000595	0.198
				Mean 0.200%
	Inner layer of test	0.1500	0.000016	0.010
	Mantle	0.1000	0.000012	(0.012)
	Gonads	0.2500	0	0

ii. Subphylum *Vertebrata*, class *Pisces*, sub-class *Elasmobranchii*, species *Squalus sucklii*.

The dog-fish was selected for examination, since I had already shown that the thyroid contained a relatively large amount of iodine, and since only in elasmobranch fishes is the thyroid found encapsuled, so that its dissection in teleosts is almost impossible without removing much adjacent tissue. The thyroid material was obtained from 82 specimens of *Squalus sucklii* caught by local fishermen in the course of one night. Sample 1 was a fair sample of the material obtained from 32 females, sample 2 from 34 females, sample 3 from 16 males, and sample 4 from 133 "pups" contained in the females. This last sample undoubtedly contained a large amount of connective tissue, removed in order to be certain that the thyroid

5 GEORGE V., A. 1915

material was obtained. The other tissues examined were obtained from two female specimens, with the exception of the testes, taken from a male specimen selected at random.

Tissue examined	Amount taken	Iodine found	Per cent. Iodine.	
Thyroid, sample 1	0.2015 g.	0.000394 g.	0.195%	
sample 2	0.2003	0.000391	0.195	
	0.1005	0.000197	0.196	Mean 0.195%
sample 3	0.1000	0.000224	0.224	
sample 4	0.0604	0.	0	
Heart	0.1000	0	0	
Pancreas	0.500	0	0	
Spleen	0.500	0	0	
Brain	0.251	0	0	
Rectal Body	0.401	0	0	
Testes	0.500	0	0	
Ovaries and Eggs	0.500	0	0	
Muscle	0.500	0	0	
Skin	0.499	0	0	
Vertebræ	0.500	0	0	
Kidney	0.499	0.000017	0.003	
	0.400	0.000012	0.003	Mean 0.003
Liver Oil	0.741	0	0	
Liver Residue	0.522	0.000015	0.003	
	0.533	trace	trace	
	0.528	0.000004	0.001	Mean 0.001
"Dog-fish oil"	1.500	0	0	

The liver residue was obtained by beating the liver at 100° C. for some time, and pouring away the clear oil. It consisted of an oily mass which could not be sampled properly (whence the varying results) and amounting to only three-elevenths of the whole. The "dog-fish oil" was a sample of the commercial oil sold in Nanaimo and used for miners' lamps. Various fish liver-oils have been reported to contain iodine [1], but in amount not detectable by the method of analysis I have employed.

[1] See for example, Stanford, *Chem. News*, 1883, **48**, 233.

The results obtained for the thyroids of *Squalus* permit a direct comparison with those for mammalian tissue in determining the relative amount of iodine, and of thyroid tissue, per kilogram of body-weight of the whole animal. The total amount of thyroid tissue obtained from 66 female fish was 1.459 grams. Ten of these fish selected at random gave an average weight of 3.8 kg. The average iodine content in the dried thyroid tissue was 0.195 per cent. The 16 male fish yielded 0.169 gram dried thyroid containing 0.224 per cent. iodine. Ten of these fish selected at random gave an average weight of 2.5 kg.

These figures may be compared with those obtained from analyses of twelve dogs (ordinary laboratory animals of no particular variety) which I have already published ([1]) in which the total weight of the dogs was 191 kg., they contained 14.33 grams thyroid tissue, containing 0.95 per cent. iodine. Hence:

Squalus sucklii (female) contains per kg. body weight 0.0058 g. dry thyroid tissue containing 0.000011 g. iodine.

Squalus sucklii (male) contains per kg. body weight 0.0042 g. dry thyroid tissue containing 0.000009 g. iodine.

Canis contains per kg. body weight 0.075 g. dry thyroid tissue containing 0.00007 g. iodine.

The figures indicate that both iodine content and amount of tissue are smaller, but of comparable order. If the figures obtained by me for *Scyllium canicula* can be regarded as comparable for body weight (I obtained the value 1.16 per cent. iodine in dry thyroid tissue ([2]), but have no data as to the weight of the fishes from which the tissue was obtained), this species would give a much closer figure to that for mammals. The cause of the difference obtained for the two species of dog-fish may be a seasonal variation (the *Scyllium* thyroids were obtained in winter), or a different diet containing less iodine, or the difference may be specific for the two species. Further work is indicated in this direction.

Discussion of Results.

In considering the results for algæ, it becomes evident that while every species examined contained iodine in detectable amount, only those of two families *Laminariaceæ*, and *Delesseriaceæ*, contained amounts of the order of 0.1 per cent. The results are in substantial agreement with those obtained by Turrentine ([3]) for many of the same species of algæ obtained further south, with the exception that many of his values are distinctly higher, in spite of the fact that his analytical method should lead to lower rather than higher figures. This is probably traceable to the fact that the specific gravity of the waters near Nanaimo is very low (due to the influx of large bodies of fresh water, such as the Fraser River), with a corres-

[1] *J. Biol. Chem.*, 1914, **16**, 472.

[2] *Biochemical J.*, 1913, **7**, 468.

[3] U. S. Senate Document, No. 190, 62nd Congress, 2nd Session, 1912, p. 220.

5 GEORGE V., A. 1915

pondingly low salinity, and a probably lower iodine content ([1]). (To the same lowered salinity may be due the total absence of *Macrocystis* in these waters, although it is common to the south, and has been reported much further north.) ([2]).

In the only species of alga in which different parts of the plant were systematically examined, *Nereocystis lütkeana*, markedly different iodine contents were observed. There seems to be no regularity in the results so far obtained, and further and more detailed work will be necessary in order to show how far variations exist throughout a single plant, in plants from the same locality, and in plants from different localities.

Balch, as the result of a few analyses of *Nereocystis* and similar forms has concluded that as a rule the stipe contains more iodine than the frond ([3]). It appears certain, both from Turrentine's figures and my own, that specimens of the same species of alga from different localities may contain differing amounts of iodine, but much further work including examination of both plants and surrounding sea-water will be necessary before any definite explanation of the variations can be put forward.

None of the sponges examined showed marked iodine content. There are no corals obtainable in the Nanaimo district. Of the types of animal life examined all except the free-floating forms and the star-fish, *Pyknapodia*, showed the presence of ·iodine in detectable amount, although in one or two cases—sea-cucumber, barnacle,—it was barely detectable. Hunter's method, employing 0.5 gram of material as in most of the above analyses, permits the detection of 0.001 per cent. of iodine with some certainty. A negative result with this amount of material indicates that the iodine is not present to an extent greater than 0.0005 per cent.

Macallum has shown that the iodine content of *Aurelia flavidula* is comparable with that of the surrounding sea-water, two litres by volume of *Aurelia* containing 0.00001 gram ([4]). His figures for the fresh tissue do not contradict mine for the dry residue, since as just mentioned, Hunter's method will not show the presence of quantities of this order.

The results obtained for the annelid worms are distinctly high. That for the inner layer of the *Diopatra* worm tube was the highest value recorded in the whole series of analyses.

The general distribution of iodine throughout the whole of the marine flora and fauna which is indicated by the results of this paper can be satisfactorily accounted for by a continuous circulation of the element in a succession of living organisms. Death and subsequent decay of a certain proportion of animals and plants would return organic and inorganic iodine to the sea-water. Such a

---

[1] During the five weeks stay at the Station, I made daily readings of the specific gravity of the water in Departure Bay. The average of 32 daily readings was 1.015, varying between the limits of 1.008 and 1.019. A few readings were taken at outside points from time to time, and these approximated to the higher value. The value for normal sea-water is about 1.03.

[2] See Setchell, U. S. Document No. 190, 1912, p. 135.

[3] J. *Industrial Chem.*, 1909, **1**, 777.

[4] J. *Physiology*, 1903, **29**, 213.

hypothesis is in line with Gautier's results for sea-water itself ([1]). He found that sea-water obtained at the surface contained no inorganic iodine, but only organically combined iodine, and iodine present in minute organisms, while the greater the depth from which the water was obtained the greater the amount of inorganic iodine it contained ([2]). His results, rigorously applied, would indicate that the algæ themselves obtain their iodine in organic form. This is perhaps not absolutely impossible, since various authors seem to have shown that algæ can assimilate organic material ([3]), including amino-acids (and as already pointed out at least part of the iodine in organic combination is in amino-acid groups), but it seems more probable that a minimal quantity of iodine reaches the inorganic stage, and is then reabsorbed by the algæ, so continuing the circulation.

A conclusion which may fairly be drawn from the data now presented is that with greater development in the organism is found greater specificity of iodine-containing tissue, until, in vertebrates, the thyroid is the only organ containing an appreciable amount. It is to be noted that in *Squalus sucklii*, the vertebrate type under examination, the only other organs in which iodine was detected were excretory organs.

Iodine has been found present in marked quantity in three different tissues in which, as far as I am aware, it has not previously been recorded. These are, the horny tube secreted by the worm *Diopatra*, the cellulose ("tunicine") test of the tunicate *Pyura*, and the external cuticle of the horse-clam *Schizothoerus*. I hope to examine these further, along with similar tissue in other species.

There are at present insufficient data for any generalisation as to the type of tissue containing iodine in relatively large amount, but it may be pointed out that the iodine in thyroid tissue is usually regarded as localised in the colloid material, which has been assumed—without experimental evidence—to consist of or to contain a globulin (thyreoglobulin), while the iodine in sponges is contained in *spongin*, a sclero-protein, that in corals in *gorgonin*, also a sclero-protein, and the organic substance of the serpulid tube is *conchiolin*, another sclero-protein. The external cuticle of *Schizothoerus* probably consists largely of a keratin, yet another sclero-protein. On the other hand, the organic material of at least one Eunicid worm (*Hyalinœcia*) appears to consist of *onuphin*, which although containing nitrogen seems closely related to dextrin or glycogen ([4]), the test of tunicates appears to consist largely of true cellulose ([5]) (which is easily caused to combine

---

[1] *Compt. rend.*, 1899, **128**, 1069.

[2] Ibid., 1899, **129**, 9.

[3] See Oltmann's "Morphologie und Biologie der Algen," 1905, Bd. 2, S. 155.

[4] Schmiedeberg, *Mitt. a. d. zool. Station zu Neapel*, 1882, **3**, 373. (*Note added to proof.* Since writing the above, I have found an observation of Mörner's (*Zeitschr. f. physiol. Chem.*, 1908, **35**, 83,) on the presence of iodine in the tubes of the worms *Hyalinœcia tubicola* and *Chœtopterus norvegicus*. The amounts are smaller, but of the same order as that I have found for *Diopatra*).

[5] Cp. Winterstein, *Zeitschr. f. physiol. Chem.*, 1894, **18**, 43.

. 5 GEORGE V., A. 1915

with iodine), and the feature markedly distinguishing the *Laminaria* from other Sea-weeds is the secretion of a mucilage also probably of a carbohydrate nature ([1]).

It seems not unlikely that careful examination of these different iodine-containing tissues may lead to the result that iodine is held in the living organism in but one or two types of organic compound. I hope to extend the work in this direction.

I wish to acknowledge my grateful indebtedness to Dr. Maclean Fraser, the Curator of the Nanaimo Biological Station, for his uniform kindness in assisting me in the work of collection and identification of the material described in this paper, to thank Mr. F. S. Collins for kindly identifying a number of algæ for me, and to thank Professors Swale Vincent and Buller for their interest and encouragement in the course of this work.

The expenses incurred in the collection and preservation of the material were defrayed by grants from the Biological Board. The expenses of the analytical work carried out at Manitoba University have been defrayed by grants through the Ductless Glands Committee of the British Association for the Advancement of Science, and (through Professor Vincent) from the Royal Society of London.

Appendix.

THE ECONOMIC VALUE OF THE PACIFIC KELPS.

The value of kelps as fertilizers has been known for a long time. In the British Islands, Norway, and the coast of Brittany they are gathered more or less extensively and spread as a manure. Along the Atlantic Coast of Canada and the New England Coast they are stated to be fairly extensively used; the torn kelp is thrown up on the shore in the Fall, and collection is rendered easy. They have been occasionally used along the Pacific Coast of the United States for the same purpose. In Japan they are extensively used for various purposes.

The fertilizing value of kelps is attributable chiefly to their potash content, and in some small part to their phosphate content. They also contain definite small quantities of iodine, although this probably does not increase their value as manures. In view of the great cost of potash fertilizers due to the increasing market and the monopoly held by the Stassfurt Syndicate, other sources of potash have been sought. The most promising of these are the giant kelp beds situated along the western coast of this continent.

The U. S. Government, realising the importance of this problem, have, during the past few years, charted out the kelp beds off their western coasts, including Alaska, to which two expeditions were sent last year for that purpose. At least two companies in California have started to extract potash from kelp, although the industry has scarcely got beyond the experimental stage ([2]).

---

[1] Very little work appears to have been carried out to determine the form in which iodine occurs in algae. Eschle (*Zeitschr. f. physiol. Chem.*, 1897, **23**, 30) showed that in *Fucus vesiculosus* and in *Laminaria digitata* the iodine was present almost completely in organic form, and considered that several different organic compounds containing iodine were present.

[2] J. *Industrial Chemistry*, 1913, **5**, 251.

In Mexico, a concession has been granted for the purpose of collecting and utilizing the kelp found floating along the western shores, and there seems possibility of commercial development here also ([1]).

Much of the information with regard to the possibilities of the kelp industry is to be found in the U. S. Senate Document, No. 190 (62nd Congress, 2nd Session, 1912), on "The Fertilizer Resources of the United States."

In this the conclusion is reached (p. 44) that the U.S. Pacific kelps could if necessary furnish per year over six million tons of potassium chloride, at present prices worth over $240,000,000, and over 19,000 tons of iodine, worth over $95,000,-000. One-sixth of these quantities could with ease be obtained, and even this would be far in excess of present requirements. This could be obtained, if the kelp were cut scientifically, without annual diminution of the size of the kelp beds. The cost of production was estimated to be covered by the value of the iodine and other bye-products, but this seems to me undoubtedly too low an estimate, since any competition would immediately lower the price of iodine (and also of potash).

Few of the algæ found along the Pacific coast can be utilized on account of the cost of collection. Of the shore forms *Fucus evanescens* and *Fucus furcatus* are found at low tide covering rocks everywhere, but they could only be collected by hand labour. Three forms of giant kelp seem particulary suitable. Far south *Macrocystis pyrifera* and *Pelagophycus porra* are found in quantity; further north the latter disappears, and yet further north, in the Puget Sound region, the principal kelp is *Nereocystis lütkeana* (bladder-kelp or bull-kelp). Each of these kelps grows in deep water, and consists of a large surface of leaves, supported by a bladder or float, which is attached by a thick stipe 40 or more feet in length to a spreading "holdfast" attached to rocks several fathoms below low water mark.

Of the three types mentioned, only *Nereocystis* will probably be found of importance economically along the Canadian Pacific coast. This plant is an annual, and could, according to Rigg ([2]), be harvested annually after the middle of July without diminution of quantity. It is found in large groves throughout the Puget Sound region. Specimens from this region contain 30 per cent. potassium chloride and 0.16 per cent. iodine. My iodine analyses for *Nereocystis* from the Nanaimo district gave similar figures, so that in all probability the potassium values are also of the same order.

The methods of harvesting this kelp, and of extracting from it its commercial products, are still in the experimental stage, although there seems no reason to doubt that the problems involved can be satisfactorily solved.

During my stay at Nanaimo last summer, I was only able to observe the kelp beds in this district for a distance of eight or ten miles on each side of Nanaimo itself. Plants of *Nereocystis* in greater or less quantity are to be seen floating wherever there is a ridge or rock running out a few feet below the sea surface. There are three fairly large beds in the area I inspected. One, in False Narrows, about eight miles south-east of Nanaimo, fills the space between Gabriola and Mudge Islands (necessitating careful navigation of the passage). It is from one and one-half to two miles long, and varies from 100 to 200 yards in width. A

---

[1] *ibid.*, 5, 338.

[2] U. S. Senate Document, No. 190, 1912, p. 43.

5 GEORGE V., A. 1915

second bed runs north east from Hammond Bay (five or six miles north of Nanaimo). It is about one and one-half miles long, and varies from 50 to 100 yards in width. The third bed covers a submerged ridge on the north side of Departure Bay; it is about three-quarters of a mile long, and from 50 to 100 yards wide. I had no opportunity of examining the beds further east and south, although, according to the charts, kelp is common in that region. The beds I saw were of medium thickness. I estimated (very roughly) that on the average there were about four plants to the square yard.

In order to obtain an approximation as to the weight of material to be obtained from such beds as those described, I weighed a number of plants of *Nereocystis* selected at random, and obtained in Departure Bay, with the following results:

The fronds of seventeen plants weighed on the average 16 oz.

The floats of seventeen plants weighed on the average 9 oz.

The stipes and holdfasts of nine plants weighed on the average 6 oz.

Samples of fronds and floats were dried, and the amount of moisture determined approximately:

5.3 g. fronds, fresh, yielded 0.57 g. dry material, approximately 11%

6.5 g. floats, fresh yielded 0.36 g. dry material, approximately 5.5%

Hence a single bed of *Nereocystis*, two miles long by 150 yards wide, and containing on the average four plants per square yard (such a bed as that in False Narrows) would yield 132 tons of dry material (neglecting the stipes, only short lengths of which would be removed by proper cutting), containing (assuming 30 per cent. potassium chloride present), about 40 tons potassium chloride, worth at $40 per ton ([1]) some $1,600. The figures utilized are all distinctly minima. My ratio for wet and dry material is distinctly less than that found by other observers. No account has been taken of the value of the iodine also obtainable. The actual weights of the plants were determined at the beginning of September, when the fronds had commenced to decay.

Further, and especially important, is the opinion of Setchell ([2]) that the degree of salinity affects the growths of these kelps. This is borne out by my observations in the Nanaimo region. The average length of nine of the plants examined was about 16 feet, while those reported on off the American coast run to 40 feet or even 70 feet ([3]). The fronds were not so large as those described in plants obtained further south. In the Nanaimo district, along a stretch of coast twenty miles in length only three beds of any size were met with. I have shown in the body of this paper that the salinity of the sea-water in this district is on the average about 1.015, instead of 1.03 as in normal sea-water. Nevertheless, if *Nereocystis* beds were scattered along the whole Pacific coast of the Dominion to no greater extent, their total economic value would be very considerable.

It seems extremely desirable that steps should be taken at an early date to investigate the extent of the beds through as great a region as possible, and especially in districts of greater salinity.

[1] Potassium "muriate," basis 80 per cent., is at present quoted in American lists at $39.07 per ton. See J. *Industrial Chemistry*, March 1914, Market Report. I have no information as to Canadian prices.

[2] U. S. Senate Document, No. 190, 1912, p. 135.

[3] *ibid.*, p. 42.

VIII.

ON SOME OF THE PARASITIC COPEPODS OF THE BAY OF FUNDY FISH.

By V. Stock, B.A.
University of Toronto.

The above field for investigation was suggested to me by Dr. A. G. Huntsman, Curator of the Marine Biological Station at St. Andrews, and I am greatly indebted to him for the kindness and assistance in collecting the material and examining the specimens.

The work was carried on between June 15 and September 9, 1912, around the Biological Station at St. Andrews and among the various islands of Passamaquoddy Bay. The parasites were collected from fish obtained by trawling, hand-lining, seining and also by visiting the various fish markets and weirs in the neighborhood. Occasionally also, excursions were made out into the Bay with the fishermen of the surrounding villages who offered every opportunity to examine the fish they caught.

1. Caligidae.

The kind of parasitic Copepods specially investigated were those belonging to the Family Caligidæ. Two species only were found *Caligus curtus* and *Caligus rapax.* Occasionally both forms were obtained on the same fish and they were found on the surface of the head, body and fins, and in the case of *C. rapax* the dorsum of the tail immediately anterior to the caudal fin appeared to be a favorite place for attachment. Only one parasite was found inside the gill cover. The subjoined table gives in brief form the general information obtained, and enables one to make comparison in regard to the parasites and the hosts from which they were collected.

In addition to the above species there were also examined:—Sculpins 123, Mummichogs 62, Sticklebacks 30, Butterfish 28, Herring 27, Smelt 23, Perch 14, Silver Hake 6, Dogfish 4, Shad 4, Mackerel 3, and also one each of Cunner, Halibut, but no Caligids were found on them.

It might be mentioned that in the above table there is not included an instance in which 190 young cod were dumped out of a weir-seine into the bottom of the boat along with a host of other struggling fish, were examined and only 3 Caligids collected from them. Another factor which should be considered in making comparisons is that in the table are included two instances, one in which 23 specimens of *C. rapax* were found on one cod, and another in which 27 of the same species were taken off a single hake—thus raising to a considerable degree the average number found in each species.

5 GEORGE V., A. 1915

It is perhaps worthy of note that in the case of *Caligus curtus* the hosts are entirely among the Gadoid fishes for both adult and chalimus stages. In this species there was practically no variation in the number found on the various forms of fish during different times of the season. *Caligus rapax* was found on a greater variety of fish and also in greater numbers on the host. As many as twenty-seven were collected from one fish, whereas the number of *C. curtus* rarely exceeded six per fish. *C. rapax* was also obtained from the Eel pout (*Zoarces anguillaris*) a host from which it has not yet been reported, This latter species was first noticed in small numbers towards the end of June, but by far the greater number were collected during the months of July and August.

Comparatively few chalimus stages were found and in many instances it was difficult to determine to which species the form belonged. The chalimus stages of *C. curtus* were obtained from the Cod and Tomcod, while those of *C. rapax* from. the Cod and Lumpsucker, chiefly from the latter, nine being collected from one specimen. Forms apparently belonging to the latter species were also collected from the Hake and Haddock. The chalimus stages were noticed particularly at two different periods—during the latter part of June and during the last week of August.

A large number of measurements were taken in order to ascertain whether there is any variation in size in the various forms throughout the season. Practically none whatever in either species was found. In the case of *C. curtus* the size of the parasite seemed to increase with the size of the host. The largest specimen of *C. curtus* obtained which was a male, was found to be 13.2 mm. and the female 11.8 mm. in length. It might be added that in the adult female only in a very few instances was the abdominal segment found to be longer than half the length of the genital segment—a marked difference from the findings of Dr. C. Branch Wilson, who in his report states the opposite to be the case. The largest specimen of *C. rapax* collected were female 6.4 mm. and male 5 mm.

In conclusion it might be added that these fish whose activity was impaired by disease or which are naturally slow in their movements appear to be particularly infested with the parasites, affording special opportunities in the chalimus stage to become attached. This is quite evident in the case of the Lumpsucker which lives among the seaweed and debris on the surface of the water and is particularly sluggish in its movements.

2. *Argulidæ.*

The fish were also examined for these parasites at the same time as the Caligids were being investigated. For the major part of the work credit must be given to Mr. N. A. Wallace who at the beginning of the season carried on all the collecting in this direction.

Only one species, *Argulus fundulus* (Kroyer) was found and this on three different hosts, *Pseudopleuronectes americanus* (Mummichog), *Heteroclitus fundulus* (Mummichog) and *Pygosteus pungitius* (Nine-spined Stickleback).

SESSIONAL PAPER No. 39b

These parasites were found attached anywhere on the surface of the body, on the gill covers and on the fins. Frequently they were completely embedded in the substance of the fin or body, resulting in a nodule showing marked inflammation.

In addition the following parasite Copepods were also found:—

Lernaea branchialis on *Gadus callarius*,

Pandorus sinuarus on *Carcharias littoralis*,

Nemesis robusta on *Carcharias littoralis*,

Chondrocanthus cornutus on *Pseudopleuronectes americanus*,

Chondrocanthus merluccii on *Merluccius bilinearis*, and the following unidentified forms:—

Chondrocanthus on the Sea Perch,

Lernæopodæ on *Raja lævis* (Barndoor Skate)

Anchorella on *Gadus callarias*, *Aeglefinus melanogrammus*, *Pollachius virens*.

Species	No. Examined.	C. curtus.			C. rapax.			Chalimus	Total No. Parasites.
		♂	♂	Total.	♀	♀	Total.		
Cod..........	154	16	79	95	6	71	77	6	178
Haddock......	103	12	32	44	10	46	56	2	102
Hake..........	168	13	32	45	5	17	22	1	68
Pollock........	38		11	11	2	3	5		16
Flounders......	122				2	5	7		
Conger Eel.....	19					1	1		-
Skate..........	95				1	9	10		10
Tom Cod......	12							2	2
Lump sucker ..	7				11	39	50	9	59
	718	41	154	195	37	191	228	20	443

IX.

SOME EXPERIMENTS ON THE FREEZING AND THAWING OF LIVE FISH.

By W. H. MARTIN, B.A.
University of Toronto.

The fishermen of the Bay of Fundy say that if, in very cold weather, a herring be thrown out on the ground and frozen so that it is apparently quite stiff, when thrown back into the water, it will swim off as soon as it thaws out again.

The following experiments were performed at St. Andrews, N.B., at the Marine Biological Station, Summer of 1913, to determine how low a temperature fish will stand and for what length of time they will survive such a temperature.

Methods.

For the experiments the species *Fundulus. heteroclitus* (the common mummichog) was used. They were easy to obtain in tide pools about St. Andrews. They are of convenient size for experiments and are wonderfully hardy: they are easily kept for several weeks in a tank, and were found to survive sudden changes of Temperature much better than any other fish used.

In the experiments a large carbide-tin was covered with felt and used as a refrigerator.

An inner tin vessel contained a mixture of ice and salt. The fish were placed in an inner jar in water or in air as required.

Results.

Experiment I. A dozen fish were put into sea water at 6°C. and the jar was lowered into the freezing mixture. The following table gives the results:—

Time.	Temperature.	Behaviour of Fish.
9.20	6° C.	All are swimming about in lively manner.
9.25	3° C.	" " " " " " "
9.32	0° C.	2 have fallen over on side—All seem to gasp for breath.
9.45	-1½° C.	All have stopped breathing and are apparently dead. Took one out and put into water at 12.5°- By 9.50 it was breathing and swimming a little. It recovered completely and lived for weeks.
9.52	-2⅛°C.	
10.03	-3° C.	
10.10	-3.5°C.	Took another out. It seems frozen stiff. Has a thin sheet of ice on it. Put into water at 21.5°. Did not recover.
10.15	-3.5°C.	All taken out and put into water at 12.5°. None recovered.

5 GEORGE V., A. 1915

Experiment II. Put 3 directly from tank (temp. 12°C.) into water at -3.5°C. Time 10.28 A.M.

At 10.33 took one out and put it into water at 13°C. At 10.39 it moved its gills and breathed for a time. Later it died, bleeding at the gills. 10.39— took other 2 out. They did not move their gills or recover.

Experiment III. Put 3 into a température of +1°C.

Time.	Temperature.	Behaviour of Fish.
11.30	1° C.	They lay on their sides in about 1 minute, but continued to breathe.
11.35	½° C.	Took one out. It at once swam around, so put it back.
11.50	-½° C.	They seem to be getting used to it, and swim a little now and then. Still on their sides however, and breathe very slowly.
12.00	-1° C.	No sign of life. Took one out and it came to life at once.
12.10	-1° C.	Took one out. Put into water at 12°C. Began to breathe in less than one minute and recovered completely.
12.30	-1° C.	Took other two out. They were dead.

Experiment IV. Done under the conditions that would exist according to the stories the fishermen tell.

Put 4 fish from water at 2°C. into dry jar at -15½°C.

Time.	Temperature.	Behaviour of Fish.
5.50	-15°C.	Put in 4 fish.
6.00	-15°C.	Put into water at 0°C. Complete recovery. It was apparently frozen stiff, like a piece of ice on the outside.
6.05	-15°C.	Took another out. It breathed but never completely recovered.
6.08	-15°C.	Took another out. It was found to be dead.
6.09	-15°C.	" " " " " " " "

Experiment V. Put 8 fish into water at -4°C. and left for 5 minutes. All seemed stiff. Took all out, and put 6 into warm water.

Cut sections through the other two. Flesh was stiff but seemed to have no ice crystals in it. The viscera were quite soft.

The 6 recovered completely.

Experiment VI. 10 fish were packed in lumps of ice in a dish so that the water drained off. They were put in refrigerator. Temperature 3°C. at 4 p. m.

At 8 a.m. the next day one was taken out and put into warm water. It recovered completely in less than one minute, and lived for days. The rest put back in refrigerator.

At 4 p.m. they were taken out and all recovered completely.

This experiment was not carried any further.

Conclusions.

From Experiments I, II and III it is seen that the fish will not survive for any length of time a temperature of -1°C. or lower.

The lower the temperature the shorter the time they will survive.

In experiment III the fish lived for 25 minutes at -1°C.

In experiment II the fish lived for 6 minutes at -3½°C.

At 0° C. and without water they survived for 24 hours and were in good condition at the end of that time.

Further experiments would be useful in solving the problem of shipments of live fish.

The fishermen's accounts are evidently partly true. Experiment IV shows that even when apparently frozen stiff they recover on being warmed, if the exposure be not for too long a time.

One withstood a temperature of -15°C. for 10 minutes, but 15 minutes proved fatal.

It seems (Exp. **V**) that even when apparently frozen stiff the viscera are not frozen at all. The body is covered with an ice coating as the water adhering to the fish freezes.

The flesh may even be quite stiff also, but there does not seem to be any freezing of the blood or flesh, but only a stiffening due to the low temperature.

SUPPLEMENT

TO THE

47th ANNUAL REPORT OF THE DEPARTMENT OF MARINE AND FISHERIES, FISHERIES BRANCH

CONTRIBUTIONS

TO

CANADIAN BIOLOGY

BEING STUDIES FROM THE

BIOLOGICAL STATIONS OF CANADA

1911-1914

FASCICULUS II—FRESH WATER FISH AND LAKE BIOLOGY

OTTAWA

PRINTED BY J. DE L. TACHE, PRINTER TO THE KING'S MOST EXCELLENT MAJESTY

1915

[No. 39*b*—1915]

PREFACE.

BY Professor EDWARD E. PRINCE, *Dominion Commissioner of Fisheries, Chairman of the Biological Board of Canada, Canadian Representative on the International Fisheries Commission, and Vice-President of the Fourth International Fisheries Congress, Washington, D. C.*

The number of papers embodying researches carried on at the three Biological Stations of Canada on the Atlantic and Pacific Coast, and at the Great Lakes Station, Georgian Bay and now completed for publication, so considerably exceeds the number which were available for each of the three preceding volumes, that it has been found necessary to divide them into two parts or Fasciculi, as pointed out in my preface to Fasciculus I. Fasciculus I consists of the papers on the Sea-fisheries and marine Biology, while the present part, the second part, now issued as Fasciculus II, includes papers treating of the Interior Fresh-water Fisheries and the Biology of the Great Lakes.

Professor B. Arthur Bensley's paper entitled "The Fishes of the Georgian Bay" is the first technical account of the fish fauna of that important part of the Lake Huron waters known as Georgian Bay, and may be looked upon as the initial systematic contribution towards a history of the fishes in the Canadian portion of the Great Lakes system. Its numerous original illustrations add greatly to its value and interest.

Dr. E. M. Walker, who was Curator of the Georgian Bay Station for several seasons, summarises his study of that important group of insects, the Odonata, which contributes either in the aquatic larval condition, or in the adult dragonfly condition, to the insect-food of fishes. Dr. Walker's eminence as a specialist gives importance to this original study which is of high scientific as well as practical interest. Taken along with Mr. W. A. Clemens' three papers on the Mayflies of the same water areas, they meet the need prominently brought before the Commission of Conservation, in January 1913, by Dr. C. Gordon Hewitt, Dominion Entomologist, who said that reliable information was absolutely necessary upon the insects and other food supplies in the waters in which fish abound, or in which fish have been introduced. Dr. Hewitt had previously brought before the Entomological Society of Ontario, a resolution expressing very strongly this need, and in the resolution it was stated that as the food of many of our important commercial fishes consists of larvæ and adult insects, a study should be made of the available or possible food supplies in the way of insect life before attempts are made at replenishing or stocking waters. Otherwise by stocking waters in which the food supply is not suitable, or cannot be made suitable, large sums of money, and considerable time and energy, will be uselessly expended, owing to the fish being planted where the food is either insufficient or of the wrong character. The resolution concluded by emphasising the necessity of more knowledge being secured as to the feeding habits and requirements of fresh-water fishes, and of the insect or other

5 GEORGE V., A. 1915

fauna and all available food supplies of the waters in which fish are living or which it is desirable to stock with fish.

As a matter of fact the Biological Board had already entered upon this field of research and Dr. Walker and Mr. Clemens have completed valuable researches on the very lines indicated, these appearing in the present volume.

The study of Insect Ecology, and the carrying on of experiments upon May-flies, and the rearing of this valuable fish food have yielded results which have direct practical bearing upon the welfare of our fish and fisheries.

Mr. A. D. Robertson, in his very detailed paper on the Mollusca of the Georgian Bay, furnishes a study similar in many ways as being a study of an important source of fish food. Sturgeon, for instance, have been found filled with the shells of many species of Mollusca such as Mr. Robertson describes, and it has been established that the spawn and the young of our fresh-water shell-fish are important as a food supply for young fishes, as well as for adults, while many of the larger bivalves have economic value owing to their producing pearls. Similarly Dr. Huntsman's able paper on the Crayfish and shrimp-like creatures of Ontario waters is really a study of fish food,—while the remarkable memoirs on a black-bass parasite (*Proteocephalus*) by Mr. Cooper and Miss Ryerson, the latter treating of leeches (*Hirudinea*), are of economic importance in relation to parasites, diseases, and enemies of fish, about which fish-culturists desire all the information that can be obtained.

Mr. White contributes a paper on a series of minute forms of fresh-water life (Lake Bryozoa) which must be also a source of food for small fishes,—while Professor MacClement and Mr. Bissonnette present botanical papers which have an intimate relation to fish studies, the plants and fungi are essential to insect life, and decaying fungi form an important *nidus* for insects, which are indeed of great moment to the fish and fisheries. Of similar interest is Mr. Klugh's paper on the Hydrophytes of Georgian Bay.

Such studies as those now collected in the present Biological Fasciculus not only indicate how fully the Great Lakes Station is carrying out the main purpose for which it was founded, (like the Marine Biological Stations), namely the benefit of the fishing industries generally, and the solution of pressing fishery problems, but all have contributed also to give an unequalled opportunity to young Biologists in the various Universities of Canada to carry on original scientific researches.

At these Stations the opportunity is offered, year by year, to all capable University students and members of University staffs, which was formerly wholly lacking, and which could only be supplied by resorting to foreign Biological Stations,—but the generosity of the Dominion Government has amply supplied the means whereby our scientific workers can carry on the highest researches, marine and fresh-water, within the limits of the Dominion and can thus contribute to our knowledge of the valuable fishery and other resources of these waters.

CONTENTS

I.

THE FISHES OF GEORGIAN BAY.

By B. A. Bensley, B.A., Ph.D., F.R.S.C.

Associate Professor of Zoology, University of Toronto

Plates I and II and six figures in the text

INTRODUCTORY.

Since the establishment of the Georgian Bay Biological Station in 1901, a number of collections representing the local fauna and flora of Go Home Bay have been brought together in successive seasons by various workers associated with the laboratory. In this way a considerable amount of information has been accumulated which it is hoped first to incorporate in separate reports on the individual groups, and afterwards to extend and correlate in such a way as to give a comprehensive view of the life of the Georgian Bay area.

The present report on the fishes of the region applies especially to the water areas in and about Go Home Bay, where extensive collections and observations have been made at all points within working distance of the laboratory. A few observations, however, have been made on the more accessible parts of the Musquash River system, which is the Georgian Bay outlet for the drainage area represented inland by the Muskoka Lakes, and at various points along the east and north shores of Georgian Bay from its southern termination at Coldwater River northward to Killarney*.

In the study of these collections prominence has been given to the identification of the various species and to their characters as shown in this region. This analysis forms the basis of the present report, but there has also been included a discussion of the factors of distribution, and various observations have been made on the breeding habits of fishes and on their food-relationships.

In the identification of certain more difficult species the writer has received assistance which is here gratefully acknowledged. To Dr. S. E. Meek of the Field Museum of Natural History, Chicago, he is indebted for the identification of darters and *Cyprinidæ*. Dr. Barton W. Evermann of the United States Bureau of Fisheries, Washington, kindly undertook the identification of the whitefishes and lake herrings. In addition, Mr. G. A. Boulenger kindly gave him permission to examine the collections of Canadian fresh water fishes preserved in the British Museum.

---

* For the purpose of making the collections described in the present paper, permission was kindly given by Mr. Edwin Tinsley, Superintendent of Game and Fisheries for Ontario to use certain nets not commonly authorized for this area, or for which special license is required.

5 GEORGE V., A. 1915

ENVIRONMENT AND DISTRIBUTION.

The information at present available concerning Georgian Bay fishes is not sufficiently extensive to permit of comparisons being made between Georgian Bay and other parts of the Great Lake system, or between the different parts of the drainage area of which Georgian Bay is the collecting basin. It is important, however, to recognize certain chief factors which may be operative in one locality or another and especially in that at present under consideration.

(1) Though forming an almost enclosed body of water, Georgian Bay is an integral part of the Great Lake system; conditions which apply to the Great Lakes will also apply to this area, except for local influences of antagonistic or modifying nature.

(2) With the exception of the North Channel of Lake Huron, and Lake Superior, Georgian Bay is peculiar in having its two principal shores underlaid by rock formations of fundamentally different type. Geological differences are the basis of topographic and environmental differences, and when pronounced, as in this area, may profoundly modify the distribution of species.

In explanation of this feature, it may be pointed out that the entire eastern and northern shore of Georgian Bay, extending from the mouth of Coldwater River, at the end of Matchedash Bay, northward to Killarney, falls within the ancient Archean area of the northern part of the province. The western and southern shore, on the other hand, including the south shore of Matchedash Bay, the Saugeen Peninsula, and Manitoulin Island, is underlaid by sedimentary strata of Silurian age. The southern part of this shore, especially in relation to Matchedash Bay, is also covered by an extensive deposit of glacial drift. The surface features of the two main shores are different in almost all respects.

(3) The Archean portion of the shore of Georgian Bay is part of an extensive eastern drainage area of which Georgian Bay itself is the common outlet. The water courses of this area are chiefly of the nature of basins, connected with other levels by rapids and waterfalls which act as barriers against upward migration. Differences in distribution have already been observed in this area, though only certain of them appear to depend on this factor.

(4) There is geological evidence that the area now occupied by this portion of the Great Lake system is smaller than in former times. The south and east shores were formerly situated at a considerable distance, respectively, south and east of the present boundaries, the water area including on the eastern side a part of the Archean district now occupied by an enormous number of more or less isolated lakes.

(5) This body of water had in former times, not only the outflow connection to the south and west as now represented by Lake St. Clair, but also temporary outlets eastward through the Trent and Mattawa valleys. The western parts of these areas are now parts of the Georgian Bay drainage (cf. Goldthwait '10).

Go Home Bay is a small indentation of the main eastern shore of Georgian Bay, lying within the Archean area, at a distance of approximately 25 miles northwestward from its southern border as recognized on the shore line by the mouth of Coldwater River. Go Home Bay is connected inland with the Muskoka Lakes through the Musquash River. The latter, beginning at Bala Falls on Lake Muskoka, flows a short distance westward, and divides into two portions, known respectively as the Moon and Musquash. The Musquash, which is the more southern branch, on approaching the main shore of Georgian Bay, expands into a lake of several miles in extent, known as Flat Rock Lake. This lake has two outlets into Georgian Bay, one through the Go Home River into Go Home Bay, the other into Georgian Bay directly at Muskoka Mills, a few miles to the south.

For the general topography of the region about Go Home Bay, reference may be made to the series of maps issued by the Department of Indian Affairs and designated as "Plans 1 to 3 of the Islands south of Moose Deer Point, Georgian Bay." The inland watercourses of the entire eastern shore are sufficiently well shown in the township map issued by the Department of Lands, Forest and Mines, Ontario. The general hydrography of Georgian Bay is given in the Admiralty Chart No. 327, and the details of the offshore water for the southern part of Georgian Bay in Chart No. 2102, designated as "Western Islands to Waubaushene."

Like other parts of the Archean area, this region is characterized by extensive exposures of the underlying bed-rock, the latter consisting for the most part of semistratified gneisses, the planes of which are inclined at a small angle to the horizontal. The rock surface is extremely uneven and eroded, and is notable for its resistant character, scanty soil accumulation, and deficiency of vegetation in all elevated or exposed positions. The surface is loosely strewn with detached rounded boulders.

These features are accountable for a variety of conditions presented by the water areas, some of which may be mentioned. Owing to the inequalities of the surface all inland depressions of a closed character tend to form water basins. The number of such basins is very large in comparison with the surface area, and they are of all kinds from small sphagnum ponds to lakes of several miles in extent. Again, the main shore lines are very irregular and sinuous. They present as a rule the character of bold rocky stretches, points, or headlands, alternating with minor indentations. The latter, owing to their unexposed nature tend to form shore swamps. Another important feature is found along the main shore of Georgian Bay, where the rock surface inclines beneath the water, leaving exposed a fringe of islands, similar in character to the mainland, and lying outside of this a more or less definite zone of naked reefs and submerged shoals. This shoal area consists of clean, smooth, but gently undulating rock, showing here and there abrupt ledges or basins containing accumulated boulders.

The inland water areas, which are perhaps more typical of the Archean formation than those of the main shore, are distinguished by three principal features. First, they are of the nature of overflow basins. If small, they are connected with lower levels by temporary rock channels, which perhaps contain water only in the spring or exceptionally rainy periods. If large, and situated on water courses, their channels are permanent, but connected with lower levels by rapids and waterfalls, and not to any extent by natural drainage inclines. Second, they show a comparatively high content of organic detritus, and are deficient in inorganic sediments, often to the point of exclusion. Third, the water itself, though free from finely divided inorganic sediments and therefore translucent, is colored in various shades from yellow to dark brown, and contains a considerable amount of finely divided organic matter in suspension. This water is of the kind commonly designated as "muskeg" water, and in some cases is opaque for depths of more than two or three feet.

Many of the smaller lakes and ponds are of the nature of shallow rock basins, the bottoms of which are occupied, often to a depth of several feet, by living and dead vegetable matter, for the most part in a suspended or semi-buoyant condition. Such areas are usually more or less filled with aquatic vegetation of the lower or higher orders, and are habitable in different degrees to various species of fishes. In the larger lakes, exposure to wind and wave action or to water currents, and the distribution of sediments made possible by greater depths, combine to produce a greater variety of environmental conditions than is possible in smaller areas. Here we find that the shores in exposed places usually consist of clean stretches of rock, while the smaller indentations, especially those connected with shore ravines are from their protected character transformed into swamps. They show the same features in general as the smaller inland lakes. In many places, where the amount of inorganic material is at all abundant, we find sand or sand and mud beaches, which are more apt to be formed where there is sufficient movement of the water to carry away the lighter organic materials.

The conditions prevailing along the main shore of Georgian Bay are similar to those of the larger inland bodies of water in respect of the alternation of bold rocky shores and shore swamps or sand beaches. There are, however, important differences, resulting from the greater degree of exposure to the action of wind, waves, and ice, and the dilution of the discolored inland water with that of the main body of Georgian Bay. On the main shore and among the shore islands and reefs, exposure to the prevailing westerly winds is naturally more direct than in inland situations, and the influence is to be seen not only in the diminished soil accumulation above water, but also in the more strenuous action of the waves on rocks and shoals. More especially, however, there are extensive movements of the entire body of inshore water, which moves in and out according to the temporary direction of the wind. In the inshore bays the difference in level often amounts to fifteen inches or more, the lower level being associated with offshore winds and the higher level with inshore winds. The movement of the entire body of water in this way produces currents in and out among the islands and assists in keeping the shore zone free from all sediments of a lighter nature. Shoal areas thus consist of cleaner rock than is found inland, and such sand beaches or channels as exist are likely to be formed of clean sand rather than a mixture of sand and mud.

The fact that the inland water courses connected with this Archean area are occupied by brownish muskeg water means in general that a large volume of this water, together with a considerable amount of organic detritus, is constantly being carried into Georgian Bay. Here it comes into contact with the clear, crystalline water of the main body of the bay, and quickly loses its identity. In general the shore water shows little of the inland or muskeg character, though it is slightly

Fig. 1. Narrows. Go Home River.

Fig. 2. Inland Swamp lake, Georgian Bay.

yellowish, and not so transparent for great depths as the purer water some distance offshore.

Referring to the factors of distribution for this particular region, it may be pointed out that no satisfactory scheme of classification can be constructed by which we may indicate a natural grouping of species with reference to environmental conditions. Each species shows in fact its own combination of factors, no two being exactly similar in habits, or especially in respect of the food supply, competitively in conflict. Some species, moreover, are intermediate in respect of certain factors, and the differences shown are more often of degree rather than kind.

If, however, we select out of the total number recognized, forty species, which are more characteristic as natural inhabitants of this region, and refer these species to the most general types of environment selected, we find that approximately 11 species are deep water fishes, 5 may be set down as shoal or rock-living fishes, 17 as inhabitants of swamp areas, and 6 as inhabitants of sand beaches. One species is characteristic of running water.

The characteristic deep water species include two species of whitefishes (*Coregonus clupeaformis* and *C. quadrilateralis*), two species of lake herring (*Leucichthys cisco huronius* and *L. harengus*), the lake trout (*Cristivomer namaycush*), the ling (*Lota maculosa*), the lake sturgeon (*Acipenser rubicundus*), the lake catfish (*Ameiurus lacustris*), and three species of suckers (*Moxostoma anisurum, Catostomus commersonii*, and *C. catostomus*). The list includes the important food-fishes, together with others of little or no value. Most species show inshore migrations for spawning purposes. In respect of the available food supply three orders are to be recognized. The two species of lake herrings feed at different levels on the plankton organisms or microscopic life of the water. Two species, the lake trout and ling are predaceous, living on smaller fishes. The remaining seven are bottom-living fishes, feeding on a variety of organisms such as molluscs, crustaceans and insects.

The extensive areas, of more or less exposed character, represented by rocky shoals and channels, are habitable to three chief species, all naturally protected and more or less predatory, including the small-mouthed bass or black bass (*Micropterus dolomieu*), the pickerel or doré (*Stizostedion vitreum*), and the rock-bass (*Ambloplites rupestris*). The former two are important game fishes, the doré also having a high commercial value. Shoal areas are especially characteristic for the black bass and rock-bass, both being only partly predatory and feeding for the most part on the crayfish which inhabit shoal areas in abundance. The doré, on the other hand is less characteristic of shoaly places, since, being almost wholly piscivorous and more or less nocturnal, it seeks its food in somewhat deeper or darker water and in places where small fishes are relatively more abundant. Two small species, finally, inhabit rocky areas chiefly for the purpose of concealment, the species being the long-nosed dace (*Rhinichthys cataractae*) and the small sculpin (*Uranidea franklini*).

The species inhabiting swamp areas of various kinds include the common pike (*Lucius lucius*), large-mouthed black bass or green bass (*Micropterus salmoides*),

5 GEORGE V., A. 1915

perch (*Perca flavescens*), sunfish (*Eupomotis gibbosus*), catfish (*Ameiurus nebulosus*), fresh water dogfish (*Amia calva*), top-minnow (*Fundulus diaphanus*), blunt-nosed minnow (*Pimephales notatus*), bream (*Abramis crysoleucas*), six species of common minnows (*Notropis cornutus, N. hudsonius, N. cayuga, N. blennius, N. heterodon, and N. atherinoides*), mud-minnow (*Umbra limi*), and brook-stickleback (*Eucalia inconstans*). This list does not include the young of the small-mouthed bass, rock-bass, or common sucker, which temporarily inhabit such areas, or any species also found on sand beaches.

Swamp areas appear to offer very favorable conditions, as indicated by the large proportion of species inhabiting them. It will be seen, however, that eleven of these species are insignificant forms, doubtless seeking the protection afforded by such situations, other places being more or less barred to them. Nothwithstanding their diminutive size, these species are relatively of great importance, since the existence of the larger, predatory forms, desirable as food or game fishes, depends at all times on an adequate food supply provided by smaller fishes.

The conditions which prevail in swamp areas are undoubtedly selected by certain species according to certain factors, but it is not clear how these factors are to be differentiated. Natural protection, provided by the shape and size of the body, as in the common sunfish and rock-bass, doubtless enable such species to occupy a situation where food is abundant more or less in the face of predatory forms. The foregoing species and also the common perch and catfish, have in addition the natural protection of spines. The environmental protection afforded by weeds or by less transparent water enables a variety of smaller fishes such as minnows, top-minnows, mud-minnows, and the young of larger types to maintain themselves also against predatory fishes. Predatory forms themselves find in swamp situations an abundant and convenient food supply. Some of them, however, as a result of the deterrent factors mentioned below, are apparently able to select this environment only within certain limits. Forms such as the green bass and pike, for example, tend to inhabit only the clearer portions of such areas, while the dogfish readily accepts the more confined situations. By way of comparison, small-mouthed bass and pickerel tend to avoid such situations entirely, or resort to them only temporarily for feeding purposes, nothwithstanding the fact that these areas contain an abundance of their favorite food.

This distinction of habitat, which also applies to many smaller species, must be based on conditions existing to a greater or less extent in swamp areas according to their more open or closed character. They possibly include excessive light, increased temperature in summer or cold in winter, deficiency of oxygen circulation pollution of the water with dissolved materials or mechanical pollution by organic detritus, stems of water plants or filamentous algae.

The fact that swamp areas of all types are present within this region makes it possible to institute comparisons as to their habitability for different species. For example, the smallest inland ponds habitable to fishes at all are as a rule occupied only by two insignificant forms, sticklebacks and mud-minnows. Somewhat larger ponds may contain in addition breams and sunfish. Swamp lakes will probably contain the latter two species, together with the commoner minnows, rock-bass,

Fig. 3. Zone of Reefs and Shoals.

Fig. 4. Shore of Station Island.

green bass, perch, and catfish, in other words those species which are more or less characteristic of ordinary swamps. Such lakes, however, will not be likely to contain pike, and will not contain small-mouthed bass or pickerel. Moreover those species characteristic of the smaller inland ponds will be present only in the more confined situations, and will likely be in the minority. Undoubtedly food supply has something to do with these differences, but it is evident from a consideration of the facts that food supply is not one of the important factors. Apart from the general questions of distribution, the matter is of some practical importance, since it involves the question of the habitability of certain smaller lakes to small mouthed bass or other game fishes and the reservation of such lakes for stocking purposes.

Sand beaches, channels, or similar clean surfaces, tend to be occupied by five species, all of which, however, are also found in the more open swamps. These are the small perch-like forms known as log-perches and darters (*Percina caprodes, Boleosoma nigrum, Etheostoma iowae*), the silverside (*Labidesthes sicculus*), and the common garpike (*Lepisosteus osseus*). The maskinonge (*Lucius masquinongy*), an important game fish, and the largest of all the inshore predaceous species, may also be included in this category, since it shows a preference for sand banks or sandy river channels.

The food supply of swamp areas and sand beaches, like that of the deep water, is of three orders, namely, (a) plankton, or microscopic organisms living on the bottom; (b) bottom organisms of a higher order, such as crustaceans, molluscs, and insects, or surface insects; and (c) smaller fishes. The smallest species and the young of all species are obliged to feed upon minute or microscopic organisms. Fishes of intermediate adult size, and also the young of large fishes at a certain period of growth, depend on crustaceans, molluscs, and insects. They show on the whole a preference for insect diet, and augment the natural supply of aquatic insects by feeding upon terrestrial insects which fall into the water. Finally, all the smaller species and the young of all larger fishes not naturally protected form a general food supply for the larger predatory types. There are no shore fishes of larger dimensions which retain the plankton feeding habit after the manner of the lake herrings in deep water.

On the whole the shallow water zone in this region does not appear to favor either the presence of a large number of species or the attainment of large size. The majority of species in which the normal adult size is not great appear to be smaller in this region than elsewhere. They may be dwarfed by some combination of environmental conditions, but the indications are that they have fewer chances of reaching the normal size. The number of intermediate and larger fishes of more or less predatory habit is eight, and the destruction wrought by these on smaller species must be enormous. The smaller species themselves have to depend for shelter on weeds, rocks, or shallows, and possibly the fact that the water is at all times transparent tends to turn the balance in favor of their natural enemies.

Certain conditions of environment which in other situations, especially in sedimentary areas, may confer advantages on certain species are here unrepresented. For example, with the exception of the silver lamprey, a parasitic form which up to the present has only been taken on fishes temporarily inhabiting

5 GEORGE V., A. 1915

running water, there are in this region no species of which this type of surrounding is characteristic. Some species, such as small-mouthed bass, pike, and pickerel, prefer running water, but all are distributed without reference to this factor. Such species as commonly inhabit running water elsewhere and are also present in this region appear to select other situations of an open character. There are in fact no permanent small streams, and no naturally flowing rivers in the entire region. It is possible also that the lack of similar advantages, such as either turbidity or exception purity, or minor conditions of food and shelter peculiar to sedimentary bottoms, may tend to restrict the development of a great variety of species in Archean waters.

Finally, we may refer to certain differences in distribution which distinguish parts of the Archean drainage area from one another or from Georgian Bay. On this question, unfortunately, detailed information is lacking, so that only fragments of evidence can be presented. Georgian Bay, for example, contains four species of fishes which are definitely known not to occur in the Muskoka Lakes, namely, the common pike, maskinonge, rock-bass, and green bass. Possibly a detailed study of these lakes would reveal the absence of other smaller species present in Georgian Bay, those named being the more conspicuous types. An important fact concerning these species is that they occur in various situations between the Muskoka Lakes and the Georgian Bay shore, and, moreover, that the first three of them occur in the Musquash River within a short distance of Bala Falls on Lake Muskoka. The absence of these species at the higher level has been attributed to the presence of waterfalls, a point which has been commented on by Meek and Clark ('02); but this factor would also affect the situation as regards many other lakes. Moreover, it is a well-known fact that certain of the inland lakes of the Muskoka and adjacent regions contain individually predominant types of game fishes, so that they are sometimes characterized as bass lakes, trout lakes or maskinonge lakes. This condition has led to the suggestion that the bass lakes have become so through the introduction of the small-mouthed bass in former times by Indians and through the tendency of this species to supplant the trout where the two species come into contact.

Whatever explanations may be brought forward either of natural barriers or of introduction by human agency, it can be shown that there are certain differences of distribution to which such causes cannot be assigned. These refer to the presence at higher levels of species not found at the lower levels. Several cases of the kind have come to light, the best example being certain species present in Muskoka Lake and absent in Go Home Bay. Muskoka Lake contains five species, namely, the speckled trout (*Salvelinus fontinalis*), found in some of the streams, the black catfish (*Ameiurus melas*), and three species of minnows (*Semotilus atromaculatus, Hybognathus nuchale*, and *Chrosomus erythrogaster*), all of which are reported by Meek and Clark ('02). These species either do not occur at all or if present do not occur naturally at the lower level of Go Home Bay. It thus appears that there are certain differentiating factors, which may be based on observed differences, such as soil content, food supply or water composition upon which the presence of certain species will be found to depend.

ANALYSIS OF THE SPECIES.

The total number of species here recognized is 48, representing 37 genera and 20 families. The number of species reported from Canadian localities, and including only fresh water forms, has been estimated by Evermann and Goldsborough ('07) at 145, representing 67 genera and 25 families. Probably more than one-third of the total number for Canada will be found either in Georgian Bay or in the streams entering it.

The majority of the families are represented by single species, the largest number of species belonging to a single family being 10 (*Cyprinidœ*). The complete list of families with the number of species representing them is as follows:—

Petromyzontidœ	2	*Umbridœ*	1
Polyodontidœ	1	*Luciidœ*	2
Acipenseridœ	1	*Pœciliidœ*	1
Lepisosteidœ	1	*Gasterosteidœ*	1
Amiidœ	1	*Percopsidœ*	1
Siluridœ	2	*Atherinidœ*	1
Catostomidœ	3	*Centrarchidœ*	4
Cyprinidœ	10	*Percidœ*	5
Anguillidœ	1	*Cottidœ*	2
Salmonidœ	7	*Gadidœ*	1

Family PETROMYZONTIDÆ.*

(Lampreys)

Representing the lower order of fish-like vertebrates (*Marsipobranchii*), animals with discoidal mouth, parasitic on other fishes. There appear to be two species represented in Georgian Bay, of which one has been collected, while the other is recognized from the evidence cited below.

Ichthyomyzon concolor, Kirtland.

(Silver lamprey)
· (*Plate 1. fig. 5*)

A number of specimens taken on pike and garpike in running water below the first falls on the Go Home River. Specimens taken during July and August are from 4 to 5 inches in length, a single specimen of 6½ inches. The length of this

---

* The classification here employed follows the plan of Jordan and Evermann ('96), but is restricted to the family divisions, ordinal reference being for the most part omitted.

5 GEORGE V., A. 1915

species is given by Forbes and Richardson ('08) for Illinois as 12 inches, and by Jordan and Evermann ('96) as 10. Go Home specimens are probably immature, none having been taken in the spring of the year.

Oral disc always expanded. Supraoral tooth bicuspid; infraoral with 7 cusps (sometimes 4 or 6). There are 11 lateral, oblique, curved rows of unicuspid teeth, of which 4 rows have their enlarged first cusps immediately lateral to the mouth. Dorsal fin continuous, with a broad notch.

For several years no specimens of the larva or *Ammocoetes* of this lamprey were discovered, but in August, 1910, a single specimen was found in a decaying submerged log at about the same place where all the metamorphosed specimens were taken. The larva is 3¼ inches in length, or approximately the size of the smallest metamorphosed specimens. The dorsal fin is continuous, this character differentiating the larva from that of the lake lamprey, the latter as described by Gage ('93) having the dorsal fin divided as in the adult.

Petromyzon marinus unicolor, DeKay.

(Lake lamprey)

This species is included provisionally. The dwarfed fresh water representative of the marine lamprey (*Petromyzon marinus*), described by Jordan and Fordice ('85), Meek ('85), and Gage ('93) occurs generally in the lakes of northern and central New York. A lamprey, evidently representing the same variety occurs abundantly in Lake Ontario, and is commonly taken by fishermen on whitefish and lake trout. In this lamprey the dorsal fins are separate, the four extraoral teeth bicuspid, the average length about 15 inches.

Though there is no reliable information as to the occurrence of lake lampreys in the upper lakes, and the whitefish and trout are practically free from lamprey marks, fishermen state that lampreys of about 15 inches in length are sometimes taken on whitefish and trout from deep water. This suggests that the lake lamprey is present in small numbers, and perhaps accidentally. The silver lamprey does not reach the size indicated, and up to the present has not been taken except in the limited area represented by the running water of the falls of the river. Fishes such as pike, on which the silver lamprey is commonly taken, are abundant elsewhere in shore waters, but do not have lampreys on them.

Family POLYODONTIDAE.

Polyodon spathula, Walbaum.

(Paddle-fish)

Naturally an inhabitant of the Mississippi valley, but occasional specimens taken in the Great Lakes. Noted here on account of two specimens reported by Nash ('08) taken near Georgian Bay waters, one at Sarnia*, the other at Spanish River on the North Channel.

* Vide, Prince E. E. Paddle-Nosed Sturgeon in Ontario. Ottawa Naturalist: Vol. XIII, No. 7, 1899.

Family ACIPENSERIDAE.

Acipenser rubicundus, LeSueur.

(Lake sturgeon)

This species was formerly abundant in Georgian Bay, but in the southern part is now rarely seen. Reported as ascending the Nottawasaga River in May and rivers of the east shore in June.

Though formerly considered of little or no value, and often in the earlier days of the Georgian Bay fishery taken from the water to be destroyed, the fish now commands a good price and a female which will yield caviare is a veritable prize to the fortunate fisherman. The quantity now taken in Georgian Bay is insignificant, the figures for 1909, as given by the Superintendent of Game and Fisheries for Ontario being 6,900 lbs. for Georgian Bay proper, and 14,155 lbs. for the North Channel.

The experiments carried on for several years in the United States ,looking to the artificial propagation of the sturgeon, yielded most promising results, though difficulty was experienced in obtaining spawn and milt at the same time. In view of the scarcity and increasing value of this fish, the matter of artificial propagation should be taken up at once by the Canadian hatcheries. There are doubtless many small details that would have to be worked out in handling fish of this size, and a suitable river must be found where the fish still ascend in numbers.

Family LEPISOSTEIDAE.

(Garpikes)

Lepisosteus osseus, Linnaeus.

(Long-nosed garpike. Bill-fish)

Not uncommon in swamps where there is more or less clean sand bottom. Probably not so abundant on any part of the typical Archean shore of Georgian Bay as in its southeastern arm, or elsewhere in sedimentary waters.

Length up to 5 feet, but the larger specimens in Georgian Bay are about 3. Body greatly elongated, slender, and little compressed. Depth 9.6 to 11.8. Head rounded posteriorly, extended anteriorly into the slender beak-like snout. Length of the head 2.9 to 3.2 in the length of the body; snout 1.3 to 1.4 in head. Eye in head 16.2 to 18.7. General coloration dark olive or greyish above, yellow or white below. Posterior part of the trunk and median fins spotted. Many of the smaller marks on the body give the impression of ink-stains run between the edges of the scales. Dorsal fin with 8 rays; anal with 8 or 9. Body covered with a hard thick armor of ganoid plates, the number of which is 9 to 11, 62 to 64, 9 or 10.

The fish is usually seen lying in the shallow water over sand bottoms and rising

5 GEORGE V., A. 1915

occasionally to the surface for air. It feeds for the most part on minnows, top-minnows and young suckers, which it captures by a sharp, quick snapping motion of the jaws. Like the fresh water dogfish, it represents an archaic type, of considerable biological interest. It is of no value otherwise, and its extermination is frequently urged on account of its destructiveness to other fishes or on account of the damage it inflicts on the nets of the fishermen.

The nests of this species have been found only on two occasions at Go Home Bay. In both cases the nests were constructed with little care in about two feet of water, and on a bottom covered with the short stems of aquatic plants. The spawning time is towards the middle of June. Young fish of from two to six inches in length are commonly taken in swamps or on sand beaches. They make practically no efforts to avoid capture. Their coloration is much more striking than that of the adult. There is a broad lateral stripe of black, and immediately above it a white band with brownish spots in its lower portion. There is a median dorsal band of dark color, and the ventral surface is occupied by a dark band containing a median white stripe. The tail is also notably different in form, the fin portion being separated for a considerable distance from a lance-shaped filamentous lobe representing the continuation of the tail proper. This delicate lobe is in the natural condition kept in almost constant motion.

Family AMIIDAE.

(Dogfish)

Amia calva, Linnaeus.

(Dogfish. Bowfin)

Present in nearly all swampy situations, but more abundant in the south-eastern arm of the bay, in the vicinity of Waubaushene, where the more extensive swamp areas doubtless provide a more congenial habitat.

Length 2 feet. Body robust forwards, compressed and gradually tapering backwards to the tail. Depth $4 \cdot 3$ to $5 \cdot 2$. Head very stout, its length $3 \cdot 5$ to $3 \cdot 8$ in the length of the body. Eye small, $8 \cdot 1$ to $11 \cdot 5$ in head. Anterior nares opening on short tubes. Coloration above and on sides dark olive green, with more or less definite darker mottlings. A black spot on the upper margin of the tail, surrounded in the male by an irregular band or ring of yellow or orange. Lower parts white or yellowish. Opercle with two fairly distinct bands of black extending backwards from the eye. Lower jaw and jugular plate with dark mottlings. Males in the breeding season have the dorsal and caudal fins greenish black, but the lower fins are bright emerald green, and have a band of green connecting them on each side of the body. In the female all the fins are dark. Dorsal fin very long with 48 (to 50) rays. Anal with 10 or 11 rays. Scales large, with more or less angular edges, 8 or 9, 67 to 69, 11 to 14.

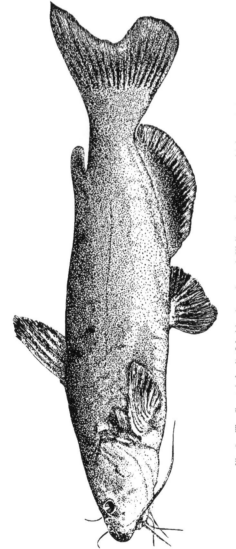

Fig. 5. The Great Lake Catfish (*Ameiurus lacustris Walbaum*). About one-third actual size.

Though a strong and vigorous swimmer, this fish is commonly seen lurking among the weeds in the shallow water of swamps, lying motionless on the bottom, or moving stealthily along by rhythmic undulations of the dorsal fin. It feeds on swamp-living fishes such as minnows, suckers, perch or the young of the game fishes, but will eat crayfish, mud-puppies, or insects.

The spawning period is from May 24th to June 1st. The eggs are deposited in large open, but fairly deep depressions, scooped out of the mud among the flag roots by the male fish. The nest is also guarded during the hatching period by the male, which at this time will be found lying motionless in the bottom of the depression, unless disturbed by the presence of minnows or other small fish on the look-out for food. Young dogfish of about two inches in length are sometimes seen in schools, swimming about in the swamps accompanied by the parent fish. It appears, however, that the young fish are taken into deeper water immediately after hatching and that as a rule they spend a considerable time in the deeper situations. The fish taken inshore are most commonly large specimens, the young in various stages of growth being unusual.

Young dogfish of two inches in length show the general features of the adult, but the darker ground markings are more conspicuous. There are three prominent lines on the side of the head, one horizontal and passing through the eye, a second passing along the upper jaw and extending backwards beneath the eye, and a third on the lower jaw. The median fins have two bands of dark color, one marginal, the other a short distance inward; also present in the adult, but obscured by the general dark coloration.

Family SILURIDÆ.

(Catfishes)

This family is represented by two species, separable as follows:—

Genus **Ameiurus.**
    a. Caudal fin rather deeply notched; anal fin with 23 to 27 rays. Catfishes of large size..lacustris.

    aa. Caudal fin at most slightly emarginate; anal fin with 21 or 22 rays. Small catfishes, usually 12 inches or less...nebulosus.

The black catfish (*Ameiurus melas*) is reported by Meek and Clark ('02) as relatively more abundant than *A. nebulosus* in Muskoka and Gull Lakes, but this species has not been recognized in Georgian Bay. The yellow catfish (*Ameiurus natalis*) is suggested by Jordan and Evermann ('96) as possibly the species referred to by Richardson ('36) as *Pimelodus coenosus*, which was described from specimens taken at Penetanguishene. This species may occur in the sedimentary swamps or streams of the region, but has not been found northward. It is probable, however, that the fish described by Richardson is *A. lacustris*.

5 GEORGE V., A. 1915

Ameiurus lacustris, Walbaum.

(Great Lake catfish)

(*Fig. 5*)

This species appears to be present along the main shore of Georgian Bay only in small numbers. It is said to be taken frequently in the Magnetewan River at Byng Inlet. A single specimen was taken near the biological station at Go Home Bay, in 1907, the weight of which was 37 lbs. In this region it occurs chiefly inland, being abundant in Flat Rock Lake and in the Musquash River immediately above the lake. The specimens taken here are commonly from 5 to 15 lbs. in weight. It may be taken in the darker water by ordinary rod-fishing during the day.

Length 2 feet or more. Body moderately elongated, the trunk very heavy forwards and laterally compressed towards the tail. Depth 4·2 to 5. Head broad and depressed, its length 3·2 to 4 in the length of the body. Eye small, 8·2 to 11 in head. Four pairs of barbels, of which the maxillaries are as long as or but slightly shorter than the head. Coloration uniformly dark ashy above, lighter below. Dorsal fin with 1 spine and 6 soft rays. Anal fin with 23 to 27 rays, its base 3·4 to 3·5 in the length of the body. Pectoral fin with 1 spine and 8 soft rays, the length of the spine 2·3 to 2·5 in that of the head. Fins all dark, except the ventrals, which are ashy at the tips.

The stomachs of several specimens examined contained nothing but crayfish. No reports are available concerning the spawning habits; but since the fish is not seen in shallow water at any time, after the manner of the smaller catfishes when spawning, it is assumed that the eggs are deposited in deep water.

Ameiurus nebulosus, LeSueur.

(Common or brown bullhead)

The small catfishes of Georgian Bay show considerable variation, but an examination of a large number of specimens indicates that there is but one species. Specimens from these waters are rarely more than 12 inches in length, and the general coloration, doubtless associated with the transparency of the water, tends to dark grey and black above with ashy shades below. The cloudy markings are present but concealed.

The fish is extremely common in all shore swamps and larger inland lakes of a swampy character, but is taken as a rule only at night. Throughout the summer the food consists almost wholly of Mayfly larvæ; for which the fish burrows in the mud of the bottom.

The following are the critical measurements of Go Home specimens indicating the reference to *A. nebulosus*. Depth of body 3·8 to 5, usually 4·1. Head 3·2 to 3·7, usually 3·4. Pectoral spine in head 2·1 to 2·5, usually 2·3. Rays of anal fin 21 or 22; length of anal base in length of body 4·1 to 5., usually 4·5 (25 specimens).

Though rarely seen in the daytime at other periods, the catfish are conspicuous objects in the swamps during the spawning season. The time of spawning varies greatly, but begins during the latter part of June and extends well into July. The nests are constructed in shallow water, often only two or three feet from the shore. In this region the fish do not make open nests as in other places, but burrow under flag roots or submerged sticks. The burrows are about eighteen inches or two feet in length, and the terminal chamber has a clean hard bottom on which the egg mass rests. The nests are chiefly made by the males, but sometimes both males and females participate in the construction. After depositing the eggs the female usually leaves the nest, the latter being then guarded by the male. As a rule he lies wholly concealed in the burrow and probably in the terminal chamber with the eggs, but darts forth when the bottom is disturbed. The eggs are hatched in a few days, and the fry gradually work their way to the outside of the burrow. Though at first unpigmented, they soon acquire a dense black coloration. For some time they move about in large schools, accompanied by the male, feeding on microscopic organisms, but later they are abandoned by the parent fish and disperse, working downwards among the bottom vegetation.

Family **Catostomidae.**

(Suckers and mullets)

Represented by two genera, separable as follows:—

> a. Air-bladder divided into three compartments. Scales in lateral line less than 50..**Moxostoma**

> aa. Air-bladder divided into two compartments. Scales in lateral line more than 50..**Catostomus**

Genus **Moxostoma.**

Moxostoma anisurum, Rafinesque.

(White-nosed sucker)

Occasionally taken in gill-nets in the southern part of the bay; more abundant towards fall. Not seen inshore in the region of Go Home Bay during the spring or summer. Many specimens were seen taken in the pound-nets at Killarney and Spanish River on the north shore. The following description is based on a single specimen, the only one thus far received for examination.

Length 15 inches (the size commonly much greater). Body heavy forwards, but moderately deep and compressed. Depth 3·2. Head short and broad, its length 3·8. Snout blunt; mouth inferior, the lips plicate, but with the folds slightly broken into papillæ. Eye 5·9. General coloration pale yellowish, darker above. Under parts and snout whitish. Dorsal fin with 16 rays. Anal with 8 rays. Scales large and coarse, 7, 44, 5.

Genus **Catostomus.**

Represented-by two species, as follows:—

a. Scales small, those in the lateral line 102 to 117. Snout elongated, conical, projecting considerably beyond the mouth.........................**catostomus**

aa. Scales larger, those in lateral line 64 to 72. Snout blunt, not projecting greatly beyond the mouth...**commersonii**

A third species, described, but perhaps wrongly, as a fine-scaled sucker is thought to occur, but no specimens have been taken.

Catostomus catostomus, Forster.

(Long-nosed sucker. Red sucker)

Commonly taken in gill-nets in deep water and in the pound-nets on the north shore. Rather infrequent inshore in the southern part of the bay. Of the young suckers common in the shore swamps all identified belong to the next species, but a few specimens of the present species have been taken in shallow water near the Giant's Tomb Island.

Length 18 inches. Body moderately elongated, rounded. Depth 4·2 to 4·9. Head rather broad and rounded behind, tapering forwards into the slender conical snout. Length of head 3·7 to 4·2. Eye 6·2 to 8·2. Mouth inferior, with thick papillose lips. Coloration uniformly dark brownish or blackish above, light below. Sides with a reddish stripe, showing only in a few of the summer specimens, but present in all males in spring. Dorsal fin with 10 or 11 rays. Anal with 7 rays. Scales small, 18 to 21, 102 to 117, 12 to 17.

In this region the fish is considered to be of no value, and is destroyed in large numbers by the fishermen.

Catostomus commersonii, Lacépède.

(Common sucker. White sucker)

Commonly taken in the gill and pound-nets. The fish probably inhabits the shallow water of shore bays, but is not seen inshore in numbers except during the spring run into the rivers, and to a certain extent near shore on the spawning beds of the rock-bass and small-mouthed black bass. Young specimens of from 2 to 5 inches in length are very commonly taken in the shore swamps, where they sometimes make up a large proportion of the seine catches.

Length 18 inches. Body rather thick and heavy forwards, moderately compressed towards the tail. Depth 4 to 4·6. Head heavy, rather broad, its length 3·7 to 4·3. Snout short, squarish at tip. Mouth inferior, with strongly papillose lips. Eye 6·3 to 8·2. Coloration grey olivaceous above, light below.

Upper parts with more or less golden reflections. Dorsal fin with 11 to 13 rays, usually 12. Anal fin with 7 or 8 rays. Scales of medium size, but rather small and crowded forwards, 9 or 10, 64 to 72, 7 to 9. Young specimens taken in the shore swamps show blackish blotches on the sides.

Like the remaining members of this family, the common sucker is a bottom-feeding fish, subsisting ordinarily on molluscs and crustaceans, but very destructive to the spawn of other fishes. The present species is abundant on the shoals where the whitefish, trout, and herring resort in the fall for spawning purposes. It also runs into the rivers, to the foot of waterfalls in the early spring, feeding on the spawn of the doré, and afterwards spawning in the same situation. It is not infrequently seen swimming lazily about in the shallow water of the swamps in June, during the spawning time of the rock-bass and black bass, and on some occasions has been observed to enter the nests of these fishes, apparently with little resistance on the part of the occupants, and leisurely devour the contents.

Family CYPRINIDAE.

(Carps and minnows)

This family is represented in Georgian Bay waters by at least five genera and ten species. One species, the common or German carp, is an accidentally introduced species for these waters. In addition there are three species, representing two more genera, the normal occurrence of which is doubtful. These are the red-bellied dace (*Chrosomus erythrogaster*) and the black-nosed dace (*Rhinichthys atronasus*), single specimens of which appear in the collections; and a species of *Semotilus* or *Hybopsis*, represented by a single specimen, too minute for identification, which was taken with specimens of *Rhinichthys* in a rock pool.

With the exception of the carp, these species are all of small size, Georgian Bay specimens appearing in general small in comparison with those of other localities southward. The largest species in the region is the redfin minnow (*Notropis cornutus*), which is commonly 3¼ inches in length, and the smallest is *Notropis heterodon*, which is barely an inch in length.

Though of small size, the *Cyprinidae* are of the greatest importance, since they form the food of larger fishes such as the bass, pike, and doré, either directly, or indirectly through the crayfish, which feed on smaller fishes to a considerable extent, and themselves form the staple food of the small-mouthed black bass and rock-bass. The *Cyprinidae* are in fact the intermediates in the range of food supply, connecting the larger fishes with the fundamental plankton food of the water, since they live very largely on small or microscopic entomostraca, blue-green and green algae, and minute insects. They are not wholly benefactors, however, for it is probable that enormous numbers of eggs of nesting fishes are destroyed in the spring of the year by the various species infesting the shore swamps. They annoy the nesting bass and rock-bass by their enormous numbers, and the temporary departure of the fish from the nest is a signal for a swift attack on the contents, which are devoured in a moment. Specimens taken under such circumstances commonly have the stomach gorged with the stolen eggs.

5 GEORGE V., A. 1915

Genus **Cyprinus**.

Cyprinus carpio, Linnaeus.

(German carp)

Abundant in the swamp waters of the southeastern end of Matchedash Bay, and reported as occurring elsewhere along the south and west shores of Georgian Bay. Very few specimens, in all probability stragglers, are reported by fishermen from any locality along the eastern and northern shores. It appears that the swamps of the Archean part of the shore are not suitable for the development of this species, which condition if true will be fortunate for the conservation of black bass and other game fishes in this region. It may be, however, that the carp is so recent an arrival in these waters that is has not had time to become distributed.

Specimens taken at Waubaushene by Capt. C. J. Swartman were chiefly of the scaled variety, but some were mirror carp, with a few large scales, and the naked or leather variety is said to be sometimes taken. The mud and sand areas of this part of Georgian Bay undoubtedly provide a suitable environment for the species, after the manner of sedimentary swamp lands elsewhere. Specimens of 10 lbs. and over are commonly taken.

Regarding the introduction of this fish into Georgian Bay waters, the general opinion is that the carp of Matchedash Bay gained access to this water through the Severn River. They are reported to have appeared in numbers about twelve years ago, at which time the fish were all small specimens of about 10 inches in length. Carp inhabit the head waters of the Severn River, Lake Simcoe, in large numbers, and the stock of this lake is thought to have been derived from specimens formerly kept in a pond near Newmarket. From this pond specimens are supposed to have escaped into the Holland River and thence into Lake Simcoe. It will be remembered, however, that the carp has had abundant opportunities to become distributed throughout the Great Lakes, and possibly those of the southern part of Georgian Bay gained access to the waters from another direction.

In the years from 1875 to 1879, the United States Fish Commission made several importations of German carp, with the object of stocking American waters with a type of fish that would thrive in waters unsuitable for other fishes and provide an abundant cheap food supply for the masses of the people. The carp were successfully bred, and were distributed in large numbers in successive years from 1880 to 1896. Between the years 1880 and 1893 several lots of carp were sent to applicants in Canada, including Mr. Samuel Wilmot, the Ontario Commission, and certain private individuals. In Ontario the fish appear to have gained access to public waters chiefly through accidents to private ponds in which they were kept.

The carp has been greatly condemned on several scores, some of which undoubtedly have a strong basis of fact. It is a bottom-living form, and produces considerable havoc in swamps, making the water muddy and rooting up aquatic plants in search of the minute molluscs which form its staple food. It is accused of polluting the water, of eating the spawn of other fishes, of driving game fishes away,

SESSIONAL PAPER No. 39b

and of destroying the wild celery of swamps frequented by ducks. It is also urged against the fish that it is a kind not acceptable as a table fish to the people at large. Such complaints, which are usually directed against the authorities responsible for the introduction of the fish, have been carefully investigated by Cole ('05), who finds on the whole in favor of the carp. There is no doubt that a great deal of foolish prejudice exists against the carp, and it is probable that when the matter is more fully understood, and especially when the necessity for a cheap class of fish food has become imperative, as it doubtless will, the work of stocking American waters will be more generally appreciated.

The carp is now firmly established in Ontario waters, and is undoubtedly there to remain, whatever attempts made be may to eradicate it. There is, however, no necessity for allowing it to increase at the present time, even in places where its presence is more or less welcome. Undoubtedly in those areas where the conservation of game fishes is in the general interest of the people, facilities should be given to fishermen to take and market the fish, or if any damage is likely to be done to smaller swamp-living fishes, to eggs of nesting fish, or to the swamp-bottoms themselves, the work should be conducted by the authorities. Some discrimination is necessary in this matter, since there are many swamp areas which sportsmen will continue to want to have recognized as game fish preserves when they are in reality suitable for little else than carp or other coarse fishes. The method of taking carp which is least objectionable from a biological standpoint on shores frequented by game fish is the use of large-meshed gill-net, set in such a way that the fish may be driven into it and the net immediately lifted. Advantage may be taken of the larger girth of the carp, permitting other fishes to escape, and no damage is done to the swamp bottoms or to their ordinary occupants. No operations of this kind should be permitted during the time that the bass or other desirable fish are on the nests if the preservation of these fish is the first consideration. It is probable that in areas like the eastern arm of Matchedash Bay, and especially on its northern side no damage worth mentioning would ensue to the game fishes if carp were permitted to be taken in the manner described at any period of the year.

The four genera of small *Cyprinidae* here recognized are separable as follows:

a. Body deep and compressed, the depth contained less than 4 in the length (except in young specimens). Abdomen compressed to a sharp edge behind the ventral fins...**Abramis.**

aa. Body at most only moderately deep, more than 4 in the length. Abdomen behind the ventral fins rounded:

b. Maxilla with a small barbel at its posterior end..............**Rhinichthys.**

bb. Maxilla without barbel.

c. Intestine considerably longer than body...................**Pimephales.**

cc. Intestine shorter than body...............................**Notropis.**

Genus Abramis.

Abramis crysoleucas, Mitchill.

(Golden shiner.   Bream)

(Plate II, Fig. 6)

Abundant, but confined to the ends of swamps and inland lakes, apparently preferring the smaller ponds where the water is dark, choked with vegetation or almost filled with bottom ooze.   Nothwithstanding its unattractive habitat the fish is one of the most striking of all the minnows, and is easily recognized by its deep flat body, which is of a bright golden coloration, and by the very oblique mouth.

Length of the larger specimens 3½ inches.   Body thin, the depth 3·5 to 4·5, relatively greater in the larger specimens.   Abdomen compressed behind the ventrals into a sharp edge.   Coloration dark olivaceous above.   Sides bright silvery with golden reflections.   A dark lateral band, conspicuous only in small specimens.   Head compressed, 3·9 to 4·3.   Mouth terminal, very oblique.   Eye 3·1 to 3·4 in head.   Dorsal fin inserted behind ventrals, with 8 rays.   Anal fin long, with 12 (sometimes 11) rays.   Scales 10 or 11; 44 to 55, 3 to 5 (usually 4). Lateral line strongly decurved, sometimes broken or irregularly connected.   Usually complete, but in some specimens, with pores only on a few anterior scales.   Intestine as long or longer than the body, 1 to 1·3.   The intestine commonly contains clean masses of green algæ.

Genus Rhinichthys.

Two species representing this genus are known to occur in Ontario waters, namely, the black-nosed dace (*R. atronasus*) and the long-nosed dace (*R. cataractœ*). Both species are reported by Meek and Clark ('02) from Hawkstone, Lake Simcoe, and from Sault Ste. Marie (*R. atronasus* being more common), but not from Muskoka Lake.

In the Georgian Bay collections there is one specimen of *R. atronasus* the presence of which may be accidental.   *R. cataractœ* occurs in limited situations as described below.

Rhinichthys cataractae, Cuvier et Valenciennes.

(Long-nosed dace)

(Plate II, Fig. 7)

The species inhabits and appears to be confined to rock-pools on exposed reefs fringing the main shore of Georgian Bay.   It is practically the only fish

SESSIONAL PAPER No. 39b

inhabiting these pools, though those open to the outside water sometimes contain other species. The fish lurks under the stones and is only taken by strategy. It is easily recognised by the spindle-like body, general dark coloration, and very small scales, or if examined closely by the minute barbel placed behind the angle of the mouth.

Length 2½ inches, the specimens commonly taken much smaller. Body spindle-shaped, not compressed. Depth 5·1 to 5·8. Color very dark olivaceous above, with black vertebral streak and dark lateral band, the latter not conspicuous except in young specimens. Sides with dark points on the scales and with more or less of fine blotching. Head long and pointed, 3·3 to 4. Eye 4·3 to 4·6 in head; in snout 1·6 to 2·4. Snout in head 2·1 to 2·6. Mouth wholly inferior, the snout projecting well beyond the tip of the lower jaw. A minute barbel behind the fleshy lobe that forms the angle of the mouth. Dorsal fin with 8 rays, inserted distinctly behind the ventrals. Anal with 7 rays. Scales minute, 12 or 13, 68 to 72, 9 to 12 (usually 10). 35 to 37 oblique rows in front of the dorsal fin. Lateral line complete, almost straight.

Genus **Pimephales.**

Of the two known species, *P. promelas* and *P. notatus*, the latter is reported by Meek and Clark ('02) as more abundant in the inland localities examined, though *P. promelas* was found at Hawkstone, Lake Simcoe, and at Trout Creek, a tributary of Lake Nipissing. Up to the present only *P. notatus* has been taken in Georgian Bay. With the exception of the red-bellied dace (*Chrosomus erythrogaster*), the natural occurrence of which in Georgian Bay is doubtful, this species is the only representative in this region of the herbivorous or mud-eating group of minnows, represented elsewhere by the species of *Campostoma*, *Hybognathus* and other genera.

Pimephales notatus, Rafinesque.

(Blunt-nosed minnow)

(Plate II, fig. 8)

With the exception of the redfin minnow (*Notropis cornutus*), this is the most abundant minnow of the region. It occurs in collections from all points from Waubaushene to Byng Inlet, in inland waters, and from the Giant's Tomb Island. Georgian Bay specimens differ in some details from those described by Forbes and Richardson ('08) from Illinois. They are rather dark, and the usual number of scales before the dorsal fin is smaller by about two rows. The intestine, described by these authors as twice the length of the body, is in the specimens examined rather shorter, the combined length of stomach and intestine, or of the intraperitoneal part of the alimentary canal, being in none equal to twice the length of the body.

5 GEORGE V., A. 1915

The minnow is easily recognized by its blunt snout, robust angular body, black lateral stripe, and crowded scales before the dorsal fin. Nearly all the scales have dark edges, giving the body a cross-hatched appearance.

Length 2⅜ inches. Body moderately elongated, but with the sides and back flattened, giving a somewhat rectangular appearance to the forward part of the trunk. Depth 4·7 to 6·2 in the length of the body. Color rather dark olivaceous above, all the scales except those about the pectoral and ventral fins with prominent dark edges. Sides dull silvery or leaden. A dark lateral stripe extending along the body and around the head, passing through the eye and the upper part of the snout; not conspicuous on the head in some specimens (spring males) on account of the dark coloration of its upper portion. A dark spot at the base of the caudal fin, and another at the anterior base of the dorsal fin, the latter spot often faint or absent. Head 4·2 to 4·5. Snout blunt, the mouth at its ventral angle, small and almost inferior. Eye 2·9 to 3·2 in head. Dorsal fin with one anterior short, swollen or club-like ray and 8 ordinary rays; situated a little behind the ventrals. Anal with 2 rudimentary and 7 developed rays. Scales 6 to 8, 42 to 49, 4 or 5; usually 7, 44, 4. Oblique rows before dorsal fin 18 to 23, usually 21, but sometimes 2 or more scales inserted between rows. Lateral line complete, slightly decurved in front, usually showing black specks above and below the pores, but the latter never conspicuous, and often very faint or absent. The length of the body is contained 1·1 to 1·9 in the length of the stomach and intestine.

The intestine commonly contains large quantities of vegetable material, for the most part green algæ in a mud-like basis, but the fish are by no means purely herbivorous. During the nesting season of the bass and rock-bass, they are commonly seen in large numbers waiting about the nests. If the latter are left for a moment the contents are quickly disposed of.

The eggs of this minnow are deposited during June and the early part of July on the under sides of stones, sticks or pieces of bark, and are watched and vigorously defended by the male fish, which at this season has the front of the head armed with 16 or 18 sharp tubercles.

Genus Notropis.

This characteristic American genus, containing in all about 100 species, is represented in this region by 6 species. The most abundant species is the redfin minnow (*N. cornutus*), which occurs everywhere in the shore swamps and in inland waters. The much smaller species, *N. blennius*, is probably next in frequency of occurrence, though more abundant in the more open swamps. Two species, *N. cayuga* and *N. heterodon*, show a tendency towards inland situations; more marked in the latter, which has been taken almost wholly in the Musquash River and in Flat Rock Lake above the first falls on the Go Home River. *N. hudsonius* is comparatively rare in the region, and appears to prefer situations where there is more sand or mud bottom. *N. atherinoides* is an extremely abundant minnow in the shore swamps in spring, but in summer appears as a rule only in small numbers.

SESSIONAL PAPER No. 39b

The six species are separable as follows:—

a. Rays of anal fin 7 or 8;

  b. Scales before dorsal in 12 to 15 oblique rows;
      c. A black stripe along the side of the body, extending through eye to end of snout;
          d. Chin white..cayuga
          dd. Chin black...heterodon

      cc. A diffuse plumbeous lateral band, only evident posteriorly. Lateral line with black specks above and below the pores.............blennius

  bb. Scale before dorsal in 18 to 20 oblique rows, a prominent black spot at the base of the caudal fin....................................hudsonius
aa. Rays of anal fin 9 to 11;

          e. Dorsal fin immediately over the ventrals. Anterior scales on sides of body rather deep and narrow.............cornutus
          ee. Dorsal fin distinctly behind the ventrals. Scales rounded normal.....................................atherinoides

Notropis cayuga, var. muskoka, Meek.

(Plate II, fig. 9)

Frequently taken in the shore swamps about Go Home Bay, and also appears in collections from Sans Souci and Pte. au Baril. It seems to prefer the less open swamps, but has not been found anywhere in abundance. The fish is easily recognized in comparison with other minnows of the region by the small crescentic markings along the sides of the body.

Specimens submitted to Dr. Meek were referred to the species *N. muskoka*, a form described by him ('99) from specimens taken in Muskoka and Gull Lakes, but with the suggestion that this form may be a variety of *N. cayuga*. In view of the intermediate characters presented by Georgian Bay specimens, the latter interpretation is here recognized.

Length commonly $2\frac{1}{2}$ inches. Body moderately elongated, only slightly compressed. Depth 4.3 to 5.3 in the length of the body. General coloration olivaceous, sometimes, in spring specimens, with a golden tinge. Scales above with prominent dark edges. Vertebral line scarcely evident. A dark line passing along the side of the body, through the opercle and snout, above the upper jaw. On the trunk this line is separated from the dark-edged upper scales by a lighter band. It is overlaid by a series of small crescentic marks, one at the base of every scale of the lateral line. Head somewhat conical 3.4 to 4 in length of body. Eye 3 to 3.7 in head. Mouth subterminal. Dorsal fin with 8 rays; anal with 7 or 8. Scales 5,34 to 37,3 or 4.15 (sometimes 16) rows of scales before dorsal fin. Lateral line incomplete, lacking pores on some of the scales. Stomach and intestine 1 to 1.3 in length of body.

5 GEORGE V., A. 1915

The species *N. muskoka* is described by Meek as differing from *N. cayuga* in the reduced size of the scales before the dorsal fin, more slender body, less blunt snout, slightly larger and more oblique mouth, and more incomplete lateral line. Georgian Bay specimens cover the range of depth variation as described by Forbes and Richardson ('08) for *N. cayuga* (4.5 to 5.2), but 5 specimens of *N. cayuga* in the British Museum collection (Silver Lake, Iowa, Meek), which have been recently examined, are much deeper (4 to 4.3), and their appearance is quite different both from the Georgian Bay specimens and from specimens of *N. muskoka*. Georgian Bay specimens commonly show 15 rows of scales before the dorsal fin, but the number is occasionally 16, and in some specimens two or three extra scales are inserted between rows. The crowded appearance is, however, not nearly so marked as in *N. muskoka*. The lateral line characters seem to be quite variable, some specimens having the lateral line almost complete, and other showing pores only on a few scales. It appears that the Georgian Bay specimens deviate in some characters from the typical *N. cayuga*, and that these characters are accentuated in the inland form. The species described by Eigenmann ('93) as *N. heterolepis*, from a single specimen taken at Qu'appelle, is, as suggested by Forbes and Richardson, referable to *N. cayuga*. The specimen is superficially much more like *N. cayuga* than are those from Georgian Bay or Muskoka Lake.

Notropis heterodon, Cope.

(*Plate II, fig. 10*)

A small species, in fact the smallest of all fishes inhabiting the region, the largest specimens being barely 1½ inches in length. It appears in collections from Go Home Bay, but probably does not occur in any numbers along the main shore. It is very abundant inland, however, a large number having been taken from Flat Rock Lake, where small specimens have been seen in millions. On account of its very small size and superficial resemblance to *Pimephales notatus*, which is abundant in the same situations, this species easily escapes detection. It is recognizable by a number of features, including a solid black lateral stripe, oblique mouth, black chin, and the small number of scales in front of the dorsal fin.

Length 1½ inches, commonly less than 1 inch. Body slender, slightly compressed Depth. 4.5 to 5. Color olivaceous, the scales above with prominent dark edges. Sides with a solid black longitudinal stripe, accentuated by overlaid specks, the anterior ones rather fainter and placed at a lower level. The stripe is continued around the head and tips the chin. Between the lateral stripe and the back there is a clear band in which the scales are not dark-edged. Head 3.4 to 4 in length of body. Mouth terminal, oblique. Dorsal fin with 8 rays; anal with 8 or sometimes 7. Scales 5, 37 or 38,3. Oblique rows before dorsal 15, sometimes 14. Lateral line developed only in front, with pores on a few scales.

Notropis blennius, Girard.

(Straw-colored minnow)

(Plate II, fig. 11)

Abundant in shore swamps, especially in the vicinity of open water. Often seen in schools containing hundreds of individuals. The species is easily recognized in the water by the short stout body and by the pale coloration, or, when examined closely, by the dark specks above and below the pores of the lateral line. The coloration on the whole is noteworthy for its lack of character.

Length 2⅜ inches. Body appearing short in comparison with its width and depth; moderately compressed, and for the most part evenly tapered at the ends, except that the ventral profile increases rapidly to the shoulder and little beyond that point. Depth 4.2 to 5.3 Coloration pale straw yellow. Scales with prominent dark edges. A narrow vertebral line, expanding in front of the dorsal fin into a more or less evident blotch. A faint broad plumbeous band, scarcely evident, along the side of the body. Lateral line decurved anteriorly, conspicuously marked out in its entire length by small black specks, one above and one below every pore. On the tail the specks tend to fuse and form small solid blocks of black. Some specimens show an extension of these lateral line specks to form faint crescentic marks as in *N. cayuga*, but the crescents are always indistinct. Lower surface of body pale. Head conical, 3.8 to 4 in body. Eye 2.6 to 3 in head. Mouth almost terminal. Dorsal fin with 8 rays; anal with 8. Scales 4 or 5,37,3. 15 oblique rows in front of dorsal fin. Stomach and intestine 1.1 to 1.3 in length of body.

The food of this species seems to be of a most general kind, the intestine containing plankton entomostraca, minute insects, and blue green or green algae, usually mixed with ingested sand-grains. Females heavy with eggs are common during the first two weeks of June.

Notropis hudsonius, DeWitt Clinton.

(Spot-tailed minnow)

(Plate II, fig. 12)

This species appears in small numbers in collections from Go Home Bay, Giant's Tomb Island, Sans Souci and Pte. au Baril, but on the whole is seldom taken. It appears to prefer solid-bottom swamps or shores such as are more characteristic of sedimentary regions. The fish is easily recognized by the pale or silvery coloration of the sides combined with the very conspicuous jet-black caudal spot.

Length 2⅞ inches. Body rather stout and laterally compressed, unlike other species of Notropis of the region, except N. cornutus, in this respect. Depth 4.2

to 4.7. Coloration in general pale yellowish, the sides silvery. A thin vertebral line. Scales of back and sides with faint dark edges. A faint plumbeous band on the side of the body showing narrower and fainter in its anterior portion. Sometimes specks above and below the pores of the lateral line but never pronounced (cf. *N. blennius*). Head short, 3.8 to 4.5 in length of body. Nose rather blunt, the mouth at its ventral angle and very slightly oblique. Eye large, 2.3 to 3.6 in head. Dorsal fin with 8 rays; anal with 8, sometimes 7. Scales 6,38 to 41,4. 16 to 19 oblique rows before the dorsal fin. Lateral line complete, decurved anteriorly. Stomach and intestine 1.1 to 1.3 in length of body.

Notropis cornutus, Mitchill.

(Common shiner. Redfin minnow)

(*Plate II, fig. 13*)

With the exception of the blunt-nosed minnow (*Pimephales notatus*) this is the most abundant minnow of the region. Represented by at least a few specimens in nearly all seine catches, and often present to the exclusion of all other species except that mentioned. It occurs in all swamps on the main shore and inland, specimens having been taken from Flat Rock Lake, Giant's Tomb Island, Waubaushene, Sans Souci, Pte. au Baril, and Byng Inlet. It is also the largest minnow in the region, though not reaching the size reported from other localities. The fish is easily recognized in the water by its somewhat deep body, slivery sides, and especially the dorsolateral gilt stripe, which is much more pronounced in this than in other species.

Length commonly to 3⅔ inches, a single specimen measuring 5 inches. Depth 4 to 4·6, the body in young specimens rather elongated, but in older ones appearing shorter and deeper. Laterally compressed, the sides quite flat. Coloration above olivaceous, with a conspicuous vertebral stripe of black. Back bordered by a gilt stripe which shows best in the water. Sides silvery, sometimes appearing blotched on account of extra pigment on groups of scales or single ones. An indistinct lateral plumbeous band, the anterior part of which is very faint and only about half the width of the posterior part. Spring males have the darker parts of the body more brilliantly expressed, and there is a bright rosy hue on the sides, especially above the pectoral fins. The lower fins are all red, and there is a flush of red on the tips of the dorsal and caudal fins and on the lower side of the head. Some males have the top of the head covered by minute tubercles. Females plain. Head 3·9 to 4 in length of body; somewhat compressed, the snout blunt. Mouth terminal, rather large and slightly oblique. Eye 3·1 to 3·8 (specimens to 3⅔ inches). Dorsal fin with 8 rays ; anal with 9. Scales 7 or 8, 41 to 43, 4 or 5. The exposed edges of the scales are very narrow and deep on the anterior end of the body at the sides, by which character alone the species would be readily recognized. 21 to 25 rows of scales in front of the dorsal fin. Lateral line complete, slightly decurved in front. Stomach and intestine 1 to 1·3 in length of body.

The large specimen mentioned above shows the body relatively much deeper (3·6 in length), and the eye relatively smaller (5 in head), its actual size being no greater than in smaller specimens.

The food of this species appears to consist largely of green and blue green algae, with some aquatic insects, and occasionally entomostraca. Specimens about to spawn have been taken as early as May 18th. During the spawning season the fish are extremely active and very tenacious of life.

Notropis atherinoides, Rafinesque.

(Shiner. Silver minnow)

(Plate II, Fig. 14)

This species occurs in small numbers in the shore swamps during the summer, but in spring is frequently seen in large schools near shore feeding for the most part on insects. It is easily distinguished from other minnows by its very slender, elongated body.

Length not usually exceeding 2⅜ inches. Body moderately compressed, very slender, the depth 5·8 to 6·9 in the length of the body. Upper part of the trunk, except for a thin triple vertebral streak, clear translucent olive, in spring deep green to almost black, bounded below by a thin gilt stripe. Sides very silvery, with a broad ground band of plumbeous shade running from the upper margin of the opercle to the base of the tail. No caudal spot. Cheek and opercle bright silvery. Spring specimens with delicate orange red spots at the bases of the pectorals and ventrals, also at the posterior end of the maxilla and above the opercle. Head 4 to 4·7, conical. Mouth terminal, somewhat oblique, the jaws more like those of larger fishes. Eye 3·2 to 3·5 in head, appearing large in some specimens. Dorsal fin with 8 or 9 rays, its anterior margin considerably posterior to a vertical line drawn at front of ventrals. Anal fin with 10 or 11 rays. Scales rounded, very lightly attached, 6,38 to 43,3. 20 to 22 rows in front of the dorsal fin. Lateral line complete, strongly bowed downwards in its anterior part.

The fish is probably the most alert and active of all the minnows, and appears to live on insects to a much greater extent.

Family ANGUILLIDAE.

(Eels)

Anguilla chrysypa, Rafinesque.

(American Eel)

Specimens of this species are reported on reliable authority to have been taken occasionally at the mouth of the Severn River and at Waubaushene at the south-

5 GEORGE V., A. 1915

east end of Georgian Bay. Since the eel spawns in the sea, and the Falls of Niagara offer an insuperable obstacle to the ascent of the young, such specimens as are taken in the upper lakes must be chance specimens that gain access through the canals.

Family SALMONIDAE.

(Whitefishes and trout)

This important family is represented in the southern part of Georgian Bay by at least three genera and five species. On the north shore an additional species is represented by the Manitoulin tullibee, recently described by Jordan and Evermann ('09) as *Leucichthys manitoulinus,* and the streams of the south and west shores contain the speckled trout (*Salvelinus fontinalis*). The latter fish also occurs in various lakes and streams inland from the eastern shore of Georgian Bay, including the streams entering Muskoka Lake. It does not appear to occur in any of the streams belonging to the Musquash River system. Speckled trout are said to have been taken occasionally in Georgian Bay, but such specimens were in all probability stragglers from the streams.

The three characteristic genera are separable as follows:

Salmoninae:
  a. Mouth deeply cleft, as usual in fishes, the articulation of the lower jaw posterior to the eye. Jaws with sharp teeth..........................Cristivomer.

  aa. Mouth not deeply cleft, the articulation of the lower jaw below or in front of the eye. Jaws weak and toothless.

Coregoninae:
  b. Mouth very small and inferior, the snout projecting beyond it......Coregonus
  bb. Mouth somewhat larger, terminal..........................Leucichthys.

Genus Cristivomer.

Cristivomer namaycush, Walbaum.

(Lake trout)

Usually taken by commercial fishermen in pound-nets or gill-nets, especially the latter. Some are taken in the summer by deep trolling, but the fish is only taken in numbers by trolling when rising to the shoals preparatory to spawning in the fall. In Muskoka Lake the fish also appear on the surface in May.

The general run of fish taken by the commercial fishermen are between 2 and 8 lbs. Very small fish, however, which would otherwise go through the nets, are sometimes captured, being entangled in the thin twine of the gill-nets by the teeth and fins. The same is true of large specimens, individuals of 20 lbs. or over, and too large to gill, being frequently taken in this way.

SESSIONAL PAPER No. 39b

The deep-bodied pale trout of the deep water of Lake Superior, known as the ciscowet, may possibly occur in Georgian Bay. Fishermen offer various reports as to very dark or pale trout, with short deep bodies, which are never taken in shallow water, and which they assume do not come inshore to spawn.

Length to 3 feet. Body elongated, moderately compressed, the depth 4 to 4·9. Head stout, with large mouth, the length of the head 3.5 to 4·1. Eye 7·3 to 9·1, in one specimen of 14 in., 5·3. Snout 3·2 to 4·1. Coloration deep grey to blackish, under parts light. Everywhere with small rounded white spots. Upper part of head and the median fins more or less vermiculated. Dorsal fin with 9 to 11, usually 11, fully developed rays. Anal with 10 or 11 rays. Scales very small. The above measurements are based on specimens of the usual run, and probably do not express the extreme variations for the species.

The lake trout is the chief predaceous species of the deep water. It feeds on herrings, young whitefish, perch or other small fishes, but has the reputation of eating almost anything that attracts its attention.

The fish is now the mainstay of the commercial fishery, the total catch of the Georgian Bay and North Channel for 1909, as reported by the Superintendent of Game and Fisheries for Ontario, being approximately 2½ million lbs., almost three times the amount of the whitefish taken during the same period, and with a value approximately three quarters in excess of that of all other species taken together. The figures of several years seem to indicate that the lake trout is withstanding the drain of the commercial fishery much better than the whitefish. There are perhaps several reasons for this. This fish is a predatory type, swimming at all levels, and thus escaping to a greater extent the operations of the gill-net fishermen. It is probable also that it is not affected to any great extent by the pollution of the bottom through lumbering operations, while the latter would be fatal to the whitefish. There is a further possibility that the artificial propagation of this fish in the Great Lakes has had a larger effect both in numbers and natural distribution than in the case of the whitefish.

Genus Coregonus.

At least two species of whitefishes occur in this region, one being the round or frost whitefish (*C. quadrilateralis*), the other the common lake whitefish (*C. clupeaformis*). They are separable as follows:

    a. Body rounded and elongated, the depth 4.8 to 5 in the length. Gillrakers few in number, 10 to 12 on the lower limb of the first arch, and short, their length about 5 in the length of the eye.........................quadrilateralis.

    aa. Body more or less compressed, elliptical, the depth 3.7 to 4.5. Gillrakers numerous, 16 to 18 on the lower limb of the first arch, and slender, their length only about 2 in eye...clupeaformis.

5 GEORGE V., A. 1915

Coregonus quadrilateralis, Richardson.

(Round or frost whitefish)

(Plate I, fig. 4)

A few specimens have been taken in shallow water in the early summer and later in the fall.   It probably exists in numbers in the deeper water, but on account of its comparatively small size and slender body it is not commonly taken in the gill-nets.

Length 14 inches.   Body elongated, somewhat cylindrical.   Depth 4·8 to 5.   Head 4·9 to 5·3 in length of body.   Eye 4·7 to 5·9.   Snout 3·8 to 4·2 in head.   Maxillary from tip of snout 4 to 4·5 in head.   Dorsal fin with 11 or 12 rays, anal with 10 or 11.   Scales 9, 88 to 91, 7 or 8.   About 32 or 34 rows of scales in front of the dorsal fin.   The sides of the body are silvery, the dorsal surface darker, brownish or sometimes bluish.   One specimen, a male taken in November, has the sides with about 7 rows of weak tubercles.

This fish is credited with the destruction of the eggs of trout and whitefish during the spawning season, and the intestines of specimens taken in the fall do contain fish eggs.   The same statement, however, may be made with reference to the lake whitefish, the fact being that both fish are bottom feeders, and in all probability they add to their ordinary diet the eggs of their own and other fishes when occasion permits.

Coregonus clupeaformis, Mitchill.

(Labrador whitefish)

(Plate. I, fig. 3)

Two kinds of large whitefishes, representing more or less separable species, but perhaps only developmental types, occur in the Great Lakes, one of them, the Labrador whitefish, or Musquaw River whitefish, having been recently recognised by Jordan and Evermann ('09) as the common whitefish of the lakes, excepting Lake Erie.   The other is the common whitefish of Lake Erie (*C. albus*).   The former species is a more or less elongated fish, of elliptical outline, and rather large and coarse head, the latter a pale, deep rather angular type, with small weak head and high nuchal elevation.

Specimens of the Georgian Bay whitefishes have been submitted to Dr. Evermann, who pronounces them fairly typical specimens of *C. clupeaformis*.

In the southern part of Georgian Bay there is a tendency on the part of fishermen to recognize two types of common whitefish, one being called the coarse-scaled, shore or shoal whitefish, the other the deep-water whitefish.   There are no whitefish inshore in the summer, and those that appear on the inshore shoals in November are recognized as shoal whitefish.   The deep-water whitefish inhabits

SESSIONAL PAPER No. 39b

the deep water during the summer, but is thought to migrate northwards to the inshore shoals for spawning. It is an interesting fact that in the southern part of the bay at least these fish do not come up on the shoals nearest their summer home. The shoal whitefish is regarded as of a poorer kind, and of inferior keeping qualities. The studies of the fish up to the present seem to lend some weight to the opinion expressed, but it is extremely doubtful if distinct races should be recognized, or indeed whether any significance should be attached to the small differences appearing locally in this species.

The following enumeration is based on 5 specimens of the shore variety or run, taken in the fall. The males have the sides with longitudinal rows of weak tubercles, the surface being distinctly rough to the touch.

To facilitate comparisons with the typical specimens recently described by Jordan and Evermann ('09), the measurements have been indicated in hundredths of the body length. Length 18 inches. Dorsal rays 11 or 12. Anal rays 11 or 12. Scales, 10, 83 to 94, 9. Gill-rakers 16 to 18. Head ·20 to ·22. Depth ·22 to ·27. Caudal peduncle, length ·07 to ·08, depth ·08. Eye, ·03. Snout ·05. Maxilla ·05 to ·06. Distance from snout to occiput ·14 to ·16. Pectoral length ·15 to ·18. Ventral length ·12 to ·15. Dorsal height ·14 to ·16. Anal depth ·10 to ·12.

The following enumeration is based on 19 specimens of whitefish taken in deep water (16 fathoms) off the Giant's Tomb Island. Dorsal rays 11 or 12. Anal rays 11 to 13. Scales 10 or 11, 79 to 93, 8 (in three specimens 9). Head ·19 to ·21. Depth ·23 to ·27. Caudal peduncle, length ·08 to ·11, depth ·07 to ·08. Eye ·03 to ·04. Snout ·05 to ·06. Maxilla ·05 to ·06. Snout to occiput ·13 to ·15. Pectoral length ·14 to ·16. Ventral length ·12 to ·14. Dorsal height ·13 to ·15. Anal depth ·09 to ·11. The gill-rakers are 16 to 18, verified in about 50 specimens.

There are several points of possible error in comparing these groups of specimens, but taking the range of variation of the first group as a basis, we find certain figures not covered by the second group, notwithstanding the large number of specimens, and indicating for the latter group slightly shorter head, greater depth, longer caudal peduncle and smaller fins. In the shape of the body the deep-water fish vary from those of elliptical form, with even dorsal profile, to those rather deep and compressed, with a considerable nuchal elevation. The head appears small, but not as in the Erie whitefish.

Measurements of the head divided into the length of the body do not appear to give the best results in comparing the size of the head in the different kinds of whitefish, the reasons being that the characters of length of head and length of body are similar or analogous. Measurements giving the proportion of head length into depth of body might, however, yield dependable distinctions. A rough trial of this proportion indicates for the 19 specimens above mentioned a proportion of ·74 to ·93. By comparison, 13 specimens of *C. clupeaformis* reported by Jordan and Evermann ('09) show roughly a proportion of ·60 to ·90, but the exclusion of two extreme specimens from the Lake of the Woods and Waubegon puts the range from ·79 to ·90. The smaller size of the head in relation to the depth in *C. albus*, from 4 specimens reported by Jordan and Evermann, is shown by the range of ·66 to ·74.

In Georgian Bay whitefish are taken by gill-nets southward and by pound-nets northward. They are occasionally taken with baited hooks. The food consists of small, sometimes minute, lamellibranch and gastropod molluscs, and small crustaceans. Specimens taken on the shoals in fall are commonly found to have eaten fish eggs, which are evidently picked up from the bottom with the usual food.

Taking Georgian Bay proper, the total catch of whitefish for 1909, as reported by the Superintendent of Game and Fisheries, was 382,392 lbs., and including the North Channel, 856,521 lbs. The statistics of a period of years show a gradual falling off in the annual catch, for which it is probable that several conditions are responsible. This matter has been discussed by the Commission appointed in 1905 by the Dominion Government to investigate the fisheries of Georgian Bay, and remedial measures are proposed. Both whitefish and trout owe any advantage that they possess in respect of escaping the nets of the fishermen to the fact that they are deep-water forms, inhabiting largely situations where complete fishing is impossible. Whitefish, however, are bottom-living types, and considering both the great amount of gill-net fishing at present carried on in these waters, and the alleged fishing of net in excess of that granted by license, it is not surprising that fishes of this kind should become less plentiful year by year. It may be pointed out also that any balance of numbers in favor of the lake-trout, as at the present time, is distinctly a balance against the whitefish, whose smaller numbers are less able to withstand the natural drain of providing through the young fish, together with lake herrings and perch, the enormous food-supply required by the lake-trout. Finally, the waters of Georgian Bay have been continuously fished for a long period of years, and little constructive work has been done in the matter of artificial propagation and distribution of whitefish in this region, a condition which it is hoped will be remedied.

Genus Leucichthys.

Specimens of the lake herrings taken in the southern parts of Georgian Bay have been examined by Dr. Barton W. Evermann, by whom they are referred to two species, one being the Saginaw Bay or Georgian Bay herring (*L. harengus*), the other the Huron herring (*L. cisco huronius*). *L. harengus* occurs in Lakes Huron and Michigan, and occasionally in Lake Erie. It is the most important element in the fisheries of Saginaw Bay, Michigan. The species was originally described by Richardson ('36) from specimens taken at Penetanguishene on Georgian Bay, but has been only recently differentiated by Jordan and Evermann ('09) from the species *L. artedi*. Only a few specimens of this type of herring have been taken, and since these are for the most part immature specimens, no analysis can be given. In general the species is close to *L. cisco huronius*, but is distinguished by the small size of the adipose fin, less cylindrical body, and grey coloration.

The Huron herring, or blueblack herring, occurs in Lakes Huron and Michigan, and occasionally in Lake Erie. A few specimens, evidently of this type, have been taken in deep water during the summer off the Giant's Tomb Island, but the fish is only seen in numbers in the southern part of Georgian Bay during the inshore run in November.

Leucichthys cisco huronius, Jordan and Evermann.

(Huron herring)

Numerous examples taken in shoal water in November. Some females distended with eggs. Almost all male fish with rough tubercles, arranged in longitudinal rows on the sides of the body, one on each scale in the row. Length 9¼ to 12 inches. Head ·20 to ·22. Depth ·21 to ·25. Caudal peduncle, length ·09 to 1, depth ·07. Eye ·04 to ·05. Interorbital distance ·06 to ·07. Maxilla from tip of snout ·07 to ·08. Snout to occiput ·14 to ·17. Ventral to pectoral ·29 to ·32. Length of pectoral contained in pectoral-ventral distance 2·03 to 2·82. Pectoral length ·12 to ·14. Ventral length ·13 to ·14. Dorsal height ·12 to ·14. Anal depth ·07 to ·09. Adipose length ·04 to ·06. Dorsal fin with 10 or 11 rays. Anal with 11 or 12 rays. Scale 8 or 9, 77 to 88, 8 or 9. 31 to 36 oblique rows before the dorsal fin.

The body is elongated, elliptical, with rather long and slender snout. Coloration of upper parts lustrous blue, the upper part of the head, maxilla and tip of the mandible dark. Lateral line almost straight.

Herrings are sometimes accused of destroying the spawn of other fishes, but there is no evidence of this in these specimens taken during the spawning time. The intestine was found to be filled with enormous numbers of minute entomostraca of the plankton, bottom materials of any kind and fish eggs being rare, and the latter probably ingested by accident.

Small meshed gill-nets operated for herrings may do considerable damage in places frequented by small whitefish of 9 inches or thereabouts in length. The taking of such fish is unlawful, and most undesirable for obvious reasons, but the regulation providing for their liberation, though evidently of preventive value, is unfortunately not very practical. The same is true of undersized whitefish taken in gill-nets of the authorized mesh for taking whitefish and trout. There are perhaps some fishermen who either cannot or do not wish to make the important distinction between adult herrings and young whitefish, and the relative numbers of small whitefish taken should be enquired into in localities where the herring fishery is permitted. While the herring fishery is admittedly valuable, it involves at least three objectionable elements, first the actual destruction of the young of larger fishes, second, the burden on the provincial authorities of inspecting for undersized fish, from the operation of small-meshed nets, and third, the removal from the waters of the food supply of the lake-trout, which should be estimated either on a basis of the amount of lake-trout taken from the water, or the damage likely to be done to small whitefish as a result of lack of abundance of herrings.

5 GEORGE V., A. 1915

Family UMBRIDAE.

(Mudfishes)

Umbra limi, Kirtland.

(Mudfish. Mud-minnow)

(Plate II, fig. 16)

Taken in the smaller inland ponds and in the muddiest parts of shore swamps. It thrives in the most uninviting puddles, in association with sticklebacks, tadpoles and newt larvæ.

Length commonly to $2\frac{1}{4}$ inches, one specimen of $3\frac{1}{2}$ inches. Body stout and compressed, caudal peduncle deep. Depth $4 \cdot 2$ to $4 \cdot 8$ in the length. Head rather heavy, its length $3 \cdot 1$ to $3 \cdot 2$. Mouth terminal, rather flattened. Eye $3 \cdot 3$ to $4 \cdot 4$. General coloration yellow or olive, but with the ground color almost obscured by dark mottlings, which form about 14 indistinct vertical bars. The sides show bluish and green reflections. A lateral stripe showing in most specimens, and a faint band through the opercle, eye, and snout. Ventral surface pale. Fins all with rounded margins, and with minute transverse striations on the rays. Dorsal placed far back near the caudal, 14 or 15 rays. Anal with 9 or 10, sometimes 8, rays. Scales rounded, 12 to 14 in oblique row from front of dorsal to anal. 34 to 36 in horizontal series. Imbedded scales on top of head, and large scales on opercles.

Family LUCIIDAE.

(Pikes)

Represented by two species, characteristic of northern waters generally; separable as follows:

> a. Cheeks scaled, opercles with scales only on the upper half. Ground color dark, with yellow or white spots on sides...............................lucius.

> aa. Both cheeks and opercles bare of scales below. Ground color light, with dark vertical or oblique bars and spots.........................masquinongy.

Lucius lucius, Linnaeus.

(Common pike)

Abundant in all places on the main shore of Georgian Bay and in the river courses. It inhabits weedy swamps and channels, where it lurks among the weeds, darting forth from time to time to capture small fishes such as black bass, rockbass, perch or minnows. Small specimens of all stages of growth are taken in the shore swamps, but are not abundant. The fish is of some commercial value in those parts of the shore where inshore net-fishing is permitted, but it is not a fish that is greatly respected by anglers. As commonly taken it is from 3 to 6 lbs. in weight, but specimens of 15 lbs. are not infrequently captured.

Length up to 3 feet. Body elongated, slender and moderately compressed. Depth 5·5 to 6·5. Head 3·3 to 3·6, rather rectangular behind, tapering forwards into the shovel-like snout which is 2·2 to 2·3 in its total length. Eye 8 to 10·4. Coloration above dark olive to black, with light irregular cross lines alternating on the two sides and connected by a wavy vertebral line, giving a somewhat reticulated appearance. These markings are obscured in dark colored specimens. Sides with longitudinal rows of white or yellow spots. Scales with V-shaped golden marks. Under parts white, except tip of mandible. Median fins yellow with dark mottlings, the paired fins more faintly marked. Dorsal fin set far back, with 18 to 21 (usually 19) rays. Anal with 15 to 17 rays. Scales small, 13 (to 15), 120 to 132, 11 or 12. Lateral line broken, and with rows of accessory pores above and below.

Young specimens taken in the shore swamps have the general coloration dark, but with rounded or oblique white markings which tend to divide the darker color into oblique bars.

Lucius masquinongy, Mitchill.

(Maskinonge. Muskellunge)

This species occurs all along the shore of Georgian Bay, though not in large numbers anywhere. It is relatively more numerous in the sand areas at the southern part of Georgian Bay, and in sandy situations in the river courses. It also occurs in many inland lakes. Though specimens of great weight are sometimes reported, the general run of fish range between 3 and 25 lbs.

The measurements here given are based on five smaller specimens, and probably do not give the complete range of variation for this region. Body greatly elongated, slender, and moderately compressed. Depth 5.7 to 6·1. Head 3·2 to 3·6 in length of body. Snout shovel-like, 2·3 in length of head. Eye rather small, 9·5 to 11 in head. Ground coloration light. Sides with brilliant dark spots, which tend to run together into vertical or oblique bars. Back and upper portion of head a beautiful deep greenish black. Under parts light. The scales exhibit bronze, gold and green reflections. The median fins are dark, with obscure spots, the paired fins plain and dark greenish. Dorsal fin with 19 or 20 rays; anal with 16 to 18 rays. Scales 15 or 16, 134 to 152, 12 to 14.

Young specimens of a few inches in length are sometimes taken in shore swamps. The coloration is different from that of the adult. There is a broad longitudinal dorsal band, usually more or less broken on the occiput; also a dorso-lateral dark band which tends to break into spots. Below the latter there is a more or less definite light stripe, followed ventrally by a series of spots. The entire ground coloration is light.

Like the common pike, the maskinonge is a predaceous type, and is very destructive to the smaller fishes and to the young of larger ones, including the game fishes. Its comparative rarity, beauty and splendid sporting qualities make it the most highly esteemed of all the fresh water game fishes.

5 GEORGE V., A. 1915

Family POECILIIDÆ.

(Killifishes)

Fundulus diaphanus menona, Jordan and Copeland.

(Menona top-minnow)

Frequently taken in somewhat weedy but rather open water near shore. The largest number of specimens taken in the gap separating the two parts of the Giant's Tomb Island, and in all probability the species favors sedimentary areas. It is the only species representing the genus or family in the region, and is one of the surface or top-minnows, interesting from their feeding habits and their value as destroyers of mosquito larvæ. The species is easily distinguished from other small fishes by its flattened wedge-like head, the top of which bears a rosette of scales, the flat tumid lips, and vertical bars of the sides.

Length 2¾ inches. Body spindle-shaped, more or less compressed posteriorly. Depth 4·4 to 5. Head 3·2 to 3·6. Eye 3·2 to 4. Dorsal fin with 12 or sometimes 13 rays. Anal with 11, sometimes 10 rays. Scales in a longitudinal row 44 or 45; in oblique row around the sides of the body from the front of the dorsal fin, 12. Lateral line inconspicuous, represented by minute rounded depressions on some of the scales. The body scales are continued over the opercle to the head, the dorsal surface of the head being scaly, with a rosette of scales on the occiput. Males have 15 to 20 vertical bars on the sides of the body, somewhat narrower than the light interspaces. Females have 12 to 16 bars, thinner, less regular and less complete, represented by rounded spots posteriorly. Dorsal surface with black blotches, sometimes almost uniformly dark. Some males have a faint horizontal mark on the dorsal fin, and one specimen taken in June has two fairly definite bars separated by a light interspace.

The characters of the Georgian Bay specimens agree for the most part with those described for the most western variety *menona* as described by Forbes and Richardson ('08), though intermediate in some respects between this and the Atlantic coast form as described by Jordan and Evermann ('96).

The food of this species consists of aquatic and terrestrial insects, minute crustacea, and occasionally small molluscs.

Family GASTEROSTEIDÆ.

(Sticklebacks)

Eucalia inconstans, Kirtland.

(Five-spined or brook stickleback)

This species, apparently the sole representative of the family in this region, occurs in a few collections, all from comparatively closed swamps and inland ponds. It appears to be rare everywhere along the shore.

Length 1¾ inches. Body fusiform, laterally compressed, with very slender

SESSIONAL PAPER No. 39b

tail. Depth 4 to 5, in one specimen 3·7. Head 3·3 to 3·8. Mouth very oblique, its aperture almost dorsal. Dorsal fin with 5 spines as a rule, but sometimes with 4 or 6, followed by 10, sometimes 9 or 11, soft rays. Anal fin with 1 free spine and 10, sometimes 9 soft rays. Ventral fins placed far forwards, with 1 spine and 1 soft ray. Between their free portions the fused pubic bones form a projecting median ridge. Body without scales or surface plates. Coloration dark olivaceous, with minute rounded clear markings on a darker ground.

Family PERCOPSIDÆ.

(Trout-perches)

Percopsis guttatus, Agassiz.

(Trout-perch)

This species is one of two representing the peculiar family *Percopsidæ*, fishes which combine the characters of the perches and salmonoids. It is reported by Jordan and Evermann ('96) as abundant in the Great Lakes. The type was described by Agassiz ('50) from specimens taken in Lake Superior, and specimens are reported by Bean ('81) from Hudson Bay. The species is also reported from Hawkstone, Lake Simcoe, by Meek and Clark ('02), but not from Muskoka Lake. Considerable interest attaches to the species in that only a single specimen has appeared in the Go Home Bay collections, this having been found floating on the surface of the water. The fish inhabits deep cold water, and may be plentiful, but up to the present has not been taken in small-meshed nets, set especially for the purpose. The following description is based on the single specimen taken.

Length 3⅝ inches. Depth 4·8. Head 3·4. Mouth slightly inferior, otherwise normal. Scales 6, 56, 8. Edges of the scales with minute teeth. Dorsal fin with 2 hard rays, the first rudimentary, and 9 soft rays. Anal with 1 hard and 6 soft rays. A small adipose fin between the dorsal and the caudal. General coloration pale, the dorsal parts with dark edges on the scales and more or less definite mottlings about the dorsal fin.

Family ATHERINIDÆ.

(Silversides)

Labidesthes sicculus, Cope.

(Brook-silverside)

(*Plate II, fig. 15*)

Commonly represented by at least a few specimens in most seinings from shore swamps. It shows a preference for localities where, in addition to aquatic vegetation, there is a considerable amount of clear sand. The largest number of specimens have been taken in the running water near the falls of the Go Home

5 GEORGE V., A. 1915

River, but enormous numbers of the young fish of scarcely more than 10 mm. in length are commonly to be found swimming in large schools outside of the main shore, either in the vicinity of the reefs or in the deep water. It is a lithe, active species, and when feeding in schools, especially towards sundown, is often seen jumping out of the water, presumably in the act of taking insects from the surface.

Length of the largest specimens 3 inches; commonly much smaller. Body very slender, little compressed, the depth 7 to 7·7 in the length. Head 4·4 to 4·8, terminating in a blunt but beak-like snout. Jaws rather narrow, and when viewed from the side bowed upwards in their middle portions. General coloration olive, the body translucent, and allowing the air-bladder and vertebral column to show through the muscles. Dorsal surface with a dark vertebral streak and with fine dark edgings on the minute scales. Sides with a silvery band, more or less underlaid by a dark line which broadens into a band on the posterior part of the body.

Dorsal fins two, the anterior consisting of 4, rarely 3, weak spines, the posterior of 12 (sometimes 11 or 13) ordinary rays. Anal fin very long, its posterior portion shallow, with 25 to 28 rays (the number reported by Forbes and Richardson ('08) for Illinois is 22 to 25). ˋScales very small and rounded, about 95 in a longitudinal row.

The food consists of minute plankton entomostraca, together with small insects, the latter including terrestrial forms which are evidently taken from the surface of the water.

Family CENTRARCHIDAE.

(Basses and Sunfishes)

This family is represented by three genera and four species, probably the most familiar of all fishes inhabiting the region, and one of them, the small-mouthed bass, important as its chief game fish.

The three genera are separable as follows:

    a. Base of the dorsal fin less than twice as long as that of the anal, the latter contained
        about 1.5...Ambloplites
  aa. Base of the dorsal fin more than twice as long as that of the anal
      b. Body very short and deep, the depth 2.2 to 2.4................Eupomotis
      bb. Body more elongated, the depth at least 2.9 and usually 3.5.....Micropterus

Genus Ambloplites.

Ambloplites rupestris, Rafinesque.

(Rock-bass)

Extremely abundant in all situations along the main shore, in the larger inland lakes in the vicinity of Go Home Bay, and in the Musquash River, though not reaching Muskoka Lake. It shows a preference for rocky ledges in the vicinity of open water, where it is commonly seen in large numbers.

Length usually 6 inches or less, specimens of 7½ inches being infrequent. Body short, deep, and compressed, the depth 2·2 to 2·4 in the length. Head 2·5 to 2·8. The general coloration varies from olive with more or less brassy reflections, in fish taken in lighter water, to almost black in fishes taken in muskeg water. Sides with rectangular blotches, more definite dorsally, and especially conspicuous in young specimens. Some of the scales below the lateral line with small dark spots, forming about 10 longitudinal stripes. A black spot on the opercle. The dorsal, caudal, and anal fins are more or less mottled or barred with pigment; lower edges of ventrals and anal black. Dorsal fin with 10 or 11 spines and 11 (sometimes 10 or 12) soft rays. Anal with 6 spines and 10 soft rays, the length of its base contained 1·5 to 1·6 in that of the dorsal. Lateral line high up on the body, and curved, the scales 40 to 46.

The food of the rock-bass consists of minnows, crayfish, and insects; the chief food depending on whether the fish is small and inhabiting swampy areas, or large and inhabiting more open shoaly places. During the period when mayflies are abundant, the smaller fish feed largely upon them, leaving their shelters after nightfall, and sucking the flies from the surface of the water.

The spawning period is for the most part during the month of June. The nesting habits are similar to those of other centrarchids. The nest is placed near shore in a swampy bay, often in only a few inches of water. It is prepared by the male fish, which usually works most energetically, fanning out the sediment with his fins, thus making a basin-like depression, clean of all debris, and of 8 or 10 inches in diameter. The female is driven into the nest and is carefully guarded until the deposition of the eggs is accomplished. During the process of spawning and fertilization the two fish lie side by side in the nest. Only a few eggs are extruded at a time, and at each period milt is extruded by the male. The operation continues for an hour or more, and at the end of the period the female leaves the nest and does not return. The eggs are carefully looked after by the male fish, which takes up a position over the nest, and every now and then sets up a fanning motion with the fins. In a few days, after the eggs are hatched, the fry gradually rise out of nest, and are soon left by the male fish to shift for themselves.

During the spawning period rock-bass nests are extremely common in the swamps. Some contain live eggs; some are empty and abandoned, and some are occupied by whitened, fungus-infested eggs which in many cases are still watched over by the male fish. The number of fish spawning at one time and the difficulty experienced by the males in getting the females into the nests, together with the lively competition for their possession sometimes results in confusion. A female for example has been observed to go alternately into two nests, and in some cases a male has been observed hopelessly trying to look after two nests, evidently undecided as to which is his own property.

The rock-bass is reported by certain authorities to reach a length of 12 inches. Possibly the decrease in number of the larger predaceous fishes, such as bass, doré, and pike, which is almost certain to take place as a result of the increase in game

5 GEORGE V., A. 1915

fishing, will put this species in a more advantageous position. At the present time, however, it is a pest to the sportsman in search of the small-mouthed bass. It inhabits the same situations, is of insignificant size and of no fighting qualities; with a propensity for biting on all occasions, regardless of experience. As a destroyer of bait intended for other fishes, it has become notorious, the more so since the supply of this commodity has now reached the dignity of a commercial enterprise.

Genus **Eupomotis**.

Eupomotis gibbosus, Linnaeus.

(Common sunfish. Pumpkinseed)

Abundant in shore swamps and inland lakes. The only species representing the brillantly colored sunfishes in this region.

Length 5¼ inches, commonly much less. Body very short, deep, and compressed, the depth 2·2 to 2·4 in the length. Mouth small. Back olive green with brassy reflections, tinges of blue color, and reddish golden spots. Below the lateral line there are wavy and more or less irregular blue lines, alternating with series of prominent reddish golden spots, the latter arranged more or less definitely into four longitudinal lines. Under parts yellow, golden, or reddish. Cheek and opercle with five blue lines, alternating with reddish golden spots. Opercular flap with a large black spot, bounded above and below by bluish, and behind by scarlet. Dorsal fin with 10 or 11 spines, followed by 11 or 12 soft rays. Anal with 3 spines and 10 soft rays, the length of its base contained 2·1 to 2·3 in that of the dorsal. Pectoral fins reach the vertical of first anal spine. Scales 40 to 45.

The food of this species consists of insects and small molluscs. The spawning period is for the most part in July, though it extends from the latter part of June to the end of August. The nests are often no more than four inches in diameter, and are placed in very shallow water near the shore. The eggs are guarded by the male fish, which at this time exhibits great courage and pugnacity in warding off enemies.

Genus **Micropterus**.

This genus is represented by two important game fishes, one being the small-mouthed bass, or black bass (*M. dolomieu*), the other the large-mouthed bass, green, or Oswego bass (*M. salmoides*). Much has been written concerning the habits of these species, their sporting qualities, and distribution, though it is unfortunate that many popular accounts do not discriminate between the two types. Not only are the two species distinct, but in a region such as this where they occur together they differ very greatly in habits, fighting ability, and in their quality as table fish, the small-mouthed bass being in every way superior.

The two species are separable as follows:

SESSIONAL PAPER No. 39b

a. Mouth very large, the posterior end of the upper jaw reaching past the vertical of the posterior margin of the eye. Scales large; 6 to 10 rows on the cheek.....
....**salmoides**

aa. Mouth smaller, the posterior end of the upper jaw reaching a point beneath the middle of the eye. Scales smaller, those on the cheek in 12 to 17 rows ....**dolomieu**

Micropterus salmoides, Lacépède.

(Large-mouthed bass. Green bass)

(Plate *I*, *Fig. 1*)

Fairly abundant in weedy swamps, swamp channels, and inland lakes. It reaches a weight of 5½ lbs. and possibly more, though as commonly taken it is from 1 to 1½ lbs. The young of all sizes are abundant in the shore swamps and inland waters. This species is said to reach a weight of 14 lbs. in the southern parts of the continent.

Body moderately elongated, the depth 3 to 3·6. Head 2·8 to 3·3. General coloration dark olive green above, white below. More or less irregular blotching dorsally. A lateral band made up of more or less disconnected blotches, very conspicuous in the young fish; sometimes also in the adult, but usually in the latter much broken, obscure or even absent. Dorsal fin with 10 spines and 12 soft rays. Anal with 3 spines and 10 soft rays, its base contained 2·6 to 3 in that of the dorsal. Scales 8 or 9, 63 to 70, 12 to 14. 6 to 10 rows on the cheek.

The food consists of smaller fishes, some specimens containing insects and crayfish. The fish is commonly taken by bait-fishing or trolling but in this region is not sought after.

The spawning period is in the early part of June. The nests are commonly placed in swamp bottoms where there is a deep deposit of detritus. For this reason the fish construct, by fanning out with the fins, huge basins, sometimes of three feet in diameter and a foot into the bottom. The eggs are watched by the male fish. After hatching, the fry gradually rise out of the nest, and begin to swim around in large schools. They are light in color and have a conspicuous lateral stripe. Like the fry of the small-mouthed bass, they are commonly found with the abdomen greatly distended from the ingested entomostracan food.

Micropterus dolomieu, Lacépède.

(Small-mouthed bass. Black bass)

(Plate *I*, *Fig. 2*)

Abundant in its favorite habitat, frequenting during the summer season rocky shoals, channels, and runways among the islands, where there is more or less clear or moving water; also pools about the openings of swamps where minnows commonly wander to and fro. Frequent in running water at the foot of waterfalls.

5 GEORGE V., A. 1915

As commonly taken in this region the fish is from 1 to 2½ lbs. in weight, specimens of 3lbs. or over being exceptional. Body moderately elongated, relatively shorter and deeper in old specimens, the depth 2·9 to 3·5. Head 3·1 to 3·4. General coloration varying from light olive green above and white below, to almost black, the difference depending on whether the fish is taken in the clear open water of Georgian Bay or from the dark water of inland localities. There are all gradations. Sides of the body with more or less irregular vertical bands, conspicuous in young specimens, but more or less obscure in older ones. Four green or dark bands on the cheek, radiating backwards from the eye and upper jaw. Fins rather light, a character by which the fish may be recognised in the water even if seen only for a moment. Young specimens have a conspicuous semicircular dark band on the tail, forming a somewhat heart-shaped figure, with a dark spot in the centre of the base of the tail. This marking is best shown in fish of two or three inches in length. Dorsal fin with 10 spines and 12 to 14 soft rays. Anal with 3 spines and 11 to 13 soft rays, the length of its base contained 2·5 to 2·7 in that of the dorsal. Scales rather small, 12 or 13, 77 to 91, 17 to 23. 12 to 17 rows on the cheek.

The staple food of the small-mouthed bass consists of crayfishes, which inhabit the rocky shoals frequented by the fish. The bass, however, shows a decided preference for minnows, and in the early part of the season when in the shore swamps, or later when in pools or channels which have swamp connections, minnows form a large portion of its food. It is an interesting fact that while swamp areas contain an abundance of minnows, the bass tend on the whole to avoid them, and in three cases in which individual fish have been enclosed by accident in such swamps they have been found dead in the water.

The spawning period is for the most part during June, though fish have been observed on the nests as late as July 20th. Towards the end of May the fish appear in the shore swamps, congregating in groups of sometimes a dozen, and basking lazily near the surface of the water, sometimes with the dorsal fin out of water. They have been observed to move out into deep water during days of colder weather and appear again later. During this early period the male fish apparently explores the shore in shallow water in search of nesting places, and having found one proceeds to put it in order. This process, as well as the deposition and care of the eggs is in all essentials as described by Lydell ('03) for the species elsewhere. The nest is constructed by the male, which is usually seen working alone. In a few cases both male and female have been observed, but the presence of the latter does not appear to be appreciated. The nest is a shallow basin of 15 or 20 inches in diameter, fanned out of the weedy or pebbly bottom, and carefully cleaned of all debris. The bottom of the nest may be of clean rock or pebble, but is more often of short stems of the aquatic plant *Eriocaulon*, which forms an ideal surface for the attachment of the eggs. It is questionable whether the female is selected before or after the construction of the nest, because she commonly remains in the deeper water some distance from shore. There are some indications, however, that in certain cases she is selected before the completion of the nest.

Before and in preparation for the actual process of spawning the male has been observed to swim out into the deeper water, and return driving the female before him. She swims into the nest, and the male circles about her, always heading her into the centre of the nest, and biting her lightly but persistently on the side of the body. If at any time she darts away from the nest, he immediately follows and brings her back. During the spawning process the two fish differ very markedly in color, the male having a uniform bronze or greenish hue, while the female has a blotched appearance, the body spots standing out strongly on a lighter ground. At the time of desposition of the eggs the body of the female is turned somewhat obliquely in the water, so that one side tends to be uppermost. Only a few eggs, perhaps 10 or 12, are extruded at a time, the extrusion being momentary, and repeated at intervals of about half a minute. It is accompanied by a trembling motion of the body and especially of the dorsal fin. The male fish lies for the most part over the female, but with the body in a slightly different direction. The milt is extruded at intervals corresponding to the periods of deposition of the eggs. After the spawning operation is completed, which may be in one half to three hours, the female either leaves the nest voluntarily or is driven out by the male. The latter then takes up a position over the eggs, fanning them from time to time, or making short excursions from the nest in pursuit of other fish that venture in the vicinity. This is maintained for the few days necessary for the hatching of the eggs. The fry, at first confined to the bottom of the nest gradually rise upwards in the water and begin to separate forming a somewhat disorganized school within a certain radius of the nest. They are watched over by the male fish for a few days and are then abandoned to shift for themselves. They are almost pure black in color, and are conspicuous objects in the water. The first food consists of the smallest of the plankton entomostraca. After the feeding process is once established they are extremely greedy, and are often found with the abdomen rounded and distended from the large amount of ingested food.

Many attempts have been made to propagate this fish by the usual artificial methods, but without success, because of the difficulty of stripping the eggs from the female. The eggs are adherent in the ovary, and under natural conditions are only extruded a few at a time. Doubtless a method of caring for the female fish could be devised by which eggs could be obtained for fertilization, but it is improbable that the number obtained in this way would be sufficient to make the work profitable. A few eggs have been fertilized in the laboratory, and eggs taken from nests have been hatched in shallow pans. The method now in common use for obtaining a supply of young fish, and the only method which promises results is that of natural cultivation in retaining ponds. This method could be applied in the Georgian Bay region by reserving for the purpose a number of the larger lakes inland from the main shore. Ponds cannot be made by excavation on the east and north shores, and the natural ponds and smaller-lakes of this region are unsuitable for this purpose in every respect.

From its wide distribution, abundance in localities not over-fished, and splendid sporting qualities, the small-mouthed bass is easily the foremost of American fresh water game fishes, and the shoal areas of Georgian Bay constitute an ideal environment for the species in a region which is most attractive to sportsmen.

5 GEORGE V., A. 1915

The habits of the fish and the methods employed in its capture in Georgian Bay have been recently described by Loudon ('10). In the southern part of the bay the bass is taken only by natural baits, and by trolling with artificial lures, but it is reported on good authority that on the north shore at McGregor Bay the fish will rise to the artificial fly. There is probably no species which is more uncertain of capture. Though at some times biting promptly and vigorously the moment the bait is in the water, at other times it is wary, or refuses with stolid indifference to respond with more than a lazy movement to anything put before it. Places which on some occasions afford fish in abundance are at other times abandoned. The fish tend to run in small groups and move about from place to place, but apparently within comparatively small areas. During the summer of 1909, one hundred fish were caught, marked with a metal tag, and returned to the water. Seven of these were afterwards taken by different persons who reported them. Those reported had been free for different periods from 4 to 30 days, but all were taken within a short distance of the place where they were liberated.

Family PERCIDÆ.

(Perches)

This family is represented in Georgian Bay waters by five genera, each with one species. One species, the pickerel or doré (*Stizostedion vitreum*) is important both as a commercial and a game fish, the others being insignificant.

The genera are separable on several technical differences, but the following analysis will suffice for Georgian Bay species.

> a. Pseudobranch gill structures on the underside of the opercle well developed; branchiostegal rays 7; preopercular bone with a spiny margin;
>   b. Sharp canine teeth on jaws and palatines...................Stizostedion
>   bb. No canine teeth...Perca
> aa. Pseudobranchs small or absent; branchiostegal rays 6; preopercular bone with its edge entire;
>   c. Premaxillaries not protractile, connected with the skin of the forehead by a median ridge;
>   d. Head broad and flat between the eyes....................Percina
>   dd. Head compressed and rounded between the eyes........Etheostoma
>   cc. Premaxillaries protractile, separated from the skin of the forehead by a transverse groove..Boleosoma

Genus Stizostedion.

The small blue pickerel, sauger, or sand pickerel (*Stizostedion canadense*), though reported by some fishermen, has not been identified in Georgian Bay, specimens taken for this species having proved in all cases to be small specimens of the ordinary doré. But one species is therefore described.

Stizostedion vitreum, Mitchill.

(Pickerel. Doré. Pike-perch. Wall-eye)

Generally abundant along the main eastern shore of Georgian Bay, though for some reason it has almost disappeared within ten years in the region of Go Home Bay. Abundant especially on the north shore, and the basis of a valuable commercial fishery in the North Channel. It is the most characteristic of the larger fishes in the darker inland waters. It inhabits the deeper places about rocky shores, projecting points, shoals or channels. It is fond of running water, and may be taken at the foot of waterfalls when other places fail. As commonly taken it is from 2 to 4 lbs. in weight, though the maximum is reported to be in the neighborhood of 20 lbs.

Body elongated and little compressed. Depth $4 \cdot 1$ to $5 \cdot 5$. Head conical, its length 3 to $3 \cdot 8$ in that of the body. Snout $3 \cdot 4$ to 4 in head. Eye large, 5 to $6 \cdot 8$ in head, the cornea whitish, giving the characteristic milky or wall-eye appearance. General coloration yellow or brassy, the upper parts both coarsely and finely blotched with black. Under parts white, yellowish or greenish. Anterior dorsal fin with dark margin and obscure dark spot at its posterior end. Posterior dorsal and caudal finely barred with dark flecks. Dorsal fin with 14 spines and 20 to 22 soft rays. Anal with 2 spines and 11 to 14 soft rays. Scales very small and closely set, 11 to 14, 80 to 110, 14 to 21.

In Ontario and westward this fish is usually and inaptly termed "pickerel". It has received a variety of names, however, of which the English term "pike-perch", French Canadian "doré" (Pic doré), or the commonplace "wall-eye" are more appropriate.

With the exception of the common pike and the maskinonge, it is probably the most predatory of all shallow water fishes. It lives on minnows, small black bass, rock-bass, sunfish, perch, and suckers, but will also take mud-puppies and crayfish. The fish is a strong, swift swimmer, and is well adapted for the capture of small fishes by its stout bony jaws and palatines, which are provided with strong canine teeth. Its large size, hard investment of closely set scales, and formidable spines defend it adequately against all natural enemies.

The doré is of considerable commercial value. In 1909 the total catch for Georgian Bay proper was valued at $4,566.00, and for the North Channel $25,-950.00. The much smaller amount taken in Georgian Bay proper is owing in part to difference in regulations, pound-nets being licensed in the North Channel, while on the eastern shore of Georgian Bay all inshore commercial fishing is prohibited.

The fish is also respected as a game fish, and though much inferior to the small-mouthed bass in sporting qualities, is quite as much in demand as a table fish on account of the firm white character of its flesh and its excellent flavor.

During the early spring, immediately after the ice disappears, the fish ascend the rivers to the spawning places at the foot of waterfalls. The eggs are deposited on sticks and stones in the running water, and are often deposited in such large

5 GEORGE V., A. 1915

masses that they probably have little chance of hatching. Considerable attention has already been given to the artificial propagation of the species, but much more could be done with very moderate expense by utilizing the various waterfalls on the river courses where the fish now spawn in abundance. In some respects the eggs are more difficult to handle than those of whitefish and trout, but, on the other hand, relatively greater results may be had with little effort and cost. The small size of the eggs permits a jar capacity of three or more times that of whitefish, and the period of operations, including the capture of the parent fish, stripping, and hatching of the eggs involves only a short season of three or four weeks according to the temperature of the water.

During the late summer the doré are said to move out into the deep water, returning again into the rivers in the following spring. In the early summer they are commonly taken by sportsmen by trolling or by bait-fishing in moderately deep water near shore, on shoals, or in channels. In clear water they bite only in the early morning or towards sundown, but in the dark inland waters they may be taken at any time of the day, though better when the light is not intense.

Genus **Perca.**

Perca flavescens, Mitchill.

(Yellow perch)

Present in all situations, except the smallest inland ponds. Probably the most abundant and generally distributed species in the region. It is taken in shore swamps, inland lakes, on shoals, and in the open water of Georgian Bay at a depth of 20 fathoms. Whether from some feature of the environment, or the presence of a large number of predaceous enemies, the fish does not reach the size that it does elsewhere. The largest specimens are about $10\frac{1}{2}$ inches in length, but the average is scarcely more than 5 inches.

Body moderately elongated, somewhat compressed, the back very convex. Depth 3·7 to 4. Head 3·2 to 3·4. General coloration yellow, light below. Sides with seven vertical dark bars. Ventral and anal fins pale yellow, bright yellow, or reddish. Specimens from the open water shoals and from deep water have the yellow of the sides replaced by grey or blackish, and the lower fins are red. Anterior dorsal fin with 12 or 13 spines. Posterior dorsal with 1 spine and 12, or sometimes 13, soft rays. Anal fin with 2 spines and 7 or 8 soft rays. Scales small and solidly attached, 6 to 8, 67 to 71, 11 to 14.

The perch is carnivorous and more or less predaceous according to its size. It feeds on small crayfish, molluscs and insects, and when large attacks smaller fishes. It seems to be more adaptive in respect of its environment than other species, and appears to be taking up the deep water area of the southern part of Georgian Bay that was formerly occupied by the whitefish. It is probably increasing rapidly in this situation, in spite of the fact that it now forms a large part of the food supply of the lake trout.

Sub-family ETHEOSTOMINAE.

(Darters)

The following three species, represent in this region an extensive division of perch-like fishes commonly designated as log-perches and darters, all fishes of diminutive size, and for the most part different in habit from the common perch and doré. They are non-predatory, living largely on minute insects and crustaceans. They inhabit sand beaches and sloping rocks in somewhat protected situations. They have the habit of lying motionless on the bottom, with the body slightly bent, and its anterior portion slightly raised on the pectoral fins, a posture which gives them an alert appearance. They move by quick jumps, rather than by swimming, the enlarged pectoral fins being used for this purpose and also for fanning up the bottom in search of food or concealment. They are characteristic of running water, but in this region have adapted themselves to a lacustrine habit.

Genus **Percina.**

Percina caprodes, Rafinesque

(Log-perch)

Usually taken on sand beaches where there is some aquatic vegetation. The fish is easily recognized by its yellow coloration, dark vertical or saddle-shaped bars, and pointed snout.

Length $3\frac{1}{4}$ inches. Body elongated, not compressed. Depth 5·7 to 6·8. Head very flat above, 3·6 to 5·2 (commonly 3·7). Snout 2·4 to 3 in head; slender and pig-like, overhanging the inferior mouth. General coloration yellow, the back and sides with 8 to 10 vertical bands or stripes, which tend to enlarge into darker spots below. Alternating with these are smaller bands or spots, either fused or disconnected with reference to the main stripes, making a somewhat irregular pattern. A definite black spot at the base of the tail. Dorsal and caudal fins barred with black or brownish. Ventral parts light. First dorsal fin with 14 or 15 spines, the second with 16 soft rays. Anal with 11 to 13 rays. Scales small, with ctenoid edges, absent before dorsal and on breast, except in young specimens, 6 to 8, 80 to 88, 15 to 17.

The food consists of minute chironomus larvæ, small amphipods, crayfish, and entomostraca.

Genus **Etheostoma.**

Only one species recognized, but some aberrant specimens suggest the character of *E. boreale,* which has been recognized by Meek and Clark ('02) as occurring in Muskoka Lake.

5 GEORGE V., A. 1915

Etheostoma iowae, Jordan and Meek.

With. the exception of the well-marked species *Boleosoma nigrum,* described below, and the doubtful species *E. boreale,* all the smaller darters, of less than two inches in length, appear to belong to a single species, identified by Dr. Meek as *E. iowae.* The specimens, however, show some variation in color pattern, especially larger specimens taken in the early part of the season.

Fairly abundant on rocks and sand beaches, but showing a more decided tendency towards the latter than *B. nigrum.* Though superficially much like the latter species, it is distinguishable in the water by the paler character of the saddle-like cross markings of the back. Spring males are easily distinguished from all other fishes by their brilliant blue and orange markings.

Length 1⅜ inches, commonly barely more than an inch. Body elongated, tapering backwards from a point in front of the dorsal fin. Depth 5 to 5·7. Head 3·4 to 4. Snout rather blunt, the mouth at its ventral angle. Premaxillaries not proctractile, joined to the forehead by a median fleshy bridge. General coloration buff, the sides with about 10 irregular blotches of cinnamon color, arranged in a bead-like series. Dorsal surface finely punctate, with 8 or 9 faint cross bars of darker color. Portions of the lateral markings sometimes tend to fuse above. A bar forwards on the snout and another downwards from the eye. Dorsal and caudal fins more or less barred. Under parts pale. The lateral line is marked out forwards as a white streak, slightly bowed upwards in its middle portion. Anterior dorsal fin with 8 or 9 spines, posterior dorsal with 10 or 11 soft rays. Anal with 2 spines and 7 or 8, sometimes 9, soft rays. Scales minute and ctenoid, 4 or 5, 55 to 60, 8 to 11. Lateral line incomplete posteriorly.

Males in the breeding season are brilliantly colored. The anterior dorsal fin has the basal two-thirds deep blue green, darker between the rays. There is a narrow band of blue at the margin of the fin, separated from the basal band by a stripe of orange. Sides with the angular cinnamon blotches very bright and alternating with greenish black spots. More or less orange at the base of the pectorals, and extended backwards by four obscure blotches to and along the base of the anal. Basal membranes of the posterior dorsal, caudal and anal with diffuse greenish.

The breeding season includes the latter part of May and June. The eggs are deposited on stones, especially in sheltered crevices, often in water of only a few inches in depth. The animals are commonly found in groups, and there is a lively competition among the males for possession of the females.

Genus **Boleosoma.**

Boleosoma nigrum, Rafinesque.

(Tesselated darter. Johnny darter)

(Plate II, fig. 17)

Abundant in rocky situations along shore, and also common on sand beaches, or in swamps where there is some clean sand bottom. Resembling the foregoing

Fig. 6. Uranidea franklini. × 1½.

species superficially, but distinguishable by the darker marks of the back, and by the W. or M-shaped flecks of the sides.

Length 2 inches or less. Body elongated, tapering backwards from the shoulder. Depth 5.4 to 6.1. Head 3·7 to 4·2. Snout blunt, the mouth at its lower angle. Premaxillaries protractile. Eyes dorsal and protruding. General coloration pale straw yellow. Back with 6, sometimes 7, cross bars of dark color. Scales more or less flecked, especially on the sides where there is a longitudinal series of W,M, or X-shaped marks. Head with a bar forwards from the eye and a spot beneath it. Spring males are dull sooty or inky black. Anterior dorsal fin with 8 to 10 spines, posterior dorsal with 12 (sometimes 10 or 11) soft rays. Anal with 1 spine and 7 to 10 soft rays. Scales 4, 43 to 48, 8 to 10. Lateral line almost complete, flexed downwards slightly in front.

Family COTTIDAE.

(Sculpins)

Represented by two genera, each with a single species. One of these is characteristic of the eastern shore, while of the other only a single specimen has been found, this from the sedimentary zone at the Giants's Tomb Island.

The two genera are separable as follows:

    a. Ventral fins with 1 concealed spine and 3 soft rays......................**Uranidea**
    aa. Ventral fins with 1 concealed spine and 4 soft rays.......................**Cottus**

Uranidea franklini, Agassiz.

(*Fig. 6*)

Found lurking under stones in shallow water, and easily recognized by its wedge-shaped body and fan-like pectoral fins. It always seeks concealment, and if dislodged from one shelter darts rapidly to another.

Length 2 inches. Body very heavy forwards, tapering backwards to the slender tail. Depth 4·7 to 5·1. Head broad, its length 2·8 to 3 and its width 3 to 4·1 in the length of the body. Eyes dorsal in position, very large and protruding. Preopercle with an abruptly hooked spine which is directed backwards and upwards. General coloration yellowish or brownish, with dark mottlings and cross blotching above and on sides. Anterior dorsal fin with 8 slender spines, posterior dorsal with 17 soft rays. Anal fin very long, with 12 to 14 soft rays. Pectorals very large and fan-like, with 15 rays. Ventrals situated forwards, very small, with 1 weak spine and 3 soft rays. Body naked. Lateral line complete.

Cottus ictalops, Rafinesque.

(Miller's Thumb)

In habit similar to the foregoing species. Probably not rare, but no specimens have been taken on the eastern shore. The species occurs throughout the Great Lakes, and is said to be especially abundant in Lake Superior.

5 GEORGE V., A. 1915

Length of the single specimen 1¾ inches. Body very robust forwards, and compressed towards the tail, Depth 4·3. Head stout and broad, its length 2·6. Eyes very large, dorsal, and protruding. Preopercular spine almost straight. Coloration dark brown or greyish above, mottled; white below. Dorsal and caudal fins finely barred with flecks of black: lower fins less so. Anterior dorsal low, with 7 weak spines. Posterior dorsal with 15 soft rays. Anal fin with 12 soft rays. Pectorals very large and fan-like, with 15 rays. Ventrals with 1 spine and 4 soft rays. Body naked, except for a few prickles behind the pectoral fins. Lateral line conspicuous anteriorly but absent posteriorly.

Family GADIDAE.

(Codfishes)

Lota maculosa, LeSueur.

(Ling. Burbot. Lake cusk)

Abundant in the deeper water of Georgian Bay, and commonly taken by fishermen in gill-nets.

Length 2 feet. Body rounded and heavy in front: greatly compressed towards the tail. Depth 5 to 7·7. Head broad and flat, its length 4 to 4·9. Snout 2·9 to 3·4 in head. Jaws and vomer with small, sharp teeth. A longer barbel below the chin, and shorter ones at the anterior openings of the nasal sacs. Eye small 6·7 to 10 in head.

General coloration olive or dark ashy above, with darker mottlings and scattered black spots. Lower parts light ashy or yellow. The general tone is darker and less yellow than in specimens from muddy waters. Anterior dorsal fin with about 10 concealed rays. Posterior dorsal very long, its base 1·9 to 2·3 in the length of the body; containing about 75 rays. Anal fin with about 68 rays, its base 2·4 to 2·7. Scales very minute and imbedded.

The ling is a voracious fish, living on perch, young whitefish, trout, herring, or on crayfishes. It is of no commercial value, is generally despised by fishermen, and is destroyed by them in large numbers. Its poor reputation is doubtless based on its slimy repulsive appearance and more or less unpleasant odor, the flesh being in reality of fair quality*.

---

* The eggs of this species were discovered in 1906 and described in a paper in the "*Ottawa Naturalist*, Mar. 1906 by Prof. Prince and Mr. A. Halkett. The egg is of a very delicate character like the pelagic floating eggs of the marine ling, cod, haddock, etc.

BIBLIOGRAPHY

1850. AGASSIZ, L.—Lake Superior, etc. Boston, 1850.

1881. BEAN, T. H.—Notes on Some Fishes from Hudson's Bay. *Proc. U. S. National Museum*, Vol. III, 1881.

1905. COLE, L. J.—The German Carp in the United States. *U. S. Dept. of Commerce and Labor, Rep. Bureau of Fisheries*, 1904.

1893. EIGENMANN, C. H., AND EIGENMANN, R. S.—Preliminary Description of New Fishes from the Northwest. *American Naturalist*, Vol. XXVII, 1893.

1907. EVERMANN, B. W., and GOLDSBOROUGH, E. L.—A Check List of the Fresh Water Fishes of Canada. *Proc. Biol. Soc. Washington*, Vol. XX, 1907.

1908. FORBES, S. A., AND RICHARDSON, R. E.—The Fishes of Illinois. *Natural History Survey of Illinois*, Vol. III, Ichthyology.

1893. GAGE, S. H.—The Lake and Brook Lampreys of New York. *The Wilder Quarter Century Book*, Ithaca, 1893.

1910. GOLDTHWAIT, J. W.—An Instrumental Survey of the Shore Lines of the Extinct Lakes Algonquin and Nipissing in Southwestern Ontario. Canada, *Dept. of Mines, Geol. Survey Branch*, Memoir No. 10, 1910.

1896. JORDAN, D. S., AND EVERMANN, B. W.—The Fishes of North and Middle America. *Bull. U. S. National Museum*, No. 47, 1896–1900.

1909. ————————A Review of the Salmonoid Fishes of the Great Lakes, with Notes on the Whitefishes of other Regions. *Bull. U. S. Bureau of Fisheries*, Vol. XXIX, 1909 (issued 1911).

1885. ———————— AND FORDICE, M. W.—A Review of the North American Species of Petromyzontidæ. *Ann. New York Academy of Sciences*, Vol. III, 1885.

1910. LOUDON, W. J.—The Small-mouthed Bass. Toronto, Hunter–Rose Co., 1910.

1903. LYDELL, D.—The Habits and Culture of the Black Bass. *Bull. U. S. Commission*. Vol. XXII, 1902.

1886. MEEK, S. E.—Additional Note on the Lamprey of Cayuga Lake. Ref. to Jordan and FORDICE *supra*.

1899. ————————Notes on a Collection of Fishes and Amphibians from Muskoka and Gull Lakes. *Field Columbian Museum, zool. series*, Vol. 1, 1895-1899.

1902. ———————— AND CLARK, H. W.—Notes on a Collection of Cold Blooded Vertebrates from Ontario. *Field Columbian Museum, zool. series*, Vol. III, 1900-1904.

1908. NASH, C. W.—Vertebrates of Ontario. *Department of Education*, Toronto, 1908.

1836. RICHARDSON, J.—Fauna Boreali-Americana, etc., pt. III, The Fish. London, 1836.

For statistics of the commercial fisheries and administration.—
(a) Annual Reports of the Minister of Marine and Fisheries, Ottawa.
(b) Annual Reports of the Superintendent of Game and Fisheries for Ontario, Toronto.
(c) Report and Recommendations of the Dominion Fisheries Commission appointed to enquire into the Fisheries of Georgian Bay and adjacent waters, Ottawa, 1908.

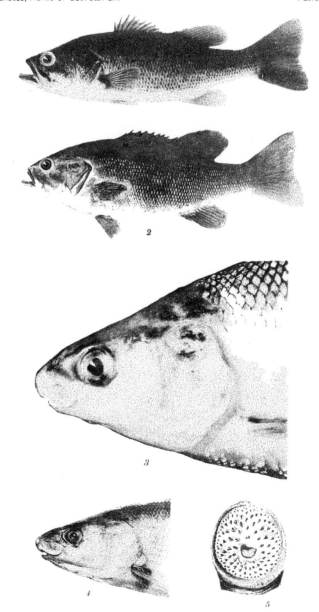

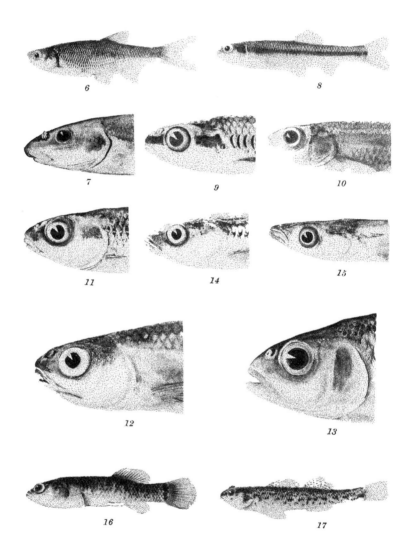

II.

NOTES ON THE ODONATA OF THE VICINITY OF GO HOME BAY, GEORGIAN BAY, ONTARIO.

By E. M. WALKER, B.A., M.B., Assistant Professor of Zoology,
University of Toronto.

(Plates III—IX, and 1 figure in text)

When I first visited the Biological Station at Go Home Bay, Ont., in June, 1907, I was struck by the great abundance of dragonflies there, and being specially interested in this group of aquatic insects, particular attention was given them during the two months that I spent there that year.

During this period an effort was made to collect both adults and nymphs of all the species native to the immediate vicinity of Go Home Bay and the Go Home River, and to determine as much as possible of their life histories, including their seasonal and ecological distribution, habits of flight, food, enemies, etc.

As practically nothing was previously known of the Odonata fauna of this locality, the preliminary work of collecting and determining species and rearing the nymphs formed the major part of the work and in this occupation the writer was ably assisted by Mr. W. J. Fraser of Toronto. A considerable quantity of material was also collected by Dr. A. G. Huntsman of the Biological Department, University of Toronto.

In 1908 I spent another period of two months at the Station, but being occupied with other work, little was added to the data already accumulated.

A third visit was paid to the Station in 1912, and as I arrived there on May 19th, nearly a month earlier than on either of the two previous occasions, and remained until Sept. 11th, I was able to add a number of observations to the seasonal distribution of some of the earlier and later species. No additions were made to the Go Home Bay fauna, but two species of *Sympetrum* previously unknown to this district were taken at the Giant's Tomb Island.

Mention is also made in this paper in the list of species, of a number of nymphs collected by Mr. R. P. Wodehouse at various other points along the shore of Georgian Bay.

PHYSICAL FEATURES OF THE GO HOME BAY DISTRICT.

Go Home Bay (Bushby Inlet) is situated on the east side of Georgian Bay about fourteen and a half miles north of Penetanguishene and its topography is typical of a large part of the eastern shore of this body of water.

The physical characteristics of this region have been described in some detail by Bensley ('14) and it will only be necessary here to refer to a few of the more salient features.

5 GEORGE V., A. 1915

The exceedingly irregular coast-line in this vicinity, with its innumerable bays, inlets and channels and its countless rocky islands and reefs, renders the region a very favorable one for the support of a varied and abundant aquatic fauna. Most of the types of environment in which dragonflies flourish are represented within a few miles of the Station Island, from the well-aerated waters of the Go Home River and the more exposed parts of the Bay to the sheltered, often shallow and marsh-bordered inlets, the shady woodland creeks and the small lakes and ponds, margined with sphagnum bogs. The shallow lagoons on the sandy beaches of Giant's Tomb Island offer still other conditions of environment.

GENERAL CHARACTERISTICS OF THE ODONATE FAUNA.

Owing to the rocky topography of the country and the scantiness of the soil the drainage of the smaller lakes and ponds, where it exists at all, is poor and the aquatic vegetation in such stations is somewhat limited in variety, while the shore plants are largely of the type that prevails on bog-soils having an acid reaction, i.e., the plants of the sphagnum-bog society. In these ponds there is an absence of some of the commonest dragonflies of the ponds in agricultural districts. Some of these species are met with in the shallow bays connected with the open water, but even here they are not the prevailing species. As examples of such species we may take *Lestes unguiculatus, Enallagma ebrium, Leucorrhinia intacta, Sympetrum rubicundulum, Libellula quadrimaculata, L. pulchella* and *L. lydia*, all abundant species in the agricultural sections of Ontario, at least in the southern part. All of these species except two have been taken at Go Home Bay, but none are very abundant and none have been taken in the sphagnum-bordered ponds. How far this scarcity is due to soil conditions and how far, in some cases, to the comparatively northern latitudes we are unable at present to say. *Sympetrum rubicundulum* and *Libellula quadrimaculata* range far to the north of Georgian Bay.

There is also an entire absence of certain regional species that breed in gentle shallow rapids with sandy or gravelly bottoms. No species of *Ophiogomphus*, e.g., has been taken in this vicinity, though Mr. Wodehouse took a nymph of a species of this genus in the Shawanaga River and I have found *O. rupinsulensis* fairly common in Algonquin Park. *Gomphus scudderi* and *Lanthus albistylus* were also taken in Algonquin Park, flying over gentle rapids, but are apparently absent from the Go Home district. They are very likely to occur on the Musquash River. Other river species common in Algonquin Park but not represented at Go Home Bay are *Agrion aequabile* and *Boyeria vinosa*.

The total absence of Cordulegasters is also worthy of note and is doubtless due to the absence of the proper conditions of environment. *C. maculatus*, an inhabitant of creeks, and *C. diastatops* of spring bogs have been taken at Port Perry, Muskoka District (Walker '06) and the former at Heyden and Searchmont, near Sault St. Marie, Ont. (Williamson, '07).

The most prominent positive feature of the fauna, as one would be led to infer from the character of the country, is the abundance of individuals of those species which develop in the well-aerated waters of the bay and the adults of which patrol

SESSIONAL PAPER No. 39b

the rocky shores or fly about over the islands and the open channels. These are all found also in the Go Home River, but characteristic river species are conspicuously lacking.

Another noticeable feature of the fauna is the great abundance in the sphagnum bogs at the edge of small lakes and ponds, of certain species that we have met with rarely or not at all elsewhere in Ontario. The most characteristic species of this group are *Nehalennia gracilis*, *Nannothemis bella* and *Leucorrhinia frigida*.

Ecological Distribution of Species.

The Odonata of the vicinity of Go Home Bay may be roughly divided into three principal ecological groups, according to the nature of their breeding-places, viz.,

Group 1:—Species inhabiting the well-aerated waters of the open bay and broader parts of the river.

Group 2:—Species inhabiting still waters, e.g., shallow bays, sluggish creeks in open marshes, small enclosed lakes and ponds.

Group 3:—Species inhabiting woodland creeks.

Two other groups might be added, namely, those species inhabiting the rapids and those breeding in the shallow sand-bottomed lagoons on the Giant's Tomb Island (Fig. 36), but no characteristic species have been found in the former, while the latter are for the most part identical with Group 2, there being but one or possibly two peculiar species.

These groups are not sharply distinguishable from one another, many species fall into more than one of them.

Group I.

These species may be further subdivided into two groups, (a) those which breed on exposed rocky shores, occurring also about the edges of currents (Figs. 26, 27, 28) and, (b) those which are inclined to occupy the lower, shallower and more sheltered parts of otherwise exposed shores (Fig. 29, 30). These sub-groups are not sharply separable, some species being equally well-placed in either.

(a)

1. Argia moesta putrida.
2. Gomphus brevis.
3.   "   lividus.
4. Dromogomphus spinosus.
5. Boyeria grafiana.
6. Basiæschna janata.
7. Macromia illinoiensis.
8. Didymops transversa.
9. Neurocordulia yamaskanensis.

5 GEORGE V., A. 1915

(b)

1. Enallagma carunculatum.
2. Hagenius brevistylus.
3. Gomphus lividus.
4.   "   exilis.
5. Basiæschna janata.
6. Nasiæschna pentacantha? (rare).
7. Epicordulia princeps.
8. Tetragoneuria cynosura simulans.

Of the species in sub-group (a) No. 2 is mainly a species of the rapids but also frequents the exposed shores of the outer islands; No. 4, is chiefly a river form, likewise occurring about the outer islands though sparingly; the others are generally distributed, though No. 5 shows a distinct preference for slightly running water, while No. 9 is most at home in the deeper waters about precipitous rocky shores or in the neighbourhood of rapids.

The species of sub-group (b), with the exception of No. 6, which is included here with some doubt, are all abundant and generally distributed.

GROUP II.

The species belonging to this group are roughly divisible into (a) those which are more characteristic of the marshy coves along the shores of the inner bays and lakes, or at the outlets of sluggish creeks (Plates VII–VIII, Figs. 31, 32) and (b) those which are partial to the edges of sphagnum bogs bordering small lakes and ponds (Plates VIII–IX, Figs. 33, 34).

(a)

1. Lestes unguiculatus (rare).
2.   "   uncatus.
3.   "   disjunctus.
4.   "   vigilax.
5. Nehalennia irene.
6. Enallagma hageni.
7.   "   calverti.
8.   "   ebrium? (one specimen).
9   "   exsulans.
10.   "   signatum.
11.   "   pollutum.
12. Ischnura verticalis.
13. Gomphus spicatus.
14. Gomphus exilis.
15. Aeshna eremita.
16.   "   elepsydra.
17.   "   canadensis.
18.   "   verticalis.

19. Anax junius.
20. Epicordulia princeps.
21. Tetragoneuria spinigera.
22.   "   cynosura simulans.
23. Dorocordulia libera.
24. Celithemis elisa.
25. Leucorrhinia frigida.
26.   "   proxima.
27.   "   intacta.
28. Sympetrum costiferum.
29.   "   vicinum.
30.   "   semicinctum.
31.   "   obtrusum.
32.   "   corruptum.
33. Libellula quadrimaculata.
34.   "   exusta julia.
35.   "   pulchella.
36.   "   vibrans incesta.

(b)

1. Lestes disjunctus.
2.   "   inaequalis (one specimen).
3. Nehalennia gracilis.
4. Enallagma hageni.
5. Gomphus spicatus.
6. Tetragoneuria spinigera.
7. Cordulia shurtleffi.
8. Dorocordulia libera.
9. Nannothemis bella.
10. Leucorrhinia frigida.
11. Libellula exusta julia.

The abundant species of sub-group (a) are Nos. 3, 4, 6, 12, 13, 14, 20, 21, 22, 24, 25, 28, 29, 31 and 34. Nos. 7, 10, 11, 16, 17, 19, 23, 30, 33 and 35 are also common, while Nos. 2, 5, 9 and 27 are not infrequently met with. The others are rare in this district, Nos. 8, 18 and 32 being doubtfully included in this group. Nos. 9, 10 and 11 form a sub-group by themselves, intermediate between Groups I and II. They seem to prefer the low parts of the shores of the river and the quiet inlets where there is a marshy tendency, but little or no growth of reeds or similar marsh vegetation. This type of habitat grades on the one hand into sub-group (b) of Group I and on the other hand into sub-group (a) of Group II, in which they have been included.

No. 32 is peculiar to the lagoons of Giant's Tomb Island.

Of sub-group (b) Nos. 1, 3, 4, 9, 10, and 11 are abundant, 9 being, however, local. Nos. 5, 6 and 11 are more characteristic of sub-group (a). Nos. 3 and 9 breed in the sphagnum bog some distance from the edge of the open water.

Helocordulia uhleri, Leucorrhinia glacialis and *L. hudsonica* probably also belong to Group II, but we have never found their nymphs.

GROUP III.

1. Agrion maculatum.
2. Ischnura verticalis.
3. Aeshna umbrosa.
4. Somatochlora williamsoni.

No. 1 is a characteristic creek and river species. It has not been taken below the "Chute" on the Go Home River, the smaller creeks in this vicinity being too sluggish to suit its requirements. No. 2 is commoner about creeks than about the swampy bays; No. 3 is essentially an inhabitant of shady creeks and ditches, while No. 4 is included here with some doubt (vide p. 85).

SEASONAL DISTRIBUTION OF ADULTS.

When we arrived at Go Home Bay on May 17, 1912, no dragonflies were abroad in the vicinity of the Biological Station, nor were any observed until we visited the Giant's Tomb Island on May 26th, when a single example of *Anax junius* was seen flying over an open bushy slope. On the 29th exuviæ of *Didymops transversa* and *Gomphus spicatus* were found on the shore of the "Pittsburgh Channel", a single example of each. During the next three days tenerals and exuviæ of *G. spicatus*, *Tetragoneuria spinigera*, and *Ischnura verticalis* were taken about Galbraith Lake. A single teneral *Leucorrhinia frigida* was also seen, while *Anax junius* and *Enallagma calverti* were both common and fully mature. On the 8th of June the first specimen of *Basiaeschna janata* emerged in the laboratory, followed by another on the 9th. By this time *Gomphus spicatus* was numerous and *Libellula exusta* was becoming common. By the 13th both of these species had become abundant and *Tetragoneuria cynosura simulans* was emerging. On the 15th the first tenerals of *Gomphus brevis* were found about the "Chute," followed next day by *G. lividus* at the Station Island. *T. cynosura simulans* was already abundant while a single specimen of *Helocordulia uhleri* was taken at Sandy Gray Falls on the Musquash River. It was also about this time that *Enallagma hageni* first made its appearance, and a few days later, on the 18th, the first young adults of *Gomphus exilis* were observed, while those of *G. lividus* were still transforming.

In 1907 these four species of *Gomphus* appeared in about the same order, but somewhat later. When we arrived on June 15th, 1907, *spicatus* was already common, but all the individuals were as yet teneral, while *G. lividus* and *brevis* did not appear until the 22nd and 23rd respectively and *G. exilis* was first found transforming on the 25th. By June 19th, *Nehalennia gracilis* and *Dorocordulia libera* had appeared and from the 22nd to the 25th (1912) three more species were added, viz., *Celithemis elisa*, *Neurocordulia yamaskanensis* and *Lestes vigilax*. On the next day the first Aeshna, *Ae. canadensis*, was recorded together with the first specimen of *Epicordulia princeps*, a species which shortly afterwards became very numerous. On the 27th *Libellula quadrimaculata*, which was first seen in 1907 on the 18th, was found in large numbers, all more or less teneral, about a rocky pond on a small island far out in the Bay. *E. hageni* and *I. verticalis* were

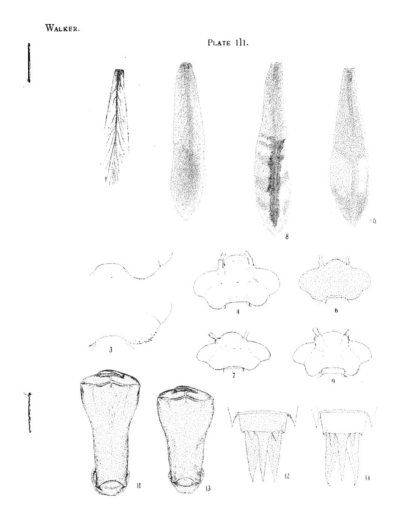

PLATE IV.

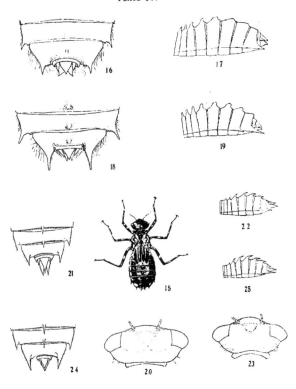

Plate V.

Fig. 26. The outer coast and islands, looking westward, Habitat of Group Ia.

Fig. 27. Islands off the outer coast with precipitous shore. Habitat Neurocordulia yamaskamensis and Argia moesta putrida (Group Ia)

Plate VI.

Fig. 28. Rapids, Musquash river. Habitat of Gomphus brevis, Boyeria grafiana Argia moesta putrida, etc. (Group Ia).

Fig. 29. Small sandy beach with boulders. Station island. Habitat of Gomphus lividus. G. exilis, Macromia illinoiensis, Didymops transversa, etc. (Groups Ia and Ib).

Plate VII.

Fig. 30.   Outlet of Galbraith lake.   A composite of habitats of Group Ia (current) and IIa (sublittoral zone).

Fig. 31.   Outer end of Galbraith lake.   Typical habitat of Group IIa.

Plate VIII.

Fig. 32. Shore of Burwash pond, showing Sphagnum-cassandra zone, with background of black spruce. Habitat of Group IIb.

Fig. 33. "Pond on Split-rock Island," showing Sphagnum-cassandra zone. Habitat of Group IIb, especially Leucorrhinia frigida and Nehalennia gracilis.

Plate IX.

Fig. 34. Mouth of small creek. Habitat of Group IIa, passing into that of Group III
in the distance.

Fig. 35. Shallow channel in the sand of Giants Tomb island. Habitat of Sympetrum corruptum.

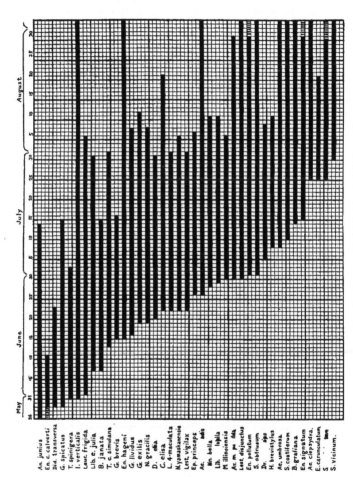

Fig. 1. Plan showing seasonal distribution of adult dragonflies.

5 GEORGE V., A. 1915

also abundant here and all three species were still transforming. By this time *E. calverti*, *D. transversa* and *T. spinigera* had about disappeared, though the last-named species was taken in 1907 on July 1 and 4. All the other species mentioned were common. During the last few days of the month several other species appeared and speedily became abundant. These were *Lestes disjunctus*, *Argia putrida*, *Nannothemis bella*, *Libellula pulchella* and *Macromia illinoiensis*. The last-named species was first observed on the 30th and during the first few days of July it came out in some numbers about the Station Island and neighbouring parts. *N. yamaskanensis* could now be taken any evening in plenty. Tenerals of *Sympetrum obtrusum* now began to appear and *Enallagma pollutum* was first noticed about some of the marshy bays. It had probably, however, been on the wing for some days as all the individuals seen were quite mature. The next species to appear in 1907 was *Dromogomphus spinosus* which was first observed in trans-formation on July 5. This species was not seen at all during 1912. On the 9th *Aeshna umbrosa* emerged in the laboratory and on the same day *Hagenius brevi-stylus* appeared on the Island, becoming common almost immediately afterwards. On the 10th the first tenerals of *Sympetrum costiferum* and one of *S. danae* were taken on the Giant's Tomb Island, but the former species did not appear at Go Home Bay until the 24th, while *danae* does not occur there at all. Mature specimens of *S. corruptum*, likewise absent from Go Home Bay were also taken at the Tomb on this date. On July 14th the last *Basiaeschna* was taken while the first *Boyeria grafiana* emerged in the laboratory, and about the same time in 1907 the first specimens of *Enallagma signatum* were seen, though these were not noticed until some time later in 1912. No other new forms appeared until the 24th when *Enallagma carunculatum* was found about the Station Island, most examples already mature, while *Sympetrum semicinctum* was added to the species of marshy habitat. The predominant species about the Island were now *Argia putrida*, *E. carunculatum*, *Epicordulia princeps*, and *Hagenius brevistylus*, while those about the marshes were chiefly *Lestes disjunctus* and a few *L. vigilax*, *Enallagma hageni*, *Nehallenia gracilis*, *Libellula pulchella*, *Celithemis elisa*, the various species of *Sympetrum* and *Leucorrhinia frigida*, though the last had become greatly reduced in numbers. On July 30, the first teneral of *Sympetrum vicinum* was noticed and after this date no new forms appéared. The Gomphi had practically gone and very few *Macromia*, *Libellula exusta* and *Dorocordulia* were to be seen. *Epicordulia princeps* had also diminished greatly in numbers and the last *Tetragoneuria cynosura simulans*, was taken on the 31st. By August 6th but little change could be noted. The Sympetrums were still emerging, most of the individuals of *S. costiferum* and *S. vicinum* being still teneral. The Aeshnas had apparently all emerged, though very few were seen in 1912. A single *Nan-nothemis* was observed on this date. On Aug. 25 and 26 *Sympetrum costiferum* and *vicinum* were abundant and a few *S. obtrusum* were seen. Many pairs were observed in copula. *Aeshna canadensis* and *clepsydra* were also fairly common and *Somatochlora williamsoni* was twice observed. Several Enallagmas were also still abroad, viz., *E. hageni*, *carunculatum*, *exsulans* and *pollutum*. *E. carun-culatum* was abundant along the shore of some of the bays and *E. pollutum* was

SESSIONAL PAPER No. 39b

also common. *Ischnura verticalis* and *Lestes disjunctus* were also observed, the latter in greatly reduced numbers.

No changes were noted after this date.

GEOGRAPHICAL DISTRIBUTION OF THE SPECIES.

Go Home Bay is situated at about the northern limit of the Transition (Alleghanian) Life Zone and its Odonate fauna thus exhibits an intermingling of Boreal and Austral elements. Many of the species range a considerable distance both north and south of this locality, being common to both the Canadian and Carolinian Zones, and therefore occurring throughout the Alleghanian (Transition) Zone. A few Carolinian forms probably find the northern limit of their geographical distribution at about this latitude, while some of the characteristic species of the Canadian Zone do not seem to occur much farther south.

The species which are generally distributed in Ontario as far or farther than the north shore of Lake Huron are the following:—

1. Agrion maculatum.
2. " aequabile.
3. Lestes unguiculatus.
4. " uncatus.
5. " disjunctus.
6. Chromagrion conditum.
7. Nehalennia irene.
8. Enallagma hageni.
9. " carunculatum.
10. Ischnura verticalis.
11. Hagenius brevistylus.
12. Gomphus lividus.
13. " exilis.
14. " spicatus.
15. Dromogomphus spinosus.
16. Boyeria grafiana.
17. Basiaeschna janata.
18. Aeshna canadensis.
19. " umbrosa.
20. Anax junius.
21. Didymops transversa.
22. Tetragoneuria spinigera.
23. " cynosura simulans.
24. Helocordulia uhleri (?).
25. Dorocordulia libera.
26. Sympetrum costiferum.
27. " vicinum.
28. " semicinctum.
29. " obtrusum.

5 GEORGE V., A. 1915

30. Libellula exusta julia.
31.   "   quadrimaculata.
32.   "   pulchella.

Some of these, such as Nos. 10, 20 and 32 are commoner in the Carolinian and Alleghanian Zones, while others, such as Nos. 18, 22, 25 and 30 are more frequent in the Canadian Zone. The others are generally distributed as far as known.

The following are Alleghanian and Carolinian:—

1. Lestes vigilax.
2.   "   rectangularis.
3.   "   inaequalis.
4. Argia moesta putrida.
5. Nehallenia gracilis. (?)
6. Enallagma exsulans.
7.   "   signatum.
8.   "   pollutum.
9. Gomphus fraternus.
10. Aeshna clepsydra.
11.   "   verticalis.
12.   "   tuberculifera. (?)
13. Nasiaeschna pentacantha.
14. Macromia illinoiensis.
15. Epicordulia princeps.
16. Nannothemis bella.
17. Celithemis elisa.
18. Leucorrhinia intacta.
19. Sympetrum corruptum.
20. Libellula vibrans incesta.

Nos. 4, 7, 8, 9, 13 and 15 range also into the Austroriparian Zone, No. 13 reaching the Gulf Strip. No. 19 is chiefly, Western (Transition and Upper Sonoran).

The following are chiefly Transitional and Canadian:—

1. Enallagma cyathigerum.
2.   "   calverti.
3. Gomphus brevis.
4. Aeshna eremita.
5. Neurocordulia yamaskanensis.
6. Cordulia shurtleffi.
7. Somatochlora williamsoni.
8. Leucorrhinia frigida.
9.   "   glacialis.
10.   "   proxima.
11.   "   hudsonica
12. Sympetrum danae.

No. 3 extends into the Carolinian Zone, No. 5 is mainly Alleghanian. The others are all more abundant in the Canadian Zone, Nos. 1, 2, 4, 6, and 11 ranging far into the Hudsonian Zone.

SESSIONAL PAPER No. 39b

NOTES ON THE SPECIES.

Calopterygidae.

1. Agrion maculatum Beauvais.

Syn. *Calopteryx maculata* (Beauv.) Burm.

A male was taken in the woods, just above the "Chute" on the Go Home River, July 7, 1907. A few others were also seen along the river shore. Another male was seen at the "Chute" on July 22, 1912. It will doubtless be found more commonly farther up the river.

The nymph has been described by Needham ('03).

2. Agrion aequabile (Say) Kirby.

Syn. *Calopteryx aequabilis* Say.

This species has not been taken in the immediate vicinity of Go Home Bay but a female was captured by Dr. Huntsman near Victoria Harbour, June 25, 1907. It has also been taken in Algonquin Park and I have an exuvia from Shawanaga River, taken by Mr. Paul Hahn, which I believe belongs to this species. The basal joint of the antennae is fully a third longer than the width of the head, this being the diagnostic character given by Needham for the nymph, that he referred to *aequabile* by supposition. In the nymph of *A. maculatum*, the basal joint of the antennae is scarcely longer than the head is wide. These two species are the only ones of this genus that occur in Ontario, so that there seems to be little doubt that the nymph referred to *aequabile* belongs to that species, particularly as the measurements are larger than those of *maculatum*, as is also the case with the adults.

My exuvia measures as follows*: Length of body 27; gills 13.5 additional; antennae 6·5; outer wing-pad 7; hind femur 10.

Coenagrionidae.

Lestinae.

3. Lestes unguiculatus Hagen.

Rare in this vicinity. A few individuals were taken in an open marsh near a small lake, on Aug. 6, 1907.

The nymph has been characterized by Needham ('03) and the writer ('14).

---

* All measurements are given in millimetres.

5 GEORGE V., A. 1915

4. Lestes uncatus Kirby.

A few specimens of this species were captured at the lagoon on the Giant's Tomb Island on July 14, 1912. A pair was observed in copula and the male was captured.

The nymph has been characterized by Needham ('03)). and the writer ('14).

Full-grown nymphs were taken in large numbers by Mr. Wodehouse from a small lake on Fitzwilliam Island, Georgian Bay, on June 29, 1912.

5. Lestes disjunctus Selys.

A very common species in all still marshy waters. It was the commonest Lestes in 1912. The first adults captured in 1907 were taken on July 23 but they had probably been on the wing for a week or more. In 1912, they were first noted on July 13.

The nymph has been described by the writer ('14).

6. Lestes rectangularis Say.

This species is rare in this vicinity. A male and two females were taken in a marshy spot on the Go Home River on July 7, 1908, and another somewhat teneral male was captured at the Giant's Tomb Island, July 14, 1912.

The nymph is described by Needham ('03) and the writer ('14).   A number of them were taken by Mr. Wodehouse from a small lake on Fitzwilliam Island, Georgian Bay, on June 29, 1912.

7. Lestes vigilax Hagen.

This was by far the most abundant Lestes in 1907 and 1908, frequenting the same stations as *L. disjunctus*, which it far outnumbered during these years. It was very much scarcer in 1912.

We have found the long slender nymphs in abundance and have bred a number of them. They have been described by Needham ('08) and the writer ('14).

The first adults taken in 1907 are dated June 28th and it was abundant in the latter half of August. In 1912, on the other hand, it was not noticed until July 16 and had practically disappeared by the end of the first week in August.

8. Lestes inaequalis Walsh.

A single male was captured while flying over a sphagnum bog on the edge of a small lake, July 3, 1907. Much further search on this and subsequent occasions failed to reveal another specimen.

This is the only Canadian record for this species.

Coenagrioninae.

9. Argia moesta putrida (Hagen) Calvert.

This is a very abundant dragonfly about rocky shores during July and August. Though common everywhere on the open bay and river, it is rather more abundant on the latter, especially in the neighbourhood of rapids. The nymphs are common under stones near the shore. We found them particularly numerous at the "Chute" in part of the river which earlier in the season had formed part of the rapids, but which later had been almost cut off from the main current as a result of the diminished flow. They are commonly associated with nymphs of *Boyeria grafiana* and *Neurocordulia yamaskanensis* and Ephemerid nymphs of the genera *Heptagenia* and *Baetis.*

Transformation takes place on the rocks near the water's edge, and the season for emergence lasts for three weeks or more. In 1907, the first teneral individuals were taken on the inner bay on June 26, but imagoes did not appear on the Station Island until nearly a week later, continuing to emerge in considerable numbers during the succeeding week. By this time many individuals on the inner bay were already pruinose. In 1912, the earliest individuals were not observed, but they were undoubtedly later in emerging than in 1907. The first specimens were taken on July 9, and a day or two afterwards pruinose individuals were seen. Tenerals kept appearing until at least the end of the first week in August. This species flies over the barer parts of the rocky shores and is the only damsel-fly met with here, except *Enallagma caranculatum*, which breeds about the Station Island and in similar places. We have repeatedly observed pairs of this species engaged in oviposition, for which act the female often selects an almost bare rock. Her peculiar habit of descending under water and remaining there for a considerable length of time, usually accompanied for a few minutes by the male is well known and need not be described in detail here.

10. Chromagrion conditum (Hagen) Needham.

Syn. *Erythromma conditum* Hagen.

A decidedly rare species in this locality. A few individuals were taken on July 3, 1907, along a small sluggish creek, bordered by a grassy marsh. We have not met with the nymph, which has been described and figured by Needham ('03).

11. Nehallenia irene (Hagen) Selys.

This usually common species is scarce in this region, but is occasionally found in shallow quiet bays in which there is a moderately abundant aquatic vegetation. The only place where it was found in any numbers was the Giant's Tomb Island, where it was common about a very shallow reedy pond in close proximity to a shallow reedy bay. The pond had probably been connected with the bay earlier in the season when the water was at a higher level. The bottom in both cases was sandy. On the day on which the insects were collected (July 29, 1908) the water of the pond must have had a temperature of 37 or 38°C., having been heated to this extent by the sun, but the nymphs of the various species of dragonflies found in it, including *N. irene*, displayed their usual activity.

The nymph of this species has been described by Needham ('03).

5 GEORGE V., A. 1915

12. Nehalennia gracilis Morse.

Very abundant on all sphagnum bogs bordering lakes and ponds; the most characteristic damsel-fly of such stations. It also occurs in smaller numbers in open marshes. It was first noticed on June 18, 1907, and was found in abundance throughout July. In 1912 it was still common on August 6, but had about disappeared by Aug. 25. Several imagos emerged in the laboratory during July.

The nymphs are found in floating sphagnum bogs and are somewhat difficult to detect. I have compared them carefully with nymphs of *N. irene* from Toronto and the only differences that I can find are the smaller size, less spinulose hind margin of the head and entire absence of spots on the gills. It is not improbable that none of these characters are constant as I had but few specimens of either species for comparison.

In *N. gracilis* the convex posterior margin of the head has only 4 to 6 slender inconspicuous, colourless spinules; in *N. irene* there are a dozen or more spinules which are somewhat coarser and blackish at base (Figs. 2, 3). Gills very slender, widest in the distal third, tapering somewhat more gradually than in *N. irene*, without any indication of spots. (Pl. III, Fig. 1).

Length of body $8 \cdot 25$-9; gills 3-$3 \cdot 75$ additional; hind wing $2 \cdot 2$-$2 \cdot 7$; hind femur 2-$2 \cdot 33$; width of head $2 \cdot 33$-$2 \cdot 4$.

13. Enallagma cyathigerum (Charpentier) Selys.

A single male was taken from an open marsh on June 21, 1907.

This is the form described by Hagen as *E. annexum* ('61). I have stated elsewhere (Walker,'12b) my belief that this form and the following are but variations of the same species, but I find on further study that this conclusion was reached too hastily; the two forms are distinct species.

14. Enallagma calverti Morse.

In 1907 only one specimen of this form was taken, on June 16, but in 1912 it was found in considerable numbers early in the season but had seemingly disappeared before the middle of June. Fully matured imagos were found in the marsh at the outer end of Galbraith Lake on June 1, many of them flying in pairs. The season for transformation was not yet over, however, for several full-grown nymphs were found, four of these emerging on June 3 and 4. Mr. Wodehouse also took a full-grown nymph at Victoria Harbour on June 1.

The nymph (Pl. III, Figs. 4, 5) is very similar in form to that of *E. hageni*, but is considerably larger with much darker gills. Eyes as in *hageni*, less prominent than in *E. signatum* and *pollutum*, the curve of the posterior median excavation of the head somewhat more flattened than that of the rather strongly convex margins on each side, the latter with a dozen or more spinules. Labium with 4 mental setae and 6 (occasionally 5) lateral setae; end-hook of lateral lobe preceded by 3 teeth of moderate size, which are preceded by 3 or 4 smaller, somewhat incurved, denticles. Gills lanceolate, widest a little beyond the middle, ventral margin straight at base dorsal margin convexly curved, apices bluntly pointed with convexly curved margins or rounded; across the middle of the gill is a distinct joint, proximad of which the margins are spinulose, the spinules of the ventral margin stronger than those of the dorsal; distad of the joint the margins are beset with a fringe of delicate hairs, much longer than those of *E. hageni*. Colour dark brownish (probably olivaceous in life), each abdominal segment except 10 with a dark lateral blotch, not seen in the exuviae; femora with a pale ring just before the apex, preceded by a dark ring.

Length of body 15·5 (exuvia)–21·5; gill 6·5–8; outer wing-pad 4·5–5; hind femur 4; width of head 3·5–3·7.

I have also bred this species at DeGrassi Point, Lake Simcoe.

15. Enallagma hageni (Walsh) Selys.

Abundant about all still waters throughout the latter half of June and July and in small numbers in August. They were first noticed about the middle of June, and were common by about June 20. They thus begin to appear about the time when *E. calverti* has nearly disappeared. On June 27 we found this species in large numbers about a small pond on an island in the open waters of Georgian Bay, about 3½ miles west of the Station Island. This island is largely bare rock but supports a thick tangle of small cedars, red and black cherry and willows in its middle part. The pond occupies a long narrow depression in the rock, emptying by a very small trickle of water at one end. The shore and bottom of this pond over the greater part of its area is bare rock, or rock covered with a thin deposit chiefly consisting of decaying vegetable matter. At a few points along the margin there are dense clumps of small reeds and at the western end, next to the outlet, is a patch of cat-tails. Owing partly to its exposed and isolated position, but few species of aquatic insects were found in this pond and these included but three species of dragonflies. These were *E. hageni, Ischnura verticalis* and *Libellula quadrimaculata*. All of these, however, were very abundant and *E. hageni* most of all. Some of the reeds were covered with their exuviæ and transforming nymphs. Most of the imagoes seen were more or less teneral, the season being apparently a little later here than on the mainland and inner islands. I also found this species in several other rock-pools on the outlying islands. It was generally the only species of Odonata present.

The nymph which has been described by Needham ('03) is exceedingly common in all swamp waters in the vicinity of Go Home Bay. Numerous specimens were also taken by Mr. Wodehouse at Matchedash Bay, Killarney and Fitzwilliam Island, Georgian Bay.

This species outnumbers all the other Enallagmas of the district taken together, at least twenty times.

16. Enallagma ebrium (Hagen) Selys.

A single specimen, a male, was taken near a small lake close to the mouth of Go Home Bay on June 20, 1907. This species is indistinguishable from *E. hageni*

5 GEORGE V., A. 1915

in the field, so that it might readily be overlooked.   I have examined hundreds of individuals in this locality, however, without finding another specimen. *E. ebrium* is very abundant at Toronto but seems to prefer ponds on a clay or alluvial soil.   Such stations are wholly lacking at Go Home Bay.

17. Enallagma exsulans (Hagen) Say.

This is one of the very abundant species of the Carolinian Zone but it is not very numerous. in the Georgian Bay region.   It does not occur about the small lakes and marshy bays but is not uncommon along the muddier parts of the river shore and along more or less shady creeks.   It also occurs occasionally on the muddier parts of the shores of sheltered inlets, but as a rule not where there is a dense growth of reeds.

The earliest capture of this species was on July 2, 1907; the latest Aug. 26, 1912.

The nymph of *exsulans* has been described by Needham ('03).

18. Enallagma carunculatum Morse.

Next to *E. hageni*, this is the commonest Enallagma in the region under discussion, but it reaches maturity later in the season than other species, the first examples noted having been seen about the 25th of July, 1907.   On Aug. 26, 1912, they were still abundant.   This species it particularly characteristic of the marshier and shallower parts of the shores of otherwise open waters.   It frequents also the narrow reed-beds which are very frequent in this region along many rocky shores of inlets and channels.   It is the only Enallagma of our fauna which develops in water that is subject to any considerable wave-action and is thus the only species that breeds about the Station Island.   The nymphs are also found at much greater depths than those of other species of this genus.   At Lake Simcoe I have found the exuviæ clinging to reeds in water five feet deep.   A description of the nymph is given by Needham ('03).

19. Enallagma signatum (Hagen) Selys.

This species, which is very abundant at Toronto, occurs somewhat sparingly at Go Home Bay, where it may be observed flying over lily-pads on sluggish creeks.   It was first noticed on July 16, 1907, but became commoner after that date.

A number of specimens of the nymph (Pl. III, Fig. 7, 8) at various stages, including full-grown examples, were collected by Mr. Wodehouse at Waubaushene, May 29, at Killarney June 24th and in a small lake on Fitzwilliam Island, June 24, 1912.   The nymph has been described and figured by Needham ('03).

19. Enallagma pollutum (Hagen) Selys.

This beautiful species is common on the river and the inner parts of the bay,

SESSIONAL PAPER No. 39b

where it frequents the marshier parts of the shore, but, like the preceding species, does not usually fly among the reeds and sedge of the marshes., but over the lily-pads and pond-weed, keeping so close to the water that it is very difficult to net. It may also be found in the more open reed-beds, where it is more easily captured.

Among the nymphs taken by Mr. Wodehouse at Waubaushene and Fitzwilliam Island are a number of specimens of an undescribed form, which is so unmistakably nearly related to *E. signatum* that we have little hesitation in ascribing it *E. pollutum.* This species is, moreover, the only Enallagma of the region except the rare *E. ebrium*, whose nymph has not been reared.*

Nymph (Figs. 9, 10);—Long and slender; eyes very prominent laterally, their postero-lateral margins forming with the sides of the head a distinct excavation. Hind angles of head with numerous slender setæ, rounded but very prominent and narrower than the median concavity. Abdominal segments 2-7 with prominent postero-lateral angles. Gills large, broad lanceolate, widest at the distal third, with a transverse median joint, basal half dark except at the base, apical half whitish or grey except a broad dark anteapical band.

Labium with 3 mental setæ; lateral setæ 5; lateral lobes, before the end-hook, with three well-marked teeth, preceded by a feebly denticulate, almost truncate margin.

Colour brown (alcoholic, probably greenish in life), sides of head and thorax with a pale longitudinal band between two dark bands, the most ventral of which passes dorso-caudad to the bases of the front wing-cases. There are usually also a few dark spots on the head and thorax. Abdomen rather dark brown, almost uniform. Legs pale, femora with a very narrow but usually well-defined dark ring at the distal fourth.

Length of body 13 (contracted) to 18 (extended); gills 5-6·5; hind wing 4·3-5; hind femur 3·5; width of head 5·23-3·4.

21. Ischnura verticalis (Say) Selys.

This ubiquitous species is not particularly abundant at Go Home Bay. It is the second species of damsel-fly to appear in the spring, being preceded only by *Enallagma calverti*. We found them in considerable numbers on June 1, 1912, on the marsh at the outer end of Galbraith Lake, where they were transforming. Nearly all the individuals seen were tenerals, while *E. calverti* was for the most part fully mature.

This species seemed to become scarcer in July, but many fresh adults appeared in August. In this district, *I. verticalis* is more frequently met with about the margins of sluggish creeks than in the marshy bays. We have not observed it in sphagnum bogs.

The nymph has been described and figured by Needham ('03).

---

*Since the above was written I have reared *E. ebrium* at Toronto. The nymph is described in Can. Ent., 46, Oct. 1914.

Aeshnidae.

Gomphinae.

22. Hagenius brevistylus Selys.

The full-grown nymphs of this large Gomphine are not infrequently met with during the first half of summer among roots and debris along the edge of the lake shore. They breed in the bay and river but do not occur in the smaller inland lakes. They do not, however, frequent the barest or most exposed parts but show a preference for the more sheltered spots, where the bottom is more or less sandy. The younger nymphs are occasionally dredged from depths of six or eight feet. Four sizes of nymphs were found, including the full-grown stage, and it would thus appear probable that the nymphal life extends over a period of three years or more.

To ensure success in rearing the nymphs of this species, the water in the breeding-jar should be kept as fresh as possible. In our first efforts, this point was not strictly observed and the two full-grown nymphs which we were attempting to rear died shortly before the usual time for emergence. They had crawled out of the water and remained out for about four days, when we replaced them in fresh water, but they soon died. They had evidently not emerged for the purpose of transformation as we had at first supposed.

The large formidable-looking imagos are first seen early in July, becoming common a few days after their first appearance. In 1907 they began to emerge on the Station Island on July 2nd, but in 1912 they were not observed until July 9th. They were still not infrequent on Aug. 10th of the latter year, but by the 20th they had nearly disappeared.

During the period of emergence and for a short time afterwards, this huge, conspicuously-coloured dragon-fly may be seen about the Station Island, flying rather slowly and within a few feet of the ground. It is fond of basking in the sun in sheltered openings in the thickets along the shore, and when disturbed it does not usually fly very far. It is also frequently seen flying swiftly over the water, close to the shore.

The food of the adult Hagenius consists chiefly, if not wholly, of other dragon-flies. We have not observed it feeding on any other kind of insect. We have taken it while devouring *Gomphus lividus, G. exilis, Neurocordulia yamaskanensis* and *Tetragoneuria cynosura simulans.*

23. Gomphus brevis Hagen.

This is the rarest of the four species of Gomphus found in the vicinity of Go Home Bay. The adults are most frequently seen in the neighbourhood of rapids, but also occur in the more exposed shores of the bay, and one was observed on South Pine Island, which lies in the open water of Georgian Bay, about 3 miles out from the coast. Another was taken on the Giant's Tomb Island. The nymphs

SESSIONAL PAPER No. 39b

inhabit the well aerated waters at the foot of the "Chute" and other stations where the imagos occur. They transform on the rocks close to the water's edge, generally early in the morning. In 1907 the first newly-emerged specimens appeared on the Station Island on June 23rd and the last adult was taken on August 14th. In 1912 they were somewhat earlier, appearing at the "Chute" on June 15th.

24. Gomphus lividus Selys.

Syn. *G. sordidus* Hagen.

This is a very abundant dragonfly, frequenting the shallower waters of the bay where the aquatic vegetation is scanty and the shores more or less wavebeaten. It is absent from the marshy bays and inland lakes and also from the steep rocky shores where the water is of considerable depth. The nymph lives in a more or less muddy or sandy bottom.

Transformation begins at almost exactly the same time as that of *G. brevis,* and usually takes place before 8 o'clock in the morning. The nymphs of these two species may be found together but on the whole those of *G. lividus* prefer quieter water than those of *brevis*.

In 1907 this species began to emerge on Station Island on June 22, becoming abundant in two or three days. In about a fortnight, however, they were nearly gone though a few females were seen as late as July 20. In 1912 the first young adults were observed on June 16 and they continued to emerge for at least 10 days, subsequently. By this time the species was very abundant and many pairs were seen in copula. By the end of the month all were mature, and shortly afterwards their numbers began to thin out, though occasional individuals were seen until the end of July. The season of flight is nearly coincident with that of the mayflies *Ephemera simulans* and *Hexagenia bilineata*, upon which they largely feed. They also devour small moths, caddis-flies, etc.

The flight of the adult males of this species is peculiar and is easily distinguishable from that of the other species of Gomphus occurring about Go Home Bay. It consists of a series of ascending and descending or dipping movements, the insect describing a series of deep curves, with the convexities downwards. These motions are not seen during the teneral state.

G. lividus is frequently captured and eaten by *Hagenius brevistylus*, but by the time the latter is common, *lividus* has already considerably diminished in numbers.

Gomphus exilis Selys.

This is the most generally distributed Gomphine of the Go Home Region, being associated with both *G. lividus* and *G. spicatus*. It is most abundant in the shallow marshy bays, but it is quite common on the Station Island, where the nymphs live in the comparatively shallow water on the south-east shore, associated with *G. lividus, Macromia illinoiensis*, etc.

5 GEORGE V., A. 1915

It is the latest of the four species of Gomphus to appear in the adult state, the first tenerals emerging a few days later than those of *G. lividus* and remaining for some time after the other species of Gomphus have disappeared. In 1907 the first adults on the Station Island emerged on June 22, continuing to appear until about the 30th, while in 1912 they were first observed on the 19th. In 1907 and 1908 they were exceedingly abundant, apparently outnumbering both *lividus* and *spicatus*; in 1912, however, they were scarcer, their numbers being distinctly smaller than those of either of these species. In 1907 a few individuals lingered as late as Aug. 12.

Though usually transforming early in the morning, this process may sometimes be observed at other times of the day.

Like its associate, *G. lividus*, this species often falls a victim to *Hagenius brevistylus*.

26. Gomphus spicatus Hagen.

This is distinctly the earliest Gomphine and one of the earliest dragonflies to appear in the adult state in the spring. When we arrived at the Station in 1907, on June 15, tenerals were already common in the neighbourhood of the small lake near the outer coast, and in 1912 they were much earlier, an exuvia having been taken on May 29th, and a very large number on the 31st. These were found floating among the reeds in the marshy outer end of Galbraith Lake. On June 15 of same year, large numbers of tenerals were flying about the same marsh and many more mature individuals in the open rocky woods nearby. On June 5 great numbers of *spicatus* were seen about one of the shallower lagoons on the Giant's Tomb Island, all apparently beyond the teneral state. By the 15th its numbers had about reached their climax in the vicinity of Go Home Bay. They did not wholly disappear until somewhat after the middle of July. The last pair in copula was noted at the Giant's Tomb Island on July 15.

G. spicatus is strictly an inhabitant of marshy places in this locality, though Kellicott ('99) states that it frequents the "borders of wave-beaten shores or rushing rivers," and Needham refers to it as an inhabitant of "all sorts of waters." These statements, especially the former, are difficult to associate with this Gomphus and it seems to me probable that Kellicott's may refer to some other species. The nymph is a very common object in dredging from the soft bottoms, consisting chiefly of rotten vegetation, of ponds and still marshy bays.

Transformation takes place close to the waterline, the nymph frequently not emerging completely. The exuviae are thus often found floating.

27. Gomphus fraternus (Say) Selys.

This species does not occur in the immediate vicinity of Go Home Bay, but it is included here on account of the capture of a male specimen by Dr. A. G. Huntsman, near Victoria Harbour (Hog Bay), June 25th, 1907.

28. Dromogomphus spinosus Selys.

On July 5, 1907, a newly-emerged male of this species with its exuvia was found on the Station Island. A few other exuviae were also found here subsequently and adults were occasionally observed though not frequently. They were apparently much more numerous on the river above the Chute, judging from the large number of exuviæ found there.

In 1912 we did not come across this species at all, but Mr. Wodehouse dredged up a number of nymphs of various stages from Shawanaga Bay and the Shawanaga River on June 9 and 13, including several full-grown ones. Those from Shawanaga Bay were taken from weedy shallow water with a sandy bottom. Usually the nymph occurs where there is little aquatic vegetation. This is the case at De Grassi Point, Lake Simcoe, where this species is the only common Gomphine. Here it lives in a bottom of very fine sand and transforms on the boulders along the shore. The imagoes fly freely over the water and often settle on passing boats and on the boulders of the shore. They may also be taken quite frequently on roads through the woods within a few hundred yards of the lake.

29. Boyeria grafiana Williamson.

This is one of the late-appearing dragonflies, August being the month in which it is most abundant. Full-grown nymphs were collected on and after June 4, and the first adult emerged in the laboratory on July 14, followed by several others during the succeeding fortnight.

Teneral adults are often found clinging to the trunks of trees or the sides of houses in the shelter of the verandah. When mature they may be seen flying up and down the lake shore, close to the water, and following a more or less regular beat. Sometimes this is limited to a little cove two or three yards across, but generally they cover a much greater distance at a time.

They are most active in the evening, but fly also during the day. In their crepuscular habits they recall *Neurocordulia yamaskanensis* which they also resemble a good deal in general appearance, especially in the dull brownish coloration. They are less swift than the latter, however, and their flight is practically restricted to a narrow littoral zone. The season of adult life extends until about the end of September.

Williamson's description of the coloration of this species does not fit the majority of specimens that we have seen, in all respects. The light markings are bright yellow at first, but become dull with age. I have never seen blue-spotted individuals such as those described by Williamson, but the colour is always distinct from that of *B. vinosa*. The fulvous tone of the wings in the latter and the dark markings at their bases are not seen in *B. grafiana*.

The dark-coloured nymphs are found rather commonly under stones, along more or less wave-beaten places or wherever there is a perceptible current. They are generally distributed along the shores of Go Home Bay and River, except in marshy places. They show a preference for the neighourhood of rapids or narrow channels wherever there is a free circulation of water. They are perhaps most nu-

5 GEORGE V., A. 1915

merous along the edges of gentle rapids, such as those above "Sandy Gray Falls" on the Musquash River. The nymphs are commonly associated with those of *Neurocordulia yamaskanensis, Argia moesta putrida* and *Basiaeschna janata.*

When ready to transform they climb upon rocks, wharves, boathouses, etc., sometimes to a height of five or six feet from the water, but often much nearer.

As the nymph of *Boyeria vinosa* was described before *B. grafiana* had been recognized as a distinct species, (Cabot, '81; Needham '01; Needham and Hart, '01), it is impossible to be certain whether these descriptions all refer to *B. vinosa* or not, but Needham's description belongs with scarcely a doubt to that species.

We have reared a number of nymphs of *B. grafiana* and collected many exuviæ as well as nymphs in several localities. We have also received a series of exuviæ of a Boyeria from the Shawanaga River, collected by Mr. Paul Hahn, which differ very slightly from those of *B. grafiana.* The latter were also found on the same river. As *vinosa* and *grafiana* are the only North American species of Boyeria and are both common in this region, there can be no doubt that the species not yet reared is *B. vinosa.*

The nymphs of these two forms may be separated as follows:

Mentum of labium 5·5 mm. long, its middle breadth scarcely less than half its length (Pl. III, Fig. 11); fourth abdominal segment without lateral spines; lateral abdominal appendages of female one-fourth to one third as long as the inferior appendages, and usually about as long as the dorsum of segment 10.....*B. vinosa.*

Mentum of labium 6·5-7 mm. long, its middle breadth distinctly less than half its length (Pl. III, Fig. 13); fourth abdominal segment generally with distinct though very small lateral spines; lateral abdominal appendages of female one-fifth to one-fourth as long as the inferior appendages and one-half to three-fifths as long as the dorsum of segment 10.................................*B. grafiana.*

B. grafiana also differs from *B. vinosa* in the slightly stouter inferior abdominal appendages which are less incurved at the tips (Figs. 12, 14) and in the slightly larger size as shown by the following measurement:

B. vinosa: Length of body 34-36·5; hind wing 6-7·5; hind femur 5-6; width of head 7·5-8.

B. grafiana: Length of body 37-39; hind wing 7·5-8; hind femur 6·6-5; width of head 8-8·5.

In coloration the nymphs of these two species are quite similar, except that the pale, wavy, dorso-lateral streak on each side of the abdomen is usually quite distinct in *grafiana* but more or less obscure in *vinosa*. In both species the depth of coloration varies considerably, usually being a rather dark brown. All the nymphs from the Go Home Bay district are very dark in colour, but the pale bands of the abdomen and legs are quite sharply defined. The most characteristic mark of Boyeria nymphs is a pale oval or diamond-shaped median blotch in the dorsum of segment 8.

Basiaeschna janata Say (Selys).

This species considerably resembles the preceding in its habits both in the nymph and adult states, but flies during the first instead of the second half of sum-

SESSIONAL PAPER No. 39b

mer. The dark-coloured nymphs cling to the under-sides of stones near the shore, and are rather more generally distributed than those of Boyeria, as they also occur in the smaller recesses or quiet places, where a few reeds and other water-plants grow. They probably feed largely on Mayfly nymphs, particularly *Heptagenia, Blasturus* and *Ephemerella*, which are abundant in places frequented by *Basiaeschna* and *Boyeria*. The exuviæ may be found, like those of *Boyeria*, on logs, wharves and boathouses, sometimes at a height of six feet, but they also occur, like those of *Anax* and *Aeshna*, on reeds.

Full-grown nymphs were found on our arrival at the Station in 1912, on May 20, and continued to be found until June 10. The first adults seen were those which emerged in the laboratory on June 8th. Adults were taken until June 25th, but were not noticed after this date. They probably were on the wing for some time afterwards, however, for in 1907 we captured specimens repeatedly until July 17th.

This species may often be seen patrolling the margins of the bay and the Go Home River, usually flying higher than Boyeria. It may also be found in the open rocky woods, a short distance from the water. It is active during the day but also flies until well after sundown.

On June 24, 1907, a female was taken in the act of carrying off a teneral specimen of *Gomphus spicatus*, upon which she was feeding.

The egg-laying habits of this species, which we have not observed at close range, have been described in detail by Needham ('01).

Aeshna eremita Scudder.

This large boreal species is quite scarce in this locality, though I found it common in Algonquin Park in 1902 and it has been taken occasionally as far south as Toronto. In the Canadian zone it is an abundant and wide-spread species, ranging across the continent and northward to the Arctic Circle.

The only adults observed at Go Home Bay were a pair taken by Dr. A. G. Huntsman on Aug. 17, 1907. A few nymphs have been taken from reed beds along the shores of ponds. Two of these taken early in August were full grown. The nymph has been described by Cabot ('81) and by the writer ('12a). The ovipositing habits have also been described by the writer ('12a).

Aeshna clepsydra Say.

Next to *Ae. canadensis*, this species has been found more frequently than any other Aeshna in the vicinity of Go Home Bay, although, generally speaking, it is one of the rarer species of the genus.

The nymph, which has been described by the writer ('12a), is occasionally dredged from reed beds along the borders of shallow ponds or bays. Two males were reared by Mr. A. R. Cooper in 1910 emerging on July 25th and 28th. Two other nymphs taken in July were nearly ready to emerge.

The adults may be taken during the latter part of July and August, flying over the reeds and sedge of their breeding-grounds or in the open woods farther

5 GEORGE V., A. 1915

away from the water.   Like most species of Aeshna they often follow the shore-line more or less closely while foraging around a bay or pond.

Aeshna canadensis Walker.

This is the most common Aeshna of the Go Home district and probably of the entire Transition Zone in Ontario.   It is also the earliest species to appear on the wing, the period of emergence commencing about June 25th and usually concluding before the middle of July.   The adults fly until the middle of September or even later.

The nymphs (Walker, '12) are very similar to those of *Ae. clepsydra* and are found under apparently precisely similar conditions.   On July 29, 1908, a number of half-grown nymphs of this species were found in a very shallow pool in the sand on Giant's Tomb Island.   This pool was close to a lagoon with which it had been connected earlier in the season.   The water was only a few inches deep and had been heated by the sun to a temperature of perhaps 37° degrees C.   The nymphs were quite active but died the following night in the laboratory, not being able to accommodate themselves to the rapid change of temperature following their removal.

Aeshna verticalis Hagen.

This species is very scarce at Go Home Bay, only three specimens, two males and a female, having been taken.   These were captured by Dr. Huntsman on August 26th and 30th, 1907.   It is a common species southward, being not infrequently met with at Lake Simcoe and at times very numerous at Toronto.   It has also been taken more or less commonly in many of the Northern and Middle States east of the Mississippi and has been recorded from Florida by Muttkowski ('10).   It is thus an Austral species, Go Home Bay being the most northerly point from which it has been taken in Ontario.

In habits it resembles the preceding species in the adult state, but first appears as a rule nearly a month later and is commonest in the latter half of August and the first half of September.

Its nymph is still unknown.

Aeshna tuberculifera Walker.

This is an insect of which the habits are quite unknown.   It is distributed from New England to Wisconsin, but is apparently nowhere common.   Only a single male has been taken at Go Home Bay, by Dr. Huntsman, on August 26, 1907.

Since the above was written this species has been reared by the writer on Vancouver Island, B.C.   The nymph will be described shortly in the Canadian Entomologist.

Aeshna umbrosa Walker.

This appears to be the most widely distributed and abundant Aeshna in North America.

SESSIONAL PAPER No. 39b

It differs decidedly from the other species of the genus, the habits of which are known, in that the nymph develops in small woodland creeks, ditches and spring-fed pools, never being found in open, weed-grown, marshy waters. The imago is also decidedly partial to somewhat shady localities and flies as readily during dull as bright weather. It flies habitually until late dusk, coursing up and down the ditches or creeks in which it breeds, or foraging in open spaces away from the water, in search of Diptera and other small insects.

On account of this type of habitat *Ae. umbrosa* is not very common about Go Home Bay. A few full-grown nymphs were taken during August in two small shady creeks, emptying into Go Home Bay and on June 10, 1912, two other grown nymphs were taken from beneath boulders in the short outlet of a small lake. These were thickly covered with brown hydras. One of them emerged on July 14th. The nymph has been described by the writer ('12a).

The imagos were not seen very often, but are by no means rare. A single female was taken on the Station Island, where it must have flown from the mainland or one of the larger islands.

Anax junius (Drury) Selys.

This common and widespread species was the first dragonfly to be seen in flight after our arrival at the Station in 1912, a single individual having been observed on the Giant's Tomb Island on May 26. On June 1 a considerable number were seen flying about the inner end of Galbraith Lake. Several couples were observed, but none actually in copula, the males adhering only by the abdominal appendages. The female of one of these couples was observed ovipositing on the under surface of water-lily pads. She remained only a few seconds at each lily-pad. A similar pair was seen at Muskoka Mills on June 31, and the male captured.

Two stragglers were taken on the Station Island, a female in good condition on June 26, '12, and a worn male on July 7. This is the latest date on which an adult of the spring brood was observed.

The nymphs are taken quite frequently with the hand-dredge and dip-net, along the marshy borders of ponds and sheltered bays, their haunts being quite similar to those of *Aeshna canadensis* and *clepsydra*.

They are not nearly so abundant here as in the vicinity of Toronto and southward.

No individuals of the fall brood had yet made their appearance at the time the Station was closed in 1912 (Sept. 11).

Nasiaeschna pentacantha (Rambur) Selys.

On August 28, 1906, Mr. W. J. Fraser found three of the strange-looking nymphs of this interesting species near Bala Falls, Muskoka, Ont. He attempted to rear them, but although easily kept in captivity they were all killed by accident. One of them was kept through the winter and brought to the station at Go Home Bay in 1907, but on crawling out of the breeding jar, probably to transform, it was accidentally crushed.

Two nymphs were found at Go Home Bay during 1907.   One of these I found clinging to my paddle while passing through the outlet of Galbraith Lake. The other was dredged from among the reeds along the edge of the "Sand Run," a shallow, sand-bottomed channel in which a more or less distinct current is usually perceptible. Following both of these captures, prolonged search was made for more specimens, but without success.   These two nymphs were kept alive until late in the winter of 1908, one of them, in the meantime, having reached the final stage, but on one unusually cold, windy night, the water in the breeding-jar, though inside the room, froze solid, and the nymphs were killed.

The only other nymph we have seen was taken by Mr. Wodehouse in a marshy bay near Waubaushene, June 1912.

The full-grown nymphs measure as follows:

Length of body 48; mentum of labium 7·3-7·5; hind wing case 10-10-5; hind femur 6·5; width of head 8·5 9.

The only adult taken in the vicinity of Go Home Bay was a fine male, captured by the writer, while flying over the marshy outlet of a small stream at the inner end of one of the sheltered bays. A colour sketch was made of this specimen, as the colours of the living insect seem not to have been recorded. The face was light grey with a slightly bluish tinge, deepening to dark brown on the frons next to the eyes. Frontal vesicle and occiput whitish, eyes brilliant blue. Thorax rather light reddish brown, the pale markings grass-green. Abdomen dull greenish black, the paler areas dull green of a somewhat bluish shade.

Go Home Bay is the most northerly locality from which this species has been obtained. It is an Austral species, being distributed as far south and west as Florida and Texas.

Libellulidae.

Corduliinae.

Didymops transversa (Say) Hagen.

With the exception of *Anax junius* this species, together with *Tetragoneuria spinigera* and *Gomphus spicatus*, is the earliest dragonfly of the sub-order Anisoptera to appear in the spring.   The first exuvia was found on the shore of one of the inner channels on May 29, 1912.   A specimen emerged in the laboratory on June 12, 1912, and in 1907 one was found emerging on June 16, the day after the Station was opened.   The latest date of emergence we have recorded is June 19, 1907.

The time of flight of this species seems to be unusually short, as none have been seen at large after June 21.   The males patrol the margins of lakes and bays, resembling Basiaeschna on the wing, but flying more swiftly.   The females are apparently secretive and are seldom seen.

The nymphs are found sprawling on the sand near the shore in clear, well-aerated water.   They are not rare about the Station Island, where the imagoes have also been occasionally taken just after transforming.   The nymphs sometimes crawl to the verandah of the dwelling-house, forty or fifty feet from the water before transformation takes place. Under the edge of the boathouse roof is another favorite spot. They may also transform on bushes. Two well-grown nymphs of this species were found in the stomach of a channel catfish (*Ameiurus nigricans* Lesueur) by Mr. A. R. Cooper. Full-grown nymphs were taken by Mr. Wodehouse at Shawanaga Bay, near Skerrevore, June 9, 1912, in "weedy shallow water, sand bottom."

Macromia illinoiensis Walsh.

The long-legged, spider-like nymphs of this dragonfly closely resemble those of the preceding species, but are somewhat larger and less distinctly marked, besides differing in the characters given by Needham ('01). Like *D. transversa* they frequent well-aerated waters, being common everywhere along the shores of Go Home Bay, except in the sheltered bays and are absent from the enclosed lakes. They generally occur among boulders on a sandy or somewhat muddy bottom. The nymphs are not infrequently seen sprawling on the surface of the sand or mud bottoms or on the stones. The exuviæ are often more or less muddy, differing in this respect from those of *D. transversa* which are always clean.

Like *transversa* they often travel a considerable distance from the water prior to the emergence of the imago. I have found exuviæ on the verandah of the dwelling-house, and under the eaves of the boathouse, on rocks along the shore a few feet or several yards from the water's edge or on tree trunks 3-6 feet from the ground.

The period of transformation commences about the end of June and continues throughout the first week in July. The first adult observed in 1907 emerged on the Station Island on June 28, while in 1912 the first individual was noticed on June 31. In about a week's time they were common about the island, flying rather low and frequently resting on the branches of trees. Both sexes appeared in about equal numbers and were easy to capture. In about a fortnight they had spread over the country and were no longer so easily obtained.

During the latter half of July and throughout most of August they may be found in sunny weather flying back and forth along the edges of woods or in small open places partly enclosed by trees. They fly swiftly, but as a rule not beyond reach of the net, and as they follow a more or less regular beat they are not very difficult to capture. Flight ceases at sundown and during dull weather.

Neurocordulia yamaskanensis (Provancher) Selys.

In the 36th Ann. Rep. Ent. Soc. Ontario, 1905, p. 69, exuviae of a Neurocordulia, referred to this species by supposition, were recorded from Algonquin Park, Ont. Shortly after the Station was opened in 1907, exuviæ of the same kind were found on the sides of the Go Home Bay Dock. On the morning following this discovery (June 28th) the dock and the steep rocks of the neighbouring shores were carefully searched for newly-transformed adults and one was finally detected

5 GEORGE V., A. 1915

with its exuvia in a crevice of a steep rocky bank. It proved to be *N. yamask-anensis.* Subsequently a number of others were found with their exuviæ on Station Island. Generally they were found between 7 and 8 a.m., but a few were taken late in the evening. Early morning appears to be the usual time for transformation. For some days adults could only be obtained in this way, but they were at last discovered by Mr. Fraser flying about the island at dusk. It was soon ascertained that their time of flight is limited to about half an hour a day, commencing soon after sundown (a little after 8 p.m.), and continuing until shortly after 8.30, after which they retire to the shelter of the trees. It is thus nearly coincident with that of the mayflies, *Ephemera, Hexagenia, Heptagenia,* etc., upon which they appear to feed exclusively.

During this short time of flight they are extremely active. They dash about erratically over the rocks among the swarms of mayflies and when one of these is captured they retire with their prey to a neighbouring tree to consume it in peace.

The majority of the individuals thus engaged are females. The males will be found at the same time flying over and within a few inches of the water close to the shore which they follow very closely. They fly back and forth in a regular beat and with extraordinary swiftness. During these flights the males apparently do not feed, but seem to be on the watch for females, for now and then a male is seen to pounce upon a female, the pair then sailing off over the water or up into the trees, where copulation takes place at rest. Except when thus seized by the males, no females were observed close to the water though plenty of them could always be seen flying over the rocks nearby.

Nymph: (Fig. 15-17) short-legged and of stouter build than most Corduliines.

Head broadly convex above and on the sides, eyes not very prominent, frontal ridge with a scurfy pubescence, the anterior margin convexly curved, hind angles of head prominent, distance between them a little greater than half the greatest width of the head; hind margin distinctly excavate.

Labium extending very slightly behind the bases of the front legs; mentum somewhat broader at the distal margin than long, the middle lobe somewhat abruptly deflexed, bluntly obtusangulate; mental setæ 9-11, the innermost 3 or 4 much smaller than the others; lateral lobes triangular, their distal margins produced into seven semi-elliptical teeth; lateral setae 6; movable hooks very slightly arcuate.

Marginal ridge of pronotum produced on each side behind the posterior angles of the head as a prominent process which is somewhat smaller than the very prominent supra-coxal processes.

Legs short, the length of the hind femora being slightly less than the width of the head.

Abdomen ovate, its greatest breadth, at segs. 6 or 7, slightly greater than two-thirds of its length; curve of the lateral margins somewhat stronger in the distal than in the proximal half; lateral spines on 8 and 9, in each case about one-third to one half as long as the corresponding segment, those on 8 strongly divergent, on 9 parallel and extending caudad scarcely or not at all beyond the tips of the appendages.

Dorsal surface rather strongly convex, dorsal hooks present on 1-9, those of the basal segment slender, nearly erect and slightly hooked, becoming gradually broader and lower caudad, and, on 7-9 reduced to scarcely more than a short ridge. Superior appendages triangular, equilateral, very slightly shorter than the somewhat divergent inferior appendages and somewhat longer than the lateral appendages.

Colour yellowish or orange brown, variegated with dark brown. Head dark brown above, generally somewhat paler in the centre and on the frontal ridge. Thorax and wing-cases variegated with pale and dark markings; femora and tibiae with two pale rings, a median and anteapical. Abdomen yellowish brown, more or less distinctly blotched with darker brown, especially on the dorsal hooks, the lateral margins and spines and the dorso-lateral scars.

Measurements: Length of body 22-24·5; hind wing, 6-7; hind femur 5-5·6; width of head 6·5; width of abdomen 9-10; mentum of labium 4.

The nymph of this species shows the following differences from that of *N. obsoleta*, two exuviæ of which I have from Lake Hopatkong, Pa., received from Professor P. P. Calvert.

Somewhat larger, more elongate and less depressed; eyes somewhat less prominent, mentum of labium a little longer and more narrowed at base, middle and hind legs somewhat less widely separated at their bases; abdomen narrower, the sides less strongly curved on the middle segments; lateral spines on segment 9 much shorter than those of *obsoleta*, in which they are fully as long as the segment and extend far beyond the tips of the appendages; dorsal hooks also less developed than in *obsoleta*, in which they form quite prominent tubercles on segs. 7-9. (Pl. IV, Figs. 16-19).

Besides the full-grown nymph we have taken specimens of two earlier instars, measuring 8 and 18 mm. in length respectively. Judging from the great difference in size between these three instars, it would seem probable that the larval period must be at least two, if not three, years long. In the youngest instar the lateral spines are relatively much longer than in the older ones.

The nymphs of *N. yamaskanensis* cling to the undersides of boulders along the more exposed shores. As the exuviæ are most commonly found on steep rocky shores, rising almost perpendicularly from the water, (Pl. V, Fig. 27) it would seem that the nymphs prefer water of considerable depth, i.e., 8 or 10 feet or more, but we have often taken nymphs of several stages in water less than two feet deep. They occur along the outer coast as well as in the river, in fact wherever the water is kept more or less constantly in motion. They are common in the vicinity of falls and rapids. One exuvia was found on a log overlying the falls at Muskoka Mills. The nymph had evidently crawled out of a comparatively quiet spot close to the swiftest part of the fall, where the water was thoroughly aerated.

The nymphs are associated with the nymphs of mayflies of the genera *Heptagenia*, *Blasturus* and *Baetis* and of the damsel-fly *Argia moesta putrida*, upon which they probably feed. I found one at the "Narrows" of the Go Home River, supporting a growth of a Polyzoan, *Plumatella* sp.

In 1912 adults emerged in the laboratory from June 23 to July 8. Their season

5 GEORGE V., A. 1915

is at its height during the second week in July and is over before the end of the month. July 23 is the latest date on which we have taken this species.

N. yamaskanensis is abundant and of general distribution in this locality and probably throughout the Muskoka and Parry Sound Districts. I have received exuviæ from various parts of Muskoka and from the Shawanaga and French Rivers, collected by Mr. Paul Hahn. They are not known north of the French River.

Epicordulia princeps (Hagen) Selys.

This large insect is very common about Go Home Bay, where it is the species most frequently observed flying far over the open water.

The nymphs live among the bottom debris of shallow bays and inlets and the larger ponds, associated with Tetragoneuria. They are quite often found clinging to the undersides of stones. None were reared in 1907 but tenerals began to appear on June 25 and in a few days became quite numerous. They appeared at the same time in 1912, the first imagoes having been observed on June 26th, this being also the date on which the first specimens emerged in the laboratory. In the first week or so of their imaginal life, they no not fly very swiftly and rest frequently so that they are easily captured, but later they wander far from their breeding-places and during the day in fine weather they seem to be in constant flight from early morning until dusk. During the evening they may be seen flying, usually rather high, in pursuit of mayflies like Neurocordulia.

This is one of the later Cordulines to remain on the wing, individuals being met occasionally as late as Aug. 6.

In specimens from Georgian Bay the dark wing markings are usually greatly reduced as compared with specimens from the Upper Austral Zone (Toronto and southward). While in some females these spots are almost as large as in southern specimens, in the great majority of both sexes they are all much smaller. The nodal spot is frequently a mere trace or it may be absent altogether, as indeed is generally the case in the males.   The apical spot is also frequently a mere trace and such individuals look much like large Tetragoneurias.

Tetragoneuria spinigera (Selys) Selys.

When we arrived at the station in 1907 (June 16) this species was already flying in considerable numbers in the open woods of the mainland near a small lake, and the season for emergence was apparently over, though that of *T. cynosura simulans* had scarcely begun. We therefore watched for the appearance of *spinigera* in 1912 as we were at the Station before the period of emergence for either species had begun.

On May 29th a single Tetragoneuria exuvia was found on the shore of Big Island and on June 1 we found large numbers of them clinging to the reeds and floating on the water in the open marsh at the outer end of Galbraith Lake. No imagos were found except a single crippled teneral with its exuvia. This was a male, however, and could be diagnosed with certainty as *T. spinigera*. Much search was made for nymphs but without success at this spot. A number of Tetragoneuria nymphs, however, were taken from beneath stones along the shore of a channel and two of these yielded female imagoes of *T. spinigera*, emerging on June 2. The other nymphs proved to be *T. cynosura simulans*. Thus, although the nymphs of these two species may be associated with each other we are inclined to the opinion that *T. spinigera* is most at home in somewhat marshier stations than those preferred by its congener.

A careful comparison was made between the exuviæ of these two species but no differences could be detected between them except that in *spinigera* the lateral abdominal appendages average slightly longer than those of *cynosura*. The difference, however, does not appear to be constant. Prof. Needham, who referred certain nymphs to this species by supposition, employed as differential characters the length and amount of divergence of the lateral spines of seg. 9. The two species discussed here are quite alike in respect to these features, which vary considerably among individuals of the same species.

The adult life of this insect appears to be rather short, July 4 (1907) being the latest date upon which it has been observed.

Tetragoneuria cynosura simulans Muttkowsky.

Syn. *T. semiaquea* (Burm.) Auctt.

In 1907 this species was exceedingly abundant. Tenerals were just beginning to appear on our arrival at the Station and by June 25 their numbers had about reached their height. Specimens were taken until July 22. In 1912 they were much less numerous and though common were not abundant. Mature nymphs were collected on May 29th, the first imago emerging on June 13, and four others on the following day. The latest date of emergence that we have recorded is June 19th and the last day on which we observed an adult is July 31. The single individual seen on this day was a female and was taken with a female of *Hagenius brevistylus* which was feeding upon it.

The nymphs of this insect are very common in sheltered bays and channels where there is a certain quantity of marsh vegetation but where the water is not stagnant. The small marshy coves which are very common along the rocky shores everywhere in this district seem to be the favorite haunts of this species. On the slender reeds which grow in such situations the exuviæ may be very numerous during the season of emergence.

I have seen half a dozen or more exuviæ on a single reed. They also frequently transform on boathouses. Full-grown nymphs are quite often taken from the undersides of stones close to the shore.

This species is most abundant about June 25. It flies everywhere on land, but is most common about the shore in sheltered places or in sunny openings in the woods. On a small, somewhat bare island just outside the outer coast we found it on June 25th, 1907, almost in swarms. They were flying about in the sunshine apparently quite aimlessly and seemed not to be feeding.

5 GEORGE V., A. 1915

Helocordulia uhleri (Selys) Needham.

This is a very rare dragonfly in this vicinity, where it has been taken but twice, both occasions on the Go Home River. The first capture was that of a male on June 23rd, 1907, taken by Mr. W. J. Fraser at the "Chute"; the other was a female taken by Mr. W. A. Clemens on the river near Sandy Gray Falls.

Cordulia shurtleffi Scudder.

This boreal species, which is common at Nipigon and probably throughout northern Ontario, is a rare insect in the Go Home Bay district, only a single imago having been captured there. This was a male, taken by the writer on July 7, 1907, in the rocky woods close to the Go Home River, just above the "Chute."

Of the nymph, which is described by Needham ('01), we have taken half a dozen specimens, all from the bottom debris of swamp waters, particularly ponds of little or no drainage.

On account of this type of habitat they are very easy to keep alive in the aquarium. Besides the nymphs from Go Home Bay one was taken from Mud Lake, Midland, and another at Killarney, Ont., by Mr. Wodehouse. Only two exuviæ were found, one dated June 16, 1907; the other has no date attached.

Dorocordulia libera (Selys) Needham.

This beautiful insect is often to be seen coursing back and forth over open marshes and sphagnum bogs, often following the course of a small stream or the edge of a pond. It is also sometimes met with in openings in woods or along their borders. It is usually seen moving rather slowly, but with rapidly vibrating wings, the body slightly tilted with the end of the abdomen uppermost. When approached it darts away swiftly, but if the collector be stationed on its regular path of flight and strikes with the net from behind, it is not difficult to capture.

The sexes occur in about equal numbers, but the females, being more retiring and more often at rest, are somewhat less frequently taken.

The few nymphs we have secured were found at the bottom of sphagnum-bordered ponds and marshy bays, such as are frequented by the imagoes.

The earliest date on which we have found the adult was June 18, 1907, and the only freshly-emerged individual that we have taken was found with its exuvia on June 27th of the same year. On June 28 they were quite numerous. Our latest capture for the Go Home district was July 30, 1912, a single male having been taken on this date, flying over a sphagnum bog on the edge of a large pond.

Somatochlora williamsoni Walker.

Like most of the Somatochloras this is a species of mainly boreal distribution, though it is not uncommon at Lake Simcoe and has once been taken at Toronto. It is not infrequently seen at Go Home Bay during August, flying rather low

SESSIONAL PAPER No. 39b

along the edge of ponds and creeks, or at a height of twenty feet or more in sunny openings in woods.

The dates on which it has been observed in flight at Go Home Bay range from July 21 (1907) to Aug. 26 (1912).

The nymph of this species has been described by Needham ('01) under the name *S. elongata*, Scudd. It has not been taken at Go Home Bay, but on Aug. 2, 1912, we found an exuvia belonging to this genus on a log at the mouth of a small forest stream emptying into the Go Home River. A similar exuvia was taken by Mr. Paul Hahn in Algonquin Park and erroneously recorded by the writer ('06) as *Cordulia shurtleffi*. These exuviæ agree with Needham's description, except in the somewhat smaller size and narrower abdomen. Width of abdomen, however, is a somewhat variable feature in exuviæ, depending much on the state of contraction, and it seems most probable that these exuviæ belong to *S. williamsoni* as this is the only Somatochlora we have observed in the vicinity of Go Home Bay.

They measure as follows (the smaller figures belonging to the Go Home specimen); Length of body 22–23; abdomen 13–15; hind femur 7–7·5; width of abdomen 7·5–8.

Libellulinæ.

Nannothemis bella (Uhler) Brauer.

This diminutive species is quite locally distributed but we have found one station where it is extremely abundant. This is a small floating sphagnum bog occupying a somewhat triangular space between two masses of rock on the edge of a small lake near the mouth of Go Home Bay. Here, in company with *Nehalennia gracilis, Leucorrhinia frigida, Lestes disjunctus* and some other less characteristic forms, it flits about among the low vegetation, settling frequently on the cotton grass, cassandra and other low plants that grow in the bog; the wings, when at rest, being bent strongly ventrad on each side of the supporting stalk.

We have not determined the time when this species begins to emerge. When first observed in 1907 on June 28, most of the males were already pruinose, though younger black individuals continued to appear for some time later. The latest capture was made at the same bog on August 6, 1912, a single male having been taken.

Careful search was made for the nymph, but without success. One exuvia, however, was found clinging to a cranberry twig, many feet back from the water's edge. The nymph had evidently emerged from the bog itself, having lived like *Nehalennia gracilis* in the water in which the sphagnum and other bog-plants were partly immersed. The nymph has been described by Needham ('01a).

Celithemis elisa (Hagen) Walsh.

This is a species of the marshes, which first makes its appearance on the wing in the latter half of June and flies until about the end of August though our latest

5 GEORGE V., A. 1915

capture bears the date August 14, 1912.   The earliest dates of its occurrence are
June 22, 1912, and June 28, 1907.

It is most often seen hovering over patches of Sweet Gale (*Myrica gale*), which
are common in the dryer parts of the open marshes of this region, especially near
the edge of the woods.

According to my observations this species does not often stray far from its
breeding-grounds, as the allied species of *Sympetrum* frequently do.

Strangely enough we have not found the nymph of this common species.   It
has, however, been bred and described by Needham ('01a).

Leucorrhinia frigida Hagen.

One of the most abundant and generally distributed of the marsh dragonflies of this district.   Though found in all the open marshes and bays it is most
abundant in the sphagnum bogs on the edges of small lakes and ponds.   Its
numbers appear to vary to some extent inversely as those of the larger dragonflies
with which it is commonly associated, e.g., *Libellula exusta julia* and *Gomphus
spicatus*.   Thus it is extremely abundant in the pond on "Split Rock Island"
(Pl. VIII, Fig. 33) where these species are absent or very rare.

The nymphs may be dredged in large numbers from the aquatic vegetation
and submerged trash along the edge of this pond and are common along the margins of all such lakes and ponds.

Teneral imagos were already common when the Station was opened in 1907
(June 16) but full-grown nymphs were still easily obtained and adults continued
to emerge for at least a week.   In 1912 the first tenerals were observed on June 1
and by the 17th were very common, though a specimen emerged in the laboratory
as late as June 24.   On August 6 this species was still fairly numerous but all the
individuals were old and pruinose.   None were noted after this date.

Needham's ('05) description of the nymph of *L. frigida*, belongs to another
species, probably *L. hudsonica* (vide infra). In a letter to the writer, he stated that
the species had not been reared but that tenerals of *L. frigida* had been found
at the spot where the exuviæ were gathered.   The nymph of *frigida*, unlike Needham's species, possesses large dorsal hooks, such as are present in all the species
of *Leucorrhinia* that have been reared.

Nymph:—(Pl. IV, Figs. 20-22).

Very similar to that of *L. intacta*, but somewhat smaller and the legs slightly
slenderer.   Head similar to that of *intacta* except in the somewhat more prominent
eyes.   Labium of similar size and form, the lateral lobes somewhat more deeply
concave within, the teeth on the distal margin obsolescent, crenate, each with a
single spinule, lateral setæ 9 or 10; mental setæ 10-13, the fourth or fifth from the
outside longest, the inner four smaller than the others.

. Abdomen broadest at seg. 6; scarcely narrowing on 7; slightly on 8; more
abruptly on 9; lateral spines on 8 one-half to three-fifths as long as the segment;
subparallel, those on 9 reaching about to the tips of the inferior appendages, their
inner margins straight and parallel.   Superior appendages somewhat less elongate

SESSIONAL PAPER No. 39b

than in *intacta*, acuminate, about twice as long as the lateral appendages and one-fourth shorter than the inferior appendages.   Dorsal hooks on segs. 3-8, larger on 3 and 4 than in *intacta*, less erect and more curved, very slender; those on 5-7 of about the same size as in *intacta* or somewhat larger and slightly more elevated, the curve of the upper margins much stronger proximally.   The apices sharp and directed straight back, reaching about the middle of the following segment; on 8 similar to those of the preceding segments, but less elevated, directed straight back

The coloration, when well marked, is so exactly similar to that of *intacta* that it seems unnecessary to describe it.   It is usually, however, rather obscure, though the legs are always distinctly banded.

Length of body 15-16; abdomen 9-10·6; hind wing 4·6-4·75; hind femur 4; width of abdomen 6-6·8; width of head 4·7-4·8.

The chief characters by which the nymph of *L. frigida* differs from that of *intacta* are thus the slightly smaller size, the more prominent eyes, the longer lateral spines on seg. 9, and the more sharply curved dorsal abdominal hooks.

Leucorrhinia proxima Calvert.

A few specimens of this species were taken in a marsh at the mouth of a small sluggish creek opening into Go Home Bay, on June 17, 1907.   It is not an uncommon species in Northern Ontario, but has not been recorded south of Go Home Bay in this province.

Its nymph is still unknown.

Leucorrhinia hudsonica (Selys) Hagen.

The adult of this northern species has not been found in this vicinity, but a number of nymphs were taken in a small marshy inlet, which we have good reason to ascribe to this form.   These nymphs are identical with two exuviæ received from Prof. Needham and erroneously referred by him ('08) to *L. frigida*.   Two nearly identical exuviæ were taken by the writer in June, 1913, at Nipigon, Ont., where *L. hudsonica* was flying in abundance, and where no other species was seen, except *L. glacialis*, whose nymph is known.   These nymphs and exuviæ are too small for *proxima* and *hudsonica* is the only other regional species whose nymph is unknown.   One of the Nipigon specimens has small dorsal hooks on segments 3, 5 and 6, the other has a single rudimentary hook on segment 4, while the Go Home Bay specimens have no trace of dorsal hooks.   In spite of these somewhat marked variations it seems almost certain that all belong to one species and that this species is *L. hudsonica*.

Ten of these nymphs were collected at Go Home Bay, five of them being full-grown.   They were collected prior to our first visit to the Station and neither date nor collector's name is known.

5 GEORGE V., A. 1915

Leucorrhinia glacialis Hagen.

A single specimen of this species was taken at Go Home Bay by Mr. J. B. Williams, on July 14, 1909.   It is more common farther north.

The nymph has been described by Needham ('01).

Leucorrhinia intacta (Hagen) Hagen.

This well-known species occurs but sparingly in the Go Home District, where it is occasionally seen in the open marshes bordering shallow bays.   It is associated in such stations with *L. frigida* which is more generally distributed and far more numerous, but we have never taken it from the sphagnum bordered ponds, where *L. frigida* always occurs.

The nymph has been described by Needham ('01.) We have not found it in this district, but have taken it in abundance at Toronto and Lake Simcoe, where it is the only species of the genus.

A single dead specimen of *L. intacta* was found in a cobweb on the small island referred to under *Enallagma hageni* and *Libellula quadrimaculata* (vide pp. 67–90).

Sympetrum danae (Sulzer) Ris.

Syn. *S. scoticum* (Donovan) Newman.

A single male of this northern species was taken from the edge of a very shallow pond in the sand on the Giant's Tomb Island, July 14, 1912.   It was a teneral and had evidently emerged on the day of its capture.   It was kept alive until the colour pattern was fully developed.

This circumpolar species is very common in Ontario north of the Great Lakes. With the exception of a single individual taken at De Grassi Pt., Lake Simcoe, the present record is the most southerly for this species in the province.

Sympetrum costiferum (Hagen) Kirby.

Our earliest captures of the adult of this species in 1907 are from the Giant's Tomb Island, July 29, 1907.   On this date a number of young individuals were flying about the shallow ponds in the sand and many exuviæ were found adhering to the reeds.   In 1912, a few tenerals were taken at nearly the same spot on July 14, but they did not appear at Go Home Bay until about a week later.   They soon became generally distributed in all the open reedy marshes bordering ponds and inlets and were often also seen away from the water.   They became quite abundant in August and were still common when the Station was closed on Sept. 11.   At this time many pairs were seen in copula.

As with most of the Sympetrums, we have neglected to rear the nymph though we are satisfied that the exuviæ referred to above and a number of full-grown nymphs of the same kind, taken at Go Home Bay and at Skerrevore, Ont., (by

Mr. Wodehouse) belong to this species.* They agree closely with Needham's ('01) description which was based on a single collapsed exuvia, except in the following particulars:—The dorsal hooks are somewhat shorter than the segments which bear them, the lateral spines of segs. 8 and 9 are also somewhat shorter than is indicated in the description, those of seg. 8 being about one-third as long as the segment, and those of 9 reaching only to the tips of the lateral appendages.

Besides these specimens, I have a number of similar but smaller nymphs, including two full-grown examples, from Fitzwilliam Island, Georgian Bay, collected by Mr. Wodehouse. Besides the smaller size these differ in the slightly shorter lateral spines of seg. 9. Specimens from Giant's Tomb Island are, however, intermediate in this character which appears to be a rather variable one. The number of mental and lateral setæ is slightly smaller in the smaller nymphs, there being 10-12 of the former and 9-10 of the latter, as compared with 13-15 mental and 10-11 lateral setæ in the larger specimens. The number of these setæ, however, seems to depend a good deal on size, and we doubt if in this case any other importance can be attached to the feature. It may be also noted in this connection that adults of *S. costiferum* vary in size with locality, specimens from Northern Ontario being distinctly smaller than those from farther south.

Sympetrum vicinum (Hagen) Kirby.

Full-grown nymphs of this species were collected towards the end of July and in early August and were found to be generally distributed along the marshy of boggy margins of still waters, their environment being similar to that of *Leucorrhinia frigida*. They were found, e.g., along the edges of sphagnum bogs as well as in shallow reed-grown waters.

The first imagos emerged on July 30 and by August 6 the pale yellow tenerals were quite common in the marshes. In the latter half of August they had for the most part acquired their bright red colour and were common everywhere. They were still numerous when the Station was closed on Sept. 11th. Many pairs were in copula at this time.

A description of the nymph is given by Needham ('01).

Sympetrum semicinctum (Say) Kirby.

This pretty species is not rare, but never appears in large numbers, as do most of the species of *Sympetrum*. Specimens were taken in open marshes adjoining shallow bays and creeks, but nothing distinctive was learned of their habits or haunts.

The nymph, which has been described by Needham ('01) was not obtained by us.

The dates of our specimens range from July 24 (1912) to Aug. 24 (1907).

---

*Since the above was written we have reared this species on Vancouver island and have verified the above determination.

Sympetrum obtrusum (Hagen) Kirby.

This common form appears considerably earlier than the other species of Sympetrum, specimens having been observed at least as early as July 1, 1912. At Lake Simcoe and southward, they appear before the end of June. The season for emergence is somewhat protracted and irregular, tenerals being seen as late as July 31. They fly until late in the season, several pairs in copula having been taken on Aug. 26th, 1912.

The adults are found in the same localities as *S. costiferum* and *vicinum*, but, as we have not found the nymph in this district we are unable to give anything distinctive as to the nature of its breeding-ground. Stray specimens of the imagos have occasionally appeared on the Station Island, where they certainly do not breed.

It is somewhat remarkable that the closely allied species *S. rubicundulum*, one of the commonest and most generally distributed of Odonata in Eastern North America, is wholly absent from the Go Home District so far as we are aware.

Sympetrum corruptum (Hagen) Kirby.

On July 14, 1912, this species appeared very unexpectedly on the low sandy eastern end of the Giant's Tomb Island. The island is divided here by a narrow channel, close to which, on the outer side, is a shallow pond or lagoon (Fig. 35). It was about the margins of the channel and lagoon, especially the former, that *Sympetrum corruptum* was observed. They were flying about from place to place, sometimes hovering over one spot, sometimes settling for a moment on the wet sand. They were so shy that it was almost impossible to get within striking distance, and more than an hour of patient effort was spent before one was secured. Two males and one female were all that were taken, all fully mature and in good condition.

This species was previously known from Ontario only by a single specimen taken at the Humber River (Walker '06). It is not known to occur east of this province, but it is common in the Prairie Provinces and also occurs in British Columbia.

The nymph has been described by Needham ('03).

Libellula quadrimaculata Linné.

This wide-spread circumpolar species is fairly common, but by no means abundant at Go Home Bay, where it frequents marshy bays and inlets. June 18 is the earliest date on which the adult was observed in 1907, while in 1912 it appeared somewhat earlier, but the exact date was not noted. It was more numerous during the latter than the former year.

A remarkable assemblage of this species was met with on June 27, 1912, on a small island in the open water of Georgian Bay, abut 3½ miles from the eastern coast. On this island, which has already been described (vide p. 67) there is a small pond filling a depression in the almost bare rock and from this pond three species of Odonata were emerging in large numbers, viz., *Enallagma hageni, Ischnura verticalis* and *L. quadrimaculata.* Along one side of the pond was bare rock and in the few clumps of small reeds that were scattered along this shore, large numbers of exuviæ of the last-named species were found. One or two emerging imagos were also noted, while resting in the bushes of a dense thicket on the opposite side of the pond, which was only a few feet wide, were scores of teneral imagos.

The unusual abundance of this species here was probably due to the lack of competition with other large species, there being apparently no others present, although I found a single dead example of *Leucorrhinia intacta* in a cobweb, which had probably developed in the same pond.

It may be noted that the season for emergence was somewhat later here than at Go Home Bay.

Full-grown nymphs of this species were also collected by Mr. Wodehouse at the French River, June 19, 1912.

Libellula exusta julia (Uhler) Ris.

. The scarcity of other species of Libellula in this region is fully compensated for by the multitudes of this form, which fly about almost every marshy bay or pond during June and July.

In the decaying organic matter at the bottom of such swamp waters, where other species of the genus are seldom found, *L. exusta julia* seems to find ideal conditions of environment, while in the ponds of agricultural districts, such as those in the environs of Toronto and Lake Simcoe, where *L. pulchella, lydia, quadrimaculata* and *luctuosa* are the prevailing species, *julia* is rare or wholly absent. It is not, however, quite uniformly distributed in the swamp waters of Go Home Bay, for in a small undrained pond on "Split Rock Island," just off the outer coast, we were unable to find the species. This pond (Plate VIII, Fig. 33) is surrounded by sphagnum bog and the aquatic vegetation is very scanty.

Full-grown nymphs were common in dredgings made on May 31, 1912, and during the week following. Imagos were first noticed on June 7 and had become abundant by the 13th. They continued so for about a month, their numbers dwindling during the last half of July until the 30th, when the last specimen was noted. A few specimens emerged in the laboratory during the latter half of June.

This dragonfly is not only common about its breeding-grounds, but also in the open rocky woods, where it takes short flights, frequently settling on the bare rocks after the manner of Gomphines. In fine still weather the males may be seen chasing each other swiftly and erratically over the water and are somewhat conspicuous objects on account of the white pruinosity of the thoracic dorsum and basal segments of the abdomen.

L. exusta julia ranges northward at least as far as Nipigon, Lake Superior.

5 GEORGE V., A. 1915

Libellula pulchella Drury.

This well-known and conspicuous dragonfly is much less common in the region under discussion than in Southern Ontario, a fact which is probably due to differences in soil and drainage conditions, (vide p. 54). Specimens are, however, quite frequently seen in the vicinity of marshy bays and in openings in the woods nearby.

In size they are not inferior to specimens from more southern latitudes.

They have been taken in the vicinity of Go Home Bay between June 28 and Aug, 9, 1912, inclusive.

Libellula vibrans incesta (Hagen) Ris.

Go Home Bay is probably near the extreme northern limit of distribution of this Austral species. It is very rare here, only two examples having been obtained. Both of these were males, not yet pruinose, and were captured at the outlet of Galbraith Lake (Pl. VII, Fig. 30) on July 15, 1907. The only other known Canadian locality for this species is Point Pelee, Lake Erie, where it is common (Walker, '06, and F. M. Root, Can. Ent., XLIV, 1912, p. 209).

Literature Cited.

1914. BENSLEY, B. A.—The Fishes of Georgian Bay, Contributions to Canadian Biology, 1911-1913, Ottawa.

1881. CABOT, LOUIS.—The Immature State of the Odonata, Part II, Subfamily Aeschnina. Mem. Mus. Comp. Zool., VIII, No. I, pp. 1–40, pls. 1–V.

1861. HAGEN, H. A.—Synopsis of the Neuroptera of North America, with a list of South American species. Smithsonian Miscel. Coll., IV, Odonata, pp. 55–187.

1899. KELLICOTT, D. S.—The Odonata of Ohio. Special Papers, Ohio Academy of Science, No. 2, viii + 116 pp., pls. 1–4.

1910. MUTTKOWSKI, R. A.—Miscellaneous notes and records of dragonflies (Odonata). Bull. Wis. Nat. Hist. Soc., VIII, No. 4, pp. 170–179.

1901a. NEEDHAM, J. G. AND BETTEN, C.—Aquatic Insects in the Adirondacks. Bull. 47, N. Y. State Mus., pp. 383–612, 36 pls.

1901b. NEEDHAM J. G. AND HART, C. A.—The dragonflies (Odonata) of Illinois, Part I, Petaluridæ, Aeschnidæ and Gomphidæ. Bull. Illinois Lab., VI, art. I, 94 pp., 1 pl.

1903. NEEDHAM, J. G.—Aquatic Insects in New York State. Bull. 68, N. Y. State Mus., Ent. 18, pp. 200–499, 52 pls.

1908. NEEDHAM, J. G.—Report of the Entomologic Field Station conducted at Old Forge, N.Y., in the summer of 1905. Bull. 124, N.Y. State Mus., 23rd Rep. State Ent., 1907, App. C, pp. 156–263, 32 pls.

1905. WALKER, E. M.—Orthoptera and Odonata from Algonquin Park, Ontario. Ann. Rep. Ent. Soc. Ont., 36, pp. 64–70.

1906. WALKER, E. M.—A first list of Ontario Odonata. Can. Ent., XXXVIII, pp. 105–110, 149–154.

1912a. WALKER, E. M.—The North American Dragonflies of the Genus Aeshna. University of Toronto Studies, Biological Series, No. 11, pp. 1–211, 28 pls.

1912b. WALKER, E. M.—The Odonata of the Prairie Provinces of Canada. Can. Ent., XLIV, pp. 253-266, 1 pl.

1914. WALKER, E. M.—The known nymphs of the Canadian Species of Lestes (Odonata). Can. Ent., XLVI, pp. 189-200, 2 pls.

1907.—WILLIAMSON, E. B.—A collecting trip north of Sault Ste. Marie, Ont., Ohio Nat., VII, pp. 129–148.

EXPLANATION OF PLATES III–IX.

PLATE III.

Fig. 1. Nehalennia gracilis.—Lateral gill.
Fig. 2. Nehalennia gracilis.—Hind margin of head.
Fig. 3. Nehalennia irene.—Hind margin of head.
Fig. 4. Enallagma calverti.—Dorsal view of head.
Fig. 5. Enallagma calverti.—Lateral gill.
Fig. 6. Enallagma bageni.—Dorsal view of head.
Fig. 7. Enallagma signatum.—Dorsal view of head.

5 GEORGE V., A. 1915

III.

THE MOLLUSCA OF GEORGIAN BAY.

By A. D. ROBERTSON, B.A., University of Toronto.

(Plates X—XII)

In 1910 the writer undertook, in connection with the work of the Biological Station, an analysis of the local molluscan fauna of Go Home Bay. This study has since been extended to include various points around Georgian Bay, but because of the labour involved in working over the material, this paper deals only with the more limited area, leaving the general distribution for future discussion. The analysis of the species is believed to be fairly complete, and special care has been taken to observe critically the specific characters and the variations, whether due to environment, age or other cause. Notice has also been taken of the food of the various forms and of the extent to which they themselves serve as food for fishes and other animals.

The collections along-shore were made by hand and hand-dredges, while in the deeper water use was made of an iron dredge, provided with a fine inner screen supported by a course outer screen. The latter method entails much labour in sorting over the material but gives excellent results.

In the identification of species the writer is indebted to Mr. Bryant Walker of Detroit and to Dr. H. A. Pilsbry and Mr. E. G. Vanatta of the Philadelphia Academy of Natural Science, who determined a number of species and confirmed the determination of others. He is also under obligation to Professor B. A. Bensley, Dr. E. M. Walker and Dr. A. G. Huntsman of the University of Toronto for much kind assistance and advice.

The environmental features of this region are of interest because it falls within the Archean area. A general account of these features is given elsewhere by Bensley ('14), but reference may be made to those which are more important from the standpoint of this paper.

1. The glaciated surface of the rock is sparingly and unevenly occupied by soil and bottom deposits, accumulating in basins and consisting chiefly of disintegrated particles of gneiss, often with high organic content.

2. The indented shore-line leaves projecting headlands of fully exposed and bare character, while the protected smaller bays form isolated swamps with usually a deep deposit of mud and much organic material.

3 The main shore gives place outwards to larger, then to smaller islands and finally to reefs and submerged shoals, with clear rock or boulder bottom.

4. The deeper waters of the larger bays and open waters have flat bottoms, consisting of mud of fairly high organic content alternating with exposed patches of the underlying rock.

5 GEORGE V., A. 1915

5. At places there are open or somewhat exposed channels with accumulations of clean sand, almost free from organic material.

6. In addition to the wave-action on the main shore and on the shoals, there is a constant flow of water in and out among the islands, giving conditions of exposure, temperature and oxygenation which are in marked contrast with those of the protected bays and especially with those of the inland ponds.

In general the species of mollusca exhibit great flexibility in their environmental relations. In many cases ecological selection is operative within broad limits, while in a few the environmental type is more or less specific. The chief factors in this selection appear to be (a) exposed or protected situation, (b) depth of water, (c) degree of aeration, (d) character of the bottom, and (e) food conditions. In the Genus *Lymnaea*, the long-spired forms occur in the stagnant bays, while the short-spired ones inhabit rocky shores. The species of *Planorbis* also occur in the swampy bays, though *P. deflectus* extends its range to the exposed rocky islands and the pools occurring on them. Most of the species of *Physa* are quite general in their distribution, but *P. integer* prefers the exposed shores. The genera *Ancylus*, *Amnicola* and *Campeloma* and the families *Unionidae* and *Sphaeriidae* occur in swampy bays and also, though much less abundantly, in inland ponds, with an extension of this range in *Amnicola*, the *Unionidae* and sometimes *Campeloma* to the sandy channels and of the *Sphaeriidae* to these channels and to sand or gravel banks in the deeper waters. *Goniobasis* is found abundantly where there are currents of clear, well-aerated water, in sand channels, on sandy beaches and on the exposed rocky shores. The genus *Valvata* is a sand-loving one, although of the two species, one, *tricarinata* is also found plentifully in weedy, muddy bays.

The inland ponds are exposed to extremes of summer and winter temperature. They are limited in the facilities they afford for migration to the deeper waters. They are poorly provided with means of aeration and are often surfeited with decaying vegetation and so afford an environment not highly favorable to molluscan life. Few forms occur and these not abundantly. Among them are *Ancylus parallelus*, *Campeloma decisum*, *Amnicola limosa* and the *Sphaeriidae*.

The protected muddy bays where these conditions are reversed afford a rich molluscan fauna with a wide range of species which includes the *Unionidae*, the *Sphaeriidae*, the genera *Valvata*, *Amnicola*, *Goniobasis*, *Planorbis*, *Ancylus*, *Physa* (with the exception of *P. integer niagarensis*) and the long-spired species of *Lymnaea* (*palustris*, *columella* and *haldemani*).

In the weedy sand-runs the same forms occur, with the exception of *Planorbis exacuous*, *P. dilatatus* and the *Lymnaea* mentioned above. In clean sand channels, free from weeds and exposed to currents, the *Unionidae*, the *Sphaeriidae* and the genera *Campeloma*, *Valvata* and *Goniobasis* occur.

The exposed rocky shores which seem to afford a scanty supply of food and an abundance of well-aerated water, yield *Lymnaea emarginata canadensis*, *L. decollata*, *L. stagnalis sanctamariae*, *Planorbis deflectus* and the various species of the *Physa*.

In the shallow island pools which are well-aerated and have a good supply of food, but which, on the other hand are subjected in some cases to destruction by drought and to severe winter conditions, *Planorbis deflectus* and *Lymnaea palustris* abound.

In deep dredging from sandy or gravelly bottoms, the *Sphaeriidae* and the genus *Valvata* are obtained.

The total number of species identified is 37, representing 14 genera in 8 families, as follows:—

I. Family LYMNAEIDAE.
    A. Genus **Lymnaea.**
        1. *Lymnaea stagnalis sanctamariae*, Walker.
        2. *Lymnaea (Galba) decollata*, Mighels.
        3. *Lymnaea (Galba) emarginata canadensis*, Sowb.
        4. *Lymnaea (Galba) palustris*, Muller.
        5. *Lymnaea (Pseudosuccinea) columella*, Say.
        6. *Lymnaea (Acella) haldemani*, (Deshayes) Binney.
    B. Genus **Planorbis.**
        7. *Planorbis (Helisoma) bicarinatus*, Say.
        8. *Planorbis (Pierosoma) trivolvis*, Say.
        9. *Planorbis (Planorbella) campanulatus*, Say.
      10. *Planorbis (Menetus) exacuous*, Say.
      11. *Planorbis, (Menetus) dilatatus*, Gould.
      12. *Planorbis (Gyraulus) hirsutus*, Gould.
      13. *Planorbis (Gyraulus) deflectus*, Say.
II. Family PHYSIDAE.
    C. Genus **Physa.**
        14. *Physa heterostropha*, Say.
      15. *Physa ancillaria*, Say.
     15a. *Physa ancillaria magnalacustris*, Walker.
     15b. *Physa ancillaria vinosa*, Gould.
      16. *Physa gyrina*, Say.
      17. *Physa integer niagarensis*, Lea.
    D. Genus **Ancylus.**
      18. *Ancylus parallelus*, Hald.
III. Family STREPTOMATIDAE.
    E. Genus **Goniobasis.**
      19. *Goniobasis livescens*, Menke.
      20. *Goniobasis haldemani*, Tryon.
IV. Family AMNICOLIDAE.
    F. Genus **Amnicola.**
      21. *Amnicola limosa*, Say.
      22. *Amnicola emarginata*, Küster.
      23. *Amnicola lustrica*, Say.
V. Family VALVATIDAE.
    G. Genus **Valvata.**
      24. *Valvata tricarinata*, Say.
      25. *Valvata sincera*, Say.
VI. Family VIVIPARIDAE.
    H. Genus **Campeloma.**
      26. *Campeloma decisum*, Say.

5 GEORGE V., A. 1915

VII.   Family UNIONIDAE.
  I.   Genus **Lampsilis**.
      27. *Lampsilis ventricosus,* Barnes.
      28. *Lampsilis luteolis,* Lamarck.
      28a. *Lampsilis luteolis rosaceus* De Kay.
  J.   Genus **Anodonta**.
      29. *Anodonta grandis,* Say.
      29a. *Anodonta grandis footiana,* Lea.
  K.   Genus **Anodontoides**.
      30. *Anodontoides ferussacianus,* Lea.
  L.   Geuns **Unio**.
      31. *Unio complanatus,* Solander.
VIII.   Family SPHAERIIDAE.
  M.   Genus **Sphaerium**.
      32. *Shpaerium simile,* Say.
      33. *Sphaerium striatinum,* Prime.
      34. *Sphaerium rhomboideum,* Say.
      35. *Sphaerium (Musculium) securis,* Prime.
      36. *Sphaerium (Musculium) partumeium,* Say.
  N.   Genus **Pisidium**.
      37. *Pisidium abditum,* Hald.

Family LYMNAEIDAE.

Represented by 13 species of *Lymnea* and *Planorbis,* together forming the third of the total number of Molluscan species.

Genus **Lymnaea**.

Of six species identified, two, *L. emarginata canadensis* and *L. Palustris* are the prevailing types. Both occur abundantly. *L. haldemani* was taken in only two situations, although it was present in numbers *L. decollata* was taken in several places, but nowhere abundantly. Of *L. stagnalis sanctaemariae* only six specimens in all were obtained. *L. columella* was frequently found, but not in numbers.

L. stagnalis sanctaemariae, Walker. The six specimens were identified as this species and variety by Mr. E. G. Vanatta. The shortened spire corresponds to the exposed situation in bare rocky channels. A light colored, transparent shell, 5 to 5½ rounded whorls with distinct suture. The aperture is broadly ovate, the edge of the lip thin, flared anteriorly; slit-like umbilicus widely open or nearly closed by the callus. (Pl. XI, Fig. 18).

L. (Galba) decollata, Mighels, occurs on rocky shores and in shallow rocky bays of outer islands. Stout, nicely formed little shell, rhomboidal in outline when viewed facing the aperture. Large body whorl, expanded aperture and short, sharp spire. Whorls 3. Color brownish horn, tinged with green, apical whorls darker, white varical thickenings on body whorl. In comparison with *L. emargi-*

nata canadensis the shell is smaller, shorter and smoother, the whorls fewer and more convex and the sutures are more impressed. (Pl. X, Fig. 7).

L. emarginata canadensis. Sowb.; very abundant on clean rocky shores, especially of the outer islands. Found also on sand and pebble bottom. Corresponding to its exposed position and in contrast to the other species, *L. palustris*, it is thick-shelled with shortened spire and is of light coloration. It is a medium-sized species (adult length 20-25mm.) and is usually recognized easily by its light horn colour and malleated surface. The spire is shorter than the aperture, the whorls well-rounded, 5-6 and the sutures distinct. The aperture is large and ovate with a somewhat flaring lip and with reddish varical thickenings usually prominent immediately behind it. A white callus spreads over the body-whorl and covers but does not close the deep slit-like umbilicus. Usually with several whitish or reddish varical thickenings. Identified as this variety by Mr. E. G. Vanatta. As it occurs at Go Home it is quite variable, especially in surface malleation, thickness of the shell, height of the whorls and length of the spire. Thinner shells show more definite malleations. Its food consists of algæ. It has been taken from the stomach of the whitefish, *Coregonus clupeiformis.* (Pl. X, Fig. 14).

L. (Galba) palustris, Muller, abundant in shallow bays on the bottom or on submerged vegetation, often on mud flats above the water's edge. It prefers moderately high temperatures. It is easily recognized by its narrow elongated form, dark colour and by the aperture which is usually shorter than the spire. Lip somewhat flared. Surface of shell variable, smoother in island pools, usually roughened by coarse lines of growth in muddy bays. Sometimes malleated. Color brown to almost black, darker in pools, often whitish due to erosion. In young, color darker and lip not flared. Distinguishable from *L. emarginata canadensis* in the more slender elongated form, narrower and shorter aperture, longer spire, darker color and distinct habitat. It feeds upon the filamentous green algæ, diatoms and desmids. Found in the stomachs of whitefish. (Pl. X, Fig. 8).

L. (Pseudosuccinea) columella, Say. Common on lower surface of lily-leaves in stagnant, muddy bays. Easily recognized by its expanded and oblique body whorl, its long aperture, expanded anteriorly, rather sharp-pointed spire and its delicate shell-structure. Lines of growth prominent. Its elongated form and especially its delicate shell are adaptations to its protected habitat. Food consists of diatoms, desmids and other green algæ. (Pl. XI, Fig. 15).

L. (Acella) haldemani (Deshayes) Binney: Found on the lower surface of lily-leaves in well-sheltered muddy bays in late summer. Observed in but two situations, both of which were removed from open water and were especially well-protected. Several specimens secured in each situation. Diligent search failed to reveal any during the early summer and nothing was found to indicate their habitat during this period. These observations agree in their main features with Kirkland's account as given by Baker ('11). Those secured were, however, considerably removed from deep water; none were observed in the approaches to the bays, neither were any secured in dredging. This is the most striking *Lymnaea* of the region. Its extremely slender form, long spire, oblique, flattened whorls, long narrow aperture, sharply angular at the posterior end and its thin transparent

39b—7½ II

5 GEORGE V., A. 1915

shell are unmistakable characters. The long spire and delicate shells are in conformity with its protected habitat. It varies in length of spire, conxevity of whorls and size and shape of the aperture and the axis is often considerably twisted. Its food consists of algæ. (Pl. X, Fig. 4).

Genus **Planorbis**.

Seven species were identified. Of these, three, *P. bicarinatus*, *P. trivolvis* and *P. campanulatus* belonging to a large-shelled group inhabiting muddy bays, possess comparatively high, sinistral shells and certain common characters in respect of the reproductive organs which will be dealt with in a subsequent paper. The others belonging to a small-shelled group with a more varied habitat possess low, flattened, dextral shells and, as far as examined, certain other characters in the reproductive organs. This group includes *P. hirsutus*, *deflectus*, *exacuous* and *dilatus*, the range of distribution of which varies with each species.

Planorbis bicarinatus, Say, occurs abundantly in weedy sand runs and weedy muddy bays; found also on rocks near the latter. Distinct and easily recognized by the two prominent angular carinae. Aperture slightly oblique, somewhat triangular, broadly rounded below, lip thin with varical thickening behind. Shell bi-concave, lower concavity with the sides interrupted by the carinae, upper smooth and funnel-like. Former apertures often evident on the body whorl as pronounced transverse ridges with darker periostracum. The possession of a broad high shell, which is carried on edge and has an aperture only slightly oblique, doubtless indicates for this form sheltered rather than exposed situations. In the young the carinae as pronounced but the aperture less oblique than in the adult. Food consists of green algæ. (Pl. X, Fig. 5).

Planorbis trivolvis, Say.; found plentifully in protected muddy bays and sometimes along the sheltered shores of the inner islands. Prefers shallow bays with comparatively high temperature. Found only in shore collections and water less than 2 feet in depth. The largest *Planorbis* of the region (Adult measurements,— width 20–30 mm, length 10–13 mm). Shell with shallow concavity above the smooth slopes of which are interrupted by the carina of about the last half of the body whorl. A deep umbilicus into which the rounded whorls disappear below. Aperture large, triangular or rhomboidal in outline, narrower above, lip thin, much flared with a varical thickening behind. One or more former apertures evident. Lines of growth coarse. Not fitted for exposed situations because of the size and shape of the shell and the vertical position in which it is carried. Varies with age. Young lighter in color, shells high and narrow, while adults are much broader than high. Recognized easily at all stages by the upper concavity of the shell. Series showing all stages readily secured. Eggs laid in flat brownish capsules on lily-leaves, sticks and even on other molluscs. Food, filamentous algæ, diatoms and desmids. (Pl. X, Fig. 6).

Planorbis campanulatus, Say.; occurs abundantly in weedy places, both muddy and sandy, up to the depth of at least three fathoms. Easily recognized by the campanulate expansion of the body whorl a short distance behind the aperture and the narrowly constricted throat just behind this. Aperture rhomboidal, narrowed above. Narrowly rounded tops of the whorls all in the same plane. Lower surface like trivolvis but narrow lower edge of the whorls more rounded and less angular. Lines of growth coarse, regular and parallel. Adapted in the same manner as the two preceding species to protected rather than exposed situations. Often distorted so that the tops of the whorls are inclined at various angles. Varies considerably in length of campanulate expansion and also in thickness of shell. Feeds on filamentous green algæ, diatoms and desmids. (Pl. X, Fig. 1).

Planorbis exacuous, Say.; occurs in protected weedy placés, never in large numbers. A well-marked species, having as distinguishing features a very sharp peripheral carina, a lens-shaped shell and small size. Whorls flattened above, broadly rounded below. Aperture triangular and very oblique. The greatly flattened shell and very oblique aperture which allow it to lie close to the surface over which it crawls would seem to adapt it to an exposed habitat, yet it was found only in protected places. Varies in color, light coloured in sandy, and brown in muddy places. (Pl. X, Fig. 3).

Planorbis dilatatus, Gould, occurs on sticks along muddy river banks and in muddy bays. Only a few obtained. Small brownish; top of shell flat; sharp peripheral keel almost level with top of shell; whorls broadly rounded below; aperture oblique; compared with *P. exacuous* it is smaller and higher in proportion, the whorls are flatter above and much more convex below and the carina is placed much higher.

Planorbis hirsutus, Gould, occurs plentifully in weedy, sandy channels and in muddy bays or in channels on smooth rocks covered with light deposit óf sediment. Easily recognized by the rough hairy shell. Shell wide and flat, having a sharp, strongly deflected peripheral keel and a very oblique aperture. Surface covered by crowded rows of hairs. Last portion of body whorl often strongly deflected. Shell varies greatly with age. A shell of about three and a half whorls is concave above and below, the aperture is only slightly oblique, the centrally-placed peripheral keel is just appearing behind the aperture, the shell is high and all the whorls are on the same level. In older shells the aperture becomes oblique, the shell becomes wide and flat, and there is a pronounced peripheral keel deflected downwards. The last whorls also drop below the level of the preceding whorls and the shell becomes saucer-shaped. (Pl. X, Fig. 2).

Planorbis deflectus, Say, the most abundant *Planorbis* species of the region, possesses a wider range of habitat than any other species, plentiful in quiet weedy bays, in weedy sand channels and in shallow dark-colored pools on islands. Occurs also on exposed shores. Recognized by its small size and rounded periphery. Aperture only slightly oblique. Color varies from yellowish on lily-leaves in bays to dark brown in dark island pools. Whorls in one plane or with last part of body-whorl deflected downwards. Periphery sometimes flattened on its upper edge, giving a peculiar sloping aspect to last whorl. Some specimens banded alternate white and dark brown.

5 GEORGE V., A. 1915

Family PHYSIDAE.

Represented by six species belonging to two genera, *Physa* and *Ancylus*.

Genus Physa.

Of this genus five species were obtained. Four of these are large and dark-colored, have short spires, thin shells and indistinct sutures and occur throughout a wide range of habitat. The fifth is small, light colored, with white varical thickenings of the whorls, has a more elongate spire and is found only on semi-exposed rocks. Although the two groups are quite distinct, the species within the first group are not so clearly differentiated. The characters upon which the species of this genus are based are exceedingly variable and the extremes of variation grade into one another so smoothly · that an attempt to verify the present classification by breeding experiments and anatomical investigation seems desirable. The writer intends to undertake the task in the near future. In the meantime the distinctions here used will be those of the literature of the genus.

Physa heterostropha, Say, occurs usually in protected situations in weedy bays or quiet rocky channels, rare. Surface smooth and shiny, without sculpture, the spire elevated and the sutures distinct. Food, diatoms, desmids and other algæ.

Physa ancillaria, Say, very abundant, almost everywhere in sheltered bays and along partially exposed shores. In the spring it may be seen collecting in vast numbers to the breeding-grounds in rocky channels and in the bays of rocky islands. Within a few days after copulation the eggs are laid in elongated capsules. A single individual may lay as many as five capsules containing in all 150-300 eggs. Spire short, sutures not so distinct as in *P. heterostropha*. Shell more robust than in that species. Surface smooth and shining, sculptured. Varies much in surface sculpture, height of spire, size and shape of aperture and number of digitations on mantle. Such malformations as forked tentacles and lobes arising from upper surfaces of foot were found. Food consists of diatoms, desmids and other green algæ. A number were found in whitefish stomachs. (Pl. XI, Fig. 19).

A variety, *magnalacustris*, Walker, with white lines on body whorl also occurs.

The variety *vinosa* occurs in sheltered bays and on partially exposed shores. Shell robust, spire short but sharp; whorls rounded and suture distinct.

Physa gyrina, Say. Not very abundant, found in sheltered bays. Differs from *P. ancillaria* in larger size, more elevated spire, more slender form and coarser surface sculpture.

Physa integer niagarensis, Lea; found on somewhat exposed rocky shores. much smaller than any of previous forms, shell much like a small reversed *L. emarginata canadensis*. Shell light horn with many white bands transverse to whorls. Shell heavy, spire elevated, apex sharp, sutures distinct and whorls rounded. Identified by Dr. Pilsbry.

Genus **Ancylus**.

Ancylus is represented by a single species.

Ancylus parallelus, Hald.; very common in sheltered bays on under sides of lily leaves and on sticks. Shell flat, pyramidal; apex ⅓ length of shell from posterior end, directed backwards and to the left; sides nearly parallel, shell narrower in front. (Pl. XI, Fig. 17).

Family STREPTOMATIDAE.

Represented by two (?) species of the Genus *Goniobasis*.

Genus **Goniobasis**.

Goniobasis livescens, Menke. Obtained abundantly where there are currents, in sand runs or along rocky shores and on the rocky shoals near the outer islands. Occurs also but not plentifully in muddy bays. Spire long and tapering, apex usually eroded away, whorls 8-9, flattened, suture not deeply impressed, distinct carina at lower edge of whorl. Aperture small and rhomboidal; prolonged anteriorly into a slight groove, closed by an operculum borne on upper side of foot. Unlike the long spired species of *Lymnaea* which inhabit sheltered situations, this form which is also long spired is well-adapted to exposed places because of its strong heavy shell. In contrast with the *Lymnaea* also, it, when detached, does not float on the water, but sinks at once into deeper water. A quite variable species. Varies in length and stoutness of spire, usually high and slender, often quite short and stout, color dark brown, shaded with green, light green or white. In young, carina well-marked; in adults, no carina on body whorls; in younger, color much darker. Feeds on diatoms and desmids. (Pl. XI, Fig. 16).

Goniobasis haldemani, Tryon; (not positively identified.) Occurred on shady beach along exposed shore. Few obtained. More slender and elongated, whorls more rounded than in *livescens*. No carina and the color white tinged with green.

Family AMNICOLIDAE.

Represented by three species, all belonging to the genus *Amnicola*.

Genus **Amnicola**.

Of the three species obtained here, *A. limosa* is the most abundant. All occur in weedy places either with mud or sand bottom. *A. limosa* is secured also on the rocky shores of even the outer islands and *A. limosa* and *lustrica* were obtained in deep dredgings. Operculate.

Amnicola limosa, Say; obtained on weeds in sand channels or muddy bays, on rocky shores and in dredging at 20 fathoms or more. Very abundant, tentacles long and constantly in motion. The jet-black eyes placed at outer bases of ten-

5 GEORGE V., A. 1915

tacles. Shell globose, whorls convex, apex usually rounded. Umbilicus small, aperture rounded. Shows considerable variation. Shell may be conic, spire elongated and apex sharp. The sutures vary in distinctness. Eggs laid in small triangular capsules on weeds, sticks, stones and even on the shells of other molluscs. (Pl. X, Fig. 9).

Amnicola emarginata, Say. Not numerous. Found with *A. limosa*. Distinguished from it by the truncated apex, the first whorl not rising above the second. Spire also more elongated than usual in *A. limosa*.

Amnicola lustrica, Say; not abundant. Occurs with other species of *Amnicola*; dredged in 20 fathoms or more of water. Compared with *A. limosa*, shell thinner, spire much more elevated, apex sharp, body whorl scarcely larger than the preceding one. (Pl. X, Fig. 10).

Family VALVATIDAE.

Two species belonging to one genus occur.

Genus **Valvata.**

Of the two species, one, *V. tricarinata* occurs abundantly. Operculate. The plume-like gills borne within the mantle-cavity.

Valvata tricarinata, Say., abundant in weedy places among islands on either sandy or muddy bottoms. Occurs on sandy bottoms even to depth of 20 fathoms. Prominent carinae, usually three. Umbilicus broad, open to apex. Whorls loosely appressed. Quite variable. One or all of the carinae may be lacking or indistinct. The order of their reduction appears to be peripheral, lower, upper. In sandy places malformations in which whorls do not touch preceding whorls occur, seen in early whorls, body whorl or intermediate ones. Found in whitefish stomachs. (Pl. XI, Fig. 21).

Valvata sincera, Say. Found only in dredgings in sandy places. Occurs at depths of up to 20 fathoms. Not abundant. Compared with *V. tricarinata* there are no carinae and whorls are more rounded. There are distinct, sharp, elevated ridges, parallel to lines of growth. In the young these ridges are finer and more closely placed. Subject in sandy places to malformations similar to those occurring in *V. tricarinata*. Abundant in stomachs of whitefish. (Pl. XI. Fig. 22).

Family VIVIPARIDAE.

Represented by one species belonging to Genus *Campeloma*.

Genus **Campeloma.**

Operculate. As family name indicates young are produced alive.

Campeloma decisum, Say; occurs abundantly in sheltered bays with soft mud bottoms and in sand channels with decaying vegetable content. Congregates in decaying lily stems and on decaying logs. Recognized easily by large, heavy greenish shell, with short spire and eroded apex, its large, broad, brownish mottled foot and its long tapering tentacles. Color varied by narrow, irregularly placed dark bands crossing the whorls. In dark water it is often rusty brown. Young lighter in colour, lip of aperture thinner and shell has numerous fine lines parallel to whorls. All stages of development from young in the uterus to the adult form are easily obtained. Feeds on decaying vegetable matter. (Pl. XI, Fig. 20).

Family UNIONIDAE.

Represented by seven species belonging to four genera.

Genus Lampsilis.

Two species of this genus are reported.

Lampsilis ventricosus, Barnes. The single specimen obtained some years ago was identified by Bryant Walker as *L. ventricosus canadensis*, Lea, conforms to descriptions of *L. ventricosus* and since Simpson (1900) includes *canadensis* in synonymy of *ventricosus* it is here designated by the latter name. Shell, thick; color yellowish, darker in front. Few faint broad radiations behind, lines of growth coarse, beaks eroded, hinge line straight, nacre white, cardinal teeth double in both valves, lateral teeth single in right valve, double in left.

Lampsilis luteolis, Lamarck. Very abundant on the slopes of deep pools in sandy channels and along sloping muddy shores. Shell much higher behind the beaks. Beak sculpture consisting of about 13 fine wavy concentric ridges. Color light or dark brown, usually with numerous, sometimes brilliant narrow green rays. Two cardinal teeth in each valve, lateral teeth double in left valve and single in the right, long, curved and lamelliform. Nacre white. Hinge line curved. Varies in periostracum which may be smooth and shining or coarsely wrinkled, in outline of shell, in color in prominence and number of rays and in cardinal teeth which are pyramidal or lamelliform. Females inflated posteriorly. Young narrower than adults. A form is common here which is large and heavily shelled, has a dark brown periostracum often with a greenish sheen towards the umbones and is coarsely and closely wrinkled at the margin of the gape. The variety *rosaceus* which has smooth reddish-brown periostracum and rosy nacre also occurs. (Pl. XII, Figs. 23, 26, 30).

Genus Anodonta.

Of this genus one species occurs.

Anodonta grandis, Say, occurs plentifully on steep slopes of sand banks in sandy channels, and also, but less abundantly in the soft mud of sheltered bays. Shell thin, smooth, inflated, hinge-teeth lacking, usually dull in colour. Beak sculpture, four or five concentric ridges with anterior and posterior loops. Varies greatly in color, sometimes dull and almost rayless; at times brilliant, with many green rays. Varies also in inflation of shell and in outline. Forms were found typical of *footiana* as well as other typical of *grandis*, s.s., but there were also many intermediates. (Pl. XII. Figs. 25, 28).

5 GEORGE V., A. 1915

Genus **Anodontoides.**

Represented by a single species.

Anodontoides ferussacianus, Lea. Plentiful in shallow sand channels and also in muddy places. Compared with *A. grandis* is smaller and much more elongated. Fine radiating sculpture at posterior of beak in addition to the five or six concentric doubly looped ridges. Color brown, tinged with green anteriorly and below and rusty brown posteriorly and above. (Pl. XII, Fig. 27).

Genus **Unio.**

Of this genus also only a single species occurs.

Unio complanatus, Solander. Very abundant in sand channels and along muddy or sandy shores of the inner islands or bays. Shells dark brown, no rays, beaks eroded, placed well forward, height behind beaks not greatly exceeding that in front, anterior end rounded, posterior tends to be angled. Ventral margin and hinge margin nearly straight, margin behind hinge curved. Teeth, both cardinal and lateral, single in right and double in left valve. Very variable, shells, narrow or broad, light or heavy; nacre white or purple. In old shells ventral margin tends to become emarginate. (Pl. XII, Fig. 29).

Family SPHAERIIDAE.

There are of this family, in this region, six species belonging to the two genera, *Sphaerium* and *Pisidium*.

Genera **Sphaerium.**

Five of the six species mentioned above belong to this genus.

Sphaerium simile, Say, occurs abundantly, buried in the sand on the slopes of deep pools in sandy channels; occurs also in the mud of sheltered bays. The largest of the family in this district. Beaks nearer anterior end of shell, inflated, closely approximated, beaks marked with coarse lines, lines of growth heavy, regular. Color brown or yellow, often brown with yellow border. Hinge line curved. Hinge slight. Varies in color. Young usually yellow, adult usually dark. Young thin, adult somewhat inflated. (Pl. I, Fig. 11).

Sphaerium striatinum, Lamarck. Abundant in sand banks in channels and in mud in sheltered bays. Shell somewhat inequilateral, beaks full, separated, lines of growth coarse with finer lines between. Beak sculpture not uniform, numerous regular coarse lines, few coarse lines irregularly placed or beak smooth. Shell thin, nacre bluish white with purple bands or patches.

Sphaerium rhomboideum, Say. Sand banks and muddy bays. Shell equilateral. Umbones depressed, approximated, marked by fine lines, lines of growth rather fine, regular; anterior slightly truncated, posterior somewhat angled below. Dark brown, narrow yellow border around margin; nacre bluish white. (Pl. X, Fig. 12).

Sphaerium (Musculium) securis, Prime. Abundant in sandy channels. Small, fragile, much higher in front of umbones which are centrally placed. Truncated behind, rounded in front, rhomboidal in outline, umbones calyculate and inflated, marked by fine concentric lines; lines of growth fine. Found in stomachs of white-fish. (Pl. X, Fig. 13).

Sphaerium (Musculium) partumeium, Say. Identified by Mr. E. G. Vanatta. Sand channels. Shell equilateral, oval in outline, large, truncated behind, color yellow, lines of growth fine.

Genus Pisidium.

One species of this genus occurs here.

Pisidium virginicum, Bourguignat. Abundant in sandy channels. Dark colored. Umbones elevated, placed posteriorly, shell heavy, brown or yellowish, truncated behind, triangular in front. Lines of growth coarse. Cardinal teeth single in right valve, inverted V-shaped; double in left; oblique, anterior narrow, posterior stout, inclined towards pyramidal. Laterals stout, double in right valve, single in left.

ARTIFICIAL KEY TO THE SPECIES IN THIS REPORT.

Since the recognition of these species is not an easy matter the following key based upon shell characters is given to facilitate their identification.

A. 1. Univalve, shell consisting of one valve.

B. 1. Non-operculate, no operculum borne on upper surface of foot and closing the aperture of shell when animal is retraced.

C. 1. Spire elevated and dextral, or flat.

Family LYMNAEIDAE.

D. 1. Spire elevated and dextral.

Genus Lymnaea.

E. 1. Spire elongated.

F. 1. Much elongated and slender, whorls very oblique, shell thin.
Lymnaea haldemani.

F. 2. Elongated but stout; dark colored.
Lymnaea palustris.

E. 2. Spire short.

F. 3. Thin-shelled, body whorl very large, whorls very oblique. Length of shell 15-18 mm.
Lymnaea columella.

F. 4. Shell large, smooth; whorls 5; length of shell 25-30 mm.
Lymnaea stagnalis sanctaemariæ.

F. 5. Shell medium sized, usually malleated; whorls 5; length of shell 20-25 mm.
Lymnaea emarginata candensis.

F. 6— Shell small, smooth; whorls 3; length of shell 10-12 mm.
Lymnaea decollata.

D. 2. Spire flat.
    Genus **Planorbis.**
    E. 3. Shell large, high and sinistral.
      F. 7. With wide shell concavity above.
        Planorbis trivolvis.
      F. 8. With narrow deep concavity above, two carinæ.
        Planornis bicarinatus,
      F. 9. With no concavity above. Expansion behind the aperture.
        Planorbis campanulatus.
    E. 4. Shell small, depressed and dextral.
      F. 10. Shell covered with bristles.
        Planorbis hirsutus.
      F. 11. No bristles on shell.
        G. 1. Peripheral keel level with the top of shell.
          Planorbis dilatatus.
        G. 2. Peripheral keel centrally placed. Shell lens-shaped.
          Planorbis exacuous.
        G. 3. No peripheral keel.
          Planorbis deflectus.
    C. 2. Spire elevated and sinistral or shell not spiral.
  Family PHYSIDAE.
D. 2. Spire elevated and sinstral.
    Genus **Physa.**
    E. 5. Shell large.
      F. 12. No sculpture on surface of shell.
        Physa heterostropha.
      F. 13, Surface sculptured, spire short, suture not impressed.
        Physá ancillaria.
      F. 14. Surface sculptured, spire more elevated, sutures impressed.
        Physa gyrina.
    E. 6. Shell small, usually whitish.
      Physa integer niagarensis.
D. 4. Shell not spiral.
    Genus Ancylus.
    Ancylus parallelus.
B. 2. Operculate, operculum borne on the upper surface of foot and
closing the aperture of the shell when the animal is retracted.
  C. 3. Spire very high, shell large, length 25-30 mm.
    Family STREPTOMATIDAE.
    D. 5. Whorls towards apex not rounded.
      Goniobasis livescens.
    D. 6. Whorls towards apex more or less rounded.
      Goniobasis haldemani.
  C. 4. Spire low or only moderately high.
    D. 7. Umbilicus narrow.

E. 7. Shell small, about 5 mm. in length.
Family AMNICOLIDAE.
F. 15. Shell globoid or low conic; apex rounded.
Amnicola limosa.
F. 16. Shell low, conic, apex emarginate.
Amnicola emarginata.
F. 17. Shell high conic, apex sharp.
Amnicola lustrica.
E. 8. Shell large and heavy. Apex usually eroded.
Family VIVIPARIDA.
Campeloma decisum.
D. 8. Umbilicus wide.
Family VALVATIDAE.
E. 9. Whorls bearing three carinae.
Valvata tricarinata.
F. 10. Whorls without carinae.
Valvata sincera.
A. 2. Bivalve shell consisting of two valves, united by a dorsal hinge.
B. 3. Shell large, one set of cardinal teeth in each valve.
Family UNIONIDAE.
C. 5. Shell heavy, bearing hinge teeth.
D. 9. Height behind beaks not greatly in excess of that in front.
Genus **Lampsilis.**
E. 11. Rays numerous and narrow.
Lampsilis luteolis.
E. 12. Rays few and broad.
Lampsilis ventricosus.
C. 6. Shell light, no hinge teeth.
Genus **Anodonta.**
D. 11. Shell high, no radiating sculpture on posterior part of beak.
Anodonta grandis.
D. 12. Shell low, elongated, radiating sculpture on posterior of beak.
Anodontoides ferussacianus.
B. 4. Shell small, two sets of cardinal teeth in each valve.
Family SPHAERIDAE
C. 7. Not trigonal in outline.
Genus **Sphaerium.**
D. 13. Beak not calyculate.
E. 13. Lines of growth regular, coarse.
F. 19. Shell usually with definite narrow yellow border and rhombic outline.
Sphaerium rhomboideum.
F. 20. Shell without definite yellow border and oval in outline.
Sphaerium simile.

E. 14. Lines of growth not regular, coarse, with numerous fine
   between.
   Sphaerium striatinum.
D. 14. Beaks calyculate.
   E. 15. Rhomboidal in outline.
   Sphaerium (Musculium) partumeium.
C. 8. Shell trigonal in outline
   Genus **Pisidium.**
   Pisidium virginicum.

LIST OF ARTICLES CONSULTED IN THE DETERMINATION OF THE FOREGOING
SPECIES.

1898. BAKER, F. C.—The Mollusca of the Chicago Area. *The Chicago Academy of Sciences,* (*Natural History Survey*); Bulletin No. III, Pt. I. 1898.

1902. BAKER, F. C.—The Mollusca of the Chicago Area, *The Chicago Academy of Sciences,* (*Natural History Survey*); Bulletin No. III, Pt. II. 1902.

1911. BAKER, F. C.—The Lymnæidæ of North and Middle America. *The Chicago Academy of Sciences:* Special Publication No. 3, 1911.

1865. BINNEY, W. G.—Land and Fresh-water Shells of North America, Pts. II and III. *Smithsonian Miscellaneous Collections:* (143, 144) vol. VII, 1867.

1870. BINNEY, W. G.—Report of the Invertebrata of Massachusetts (A. A. Gould, edited by Binney) Boston, 1870.

1901. CRANDALL, O. A.—The American Physæ, *Nautilus,* 1901.

1905. DALL, W. H.—Land and Fresh-water Mollusks. *Harriman Alaska Expedition,* vol. XIII, New York, 1905.

1882. LATCHFORD, F. R.—Notes on the Ottawa Unionidæ, *Transactions Ottawa Field Naturalists' Club,* No. 3, Ottawa, 1881–82.

1865. PRIME, T.—Monograph of American Corbiculidæ. *Smithsonian Miscellaneous Collections.* (145) vol. VII, 1867.

1858. SAY, T.—The Complete Writings of Thos. Say on the Conchology of the United States. Edited by W. G. Binney, New York, 1858.

1900. SIMPSON, CHAS. T.—Synopsis of the Naiades or Pearly Fresh-Water Mussels. *Proceedings of the U. S. National Museum:* vol. XXII, Washington, 1900.

1873. TRYON, G. W., JR.—Land and Fresh-water Shells of North America, Pt. IV, Smithsonian Miscellaneous Collections (253) vol. 16; Washington 1873.

SESSIONAL PAPER No. 39b

EXPLANATION OF PLATES.

PLATE X.

1. Planorbis campanulatus, Say, x2½.
2. Planorbis hirsutus, Gould, x3.
3. Planorbis exacuous, Say, x3.
4. Lymnæa haldemani, (Deshayes) Binney, x3.
5. Planorbis bicarinatus, Say, x2.
6. Planorbis trivolvis, Say, x1½.
7. Lymnæa decollata, Mighels, x3.
8. Lymnæa palustris, Muller, x2.
9. Amnicola limosa, Say, x1½.
10. Amnicola lustrica, Say, x1½.
11. Sphærium simile, Lamarck, x1½.
12. Sphærium rhomboideum, Say, x1½.
13. Sphærium (Musculium) securis, Prime, x3.

PLATE XI.

14. Lymnæa emarginata canadensis, Sowb, x2.
15. Lymnæa columella, Say, x3.
16. Goniobasis livescens, Menke, x2.
17. Ancylus parallelus, Hald, x3.
18. Lymnæa stagnalis sanctaemariae, Walker, x2.
19. Physa ancillaria, Say, x3½.
20. Campeloma decisum, Say, x1¼.
21. Valvata tricarinata, Say, 3½.
22. Valvata sincera, Say, 3½.

PLATE XII.

23. Lampsilis luteolis, Lamarck, x¾.
24. Series, lamellar to pyramidal teeth in Lampsilis luteolis, Lamarck, x¾.
25. Anodonta grandis, Say, x¾.
26. Lampsilis luteolis, Lamarck, x¾.
27. Anodontoides ferussacianus, Lea, x¾.
28. Anodonta grandis, Say, x⅘.
29. Unio complanatus, Solander, x¾.
30. Lampsilis luteolis, Lamarck, x¾.

PLATE X

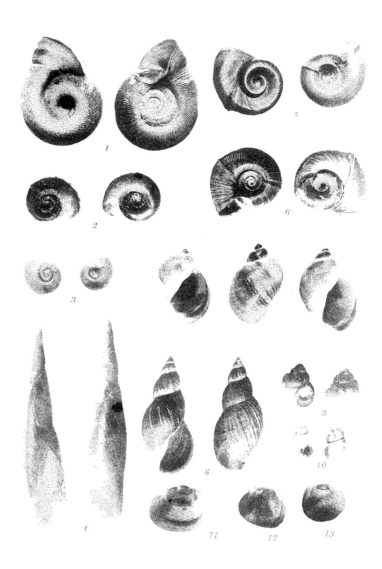

PLATE XI

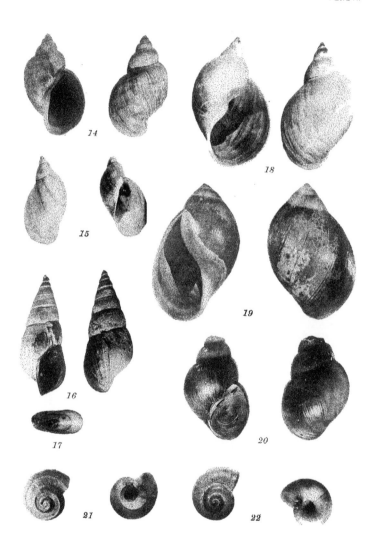

PLATE XII

23

27

24

28

25

29

26

30

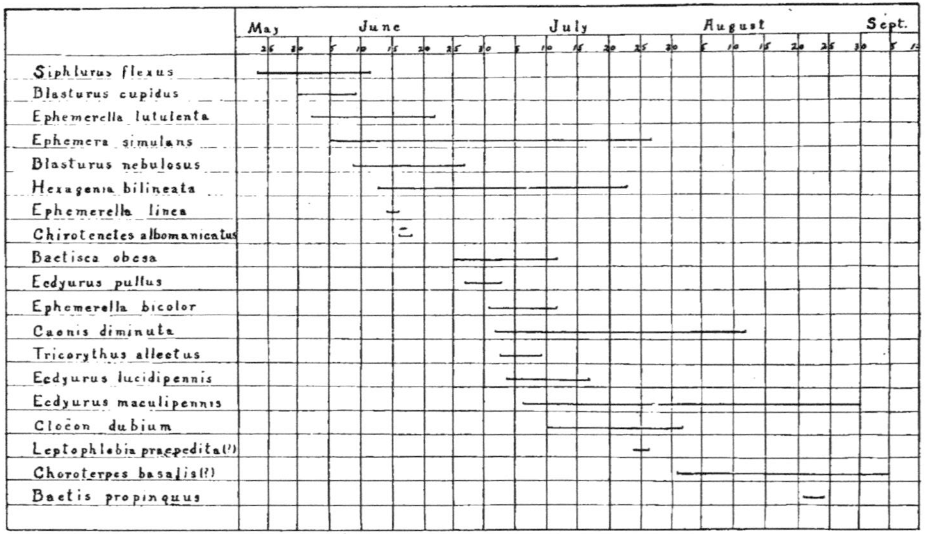

Fig. 1. Plan showing seasonal distribution of adults of Ephemeridæ.

IV.

REARING EXPERIMENTS AND ECOLOGY OF GEORGIAN BAY EPHEMERIDAE.

By W. A. CLEMENS, Department of Biology, University of Toronto.

(Plates XIII and XIV and 1 figure in the text)

The results given in the present paper are based upon a series of observations on the distribution and life histories of various species of this family, which were begun on the advice and under the supervision of Dr. E. M. Walker. Owing to the very imperfect knowledge of these species as they occur in Canadian localities, it was considered desirable to make collections of the local forms occurring in the vicinity of the Biological Station and to conduct breeding experiments to determine the identity of nymphs and imagos, and discover the time of emergence. These insects, as is well-known, are an important source of fish food. In view of the comparative abundance of the species of *Heptagenia* occurring in this region, however, and the exceptional facilities for their study, it was decided to deal with these species in a separate paper which appears elsewhere.

The life histories of comparatively few North American forms, comprising in all about 31, out of a total number of about 114, have been described. The first was that of *Baetisca obesa* Say, by Walsh in 1864. In 1901, Professor J. G. Needham reared and described six species; in 1904 he published the life histories of 11 more, and since then 2 others. In 1903, Mr. Edward Berry described the life histories of 3 forms and in 1911 Dr. Anna Morgan described 8.

The particular species are as follows: Needham (1901, 1904) *Heptagenia pulchella* Walsh; *Baetis pygmea* Hagen; *Siphlurus alternatus* Say; *Caenis diminuta* Walker; *Hexagenia variabilis* Eaton; *Ephemera varia* Eaton; *Chirotenetes albomanicatus* Needham; *Ameletus ludens* Needham; *Choroterpes basalis* Banks; *Callibaetis skokiana* Needham; *Ephemerella bispina* Needham; *Tricorythus allectus* Needham; *Leptophlebia praepedita* Eaton; *Heptagenia interpunctata* Say; *Ecdyurus maculipennis* Walsh; *Polymitarcys albus* Say; (By W. E. Howard); *Ephemerella dorothea* Needham; *Potamanthus diaphanus* Needham; Berry (1903); *Leptophlebia americana* Banks; *Blasturus cupidus* Say; *Callibaetis ferrugineus* Walsh.

Morgan (1911) *Ephemerella cornuta* Morgan; *Ephemerella rotunda* Morgan; *Ephemerella serrata* Morgan; *Ephemerella lata* Morgan; *Ephemerella tuberculata* Morgan; *Ephemerella deficiens* Morgan; *Ephemerella plumosa* Morgan; *Ephemerella spinosa* Morgan; *Iron fragilis* Morgan; *Epeorus humeralis* Morgan.

As for Canadian forms, L'Abbé L. Provancher, in 1877, recorded the following from Quebec; *Ephemera simulans* Walk.; *Hexagenia bilineata* Say.; *Heptagenia terminata* Walsh; *H. canadensis* Walker; *H. quebecensis* Provancher; *Siphlurus alternatus* Say, *Baetis rubescens*, Hagen. In the Monograph of Eaton, 1888, are described the imagos of 21 taken in Canada. The following is a list of the species recorded and the localities from which they were taken. Those marked with an asterisk are recorded from Canada only:

> *Polymitarcys albus* Say; Winnipeg River.
> *Emphemera guttalata* Pict.; Quebec.
> *Ephemera simulans* Walk.; St. Martin's Falls, Albany River.
> *Blasturus cupidus* Say; Nova Scotia.
> *Blasturus nebulosus* Walk.; St. Martin's Falls, Albany River.
> *Ephemerella walkeri* Eaton; St. Martin's Falls, Albany River.
> *Ephemerella invaria* Walker; St. Martin's Falls, Albany River.
> *Baetis rubescens* Hag.; Quebec.
> *Baetis pygmeus* Hag.; St. Lawrence River.
> *Centroptilum luteolum* Müller; St. Martin's Falls, Albany River.
> *Callibaetis hageni* Etn.; Puget Sound.
> *Callibaetis ferrugineus* Walsh; Quesnel Lake, B.C., and Vancouver Island.
> *Siphlurus alternatus* Say; North West Territory and Quebec.
> *Siphlurus bicolor* Walker; St. Martin's Falls, Albany River.
> *Rhithrogena vitrea* Walker; St. Martin's Falls, Albany River.
> *Heptagenia canadensis* Walker; Canada.
> *Heptagenia verticis* Say; St. Martin's Falls, Albany River.
> *Heptagenia luridipennis* Burmeister; St. Martin's Falls, Albany River and St. Lawrence.
> *Heptagenia vicarius* Walker; St. Lawrence River.
> *Heptagenia quebecensis* Prov.; Quebec.
> *Heptegenia basalis* Walker; Lake Winnipeg.

Specimens of many of these are in the British Museum, London, England. These were probably only casual captures and would seem to indicate a rich fauna in our northern inland waters.

I commenced collecting nymphs on May 25 and continued until September 6. The area covered was within a radius of about five miles of the Biological Station Island. Collections of nymphs were made in localities as varied as possible, such as along open shores, in quiet bays, quiet streams, rapids, above and below waterfalls, pools, ponds, lagoons, and in water from fifteen to forty-five feet deep.

The chief method of collecting was that of picking up stones along the shores from water three inches to two feet deep, and picking off the nymphs clinging to them with a pair of forceps, or lifting off the nymphs with the blade of a pocket knife. The dipnet was used in some localities and for deep water a dredge was dropped from the stern of a gasoline launch.

Each collection of nymphs, as it was brought in, was carefully examined under the binocular microscope and the species separated. A number of each species were then transferred to breeding jars and the remainder were killed and preserved in 70% alcohol. Glass battery jars were arranged on the centre table of the laboratory and each fitted up as nearly as possible to the conditions in which the

SESSIONAL PAPER No. 39b

nymphs were found. For instance, for most of the nymphs of the genus *Heptagenia* which for the most part inhabit the swift water, a mixture of earth and sand was placed in the bottom of the jar and a couple of stones to which the nymphs could cling. Sticks were placed in the jars for the nymphs to crawl out upon when ready to emerge and a constant stream of fresh water supplied. For the *Hexagenia* nymphs, which were taken from deep water, the jar was partly filled with mud, which was dredged up in the locality from which the nymphs were taken. This was for the nymphs to burrow in. Only a trickling stream of water was necessary. *Blasturus* and *Caenis* nymphs did not require running water, as they were taken for the most part in ponds, pools and pot-holes in which the water was often almost stagnant. However the water in the jars was changed every day or so. Some dead leaves and twigs were placed in the bottom of the jar, to imitate the natural conditions.

Usually the stones placed in the jars were covered with algal forms upon which the nymphs could feed, but often algal material scraped from the stones was added.

Wire cages were placed over the jars to catch the subimagos as they emerged. It was impossible to set up breeding cages in the open on account of the changes of level of the water in Georgian Bay and because of waves produced by winds, or passing boats. Go Home River was too far from Station Island to be available.

When the subimagos appeared they were transferred to other vessels, where they were kept in an atmosphere very slightly humid and out of the direct sunlight, until their final moult. The imagos were killed with potassium cyanide and then preserved dry or in 70% alcohol. The final nymph slough and the subimago exuvia were both preserved for future reference.

In this way about 180 specimens were bred out. Altogether there were taken 29 species belonging to 16 genera.

The following are the genera represented:

Sub-family *Ephemerinae*
    1. *Hexagenia.*
    2. *Ephemera.*

Sub-family *Heptageninae*
    1. *Heptagenia.*
    2. *Ecdyurus.*

Sub-family *Baetinae.*
    1. *Baetisca.*
    2. *Leptophlebia.*
    3. *Blasturus.*
    4. *Choroterpes.*
    5. *Ephemerella.*
    6. *Drunella.*
    7. *Caenis.*
    8. *Tricorythus.*
    9. *Chirotenetes.*
   10. *Siphlurus.*
   11. *Baetis.*
   12. *Cloëon.*

Dr. Anna H. Morgan was kind enough to identify a number of species for me.

5 GEORGE V., A. 1915

Hexagenia bilineata Say.

(Pl. XIII, Fig. 1).

Nymphs of this species were first taken on June 6, 1912, by dredging in water 15 to 45 feet deep. The bottom was very muddy. These were taken to the laboratory and about ten were placed in a breeding-jar, ¾ filled with soft muck. The nymphs immediately began to burrow, using their fore-legs to displace the mud. They were able to bury themselves in a remarkably short time. At first the gills were left partly exposed and the position of the creatures could be detected by the waving motion of these in the thin mud. They remained this way for a short time, but later on only the round openings of their burrows could be seen.

The first subimago to emerge from the breeding-jar was on July 3, and others followed during July and August. One nymph was still alive in the jar when I stopped my work on September 9th. On June 13th the first subimago was captured at large and from this on a few subimagos and imagos were taken at various times, but not until June 28th did they appear in large numbers. On this date about dusk, a large number of females were discovered flying up and down a long narrow channel between an island and the mainland. They dipped down frequently to deposit their eggs and many fell victims to hungry fish. For a couple of weeks after this, this species appeared in immense numbers. They commenced their flight about three-quarters to half an hour before dark and swarmed about the tree-tops, forty feet high. None were observed after July 23rd. On July 12 I caught a female just after copulation and held her over a jar of water, touching her abdomen to the water occasionally and she deposited a large number of eggs. The water was changed from time to time to keep it from becoming stagnant, and on August 17 a number of very small nymphs appeared. This was a period of thirty-six days.

Description of nymph. Length of body 30-35 mm.; setæ 13-15mm.; antennae 5-6mm. Head yellowish with the dorsal surface between ocelli and between eyes entirely brown, or in some cases lighter along median line and posterior margin. Antennae very hairy at joints of basal halves, while apical halves are entirely bare and become very slender. Margin and base of frontal piece hairy. Clumps of hairs between eyes and bases of antennae, in front of lateral ocelli and posterior to eyes. Mandibular tusks, ¾ length of antennae, upcurved, brown at tips, and with three longitudinal rows of hairs. Prothorax brown for the most part dorsally. Each abdominal segment has a large almost triangular brown area with two light areas within it. These light areas often reduced to mere stripes. Ventrally on segments 6 to 8 a faint median longitudinal dark streak, while on 9th segment there are two lateral streaks. Setæ of about equal length and very heavy at joints for entire length. Gills and legs of the usual Hexagenia type.

Ephemera simulans Walker.

For some inexplicable reason I was unable to find *Ephemera* nymphs at Go Home Bay, although the imagos were very abundant and the shore was strewn with the nymph sloughs. Dredging failed to bring them up, although *Hexagenia* nymphs were dredged up almost everywhere in Go Home Bay. However, Mr.

SESSIONAL PAPER No. 39b

R. P. Wodehouse kindly gave me a number of specimens which he took at Shawanaga Bay, about fifteen miles north of Parry Sound on June 9 in 2 to 8 feet of water; some from the south east shore of Manitoulin Island, June 26th, in water two to five feet, and at Waubaushene on May 31 in 6 to 9 feet of water. Nymph sloughs were taken at Go-Home Bay from June 24 to July 9.

The first imago of this species was taken on June 5th at Giant's Tomb Island, 4 miles south west of Station Island, but none were taken at Station Island until June 21. After this date they became very abundant and remained so until July 27th. The males occurred in fairly large swarms all along the shore. They maintained their position in the air by a dancing motion, at a height of 10 to 35 feet. They appeared shortly before 8 'clock in the evening and continued until dark. When a female appeared among them quite a commotion was noticed. The successful male flying up beneath the female would grasp her around the prothorax with his fore-legs, and, bending up his abdomen, would put his forceps around her abdomen. His setæ usually aided him in securing and maintaining his hold, by being bent up over the female's body. The couple would then go off on a gradual downward slant toward the water, before reaching which the male would disengage himself and fly back to the swarm, while the female would fly out over the water close to the surface and soon begin depositing her eggs, by skimming the water with her abdomen. A peculiar thing was noticed, namely, that the male *Ephemera* frequently attempted copulation with the male *Hexagenia* evidently being deceived by the colour.

Heptagenia.

This proved to be a very abundant and interesting genus and is treated separately elsewhere. The nymphs of eight species were taken and the imagos of all of them reared, three of which proved to be new species. The life histories of none of these have been previously described. Besides these eight, Mr. R. P. Wodehouse gave me several nymphs of another species which he discovered along the east shore of Manitoulin Island, June 26th, 1912. These were not bred, so the species has not been determined.

Genus **Ecdyurus.**

Ecdyurus maculipennis Walsh.

(Pl. XIII, Fig. 2).

The nymphs were quite widely distributed, being common along open stony shores and in rapids. They were taken as follows:

(1) At Station Island, on July 2.

(2) At Giant's Tomb Island on July 14, in a large stony bay commonly called the "Gap," on the west side.

(3) On August 19th at the South Watcher Island, 6 miles from the mainland. This island is about 3 acres in extent and composed entirely of loose stones, with a clump of small poplar, willow and alder trees in the centre, and was the breeding-ground of hundreds of gulls.

5 GEORGE V., A. 1915

(4) In the rapids above Sandy Gray Falls, on August 23rd.

The imagos of these collections emerged on July 6th, 17, August 23 and 30th respectively.   Only a few imagos were taken at large.

Ecdyurus lucidipennis Clemens*

Male imago: (Pl. XIII, Fig. 3).

Measurements: Body 6 mm.; wing 7 mm.; fore-leg 6·5 mm.

Face very slightly obfuscated.   Dorsal surface of head dark brown or reddish. Notum dark brown; sides of thorax and ventral surface light yellow.   Dorsum of abdomen a blackish brown and venter considerably lighter.   Penis lobes and bases of forceps yellow.   Forceps tinged with black.   Setæ: basal half slightly tinged with black, minutely hairy.   Fore femora dark, middle and hind yellowish. Wings hyaline; longitudinal veins slightly dusky, especially costa and subcosta; cross veins entirely colourless.

Female imago:

Measurements: Body 6 mm.; wing 7·5; fore-leg 4.

Thorax and abdomen lighter in colour than male.

Nymph:

Measurements: Body 7-8 mm.; setæ 3-4 mm.

Head brown with numerous light spots, chief of which are 6 along anterior margin; 2 lateral to each antenna, 4 elongated ones between antennæ and 2 small round spots anterior to these latter.   Thorax lighter brown with numerous light areas.   Anterior part of each abdominal segment brown.   Four light spots along anterior margin, one large spot at each lateral margin and 3 along posterior margin. Setæ of about equal length and fringed with hairs; middle one slightly smaller in size than lateral ones.   Femora flattened, fringed with spines along anterior margin and with hairs along the posterior; rather light in colour with two zigzag brown marks about middle and brown areas at distal and proximal ends.   Tibiæ banded about the middle with brown.   Tarsi with distal and proximal ends dark.

Nymphs of this species were collected at Station Island, July 1, and at Giant's Tomb Island, July 14th.   Imagos were reared from these collections on July 4 and July 17 respectively.

Ecdyurus pullus Clemens†

Male Imago: (Pl. XIII, Fig. 4).

Measurements: Body 10-11 mm.; wing 11 mm.; setæ 22 mm.; fore-leg 11-12 mm.

*Clemens, '13, p 329.
†Clemens, '13, p. 330.

Face pale, slightly tinged with brown along the carina. Dark brown on dorsal surface of head between eyes. Pronotum dark brown; mesonotum lighter; a dark brown line on each side of prothorax, extending forward from base of fore wing; other dark brown marks at bases of wings and legs. Dorsal surface of abdomen dark brown, somewhat lighter laterally toward anterior margin. Ventral surface light in colour. Genitalia of usual *Ecdyurus* type. Legs light in colour, dark at joints. Tarsi of fore legs in order of increasing lengths 1, 5, 4 (3 and 2) equal. Wings with longitudinal and cross veins brown, and very slightly darkened in apical costal region.

Nymph:

Measurements: Body 12 mm.; setæ 15.

Head brown with a colourless area on each side from eye to lateral margin of head and 3 light dots between eyes; slightly fringed with hairs along anterior margin. Pronotum somewhat lighter in colour than head, colourless areas along anterior and lateral margins and a light area about the middle of each half of pronotum. Mesonotum darker with numerous light spots. Each segment of abdomen brown; 1-8 have 6 light spots; on segments 4-8 the 2 near the median line are fused, forming a large, almost rectangular spot; segment 9 with only 4 light spots; segment 10 entirely brown. Gills comparatively small; lamellæ oval. Setæ of about equal size, with each 2 alternate segments brown; sparsely fringed at joints; outer margins of lateral ones not fringed. Femora stout and flattened, brown in colour; lighter at distal and proximal ends and 2 or 3 irregular light areas toward middle; covered with minute spines and fringed along posterior margin with hairs. Tibiæ alternately light and dark banded, fringed along both anterior and posterior margins. Tarsi brown with proximal tips colourless. Ungues double on each leg; the large one well curved; the other small and lateral to the large one.

The nymphs were collected along the very stony shores of islands three and four miles out in the open bay, from June 23 to July 6. Imagos were reared on July 2 and a few captured June 27th.

In the key to the genera of Mayflies of North America by Professor Needham in Bulletin 86, New York State Museum, there is a slight error in the separation of the genera *Ecdyurus* and *Heptagenia*. In *Ecdyurus* the basal segment of the male fore tarsus is shorter not longer than the fifth segment and the second and third segments of equal lengths. In *Heptagenia* the basal segment of male fore tarsus is longer than the fifth segment and the second and third segments may be equal or unequal.

Baetisca obesa Walsh.

This very interesting nymph was taken in only two localities. The one was along the north east shore of Giant's Tomb Island. This shore is quite sandy with numerous small stones and deepens very gradually. The nymphs were abundant here May 26, clinging to the stones in water from 3 to 15 inches deep. Some

5 GEORGE V., A. 1915

of these were put in breeding jars, but did not emerge until July 13. On July 14 I visited this place again but could not find a single specimen, nor any sloughs along the shore. The other locality was the south east shore of Station Island, but the nymphs were not abundant. Only one imago, a female, was captured.

Leptophlebia (?) praepedita Eaton.

The only representative of this genus was a single almost mature nymph taken on July 21st in quiet water at the side of an old lumber chute. I was unsuccessful in breeding it and so am doubtful as to the species. It agrees with the description by Professor Needham, Bulletin 86, N.Y. State Museum, but this description is rather more generic than specific.

Genus Blasturus.

Blasturus cupidus Say.

This is an early species. Nymphs were first taken May 23. Subimagos appeared May 31 and transformed next day. The imagos were never very abundant and were captured around Station Island only. The last observed was June 9.

A small nymph collected May 31 was observed to be filled with small oval brownish bodies. These, upon dissection by Mr. A. R. Cooper, were found to be a trematode of the genus *Halicometra* and its eggs. Another nymph taken some time afterwards was also discovered to be parasitized.

Blasturus nebulosus Walker.

The nymph and imagos of this species were first taken June 9, on a small bare granite island, a short distance out in the open bay. On the top of this island were numerous pot holes of all sizes filled with water, and in these, under loose pieces of rock and some rubbish, the nymphs were very abundant, having tadpoles, chironomid larvæ and water beetles for associates. Many were covered with *Vorticella*. Several nymphs were seen to crawl out of the water and transform on the rock. Subimagos were clinging to the sides of the rocks in sheltered places while a few imagos were flying above the pools.

This species was again taken on June 27th on an island 5 miles from the mainland. This island had an area of about 3 acres and was almost smooth bare granite. On top was a pretty lagoon margined with water plants, shrubs and a few small trees. Imagos of *B. nebulosus* were dancing over this pond in the sunlight about 3 p.m., matings frequently occurring. A few nymphs were taken from the lagoon.

Up to the present time I have not been able to find any difference between the nymphs of these two species, but am adding a description of the nymph of *Blasturus nebulosus*.

SESSIONAL PAPER No. 39b

Nymph:

Measurements: Body 9·5 to 10 mm.; setæ, 7-10 mm.

General colour blackish brown.   Head with a dark area behind middle ocellus and between lateral ones; black, scroll-like markings between the eyes. Prothorax has a small light spot on each side, close to median line and near anterior margin; posterior to this and, farther from the median line is another larger oval light spot.   Lateral to this is an elongated light area, beyond which is the light rounded lateral margin of the prothorax.   Abdomen is blackish brown, with light brown markings.   Segments 5 or 6 to 10 have a light median longitudinal stripe.   On each segment is a slightly elongated incurved small light spot on each side of median line toward the anterior margin of the segment; posterior to this and more lateral is a larger round light area, which disappears usually on segments 8, 9 and 10.   Ventral surface is light brown with three faint dark longitudinal lines, one median and two lateral.   On each side of the median line in each segment is a very small, white oblique line near anterior margin and posterior to this is a small, white dot.   Median seta shorter, slenderer and lighter in colour than the lateral ones.   All fringed with hair at joints.   Legs light brown.   Posterior margin of tibia and tarsus fringed with hairs; anterior margin of femur fringed with spines, while anterior margin of tibia and tarsus have numerous serrated teeth.   Inner margin of ungues with a row of teeth for its entire length.

Choròterpes (?) basalis Banks.

This is a late summer form.   When I was beginning to think I had exhausted the collecting ground, I discovered this form in a small creek which formed the outlet of a chain of small lakes and which I had not visited for a month and a half. Large numbers of the nymphs were found here, July 30, clinging to stones in the quiet water.   The next day several imagos emerged.   As late as September 5th mature nymphs could be found.   On July 31 a few nymphs were taken at Station Island and imagos on August 19th.

This later appearance of imagos at Station Island was noted also in the case of *Heptagenia tripunctata*.   Mature nymphs of this species were taken in this creek May 31 and imagos emerged June 2, whereas no imagos appeared at Station Island until June 11th.   This was probably due to the lower temperature of the water of Georgian Bay.

Genus **Ephemerella**.

Ephemerella lutulenta Clemens.[*]

Male imago:

Measurements: Body 8-9 mm.; wing 10 mm., setæ 12-14; fore-leg 8.

Face dark brown; a spotted reddish gray streak down carina and 2 similar

[*]Clemens, '13, p. 335.

5 GEORGE V., A. 1915

lateral streaks from it to the base of antennae. Thorax dark reddish brown. Abdomen blackish brown; segments 9 and 10 slightly lighter in colour. Venter pale. Posterolateral margin of 9th segment produced into spines. Forceps pale with tips brown. Setæ reddish brown towards base but becoming pale toward tip; joinings brown. Legs greenish yellow, ungues brown. Segments of fore tarsi in order of increasing lengths 1, 5, 4, 3, 2; 1 very small; fore femur about 5/6 length of fore tibia. Wings entirely clear.

Female imago:

Measurements: Body 9-10 mm.; wings 10; setæ 10-12; fore-leg 5.

Quite similar to male. Posterolateral projection of 9th abdominal segment not as long as in male. Ninth segment ventrally produced posteriorly into a truncated triangular plate, with end emarginate.

Nymph:

Measurements: Body 10-11 mm.; setæ 6-7.

A large species, with colour varying from a dirty brown to a deep blackish brown, often of a granular appearance. Body and legs hairy. Head with a pair of occipital tubercles of varying size; in the male sometimes obscured by the developing eyes of the imago. Pronotum rectangular. Abdominal segments 2-9 produced laterally into flat spines; none on segment 1, minute on 2, increasing in size to the 9th; none on segment 10. A double row of spines on dorsal surface, very minute on segment 8-10, large on 1-7. On venter 6 small black dots on each segment, sometimes very faint. Rudimentary gills on segment 1; gill on segments 4-7; a large jointed elytroid gill cover 1·5 mm. in length. Femora stout, brown in colour with numerous round white dots and several irregular light areas. Tibiæ with median brown band, distal ends light, proximal ends dark. Tarsi about same length as tibiae and with proximal half dark and distal half light. Claw with numerous pectinations. Setae well fringed with hairs along middle, almost bare at base and tip. Each 2 alternate segments brown.

The nymphs were taken almost everywhere about Go Home Bay from May 29th to June 19th. Mr. R. P. Wodehouse has also given me specimens from various places around Georgian Bay including Shawanaga Bay, Pentecost Island, French River, Sturgeon Bay.

Ephemerella lineata Clemens.[*]

(Pl. XIII, Fig. 5).

Female imago:

Measurements: Body 9 mm.; setæ 14; wing 10·5 mm.

Very similar to female of *E. lutulenta* but has a distinct rusty brown median longitudinal stripe on dorsal surface of abdomen. In a fresh specimen the stripe would probably extend over the thorax and thus correspond to the stripe of the nymph.

[*]Clemens '13, p. 336.

Nymph:

Measurements: Body 10 mm.; setæ 6 mm.

Slightly smaller than *E. lutulenta*, but very similar in colour, except that there is a dorsal median longitudinal white stripe from the interior margin of pronotum to the posterior margin of 10th abdominal segment. This stripe lies between the double row of spines on the abdomen. Occipital tubercles slightly longer than those of *E. lutulenta*.

The nymphs of this species were not very abundant and were found in about the same localities as *E. lutulenta* from June 3 to July 9. My bred specimens are dated June 14th and June 15th. I was unsuccessful in rearing a male.

Ephemerella bicolor Clemens.*

(Pl. XIV, Fig. 1).

Male imago:

Measurements: Body 5–6 mm.; wing 6mm.; setæ 8–9; fore-leg 6.

A small wholly brown species. It is very similar to *E. lutulenta*, in form and structure and apparently there are no satisfactory characters by which to distinguish it, except its size.

Female imago: slightly larger than male.

Nymph:

Measurements: Body 6–6·5mm; setæ 3mm.

These nymphs show a great variation in colour pattern. The light coloured specimens are of a dirty white colour with brown markings. Head for the most part brown, slightly paler towards posterior margin. Sides of pronotum brown; anterior margin of mesonotum brown and a brown area at posterior margin between the wing pads. Anterior halves of abdominal segments 2 and 3 brown and slight marks on 4th segment; brown areas on 6 and 7 about the median line, and on segment 9, there are 2 small brown dots at anterior margin and a rather semi-circular brown band posteriorly. Some specimens are almost entirely brown and between these two extremes the amount of brown and white varies. A few specimens, especially females, show a slight indication of tubercles but they are never large as in the preceding species. A double row of spines on abdominal segments 1–7. Posterolateral margin of 3–9 produced into broad flat spines. Gills on segments 4–7, covered by a large jointed elytra. Setæ light brown basally, becoming paler distally; well-fringed with hairs; joints brown. Legs rather small; femora stout; colour for the most part brown, divided into 2 areas; the proximal one large and contains a rectangular white spot; the distal one smaller and contains a perfectly round white dot. Tibiae brown at proximal end and a brown band near distal end. Tarsi with a brown band toward proximal end; claws dark and pectinated.

The nymphs were everywhere abundant, especially along the open shore of Station Island. I have them also from Rattlesnake Harbour, Gray Island, Giant's Tomb Island, and Musquash River. The dates are from June 3 to July 9. Imagos were captured and reared from July 1 to July 12th.

---

*Clemens, '13, p. 336.

5 GEORGE V., A. 1915

Genus Drunella.

I have two nymphs of this genus, identified for me by Dr. Morgan, but as I have not reared any imagos, I think it advisable not to describe the nymphs at the present time.

Caenis diminuta Walker.

This little nocturnal species came to the lamp in the reading room for the first time on July 2, and was taken as late as August 12th.

The nymphs are quite abundant in shallow, almost stagnant pools and lagoons from June 5 to July 30. I have them from various places around Georgian Bay.

Tricorythus allectus Needham.

The nymph was dredged up from a slightly sandy bottom in water 5 to 15 feet deep on Sept. 3. They were not reared, but imagos were taken July 3 and 9.

Chirotenetes albomanicatus Needham.

On June 16 I found a nymph slough at Sandy Gray Falls on the Go Home River but was unable to find either nymphs or imagos. I did not get up to the falls again until August 23 and then found the numerous small nymphs of the next generation.

Siphlurus flexus Clemens.*

Two beautiful *Siphlurus* nymphs were taken early in the season but both died before time of emergence. The first was found May 25th in the bottom of a canoe when some water was being emptied from it. The other was found June 3 beneath a stone in about one and a half feet of water along the open exposed shore of Station Island. Quite a number of imagos, apparently *Siphlurus*, were captured about this time and it seemed quite probable that they were the same species as the nymphs; and I think I have proved this quite conclusively by the wing venation. The wing of the imago has a very characteristic bend in Cubitus 2 at the base and the wing pad of the nymph shows this bend very distinctly. Again, the imago apparently has claws like an *Ameletus*, the two on one leg being unlike, and this can be made out in one nymph distinctly, due to the nymph dying just when about to emerge.

Male imago:

Measurements: Body 13–14 mm.; wing 12–13; setæ 23–24; fore-leg 12–13.

Head blackish brown except lower part of face, which is hyaline, tinged with brown; eyes large, meeting dorsally. Notum blackish brown. Sides of thorax marked irregularly with white. Abdominal segments 1, 8, 9 and 10 dark, segments 2–6 lighter in colour; these are light toward anterior margin and brown toward posterior; in the median line the brown is dark and forms a triangular area, the

---

*Clemens, '13, p. 338.

apex extending almost to the anterior margin; from the anterior margin in the median line, 2 bands arise, composed of black dots, which pass backwards curving outwards and ending near the base of the triangular brown area; between this line and the triangular area is a light brown oval area; segments 7–10 almost entirely blackish brown dorsally, but 7 and 8 have triangular white areas on sides, and 9 a slight indication only; segment 10 has sides of dorsum white, ventrally segment 1 dark brown; remainder white with brown markings; segment 2 has 2 brown spots, 3 with 2 smaller brown spots and a slightly reddish area at anterior margin in median line; on 4 and 5 the brown spots become smaller and the reddish area larger; segment 6 the reddish area is elongated to the posterior margin; segments 7 and 8 have a median longitudinal brown line, thickened about the middle, and 2 dots of unequal size on each side of it; segment 10 brown except for a lateral white streak on each side. Forceps white; 4 jointed; setæ white with brown joints, minutely pubescent. Fore-legs brown; femur with a light area near distal end, lateral to which is a dark brown band; tarsi with segments 1, 2 and 3 about equal in length, 4 slightly shorter, and 5 about half the length of 4. Hind legs lighter in colour than fore; a brown band on femur is distal half; tibia with a brown band about middle; tarsus light but brown at joints; joint between tibia and tarsus 1 not distinct. Claws unlike. Wings with brown neuration; costal cross-veins and others towards base of wing margined more or less with brown; slightly clouded in apical costal area; a heavy brown cloud at bulla; often a small cloud at bifurcation of median vein; cubitus 2 strongly bent at base. Hind wing with a large brown cloud at base.

Nymph:

Measurements: Body 15 mm.; setæ 5 mm.

I have two of these graceful nymphs, a male and a female, both mature, but unfortunately both died when just about to emerge. On this account it is difficult to describe the colour pattern as the body of the subimago shows through the nymph skin.

Head vertical; body curved. Posterior lateral margins of abdominal segments 1–9 produced into spines. Dorsal colour pattern distinct on segments 9 and 10 only; 9th segment pale with a short brown median longitudinal stripe, commencing at anterior margin; on each side of this is a short stripe of about the same length, but placed more posteriorly; lateral to this again is a large brown area, roughly triangular, apex at posterior margin, base at anterior; at lateral margin slightly below middle line is a small brown spot; on 10th segment is a median brown longitudinal stripe with 2 dots on each side of it. Ventral surface of abdomen white with 3 longitudinal brown stripes, one median and 2 lateral. Gills on segments 1–7; double on 1, 2 and 3. Three setæ of equal length; lateral ones fringed with hair on inner margins only except towards tips; in these specimens the lateral setæ are brown, lighter towards tips, while the median one is whitish; setæ banded toward distal end with brown. Legs pale; femur with proximal end brown and a brown band beyond middle; tibia with a brown band about the middle; tarsus with brown band towards proximal end; fore tarsus much longer

5 GEORGE V., A. 1915

than fore tibiae; fore tarsus only slightly longer than hind tibiae; fore claw rather short, broad and bifid at tip; hind claws about twice length of the fore and very pointed.

Imagos were captured on the following dates; May 23, May 26 and June 12th. On the latter date a swarm or 12 or 15 individuals was observed flying off the west shore of Island Station from 12 to 20 feet from the surface of the water at 5.30 p.m. About 8 of these were taken.

Baetis propinquus Walsh.

The imago is described in Eaton, but my specimens do not show the subopaque area between the 2 nervures of the hind wing.

Nymph:

Measurements: Body 6 mm.; setæ 2.

Face vertical, mostly brown in colour; on dorsal surface of head on each side of median line is a row of irregularly shaped light spots.   Notum brown with various light areas.   Dorsum of abdomen for the most part brown; segments 2-4 brown with a light area in each half of segment and colourless margins; on segment 4 there is also a light area in median line; segment 5 quite light in colour; segment 6 brown with a light area along anterior margin and 2 faint ones posterior to it; segments 7 and 8, each with two rather large pale areas in posterior half; segment 9 almost entirely pale; segment 10 slightly brown, especially along posterior margin; on each side of the brown segments there are 2 small faint, pale, oblique, slightly curved streaks and a pale dot posterior to each.   Ventrally the joinings of segments brown. Setæ slightly tinged with brown, with tips darker brown and a brown band beyond the middle; lateral setæ fringed on inner sides only. Legs pale; femora banded with brown about middle; tibiæ and tarsi darker toward distal ends; each claw with a lateral row of pectinations.

Nymphs of this species were taken at Go Home Bay from June 14 to July 22; on August 19 large numbers of them were discovered in a little bay of a small bare island about three miles out in the open.   This rock was the home of numerous gulls and hence is commonly called "Rookery Island."   The nymphs were mature and imagos emerged on August 21 and 22

Cloëon dubium Walsh.

The imagos I have agree with the description in Eaton, except that the intercalar veins are single, not in pairs.   Probably the description is in error as the genus *Cloëon* typically has the intercalar veins single.

Nymph:

Measurements: Body 4-4·5 mm.; setæ 1·5.

Face vertical with 2 large pale areas above antennæ; between eyes a large pale area partly divided into 2 and containing 2 brown stripes. Notum brown with irregular light areas. Dorsum of abdomen brown except lateral margins which are colourless; on each segment there are 2 small oblique pale streaks and 2 round dots posterior to the streaks. Setæ pale with brown band toward distal end; lateral setæ fringed on inner sides only. Gills double, apparently on segments 1 and 2 only; broader than gills of *Bætis;* a main trachea in each, slightly to outer side and branchlets on inner side only. Legs pale; femora banded with brown in distal half; tibiæ and tarsi brown toward proximal ends; claws comparatively long, sharp-pointed, and not pectinated.

The nymphs were not very abundant; my collections date from July 30 to Aug. 12.

Imagos were reared July 30 and August 2. Adults were quite numerous at Station Island about July 10, flying in small swarms along the shore, at a height of from 10 to 15 feet. They appeared about 7·45 in the evening.

This paper and the following one on the genus *Heptagenia* contain the results of but a few months collecting and rearing. The complete life histories of 9 new species were secured and the hitherto unknown nymphal stages of 9 other species determined by rearing. Besides a few observations on the habits of several species have been recorded. The results may be taken as an indication of the richness of our inland waters in aquatic insect life.

I am adding a diagram showing the length of time imagos of these species were seen, captured or bred. I find in a number of instances that the dates are somewhat later than those given for the same species at Fall Creek, Ithaca, New York.

REFERENCES.

1877. L'Abbé L' Provancher.—Faune Entomologique du Canada, et Particulièrement de la Province de Québec. Vol. II.

1888. Eaton, Rev. A. E.—A Revisional Monograph of Recent Ephemeridæ or Mayflies. *Trans. of the Linnæan Soc., Second Series, Vol. III, Zoology, London, 1883.*

1901. Needham, J. G.—Aquatic Insects in the Adirondacks. *New York State Museum, Bulletin* 47, 1901.

1904. Needham, J. G.—Mayflies and Midges of New York. *New York State Museum, Bulletin* 86, 1904.

1911. Morgan, Anna H.—Mayflies of Fall Creek. *Annals of the Entomological Society of America, Vol. IV, No. 2, 1911.*

1913. Clemens, W. A.—New Species and New Life Histories of Ephemeridæ or Mayflies. *Can. Entomologist,* Vol. XLV, Nos. 8 and 10.

MARINE AND FISHERIES

EXPLANATION OF PLATES.

PLATE XIII.

Fig. 1. *Hexagenia bilineata* Say.
Fig. 2. *Ecdyurus maculipennis* Walsh.
Fig. 3. *Ecdyurus lucidipennis* Clemens.
Fig. 4. *Ecdyurus pullus* Clemens.
Fig. 5. *Ephemerella lineata* Clemens.

PLATE XIV.

Fig. 1. *Ephemerella bicolor* Clemens.
Fig. 2. *Bœtis propinquus* Walsh.
Fig. 3. *Cloëon dubium* Walsh.
Fig. 4. Venation of wing pad of *Siphlurus flexus* Clemens.
Fig. 5. Wings of *Siphlurus flexus* Clemens. '
Fig. 6. Fore-claw of nymph of *Siphlurus flexus* Clemens.
Fig. 7. Fore-claws of imago of *Siphlurus flexus* Clemens.

PLATE XV

Fig. 1

Fig. 2

Fig.3

Fig.4

Fig.5

PLATE XVI

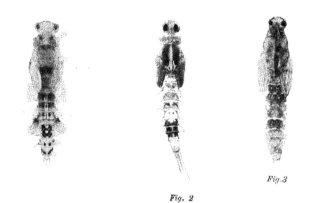

Fig. 2

Fig.3

Fig. 1

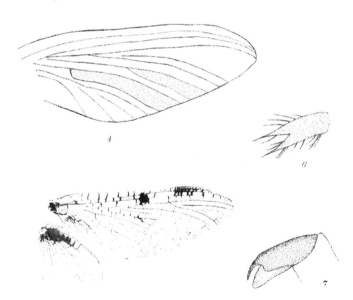

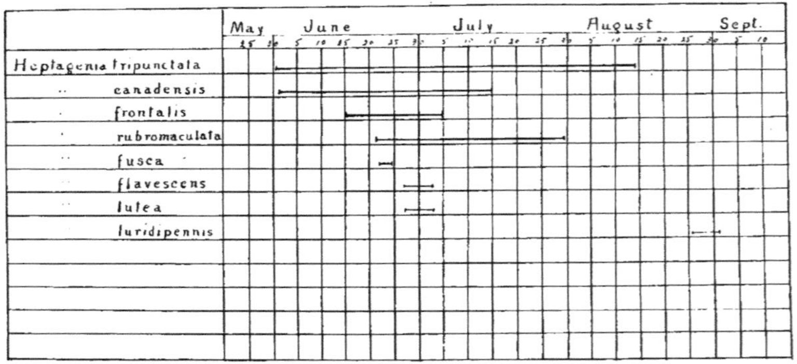

Fig. 1. Plan showing seasonal distribution of *Heptagenia*.

V.

LIFE-HISTORIES OF GEORGIAN BAY EPHEMERIDAE OF THE GENUS HEPTAGENIA.

By W. A. Clemens, B.A., University of Toronto.

(Plates XV—XVIII, and 1 figure in the text)

In the preceding paper on the mayflies or Ephemeridæ of Georgian Bay, the consideration of the genus *Heptagenia* was omitted for reasons there stated. The present paper is based on the results of observations and breeding experiments in connection with this genus covering a period of slightly over three months during the summer of 1912.

The genus *Heptagenia* is a comparatively large one as it occurs in America. Rev. A. E. Eaton in his Monograph of Recent *Ephemeridæ* made a summary of the then known North American species, amounting in all to at least 13, some of them, however, having been referred tentatively to the genus *Ecdyurus*. In 1910 Mr. Nathan Banks described 4 new species, making a total of 17 species recorded from America. Up to the present the nymphs of only 2 of these have been described namely. *Heptagenia pulchella* Walsh, and *H. interpunctata*, Say, both by Professor J. G. Needham in 1901 and 1904 respectively. In this paper are given the descriptions of the nymphs of five more as well as descriptions of the nymph and imagos of 3 new species.[*]

The nymphs of this genus inhabit swift water for the most part: clinging close to the sides and bottoms of stones. They are adapted to this life by reason of many interesting specializations, chief of which are, much-flattened bodies, flaring margins to head, spreading legs with flattened femora, pectinated claws, gills dorsally placed in an overlapping series and spreading setæ. A few species, however, are common in quiet water, notably *Heptagenia canadensis*, and *H. frontalis* while *H. tripunctata* was found to be everywhere abundant. The nymphs are quite active, for when a stone is lifted from the water they scurry over its surface usually seeking the lower side. The clinging habit was frequently demonstrated when quite a number were brought in and put in a vessel of water without a stone or stick for them to cling to; not having anything else, they would begin clinging to each other and soon would all be in a single mass. As for food, being herbivors, they usually find abundance of various algal forms on the stones to which they cling.

A *Heptagenia* completes its life-cycle in a year. The egg is deposited in the water and hatches in about 40 days. The remainder of the mayfly's life is spent in the water as a nymph with the exception of a short aerial life of from 2 to 4 days as a subimago and imago. As the time of emergence approaches, the nymphs probably migrate to the quieter water. I have not observed a *Heptagenia* emerge

---

[*]Since the above was written these new species have been described by the writer (Clemens) '13.

5 GEORGE V., A. 1915

in the open, but in the laboratory they were observed to crawl up the sticks placed
in the breeding-jar for the purpose and transform just above the water-level.
The subimago stage generally lasted a day, but in the early part of the season it
quite frequently lasted 3 days, and in a couple of instances 4 days. No doubt
this time would have been shortened had the subimagos been out of doors. The
imagos never appeared in large swarms as in the case of *Ephemera* and *Hexagenia*,
but a swarm would consist of perhaps 50 to 100 individuals. They would begin
their flight from three quarters to one half an hour before dusk, dancing up and
down in their rhythmic manner at a height of from 12 to 20 feet. On calm evenings they could be found in numerous swarms all along the shore of the island,
but on windy evenings would congregate on the lee side. The females of all the
species observed at Station Island deposited their eggs by skimming the surface
of the water and brushing off the eggs as they appeared from the openings of the
oviducts. The earliest species was *H. luridipennis*, mature nymphs of which
were taken on the afternoon of May 31, and one subimago emerged the same
afternoon. The last was *H. luridipennis*, the imagos emerging Sept. 2, from
nymphs collected August 23rd.

The following are the generic characteristics of the *Heptagenia* nymph:

Body flattened; head orbicularly rounded with flaring margins; eyes dorsally
placed; postero-lateral angles of abdominal segments produced into spines; femora
flattened; gills on segments 1 to 7, placed dorsally in an overlapping series and in
life move in waving undulations; lamellae oblong or oval pointed, the 7th small
and lanceolate. Branchial filaments bifid and united basally into a flat triangular plate. Setæ from one to one-and-one-half times length of body, spreading,
fringed with hairs at the joints of the segments. Mouth parts—labrum with
width nearly twice the length and a row of short spines along the ventral surface,
just inside the anterior margin. Anterior margin densely fringed with hair.
Mandibles rather triangular in shape; fangs two in number, the exterior one of
the right mandible stouter than inner and separated along inner edge; the inner
fang bifid at tip. Mandible fringed with hair along the exterior margin. Lacinia
of first maxilla externally rounded, the anterior part being beset with spines and
hairs. The internal margin beset with a very dense even row of hairs and fine
bristles and several spines at upper corner. Palpus 3-jointed, basal one small,
middle one stout, distal longer and more slender ending in a curved tip; a row of
short spines near the apex. Palpus hairy along outer and inner margins.

Labium with two pairs of lobes. The outer oval and densely covered with
hairs; the inner more slender, more pointed and incurved; also hairy. The anterior end of the distal segment of the palpus densely beset with long hairs and
sharp pointed projections with teeth along inner sides, somewhat resembling a
rake. Beneath this crown is a chitinized ridge. Hypopharyax with a triangular
tongue; paraglossae extend outwards with ends curved slightly backwards.

Generic characters of the imago:

Fore leg of male as long or slightly longer than body. The lengths of male
fore tarsi arranged in order of increasing lengths are 5, 1, 4, 3, 2; 3 and 2 equal

in some forms. Eyes simple; large, especially in male. Antennae short, 1 to 1·5 mm. Setæ 2½ to 3 times length of body; segments of basal half alternately darkened; minutely pubescent. Penis lobes extend backwards and outwards, almost L-shaped. In some forms the lateral extension lacking and hence rather oblong in shape. Stimuli adjacent between the lobes.

The species which I have taken fall into two distinct groups:—

In the first group, consisting of *H. tripunctata, H. luridipennis, H. flavescens, H. rubromaculata, H. fusca, H. Lutea*, the nymphs are characterized by having the lamellae of the gills oblong, claws usually pectinated, the distal segment of maxillary palpus thickest about its middle and with a small tuft of hair near its end. The body is much flattened and the colour more or less brown. The male imagos have the penis-lobes L-shaped and the 2nd and 3rd tarsal segments of fore leg are equal, while the fourth is about 4/5 the length of the 2nd.

In group 2, consisting of *H. canadensis, H. frontalis* and a third undetermined species represented by the nymph only, the nymphs have the lamellae of the gills oval and produced distally into a sharp point; claws are not pectinated, the distal segment of maxillary palpus thickest toward distal end and the tuft of hair is larger than in group 1; usually there are more teeth along anterior margin of lacinia. The body is less flattened and more reddish or yellowish. The male imagos have the penis lobes oblong, not L-shaped and the 2nd and 3rd tarsal segments not quite equal, while the 4th segment is about ½ the length of the second.

Key to Male Imagos

A. No black spots or bands on face below antennae, Group 1.
  B. Very pale species.
      C. Notum ferruginous, stigmal dots distinct............H. flavescens.
      C.C. Notum lighter; no stigmal dots........................H. lutea.
  B.B. Dark species.
      D. Large, entirely brown species.
          E. Thorax with a broad dark median stripe or two narrow stripes close together..................................H. verticis.
          E.E. Thorax without dark median stripe.................H. fusca.
      D.D. Not entirely brown
          F. Two very small dots on median carina between antennae ....H. luridipennis.
          F.F No dots on median carina. Thorax and top of abdomen dark.
              G. Reddish area in pterostigmatic space of wing. ....H. rubromaculata.
              GG. Without reddish area in wing...........H. luridipennis.
AA. Two black spots or bands on face below antennae, Group II.
          H. A black band on face below antennae, a dark dash in wing. Abdomen dark.........................H. canadensis.
          HH. A black spot on face below antennae, no dash in wing, abdomen lighter...........................H. frontalis.

39b—9½ II

5 GEORGE V., A. 1915

Key to Nymphs.

A. Gills oblong, Group I.
   B. Nymphs entirely brown without a distinct dorsal colour pattern.
      C. An inverted dark U-shaped mark on ventral surface of 9th segment and a dark spot on ventral surface of 8th. Dorsal surface of body has a smooth appearance..............................H. flavescens.
      CC. A row of dark mushroom-shaped marks along ventral surface and a rectangular dark mark on 9th. Dorsal surface has a rather granular appearance and lateral margins of body quite hairy. H. rubromaculata.
   BB. Nymphs not entirely dark brown and have a distinct colour pattern.
      D. Ventral surface of abdominal segments banded with dark bands along posterior margins.
         E. Broad dark bands at posterior margin of each segment on dorsal surface...H. fusca.
         EE. Dark bands at posterior margins of segments 7, 8, 9, and 10, not as broad as preceding species and a more elaborate colour pattern..H. lutea.
      DD. Ventral surface not banded.
         F. Two rows of black dots along ventral surface of abdomen.
         ....H. luridipennis.
         FF. No dots..................................H. luridipennis.
AA. Gills oval and pointed, Group II.
   G. Two light longitudinal stripes on dorsal surface of abdomen close to median line.
      H. Stripes fairly uniform for entire length. Reddish species..H. canadensis.
      HH. The stripes not of uniform width, very wide on 8th segment, very narrow on 5, 6, and 7 so that darker intermediate parts have oval shapes. Lighter species.......................H. frontalis.
   GG. Dorsal surface of abdomen has appearance of 3 longitudinal dark stripes. Colour greenish yellow................H. (undetermined species).
   Mr. Nathan Banks kindly identified the imagos for me so far as possible.
Description of Species.

Heptagenia flavescens Walsh.

(Pl. XV, Figs. 4–5).

Male imago (Description taken from Monograph of Eaton, '88):
   Measurements: Body 9+mm.; wings 11+mm.; setæ 27-38 mm.
   Yellowish. Eyes bright greenish yellow during life. Notum ferruginous, sometimes verging upon piceous. Dorsum of abdomen ferruginous, darker at the tips of segments 2-7 and with a pair of subobsolete pale vittae at the base of each of them; venter pale greenish in segments 2-7 or 8. Setæ whitish; the joinngs fuscous. Fore leg pale ferruginous with a median and a terminal band on the femur, the tip of the tibia and the tarsal joinings and tips fuscous. Hinder legs

SESSIONAL PAPER No. 39b

yellowish with the tips of the femora fuscous and the tarsal joinings and tips a little cloudy. Fore wing hyaline with a pale ferruginous cloud in the pterostigmatic region; neuration fuscous, excepting the basal $\frac{2}{3}$ of the costa, subcosta and radius which are yellowish; the thickening at the bulla of the subcosta about 0·5 mm. long, is more or less obfuscated.

Female imago:

Measurements: Body 8 mm.; wing 10 mm.

Face clear; thorax yellow, slightly tinged with brown. Abdomen yellow; segments banded along posterior margin dorsally with black; stigmal dots marked. Femora with median and distal ferruginous bands. Most of the longitudinal veins of the fore-wing colourless; cross veins heavy and brown; very slightly clouded in the apical costal space; venation of hind wing almost colourless.

Nymph:

Measurements: Body 8–9 mm.; setæ 10–13 mm.

Head brown, very slightly covered with light dots; a light spot above each ocellus; a small light dot on each side of median ocellus; an irregular light area anterior and lateral to each eye. Pronotum brown with two light spots on each side; part of lateral margin. Mesothorax similar in colour to prothorax. Abdomen of a uniform brown colour dorsally, having a smooth appearance; lighter ventrally with a semicircular brown band on 9th segment and a median brown spot on 8th. Spines of lateral edge short. Setæ banded usually 3 dark and 1 light; sparsely fringed, usually only at base of light segment. Femora much flattened, brown and dotted with light spots, and having 3 irregular light bands; covered dorsally with small spines and posterior margin fringed with hairs and spines. Tibia with median and distal light bands. Tarsus tipped with white. Claws with two pectinations.

The nymphs of this species were taken up the Go Home River on June 16, 1912, immediately above Flat Rock Falls, where the water was flowing swiftly but smoothly. The nymphs were clinging to stones in water 1 to 1½ feet deep, not far from the shore. On this date they were also found just below Sandy Gray Falls in swift rough water, but close to shore. I was successful in rearing two specimens, but one escaped from the cage over the breeding-jar and could not be found. The remaining one was a female and hence to make the description as complete as possible, I have inserted the description of the male from Eaton. The dates of the two emergings were June 27 and July 3.

Heptagenia lutea Clemens.

(Pl. XV, Fig. 2).

Male imago:

Measurements: Body 9–10·5 mm.; wing 10·5 mm.; setæ 20 mm.; fore-leg 10 mm.

This is a light coloured species, slightly reddish on face below antennae; between ocelli and eyes, reddish brown. Thorax almost whitish yellow dorsally; light yellowish brown laterally; dark area on each side of pronotum; slight red and brown markings, beneath bases of fore and hind wings. Each of the abdominal segments 1–8 banded dorsally at posterior margin, remaining parts of these segments being almost white; segments 9 and 10 entirely reddish brown; stigmal dots not marked; wings clouded in pterostigmatic space, a few cells reddish, femora with median and apical bands; tibia-tarsal and tarsal joints black; fifth tarsi and ungues dark.

Female imago:

Measurements: Body 11 mm.; wing 12 mm.; setæ 22 mm.

Abdomen more yellowish than male.

Nymph:

Measurements: Body 10 mm.; setæ 13–16 mm.

Head light brown in colour and dotted with light dots; light areas over ocelli; another at posterior margin of head in median line and a larger one lateral to each eye. Pronotum with a broad colourless lateral margin, remainder light brown numerous irregular light spots. Abdomen darker dorsally and with a rather complicated colour pattern. First segment light with 2 brown areas at side; 2nd, a narrow brown band along posterior margin and 5 brown areas and four light ones placed alternately; 3rd almost entirely dark with a few light dots; 4th with 2 dark spots in posterior lateral angles of segment; also a large dark area in centre of segment with a light area within it; 5th with a dark spot in each posterior lateral angle as in preceding segment; a dark band along posterior margin; 2 light areas surrounded with brown and a dark spot in centre of each; 6th almost entirely brown except for two light areas in anterior lateral angles; 7th with 2 large light areas with a brown dot in each toward inner side; 8th an irregularly light and dark coloured segment; 9th has a narrow brown band along posterior margin and a dark longitudinal stripe in median line; 10th almost entirely dark. Ventrally, the lateral and posterior margins of segments 2–8 dark; segment 9 with two large brown spots. Setae greenish; basal half well fringed at joints, distal half with each two segments alternately light and dark and few hairs at joints. Legs, femora with alternately light and dark irregular bands and covered with minute spines dorsally; posterior margin fringed with hairs; anterior margin also fringed but hairs shorter. Proximal end of tibia dark and has a dark band slightly beyond middle. Tarsi with a reddish brown band, very near proximal end. Claws with two pectinations.

The nymphs were quite abundant along the open shore of Station Island and west of it, my collections dating from June 3 to July 2. Besides I have a few from a small waterfall on the Musquash River, 3 miles south of Go Home Bay, taken June 30, and 3 small nymphs from Sandy Gray Falls, Aug. 23. Imagos emerged from June 27 to July 3.

Heptagenia fusca Clemens.

(Pl. XVI, Fig. 1).

Male imago:

Measurements: Body 10 mm.; wing 13; setæ 26.

No marks on face; ocelli almost in a straight line, middle one the smallest. Pronotum brown, slightly darker along the median line. Mesothorax uniformly brown. Abdomen with posterior ⅓ of each segment of same brown colour as thorax and projections from this band anteriorly in the median line, almost forming a continuous longitudinal stripe on the abdomen; the band widens slightly laterally also; remaining portions of each segment somewhat light brown; ventrally very slightly banded. Forceps and penis lobes of usual form. Femora banded in middle and at distal end. Wings large; costa, subcosta and radius light in colour while remainder of longitudinal and the cross veins brown. No cloud in pterostigmatic space.

Female imago:

Measurements: Body 10–12 mm.; wing 14 mm.; setæ 18 mm.

Quite similar to male, except that abdomen is considerably darker.

Nymph:

Measurements: Body 12–14 mm.; setæ 15–20; antennae 3 mm.

Head brown, dotted with light spots; usually 3 light areas at posterior margin between eyes and 2 lateral to each eye; anterior margin well fringed with hairs. A light longitudinal median line on pronotum; 2 light areas on each side and lateral margin colourless; remainder of pronotum brown with small light dots. Posterior ⅓ of each abdominal segment 6–10 almost black; segments 1–6 brown; the remainder of each segment varying from light brown to greenish yellow; ventrally posterior ⅓ of each of the segments 2–8 brown; 9th segment has 2 dark areas laterally. Femora light brown on upper surface with a few lighter areas and covered with minute spines dorsally; posterior margins fringed with hairs; proximal end of tibia dark brown and its third ⅓ dark; proximal half of tarsus dark. Setæ well fringed with hairs at the joints.

While on a canoe trip up the Go Home River, June 16th, I collected a number of the nymphs of this species just below Sandy Gray Falls. The only imagos I have are the ones reared from this collection. The dates of emergence are June 23rd and 24th.

This species is close to *H. verticis* but lacks the dark median stripe on the thorax, and does not show the slightest trace of a dash in the wing under the bulla.

Heptagenia tripunctata Banks.

(Pl. XV, Fig. 1).

Male imago:

Measurements: Body 9–11 mm.; wings 12–13 mm.; setæ 25-35 mm.; fore-legs 12-14 mm.

Two small dots on median carina, slightly below level of antennæ. Thorax brown; on pronotum a dark spot at anterior margin in median line, sometimes divided into two by a fine light line; two small dark spots just posterior to these; an oblique dark streak on each side of pronotum; a brown stripe on coxa of foreleg and extending up on side of prothorax.

Short dark stripes at bases of fore and hind wings. Abdominal segments 1 to 7 paler than rest of body; segments 8-10 dark, similar to thorax; three dots on dorsum of each abdominal segment at posterior margin; stigmal dots well marked. Setæ with alternate joints of basal half dark. Femur of fore-leg darkened at both ends and with a median band. Tibia-tarsus joint dark. Yellowish in apical costal space of wing and a reddish area in pterostigmatic space.

5 GEORGE V., A. 1915

Female imago:

Measurements : Body 10-12; wings 14-16; setæ 22-25.

Nymph:

Measurements: Body 11-14 mm.; setæ 12-16.

Head deep brown, occasionally almost black dotted with light spots; three light areas along anterior margin of head and one at posterior margin between eyes. Pronotum similar in colour to head with light dots and about 5 larger light areas on each side; lateral margin with a light area which extends in some distance. A light area in antero-lateral angle of mesothorax. Femora stout with 5 irregular light areas; small spines very numerous; posterior margins fringed with hairs. Tibiæ with 2 dark and 2 light areas, arranged alternately. Abdomen similar in colour to head and thorax; a light area on segments 4 and 5 containing a small triangular dark area at anterior margin of segment 5, lateral to which are 2 dark dots; another light area on segments 7, 8, 9 and 10 containing 2 dark dots on 8 and 2 on segment 9; usually the 3 dark spots at posterior margin of the segments of the abdomen of the imago can be distinguished; ventrally two longitudinal rows of dark dots, increasing slightly in size toward posterior end; segment 9 usually with 2 pairs, the anterior pair small, posterior pair larger. Setæ with alternate dark and light areas. Gills have the lamellæ slightly rounded at distal end.

This species was by far the most abundant at Go Home Bay. The nymphs were found in almost every locality where there was a stone to cling to, except, of course, in stagnant water, and could be taken at any time during the three months. The first specimens bred emerged May 31, but the first capture was not made until June 11. On this date a small swarm of about 20 individuals was discovered about 8.15 p.m., flying from 10 to 20 feet high, facing north. One female and several males were taken. Soon after this they became very abundant and remained so until about July 5th. The last specimen bred is dated Aug. 13.

Heptagenia rubromaculata Clemens.

(Pl. XVI, Fig. 2).

Male imago:

Measurements: Body 8 mm.; wing 8 mm.; setæ 17 mm.; fore-leg 7 mm.

No marks on face; darker spot at posterior margin of head between eyes. Thorax dark; median longitudinal dark stripe on pronotum; dark brown stripe on coxa of fore-leg and extending up the side of prothorax. Abdominal segments 1-7 lighter; 8-10 dark, similar to thorax; each segment banded at posterior margin; stigmal dots distinct; wing has a reddish area in pterostigmatic space.

Female imago:

Measurements: Body 9-9·5 mm.; wings 13-14 mm.; setæ 15-22 mm.

Often slightly reddish on face beneath antennæ. Dark brown on dorsal surface of head behind ocelli. Abdomen varies from a reddish to a yellowish colour in dried specimens.

Nymph:

Measurements: Body 9-10 mm.; setæ 10 mm.

Head dark brown, dotted with minute light spots. Pronotum similar in colour to head; two light areas on each side, the outer one sometimes joined to the light margin. Abdomen dark brown with a granular appearance; sometimes a faint broad, dark longitudinal streak can be made out with 2 dots on each side of it on each segment excepting 9 and 10; ventral surface lighter with a median row of irregular dark spots and lateral rows of small dots or lines, the median dots sometimes broken up so that only 4 or 5 small dots remain in their place. On segment 9 the markings are usually jointed, forming roughly three sides of a square. Femora with 4 irregular dark bands; both posterior and anterior margins very hairy. Claws pectinated. A very hairy species, having anterior margin of head, sides of thorax and abdomen very hairy.

This nymph was first taken on June 15 in what is commonly called the Narrows, near the mouth of the Go Home River. The water here had a well-marked current, but scarcely swift. On June 30 I found them very numerous in the very swift water of a rapid near the mouth of the Musquash River. Nearly a month after this on July 20 and 22 I discovered mature nymphs at an old lumber chute on the Go Home River in fairly swift water.

Imagos were bred from the nymphs taken at the Narrows on June 22nd and 25th; at the Musquash rapids from July 3 to 5th; at the chutes July 24-29th. No imagos were captured.

Heptagenia luridipennis Burm.

Male imago: (Pl. XV. Fig. 3).

Measurements: Body 7-8 mm.; wings 8 mm.; setæ 2 -22 mm.; fore-leg 8 mm'

Face clear; slight dark marks at posterior margin of head between eyes. Median longitudinal stripe on pronotum; sides brown; mesonotum dark brown; brown area in front of base of middle leg. Segments 1-7 of abdomen light coloured segments 8·9 and 10 dark, similar to thorax, narrow black bands along posterior margins of the segments; stigmal dots distinct. Apical costal area of wing not distinctly darkened and no reddish coloured area.

Female imago:

Measurements: Body 9 mm.; wings 10 mm.

Nymph:

Measurements: Body 7-8·5 mm.; setæ 10-14 mm.

Head brown with light dots; anterior margin fringed with hairs. Prothorax similar in colour to head; on pronotum a light spot on each side of median line; lateral to this another larger one and lateral to this again another which extends to the lateral margin. Abdomen similar in colour to prothorax; a row of black dots on each side corresponding to the stigmal dots of imago; segment 3 for the most part light with a round brown spot in median line and with two short projections laterally; segment 4 with a small triangular brown spot in median line

5 GEORGE V., A. 1915

with base to anterior margin while the apex meets a large brown area leaving a small light area on each side of the triangle; lateral to the brown area is a light one and lateral to this is a triangular dark spot in the posterior angle of the segment; segment 5 much like the 4th; segment 6 entirely dark except for two small spots at anterior margin and 2 toward lateral margin; segment 7 with a triangular dark spot in median line with base to anterior margin and apex reaching about middle of segment; on each side of triangle 2 dark spots; segment 8 similar to the 6th; segment 9 irregularly marked; roughly, it is dark with a darker median longitudinal stripe, 2 light spots on each side and another at lateral margin; segment 10 entirely dark; ventrally 2 dark spots at lateral margins of 9th segment just beside the lateral spines of that segment; sometimes a triangular spot in the median line. Setæ with basal half fringed with hairs at joints.

The nymphs of this species were the last to be taken. On August 23rd I found them in a rapid just above Sandy Gray Falls, about 5 miles from Station Island. I was successful in rearing quite a number of these dating from August 28th to Sept. 1.

Heptagenia canadensis Walker.

(Pl. XVI, Fig. 4).

Male imago:

Measurements: Body 8–9 mm.; wings 9 mm.; setæ 20–22 mm.; foreleg 9 mm.

A dark species; a black band on face below each antenna; dark reddish brown between lateral ocelli; a small black dot close to inner margin of each eye; posterior margin of head with a narrow black line thickened at median line. Prothorax brown; short dark band along posterior margin of pronotum, the ends of which turn obliquely across the side of pronotum. Mesothorax rich brown dorsally, sides lighter, oblique dark stripes at base of fore and hind limbs; dorsum of abdomen black; slightly darker in median line; and a lighter area in each segment on either side of it. Posterior margins of segment 1–9 margined with black; 10th segment lighter; stigmal dots obscured by the black colour of the abdomen; penis lobes rather oblong in shape; setæ almost white, tinged with black; joints darker. Femur of fore-leg almost yellowish tinged with black and having median and distal black bands. Tibia lighter; tibiotarsal joint black; tarsi tinged with black; lengths of segments arranged in increasing order 5, 1, 4, 3, 2; the 2nd slightly longer than the 3rd. Wings with a dark dash and numerous cross-veins margined with black between the dash and base of wing; terminal margin of hind wing slightly tinged with black.

Female imago:

Measurements: Body 9–10 mm.; wing 12 mm.; setae 15 mm.

Abdomen very reddish, often blackish red.

Nymph:

Measurements: Body 11 mm.; setæ 15 mm.; antennae 3·5 mm.

Head reddish brown in colour; a small dark area immediately in front of each antenna; and another about the same size in front of each eye; a black dot behind each lateral ocellus; a light area in front of median ocellus, and a larger light area

SESSIONAL PAPER No. 39b

between each lateral ocellus and eye. Another lateral to each eye along margin of head; mouth parts of the type belonging to group 2.

Pronotum reddish brown with a dark and an approximate light area in each lateral half; margin colourless.

Abdomen darker than thorax; each segment with four light longitudinal streaks, 2 near median line and the other two near lateral margin ; black dots, corresponding to the stigmal dots just inside of lateral light streaks. Ventrally the abdomen is almost white, each segment has two light brown lateral streaks, while 9th has its lateral and posterior margins margined with light brown. Short lateral spines at posterior lateral angles of segments 8 and 9. Setæ of equal length; light brown; joints fringed with hair. Gills oval and pointed, femur of fore-leg light brown with 4 light areas. Two small ones towards anterior margin and two large towards posterior. Distal end light coloured. Femora of hind legs with fewer pale-markings. Tibae alternately banded with brown and white. Tarsi have very broad median bands, legs slightly hairy along posterior margin.

This species was the second most abundant at Go Home Bay. The nymphs were taken from May 25th to June 31st in various localities, but never in swift water, the usual place being quiet bays. On Sept, 5th some small nymphs were found in a small creek which were evidently the next generation.

Almost mature nymphs were taken in this creek on May 31. The first bred specimen is dated June 1, and the last July 4th. Imagos were very abundant at Station Island from June 25th to July 15th.

Heptagenia frontalis Banks.

(Pl. XVI, Fig. 3).

Male imago:

Measurements: Body 7–8 mm.; wings 9 mm.; setæ 18–20 mm.; forelegs 7 mm.

Much like *H. canadensis* but lighter in colour, face yellowish a black dot on face below each antenna; a smaller black dot near inner margin of eye; pronotum light brown with a black streak on each side. Mesonotum rich brown; sides of thorax whitish yellow, segments 1–7 of abdomen very light, with posterior margin black; 8–10 reddish dorsally; stigmal dots distinct; setæ white. No dash in wing and cross veins not margined. Femora yellowish with black median and apical bands.

Female imago:

Measurements: Body 8–9 mm.; wings 10 mm.; setæ 15 mm.

Head and thorax light yellow ; dots on face beneath antennae almost forming bands; sometimes a black dot at lateral margin of each side of pronotum, usually a few cross-veins margined with black on the wing.

Nymph:

Measurements: Body 9–10 mm.; setæ 9–10 mm.

Head yellowish brown in colour; three almost round light spots along anterior margin of head; a light area in front of each ocellus; usually a light area along median line between eyes and 2 smaller ones lateral to this along posterior margin

5 GEORGE V., A. 1915

of head.   A black dot below each antenna, in front of each eye and near inner margin of each eye.

Thorax lighter in colour than head; on each side of pronotum, near median line is a small light spot; just lateral to this is a triangular dark spot and lateral to this again is another light area.   In anterior angle of pronotun is an oval light spot.   Along posterior margin extending some distance on either side of median line is a broad light band, which is connected by a light longitudinal stripe along median line of mesonotum, to a large irregular light area on the mesonotum.

Abdomen usually light yellowish brown; the colour pattern, roughly, has the appearance of a broad light band along median line in which, on segments, 5, 6, and 7 are oval dark areas, in 8 a narrow stripe and in 9 a round dark area in each segment, on either side of this broad light band is a short light stripe; ventral surface almost white with 2 lateral light brown longitudinal stripes on segments 1–9; a broad band across 9th along posterior margin, joining the 2 lateral stripes. Segments of setæ alternately light and brown.   Legs pale, colour pattern similar to *H. canadensis*.

This species was not nearly so abundant or wide-spread as *H. canadensis*. Nymphs were taken in quite similar localities and at about the same time.   They were taken from June 15 to July 2, and imagos reared from June 26 to July 4.

Heptagenia sp. indet.

(Pl. **XVI**, Fig. 5).

Nymph:

Measurements: Body 10–11mm.; setæ 12–13 mm.

Head light brown; sometimes 3 light areas along entire margin but frequently middle one is lacking and the 2 lateral ones are connected with the light margins lateral to eyes.   An almost black spot in centre of each half of pronotum; around this is an irregular light area, exterior to which is a brown area.   Abdomen whitish yellow with 5 longitudinal yellowish brown stripes in each segment 1-8.   Setæ light greenish yellow; joints abundantly fringed with hairs.   Legs yellowish brown in colour; pattern similar to the 2 preceding species.

Mr. R. P. Wodehouse kindly gave me these nymphs which he collected along the east shore of Manitoulin Island on June 26th, 1912.   As they were not reared the species cannot be ascertained at present.

REFERENCES.

1888. EATON, REV. A. E.—A Revisional Monograph of Recent Ephemeridæ or Mayflies. *Trans. of the Linnæan Society:* Second Series, vol. III, Zoology; London, 1888.

1901. NEEDHAM, J. G.—Aquatic Insects in the Adirondacks. *New York State Museum Bulletin,* 47, 1901.

1904. NEEDHAM, J. G.—Mayflies and Midges of New York. *New York State Museum Bulletin,* 86, 1904.

1910. BANKS, NATHAN.—Ephemeridæ of the Genus Heptagenia. *Can. Entomologist,* vol. XLII. No. 6, 1910.

SESSIONAL PAPER No. 39b

1911. MORGAN, ANNÀ H.—Mayflies of Fall Creek. *Annals of the Entomological Society of America,* vol. IV., No. 2, 1911.

1913. CLEMENS, W. A. — New Species and New Life Histories of Ephemeridæ or Mayflies. *Can. Entomologist,* vol. XLV., Nos. 8 and 10.

EXPLANATION OF PLATES.

PLATE XV.

HEPTAGENIA NYMPHS.

Fig. 1. *Heptagenia tripunctata* Banks.
Fig. 2. *Heptagenia lutea* Clemens.
Fig. 3. *Heptagenia luridipennis* Burm.
Fig. 4. *Heptagenia flavescens* Walsh.
Fig. 5. *Heptagenia flavescens,* ventral view.

PLATE XVI.

HEPTAGENIA NYMPHS.

Fig. 1. *Heptagenia fusca,* Clemens.
Fig. 2. *Heptagenia rubromaculata* Clemens.
Fig. 3. *Heptagenia frontalis* Banks.
Fig. 4. *Heptagenia canadensis* Walker.
Fig. 5. *Heptagenia,* undetermined.

PLATE XVII.

MOUTH-PARTS AND GILL OF NYMPH OF *H. LUTEA* CLEMENS, AND GENITALIA OF *H. TRIPUNCTATA* BANKS.

Fig. 1. Left maxilla.
Fig. 2. Labium.
Fig. 3. Labrum.
Fig. 4. Hypopharynx.
Fig. 5. Left mandible.
Fig. 6. Gill.
Fig. 7. Genitalia.

PLATE XVIII.

MOUTH PARTS AND GILL OF NYMPH OF *H. CANADENSIS* WALKER, AND GENITALIA OF IMAGO.

Fig. 1. Left maxilla.
Fig. 2. Hypopharynx.
Fig. 3. Labrum.
Fig. 4. Labium.
Fig. 5. Left mandible.
Fig. 6. Gill.
Fig. 7. Genitalia.

PLATE XVII

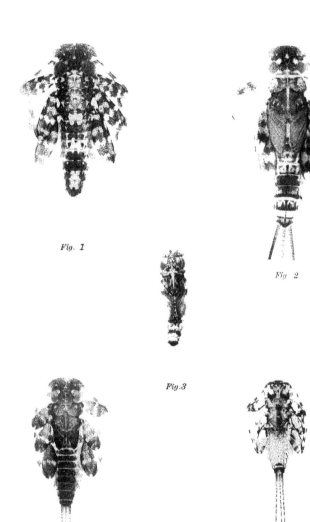

Fig. 1

Fig 2

Fig. 3

Fig. 4

Fig. 5

PLATE XVI

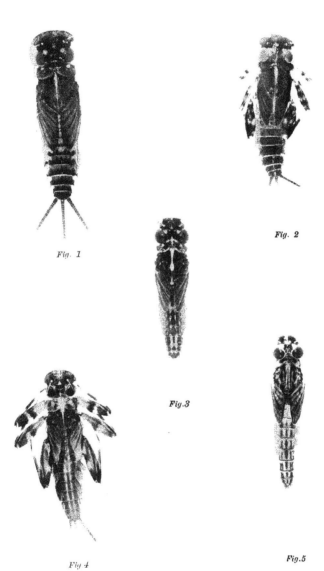

Fig. 1

Fig. 2

Fig.3

Fig 4

Fig.5

PLATE XVII

PLATE XVIII

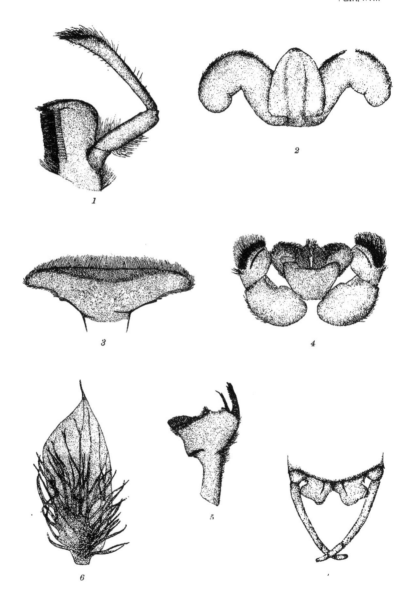

1

2

3

4

5

6

VI.

THE FRESH-WATER MALACOSTRACA OF ONTARIO.

By A. G. Huntsman, B.A., M.B., University of Toronto.

(13 figures in the text)

The greater part of the material upon which this paper is based was collected at the Georgian Bay Biological Station. As it seemed desirable to bring together records of all the species known to occur in the region of the Great Lakes, the scope of the paper was extended beyond the Georgian Bay region and material from other localities in Ontario was studied.

The scope of the paper may be considered as the Canadian part of the region of the Great Lakes, which is practically limited to the Province of Ontario. Only a small part of our waters has as yet been examined, and the following list of species cannot be considered as exhaustive, but it may be noted that few species have been added to the list of Malacostraca reported from the Great Lakes in "The Fresh-water Crustacea of the United States" published by S. I. Smith in 1874, and probably very few remain to be added.

As the literature necessary for the identification of species is more or less scattered and inaccessible, it has been deemed advisable to include keys for the determination of the species, together with figures of the principal parts useful in diagnosis so that this account may serve as a basis for future work.

Much work remains to be done to determine the distribution of the various species. The localities from which specimens have been obtained, are given, but no systematic collecting has been done in any part of the Province with the exception of Georgian Bay. Doubtless the majority of the species occur throughout the entire region, wherever suitable habitats are to be found.

The importance of the Malacostraca in connection with our fresh-water fisheries can scarcely be overestimated. They form the chief element in the food of many of our food-fishes. Their large numbers, their free-living habits and their general edibility render them particularly suitable as fish food. It is very desirable to learn more of their life-histories, habits, food, etc., so that their numbers may be increased or extralimital species that are desirable may be introduced. The practical value of such work would be very great, as the lake area of the Dominion is exceedingly large and able to support an immense number of fish. The probability of a successful issue of such researches is greater in the case of fresh-water than in that of marine forms, owing to the fact that the various conditions can be much more readily controlled in closed-in bodies of water.

5 GEORGE V., A. 1915

Some of our Malacostraca are large enough to serve as food for man. The crayfishes and shrimps are marketed in many of the American states, but in Canada little use has been made of them. Our crayfishes are quite large, but our shrimps are small. A species of shrimp that occurs in Ohio could doubtless be introduced here very readily.

The Malacostraca are also of importance as serving as intermediate hosts for many parasites which occur in fishes.

The greater part of the material that I have examined was collected in the summer of 1912 by Mr. R. P. Wodehouse at various points in the Georgian Bay. To Dr. E. M. Walker, Mr. A. R. Cooper and Mr. A. D. Robertson I am indebted for material from the Georgian Bay and from other points in Ontario. I have indicated the sources of my material in the following way,—from Mr. R. P. Wodehouse—(Wo), from Dr. E. M. Walker—(Wa), from Mr. A. R. Cooper—(C), and from Mr. A. D. Robertson—(R).

Key to the Orders.

A_1. Eyes sessile (Fig. 1). First thoracic segment fused with head. Remainder (seven in number) free, with large appendages.

    B_1. Body compressed dorso-ventrally. Branchiae on abdominal appendages...Isopoda.

    B_2. Body laterally compressed. Branchiae on thoracic appendages (Fig. 3, br)...Amphipoda.

A_2. Eyes pedunculated. The majority of the thoracic segments fused with the head to form dorsally a carapace (Fig. 5).

    C_1. Thoracic legs similar and biramous (Fig. 6). Several of the posterior thoracic segments not fused with the carapace...............Mysidacea.

    C_2. Posterior five pairs of thoracic legs uniramous and large, anterior three biramous and small. Not more than one thoracic segment free from carapace.
....Decapoda.

Order ISOPODA.

For North America, this group has been monographed by Miss Harriet Richardson (see Bibliography). Only two fresh-water species have been reported from the region of the Great Lakes. They belong to the family Asellidae. For Canada no records have been published. The Isopods are to be found crawling about in shallow water in a variety of situations (among weeds, under stones, etc.).

Key to the Genera.

A_1. Mandible with palp (Fig. 1, d).....................................*Asellus.*

A_2.   "   without palp (Fig. 1, b)..........................*Mancasellus.*

SESSIONAL PAPER No. 39b

Asellus communis Say.   Fig. 1, c, d.

Harger in Smith, 1874, p. 657; Richardson, 1905, p. 420.

Abundant nearly everywhere in shallow water among weeds and frequently found in stagnant pools.   Richardson records it from Massachusetts to Michigan on the north.

Localities.—Georgian Bay: Go Home, Fitzwilliam Id. (R); Waubaushene, Go Home, Shawanaga, French River, Fitzwilliam Id. (Wo).   Lake Ontario: Toronto.

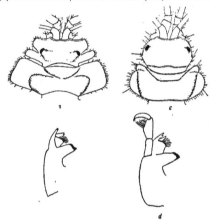

Fig. 1.

This species differs from other members of the genus in having the head broadest about its middle, the uropods about as long as the last segment, distinct epimera on all the thoracic segments and the palmar margin of propodus of first gnathopod with one or two teeth.   Length up to 11 mm. (15 mm. Harger).

The specimens that I have examined do not agree perfectly with the description of *Asellus communis* given by Richardson.   They are in many respects intermediate between her descriptions of that species and of *Asellus intermedius*. With the latter my specimens agree in having a distinct lobe at the posterolateral angle of the head, and the antennule extending to the middle of the last segment of the peduncle of antenna.   According to Richardson, the flagellum of the antennule is composed of fourteen articles in *communis* and of nine articles in *intermedius*.   In my specimens they vary in number from nine to twelve.   The inner branch of the uropods is sometimes the same width as the outer as described for *intermedius* and sometimes twice as wide as the outer as described for *communis*.   It is to be doubted whether these are distinct species.   Richardson copied Smith's figure of *Asellus communis*, which does not agree with her description of that species but agrees more nearly with her description of the other species.   Until the matter is settled by further study, I consider it best to use Say's name, which is the older.

5 GEORGE V., A. 1915

Mancasellus tenax (Smith).   Figs. 1, a, b; 2.

Asellus tenax Smith, 1871, p. 453.
Asellopsis tenax Harger in Smith, 1874, p. 601.
Mancesellus tenax Underwood, 1890, p. 359; Richardson, 1905, p. 415

This species is less abundant than the preceding one and is more restricted to open and pure water, although one record is from a small inland lake. It has been reported as ranging from Lake Superior to the Detroit River. This range must be extended to Lake Ontario. It is probably found throughout the entire region of the Great Lakes. It has been found in as deep water as 30 fathoms (Smith).

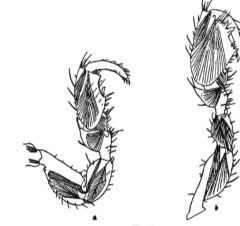

Fig. 2.

Localities.—LAKES SUPERIOR and HURON (Smith).

GEORGIAN BAY: Sydney Bay (Wiarton), Go Home (R); Sturgeon Bay, Shaw-anaga, Tamarac Bay (Manitoulin Id.), Fitzwilliam Id. (Wo).

LAKE ONTARIO: Toronto Island.

Stomachs of *Coregonus clupeaformis* (R), *Perca flavescens* (Forbes).

In addition to the generic difference given in the key, this species is readily distinguished from the preceding by the extended truncate epimera (not separate from the segments) and by the head being much broader than long and with a deep incision on each side (Fig. 1, a). Length up to 11·5 mm. (13 mm., Harger).

From other species of the genus this differs in having deep lateral incisions in the head, antennae about half the length of the body and the uropods about two-thirds the length of the last segment.

Two subspecies or varieties are distinguished, *M. tenax tenax*, the typical and commoner variety and *M. tenax dilata* (Smith) 1874, p. 661. The latter differs from the former in being broader, in having more joints in the flagellum of the antennule, and in having three teeth (the middle one largest) on the palmar margin of the propodus of the first gnathopod of the adult male (see Fig. 2). I am unable to definitely separate these, either as to the width of the body, number of joints in flagellum of the antennule or in the shape of the first gnathopoda of the male. The extremes in the conditions of the gnathopoda are shown in Fig. 2. In some lots of specimens, one or other form appears to predominate and in others all gradations between the two extremes are to be seen. The typical *dilata* I have seen only from the north end of the Georgian Bay (Fitzwilliam Id.). Smith's specimens came from the Detroit River.

Order AMPHIPODA.

Miss Weckel (see bibliography) has recently given an account of the fresh-water species of this group occurring in North America. Six species have been reported from the region of the Great Lakes, although only three have actually been recorded from Canada. The Amphipods occur at practically all depths, either crawling about among debris or swimming freely near the bottom. Only rarely do they venture out into the open water.

Key to the Genera.

A_1 Last thoracic leg shorter than preceding one and with basal joint large and leaf like. (Fig. 3)..*Pontoporeia.*

A_2 Last thoracic leg longer than preceding one and with basal joint little larger than that of preceding one.

    B_1 Telson cleft to base. Third uropod biramous, rami nearly equal (Fig. 4, b).
....*Gammarus.*

    B_2 Telson notched. Third uropod biramous but inner ramus rudimentary (Fig. 4, c)...*Eucrangonyx.*

    B_3 Telson entire. Third uropod uniramous (Fig. 4, d)...........*Hyalella.*

Pontoporeia hoyi (Stimpson Mss.) Fig. 3.

P. affinis Smith, 1871, p. 452.
    "   "   Nicholson, 1872, p. 501.
" *hoyi*   Smith, 1874, p. 647.
    "   "   Weckel, 1907, p. 26.

This species occurs in abundance on muddy or gravelly bottoms at various depths down to 169 fathoms (Smith). In Lake Superior, according to Smith, it is found in as shallow water as 4 fathoms. It is the same at the north end of Georgian Bay, where it was dredged last summer (1912) by Messrs. Robertson and Wodehouse in Rattlesnake Harbour, Fitzwilliam Id. In this harbour many whitefish are caught in pound nets, and they doubtless feed upon this species in

5 GEORGE V., A. 1915

the harbour. In the southern end of Georgian Bay, I do not know of it being taken in shallower water than about 20 fathoms and, in Lake Ontario, Nicholson did not obtain it in shallower water than about 30 fathoms.

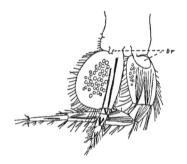

Fig. 3.

Localities.—Lakes Superior and Michigan (Smith).

Georgian Bay: Near Pine Ids., 20 fathoms (Wa); East of South Watcher Id., 20 to 25 fathoms; Rattlesnake Harbour, Fitzwilliam Id., 4½ fathoms (R and Wo).

Lake Ontario, near Toronto, 30 to 40 fathoms (Nicholson).

Stomachs of whitefish from Lakes Superior and Michigan (Smith), from Georgian Bay (Wa, C, R); *Uranidea formosa* (C) from Port Credit, Lake Ontario.

In addition to the differences given in the key, this species can readily be distinguished from our other Amphipods by the rudimentary condition of the 'hands' of the second gnathopods. Length up to 8 mm.

Pontoporeia filicornis (Stimpson Mss.)

Smith, 1874, p. 649; Weckel, 1907, p. 28.

A single specimen was dredged by Stimpson in Lake Michigan near Racine in 40 to 60 fathoms. It has not since been found. It is distinguished from the preceding species chiefly by the very long antennæ and antennulæ, which are as long or longer than the body.

Genus Gammarus.

Key to the Species.

A₁ Terminal joint of outer ramus of last uropod without long plumose hairs on outer margin. . . . .*G. fasciatus.*

A₂ Terminal joint of outer ramus of last uropod with long plumose hairs on outer margin. . . . .*G. limnaeus.*

Gammarus fasciatus Say.   Fig. 4, b.

(?) *Gammarus* sp. Nicholson, 1873, p. 500.
Gammarus fasciatus Smith, 1874, p. 653, Weckel, 1907, p. 40.

Generally distributed in shallow water, under stones and among weeds, etc. It is probably our commonest Amphipod, although not found in as large numbers as is *Hyalella*.   It is found from Maine to Wisconsin on the north according to Smith.

Localities.—LAKES SUPERIOR and MICHIGAN (Weckel).

GEORGIAN BAY: Waubaushene, Rattlesnake Harbour (Fitzwilliam Id.) (Wo); McGregor Bay (Wiarton) (R).

LAKE ONTARIO: Toronto; Coburg (Wa).

NIAGARA FALLS (Weckel).

Stomachs of Black Bass (Forbes).

The characters of this species have been sufficiently indicated in the keys. Length up to 15 mm.

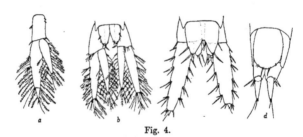

Fig. 4.

Gammarus limnaeus Smith.   Fig. 4, a.

G. lacustris Smith, 1871, p. 453.
G. limnaeus Smith, 1874, p. 651; Weckel, 1907, p. 42.

This species is much less abundant than the preceding and occurs with it. According to Weckel, it ranges from Maine to Utah.

Localities.—LAKES SUPERIOR and MICHIGAN (Smith and Weckel).

GEORGIAN BAY: Rattlesnake Harbour (Fitzwilliam Id.) (Wo).

Stomachs of Trout (Smith).

I have been unable to separate this species from the preceding, except in regard to the presence or absence of bristles on the sides of the terminal segment of the outer ramus of the last uropods.   The number of joints in the secondary flagellum of the antennule varies from 2 to 4 in *G. limnæus* and from 3 to 6 in *G. fasciatus*.   The other differences given by Weckel are only differences of degree and not easily applied.   Length up to 20 mm.

5 GEORGE V., A. 1915

Eucrangonyx gracilis (Smith).   Fig. 4, c.

Crangonyx gracilis Smith, 1871, p. 453; 1874, p. 654.
(?) *Crangonyx* (?) sp. Nicholson, 1873, p. 501.
Eucrangonyx gracilis Weckel, 1907, p. 32.

This species does not appear to be very common.   It is found in shallow water among weeds and down to 13 fathoms (Smith).   According to Weckel it ranges from Rhode Island to Wisconsin.

Localities.—LAKES SUPERIOR, MICHIGAN and HURON (Smith and Weckel). GEORGIAN BAY: Go Home (R); Shawanaga (Wo).

BOND LAKE (near Toronto).

Stomach of Mud-minnow (*Umbra limi*) (Forbes).

This species is well characterized by the features mentioned in the keys and by the figure.   Among other things it can be distinguished from the two species of *Gammarus*, which it very much resembles, by the absence of stout bristles on the dorsal surface of the abdomen and by the structure of the secondary flagellum of the antennule, which consists of two joints, the last one very short.

Length up to 18 mm.

Hyalella knickerbockeri (Bate).   Fig. 4, d.

H. dentata Smith, 1874, p. 645.
H. knickerbockeri Weckel, 1907, p. 54; Jackson, 1912.

This species is extremely abundant among weeds in shallow water, both in the Georgian Bay and in Lake Ontario.   Smith reports it from Maine to Wisconsin on the north.

Localities.—LAKES SUPERIOR and MICHIGAN (Smith and Weckel).

GEORGIAN BAY: Go Home (C and R); Matchedash Bay (R); Waubaushene, Shawanaga, French River, Killarney, Tamarac Bay (Manitoulin Id.), Fitzwilliam Id. (Wo).

LAKE ONTARIO: Toronto.

Stomachs of the following fishes according to Forbes (1888),—*Perca flavescens, Percina caprodes, Micropterus dolomieu, Eupomotis gibbosus, Lepomis pallidus, Ambloplites rupestris, Aphredoderus sayanus, Fundulus diaphanus, Notropis cornutus, N. heterodon, Ictalurus punctatus, Ameiurus natalis, A. nebulosus, Amia calva* and *Polyodon spathula*.

This species is easily recognized by the exceedingly broad and clumsy hands of the second gnathopods of the male, by the absence of a secondary flagellum on the antennule and also by the spines projecting backward from the middle of the posterior margin of each of the first two abdominal segments.   Weckel includes in this species forms without these spines.   In my material I have not seen any individuals without them.   Length up to 7 mm.

Order MYSIDACEA.

Of this group only a single species occurs within our limits.

Mysis relicta Loven. Figs. 5 and 6.

Mysis oculata var. *relicta* Sárs, 1867, p. 14.
" *relicta* Smith, 1874, p. 642.

This species swims about in shoals near the bottom in rather deep water (from 4 to 148 fathoms, Smith) in probably all our lakes. It forms a large part of the

Fig. 5.

food of many of our fishes: It has been reported by Smith from Lakes Superior and Michigan. To these I can add the Georgian Bay and Lake Ontario. It also occurs in the Scandinavian Lakes and in Ireland.

Localities.—LAKES SUPERIOR and MICHIGAN (Smith).

GEORGIAN BAY: Near South Watcher Id., 20 f., sand (Wa).

LAKE ONTARIO: Near Port Credit (C).

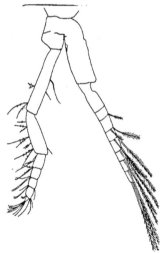

Fig. 6.

Stomachs of whitefish, Lake Superior (Smith), and of herring, *Clupea aestivalis*, in Lake Ontario (C).

The identification of the *Mysis* of our lakes with that occurring in the Scandinavian lakes seems at first sight very surprising. It is impossible for it to have been transported from one place to the other. This makes it practically certain that they have both been derived independently from one of the marine species, probably *Mysis oculata*. Smith, after comparing the American with European specimens, states that he is unable to find any differences. I have had no European specimens for comparison, but a comparison with Sars' account (1867), has shown me no differences, with the possible exception of the third and fourth pleopods of the male. But as these appendages differ in different individuals from our lakes and as the figures of Sars are evidently from a somewhat immature male, I hesitate to ascribe any importance to the differences noticed. A careful study of an abundance of material may yet show that our species is distinct from the European. Length about 16 mm.

Order DECAPODA.

The forms belonging to this group are larger than those of the other groups. There are a single shrimp and eight species of crayfishes from the region of the great lakes.

Key to the Genera.

A$_1$ Third from last pair of thoracic limbs not chelate (provided with pincers).
....*Palaemonetes.*
A$_2$ Third from last pair of thoracic limbs chelate................*Cambarus.*

Palaemonetes paludosa (Gibbes). Fig. 7.

Hippolyte paludosa Gibbes, 1851, p. 197.
Palaemonetes exilipes Stimpson, 1871, p. 130; Smith, 1874, p. 641.
Palaemonetes paludosa Kingsley, 1878, p. 97; Underwood, 1890, p. 374.
This is our only large shrimp. It is found swimming about in the bays and rivers of the Lake Erie drainage area. It has not previously been recorded from Canada.

Fig. 7.

Localities.—Detroit River and Sandusky Bay (Lake Erie) (Smith). Welland River (Wa).
Stomachs of *Perca flavescens, Apomotis cyanellus,* and *Ameiurus natalis* (Forbes, 1888).

SESSIONAL PAPER No. 39b

This species is readily distinguished from our other Crustacea by its laterally compressed, dentate rostrum (see figure), by the sharp bending of the abdomen at the third abdominal segment and by the enlarged lateral plates of the second abdominal segment. There are from 7 to 9 teeth on the dorsal edge of the rostrum and from 1 to 3 on the ventral edge. Length up to 38 mm.

Genus *Cambarus.*

All our crayfishes belong to the genus *Cambarus.* One species of the allied genus *Astacus* or *Potamobius* occurs on the coastal slope of British Columbia.

We have two monographs of the crayfishes of North America, one by Hagen (1870), and another by Faxon (1885). Ortmann (1905) has given the most recent revision of the group.

The crayfishes are bottom forms, living altogether in shallow water, not descending deeper than a few fathoms. For the most part they shelter themselves during the day under stones, plants, etc. or in holes excavated in the mud.

Eight species are properly referable to our region. Four of these have already been reported from Ontario. and to these I can add two. The other two I have not seen.

Outside of Ontario, *C. bartonii* has been reported from New Brunswick and Quebec, and *C. virilis* from several points in the middle west (Lake Winnipeg, Saskatchewan River, Red River).

Key to the Species.

A₁ First abdominal appendages of male hooked (Fig. 8, e–f). Rostrum without lateral teeth (Fig. 9, d, e—f).

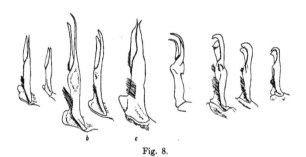

Fig. 8.

B₁ Areola of moderate width (Fig. 9, e)......................*C. bartonii.*
B₂ Areola nearly or quite obliterated in the middle (Fig. 9, d).
  C₁ Dactyl of first legs excavate at base on outer side (Fig. 11, a).*C. fodiens.*
  C₂ Dactyl of first legs not excavate at base (Fig. 11, b).........*C. diogenes.*

5 GEORGE V., A, 1915

A$_2$ First abdominal appendages of male styliform and curved (Fig. 8, d). Rostrum
without lateral teeth or occasionally with. *C. immunis.*
A$_3$ First abdominal appendages of male styliform and straight (Fig. 8, a, b, c
Rostrum with lateral teeth (Fig. 9, a, b, c).
  D$_1$ Sides of carapace with many teeth (Fig. 9, c). . . . . *C. limosus.*
  D$_2$ Sides of carapace with only one distinct tooth on each side (Fig. 9, a, b).
    E$_1$ Areola of moderate width (Fig. 9, a). First abdominal appendages of
    male with short tips (Fig. 8, a). . . . . *C. propinquus.*
    E$_2$ Areola rather narrow (Fig. 9, b). First abdominal appendages of male
    with long, tapering tips (Fig. 8, b).
      F$_1$ Sides of rostrum straight. . . . . *C. virilis.*
      F$_2$ Sides of rostrum concave . . . . *C. rusticus.*

Having examined only five of the eight species, I have not been able to devise
a key for all the species, that would be applicable to both sexes. The five species
can be readily distinguished from each other by characters of the carapace or
chelipeds, as shown in figures 9 and 10. The following keys may be found useful
and include all the species* that have been found within our borders but not all
that will probably be found to occur.

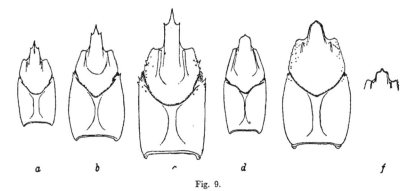

Fig. 9.

Key based upon the characters of the **carapace.** (Fig. 9).

A$_1$ Rostrum with lateral teeth.
  B$_1$ Numerous spines on sides of carapace. . . . . *C. limosus.*
  B$_2$ Only one pair of spines on sides of carapace.
    C$_1$ Areola broad (about 2 mm) . . . . *C. propinquus.*
    C$_2$ Areola narrow (about 1 mm). . . . . *C. virilis.*
A$_2$ Rostrum without lateral .teeth.

*Except *C. immunis.*

SESSIONAL PAPER No. 39b

D_1 Areola nearly or quite obliterated. . . . .*C. fodiens* and *diogenes.*
D_2 Areola broad.
  E_1 Rostrum nearly square. . . . .*C. bartonii bartonii.*
  E_2 Rostrum oblong. . . . .*C. bartonii robustus.*

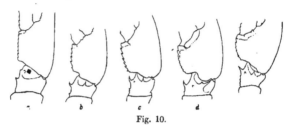

Fig. 10.

Key based upon the characters of the large **chelipeds.** (Fig. 10).

A_1 Carpus without ventral median anterior tooth. Inner border of hand or pro-
  podus straight. . . . .*C. propinquus.*
A_2 Carpus with ventral median anterior tooth. Inner border of propodus curved.
  B_1 Two teeth on ventral margin of joint between dactyl and propodus.*C. virilıs.*
  B_2 No teeth on ventral margin of that joint . . . .*C. bartonii.*
  B_3 One distal tooth on ventral margin of that joint. Dactyl not excavated at
    base. . . . .*C. limosus.*
  B_4 One median tooth on ventral margin of that joint. Dactyl excavated on
    outer side near base. . . . .*C. fodiens.*

Cambarus bartonii (Fabr.) Fgs. 8, e; 9, e, f; 10, d; 12, c.
i
Hagen, p. 75; Faxon, p. 59; Ortmann, p. 120.
    This is perhaps our commonest crayfish. It is found under stones, etc. in
running or open water, often at considerable depths. According to Abbott (see
Faxon) it sometimes burrows in muddy banks. It has been reported from St.
John, N.B., to Lake Superior on the north.
    Localities. — LAKE SUPERIOR (Hagen). Searchmont (Algoma District),
(Williamson).
    GEORGIAN BAY: Giant's Tomb, Go Home, Shawanaga, Bustard Ids., French
River (Wo); Go Home from fish nets (Wa); Copperhead Id.
    GRANT RIVER (Brant Co.), WELLAND RIVER.
    NIAGARA (Hagen).
    HUMBER and DON RIVERS near Toronto (Hagen).
    IROQUOIS (C).
    Length up to 100 mm. Antennae frequently longer than body. Rostrum
without lateral teeth, in shape from nearly square to rather long rectangular. Areola
from 1/7 to 1/10 the width of the carapace. Only one distinct spine on each side

5 GEORGE V., A, 1915

of carapace, although there are numerous tubercles on each side, particularly anteriorly. In large specimens, the fingers of the large chelipeds are relatively very long, narrow and curved. The propodus or hand may be even more than two-thirds as long as the body. The annulus ventralis of the female (Fig. 12, c) has a very small excavation which is almost in the middle line. It is sometimes on the right side (lower figure) and sometimes on the left (upper figure). This recalls the dimorphism that has long been known to exist in the males and that was first described by Hagen and that affects the first pair of abdominal legs. Whether this dimorphism in the female is strictly comparable with that in the male may be doubted. It is more like the *inversio viscerum* that occasionally occurs in many animals. I have observed it in other species of *Cambarus* although not in as well marked a state as in this species.* The specimens show roughly about equal numbers of the two kinds. The dimorphism shown in the first abdominal appendages of the males of this species is represented in Fig. 8, e.

This species has a number of varieties. From western Ontario I have seen only the form known as *robustus* (Fig. 9, e). From Iroquois in eastern Ontario, I have received specimens both of *robustus* and of the tpyical *bartonii*, with a square rostrum. The latter appears to mature at a much smaller size. A male, 23 mm. long, has the first abdominal appendages well developed and extending forward between the fourth pair of thoracic legs. In a male *robustus*, 33 mm. long, the first abdominal appendages are small and rudimentary, not extending in front of the fifth thoracic legs. Hagen considered these two forms as distinct species but Faxon subsequently united them into one. It is probable that further study will show that they are distinct.**

Cambarus fodiens (Cottle). Figs. 8, f; 9,d; 10,e; 11,a; 12,e.

Astacus fodiens Cottle, 1863, p. 216.
Cambarus argillicola Faxon, 1885, p. 76.

This appears to be the common burrowing crayfish in Ontario. It is found in swamps, etc. which become dry in the summer. At this time it retreats to its burrows, the mouths of which are surmounted by the so-called 'chimneys' which are formed by pellets of mud.

Cottle records it from Ontario, but does not give the locality. A few years previous to the time of publication of his article, he was residing at Woodstock. Faxon records it from Toronto and Detroit. The only adult specimens I have seen were given to me by Dr. E. M. Walker, who had received them from a student but without any record of the locality. Strathroy (H. B. Sifton).

Length from 60 to 70 mm. (76 mm. according to Faxon). Rostrum without lateral teeth, similar to that of *Cambarus bartonii robustus* but without thickened margins and with the tip well bent down. Carapace practically without lateral spines or tubercles. Depth of carapace (except in young individuals) equal to or

---

*Andrews (Proc. Bost. Soc. Nat. Hist., vol. 32, 1906, p. 477) found it in four out of five species examined, and thinks it may be general in *Cambarus*.
**Ortmann in Williamson (1907) reports the typical *bartonii* from near Lake St. John, Quebec, as well as from Searchmont, Algoma District.

greater than breadth (in *C. bartonii* it is only two-thirds of breadth).   Areola not entirely obliterated at any point, but nearly so.   The excavation at the base of the movable finger of the large chelae enables one to readily recognize this species.

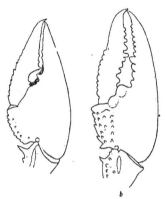

Fig. 11.

Faxon apparently had no knowledge of Cottle's article.   The description of Cottle leaves no doubt as to the identity of his species with that of Faxon.   Cottle gives an account of the burrowing habit and of the duration of the spawning period.

Cambarus diogenes Girard. Fig. 11, b.

C. obesus Hagen, p. 81.

C. diogenes Faxon, p. 71.   Ortmann, p. 120.

This is the common burrowing form of the eastern United States.   Hagen records it from Lakes Erie and Ontario and Faxon from Detroit.   It has not yet been found within our borders but probably occurs.

Length up to 111 mm. (Faxon).*   Rostrum without lateral spines, areola obliterated in the middle, first abdominal appendages of male hooked, movable finger of large chelae without excavation at base.

Cambarus immunis Hagen. Fig. 8, d.

Hagen, p. 71; Faxon, p. 99.   Ortmann, p. 113.

This species is reported by Faxon as being found concealed among weeds in muddy pools and ditches connected with the Detroit River.   Localities:— Twenty-mile creek near Tintern, Lincoln County.

Length up to 3·2 inches.   Rostrum usually without lateral spines.   Areola narrow.   Movable finger of large chelae usually excised near base on outer side.   First abdominal appendages of male styliform and curved.

*124 mm. (Williamson).

5 GEORGE V., A, 1915

Cambarus limosus (Raf.)   Figs. 8,c; 9,c; 10,c; 12,d.

C. affinis. Hagen, p. 60; Faxon, p. 86.
C. limosus Ortmann, p. 107.

This appears to be one of the commonest and largest species of the eastern United States and is the one usually sold in the markets, according to Faxon. According to Abbott it is mostly found in the rivers under flat stones in deep water.

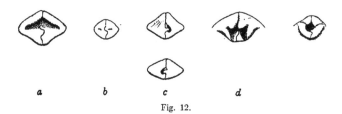

a   b   c   d

Fig. 12.

Localities.*—LAKE SUPERIOR (Faxon).
LAKE SUPERIOR and NIAGARA (Hagen).
IROQUOIS, Ontario (C).

Length up to 120 mm. Areola of moderate width, about 1/12 width of body. Rostrum with well developed lateral spines and a rather long narrow tip. Numerous spines on sides of carapace, chiefly in front of cervical groove and along its posterior margin. First abdominal appendages of the first form of male with tapering tips, the extremities being bent away from each other. Annulus ventralis of female with a transverse sinus behind and two projections in front, one on either side of a median depression.

Cambarus propinquus Girard.   Figs. 8,a; 9,a; 10,a; 12,b; 13,a.

Hagen, p. 67; Faxon, p. 91; Ortmann, p. 112.

This is our smallest species. It is generally distributed over the whole region. On the whole it keeps nearer to the shore than the other open water species and is at times found in quite stagnant water (small inland lakes along shore of Georgian Bay).

Localities.—LAKE SUPERIOR (Hagen): St. Mary's River, Heyden, Searchmont (Williamson).

GEORGIAN BAY: Sturgeon Bay, McCoy Id., Shawanaga, Bustard Ids., French River, Killarney, Tamarac Bay (Manitoulin Id.), Fitzwilliam Id. (Wo); Go Home, Santa Gre, Shawanaga, Tamarac Bay (R); Collingwood, Go Home (Wa).

WALKERTON (I. A. Sinclair), GRANT RIVER (Brant Co.), WELLAND RIVER (C. O. E. Kister).

---

* Ortmann considers that the records of Faxon and Hagen are incorrect, and that this spe cies does not occur in the lake region. I did not myself collect the Iroquois specimens.

DETROIT RIVER, ST. CLAIR RIVER, (Faxon).
STRATHROY (H. B. Sifton).
NIAGARA (Hagen).
LAKE ONTARIO (Girard).
Toronto (Faxon), IROQUOIS (C).

Fig. 13.

Stomachs of *Micropterus dolomieu* (C), *Lota maculosa, Micropterus dolomieu* (Forbes).

Length up to 65 mm. Usual length of adult individuals, from 40 to 50 mm.

Rostrum with lateral spines. A single spine on each side of carapace. Areola broad, from ⅛ to ¼ of width of body. No teeth on anterior border of third joint of large legs (Fig. 13,a). No middle anterior spine on ventral side of carpus ol large legs. Inner border of hand or propodus quite straight. First abdominaf appendages of male with short straight tips. Annulus ventralis of female without sulcus or processes.

Cambarus virilis Hagen. Figs. 8,b; 9,b; 10,b; 12,a; 13,b.

Hagen, p. 63; Faxon, p. 96; Ortmann, p. 113.

This species appears to be quite abundant in the Georgian Bay but not in Lake Ontario. Like the last species it occurs in open water and also in rather stagnant pools, and in depth down to 8 fathoms.

Localities.—LAKE SUPERIOR (Hagen).

GEORGIAN BAY: Waubaushene, Giant's Tomb, Go Home, McCoy Id., Shawanaga, Bustard Ids., Killarney, Tamarac Bay, (Wo); Go Home, (Wa); Wiarton Collingwood, Bustard Ids., Killarney, (R).

LAKE ROSSEAU.

TORONTO (Hagen); Sandy Lake (Ortmann).

Stomachs of *Micropterus dolomieu* (C), *Perca flavescens, Anguilla chrysypa* and *Amia calva* (Forbes).

Length up to 90 mm. (as much as 6¾ in. according to Bundy). Usual length of mature specimens, from 60 to 75 mm.

Rostrum with lateral spines. A single spine on each side of the carapace. Areola narrow (1/21 to 1/15 of the width of the body). Several teeth along anterior border of third joint of large legs (Fig. 13, b). A middle anterior spine on ventral side of carpus of large legs and occasionally a small accessory one between this spine and the inner spine of the carpus (Fig. 10, b). First abdominal appendages of male with long somewhat curved tips. Annulus ventralis of female with a deep and broad transverse sinus.

Cambarus rusticus Girard.

Hagen, p. 71; Faxon, p. 108.

This species has been reported by Hagen from Lake Superior and by Faxon from Lake Erie. It will doubtless be found within our limits.

Length up to 73 mm. Rostrum with lateral spines, its margins concave. Areola narrow. First abdominal appendages of male with long, straight or somewhat curved tips.

LIST OF REFERENCES.

For further bibliography the reader is referred to the following papers,— that of Underwood for general bibliography prior to 1885, that of Richardson for articles on the Isopoda, that of Weckel for articles on the Amphipoda, and that of Faxon for articles on the crayfishes.

COTTLE, T. J.—On the Two Species of Astacus found in Upper Canada. *Canad. Journ.*, n. ser., vol. VIII, p. 216. 1863.

FAXON, W. A.—A Revision of the Astacidæ. *Mem. Mus. Comp. Zool. Harv.*, vol. X, no. 4, 1885.

FORBES, S. A.—On the Food Relations of Fresh-Water Fishes: a Summary and Discussion. *Bull. Ill. Lab. Nat. Hist.*, vol. II, p. 475. 1888.

GIBBES, L. R.—On the Carcinological Collections of the United States, etc. *Proc. Am. Ass. Adv. Sc.*, vol. III, p. 167. 1851.

HAGEN, H. A.—Monograph of the North American Astacidæ. Ill. *Cat. Mus. Comp. Zool. Harv.*, no. III. 1870.

JACKSON, H. H. T.—A Contribution to the Natural History of the Amphipod, Hyalella Knickerbockeri (Bate). *Bull. Wisc. Nat. Hist. Soc.*, vol. X. p. 49. 1912.

KINGSLEY, J. S.—Notes on the North American Caridea, etc. *Proc. Acad. Phil.*, vol. XXX, p. 89. 1878.

NICHOLSON, H. A.—Contributions to a Fauna Canadensis, etc. *Canad. Journ.*, n. ser., vol. XIII, p. 490. 1873.

ORTMANN, A. E.—The mutual affinities of the species of the genus, Cambarus, &c. *Proc. Amer. Phil. Soc.*, vol. XLIV, p. 91, 1905.

RICHARDSON, H.—Monograph on the Isopods of North America. *Bull. U. S. N. Mus.*, no 54. 1905.

SMITH, S. I.—The Crustacea of the Fresh Waters of the United States. *Rep. U. S. F. Comm.*, pt, 2 for 1872–1873, p. 637. 1874.

SMITH, S. I. AND VERRILL, A. E.—Notice of the Invertebrata dredged in lake Superior, etc. *Am. Journ. Sc.*, ser. 3, vol. II, p. 448. 1871.

SESSIONAL PAPER No. 39b

STIMPSON, W.—Notes on North American Crustacea. No. III. *Ann. N. Y. Lyc. Nat. Hist.*, vol. X. p. 92. 1871.

UNDERWOOD, L. M.—List of the Described Species of Fresh-Water Crustacea from America North of Mexico. *Bull. Ill. Lab. Nat. Hist.*, vol. II, p. 323. 1886.

WECKEL, A. L.—The Fresh-Water Amphipoda of North America. *Proc. U. S. N. Mus.*, vol. XXXII, p. 25. 1907.

WILLIAMSON, E. B.—A Collecting Trip North of Sault Ste. Marie, Ontario. *Ohio Naturalist*, vol. VII, No. 7, p. 129, 1907.

VII.

NOTES ON THE HIRUDINEA OF GEORGIAN BAY.

By Miss C. G. S. Ryerson, B.A., University of Toronto.

Some time ago Dr. E. M. Walker placed in my hands, for the purpose of identification and morphological study, the entire collection of Hirudinea belonging to the Georgian Bay Biological Station. This collection contained numerous specimens obtained in different seasons and particularly the special collection made by Mr. R. C. Coatsworth in 1910. This collection was accompanied by extensive field notes which were kindly placed at my disposal by Mr. Coatsworth, and which have been freely used. Useful information has also been supplied by Mr. A. D. Robertson. In several cases it has been difficult to make satisfactory determination on account of lack of information on anatomical features, a study of which is now in progress.

Hitherto, collections of leeches have been made chiefly in the lakes and rivers of the United States and little work has been done in Canadian waters. Bristol (1899) in his paper on the Metamerism of *Nephelis*, mentions having received several specimens from Toronto. Verrill (1872) found *Ichthyobdella punctata* in Lake Superior. Nicholson (1872) in his "Contributions to Fauna Canadensis" describes four species from Lake Ontario. Moore (1905) in his paper on "Hirudinea and Oligochæta of the Great Lakes," describes species which, except for two parasitic forms are common around Toronto.

There appear to be four families represented in the Georgian Bay region, namely, the *Glossiphonidæ*, *Hirudinidæ*, *Erpobdellidæ*, and *Ichthyobdellidæ*. Judging from the field notes, there is a great uniformity in the environmental conditions of the various species. Whether free-swimming forms such as the *Hirudinidæ* and *Erpobdellidæ* or less active creeping forms such as the *Glossiphonidæ*, leeches, as a rule, choose sheltered places where their movements will not be hampered by the motion of the water. Further in such places are found snails, oligochætes and other invertebrates which form the food of the majority of species. Semi-permanent parasites such as *Placobdella parasitica* or the *Ichthyobdellidæ* accommodate themselves to the environment of the host, but during the breeding season retire to the shelter of plants or stones.

The following is a list of the species herein described:

    I.   Family Glossiphonidæ.

        A.   Genus **Glossiphonia,** Johnston.

            1.   *G. stagnalis,* Linn.

            2.   *G. fusca,* Castle.

            3.   *G. nepheloidea,* Graf.

            4.   *G. heteroclita,* Linn.

            5.   *G. complanata,* Linn.

B. Genus **Placobdella**, Blanchard.
    6. *P. parasitica*, Say.
    7. *P. rugosa*, Verrill.
    8. *P. montifera*, Moore.
    9. *P. phalera*, Graf.
    10. *P. picta*, Verrill.
II. Family HIRUDINIDÆ.
  C. Genus **Macrobdella**, Verrill.
    11. *M. decora*, Say.
  D. Genus **Haemopis**, Savigny.
    12. *H. marmoratis*, Say.
    13. *H. grandis*, Verrill.
III. Family ERPOPDELLIDÆ.
  E. Genus **Erpobdella**, Blainville.
    14. *E. punctata*, Leidy.
  F. Genus **Nephelopsis**, Verrill.
    15. *N. obscura*, Verrill.
IV. Family ICHTHYOBDELLIDÆ.
  G. Genus **Piscicola**, Blainville.
    16. *P. milneri*, Verrill.
    17. *P. punctata*, Verrill.

Family GLOSSIPHONIDÆ.

Leeches of small to medium size, rather short and broad; caudal sucker usually broad and flat; the somites of the middle region of the body usually of three annuli and in most species sensillæ and cutaneous papillæ, in some species several series. Eyes 1–4 pairs, situated close to the median line. The first may be compound, the others simple. From the mouth situated in the oral sucker the pharynx passes backwards and forms a sheath for the protrusible proboscis. This is succeeded by the œsophagus and the crop. The crop possesses from one to ten pairs of lateral cæca and the stomach four pairs. In the literature of the group this family is characterized by the possession of salivary glands, but five species possess in addition to these a pair of œsophageal glands (Hemingway, 1908). The eggs and young of these forms are carried on the ventral surface of the body of the parent. The adult individuals are usually found clinging to fish or frogs, sticks or stones.

Genus Glossiphonia, Johnston.

Moderately depressed or elongated, tapering more or less toward the anterior end. Eyes 1–3 pairs, simple; cutaneous papillæ present in some species but usually not conspicuous. Pharyngeal glands diffuse; gastric cæca 1–7 pairs, not greatly branched. Sperm ducts forming long open loops. Chiefly free-living forms.

SESSIONAL PAPER No. 39b

Glossiphonia stagnalis, (Linn.) Johnston.

Hirudo bioculata, Bergmann (1757).
Hirudo stagnalis, Linnæus (1758).
Clepsine modesta, Verrill (1872).
Helobdella stagnalis, Blanchard (1896).

In the collection there are fifty-seven adult specimens, together with a number of young which appear to belong to this species. Size small, usually not exceeding an inch in length when extended. The crop, when empty, shows one pair of long posterior cæca, which lie alongside the stomach. When the crop is full, five or six pairs of cæca may be visible, but usually only three or four pairs can be seen. At the twelfth annulus there is a small brown chitinous plate on the dorsum, which marks the position of the nuchal gland; this plate is usually visible to the naked eye and furnishes a means of distinguishing the species. The simple eyes in the fourth somite of this leech correspond to those of *G. nepheloidea*. The color is generally white to semi-transparent, but some specimens are flecked with grey.

In specimens collected in the vicinity of Toronto last autumn a pair of conspicuous white spots, close to the dorsal median line were observed. These, on closer examination, proved to be the atria showing through the transparent body-wall. These spots were not observed in the Georgian Bay specimens, but since they become less conspicuous after preservation, it is probable that their absence is due to the effects of the preserving fluids, or possibly there is a difference in respect of the season of the year.

In one instance a leech of this species was found attached to a tadpole in a pool where tadpoles were numerous. In other instances, specimens were taken in dredgings from swampy bays or from under stones or again on aquatic plants.

Glossiphonia fusca, Castle.

Clepsine papillifera, var. *lineata*, Verrill (1874).
[Not *Hirudo lineata*, O. F. Müller (1874)].
Glossiphonia lineata, Moore (1898).
Glossiphonia fusca, Castle (1900).

Twenty-one specimens have been identified as belonging to this species. Size from 5 to 13 mm. in length and from 1·5 to 3·5 mm. in breadth; crop of five or six pairs of caeca, not greatly branched. Shape very similar to that of *G. stagnalis* but not so capable of extension. The color varies from yellowish grey to almost white. On the dorsal surface there are usually three to five series of rather prominent papillae. Along the line of the papillae there is a more or less complete longitudinal white band and the papillae are frequently tipped with black. The simple pair of eyes are located on the fourth annulus or in the furrow between the third and fourth. Several specimens were taken from the lower sides of sticks.

5 GEORGE V., A. 1915

Glossiphonia nepheloidea, Graf.

Clepsine nepheloidea, Graf (1899).
Glossiphonia elongata, Castle (1900).   •

The collection contains but three specimens identified as this species. They are rather elongated and worm-like in form with weak suckers. In size similar to *G. stagnalis*, but capable of greater extension. The single pair of crop diverticula are shorter than those of *G. stagnalis*. The single pair of eyes are large and widely separated. Color light brownish to white.

These specimens were obtained from dredgings at a depth of from one to six feet in a soft muddy bottom.

Glossiphonia heteroclita, Linn.

Hirudo heteroclita, Linn. (1761).
Hirudo hyalina, O. F. Müller (1774).
Clepsine hyalina, Moquin Tandon (1826).

Four specimens identified as this species were from 5 to 11 mm. in length and from 1·5 to 3·5 in breadth. In shape they are similar to *G. stagnalis*, but not so extensible. In color they are white to semi-transparent, which enables one to distinguish the six pairs of gastric caeca when filled with blood. The three pairs of eyes, arranged in two parallel lines, close to the median line, show clearly against the white background.

The specimens were taken from the lower sides of stones along the shore of a small bay and in a pool.

Glossiphonia complanata, (Linn.) Johnston.

Hirudo complanata, Linn. (1758).
Clepsine elegans, Verrill (1874).
Clepsine patelliformis, Nich. (1872).

Thirty-six specimens have been referred to this species. In size, they vary from 10–16mm. in length and from 2 to 6·5 mm. in breadth, though larger specimens have been found. Individuals of this species are larger than those of the species already described. The margins are thick and the head is not distinctly widened. The three pairs of eyes are usually situated on annuli 2, 3, and 4. The second pair, largest and farthest apart, is said by Castle (1900) to correspond to the single pair of eyes in *G. stagnalis*. Seven pairs of gastric caeca. The ground color varies from brown to bright green dorsally, paler ventrally. On the dorsal surface are two brownish lines running from just behind the eyes backward. These lines are usually more or less interrupted by whitish spots metamerically arranged. Elsewhere on the dorsal surface appear series of white or yellow flecks. On the ventral surface there is also a pair of longitudinal brown lines but rather paler. Dorsal cutaneous papillae are present but are not conspicuous.

An examination of serial sections revealed a pair or tubes which come into view several microns posterior to the female genital pore. These tubes appear to bend on themselves, the outer arm ending blindly at about the point where the first pair of gastric caeca appear. The other arm passes backward and appears to connect posteriorly with the seminal duct. This blind end may be merely the anterior end of the outer arm of the seminal loop. Also between the genital pores, appears the end of a loop, similar in structure to the oviducts, which passes backward and connects with the oviducts posterior to the female pore. The point of attachment is approximately that of the caecum attached to the oviduct of *P. montifera* (Moore 1912).

These specimens were obtained from dredgings in a channel from 3 to 5 fathoms deep, from under shells, stones and logs in small inlets or in pools.

Genus **Placobdella,** Blanchard.

The collection affords five species belonging to this genus. Form more or less broad and flattened, crop with seven pairs of caeca which are usually more or less branched. As commonly described, the species of this genus possess compact salivary glands, but in three of the five species these glands were found to be diffuse. In all the species, however, there was observed a pair of glands opening into the oesophagus similar to those mentioned by Hemingway (1908) as occurring in *Placobdella pediculata*. These glands are lined with columnar epithelium, whereas the salivary glands are unicellular. Another generic feature is the absence of a seminal loop, but, in one species, viz., *P. picta*, the seminal duct was observed to form a distinct loop.

Placobdella parasitica, (Say) Moore.

Hirudo parasitica, Say (1824).
Glossiphonia parasitica, var. *plana,* Castle (1900).
Placobdella parasitica, Moore (1901).

Sixty-five specimens of the collection have been assigned to this species. In size they vary from 8 to 60 mm., in length and from 1·5 to 18 mm. in breadth. The color varies from dark or light brown to deep green with a series of yellow markings along the margins and a yellow vitta in the dorsal median line. This vitta may reach the length of the body, expanding at intervals of about three somites or it may be confined to a few somites at the anterior end of the body. Some specimens have an intermediate series of yellow spots. The ventral surface is striped longitudinally with light and dark. Dorsally there are three series of more prominent cutaneous papillae and several series of smaller ones. These papillae are always smooth, though inconspicuous in some specimens. The oesophagus is long and looped and at the anterior end of the first loop are the long oesophageal glands.

The greater number of specimens were found attached to turtles of various kinds, *Clemmys guttatus, Aromochelys odoratus, Chelydra serpentina* and *Chrysemys picta.* One was found on a perch, another on a *Macrobdella decora* and numbers were taken from the lower sides of stones, sticks, etc., in small bays or lakes.

5 GEORGE V., A. 1915

Placobdella rugosa, (Verrill) Moore.

Clepsine ornata, var. *rugosa*, Verrill (1874).
Glossiphonia parasitica, var. *rugosa*, Castle (1900).
Placobdella rugosa, Moore (1901).

Thirty-five specimens possess the general form and coloration characteristic of this species. Sizes from 10 to 59 mm. in length and from 3 to 17 mm. in greatest diameter; in shape similar to *P. parsitica*, broad, flat and blunt at the anterior end. In color also these two species are alike except that in *P. rugosa* the contrasts are rather less striking. In *P. rugosa* there is usually an interrupted, dark, dorsal median band. The cutaneous papillae on the dorsal surface are arranged as in *P. parasitica* but the surface of these is especially rough, hence the specific name *rugosa*. The oesophageal glands are in this species also, long, blind tubes.

The collections were made from dredgings, in small lakes and bays, from under sticks and stones in pools or ponds or from the turtles *Chelydra serpentina* and *Chrysemys picta*.

Placobdella montifera, Moore.

Not *Clepsine carinata*, Diesing (1858).
Clepsine papillifera var. *carinata*, Verrill (1874).
Hemiclepsis carinata, Moore (1901).

Among the leeches collected at Georgian Bay, there are twenty-six specimens with the discoidal head and three prominent series of papillae characteristic of . this species. Shape more slender and less flattened than most of the species of this genus. In size varying from 5 to 21 mm. in length and from 1·5 to 5 mm. in breadth. The ground color is greenish or brown. A darker band is frequently to be seen in the dorsal median line. Along the margin a yellowish band may be seen and the papillae may be tipped with yellow.

Moore (1912) describes compact salivary glands for this species but the dissection of the specimens showed that these glands are diffuse rather than compact.

An interesting similarity has been observed between the position and the structure of the oesophageal glands in *P. montifera* and *P. pediculata*. In one specimen, a tube-like body is attached at the male pore, but it has not been found possible to explain its presence.

The majority of the specimens were collected from the lower sides of logs, stones, or shells of molluscs. Several were obtained by dredging at a depth of six fathoms. One specimen was found clinging to a garpike (*Lepisosteus osseus*) and another to a sunfish (*Eupomotis gibbosus*).

Placobdella phalera, Graf.

Seven specimens showing the markings characteristic of this species were collected. In size varying from 10 to 18 mm. and from 2 to 6 mm. in breadth. Body broad and flattened, tapering to a rather slender anterior end. Colour brown with a striking greenish tinge. From the anterior end backward to about the seventh annulus, the dorsal surface is yellow and a yellow band passes around the body in the region of the eleventh or twelfth annuli. Along the margin of the body are yellow spots on the tips of the first and third annuli of the somite. In some specimens there is a median dark brown line interrupted in some cases by yellow patches. There are, usually, three series of papillae, but, in one specimen, five were observed.

As shown by dissection, there is a long looped oesophagus with a short pair of oesophageal glands connected with it. The diffuse salivary glands attached at the base of the proboscis, are of large size and stretch outward and backward, not forward as in other species. In one specimen the stalk of the posterior sucker was found to be quite long. In another specimen small bodies which appeared to be spermatophores were found attached to the body.

The specimens collected were taken from dredgings at a depth of from one to six feet and from the lower sides of stones or clam-shells on a sandy shore.

Placobdella picta, (Verrill) Moore.

Clepsine picta, Verrill (1872).

In the collection there is only one specimen answering to the description of this species. This specimen measures 29 × 5 mm. and is broad and flat in shape. The dorsum is marked with numerous longitudinal lines of deep green. Verrill describes this species as having a marginal series of yellow spots on the tips of the first and third annuli of the somite. In the preserved specimen the presence of these spots cannot be observed with certainty and the notes give no information on this point. The colour of the ventral surface is a flecked green. In the living specimen numerous papillae were observed on the dorsum.

There are diffuse salivary glands present and the oesophageal glands are long and similar in form to those of *P. rugosa* or *P. parasitica*. The oesophagus is almost straight, not looped as in the other species of this genus. The seminal duct forms a long loop connecting with the testicles anteriorly, much as in the species of *Glossiphonia*.

The single specimen of this species collected was found on the lower side of a clam shell on a sandy bottom.

Family HIRUDINIDAE.

Distinguished by the presence in most species of five pairs of eyes, a five-ringed somite, three toothed jaws and a large mouth occupying the entire oral sucker. There is no protrusible proboscis.

These leeches are free swimmers and subsist upon the blood of animals or upon weaker invertebrates.

Genus **Macrobdella**, Verrill.

Characterized by large size and the presence of metamerically arranged black and red spots on the dorsum.

5 GEORGE V, A. 1915

Macrobdella decora, (Say) Verrill.

Hirudo decora, Say (1824).
Hirudo decora, Leidy (1868).
Macrobdella decora, Verrill (1872).

Thirty-six adults and several young specimens in the collection possess the markings and general form peculiar to this species. In size, there is considerable variation, the largest specimen being 120x18 mm. The body is rather long and narrow, soft and limp. Dorsally the ground color is some shade of olive green, with conspicuous red and black dots metamerically arranged. The male and female genital pores are separated by five annuli and posterior to the female pore appear four openings which are arranged in a quadrate figure. These are the openings of the copulatory glands.

This form was usually found free in channels, ponds or bays and also clinging to sticks or stones.

Genus Haemopis, Savigny.

Among the *Hirudinidae* examined there are two species of leeches with the mottled or sooty gray colour characteristic of this genus. There is no appearance of metameric arrangement in the blotches. Especially in the contracted specimens, there is a noticeable angle in the posterior half of each annulus.

Haemopis marmoratis, (Say) Moore.

Hirudo marmorata, Say (1824).
Aulastomum lacustri, Leidy (1868).
Haemopis marmoratis, Moore (1901).

Twenty-one specimens have been assigned to this species. Size 40 to 90 mm. in length and 8 to 11 mm. in breadth. The colour in a majority of cases is dark and mottled or almost black, but in a few instances the ground color tends toward light grey. The angle in the posterior half of each annulus is quite prominent in this species and the body is more rounded at the margins than in *H. grandis,* the other species collected in this district.

These specimens were taken from the lower sides of stones in small bays or channels or from water plants.

Haemopis grandis, Verrill.

Semiscolex grandis, Verrill (1874).

Six specimens were identified as belonging to this species. Size large, 100 to 160 mm. long and 20 to 30 mm. wide, in the contracted condition. Colour dorsally slightly mottled, greenish-grey, ventrally plain. Lateral angle sharp; the male genital pore, twenty-four rings posterior to the mouth.

Dissection showed eleven pairs of testes, though ten pairs appears to be the more usual number (Moore, 1912). The gastric caeca in the specimen dissected were much larger in both dimensions than in the small *H. marmoratis.*

One specimen was obtained by dredging in the French River at a depth of twenty-five or thirty feet. The others were found in small lakes and among the islands. This leech is reported to have been seen feeding upon dead fish, but I have not been able to obtain positive information on this point.

Family ERPOBDELLIDAE.

The medium size, long, slender form and firm muscular body distinguish this family from others. The presence in the collection of one hundred and eighty specimens in the two species of this family shows that the group is well represented in the Georgian Bay region.

Genus **Erpobdella**, Blainville.

The representatives of this genus are slightly depressed in the posterior region of the body and rounded anteriorly. The five annuli of the complete somite are approximately equal in length. In some specimens the fifth annulus was slightly enlarged and showed signs of division, but dissection revealed the form of sperm duct characteristic of this genus.

Erpobdella punctata, (Leidy) Moore.

Nephelis punctata, Leidy (1870).
Erpobdella punctata, Moore (1901).

Of all the leeches in the collection, this species appears to have been the one most commonly taken. Specimens were obtained by dredgings in sandy channels or muddy bays and, along every sheltered pebbly shore either the leech itself or its cocoons were to be found on the lower sides of sticks and stones. The variati ons in color are considerable. Some specimens are light or dark brown with practically no markings while others show a series of dark flecks or dots on either side of the median line.

While examining these specimens I noticed that a considerable number possess four pairs of eyes, two pairs in somite II and also two pairs in somite IV. The usual number is three pairs (Moore 1901).

Genus **Nephelopsis**, Verrill.

Size large, body much depressed posteriorly. All annuli of complete somites more or less distinctly subdivided.

Nephelopsis obscura, Verrill.

Seventy-four specimens, large and small, have been thus identified. In size, these vary from 13 to 42mm. in length and from 3 to 5 mm. in breadth. The leech is evidently of the *Herpobdellidae*, but the greater depression of the body posteriorly and the greater diameter at that point mark it as distinct from *E. punctata*. The margin of the body is sharp and may tend upward in the preserved specimen.

5 GEORGE V., A. 1915

The color is usually light, mottled grey but in some the dorsum is blotched with dark pigment.

Cocoons similar to those described by Verrill (1872) were found on the lower sides of stones in a pool beyond the reach of the waves.

Family ICHTHYOBDELLIDAE.

This family is represented in the collection by forty-two specimens apparently belonging to at least two species. In the one type there is a slender, rounded body and large explanate suckers; in the other the suckers scarcely exceed the body in diameter while the rounded, slender body tapers toward the anterior end. Both forms possess a protusible proboscis and are parasitic on fishes.

Genus Piscicola, Blainville.

For the determination of these forms Verrill's paper (1872) was used and two species have been identified with his genus *Ichthyobdella.*

An examination of the external features of the one species would lead one to suppose that it belongs to this genus. The form is slender and rounded, the suckers large and explanate. So far as can be ascertained, fourteen annuli constitute a complete somite. No papillae or sensillae are in evidence. There are two pairs of eyes widely separated on the base of the head.

Where information is given the specimens obtained at the Biological Station were found clinging to lake trout (*Cristivomer namaycush*), but the same form has been found free in the waters of Lake Ontario.

Piscicola milneri, Verrill.

Ichthyobdella milneri, Verrill (1872)

In size this leech varies from 12 to 35 mm. in greatest diameter. The body slender and rounded, tapering toward the anterior end. There are two pairs of eyes plainly visible. The anterior pair are larger and farther apart than the posterior pair. The suckers are two or three times as wide as the body and are deeply cupped and excentrically attached. In the lateral region of the body seventeen pairs of vesicles were observed. The color is deep yellow with a symmetrical pattern in brown. There are four longitudinal yellow bands, dorsal median, lateral and ventral median. The brown color in the form of irregular pigment cells, is laid down in twelve longitudinal lines which are arranged in four groups of three, each group alternating with the yellow bands.

In each group of brown lines the uppermost line is more or less broken, showing a tendency to a series of heavy brown metameric bars. On the posterior sucker twelve dark brown eye-spots were observed. Verrill speaks of a tinge of green, but this was not observed in any of the specimens in the collection. The absence of this color, however, may be due to the effect of the preserving fluids.

These specimens were taken on lake trout (*Cristivomer namaycush*).

Piscicola punctata, Verrill.

In size these specimens vary.from 15 to 30 mm. in length and from 2 to 3mm. in greatest diameter. The form is rounded and slender, and the division of the body into anterior and posterior portions is evident. The suckers are slightly explanate but do not exceed the body in diameter nor are they so deeply cupped as in *P. milneri*. The separation of the suckers from the body is not so well defined as in most Piscicolas.

In these specimens one pair of eyes has been observed but the number of annuli in a complete somite has not been determined.

At about the anterior end of the middle third of the body is an area covering apparently seven annuli. This area has a more or less swollen, porous appearance and contains the genital pores. This region evidently answers to the description of a clitellum, although a distinct clitellum is described as absent in this family (Moore, 1912).

An examination of a dissected specimen reveals the presence of a protrusible proboscis, an oesophagus and a moniliform crop. Attached at the base of the proboscis are a number of whitish bodies irregular in shape and provided with long white "ducts". These are, in all probability, the diffuse salivary glands. Attached to the oesophagus about midway is a pair of bodies which resemble the oesophageal glands of *Placobdella montifera*.

The ovaries appear in the form of two elongated sacs. There are five pairs of testes. Attached in the region of the "clitellum" are four thick layers of tissue just beneath the layer of longitudinal muscle. These bodies, probably the clitellar glands, pass backward toward the posterior end of the body.

These specimens were found clinging to rock bass (*Ambloplites rupestris*).

BIBLIOGRAPHY.

1899. Bristol, C. L.—The Metamerism of Nephelis. *Journal of Morphology*, vol. XV.

1900. Castle, W. E.—Some North American Fresh-Water Rhynchobdellidæ and their Parasites. *Bull. Mus. Comp. Zool.*, vol. XXXVI, No. 2.

1898. Moore, J. Percy.—The Hirudinea of Illinois. *Bull. Ill. State Lab. Nat. Hist.*, vol. V.

1905. Moore, J. Percy.—Hirudinea and Oligochaeta Collected in the Great Lakes Region. *Bull. U. S. Bur. Fish.*, Vol. XV.

1899. Moore, J. Percy.—The Leeches of the U. S. National Museum. *Proc. U. S. Nat. Mus.*, vol. XXI.

1912. Moore, J. Percy.—The Leeches of Minnesota. *Geological and Natural History Survey of Minnesota.* Zoological series No. 5.

1872. Verrill, A. E.—Synopsis of North American Fresh-water Leeches. *Rep. U. S. Fish Comm.* (Refers to *Amer. Journ. Sc.*, vol. III).

1891. Whitman, C. O.—Description of *Clepsine plana*. *Journal of Morphology*, vol. IV.

VIII.

CONTRIBUTIONS TO THE LIFE HISTORY OF PROTEOCEPHALUS AMBLOPLITIS LEIDY.

A Parasite of the Black Bass.

By A. R. Cooper, M.A., University of Toronto.

(Plates XIX—XXI)

During the summer of 1909 the writer began a systematic study of the parasites infecting fresh-water fishes of the Georgian Bay region. In the course of this work it was noticed that the visceral organs of the small-mouthed black bass were greatly infected with the plerocercoids of some species of Proteocephalus. Up to that time Leidy's description of *Tænia micropteri* was the only reference to plerocercoids found in the bass, so that it was thought that these were individuals of that species. Furthermore, there appeared to be a close resemblance between the scolex of this form and that of *P. ambloplitis* Leidy, which was found in the intestinal tract of the same host, consequently a comparative study was undertaken to find out whether the resemblance was sufficient to warrant the view that the former was a larval stage of the latter. In order to ascertain the local distribution of the infection, adult hosts ranging in length from 22-23 cm., were taken in different localities around the Lake Biological Station on Georgian Bay, from the outlying islands and reefs some miles from shore inwards to the inland lakes and the Go Home River. The present paper is devoted chiefly to a description of certain stages of these plerocercoids and their identification with *P. ambloplitis*, but a number of observations on the life-history of this species have also been appended.

As a rule, bass of small size caught inshore are not greatly parasitized by *P. ambloplitis*, only occasionally is a young one found to contain a number of individuals of this species. Large bass, on the other hand, are invariably much infected. It is probable that the harboring of even a dozen or more adult specimens of this worm would have no noticeable effect on the fish in view of the presence of scores and even a hundred or more of echinorhynchi which are found in the pyloric cœca and intestines of every adult bass one examines. Of a small lot of bass caught near a group of islands lying about three miles from the mainshore, three, averaging 26 cm. in length, were examined for parasities, and in only one of these were adult specimens of *P. ambloplitis*, to the number of nine, found in the stomach. On the other hand the plerocercoid above-mentioned, which will be called *P. micropteri* Leidy (LaRue, '11) was well represented. In ten bass from twenty-one to to twenty-nine centimetres in length, only three harbored adults (*P. ambloplitis*) namely, two, each ten centimetres long when extended, in the first bass; two, thirty-three and ten centimetres, respectively, in the second; and three much smaller

5 GEORGE V., A. 1915

in the third. As for the bass taken up the river, no adult tape-worms were found, yet the whole aspect of the parasitic fauna of these fish otherwise presents practically no differences from that of the hosts procured farther out among the islands off the shore.

So far as the influence of seasonal changes on the presence of adult individuals of this parasite is concerned, everything seems to depend on the food-supply and its alteration. In the late spring and early summer, when the bass are inshore spawning, the food appears to consist almost entirely of minnows which are then very plentiful; later the diet is restricted to crayfish. There is, however, a variation in the proportions of these two kinds of food from season to season, a variation which obviously depends on the numbers to be found by bass on the feeding-grounds, but which has a distinct influence on the presence of cestodes in the host. Again, the earliest fish to come in for spawning in June harbor comparatively few adult tapeworms, while later, about the middle of July, more are met with. This points to a rapid growth from the oncosphere stage, as has been noted by different authors for other species.

The Occurrence of the Plerocercoid in the Host.

In 1887 Leidy described under the name of *Tænia micropteri* a plerocercoid which he found in the body-cavity of the black bass, *Micropterus nigricans*, (the green or bayou bass, now called *M. salmoides* Lacépède, but since his description was based on external features only, it is now of comparatively little value. However, it is evident that this worm is the larval stage of some species of Proteo-cephalus (LaRue, '11). Furthermore, Leidy's description of the scolex: "head large compressed spheroidal, with four subterminal, spherical bothria and a papillaform, unarmed summit; neck none..........," is so suggestive of the plerocercoids here shown in Figs. 4 and 6, that in spite of the fact that no specimens were found in the few adults of *M. salmoides* examined, and that, to my knowledge, Leidy's original specimens have not been studied in serial sections, I feel justified in concluding that, in all probability, *P. micropteri* and the plerocercoid described below belong to the same species.

A number of hosts were dissected, and all the visceral organs excepting the air-bladder and the heart were found to be infected. The following table shows to what extent this occurs taking into consideration only those plerocercoids which could be seen with the unaided eye in nine specimens of the host species:

Number.	Length in cms.	Stomach.	Intestine.	Liver.	Ovaries.	Testes.	Mesenteries and cœlomic cavity.	Spleen.	Cœca.	Kidneys.
1	23.7	......	2	......	10	......	......	......	......	......
2	32.8	1..	......	......	......	......	......	......	......	......
3	26.2	1	6	......	17	......	2	1	2	9
4	?	......	2	7	......	9	14	......	5	......
5	29.6	......	1	5	10	......	......	1	......	......
6	21.8	......	......	1	......	1	2	......	......	......
7	25.9	......	......	......	......	......	......	2	......	......
8	22.5	......	......	......	......	......	......	2	......	11
9	25.0	......	......	5	11	......	......	......	......	......

Above Table shows the occurrence of the plerocercoid in visceral organs of nine specimens of *M. dolomieu*.

From this it is seen that there is considerable variation in the numbers of the plerocercoid infecting the different organs: there is also a variation in their size. Those found in the *stomach* are very few in number and quite small. The *intestine*, on the other hand, harbors most of the plerocercoids found in the alimentary tract, their size ranging from 0·5 cm. to the adult condition (vide infra). Most of those found in the *liver* (Pl. XIX, Figs. 2, 3, 4 and 5) which, like the livers of most fishes harboring larval cestodes is much infested, average about 1 cm., the limits being from less than 1 mm. to 2 or 3 cms. as dissected out without the use of a lens or dissecting microscope. The smaller specimens are more cylindrical and compact in their structure than are the larger ones, the latter being, as Leidy describes them, "soft and white." The plerocercoids found in the *ovaries* and *testes* are somewhat flattened behind the constriction between the scolex and the body, soft and distended as if well provided with nutriment, that is, the constriction itself is deeper and the apex of the scolex is also better developed than in those found in the other of the visceral organs (Pl. XIX, Fig. 6). The scolex is attached to the outer wall or stroma of the gonad, while the body lies free away among the eggs or sperms, as the case may be, thus surrounded with a rich nutritive medium. The presence of such a food supply doubtless accounts for the greater diameter, the length remaining more nearly the same for similar stages of development. Furthermore the plerocercoids found in the gonads are on the average much larger than those found in the other viscera, another point which illustrates the influence of the surrounding tissue on the growth

5 GEORGE V., A. 1915

of the worm. The few plerocercoids which are found on the mesenteries and in the cœlomic cavity average about 1 cm. in length and resemble those found in the ovaries and testes in that they are more distended than specimens from the alimentary tract. The presence of these will be discussed below in connection with the transference of the oncosphere and its further development. A number of cases were met with in which plerocercoids about 1·5 cm. in length were protruding into the coelomic cavity through apertures in the intestinal wall. Similar apertures are often caused by the probosces of echinorhynchi, and in one case a plerocercoid was found protruding from one of them together with one of these parasites. Again, in a small number of cases larvæ were found with their scolices imbedded in the stroma of the ovaries while their bodies were lying in the cœlomic cavity. Fish No. 2, in the table, harbored only one larva whose scolex was imbedded in the wall of the stomach, the body, about twice as long as the diameter of the scolex, remaining suspended in the lumen of the tract. The scolex was surrounded by a cavity, a little larger than itself, whose diameter was that of the thickness of the wall of the stomach less a thin outer membrane separating the cavity from the cœlome of the host; and in this space were the remains of the stomach wall in a comminuted state much resembling digestive débris. In the wall of the duodenum near the pylorus of fish No. 5, there was a similar cavity containing a plerocercoid about 2 or 3 mm. in length, with its suckers invaginated, which condition will be seen below to be normal for specimens of that size. These two cases could be explained by the development of the oncosphere which had not burrowed far into the wall of the alimentary tract, and perhaps the others could be dealt with in a similar way, but the evidence, though quite meagre, seems to point to an active boring by the larva. In this connection, several authors have recorded the wandering of larvæ in the tissues of the host and in the cœlomic cavity. Those found in the *spleen* are quite like specimens taken from the liver of the host. The *kidney*, on the other hand, is infested with small spherical forms with their scolices invaginated as shown in Figs. 1b and 1c.

External Features of the Plerocercoid.

The larvae are found with or without the scolex or sucker-bearing portion evaginated. In the very young forms, (Pl. XIX, Figs. 1a, b and c) the suckers are constantly invaginated, but when a length of about 1 mm. has been reached the suckers are found evaginated. From that time until a length of 6 or 7 mm. is attained (Pl. XIX, Fig. 5) they may be found in either condition depending on the location in the host and the manner of preserving or fixing. From observations of a number of plerocercoids of all sizes from different visceral organs it may be concluded that the sucker region remains permanently evaginated after a length of about 10 mm. has been reached. However, there are exceptions, as many specimens much longer are found with the scolex in the former condition. For example, the scolices of those found in the gonads of the host are protruded where the length of the body ranges from 4–40 mm., the latter being the length of the largest specimen I have yet found. When a fixing fluid is applied to small specimens whose suckers are temporarily evaginated, there is often a sudden invagination of the scolex, while the converse is the case with somewhat older specimens. Both actions are apparently due to the instability of the conditions.

In young specimens where the scolex is only temporarily protruded, the anterior end bearing the organs of adhesion is somewhat cone-shaped with the base resting squarely on the anterior end of the body proper, as shown in Figs. 3 and 4. This structure is also to be seen in the large plerocercoids found in the gonads of the host, but from the size and greater development of the end-organ, which occupies a large space in the apex of the scolex, together with the well-nourished condition of the body, it is obvious that the neck, if the term may be used, is almost obliterated (Pl. XIX, Fig. 6). In larvae with the scolices permanently extended (Pl. XIX, Fig. 5) there is a well-defined neck, while the scolex is shaped like two truncated pyramids placed base to base, thus very closely resembling the scolex of the adults of *P. ambloplitis* as described by Benedict ('00). The body of the worm varies from the oval shape seen in Figs. 1, a, b, and c, Pl. XIX, through the elongated oval or elliptical outline of the older invaginated specimens, (Pl. XIX, Fig. 2) to the cylindrical form as shown in Figs. 4 and 5, Pl. XIX. Later when segmentation commences, the body is quite torulous. The flattening is well marked in those found in the gonads of the host some time before the development of the rudiments of the male reproductive organs shows that segmentation has commenced.

After the suckers have become permanently everted they are seen to undergo movements which may be observed at will when the animals are placed in tepid normal saline solution. These movements are rather indefinite and spontaneous at first, but as the plerocercoid develops they become more apparently purposive, and still later they are identical with those observed in adult specimens of *P. ambloplitis*. When the worm is not attached to the bottom of the receptacle, the suckers grope around here and there through the solution, being alternately protruded and withdrawn in diagonal pairs, while the whole scolex moves slowly to the right or left or occasionally rises from the bottom. The apex does not take part in these movements. Sometimes two adjacent suckers attach themselves firmly to the bottom of the vessel while the two free ones protrude and retract alternately. Again the worm may move along slowly, by alternately freeing and reattaching the two lower suckers while the other two continue with the groping movements. When this takes place the body is drawn along the distance travelled, generally not more than the width of the scolex or the distance between the centres of the adjacent suckers, by a bead-like contraction commencing near the scolex and travelling slowly towards the posterior end of the body. Occasionally all four suckers are used for attachment, and then the only movements to be seen are the contractions which follow one another slowly backwards. After a few seconds of attachment in this manner, the two anterior suckers are raised and the motions are resumed as described above.*

* A similar movement observed first by Batsch and later by Kraemer for *Taenia* (*Proteocephalus*) *torulosa* Batsch was described as "paarweise."

5 GEORGE V., A. 1915

In the smaller forms, e.g., those shown in Fig. 1c, the movements are confined to irregular contractions of small amplitude of the whole body in a longitudinal direction.

Anatomy of Larvae of Different Sizes.

The smallest specimen investigated by means of serial paraffin sections measured 0·29 mm. in diameter by about 0·25 mm. in length. The suckers show narrow spindle fibres, two zones of nuclei and circular muscle-fibres on the inside and outside of the spindles, all characteristic of the adult *P. ambloplitis* as described by Benedict. The invagination chamber is large and contains mucus. The measurements are considerably less than those given below for an older larva, the end-organ being 0·058 mm. in diameter and the suckers 0·084 mm. The former is essentially similar in structure to that of the older plerocercoids. A few nuclei are found lying within the basement membrane, and there is found a cross of large muscle-fibres in the parenchyma behind the organ. This parenchyma is very loose and open, especially immediately behind the end-organ (here situated more posteriorly than the suckers, since the scolex is invaginated). Longitudinal muscle-fibres are few, but there are many nucleated anlagen in that area. The cuticula is thin and the cuticular muscles are poorly differentiated. The caudal vesicle gives off two main branches. Parenchyma cells surround these branches, as described below and extend for some distance out on secondary branches, thus suggesting the origin of the excretory vascular system (cf. Braun, '94–'00). Only a few parenchymatous spaces are to be observed, and the connections between them and the branchlets of the caudal vesicle are not evident.

Later the parenchymatous cells grow and take on a more definite stellate appearance showing their fine processes distinctly while the muscle-fibres become more strongly developed from the anlagen in the parenchyma.

Larva, 0.7 mm. in length, Pl. XIX, Fig. 1c.

At this stage of the development the cuticula measures 8μ in thickness, just 1μ less than that given by Benedict for the adult *P. ambloplitis*. The tube leading from the invagination chamber to the exterior has a diameter of 48μ including the cuticula itself, which is here deeply incised. Around this tube the circular muscles are well developed while the longitudinal fibres are very numerous and quite large. The suckers are 110μ in diameter and show at their centres spindle fibres 32μ long, the rest of the musculature being well-developed in plerocercoids of this size. The crenulated cuticle lining the cavities of the suckers is 3μ in thickness. The end-organ, which, so far as structure is concerned, seems to be as well-developed as that of the plerocercoid described below, has a diameter of 0·150 mm. and a length of 0·135 mm., which measurements show that it is proportionately much larger than in an older plerocercoid (vide infra). The caudal vesicle, 60μ in length, is forked for a distance of 15μ, and this forked portion is lined with a continuation of the cuticula applied to the inside of the vesicle itself. Small absorptive cells are grouped around the vesicle in the typical manner, but the cuticular muscles are here poorly developed. Proceeding from the forked portion of the vesicle are two main longitudinal excretory vessels, each 10μ in diameter, with very thin but distinct walls. These vessels course slightly backwards before passing forward where they connect with a meshwork of vessels of the same size situated in the scolex region; but on account of the invagination of the scolex the latter are directed posteriorly again. Some distance in front of the caudal vesicle the beginnings of the second excretory vessels may be seen in the parenchyma as a very small tube running along the larger vessel on each side and gradually diverging from it as far anteriorly as the latter can be traced. That part of the anterior anastomosis of the excretory vessel mentioned above, which is closely associated with the organ-end and invaginated suckers, is circularly disposed as are the parenchyma cells, owing to the compression due to invagination; at a later stage when the scolex is permanently everted, they are more loosely arranged.

Plerocercoid, 2.9 mm. in length, Pl. XIX, Fig. 4.

The plerocercoid of th s size shows practically all of the structures found in the older specimens, so that it will be described somewhat at length.

Masculature of Scolex.—At a depth of about 15μ from the apex of the scolex general oblique muscle fibres are found coursing from the lateral walls to the dorsal and ventral surfaces, thus forming a rhomboid whose diagonal axes lie in the coronal and sagittal planes of the animal. These also surround the end-organ and its opening quite like similar fibres described by LaRue ('09). As seen in Pl. XX, Fig. 7, most of them are attached to the wall of the scolex near the edges of the suckers but some end in the parenchyma before the sucker is reached. They can be traced backward from the tip to a distance of about 150μ beyond which they remain as vestiges only, attached to the indentations between the suckers, Pl. XX, Fig. 8; and, furthermore, the farther back one traces them the fewer are those fibres which run between the suckers and the end-organ. This shows that from their points of attachment on the wall of the scolex the fibres curve forward towards the apex, which is well shown in longitudinal sections.

No "muscle-cross" due to the crossing of rhomboid fibres, with fibres running dorso-ventrally and laterally, connecting opposite structures, as described by LaRue for *P. filaroides*, can be made out in this region of the scolex, for here is situated the very large end-organ (Pl. XX, Fig. 8). It is surrounded by a thick mat of circularly arranged fibres which do not appear to run transversely or dorso-ventrally in any part of their course.

At a depth of 140μ transverse sections of the flared ends of the "diagonal muscle cross" may be seen between the inner walls of the suckers and the wall of the end-organ. Farther on these ends are cut more obliquely and converge towards the end-organ fast diminishing in size as the sections go farther back, until at a level of 230μ the end-organ is just passed and the muscle cross itself is seen very distinctly (Pl. XX, Fig. 10). In this section the flared ends of the two crossing bundles and their narrowed centres are quite characteristic (LaRue); the fibres are, however, more numerous than those of *P. filaroides*, according to LaRue's figure and each

5 GEORGE V., A. 1915

bundle is about 35μ at its widest.   This muscle-cross can be traced for 20μ farther. Just before it disappears its fibres become closely arranged in the centre of the section, but the flared ends may pass a little farther back if any part of a sucker remains past the decussation. From this and the appearances in longitudinal sections passing through diagonally opposite suckers it is seen that the relation between the end-organ and the muscle star is that of a body suspended in a sling; contraction of the fibres would obviously protract the apex of the scolex both by the retraction of the suckers and the protrusion of the end-organ.

Just before the posterior end of the end-organ is reached straggling fibres coursing dorso-ventrally and laterally appear in four groups in the areas bounded by adjacent suckers and the walls of the end-organ.   Farther back these elongate centrally and mingle with the descussation of the diagonal fibres before the latter disappear (Pl. XX, Fig. 10) the double crossing forming Riggenbach's "Muskelsterne". They are rather loosely arranged, are quite narrow as compared with those of the diagonal group, and continue posteriorly to the caudal vesicle, around which a few may be found; they are the dorso-ventral and lateral muscles of the adult strobila (Fig. 12). Benedict in his paper on *P. ambloplitis* describes them as originating from cells which may be situated anywhere within the longitudinal muscles of the plerocercoid.   The cell itself is spindle-shaped, has a large nucleus which fills up most of the body of the cell and sends off fibres at least in two opposite directions.   Other fibres crossing these muscle-cells near the centre give the appearance of as many as four originating from one cell. The fibres themselves run out into the cortical parenchyma well towards the absorptive cells.

Longitudinal Muscles.—The longitudinal body-muscles are quite prominent and situated about three-eighths of the length of the shorter radius from the cuticula, (Pl. XXI, Fig. 12); towards the ends of the major axis of the more or less elliptical cross-section they lie relatively nearer the latter.   The fibres themselves cannot be said to be regularly arranged in groups as described by Benedict for the adult, yet here and there two to four and sometimes more are somewhat isolated from their fellows.   Towards the posterior end of the plerocercoid they approach the centre, but at about the anterior end of the caudal vesicle they fall off considerably in number.   Anteriorly most of them after passing the neck constriction break up into four groups each of which is attached to the posterior half of a sucker. A very few fibres, however, pass by the suckers and become lost in the parenchyma around the equatorial region of the end-organ.

Parenchyma.—In the area enclosed by the longitudinal muscles the parenchyma is in the form of an open mesh-work of very fine fibres and cell-processes. The nuclei of these cells are scattered irregularly throughout the area and are easily confused with nuclei of dorso-ventral and lateral muscle fibres.   The cortical parenchyma, on the other hand, is more compact, the cells being arranged roughly in a radial manner.   Throughout the parenchyma, more especially in the medullary portion, very many comparatively large spheroidal spaces are to be seen.   These may reach a diameter of 15μ. While the fixing of fresh material with glacial acetic acid demonstrates the presence of much calcareous matter in the parenchyma by effervescence and the passing out of gas-bubbles through the cuticula, it cannot be concluded that these spaces are filled with chalk bodies; it is quite probable that they accommodate oil-globules (LaRue). Furthermore, in plerocercoids from the ovaries or testes of the black bass where they are richly supplied with food, these spaces are very numerous, quite large and crowded closely together. It is doubtless their distension with fat which causes the well-nourished appearance of these larvæ as mentioned above.

Subcuticula or Absorptive Cells.—The subcuticular cells are quite granular in consistency and possess comparatively large nuclei which stain deeply with Heidenhain's iron-haematoxylin stain (Pl. XXI, Figs. 12 and 13). Centrally they are more or less abruptly attenuated, thus not proceeding far into the cortical parenchyma with processes of whose cells they mingle and anastomose. Including these attenuations as far as they may be distinctly traced with a magnification of 450 diameters, the cells average 20μ in length. Their peripheral ends are truncated, slightly expanded and apparently closely applied to the outer circular layer of muscles, while the longitudinal cuticular muscle fibres penetrate their broad bases some little distance from the latter. However, in gaps in the layer of circular muscles the absorptive cells are seen to proceed farther out as fine processes which can be distinctly traced as such into the cuticula for at least one-third of its thickness. Furthermore, it is quite likely that these processes proceed farther out, perhaps as far as the boundary between the two layers of the cuticula, as described below, but the highest powers used did not show this positively. A study of better sections with various kinds of fixations would doubtless much elucidate this problem which has occupied the attention of so many workers during the past.

Cuticular Muscles.—The cuticular muscles are quite typical in their structure and arrangement and closely resemble those figured by Benedict for the adult *P. ambloplitis* They are shown in various figures, especially in Pl. XXI, Fig. 13.

Cuticula.—By the use of the iron-haematoxylin stain the cuticula is resolved into two distinct layers, the outer of which takes no stain as compared with the inner. The latter (Pl. XXI, Fig. 13 cu″) is about four times as quick as the former, and takes the stain better in·its outer parts. But in deeply stained series the inner portions show the structure described above under the subcuticula. In the middle third of the cuticula, which takes the stain well, what appear to be fine processes from the absorptive cells become arranged in a more or less parallel manner and extend to the boundary between the two layers where a layer of comparatively large granules, quite regularly arranged, is plainly to be seen. Beyond this the cuticula appears to be quite homogenous with the highest powers of magnification available. The parallel processes, however, are identified more by small spindle-shaped granules placed along their courses than by the parts of what must be canals between these enlargements. Thus it seems that the outer homogeneous layer of the cuticula is something quite different from the inner layer although it takes a transparent counter-stain like Orange G. to the same extent as the latter. Concerning the significance of these layers, the extent of the present work will not permit the making of any definite statements. It seems, however, that the external layer of the cuticula is a definite structure and not something added from the outside since it is of uniform thickness, excepting where broken by injury, and has a

· 5 GEORGE V., A. 1915

definite outer boundary which is at least optically different from the rest of the layer.

Nervous System.—In plerocercoids of this size the nervous system is quite well developed and essentially the same as that described for the adult by Benedict. The nerve ring is found at a depth of about 120μ from the apex, but it is quite thin especially where it passes between the large end-organ and the suckers which are quite close together at this level (Pl. XX, Fig. 8). At the points where the large nerves supplying the suckers are given off the nerve ring is swollen to form ganglia. From the ring two somewhat flattened cords course posteriorly to supply the body of the plerocercoid. In the anterior part of the body they are situated in the cortical parenchyma, while in the posterior region they approach the centre somewhat and lie in the band of longitudinal body-muscles just outside the excretory vessels.

Excretory System.—The excretory system at this stage is characterized by the presence of a large number of flame cells and two longitudinal vessels, connecting anteriorly with a meshwork of fine tubes surrounding the suckers and end-organ. These two vessels are unequal in size, nor are their courses and connections similar. The larger, averaging between 5 and 8μ in diameter, has thin walls and gives off a large number of branches whose diameters are quite as great as that of the main vessel. These branches are distinguishable as vessels with walls for very short distances only, since they soon fuse with parenchymatous spaces in a complicated manner. Here and there branches can be seen running from this vessel to the periphery, narrowing as they approach the cuticle and eventually piercing it by apertures much smaller than the diameter of the main vessel. These, however, are not as numerous as might be expected from the development of the main tube itself. The other vessel is from one-quarter to one-third the size of the larger and pursues a straight course; on the other hand, it has thicker walls in which prominent nuclei are to be seen. While its origin in the region of the suckers is more easily made out than that of the larger vessel, posteriorly it becomes so constricted at different levels that it all but disappears from view; near the caudal vesicle, and just before joining the latter, it bends forward and inward behind the anterior end of the vesicle and opens by an aperture quite separate from its fellow of the opposite side.

On each side of the plerocercoid the two excretory vessels are situated just within the longitudinal body muscles, about 35μ apart and on a line inclined at various angles to the perpendicular to the longitudinal axis of the transverse section, the smaller vessel constantly lying nearer the centre of the section.

The caudal vesicle is 70μ long and 10μ in diameter, including the cuticular lining. The lumen itself is somewhat stellate in shape owing to deep incisions and folds in the cuticula. The absorptive cells follow the cuticula from the outer wall of the worm throughout its whole length, while the cuticular muscles are well developed as far as the apertures of the excretory vessels. As this place is approached the longitudinal fibres diverge and become lost in the parenchyma; likewise the absorptive layer suddenly disappears.

Flame-cells are very numerous and comparatively large at this stage. They are found to be confined to an area around the excretory vessels as described by

LaRue for the species *P. filaroides;* Pl. XXI, Fig. 14, shows a typical group of flame-cells connected to the smaller excretory vessel (vide supra). While it was rather difficult to make out the exact point where the common duct emptied into the longitudinal excretory vessel, it could be seen that more flame-cells poured their excretions into the smaller vessel than into the larger and more irregular of the two. The stellate appearance at the ends of the flame-cells in this figure seems to be due to contraction of the cell-body and the staining of numerous radiating strands in its protoplasm, which do not appear in sections stained lighter. The parts of the flame-cells itself are seen in Pl. XXI, Fig. 15. The outlines of the cell-body are difficult to discern, but they are quite irregular, as shown, the protoplasm being prolonged into many processes of different lengths. Very little structure can be detected in the protoplasm, but it is evident that vacuoles observed by various authors are present. The nucleus is comparatively large and stains deeply. The ciliary flame is rather large and attached at its proximal end to a very deeply staining body situated close to the nucleus, doubtless the basal granules of the individual cilia massed together, which, by the way, are difficult to separate optically. The conical cavity in the cell which accommodates the ciliary flame is easily seen and has peculiar elongated thickenings in its walls, much resembling elongated nuclei but which show practically no structure. The cavity itself is directly continous with the very thin-walled, homogenous canaliculus which connects it with those of its neighbours to the longitudinal excretory vessels.

End Organ.—In his description of the scolex of *P. ambloplitis* Benedict makes the following statement: "Directly beneath the apex of the scolex is a sac of cuticular structure enclosing a small number of circular masses, closely packed together. The masses seem to be of a calcareous nature and are penetrated by numerous fine canals. No connection whatever could be traced between this sac and any outside system, although the excretory ducts form a thick network around it." Longitudinal sections of the two scolices of sexually mature specimens showed two conditions of this end-organ (Pl. XXI, Figs. 16 and 17) which are very suggestive. In both cases the organ, although not separated from the surrounding parenchyma by a clear zone as in LaRue's account of *P. filaroides*, is quite distinct from the latter. Evidently Fig. 16 represents a younger stage than does Fig. 17. In the latter it is to be noted that the whole of the central tissue has lost its structure, remaining as so much connective tissue, irregular muscle-fibres and other deeply-staining bodies; the granular nature of the organ is more evident and the material seems to be arranging itself into definite areas, doubtless to form the calcareous bodies above mentioned. The connection between the cuticula and the organ is much less evident; the muscular bounding fibres, which are continuations of the longitudinal muscular layer of the body of the worm, are losing their connections with the musculature beneath the cuticula; in fact the whole organ and its surroundings seems to be in a degenerate state. On the other hand, Pl. XXI, Fig. 16, presents what one might consider a more functional structure. Although there is no direct aperture through the cuticula of the scolex connecting the organ with the exterior, it is quite evident that at a slightly earlier stage such might be found. The muscular boundary is more definite, and the contents of the organ,

5 GEORGE V., A. 1915

namely, peculiar basal cells whose free parts project into a fibrous meshwork in the spaces of which there is to be found a fluid with very fine granules, more nearly approach the condition about to be described.

In the plerocercoid described above (Pl. XIX, Fig. 4) the end-organ extends 220μ from the apex. Its cross section throughout the series is somewhat elliptical, the major axis measuring at its greatest 290μ and the minor 196μ, diameters 229 and 230μ, thus presenting a more nearly spherical outline. The organ has a thick wall composed of two layers. The outer and thicker is made up of comparatively large muscle-fibres running in a general circular direction and intermingled with longitudinal fibres which constitute the inner layer. These fibres fuse with the cuticular musculature at the anterior end of the end-organ. Next towards the centre of the organ comes a very thin basal membrane much resembling the cuticula on the exterior of the plerocercoid and continuous with that lining the invagination chamber of the apex of the larva (Pl. XX, Fig. 9). The organ opens to the exterior by an aperture 29μ in diameter and circular in shape. It is lined with the cuticula from the surface of the worm, which continues down into the lumen of the organ for about half its diameter as a thin-walled tube, perforated freely, more especially as it nears the centre of the organ, by wide irregular openings. This tube is supported by numerous radiating strands of tissue attached to short, conial and wedge-shaped processes from the cells situated on the basement membrane. In most series of plerocercoids of this age these radiating filaments disappear at the posterior end of the organ as distinct connections between the basal cells and the central tube, leaving only scattered pieces lying in radial directions from the latter. The general arrangement is best seen in transparent preparations of whole plerocercoids; in these the strands all appear to emanate from the aperture of the organ. The basal cells are very irregular, granular, highly-stainable and have large nuclei, themselves readily taking the stain. The processes both free and attached to the central tube are bathed in a fluid filling the organ, which is very fine and granular in consistency and stains very deeply with Heidenhain's iron-haematoxylin. In some series a clear area surrounding the inner end of the central tube shows where some of the material has been expelled from the organ, since in longitudinal sections a band of material is often found protruding through the aperture to the exterior.

LaRue discusses this end-organ at some length in *P. filaroides*, among other things mentioning its occurrence in the plerocercoid found in *M. dolomieu*, in all probability that being dealt with in this paper. Apart from this species the end-organ has been described only in Riggenbach's *P. sp.?*, *P. Lonnbergii* Fuhrmann, and in *P. ambloplitis* Leidy, the latter by Benedict.

As to the function of the organ, if it has any definite function, the extent of my studies will not permit me to give anything further than suggestions. From its early disappearance in *P. filaroides* and its great development in this plerocercoid, one would be inclined to conclude that it functions only in the larval stages, since obviously the organ as found in the adult is functionless at least so far as the external surroundings of the plerocercoid are concerned. Unfortunately I have not at hand a complete series from the plerocercoid to the adult stage, the oldest specimen of the former condition which shows signs of segmentation being only 39 mm. long. In this specimen (Pl. XIX, Fig. 6) the apex of the scolex is very prominent and is occupied almost wholly by the end-organ which is somewhat flattened dorso-ventrally as is the scolex itself and measures 426 by 360μ in the cross-section by 380μ in length. Furthermore, the basal cells are represented by only small remains with here and there short processes, and widely-separated radiating pieces represent the strands connecting them with the central tube of the organ. The aperture is relatively quite large. The contents show larger granules, while only that part at the posterior end of the tube seems to be very fluid. When fresh material is fixed it is a common occurrence to see a short, thick, viscid stream of liquid oozing from the apex of the scolex. As soon as this material comes in contact with the fixing-fluid, it coagulates, thus demonstrating its protein nature. Microchemical tests show that it contains lime-salts, while the basal cells are likewise rich in calcium. From this it would appear that the basal cells secrete the material found in the cavity of the organ, but whether the material is for digestion in connection with the boring action which some authors attribute to the plerocercoids, for adhesion or merely represents the remains of a much-altered rostellum, perhaps in connection with excretion, must remain conjectural until further study throws more light on the subject.

The study of plerocercoids intermediate in length between that just described and the 39 mm. specimen mentioned above, showed that besides the general growth and differentation of all of the tissues there is particular development in the cuticula, end-organ and excretory vessels.

The cuticula as a whole gets much thicker while its external layer becomes relatively thinner. The end-organ grows comparatively rapidly until it occupies almost the whole of the apex of the plerocercoid (Plate XIX, Fig. 6). Its degeneration into the calcareous bodies of the adult scolex must take place very quickly, as has been demonstrated for *P. filaroides* by LaRue. Unfortunately I have not yet procured specimens showing this degeneration.

In the 39 mm. plerocercoid the excretory vessels are three or four on each side in the neck region. One pair lying in a "median frontal plane" (Benedict) are the largest and most regular of them all; few branches are given off from them in the scolex where they gradually diminish in size and disappear near the apex. Another pair giving off many branches, a large number of which go to the exterior, lies in a sagittal plane on each side of the body just outside the first vessel but within the longitudinal body-muscles, thus forming the base of a triangle whose apex is the largest vessel. Other large vessels in the region of the scolex are merely branches, but some run parallel to the main vessels for considerable distances, and one may develop into a fourth vessel. This latter statement refers especially to one seen outside the longitudinal muscle zone, about half-way between it and the cuticula. This arrangement of vessels is also found in specimens only 10 mm. long, where even a fifth vessel may be seen running parallel to the others for a short distance. However, when these vessels are traced backward, they all, excepting the smaller pair in the median frontal pair unite to form the single pair of large vessels, evidently ventral in position, which course irregularly backwards and unite with the caudal vesicle.   The smaller, median frontal pair are the smaller vessels described above for a shorter plerocercoid which becomes lost in the parenchyma around the end-organ forward and the caudal vesicle posteriorly, thus exactly coinciding with Benedict's median frontal pair, excepting that this writer did not see the posterior connections.   The large size of this pair as described above for the 39 mm. larva must be due to some physiological condition or individual variation since they are not thus distended in the 25 mm. specimens.   An important point to be noticed in connection with the development of the excretory vessels is that the posterior end of the plerocercoid remains in a primitive condition while the anterior end specializes; and the development of the other parts bears out this statement.

The evidence given above appears to establish the idea of the identity of this plerocercoid and *P. ambloplitis*, more especially with regard to the following points:
(1) The excretory vessels of advanced stages of the former are identical with those of the latter;
(2) Measurements of the cuticular structures and the parts of the suckers are the same in both forms, relatively speaking;
(3) The movements of the suckers during life are identical;
(4) The nervous system of *P. microperi* is essentially the same as *P. ambloplitis;*
(5) The stages in the development of the end-organ, although not complete, suggest a continuity between the two forms.

The Intermediate Hosts.

Our knowledge of the development of the genus Proteocephalus (Ichthyotaenia) dates as far back as 1878 (Gruber).   Since then data have been added from time to time, so that only now are we getting a general idea of the whole process.   Gruber found several stages of a plerocercoid in *Cyclops breviciudatus,* which he believed to be those of *Proteocephalus (Taenia) torulosa* Batsch. Zschokke ('84) found the unsegmented larva of *P. longicollis* Rud. in the liver of *Salmo umbla* in which the adults were found, and what he called the larva of *P. torulosa* in *Coregonus fera* in the month of January, in the intestine of *Lota vulgaris* in the month of February, and in *Alburnus lucidus* in March.   These observations in the light of present knowledge suggest a comparatively simple lifehistory in that the larvæ may develop from the oncospheres in the final host and in a short period, as found by LaRue ('09).   The former was also observed for *P. longicollis* by von Linstow ('91).   Riggenbach ('96) describes the plerocercoid found in the parenchyma of the scolex of *Corallobothrium lobosum* Rigg., which closely resembles that of *P. ambloplitis* found in the black bass and other freshwater fishes, but he gives no suggestions as to its adult existence.   In the section, entitled, "Development," he merely mentions work contained in some of the above references, after saying that "on the development of the uterine eggs, as well as

the early stages of the Ichthyotaeniae almost nothing has yet been published."
Schwarz ('08), in speaking of the development of the reptilian Ichthyotaeniac, takes
Gruber's observations as his basis and proceeds to elucidate the infection of rep-
tiles through Cyclops and the aquatic habits of the hosts concerned. Fuhrmann
('03) considers the larvæ found in the livers of Salmonids and Percids by von
Linstow, von Siebold and Zschokke as strayed larvæ having mistaken hosts and
having then taken a particular aspect. This view would explain the fate of numbers of the plerocercoids of *P. ambloplitis* found encysted in young and old bass.
Unless the bass were eaten by larger fish such as *Amia, Lepisosteus, Esox* or
Salvelinus in which they could develop as in a second final host they would surely
disintegrate eventually. Fuhrmann showed by his infection experiments that
the intermediate hosts of the Ichthyotæniæ were the one or the other of the copepods found in the plankton used as food. LaRue's infection experiments with
Chironomus larvæ, *Daphnia, Cyclops, Notonecta,* some larvæ of the *Dytiscidae,*
tadpoles of *Rana catesbiana,* besides the Salamander (*Amblystoma tigrinum*)
itself proved failures as did those of Schneider ('04). However, he furnishes
conclusive evidence, "first, that the encysted plerocercoid (of *P. filaroides*) is
the larval form of the cestode found in the same host; second, that the period
of development from the plerocercoid after ingestion is short."

In connection with the present study only a few infection experiments were
tried, but they gave no results; it was found very difficult to keep the young of
M. dolomieu alive and unmolested,—they are very sensitive to change of environment—whereas the young of *M. salmoides* are easily kept in captivity. However,
a thorough series of dissections was carried out with bass of all sizes from those
taking their first food after hatching, about 8 mm. in length, to adults. This
resulted in a very fair acquaintance with the great variety of food-forms of *M.
dolomieu* as found on the eastern shore of Georgian Bay, but the points elucidating
the life-history of *P. ambloplitis* were rather few in number.

Plerocercoids were first found in hosts about 40 mm. in length and between
that and 50 mm. the infection was not considerable. The organs infected were,
first the liver, then the alimentary tract and coelome. Observations on methods
of infection pointed, first, to the direct development of the oncospheres accidentally
introduced (autoinfection), their subsequent transference by means of the blood
stream (Braun), and the boring of the oncospheres themselves; second, to infection
from invertebrate food-forms such as *Sida, Daphnia, Chironomus* larvæ and *Coriza;*
and thirdly, to infection from minnows and young perch which constitute a part
of the food and from which large numbers of very small plerocercoids closely
resembling the youngest stages described above are freed in the stomachs of the
bass. This latter method is borne out by the fact that no tape-worms were found
in those fish examined in the early fall of 1910 when the food taken was composed
wholly of crayfish, whereas a comparatively heavy infection was found with those
taken on the outlying reefs and islands where minnows constitute the bulk of the
food. Thus the evidence points to *P. ambloplitis* having at least two intermediate
hosts, the first, some unknown species of aquatic arthropod, and the second, either
different species of minnows, small perch or the final host itself.

5 GEORGE V., A. 1915

Our knowledge of the identity of the first intermediate host of the genus Proteocephalus is confined to Barbieri's paper on *P. agonis* Barb.   Although he did not absolutely prove his hypothesis, he collected sufficient evidence to lead one to feel justified in concluding that *Bythotrephes* and *Leptodora* are the forms in which the oncosphere of that species develops into the very young plerocercoids.

The Egg.

Up to the present the egg of the genus Proteocephalus has been described for only a few species, but the descriptions all show that it consists of a six-hooked embryo or oncosphere surrounded by three membranes.   The outer or first membrane is very variable in shape and size while the other two are constant in mature eggs, that is, in eggs showing three pairs of hooks.   The third or innermost membrane is difficult to differentiate in whole specimens since it is so thin and so closely applied to the embryo.

The egg of *P. ambloplitis* is shown in Pl. XX, Fig. 11.   It is to be seen that the first membrane varies from a nearly spherical shape to that seen in e, which is rarely found.   These extreme variations appear in eggs all procured from a single ripe proglottis, but those shown in Figs. 11 a, c and d are commonest.   In fact, apparently all the eggs in most ripe proglottides possess these peculiar dumb-bell shaped outer membranes, thus leading one to consider their structure as characteristic of the species.   At any rate, such appendages do not appear in the eggs of any other of the several species of Proteocephalus I have examined.   On the other hand the rest of the egg is quite typical.   As suggested in Fig. 11d and shown in the extreme in e, these characteristic swellings of the outer membrane are not quite in line with the longitudinal axis of the egg.   In such eggs as shown in Figs. 11 a, c and d, from which living oncospheres can be expressed, the outer membrane varies in length from 55μ to 75μ.   The second membrane, however, is more constant in diameter, varying only from 24μ to 27μ.   The third membrane is not easily seen in the intact egg, but parts of it appear after the oncosphere has been pressed out.   The granular layer between the second and third membranes as seen in optical sections is quite uniform in thickness, about one tenth of the diameter of the second membrane, and is composed of fine granules and spherical yellowish globules scattered about so as to leave irregular, often circular, clear areas through which one can see the oncosphere.   The largest of these small granules are, however, apparently identical with the smallest globules, so that the whole suggests fat droplets of various sizes.

The oncosphere may be easily pressed out under a glass-cover from the central parts of the egg to either of the expansions of the outer envelope, that is, to the space between the outer and second membranes.   There it is seen to move vigorously, the hooks acting in a manner very similar to that described by LaRue for *P. filaroides*.   A pressure which is not quite sufficient to cause the oncosphere to escape from the second membrane almost invariably stimulates it to begin its movements *in situ*.   These take place outside of the egg-membranes at the rate of about fifteen per minute.   From the 10μ sections of ripe proglottides stained in Heidenhain's iron-haematoxylin the oncosphere was seen to be made up of a number of cells packed closely together, whose boundaries were obscure while their nuclei were deeply stained. In living oncospheres each hook was observed to be imbedded in a cone of homogenous material, the apex of which surrounded the proximal end, slightly swollen in this species, while the base at the surface of the oncosphere was about three times the diameter of the distal end of the main shaft of the hook. The tips of the hooks appeared to protrude from the surface of the onocsphere, especially during the phase of separation of the former.

All of these observations and measurements were made from fresh material in normal physiological saline solution, so that the various stages in the swelling of the outer envelope were not due to osmotic action; this takes place apparently within the uterus as a stage in the development of the egg.

March 3, 1913.

WORKS CITED.

1878. GRUBER, A.—Ein neuer Cestoden-Wirth. *Zool. Anz.*, Vol. 1, 1878, p. 74.

1884. ZSCHOKKE, T.—Recherches sur l'organisation et la distribution zoologique des vers parasites des poissons d'eau douce. *Archives de Biologie*, Tome V, pp. 153-241.

1891. VON LINSTOW, O.—Ueber den Bau und die Entwicklung von Tænia longicollis Rud· *Jenaische Zeitschrift f. Naturw.*, Bd. XXV, n. f. XVIII, pp. 565-576.

1892. KRAEMER, A.—Beiträge zur Anatomie und Histologie der Cestoden der Süsswasserfische· *Zeit. f. Wiss. Zool.*, Bd. LIII, pp. 647-722, Pl. XXVII, XXVIII.

1896. RIGGENBACH, E.—Das Genus Ichthyotænia. Inaugural Dissertation, Geneva; *Revue Suisse de Zool.*, Bd. IV, pp. 165-276, 3 plates.

1900. BENEDICT, H. M.—Structure of Two Fish Tapeworms from the Genus Proteocephalus, Weinland, 1858. *Journal of Morphology*, Vol. XVI, pp. 337-368, 1 pl.

1900. BRAUN, M.—Cestodes, in *Bronn's Klassen und Ordnungen des Thierreichs;* Vermes, Band IV Abt. 1 b.

1903. FUHRMANN, M. O.—L'évolution des Ténias et en particulier de la larve des Ichthyoténias. *Archives des Sciences Physiques et Naturelles*, Vol. 16, 4th Period, pp. 335-337.

1904. SCHNEIDER, G.—Beiträge zur Kenntnis des Helminthen-fauna des Finnischen Meerbusens, *Acta Soc. pro Fauna et Flora*, Fenn. XXVI, No. 3, pp. 1-34.

1908. SCHWARZ, R.—Die Ichthyotænien der Reptilien und Beiträge zur Kenntnis der Bothriocephalen. Inaugural Dissertation, Basel.

1909. LaRUE, G. R.—A new Cestode. *Trans. Amer. Micros. Soc.*, Vol. XXIX, No. 1. pp. 17-46.

1909. BARBIERI, C.—Ueber eine neue Species der Gattung Ichthyotænia und ihre Verbreitungsweise. *Central. f. Bakt. Parasit. u. Infckt.*, Bd. XLIX, Heft 3, pp. 334-340.

1911. LaRUE, G. R.—A Revision of the Cestode Family Proteocephalidæ, *Zool. Anz.*, Vol. 38, pp. 473-482.

5 GEORGE V., A. 1915

EXPLANATION OF FIGURES.

All drawings, unless otherwise mentioned, were drawn to the scale indicated with the aid of an Abbé camera lucida.

Abbreviations.

ac.	Absorptive cell.	lcm.	Longit. cuticular muscles.
bac.	Basal cells.	m.	"Muskelsterne".
bm.	Basal membrane.	mw.	Muscular wall of end-organ.
cu.	Cuticula.	n.	Neck.
cu'.	Outer layer of cuticula.	ng.	Nerve ganglion.
cu".	Inner layer of cuticula.	nac.	Nuclei of absorptive cells.
ccm.	Circular cuticular muscles.	om.	Outer membrane.
ct.	Central tube.	on.	Oncosphere.
eo.	End-organ.	pac.	Parenchyma cells.
eeo.	Entrance to end-organ.	par.	Parenchyma.
exv.	Excretory vessels.	sm.	Second membrane.
gr.	Granular material.	weo.	Wall of end-organ.
lbm.	Longitudinal body muscles.	ym.	Yolk mass.

PLATE XIX.

Figs. 1, a, b.—Plerocercoids from kidney of host, × 33.

Fig. 1 c.—Small plerocercoid from liver of host, × 33.

Figs. 2, 3, 4, 5.—Plerocercoids from liver and intestine of host, × 33.

Fig. 6.—Plerocercoid from gonad of host, × 33.

PLATE XX.

Fig. 7.—Transverse section through a 2·9 mm. plerocercoid, 30µ from the apex, × 105.

Fig. 8.—Transverse section of same, 130µ from apex, showing end-organ and suckers, × 130.

Fig. 9.—Longitudinal section through end-organ of another specimen, showing structure × 130.

Fig. 10.—Transverse section through the 2·9 mm, plerocercoid 240µ from apex, showing "Muskelsterne", × 130.

Fig. 11.—Egg, showing structure and various forms of the outer membrane, *a* × 700; others × 350.

PLATE XXI.

Fig. 12.—Transverse section through body of plerocercoid, showing general structure: semi-diagrammatic, × 130.

Fig. 13.—Part of a transverse section through the body of a plerocercoid, showing details of cuticular structures, × 600.

Fig. 14.—A group of flame-cells in connection with one of the excretory vessels of the median frontal pair, × 1,000.

Fig. 15.—A flame-cell greatly magnified, showing structure, × 2,000.

Fig. 16, 17.—Longitudinal sections of different conditions of the end-organ in adult specimens of P. ambloplitis, × 330.

PLATE XIX

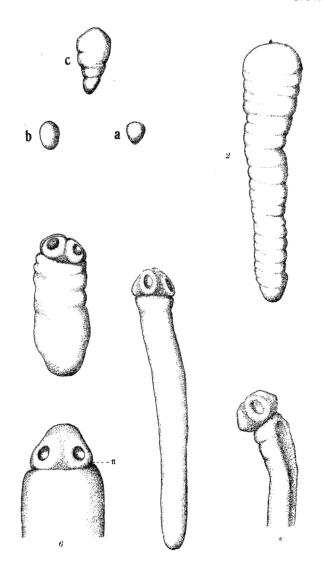

PLATE XX

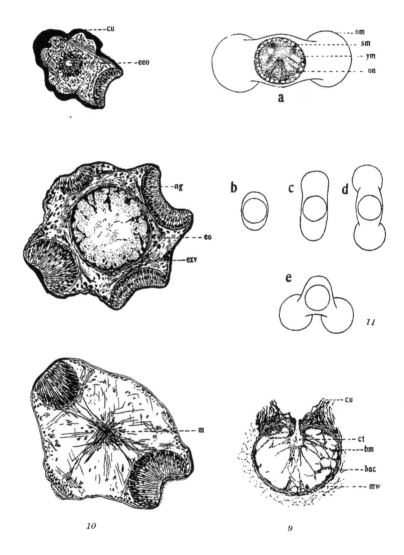

PLATE XXI

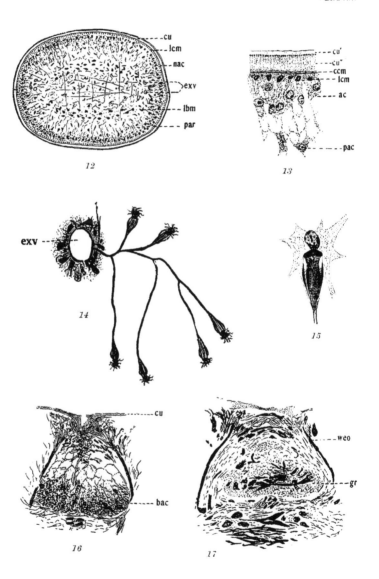

12

13

14

15

16

17

IX.

BRYOZOA OF THE GEORGIAN BAY REGION.

By H. T. White M.A., High School, Sudbury, Ont.

The work upon which this paper is based was carried out at the Georgian Bay Biological Station during the summers of 1911 and 1912, under the direction of Dr. B. A. Bensley and Dr. E. M. Walker. I have been concerned chiefly with collecting and identifying the species and with noting the habitats and variations shown.

At Go Home Bay, nine species and one variety were identified. This is double the number reported from any other locality in North America. The Bryozoa are pretty well distributed around the Georgian Bay, and most of the species are found wherever suitable places occur. The relative abundance varies with the season

It was found necessary to introduce certain changes in the classification of the Plumatellas, as given by Kraepelin (1887) and Davenport (1904). Otherwise the classification of those authors has been followed. The nomenclature has been changed from that of the authors quoted, in accordance with the law of priority.

The changes in the genus *Plumatella* were deemed necessary, because there were as great differences between varieties of a species as between different species. For that reason, *Plumatella polymorpha* Kraepelin has been divided into *P. repens*, *P. fungosa*, and *P. appressa*. New variations in some of the characters of the species have been noted.

Comparatively little has been published concerning the Bryozoa of Canada. In 1855, Goadby and Bovell published notes concerning a 'Plumatella' from Rice Lake, Ontario. It evidently was *Pectinatella*. In 1880, Thomas Hincks published some notes made by his father on 'a supposed Pterobranchiate Polyzoan' collected in the Humber River near Toronto. According to Osler, this may have been *Pectinatella*. In 1883, Prof. William Osler, then at McGill University, gave an account of a number of Bryozoa from Canada. He records *Cristatella* from several points in Quebec, *Pectinatella* from Quebec and Ontario, and *Plumatella arethusa*, *P. vitrea* and *P. diffusa* (probably = *P. repens*, *P. punctata* and *P. emarginata*, respectively) from various localities.

Paludicella articulata (Ehrenberg). (=*ehrenbergii* auct.)

This is quite inconspicuous and may easily be overlooked. It occurs at Go Home, Skerryvore, French River, Killarney and Waubaushene.

Habitat very varied; under stones in rapid streams or fairly exposed shores, or more protected places, e.g. bays and ponds. In the latter it is found under waterlily leaves, or sticks. June to September. Common both in 1911 and 1912.

Fredericella sultana (Blumenbach).

The colonies are all small and the statoblasts few. The colonies present about the same appearance throughout the season, as in the case of *P. articulata*. They

5 GEORGE V., A. 1915

do not appear till about June and remain throughout the summer. It occurs at Go Home, Parry Sound, Skerryvore, French River, Killarney, Manitoulin Island,. and Waubaushene. It has also been found at Brantford and Sudbury.

Habitat.—Found in the same places as *P. articulata*, on the under sides of stones in streams or along exposed shores, or under sticks, water-lily leaves, etc., in bays and ponds. The abundance was about the same in 1911 and 1912.

Genus *Plumatella*.
Key to species.
A₁ Colony with vertical as well as horizontal branches.
   B₁ Cuticula thick and brown, with a keel that broadens at the aperture.
       Free statoblasts elongated; proportions 1: 1·53 to 1: 2·8.   *P. emarginata*.
   B₂ Cuticula thick and colorless; colony robust; zooecia in groups; keel
       absent; free statoblasts nearly circular.   *P. repens*.
   B₃ Cuticula colorless to brown; tubes elongated, often pendant; may be
       keeled and emarginate; free statoblasts nearly circular.   *P. fungosa*.
A₂ Colony with horizontal branches only (rarely vertical).
   C₁ Cuticula colorless to brown; tubes usually with a clear, longitudinal
       band; depressed and closely adherent to the substratum (usually). Free
       statoblasts nearly circular.   *P. appressa*.
   C₂ Cuticula delicate, colorless to white; mouth cone elevated, often wrinkled
       and speckled with white. Free statoblasts nearly circular. *P. punctata*.
Plumatella emarginata Allman. (=*princeps* Kraepelin). var. *emarginata*.

Tubes openly branched, repent, with short lateral branches, antler-like. Statoblasts always few, but more abundant in older colonies. Although 1ather well distributed about Georgian Bay, this form is nowhere very abundant, and the colonies are rather small. It is found at Go Home, Parry Sound, Skerryvore, French River, Killarney, Manitoulin Island, Fitzwilliam Island, Tobermory, and McGregor Harbor. Specimens have been collected also at Brantford.

Habitat varied; frequently under stones along rather exposed shores, but also under stones or sticks in bays or in running water. Appears June to September, more abundant in July than earlier. Colonies become darker with incrustations, but modifications with the season not great. More abundant in 1912 than in 1911 at Go Home.

Plumatella repens (Linn.) (=*P. polymorpha*, var. *caespitosa* and *repens* Kraepelin).

This is the most variable of all the species found in Georgian Bay. The colony starts from an embryo of usually two individuals, and single tubes develop, branching somewhat openly. At this stage it is much like *P. punctata*, var. *prostrata*. Later the zooecia are found in groups, and the colony is caespitose. If the area of the substratum is very limited, a dense mass, half or three-fourths of an inch thick is formed with only the apertures rising free. In some cases the tubes remain scattered, or are intertwined rather than bunched. It is very probable that this last variation is the *P. polymorpha*, var. *repens* of Kraepelin. Intermediate stages may readily be found. It is, of course, found in the same places as the more usual variety. .The statoblasts of this species are very abundant in well developed colonies. It has been found at Go Home, Skerryvore, French River, Killarney, Fitzwilliam Island, and Tobermory.

Habitat.—This species is found chiefly under the leaves of water-lilies and other plants, on sticks, stones, and old iron, in ponds and sheltered bays, sometimes exposed to the direct sunlight. The colonies appear first from the middle to the end of May. The first colonies are found chiefly on twigs and bark, since the leaves are not yet developed. They do not seem to avoid the light and the twigs offer almost no shade. About a month later the colonies are fully developed, and soon afterwards disappear. At this stage the statoblasts are exceedingly abundant, and numerous embryos may be seen swimming about. These soon develop into small colonies, and during the latter half of July and in August these are in many places found in great abundance under water-lily leaves.

Brown bodies are very abundant in older colonies, especially of the first generation, shortly before it disappears. The branches of the colony usually contain *Chironomus* larvæ, which devour the cuticula. In some cases swarms of unicellular organisms may .be seen in the zooecia. This species was common at Go Home, both in 1911 and 1912.

Plumatella fungosa (Pallas), (=*P. polymorpha*, var. *fungosa* Kraepelin).

The statoblasts of this species are more elongated than those of *P. repens*. Davenport gives the limit of the varieties of *P. repens* as 1 : 1·5; but specimens from Georgian Bay are often more elongated, the proportions being as high as 1: 1·65. The lower limit of the statoblasts of *P. emarginata* is given as 1 : 1·53, thus overlapping with this species; but this does not prevent the identification of the species by means of the statoblasts, since many of those of *P. fungosa* are quite round, specimens with the proportions of 1 : 1·2 being found.

This species occurs at Waubaushene, Go Home, Skerryvore, French River, Killarney, Club Island, Tobermory, and McGregor Harbor.

The habitat of this species is on leaves of pond weeds, water-lilies and sticks. It coats leaves of pondweeds (*Pontederia*) and is thus somewhat exposed to sunlight. It is found in still water or only moderately exposed to waves. Brown bodies and statoblasts very abundant in older colonies. Like *P. repens*, this species is sometimes found in dense masses, with strings of tubes extending out two and a half inches, or more. They are found in almost incredible numbers during the latter part of July in Matchedash Bay, near Waubaushene, coating the pondweeds which clog the bay. From July to September. Not rare in 1911 or 1912, but found in a number of places in 1912, where they were absent in 1911. The form is rather constant throughout the season, except for the changes due to crowding.

Plumatella appressa Kraepelin. (=*P. polymorpha* var. *appressa* Kraepelin).

Cuticula transparent to brown, coriaceous; tubes flattened, closely adherent to the substratum, and seldom rising from it. There is usually a clear longitudinal band, or a low keel. The branching is angular. The fixed statoblasts are abundant in this species, and may be seen adhering to the under sides of the flat

5 GEORGE V., A. 1915

stones, which are the favorite resort, long after the colony has disappeared. The free statoblasts are nearly circular. Statoblasts and brown bodies very abundant in mature colonies.

This species is the most abundant and widespread of the Bryozoa of Georgian Bay, being found all around the bay along exposed shores.

The habitat is chiefly under flat stones along rather exposed shores and in rapid streams, but sometimes in more protected places, and then it may rise from the substratum, the tubes become more rounded and intertwining, and the longitudinal clear band often be absent. From the end of May till September. The form is rather constant throughout the season.

Plumatella punctata Hancock.

Var., *prostrata.* Stock repent and open, forming long hyaline tubes that give rise to only a few, likewise repent, lateral tubes. This was found at Go Home, Skerryvore, French River and Killarney. Outside of Georgian Bay it has been found in several places, Brantford, Aurora, and Sudbury.

The habitat is under stones or sticks in running water, or along more or less exposed shores. At the chute in the Go Home River it occurs associated, or even intertwined with *Paludicella articulata* and *Fredericella sultana.* Brown bodies and statoblasts are not abundant. The colonies vary but little with the season. Common in both 1911 and 1912.

Pectinatella magnifica Leidy.

The colonies are conspicuously marked with white bodies, situated at the outer ends of the mouth cones and near the ends of the lophophores. The latter is due to the habit of the animals of flexing the lophophores so that they touch the white body on the mouth cone. Part of the substance adheres to them.

This species was found at Go Home Bay and French River. It was not abundant.

The habitat is chiefly under sticks, stones, logs, etc., sometimes under water-lily leaves. It mostly lives in sheltered bays, ponds, or slow streams. July to September. More abundant and widespread in 1912 than in 1911.

Cristatella mucedo Cuvier.

This species found above the chute Go Home River and sparingly at Tobermory.

Its habitat is on or under logs, sticks, or sometimes water plants in slow-flowing water. It does not always avoid the light. The statoblasts are abundant. The colonies do not change greatly with the season, but may disappear very quickly. Abundant for a short time in 1911 and 1912.

REFERENCES.

DAVENPORT, C. B.—Report on the Fresh-Water Bryozoa of the United States. *Proc. U. S. N. Mus.*, vol. XXVII, p. 211. 1904.

GOADBY AND BOVELL, J.—Passing Visits to the Rice Lake, Humber River, Grenadiers' Pond, and the Island. *Canad. Journ.*, vol. III, p. 201. 1855.

HINCKS, T.—On a supposed Pterobranchiate Polyzoan from Canada. *Ann. Mag. Nat. Hist.*, March, 1880.

KRAEPELIN, K.—Die Deutschen Süsswasser Bryozoen. Eine Monographie. I. Anatomisch-systematischer Teil. *Abhandl. Naturwissen. Ver. Hamburg*, Bd. X, No. IX. 1887.

OSLER, W.—On Canadian Fresh-water Polyzoa. *Canad. Natur.*, n. ser., vol. X, p. 399. 1883.

X.

PRELIMINARY REPORT ON THE PLANTS OF GEORGIAN BAY.

A CONTRIBUTION TO THE BIOLOGY OF THE GEORGIAN BAY WATERS.

By W. T. MacCLEMENT, M.A., D.Sc.
Professor of Botany, Queen's University, Kingston.

With an added List of Algæ collected and determined by A. B. Klugh, M.A., Lecturer on Botany, Queen's University, Kingston.

During my stay in 1911 at the Dominion Biological Station, Go Home Bay, Georgian Bay, I hoped to begin a study of the fungus attacking fish-eggs in the vicinity of the Station. I was unable to reach the Station until June 19th, the first summer trip of the passenger steamer from Penetang. I found that by that date all the Black Bass had left their eggs, which had either hatched, decayed, or been in some other way hidden or destroyed. Rock Bass,—*Ambloplites rupestris* —were still protecting their eggs. I collected and preserved all I could find of these, as in every case they were attacked by a fungus. This I was able to identify as a '*Saprolegnia*,' probably '*mixta*.' On most of the lots of these eggs the fungus had reached the zoopore stage, but I have been unable to discover any *Saprolegnia* oospores on them. I gathered a good many facts regarding this fungus and its distribution and its attacks on fish and fish eggs, but this should accompany an account of its conditions and effects in our waters, such as I hope to be able to prepare at some future time after a study of these waters earlier in the summer.

Disappointed in my hope of studying water-moulds I turned my attention to the green water-plants of Georgian Bay. So far as I can learn little work has been done in this field, and no report upon them published. Dr. Bensley who was Curator of the Station, informed me of the desirability of gathering materials for a complete biological survey of Georgian Bay, on account of the close relation of these facts to the fishing industry.

Accordingly I gave my time to the collecting and determining of the littoral and plankton flora, of the waters within convenient reach of the Station. Incidentally I collected and classified all the fleshy fungi I found—some thirty-five species in 1910 and 1911. My list of plants must be considered as preliminary, as many common genera were not in fruit at the time I collected them and could not be identified. Also in such a group as the Diatoms my identifications are only of the well marked species.

My assistant, Mr. Klugh, spent May and June on the west side of Georgian Bay in the vicinity of Colpoy's Bay, and at my suggestion studied the algæ found there. I am inserting his list to supplement my own.

5 .GEORGE V., A. 1915

I believe we shall find the flora of Georgian Bay quite as luxuriant as that of Lake Michigan, or Lake Erie, and possibly approaching that of Lake St. Clair, although the conditions are quite dissimilar from those reported by Thompson, Snow, and Pieters. This work is valuable from the purely scientific, as well as from its economic side, because of our lack of knowledge of the distribution of the fresh-water algæ of Canada.

The list of water plants now presented is the result therefore of a few weeks' work at the Biological Station at Go Home Bay on the south east shore of Georgian Bay, during August and September, 1911.

A study of algæ is especially important in connection with those waters which are the spawning ground and nurseries of the food fish, of which Lake Huron furnishes so large a supply. The innumerable islands, points and inlets along the east shore of Georgian Bay seem to furnish almost ideal conditions for the development of fish life. If we can show that the microscopic creatures are present which form the first food of the fry, and that for these minute animals there is an adequate quantity of the still more minute plants on which they feed, —we shall have gone far toward furnishing a basis for the expectation that scientific methods of conservation and propagation will renew the copious supply of fish for which these waters were once famous.

The chain of life which begins with the unicellular algæ and ends with man, has been often demonstrated. The one-celled plants convert the non-living substances—atmospheric gases and water with its dissolved salts—into the lowest form of living matter. Mingled with these are many forms, so lacking in definite characters that so far it has been found impossible to decide their affinities. They constitute the *Protista*, possibly neither plant nor animal, but of the common structure from which both branches of life have developed. The quantity of unicellular plants per unit volume of water decides the quantity of the Protozoa, Rotifera and Crustacea which may inhabit the waters. These latter are known to serve as the chief if not the only food of the young and small fish. Favorable conditions of shelter and food are indispensable to the growth and rapid development of the young food fish. We are therefore quite safe in deciding that a prime biological condition for a plentiful fish fauna is the presence of an abundant growth of microscopic plants.

The surroundings most favourable for the growth of the more minute algæ are quiet waters, sunlight, and a plentiful growth of larger plants such as Chara, Potamogeton, Elodea, Utricularia, and Myriophyllum, as bottom and shore growths. These larger plants serve as shelters and homes for the minute forms, and wherever the former are absent, we cannot expect the latter to be abundant.

The prevailing westerly winds give such an eroding power to the water washing the islands and eastern shore of Georgian Bay that only in the deeper inlets and sheltered bays and river mouths can we find conditions suitable for shore growths of the larger plants. The steepness of the gradient at which the crystalline rocks forming the shore enter the water, seldom permits of an extended submerged terrace of proper depth for the anchored society of plants. Hence only in a few places, and those more or less remote from the open bay, can we find littoral zones characteristic of such quiet shallow waters as Lake St. Clair.

Beds of Chara—that most important breeding ground for minute algæ,—were seldom found. The following lists are far from being exhaustive even of the small district investigated.

Not expecting to undertake a study of the green plants, my supply of reference authorities was far from complete, and necessitated much drawing and recording of measurements for future reference. Only such specimens as' lent themselves by good condition, reasonable size, and characteristic marks, were determined. Many non-fruiting forms of the higher algæ were passed over unidentified. Subsequent observation at a different season may add very considerably to every group.

SHORE AND BOTTOM FLOWERING PLANTS.

From the moist border outward about the following order may be found but never all at one place:

Gratiola aurea Muhl.
Utricularia cornuta Michx.
Gerardia purpurea L.
Isoetes echinospora var. muricata. Engelm.
Ranunculus flammula L. var. reptans L (Meyr).
Lobelia Dortmanna L.

The above are usually, but not always, in the water.

Juncus Balticus Willd. var. littoralis Engelm.
Typha latifolia L.
Eriocaulon articulatum (Huds) Morong.
Sparganium eurycarpum. Engelm.
Sagittaria latifolia. Willd, forma diversifolia Engelm.
S. graminea Michx.

These are found usually in water less than one foot deep.

Pontederia cordata L.
Scirpus hudsonianus (Michx) Fernald.
Nymphaea advena Ait. var. variegata (Engelm).
Nymphoides lacunosum (Vent) Fernald.
Brasenia Schreberi Gmel.

The above are found in water up to three feet in depth.

Utricularia vulgaris L. var. americana.
U. minor L.
Ceratophyllum demersum L.
Myriophyllum spicatum L.
Elodea canadensis Michx.
Valisneria spiralis L.
Potamogeton heterophyllus Schreb; forma myriophyllus (Robbins) Morong.

5 GEORGE V., A. 1915

Potamogeton pectinatus L.
No doubt other Potamogetons are to be found.

Chara and Nitella were found only in water less than one foot in depth, but may occur at greater depths.

CYANOPHYCEAE.

Chroococcus turgidus (Kutz) Naeg.
C. turicensis (Naeg) Hansg.
> Both of the above are frequent in washings from submerged plants collected in muddy bays.

Gleocapsa sp. In washings from submerged moss from Go Home River at Chute.
Aphanocapsa Grevillei (Hass) Rab.
Aphanothece pallida (Rab).
> On Chara from Louden's Bay.

A. stagnina (Spring) A. Br.
Gonphosphaeria aponina Kuetz.
Clathrocystis aeruginosa. (Kuetz) Henfrey. In floating plankton.
Coelosphaerium Kuetzingeanum Nag.
> In all surface collections made with plankton net in quiet waters.

Merismopedium glaucum (Ehren) Nag.
> In shallow bays of warm water.

Eucapsis alpina Cl. & Sh.
Oscillatoria limosa Agardh.
> In surface plankton in steamer channel.

Lynbya sp. in scrapings from submerged stones in Gap, Giant's Tomb.
Scytonema Naegelii Kg. (Tolypothrix penicillata) (Agardh) Thuret.
> In scrapings from rocks. Fraser's Channel.

S. crispum Bornet.
> Plentiful, in scrapings from submerged rocks.

Nostoc comminutum Kutz.
> Common in surface collections with the next.

Anaboena flos aquae. Kutz.
Dichothrix horsfordii Barnet.
Rivularia dura, Roth.
> In scrapings from rocks, Fraser's Channel.

R. echinulata (Smith) Barnet.
> On culms of Scirpus. Go Home River near the Chute.

Stigonema mamillosum Agardh.

CHLOROPHYCEAE

Volvocales.

Chlamydomonas pulvisculus Ehrb
> Common in surface plankton with the two following.

Botryococcus Braunii Kutz.
Tetraspora lubrica (Roth) Agardh.
Pandorina morum (Mull) Bory.
    In washings from a submerged moss from Go Home River near the chute.

Protococcales.

Pleurococcus vulgaris. Meneg.
    In scrapings from back of a large Snapping Turtle.
Selenastrum acuminatum Lagerh (Conn & Webster's Fresh Water Algæ of
.Conn). In washings of Nitella from Louden's Bay.
Palmodictyon viride, Kutz—with above.
Scenedesmus quadricauda (Turp) Breb.
S. obliquus (Turp) Kutz.
S. bijuga (Turp) Wittr.
    With the next in surface plankton.
Rhaphidium convolutum (Corda) Rabenh.
Schizochlamys gelatinosa A Br.
Tetracoccus botryoides West, as described in West.
Coelastrum cambricum Archer.
C. sphaericum Nag.
    These are frequent in collections.
Dimorphococcus cordatus Wolle.
    Plentiful.
Pediastrum Boryanum (Turp) Meneg.
P. tetras. Ehrenb.
P. Ehrenbergii A. Br. An unsymmetric specimen noted.
P. sp.
    A symmetrical non-clathrate form of 64 cells, bearing on the outer margin
    slender projections, each with a well-marked capitellate termination.
    Very similar to the portion of P. glanduliferum Benn. as figured by West.
    Found in washings from Limnea and other crustacea.

Confervales.

Confervaceae.
Ophiocytium capitatum Wolle
    In surface plankton but infrequent.
Characium heteromorphum (Reinsh) Wolle.
    Attached to Œdogonium.
Chlorobotrys regularis (West) Bohlin.
    In washings from Nitella.
Dictyosphaerium Ehrenbergianum, Nag.
D. reniformis, Bulnh.
    Both of these frequent in collections.

5 GEORGE V., A. 1915

Chaetophoraecae and *Oedogoniaceae.*

Chaetosphaeridium globosum (Nordst) Klebahn.
In scrapings from rocks in Fraser's Channel.
Œdogonium fragile Wittr.
Œ. crispum Wittr.
Bulbochaete monile Wittr. & Lund.
With Œdogonium from near the chute in Go Home River.
Bulbochaete sp. In washings from Utricularia purpurea.

Coleochaetaceae.
Coleochaete soluta. (Breb) Pringsh.
On submerged culms of Scirpus in Go Home River below the chute, and
plentiful, in scrapings of submerged rocks in Fraser's Channel, August
23rd, bearing oogonia many of which were brownish at that date.

Conjugales.

Mougeotia calcarea (Cleve) Wittr. On Island 218, two miles north of Go
Home Bay. M. genuflexa, Agardh.
Desmidaceae.—
Penium oblongum D. By.
P. rupestre Kg. Common in washings of submerged moss.
Closterium striolatum Ehrb.
var. intermedium.
Cl. parvulum, Naeg.
Cl. Dianae Ehrb. Frequently found.
Cl. pronum, Breb.
Several other species undetermined.
Cosmarium moniliforme, Ralfs.
Cos. sexangulare, Lund.
Cos. orbiculatum, Ralfs.
Cos. perforatum, Lund.
Cos. pyramidatum, Breb.
Cos. Meneghinii Breb—plentiful in washings from Fontinalis.
Cos. Nordstedtii Delfs.
Cosmarium sp. agreeing with description and figure of Cos. Eloiseanum Wolle,
but lacking the granular tumors.
Docidium Baculum Breb.
Pleurotaenia Trabecula (Ehrb) Nag.
P. crenulatum (Ehrb) Rab.
Xanthidium cristatum (Breb) Ralfs.
X. antilopeum (Breb) Kg.
var. Minneapoliense Wolle.
X. fasciculatum (Ehrb) Ralfs.
Staurastrum dejectum Breb.

St. ophiura Lund. var. tetracerum Wolle.

St. " " " pentacerum "

St. odonatum, Wolle.

Euastrum elegans Kg.

E. magnificum Wolle.

E. ventricosum Lund.

Micrasterias furcate (Ag) Ralfs.

M. pseudo-furcata, var. Minor(?) Wolle.

M. lanticeps, Nord. common.

M. crux Melitensis (Ehrb) Hass.

Diatomaceae

Acnanthes exilis Kg.

Asterionella formosa var. gracillima, V.H.

In surface plankton in steamer channel at entrance to the Bay.

Amphora ovalis Kg.

Cocconema lanceolatum Ehrb.

Cosinodiscus lanceolatum Ehrb.

Cosinodiscus plentiful in, and characteristic of dredged material from east edge of sand beach Giant's Tomb.

C. lacustris, from inner bay.

Craspedodiscus microdiscus Ehrb. (?)

Denticula lauta Bail.

Encyonema gracile Rab.

Epithemia turgida Kg.

E. argus Kg.

Fragilaria—Ribbons of acute-pointed individuals frequent in scrapings from submerged rocks and in floating plankton in steamer channel.

Gonphonema geminata Ag.

Melosira granulata Bail.

Navicula viridis. Kg.

In ribbons of 100 individuals among decaying Zygnema.

Stauroneis Phoenocenteron Ehrb.

Surirella elegans Ehrb.

Synedra ulna var. splendens.

Tabellaria fenestrata (Lyng) Kg.

T. flocculosa (Roth) Kg.

Terpsinoe Musica Ehrb.

Through the kindness of Mr. C. S. Boyer, of Philadelphia, one of the authorities on Diatoms, I am able to add the following, identified from the material I collected in the immediate vicinity of Go Home Bay:—

Amphora ovalis (Bréb) Kütz.

Anomœoneis serians Bréb.

Cyclotella striata Kütz.

Cymbella cuspidata Kütz.

C. gastroides Kütz.
C. cistula (Hempr.) Kirchn.
C. lanceolata (Ehr.) Kirchn.
Cymatopleura elliptica Itm. Sm.　Rare form.
Diploneis elliptica (Kütz.) Cl.
Eunontia gracilis (Ehr.) Rab.
E. major. (Itm. Sm.) Rab.
　　　　　var. impressa.
E. formica (Ehr.)
Frustulia vulgaris Thw.
Gonphonema constrictum Ehr.
G. capitatum Ehr.
G. acuminiatum. f. coronatum (Ehr.) Rab.
Melosira granulata.
Meridion intermedium var constrictum H. L.S.
Nitzchia amphioxys (Ehr.) Itm. Sm.
Navicula pseudo-bacillum.　Grun.
Neidium iridir (Ehr.) Cl.
Pinnularia divergens.　Ralfs.
P. nobilis Ehr.　Also varieties.
P. tabellaria Ehr.
Stauroneis gracilis.　Itm. Sm.
Surirella splendida Itm. Sm.　Also varieties..
Synedra danica Kiltz.

Batrachospermum moniliforme. Roht.—though not a green alga.—should be mentioned. It was found attached to timbers of a rude wharf.

The following named algæ were collected and identified by my colleague, Mr. A. B. Klugh, during May and June, 1911. The collections were made at various points as indicated in the notes, but all along the Georgian Bay shore or in the waters immediately tributary to the Bay.

Chroococcus turgidus, Naegeli. Bog, Mud Lake, near Colpoy's Bay, June 7, Marsh Oliphant, June 14.

Microcystis marginata, Kuetzing. Floating among other algæ at windward shore of Sky Lake, near Oliphant, May 28; Bog, Mud Lake, Near Colpoy's Bay, June 26.

Merismopedium glaucum, Naegeli, Plankton, Pool on the Commons, Colpoy's Bay, May 8, 1911; Swale, Colpoy's Bay, May 20; Pool, McGregor's Harbour, Cape Croker, May 30; Shore of Lake Huron at Oliphant, June 14; Sky Lake May 28.

Oscillatoria tenuis, Agardh. Damp places on rock. Colpoy's Bay, May 11.

Oscillatoria subtilissima, Kuetzing. Damp places on rock, Colpoy's Bay, May 11.

Oscillatoria formosa, Bory. On timber in a small stream near Colpoy's Bay, May 27.

SESSIONAL PAPER No. 39b

Nodularia paludosa, Wolle. Swale near Colpoy's Bay, May 20: Swamp, Golden Valley, June 1.

Anabaena torulosa, Lagerheim, Swale, Colpoy's Bay, May 20; Swamp, Golden Valley, June 1.

Stigonema mamillosum, Agardh. Bog, Mud Lake, near Colpoy's Bay, June 7.

Calothrix parietina, Thuet. Damp place on limestone rock, Colpoy's Bay, May 11.

Ophiocytium cochleare, A. Braun. Swale, Colpoy's Bay, May 20; Swamp, Golden Valley, June 1; Swamp near Boat Lake, June 16; Ditch, Oliphant, June 14.

Ophiocytium parvulum, A. Braun. Swale, Colpoy's Bay, May 20; Swamp Golden Valley, June 1; Bog, Mud Lake, June 7; Swamp near Boat Lake, June 16; Ditch, Oliphant, June 14; Pool, Hope Bay, June 8.

Ophiocytium gracilipes, Rab. Scarce, in a collection from a marsh on the Cape Croker road, May 30.

Conferva bombycina, Agardh. Swamp, Golden Valley, June 1; Swamp, Mar road, June 5. Diteh, near Boat Lake, June 16; Stream in sandy shore, Oliphant, June 14.

Zygnema leiospermum, De Bary, common near mill at Lake Isaac, June 5.

Spirogyra catenaeformis, Kuetzing, Bog, Mud Lake, June 26.

Spirogyra varians, Kuetzing. Common with abundant zygospores, in ditch near Wiarton, May 5.. By May 26 it had completely disappeared though the ditch still contained plenty of water. Scarce in a swamp near Boat Lake, June 16, zygospores present.

Spirogyra orthospira, Naegeli. Small stream from spring, Oliphant, June 14, in all stages of conjugation; Ditch, Colypoy's Bay, June 23.

Spirogyra weberi, Kuetzing. Common in pools in sand of shore of Lake Huron at Golden Valley, in all stages of conjugation, June 1; Ditch, Lion's Head, June 8, just beginning conjugation; in small stream in sandy shore at Oliphant, June 14, spores mature; Swamp, Adamsville, June 8.

Spirogyra insignis, Kuetzing. Ditch, near Wiarton, June 4, spores nearly mature.

Mougeotia genuflexa, Agardh. Common in a small marsh near Purple Valley, May 30, very sparingly fruited; Swamp, Golden Valley, June 1.

Mougeotia scalaris, Hassall. Near mill, Lake Issac, June 5; Pool in swamp at Mud Lake, June 6.

Mougeotia viridis, Wittrock. Common in swale, Colpoy's Bay, May 20.

Chlamydomonas communis, Snow. Abundant in a collection from a swamp on Mar road, June 5.

Chlamydomonas globosa, Snow. Common in pools and swamps throughout the Peninsula.

Pandorina morum, Bory. In small marsh at Sky Lake, May 28; in marsh on Cape Croker road, May 30; Swamp, Golden Valley, June 1; abundant in a collection from a swamp on Mar road, June 5.

Tetraspora lubrica, Agardh. Common in a stream in a pasture, Colpoy's Bay, April 30th; in pools along a bush road, near Mar, May 10. In a stream between Colpoy's Bay and Purple Valley, May 27.

Chlorococcum humicola, Rabenhorst. Common under dripping water.

5 GEORGE V., A. 1915

Characium naegelii A. Braun.   Common on other algæ, particularly on Conferva bombycina throughout the peninsula.

Characium ambiguum, Hermann.   On Conferva bombycina in swale near Colpoy's Bay, June 20th.

Rhaphidium falcatum, Cooke.   Swamp, Mar road, June 5; Ditch, near Boat Lake, June 16; Pool, Hope Bay, June 8; Shore of Lake Huron at Oliphant, June 14.

Rhaphidium falcatum aciculare, Hansgirg.   Swale, Colpoy's Bay, May 20; Pool near Colpoy's Bay, May 30; common in swamp near Golden Valley, June 1.

Nephrocytium agardhianum, Naegeli.   Swamp on Mar road, June 5.

Tetraedron minimum, Hansgirg.   Pool, Hope, June 8; small stream, Oliphant, June 14.

Scenedesmus bijuga, Wittrock.   Pool, Hope Bay, June 8; Pond on Commons, Colpoy's Bay, May 11

Scenedesmus obliquus, Kuetzing.   A common plankton form throughout the Peninsula.

Scenedesmus quadricauda, Brebisson.   A common plankton form throughout the region.

Scenedesmus quadricauda abundans, Kirchener.   Pool, McGregor's Harbour, Cape Crocker, May 30; Ditch, near Boat Lake, June 16.

Coleastrum proboscideum, Bohlin.   Swale, near Colpoy's Bay, June 5; Marsh, Oliphant, June 14.

Sorastrum spinulosum, Naegeli.   Scarce, in collection from a pool at Hope Bay, June 8.

Hydrodictyon reticultaum, Lagerheim.   Forming a sheet over the surface of a large pool at edge of swale near Colpoy's Bay, June 5.

Pediastrum boryanum, Meneghini.   A very common plankton form throughout the Peninsula.

Pediastrum tetras, Ralfs.   Scarce, in collection from a marsh at Oliphant, June 14; Pool, Hope Bay, June 8.

Ulothrix aequalis, Kuetzing.   This species and Ulothrix zonata are the commonest filamentous forms on the rocks of the shores of Georgian Bay.   They occur in patches consisting of one species only.   Gametes were mature on April 30.

Ulothrix zonata, Kuetzing.   Common, on rocks along shores of Georgian Bay; fruiting on May 7.

Oedogonium capilliforme, Kuetzing.   Swale, Colpoy's Bay, June 5.

Chaetosphaeridium globosum, Klebahn.   On Oedogonium capilliforme in swale, Colpoy's Bay, June 5.

Chaetophora elegans, Agardh.   Forming globular gelatinous masses about 5 mm. diameter on stones in a pool on the Cape Croker road, May 30; forming light green spheres from extremely minute size up to 1 mm. diameter on sticks at edge of a willow swale near Colpoy's Bay, June 5.

Chaetophora incrassata, Hazen.   Attached to a log in a ditch, near Wiarton, May 12; common on stones at bridge over Patanelly River, near Mar, June 1.

Stigeoclonium lubricum, Kuetzing.   Common in a little stream from a spring near Wiarton, May 5.

Draparnaldia acuta, Kuetzing. In pools with Tetraspora lubrica on a bush road near Mar, May 10; Stream near Wiarton May 19; Stream near Golden Valley, June 1.

Draparnaldia glomerata, Agardh. Swale, Colpoy's Bay, May 20: Swamp, Golden Valley, June 1.

Pleurococcus vulgaris, Meneghini. Common on trees, walls, etc.

Tretepohlia aurea, Martius. Scarce on limestone rocks in Populus-Thuja scrub along Mar road, June 20: forming bright orange velvety cushions from 1 to 2 dm. in extent; forming light orange-colored patches on rocks along the shore road at Colpoy's Bay.

Cladophora callicoma, Kuetzing. Scarce in stream at Colpoy's Bay.

Vaucheria sessilis, D. C. Common in swale along Wiarton road, oospores not yet mature, June 23.

Vaucheria geminata racemosa, Walz. Swamp near Boat Lake, June 15.

XI.

LIST OF GEORGIAN BAY FLESHY FUNGI AND MYXOMYCETES.

By T. H. Bissonnette, M.A., Queen's University, Kingston.

The following is a preliminary list of fleshy fungi which I collected and classified August and September, 1912. In the work of collecting and classifying Miss Penson and Mr. Woodhouse preceded me at the Biological Station, and this report combines the work of the three of us.

N.B.—Only where the species were classified are any entered. Almost all the genera found are included, but only those species are included which were determined and duly classified.

Agarics.

1. Amanita phalloides No. 35 Aug. 3/12 Wishart's Bay.
　　A.　　muscaria　"　77 Aug. 7/12 Long Bay.
　　A.　　Frostiana　"　176 & 192.　Aug. 25/12 Galbraith's Creek.
　　　　　　　　　　　　　　　　Aug. 27/12 Loudon's Bays.
　　A.　　mappa　　"　130 Aug. 25/12 Galbraith's Creek.　　.
2. Ananitopsis strangulata No. 44. Aug. 3/12 Wishart's Bay.
　　A.　　vaginata fulva No. 2, Aug. 2/12 Creek near Chute.
　　A.　　vaginata livida No. 196 Aug. 30/12 Meuller's Bay.
3. Lepiota acutesquamosa No. 143 Aug. 21/12 Creek near Chute.
　　L.　　asperula　　"　218 Sept. 3/12 Laforge's Wood.
　　L.　　granulosa　　"　227 Sept. 4/12 Long Bay.　　.
4. Armillaria mellea No. 121 & 14 Aug. 19/12 Laforge's Wood &
　　　　　　　　　　　　　　　　Aug. 2/12 Creek near Chute.
5. Tricholoma　　No. 134 Aug. 20/12 Fenton's Bay.
　　T.　sejunctum　"　149 Aug. 21/12 Creek near Chute.
　　T.　album　　"　166 Aug. 23/12 Skidway above Sandy Gray.
　　…………　　"　9 Aug. 2/12 Creek near Chute.
6. Clitocybe　No.　24 July 20/12 & Aug. 3/12 Creek near Chute.
　　"　　　"　60 Aug. 4/12 Sandy Gray Falls.
　　"　　　"　118 Aug. 16/12 Giant's Tomb.
　　C.　infundibuliformis No. 122 Aug. 19/12 Laforge's Wood.
　　C.　phyllophila　　"　127 Aug. 19/12 Laforge's Wood.
　　C.　clavipes (media?)　"　131 Aug. 19/12 Laforge's Wood.
　　C.　subditopoda　　"　197 Aug. 30/12 Meuller's Bay.
　　C.　media　　"　240 Sept. 6/12 Galbraith Lake.
7. Cantharellus, cinnabarinus No. 57 Aug. 4/12 Sandy Gray Falls.
　　C.　infundibuliformis　　"　58 Aug. 4/12 Sandy Gray Falls.
　　C.　aurantiacus　　"　59　"　"　"　"

5 GEORGE V., A. 1915

C. brevipes   No. 106 Aug. 12/12 Giant's Tomb.
C. cibarius   " 150 Aug. 21/12 Creek near Chute.

8. Myctalis,—
9. Lactarius indigo No. 20 Aug. 2/12 Creek near chute.
     L.   piperatus No. 68 Aug. 4/12 Sandy Gray Falls.
     L.   resimus   " 164 Aug. 23/12 Skidway above Sandy Gray Falls.
     L.   cinereus   " 175 Aug. 25/12 Galbraith's Creek.
     L.   regalis   " 194 Aug. 28/12 Sandy Gray Falls.
10. Russula virescens   "   39 & 142 Aug. 3/12 Wishart's Bay.
     R.   alutacea   " 169 Aug. 23/12 Skidway above Sandy Gray Falls.
     R.   emetica   " 188 July & Aug. Everywhere.
     R.   rubra   " 189 Aug. 22/12 Loudon's Bay.
11. Hygrophorus miniatus No. 94, Aug. 7/12 Long Bay Aug. 27, Loudon's Bay No. 187.
12. Pleurotus sapidus No. 112 Giant's Tomb, Aug. 12/12.
     P.   ostreatus No. 161 Aug. 23/12 Skidway above Sandy Gray.
     P.   petaloides No. 205 Sept. 1/12 Giant's Tomb.
13 Collybia radicata No 3, Aug 2/12 Creek near chute.
     C.   familia (Marshall) No. 53, Aug. 4/12 Sandy Gray Falls.
     C.   velutina No. 79 Aug. 7/12 Long Bay.
     C.   zonata, No. 141 Aug. 21/12 Creek near chute.
     C.   myriadophila No. 89 Aug. 7, Long Bay.
     C.   confluens, No. 178, Aug. 25/12 Galbraith's Creek.
14. Mycena galericulata No. 177 Aug. 25/12 Galbraith's Creek.
     M.   Leaiana   No. 180 Aug. 25/12 Galbraith's Creek.
15. Omphalia companella No. 10 & 140 Aug. 2/12 Creek near chute.
                                         Aug. 21/12.
16. Marasmius rotula, No. 55, Aug. 4/12 Sandy Gray Falls.
             velutipes No. 12/12 Laforge's Wood.
             siccus No. 160 Aug. 23/12 Skidway above Sandy Gray.
17. Xarotus.
18. Heliomyces.
19. Lentinus.
20. Panus—strigosus (?) No. 183 Aug. 19/12 Giant's Tomb.
21. Trogia crispa No. 199 Sept. 1/12 Giant's Tomb.
22. Schizophyllum commune No. 206, Sept. 3/12 Station Island and elsewhere.
23. Lenzites separia No. 193 Aug. 28/12 Portage from Sandy Gray to Flat Rock
        Rock Lake, in Woods.
24. Volvaria.
25. Annularia.
26. Pluteus cervinus No. 113, Aug. 12/12 Giant's Tomb.
27. Entoloma rhodopolium, No. 146, Aug. 21/12 Creek near chute.
28. Clitopilus prunulus No. 167 Aug. 21/12 Creek near chute.
     C.   Noveboracensis No. 204 Sept. 1/12 Giant's Tomb.
     C.   abortivus No. 152 Aug. 21/12 Creek near chute.
     C.   orcellus (?) No. 67, Aug. 4/12 Sandy Gray Falls.

29. Claudopus nidulans (?) No. 182 Aug. 25/12 Galbraith's Creek, Aug. 19 Giant's Tomb.
30. Eccilia.
31. Leptonia.
32. Nolanea.
33. Pholiota dura (dwarf) No. 212, Sept. 1/12 Giant's Tomb.
    P.   squarosa No. 253 Sept. 8/12 Laforge's Wood.
34. Cortinarius alboviolaceous No. 167 Aug. 23/12 Old skidway above Sandy Gray Falls.
    C.   armillatus No. 213 Sept. 3/12 Laforge's Wood.
    C.   cinnamoneus No. 243, Sept. 7/12 Burwash Lake.
35. Flammula.
36. Inocybe.
37. Hebeloma glutinosum No. 148 Aug. 21/12 Creek near chute.
38. Paxillus.
39. Crepidotus versutus No. 165 Aug. 23/12 Skidway above Sandy Gray.
    C.   mollis   " 219 Sept. 3/12 Giant's Tomb.
40. Tubaria.
41. Naucoria hamodryas No. 128; Aug. 19/12 Laforge's Wood.
42. Pluteolus.
43. Galera crispa or laterita No. 181 Aug. 25/12 Galbraith's Creek.
44. Bolbitius.
45. Chitonia.
46. Pilosace.
47. Agaricus sylvaticus No. 97, Aug. 12/12 Giant's Tomb.
48. Stropharia.
49. Hypholoma perplexum, No. 99, Aug. 12/12 Giant's Tomb.
50. Deconica.
51. Psilocybe, spadica, No. 168, Aug. 23/12 Skidway above Sandy Gray Falls.
52. Psathyra—No. 4, Aug. 2/12 Creek near chute.
53. Coprinus atramentarius. No. 12 Aug. 2/12 Creek near chute.
    C.   ovatus No. 173 Aug. 25/12 Galbraith's creek.
54. Gomphidius.
55. Psathyrella.
56. Panaeolus.
57. Chalymatta.
58. Anellaria.

                    Polyporaceae—Pored Fungi.

59. Boletus scaber No. 64, Aug. 4/12 Sandy Gray Falls.
            subtomentosus No. 65, Aug. 4/12 Sandy Gray Falls.
            felleus No. 66 Aug. 4/12 Sandy Gray Falls.
            versipellis No. 132 Aug. 20/12 Fenton's Bay.
            edulis No. 144 Aug. 21/12 Creek near chute.
            americanus No. 186, Aug. 27/12 Loudon's Bay.

5 GEORGE V., A. 1915

Boletus edulis clavipes No. 145 Aug. 21/12 Creek near chute.

chrysenteron No. 162, Aug. 23/12 Skidway above Sandy Gray Falls.

60. Strobilamyces strobilaceous No. 114, Aug. 16/12 Giant's Tomb.
61. Boletinus pictus, No. 242, Sept. 7/12 Burwash Lake.
62. Fistulina.
63. Polyporus velutinus, No. 34, Aug. 2/12 Creek near chute.

fomentarius No. 75, Aug. 4/12 Sandy Gray Falls.

picipes (or elegans), No. 88, Aug. 7/12, Long Bay.

frondosa No. 110, Aug. 12/12 Giant's Tomb.

applanatus No. 129, Aug. 19/12, Laforge's Wood.

pubescens No. 163, Aug. 23/12 Skidway above Sandy Gray.

cinnabarinus No. 179, Aug. 25/12 Galbraith's Creek.

resinosus, No. 182, Aug. 25/12   "   "

12439 Gal 87 Dept P P & S Vol 2 M. Curran
64. Polystictus biformis No. 42, Aug. 3/12, Wishart's Bay.

versicolor No. 133, Aug. 20/12, Fenton's Bay.

perennis, No. 216, Sept. 3/12, Laforge's Wood.

pergameus, No. 136, Aug. 19/12   "   "

hirsutus, No. 209, Sept. 1/12, Giant's Tomb.

65. Fomes leucophaeus, No. 129, Aug. 19/12, Laforge's Wood.
66. Trametes.
67. Merulius.
68. Daedalea quercina No. 37, Aug. 3/12, Wishart's Bay.

confragosa No. 185 Aug. 27/12, Loudon's Bay.

ambigua No. 195, Aug. 29/12, Meuller's Bay.

unicolor, No. 248 Sept. 8/12, Laforge's Farm.

69. Favolus canadensis, No. 137, Aug. 19/12, Laforge's Wood.

alveolatus, No. 40, Aug. 3/12, Wishart's Bay.

70. Cyclomyces.
71. Glaeoporus.

Hydnaceae—Fungi with Teeth.

72. Hydnum coralloides No. 100, Aug. 12/12, Giant's Tomb.

caput-ursi No. 101, Aug. 12/12   "

caput medusae No. 102, Aug. 12/12   "

spongiosipes, No. 125, Aug, 19/12, Laforge's Wood.

septentrionale, No. 126, Aug. 19/12   "   "

adustum, No. 135, Aug. 19/12   "   "

pulcherrimum No. 251, Sept. 8/12   "   "

73. Irpex.
74. Phlebia.
75. Grandinia.

Thelephoraceae—Smooth hymenium or wrinkled.

76. Craterellus.
77. Corticium sambucum No. 191, Aug. 27/12, Loudon's Bay.
78. Thelophera.

79. Stereum.
80. Hymenochaetc.

Clavariaceae, Coral Fungi.

81. Sparassis.
82. Clavaria. flava No. 11, Aug. 2/12, near chute Aug. 4, Sandy Gray Falls.
       stricta No. 156, Aug. 21/12, Creek, near chute.
       pistillaris, No. 157, Aug. 23/12, Sandy Gray point—small form.
       cristatum No. 159, Aug. 23/12, Skidway above Sandy Gray Falls.
       aurea, No. 203, Sept. 1/12, Giant's Tomb.
       mucida, No. 174, Aug. 25/12, Galbraith's Creek.
       pistillaris, No. 201, Sept. 2/12, Giant's Tomb, large form.
83. Calocera.
84. Typhula.
85. Lachnocladium.

Tremellini.

86. Tremella.
87. Tremellodon gelatinosum Sept. 8/12, Laforge's Wood.

Ascomycetes—Sac Fungi.

88. Morchella.
89. Verpa.
90. Gyromitra.
91. Helvella.
92. Hypomyces.
93. Leptoglossum luteum No. 70, Aug. 4 and Aug. 23/12, Sandy Gray Falls.
94. Spathularia.
95. Leotia-lubrica No. 115, Aug. 15/12, Wishart's Bay.
                No. 184, Aug. 27/12, Loudon's Bay.
96. Peziza nebulosa No. 74, Aug. 4/12 Sandy Gray.
      badia, No. 87, Aug. 7/12, Long Bay.
      scutellata, No. 138, Aug. 19/12, Laforge's Wood.
      semitosta, No. 139, Aug. 19/12,   "   "
      hemispherica, No. 155, Aug. 21/12, Creek near Chute.
      repanda, No. 249, Sept. 8/12, Laforge's Wood.
      aurantea, No. 252, Sept. 8/12,   "   "
97. Urnula.
98. Helotium.
99. Bulgaria inquinans, not preserved.

Nidulariaecae—Bird's nest Fungi.

100. Cyathus.
101. Crucibulum.
102. Nidularia pisiformis 116, Aug. 16/12, Giant's Tomb.

Basidiomycetes.

103. Phallus duplicatus, No. 98, Aug. 12/12, Giant's Tomb.

5 GEORGE V., A. 1915

104. Lysurus.
105. Mutinus.
106. Calvatia maxama No. 1, Aug. 2/12, Creek near Chute.
107. Lycoperdon gemmatum No. 120, Aug. 19/12, Laforge's Wood.
         pyriforme (log form), No. 120, Sept. 1/12 Giant's Tomb.
         pyriforme ground form 211, Sept. 1/12   "   "   larger.
108. Bovista pila—No. 119, Aug. 19/12, Laforge's Farm.
109. Bovistella.
110. Scleroderma aurantium, No. 158, Aug. 23/12, Old Skidway above Sandy Gray.
111. Catastoma.
112. Secotium.
113. Polysaccum.
114. Mitremyces.
115. Geaster triplex No. 90, Aug. 23/12, old skidway above Sandy Gray.

116. Cordyceps.   Sphaeriaeceae.
117. Claviceps.
118. Xylaria polymorpha No. 15, Aug. 2/12, Creek near chute and elsewhere.
         polymorpha spatularia, No. 73, Aug. 4/12, Sandy Gray Falls.

Myxomycetes.

119. Stemnotis fusca No. 29, Aug. 12/12, Creek near chute.
120. Calcareous Myxomycete, on sticks. Not identified.
     Creamy one, also Tycogala, small red, Aug. 2, No. 30. Creek near chute.

     I may add that the method pursued in this mycological research was as follows:—

     Each afternoon, when possible, baskets and bottles for small specimens were taken, and we visited the hardwood bushes of the neighborhood and collected specimens till dark or nearly so. On our return to the Laboratory, we arranged our collections upon white paper in such a manner that spores would be shed over night and this means of identification utilized. The following morning we identified as many as possible and took description of those we could not identify, after which we preserved all but the Boleti and the Bracket fungi in 4% formalin in jars, tagged with numbers corresponding to our list, which contained either the species or the description and date and place where the species was found. Bracket fungi we preserved dry in cigar boxes or, when large, in large boxes. Boleti, we preserved in a mixture of 4% formalin and 30% alcohol, as they broke down in 4% formalin. In many cases one day's collecting furnished material for two or more days' work in classification. Where trips were longer, a whole day was required for collecting. The collecting covered the area around the station near the shores, within a radius of 7 miles among the islands and about 7 miles up the Go Home river. In no cases were we able to take long trips inland because of the roughness of the country and the difficulty of finding our boat on our return to the shore.

XII.

NOTES ON THE AQUATIC PLANTS OF GEORGIAN BAY.

By A. B. KLUGH, M.A., Queen's University, Kingston.

During the latter part of August and the first three weeks of September, 1912, I was engaged in a study of the Hydrophytes of Georgian Bay in connection with the Biological Station at Go Home Bay, Muskoka. In company with Mr. A. D. Robertson, M.A., I made a trip round the Bay, collecting along the shore at numerous points. A full report upon the Algæ collected on this trip is given in my paper in "*Rhodora.*," Vol. 15, No. 173 (May, 1913), in which a new species of *Rivularia* is described, and twelve species of Algæ new to Canada are recorded.

The following are the records of the Pteridophytic and Spermatophytic Hydrophytes observed on the trip referred to.

PTERIDOPHYTA.

Isoetes macrospora. Dur. Growing on muddy bottom in eight feet of water near the mouth of the Shawanaga River Aug. 26.

Isoetes echinospora braunii Engelm. Growing submerged near shore in the Shawanaga River Aug. 27. Submerged in a foot of water at Killarney Sept. 4. Submerged, in Collins' Inlet Sept. 3.

SPERMATOPHYTA.

MONOCOTYLEDONEÆ.

Sparganium simplex, Huds. In lake off Shawanaga River, Aug. 27.

Potamogeton natans, L. Severn River, Aug. 21. At mouth of Shawanaga River, Aug. 27.

Potamogeton heterophyllus, Schreb. Waubaushene, Aug. 27. In lake off Shawanaga River, Aug. 27.

Potamogeton epihydrus cayugensis, Benn. In the French River at foot of the last rapids, Sept. 1.

Potamogeton lucens, L. At mouth of Shawanaga River, Aug. 26.

Potamogeton perfoliatus, L. Severn River and covering large areas of water in Georgian Bay at mouth of Severn River, Aug. 21. Near mouth of Shawanaga River, Aug. 26. French River, Sept 1. Killarney, Sept 4.

Potamogeton zosterifolius, Schw. Near the mouth of the Shawanaga River, Aug. 26. Waubaushene, Aug. 21.

Potamogeton pusillus tenuissimum, Mertens and Koch. In a very small almost cut-off inlet in one of the Bustard Islands off French River. Aug. 30.

5 GEORGE V., A. 1915

Potamogeton dimorphus, Raf. In lake off Shawanaga River, Aug. 27. In little almost cut-off inlet in one of the Bustard Islands off French River, Aug. 30. In pool at the foot of the last rapids on the French River, Sept. 1.

Potamogeton pectinatus, L. Matchedash Marsh, Waubaushene, Aug. 24. Killarney, Sept. 4.

Najas flexilis, Rostk & Schmidt. Waubaushene, Aug. 21. In lake off Shawanaga River, Aug. 27. In the Shawanaga River, Aug. 26. In pool at foot of the last rapids on the French River, Sept. 1.

Elodea canadensis, Michx. Waubaushene, Aug. 21. In pool at foot of last rapids on the French River, Sept. 1.

Vallisneria spiralis L. Waubaushene, Aug. 21. In pool at foot of last rapids on the French River, Sept. 1.

DICOTYLEDONEÆ.

Ceratophyllum demersum, L. Waubaushene, near Canary Island, Aug. 21.

Nymphaea advena, Ait. Severn River Aug. 21, In the Shawanaga River, Aug. 26. In lake off the Shawanaga River, Aug. 27. In pool at foot of last rapids on the French River, Sept. 1. In Collin's Inlet, Sept. 3.

Castalia odorata, Woodville & Wood. In lake off the Shawanaga River, Aug. 27. In pool at foot of last rapids on the French River, Sept. 1. In Collin's Inlet, Sept. 3.

Brasenia schreberi, Gmel. In lake off the Shawanaga River, Aug. 27.

Callitriche autumnalis, L. In little almost cut-off inlet in one of the Bustard Islands off French River, Aug. 30.

Hypericum boreale, Bicknell. Submerged form; in the Shawanaga River, Aug. 27.

Myriophyllum spicatum, L. In the Severn River, Aug. 21. At the mouth of the Shawanaga River, Aug. 26. In little almost cut-off inlet in one of the Bustard Islands off French River, Aug. 30. In pool at the foot of the last rapids on the French River, Sept. 1.

Myriophyllum heterophyllum, Michx. In pool at foot of the last rapids on the French River, Sept. 1.

Nymphoides lacunosum, Fernald. In lake off the Shawanaga River, Aug. 27.

XIII.

ENTOMOSTRACA OF GEORGIAN BAY.

By G. O. SARS. Professor of Zoology, Christiana University, Norway.

The following is a list of Entomostraca occurring in a series of surface tow-nettings made by Dr. E. M. Walker in the summer of 1907, at the Georgian Bay Biological Station, Go Home, Georgian Bay, Lake Huron.

CLADOCERA.

1. *Holopedium gibberum*, Zaddach.
    Very common in all the samples.
2. *Sida crystallina*, Müll.
    Occasionally from the bottom.
3. *Daphniella brachyura*, Lievin.
    Not infrequent in some of the samples.
4. *Daphnia hyalina*, var. *oxycephala*, G. O. Sars.
    This form was recorded by the present author in 1890 as a variety of *D. galeata*, G. O. Sars. I now, however, regard it as more properly belonging to the species D. *hyalina*, Leydig, with which also Herrick has identified it.
    Occasionally in most of the samples.
5. *Hyalodaphnia retrocurva*, var. *intexta*, Forbes.
    This form has erroneously been identified by Herrick with *H. kahlbergensis*, Schœdler, which is a variety of a quite different species, viz., *H. cucullata*, G. O. Sars. The present variety exhibits in its general appearance and particularly in the shape of the head, a strong resemblance to *H. cederstræmii*, Schœdler, which is a variety of *H. cristata*, G. O. Sars. It differs, however, among other things, in the more obtuse rostrum and in the presence of a well-devel oped natatory seta on the first joint of the lower or inner ramus of the antenna, this seta being wholly absent in *H. cederstræmii*.
    Rather common in all the samples.
6. *Ceriodaphnia scitula*, Forbes.
    The most conspicuous character distinguishing this species is the peculiar shape of the fornix, which is produced above the bases of the antennæ on each side to a rather large gibboiform prominence.
    Not infrequent in some of the samples.
7. *Bosmina longirostris* (Müll.) var.
    This is a very small variety, especially distinguished from the type by the somewhat longer shell-spines, which, moreover, exhibit each one or two well-marked serrations not found in the usual form.
    Not infrequent in some of the samples.

5 GEORGE V., A. 1915

8. *Polyphemus pediculus,* Müll.
   Rather abundant in most of the samples.
9. *Leptodora hyalina,* Lilljeb.
   Likewise rather common.

COPEPODA

10. *Epischura lacustris,* Forbes.
    Very abundant in all the samples.
11. *Diaptomus oregonensis,* Lilljeb.
    Likewise abundant in most of the samples.
12. *Diaptomus minutus,* Lilljeb.
    Together with the preceding species, but less abundant.
13. *Cyclops brevispinosus,* Herrick.
    This form is closely related to *C. robustus,* G. O. Sars, and is perhaps the same species.
    Only a few specimens observed, apparently dredged from the bottom.
14. *Cyclops thomasi,* Forbes.
    This form has been identified by Dr. Schmeil and some other authors with *C. pulchellus,* Koch (=C. *bicuspidatus,* Claus). It is however, as I have convinced myself, a well-defined species.
    Not infrequent in some of the samples.
15. *Cyclops edax,* Forbes.
    Nor in the case of this species has its validity been admitted by Dr. Schmeil, who regards it as only a variety of *C. leuckarti,* Claus. I find it, however, to be quite certainly specifically distinct though nearly allied to that species.
    Common in most of the samples.

OSTRACODA.

16. *Cyclocypris serena,* Koch.
    Occasionally from the bottom.

Lightning Source UK Ltd.
Milton Keynes UK
UKHW012349210119
335965UK00006B/137/P